6G Migrates from Communication to XaaS

Guangyi Liu · Fei Qin · Jianhua Zhang ·
Chengjun Sun · Jianjun Wu · Xiangyang Duan ·
Liyu Cai · Shaohui Sun · Ning Yang

6G Migrates from Communication to XaaS

中国工信出版集团
電子工業出版社
PUBLISHING HOUSE OF ELECTRONICS INDUSTRY
http://www.phei.com.cn
Springer

Guangyi Liu
China Mobile Research Institute
Beijing, China

Jianhua Zhang
Beijing University of Posts
and Telecommunications
Beijing, China

Jianjun Wu
Future Mobile Communication Forum
Shanghai, China

Liyu Cai
Nokia Shanghai Bell Co., Ltd.
Shanghai, China

Ning Yang
Standardization and Research Department
OPPO
Beijing, China

Fei Qin
vivo Communications Research Institute
Beijing, China

Chengjun Sun
Beijing Samsung Telecom R&D Center
Beijing, China

Xiangyang Duan
ZTE Corporation
Shenzhen, Guangdong, China

Shaohui Sun
CICT Mobile Communication Technology
Co., Ltd.
Beijing, China

ISBN 978-981-96-7578-4 ISBN 978-981-96-7579-1 (eBook)
https://doi.org/10.1007/978-981-96-7579-1

Jointly published with Publishing House of Electronics Industry
The print edition is not for sale in China (Mainland). Customers from China (Mainland) please order the print book from: Publishing House of Electronics Industry.

Translation from the Chinese Simplified language edition: "6G: 从通信到多能力融合的变革" by Guangyi Liu et al., © Publishing House of Electronics Industry Co., Ltd 2023. Published by Publishing House of Electronics Industry Co., Ltd. All Rights Reserved.

This Springer imprint is published by the registered company Springer Nature Singapore Pte Ltd.
The registered company address is: 152 Beach Road, #21-01/04 Gateway East, Singapore 189721, Singapore

Foreword

Human society is undergoing the fourth industrial revolution, which is primarily driven by the convergence and innovation of digitalisation, Information and Communication Technology (ICT), and Artificial Intelligence (AI) technologies. Among them, ICT plays a crucial role in the evolution of human society towards a new stage of digital intelligence. The fifth-generation mobile communication system (5G) opens up a new paradigm for the interconnection of people, machines, and things by organically integrating ubiquitous communication, computing, and control (UC^3) capabilities. The sixth-generation mobile communication system (6G) will further extend the boundary of communication from the physical world to the digital world, reshaping the world by providing instant, efficient, and intelligent "hyper-connections" between the physical and digital worlds. This trend will open a new chapter in mobile communications.

From the perspective of industry research, 6G is both an inheritance and development of mature technologies in past mobile communication networks, and a revolution of past mobile communication network systems. On the one hand, to meet the future society's demand for capacity growth in mobile communication networks, 6G will inherit mature technologies in 5G, such as Orthogonal Frequency Division Multiplexing (OFDM), large-scale antennas, Low-Density Parity-Check (LDPC)/Polar coding, etc., and will optimise and improve these technologies to further enhance the communication transmission efficiency of the network and adapt to the needs of new 6G application scenarios. On the other hand, the application practice of 5G in vertical industries has proven that traditional communication connection capabilities alone can no longer fully meet customer needs. Customers need turnkey solutions that integrate communication with big data, cloud computing, AI, etc. Therefore, 6G will be a new generation of mobile information systems that transcend traditional mobile communication categories. By integrating communication with sensing, big data, computing, AI, security, etc., it will further expand the boundaries of network capabilities, better empower more differentiated and fragmented application scenarios, and achieve "personalised and customised" service providing. Therefore, cloudification and platformisation of end-to-end networks will be an important trend in the development of 6G networks.

As a new generation of digital information infrastructure and the main channel for the future development of the global mobile communication industry, 6G has become a new strategic high ground for competition among major countries. Multiple countries and regions around the world and international organisations are promoting the formulation of 6G visions and requirements. The United States is fully preparing for 6G through the "Next G Alliance", and the European Union, China, South Korea, and Japan have also successively released 6G action plans, proposing their respective technology evolution roadmaps. The community has made significant progress in areas such as cutting-edge technology research, patent layout, and ecosystem construction.

With the advancement of the International Telecommunication Union's 6G vision and technology trend research, the research on 6G has entered a critical window period. 3GPP have start its SA1 study of scenarios and requirements on 6G in June 2024 and RAN study of scenarios and requirements on 6G in December 2024. The first release of 6G specification is expected to be completed by June 2029 in 3GPP, and 6G network will be commercial ready by 2030. From the development experience of past mobile communication industries, the development of 6G still faces many challenges, such as the lack of break through in fundamental theories, the lack of "killer" application and its necessary supporting ecosystem, and the incomplete open-source ecosystem. Therefore, from theory, technology to application ecology, 6G still needs further exploration. It is urgent for industry and academia to join hands, seize opportunities, face difficulties, and carry out coordinated research on potential key technologies of 6G, prototype verification and field trials, and cultivation of industry and application ecology.

This book is written by the top technical experts in the field of mobile communications in China. They have deeply participated in the work on the research, standardisation, industrialisation, and application in the field of 3G, 4G, and 5G mobile communications globally. They have rich practical experience and have carried out fruitful work in 6G research. This time, the research results are compiled into a book, which will surely broaden the research ideas of the industry, provide useful references and new ideas for the industry's 6G research, jointly promote the formation of 6G industry consensus and global consensus, jointly promote global unified standards and unified ecology, and make positive contributions to the health and sustainable development of 6G and the entire mobile communication industry.

Zhang Ping
Fellow of the Chinese Academy
of Engineering
Beijing, China

Preface

With the large-scale commercial use of 5G underway, universities and research institutions worldwide are turning their attention to the research of the next generation of mobile communication technology. Major companies in the industry have also released their own 6G white papers, explaining their different understandings of the future 6G. Various mobile communication markets and industrially developed countries have introduced their 6G development plans, hoping to take the lead in the future development of 6G and boost the development of their own industries, economies, and societies. The most aggressive of these is the South Korean government, which claims it will be the first in the world to commercialise 6G by 2028, while Japan plans to do so by 2030. From the current situation, China's overall 6G layout is basically in sync with the world. In June 2019, under the guidance of the Ministry of Industry and Information Technology, the China Academy of Information and Communications Technology took the lead in establishing the IMT-2030 research group, organizing domestic universities and enterprises to carry out 6G technology research; in November 2019, the Ministry of Science and Technology officially established the 6G promotion group and the overall expert group, guiding the Ministry of Science and Technology's subsequent 6G R&D planning, and the Ministry of Science and Technology also launched a batch of 6G special research project.

From the existing 6G white papers, it can be seen that there are significant differences in the early understanding of 6G in the community. For example, one view is that 6G is "5G + satellite communication", believing that 4G and 5G are sufficient, as long as the coverage is resolved; another view is that 6G is terahertz (THz), because 6G needs to provide peak rates 10~100 times higher than 5G, requiring continuous larger bandwidth, and the possible larger bandwidth may come from the terahertz frequency band. Therefore, how to define 6G is a controversial issue in the community. As research deepens, some views in the industry gradually converge: in terms of 6G vision, the mainstream view converges on "digital twins, ubiquitous intelligence", although everyone's terminology and expressions are different; in terms of 6G technology requirements, in addition to further enhancing traditional communication capabilities, 6G needs to build new capability dimensions, achieving everything as a service (XaaS); in terms of wireless access technology, 6G will fully inherit and develop existing 5G technology and carry out scenario-based optimization design; in terms of network architecture, the boundaries of the Radio Access Network (RAN) and the core network will become more blurred, lightweight, distributed core networks are the trend, and the integrated design of communication with sensing, computing, artificial intelligence (AI), big data, security, and other capabilities is both an opportunity and the biggest challenge for 6G; in terms of network service methods, 6G will shift from the traditional "use what the network have" mode to a "service providing on demand" mode; in terms of network operation and management, 6G will shift from the existing "automation" to "digital twins and native AI-enabled high autonomy".

With the definition of the International Telecommunication Union's 6G vision and requirement indicators, and the commencement of 6G technology trend research, 6G research has entered a critical window period for core technology breakthroughs. The necessary technical convergence and consensus building are beneficial for the industry to form a joint force, accelerate technological breakthroughs, and later land in standardization. The authors of this book are all distinguished experts who have long been engaged in the research, standardization, and industrialization of 3G, 4G, and 5G mobile communication technologies and are also the heads of 6G research in major companies in China. Through this book, we hope to share with readers our definition of 6G from an industry perspective, as well as our deep thinking and understanding in the research of key technologies and network architecture design, to promote the formation of technical consensus and the convergence of technical directions, accelerate breakthroughs in relevant technical directions of 6G, provide key technical support for subsequent standardisation, and promote the full realization of the 6G vision.

This book first reviews the development history of mobile communication systems, starting from the experience and challenges of 5G development, outlines the driving forces for 6G development and evolution, summarizes and collates the research progress and achievements of major institutions, organizations, projects, and enterprises engaged in 6G research globally, and looks forward to how to define 6G from the perspectives of vision, demand, frequency, network architecture, etc.; then, focusing on the optimization of capacity, efficiency, coverage, etc., of traditional communication networks, it introduces several most representative technical directions in the 6G wireless field, including distributed multiple-input multiple-output (MIMO), intelligent reflective surface (RIS), AI-enabled air interface, advanced full duplex, orthogonal time-frequency space modulation, as well as the integration of communication and sensing, zero-power communication, etc.; finally, from the perspective of the entire network architecture, it focuses on the integration of air, space and ground, as well as the design of 6G multi-capability integrated mobile network architecture.

The views in this book represent only the personal views of the authors and do not represent the views of any company. The book is compiled by Guangyi Liu, Chaps. 1–3 are written by Guangyi Liu and Jianhua Zhang, Chap. 4 is written by Jianhua Zhang, Guangyi Liu, Na Li and Juan Deng, Chap. 5 is written by Guangyi

Liu, Jianhua Zhang, Juan Deng and Na Li, Chap. 6 is written by Na Li, Chunjing Yuan, Guangyi Liu, Chap. 7 is written by Nuan Song, Hao Liu, Yan Zhao, Dongxu Zhang, Tao Yang and Liyu Cai, Chap. 8 is written by Xiangyang Duan, Yajun Zhao, Jianwu Dou, Lin Peng, Yijian Chen and Mengnan Kan, Chap. 9 is written by He Wang, Liying Zhou, Lei Wang, Chengjun Sun, Ning Yang, Wenqiang Tian, Jia Shen, Yang Xu, Jingran Chen and Boren Guo, Chap. 10 is written by Bin Yu, Hengzhu Chi, Di Su, Chen Qian, Peng Lin, Shihai Shao and Chengjun Sun, Chap. 11 is written by Pu Yuan, Bule Sun, Fei Qin, Wenqiang Tian, Jia Shen and Ning Yang, Chap. 12 is written by Jian Yao, Jianzhi Li, Shengli Ding, Yannan Yuan, Dajie Jiang and Fei Qin, Chap. 13 is written by Wei Huang, Rongling Jian, Huan Li, Junjie Tan, Dajie Jiang, Fei Qin, Ning Yang, Weijie Xu, Shengjiang Cui, Chuanfeng He, Zhisong Zuo, Rongyi Hu and Lu Gan, Chap. 14 is written by Shaohui Sun, Shaoli Kang and Bo Han, Chap. 15 is written by Jianjun Wu, Chenghui Peng, Xueqiang Yan, Shaoyun Wu, Wenliang Liang, Mingyu Zhao, Jun Wang, Zhe Liu, Weijun Xing, Fei Liu and Donghui Wang. At the same time, thanks to the editor Li Shulin of the Electronic Industry Press for his planning and meticulous editing of this book.

The writing of this book was supported by the Key R&D Program of the Ministry of Science and Technology "Multimodal Network and Communication" 6G Special Project "6G Network Architecture and Key Technologies (2020YFB1806800)", and the Beijing Municipal Science and Technology Commission and Zhongguancun Science Park Management Committee New Generation Information Communication Technology Innovation Project "6G New Air Interface Technology Test Verification Platform Development" support.

This book can be used as a reference book for 5G practitioners and 6G R&D personnel, as well as a reference book for undergraduate and postgraduate students and also as a reference book for enterprises to layout future mobile information application-related businesses.

Beijing, China Guangyi Liu
Beijing, China Fei Qin
Beijing, China Jianhua Zhang
Beijing, China Chengjun Sun
Shanghai, China Jianjun Wu
Shenzhen, China Xiangyang Duan
Shanghai, China Liyu Cai
Beijing, China Shaohui Sun
Beijing, China Ning Yang

Contents

Chapter 1
The History of Mobile Communication

Mobile communication systems were born in the 1970s, but it was not until the 1990s that they became widely popular with the advent of the Global System for Mobile Communications (GSM), which is still widely used today. Mobile communication is a highly complex industry that integrates numerous advanced technologies. It has evolved from regional standards to global standards, and the widespread adoption of mobile communication has been facilitated by the establishment of these global standards. By setting unified standards, interoperability between different manufacturers' equipment, different network elements, different terminals, and different operators and users can be achieved. Through sharing, the costs of research and development and equipment have been significantly reduced on a global scale, making what was once a luxury accessible to ordinary households. Looking at the history of mobile communication development, the update of mobile communication technology roughly follows a "decade per generation" pattern, from key technology research, standard defining, product development, large-scale network construction, to the commercialisation of network services, each step is a challenging process. The development of mobile communication technology usually follows the pattern of "use one generation, deploy one generation, research one generation". For example, although we are currently using 4G on a large scale, we are also deploying and commercializing 5G on a large scale, and we have already started the research of 6G. This chapter briefly reviews the development history of mobile communication systems and the contributions China has made to the global development of mobile communications.

G. Liu et al., *6G Migrates from Communication to XaaS*,
https://doi.org/10.1007/978-981-96-7579-1_1

1.1 The Development History of Mobile Communication Systems

The development of mobile communication systems has gone through several major stages. The mobile communication industry generally uses "generations" to distinguish between them. Each generation of mobile communication systems has distinct technological and feature differences, catering to different business and application scenarios, and generally follows a "decade per generation" development pattern. The development history of mobile communication systems is shown in Fig. 1.1.

The mobile communication systems that have appeared so far mainly include the first generation mobile communication system (1G), the second generation mobile communication system (2G), the third generation mobile communication system (3G), the fourth generation mobile communication system (4G), and the fifth generation mobile communication system (5G). From the perspective of changes in service and capabilities offered by the communication system, the development of mobile communication systems can be divided into four stages: ubiquitous communication, ubiquitous information, ubiquitous perception, and ubiquitous intelligence.

1G mainly supported voice communication on the move, extending traditional fixed telephony to mobile telephony; 2G solved the quality and capacity issues of voice communication, enabling large-scale global application of mobile communication, and also initiated the non-real-time communication mode of text messaging, with the ability to make calls anytime, anywhere, truly achieving ubiquitous communication.

3G further increased the capacity of communication, while also promoting the emergence of broadband communication, especially in the later stages of 3G, with the advent of smartphones, high-speed mobile data communication became an urgent need; 4G solved the problem of high-speed mobile data communication, making great

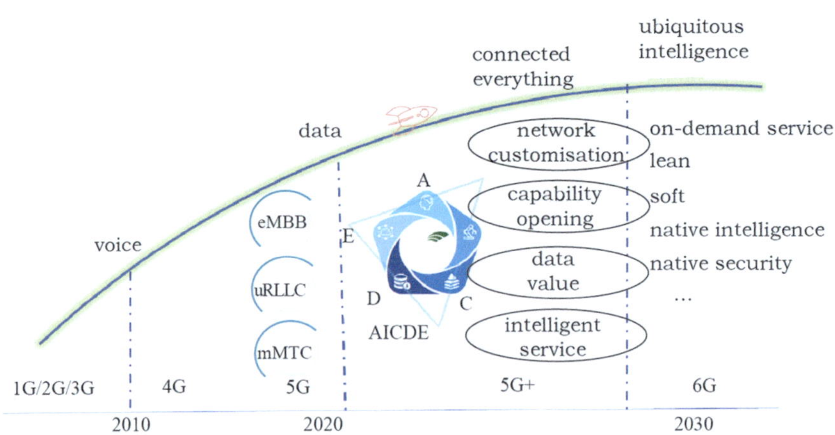

Fig. 1.1 The development history of mobile communication systems

strides in communication quality, capacity, and efficiency, and in the later stages also triggered the application needs of the Internet of Things (IoT), leading to unprecedented prosperity in information consumption, achieving ubiquitous information, with smartphones becoming an important entry and platform in people's daily lives, essentially achieving "with one device in hand, the world is at my fingertips".

5G has expanded the scope of mobile communication in an unprecedented way, aiming to achieve interconnectivity of everything, and for the first time, formally extending the scope of mobile communication to application scenarios beyond human use. It intends to cover typical application scenarios such as enhanced Mobile Broadband (eMBB), ultra-Reliable and Low Latency Communications (uRLLC), and massive Machine Type Communications (mMTC), striving to extend human perceptual capabilities to everything, achieving ubiquitous perception, and bringing a brand new experience of "information at will, everything at your fingertips". Since 2019, 5G has begun to be deployed on a large scale globally, and 5G will inevitably accelerate its integration with cloud computing, big data, artificial intelligence, edge computing, etc., bringing ubiquitous information and perception through network customisation, capability exposure, data value, and intelligent services, accelerating the digitisation of the entire society.

With the large-scale commercial application of 5G network, the mobile communication industry has begun to turn its attention to the next generation of mobile communication systems, that is, the sixth generation mobile communication system (6G). The development vision of 6G is "digital twin, ubiquitous intelligence", promoting the whole society to fully move towards digital twin, achieving prediction and change of the future through the mapping and interaction between the digital world and the physical world, and empowering the entire society's intelligent upgrade through ubiquitous intelligence, greatly improving the efficiency of the entire society's operation and governance, improving the efficiency and quality of people's life, work and production, promoting the development of human society into a new stage, further realising human self-liberation, and promoting the pursuit and realisation of self-value. To adapt to the development needs of "digital twin, ubiquitous intelligence", serving more fragmented and differentiated new application scenarios, 6G needs to have the characteristics of on-demand service, lean network, soft network, native intelligence, native security and digital twin, achieving ubiquitous connection, ubiquitous computing, ubiquitous sensing, ubiquitous AI, and omnipresent security.

1.2 China's Contribution to the Global Mobile Communication Industry

Cellular mobile communication technology has gone through the evolution from 1 to 5G, as shown in Fig. 1.2. In the early stage of the development of China's mobile communication industry, the entire industry lacked funds, talent, and modern management systems, and information communication services were only scarce

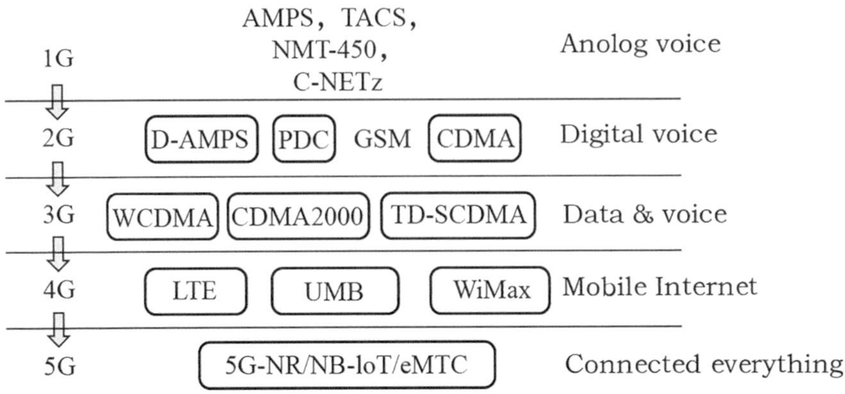

Fig. 1.2 Evolution from 1 to 5G [1]

resources enjoyed by a few people. Looking back at the development history of China's mobile communication industry, we have successively experienced the development process of "1G blank, 2G follow, 3G breakthrough, 4G parallel, and 5G lead". China's mobile communication industry has made a huge contribution to the development of the global mobile communication industry. After decades of unremitting efforts, China's mobile communication industry has achieved leapfrog development in communication capabilities, user scale, and technical level through digestion and absorption, technological innovation, standard breakthroughs, service and application innovation. China has become the world's largest mobile communication market, and despite various shortcomings, it also has the world's most complete mobile communication industry chain and has stepped into the world's leading level in mobile communication network construction.

1.2.1 1G

In November 1987, prior to the Guangzhou Sixth National Games, China launched its first large-scale cellular mobile communication system, marking the beginning of the application of mobile communication technology in our country. At that time, China's mobile communication industry was completely in a blank period, with no technology, no talent, and the entire mobile communication network and terminal equipment, network construction and maintenance were entirely dependent on foreign equipment vendors. The high cost meant that mobile communication services could only serve as luxury items for a very small number of people. During this period, under the backdrop of economic development far behind developed countries, our market demand for the mobile communication industry had not yet been awakened, and there was a huge gap in the mastery, understanding, and application of technology compared to developed countries. From mobile communication equipment, technology to the operation of mobile communication, all were introduced from the West countries. China used its untapped vast market to exchange for advanced mobile communication technology from the West countries, constantly digesting and absorbing it. Later, companies like Huawei and ZTE, which caused waves worldwide, also began to grow slowly among many small companies in Shenzhen.

1.2.2 2G

The application and development of 1G achieved a breakthrough in mobile communication from 0 to 1, but it also exposed a series of problems, such as poor call quality, low capacity, expensive equipment, and the lack of unified standards and incompatible systems of "eight systems from seven countries", which affected the widespread use of mobile communication and global roaming. Therefore, European countries jointly formulated the GSM technical specifications around the new generation of mobile communication systems, which better solved a series of problems faced by 1G from a technical perspective and successfully promoted it globally. The GSM technical specifications began to be deployed on a large scale in China from 1996 and became mainstream, sparking a boom in the development of mobile communication. In addition, Qualcomm in the United States, based on the spread spectrum technology used in military applications, developed the IS-95 technical standard based on CDMA, which was first commercially successful in the US, Japan, Korea, and other countries, and officially launched in China in 2000 through the opportunity of WTO negotiations, with China Unicom carrying out large-scale construction and operation. The right timing, correct technology selection, and huge potential market led to the rapid development of China's mobile communication industry in the following years, with the number of users doubling year by year, and the network expansion speed was astonishing, becoming the country with the most mobile communication users in the world. During this period, although technology introduction was still the main focus, driven by the high market profits of mobile communication equipment, through digestion, absorption, and learning, China's mobile communication industry also began to give birth to a number of local mobile communication companies, such as Julong, Datang, Jinpeng, ZTE and Huawei network equipment manufacturers, as well as Bird, Haier, TCL and other mobile phone manufacturers, began to accumulate strength in the low-end equipment market, but due to the monopoly of technical standards, the development of local companies in China was difficult.

1.2.3 3G

At the start of the third-generation mobile communication system, the European and American mobile communication industries, based on their own technical accumulation and advantages, respectively proposed two independent candidate technical standards, Wideband Code Division Multiple Access (WCDMA) and CDMA2000. After the introduction and digestion and absorption of 1G and 2G, China's mobile communication industry has accumulated certain technology and strength, and is eager to propose its own 3G standard to break the monopoly of technology and products. Therefore, based on the Synchronous Code Division Multiple Access (SCDMA) technology of wireless local access, Datang Telecom formed the Time Division-Synchronous Code Division Multiple Access (TD-SCDMA) system solution. After intense debate, the Science and Technology Department of the Ministry of Posts and Telecommunications finally agreed to submit TD-SCDMA as a 3G candidate technology to the International Telecommunication Union Radiocommunication Sector (ITU-R). Therefore, in 1998, during the process of the International Telecommunication Union (ITU) soliciting 3G standard proposals globally, Datang Telecom represented China for the first time to submit a 3G candidate technology proposal - TD-SCDMA to the ITU, which was accepted as one of the 15 candidate technologies. After the hard work of the Chinese government and enterprises, China's original TD-SCDMA technology solution and the TD-CDMA proposed by Germany were harmonized, and after a certain parameter harmonization with WCDMA, the final TD-SCDMA technology solution was formed, which was finally accepted as the 3G technology standard recognised by ITU along with WCDMA and CDMA2000, and then detailed standard protocols were developed in 3GPP, thus achieving the first international standard breakthrough in the history of China's mobile communication industry. Due to various reasons, TD-SCDMA did not receive support from other countries, and foreign companies supporting TD-SCDMA technology were few and far between, with only Germany's Siemens actively investing and participating in the product development of TD-SCDMA. In the process of developing TD-SCDMA technology, in order to integrate various forces and achieve overall industrial development, China established the TD-SCDMA Industry Alliance and TD-SCDMA Technology Forum. In order to realise the industrialisation and large-scale commercial use of TD-SCDMA, China's mobile communication industry began to attempt the construction of end-to-end industry ecosystem and the research and development of end-to-end products, with more and more companies participating, gradually establishing a mobile communication ecosystem from antennas, access network equipment, core network equipment to chips, instruments and terminals. In December 2003, at Datang Mobile, a milestone first mobile call of the TD-SCDMA/LCR full system was made. In January 2006, the Ministry of Information and Industry Technology (MIIT) officially established and promulgated TD-SCDMA as the standard for the Chinese communication industry, and finally during the 2008 Beijing Olympics, China Mobile commercialized TD-SCDMA.

There is a great deal of controversy in the industry about whether TD-SCDMA is successful. Objectively, TD-SCDMA is used only by China Mobile globally, and in terms of both industry scale and application effect, TD-SCDMA is not as good as WCDMA and CDMA2000. However, it is worth noting that TD-SCDMA started late and had greater development difficulties, but after the industry's hard and extraordinary efforts, its many unique advantages in technology are evident—higher spectrum utilisation, more suitable for supporting mobile Internet business, many technologies represent the direction of mobile communication technology development, especially TDD and smart antenna technology gradually becoming the mainstream direction of subsequent 4G and 5G development. TD-SCDMA has enabled Chinese companies to fully participate in the international standardization, learn the rules of international standard organisations, accumulate rich experience, and cultivate a large number of excellent standard representatives and talents. The industrialisation of TD-SCDMA has also enabled China's mobile communication industry to start a comprehensive layout of end-to-end industry ecology, and the lack of support from foreign companies has provided an excellent time window for domestic companies to grow. Through the practice of TD-SCDMA, domestic companies have cultivated a large number of engineering and technical personnel, explored and accumulated rich experience in the construction and cultivation of industry ecology, and initially constructed an end-to-end industry ecology. In addition, the development of TD-SCDMA has layed the foundation for Chinese companies to play a important role in the subsequent development of 4G and 5G.

1.2.4 4G

Although China proposed its own TD-SCDMA standard in the 3G era, achieving a breakthrough in standards and implementing large-scale commercial use with the full support of China Mobile, globally, only China Mobile deployed the TD-SCDMA network. The maturity and robustness of the TD-SCDMA ecosystem could hardly compete with the global scale effects of WCDMA and CDMA2000. The types and quality of terminals were far inferior to WCDMA and CDMA2000. In order to increase user stickiness and improve mobile broadband data service capabilities, China Mobile built more than 5 million Wi-Fi wireless access points (Access Point, AP) nationwide. However, due to poor user experience, it could not prevent a large number of users from leaving. As TD-SCDMA went commercial amidst controversy, the global microwave access interoperability (World Interoperability for Microwave Access, WiMax) also began to emerge in China initially applied in many broadband access scenarios, showing great potential. Therefore, in the 3G era, the development of the mobile communication industry clearly showed two branches, Frequency Division Duplex (FDD) and Time-Division Duplex (TDD). FDD (including WCDMA and CDMA2000) was in an absolute monopoly position, while TDD (only TD-SCDMA) was just starting and was in a very weak position. Looking at the industry situation at that time, WCDMA was the world's largest, CDMA2000 was second, WiMax was growing steadily, and TD-SCDMA was the weakest. 3GPP began the study and standardization of Long Term Evolution (LTE) in 2005, and the initial standard included one FDD mode and two TDD modes. One of the TDDs was fully compatible with WCDMA TDD Type 1 [3GPP calls it High Code Rate (HCR)], as shown in Fig. 1.3 for the 3GPP LTE Type 1 TDD frame structure; and the other TDD was fully compatible with TD-SCDMA [3GPP calls it Low Chip Rate (LCR)], as shown in Fig. 1.4 for the 3GPP LTE Type 2 TDD frame structure. At the same time, 3GPP2 also launched the technical standard for the subsequent evolution of CDMA2000—the Ultra Mobile Broadband (UMB) system, which includes FDD and TDD modes. Therefore, by the 4G era, the inherently weak TDD faced a choice of four technical routes, leading to further industry differentiation and fragmentation, which added to the difficulties of TDD industry development and made operators and manufacturers even less confident in the development of TDD. How TD-SCDMA develops and evolves has become the focus of attention in China's mobile communication industry, and there is also a lot of debate in the industry. Route 1: It should evolve towards WiMax, because it is supported by many companies, and also because "WiMax is a technology standard supported by the United States, and the technology in the United States was the most advanced at the time". Route 2: China should stick to its own standards. 3G has its own standards, and 4G should also have its own standards, so Type 2 TDD should be chosen. Route 3: We should return to the international mainstream standards, choose Type 1 TDD to share the global industry scale, and provide users with high-quality and cost-effective terminals and services.

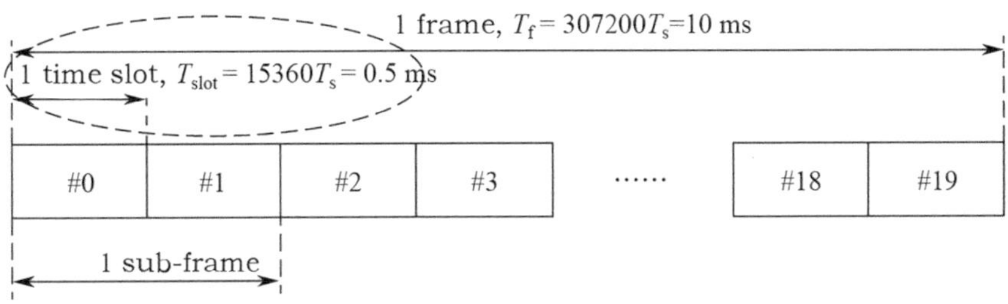

Fig. 1.3 3GPP LTE Type 1 TDD frame structure

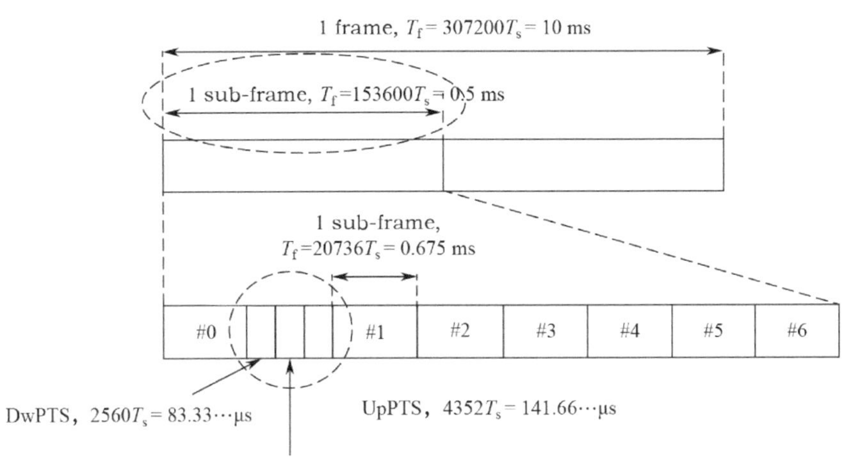

Fig. 1.4 3GPP LTE Type 2 TDD frame structure

Starting from the perspective of strengthening TDD, Chinese mobile communication industry has carved out its own path. In 2007, on the eve of the freezing of the first version of the LTE standard, China Mobile together with Vodafone initiated a discussion on TDD frame structure harmonization and optimisation in 3GPP, hoping to reduce the TDD system options, optimise TDD performance while coexisting with TD-SCDMA. China Mobile first initiated relevant technical discussions in China, and finally, with the support of Huawei and ZTE, a consensus was reached among major Chinese companies. The research on the integration of TDD and FDD began in 3GPP. Finally, with the promotion of China Mobile, Vodafone, and Verizon, through the hard work of 3GPP major companies, this work received the support of more than 30 major companies in 3GPP, including Ericsson, Nokia, Qualcomm, Alcatel-Lucent, and Nortel, and finally reached a consensus on the new frame structure at the Jeju meeting in South Korea in November 2007, as shown in Fig. 1.5. It not only takes into account the basic characteristics of TD-SCDMA, maintains backward compatibility with TD-SCDMA, but also further optimises performance, while also keeping the basic parameters of LTE's FDD and TDD consistent, laying the foundation for the subsequent integration and development of FDD and TDD industries. By this point, 3GPP finally achieved the integration of FDD and TDD in the LTE standard just as the first version was about to be frozen, laying the technical framework foundation for Time Division Long Term Evolution (TD-LTE). The Science and Technology Department of MIIT thus identified TD-LTE as the subsequent evolution technology of TD-SCDMA, basically clarifying that the technical path for the subsequent 4G development of China is LTE. In the subsequent standardization of the LTE, China Mobile, as the largest potential user of TD-LTE, joined forces with 3GPP companies to accelerate the formulation and improvement of the TD-LTE standard, achieving the synchronised development of TDD and FDD standards.

In the selection process of the next-generation mobile network technology by the global operator alliance NGMN (Next Generation Mobile Networks), China Mobile,

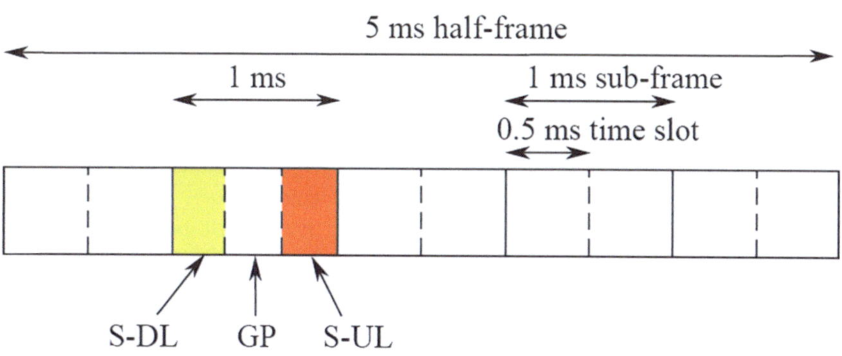

Fig. 1.5 New frame structure after LTE FDD/TDD fusion

Vodafone, and Verizon, together with industry partners, successfully promoted TD-LTE to become the only TDD choice, thereby establishing the international mainstream status of TD-LTE in one fell swoop. In the subsequent industrialisation process, China Mobile fully summarised the experiences and lessons from the development of the TD-SCDMA industry, guided by commercial needs and driven by market scale, attracted and joined global industry partners to jointly build the TD-LTE global ecosystem. With a series of demonstrations and promotions such as the Barcelona Exhibition and the World Expo, the industrialisation progress of TD-LTE was continuously accelerated. Through the construction of the TD-LTE global industry promotion platform - the Global TD-LTE Initiative (GTI), the international influence of TD-LTE was continuously enhanced, and it gradually caught up with the pace of FDD LTE industry development, achieving industry synchronisation. In addition, the Chinese regulator actively planned the entire 2.6 GHz as TDD, which has constructed a distinct spectrum advantage, providing a successful example for the TDD spectrum planning of other countries and making an important contribution to the development of the global TDD industry.

Eventually, through technical trials, large scale trials, and expanded large scale trials, following the commercial use of TD-LTE by SoftBank in Japan, China began large-scale commercial use of TD-LTE at the end of 2013. Although China Unicom and China Telecom were not enthusiastic about the development of TD-LTE, the millions of base stations built by China Mobile in just a few years, with network coverage and quality consistently at the world's advanced level, have enabled the rapid popularisation of TD-LTE applications and injected a huge impetus into the development and growth of China's TDD industry. Huawei, ZTE, and Datang account for more than 70% of the base stations in China Mobile's TD-LTE network, and their share of base stations in the international TD-LTE network market also exceeds 60%, causing the market share of Huawei and ZTE's mobile communication products to quickly rise to the top four globally. The booming development of the 4G market has also created a number of leading smartphone manufacturers. Huawei, OPPO, vivo, Xiaomi, ZTE, Lenovo and other mobile phone brands are among the top ten in the world in terms of shipment volume, with an unprecedented richness of high-end, mid-range, and low-end smartphone categories, bringing about an unprecedented prosperity of mobile internet applications. Applications such as Taobao, WeChat, TikTok, Weibo, and Alipay have had a huge impact globally, and China has quickly become one of the most developed countries in e-commerce. In the development of 4G, as of the end of October 2019, the scale of 4G users reached 1.269 billion, and the number of 4G base stations accounted for more than half of the world.

Therefore, whether from the perspectives of technology, standards, industry, or network application scale, Chinese mobile communication industry has reached the world's advanced level, achieving synchronous development with developed countries.

1.2.5 5G

The development of 4G has fully demonstrated to China the driving and promoting role of information consumption on economic development, bringing about earth-shaking changes to social development, and bringing a golden period to the development of Chinese mobile communication industry. Chinese mobile communication industry, in the development of TD-LTE, has accumulated strength, experience, and the ability to make greater contributions to the development of the global mobile communication industry in terms of technology, standards, and industrialisation. Therefore, looking forward to the commercial use of 5G in 2020, Chinese mobile communication industry has actively invested in the new round of industrial innovation.

Through systematic analysis of the development trends of existing networks, vertical industries and emerging applications, and in-depth exploration of industry application needs, in May 2014, the IMT-2020 (5G) Promotion Group issued the "5G Vision and Requirements White Paper" [2], Chinese companies proposed the 5G vision of "information at will, everything at your fingertips" [3] and designed the famous 5G requirements flower, comprehensively defining the various objective and subjective technical indicators of 5G, eight of which were adopted by ITU-R, which guided the subsequent 5G technical study and standardization very well.

In terms of key technologies, around the IMT-2020 Promotion Group, Chinese companies have carried out in-depth research and verification in the directions of large-scale antennas (also called 3D-MIMO), non-orthogonal multiple access, new modulation and coding, full duplex, new network architecture, etc., especially the world's most advanced 128 antennas and 64 ports 3D-MIMO was realised in TD-LTE's evolution as early as 2015 [4]. Commercial products in TD-LTE networks are widely deployed, fully utilising the channel reciprocity characteristics of TDD. Without changing existing standards and terminals, 16 data stream spatial multiplexing transmission can be achieved through software and hardware updates of the base station. This brings a twofold increase in spectral efficiency compared to commercial 8-antenna TD-LTE base stations, effectively addressing the challenges of power consumption, cost, weight, and size faced by large-scale antenna arrays in cellular networks, establishing a solid leading advantage for the subsequent industrialisation of 5G base stations.

In terms of standardisation, Chinese companies actively participate in the discussion and formulation of 5G standards [5] and makes a significant contribution to the timely and high-quality completion of the entire 5G standard. The statistics of project initiation led by major companies of 5G NR Rel-15 are shown in Fig. 1.6.

In June 2018, 3GPP officially announced the release of the Rel-15 standard. China Mobile, in conjunction with dozens of companies worldwide, issued a joint press release expressing support for 5G SA (standalone networking). This fully demonstrates that the contribution of Chinese companies to the 5G standard has been recognized by all 3GPP players.

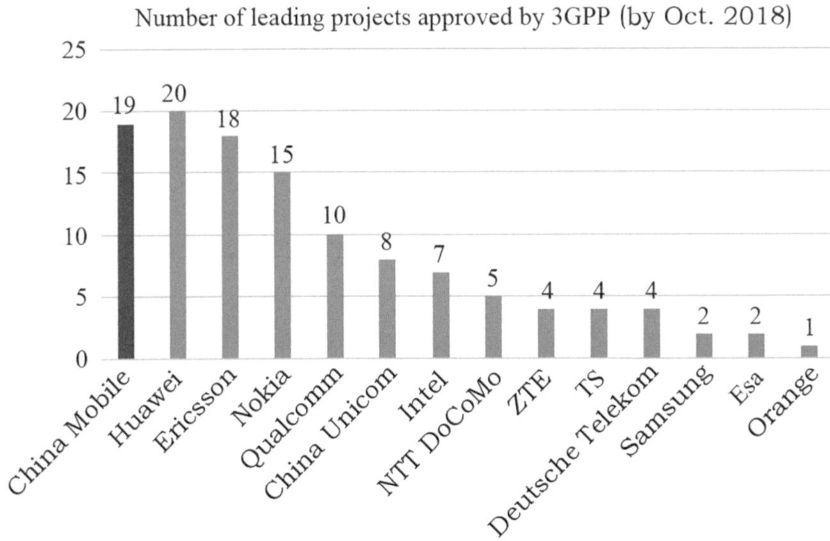

Fig. 1.6 5G NR Rel-15 project initiation statistics led by major companies

Since 2014, countries have started to invest in 5G technology research. At this time, Chinese companies represented by Huawei and ZTE have grown in strength, relying on the vast market and the ability to manufacture high-quality products. At the same time, the Chinese government fully recognises the strategic value of 5G, actively supports 5G R&D and business cultivation, and gradually leads the world in R&D. As of August 2019, China's achievements in the 5G field are globally leading. The first batch of companies to launch 5G mobile phones globally are mainly domestic companies such as Huawei, ZTE, Lenovo, OPPO, vivo, Xiaomi, etc.

Under the guidance of the "promote the service cultivation by building the network first" 5G development strategy, the MIIT and the four major operators officially announced the commercial use of 5G in November 2019. As of June 2022, 200 operators in 78 countries and regions around the world have commercially launched 5G, and Chinese operators have taken the lead in achieving commercial launch of 5G standalone (SA). At present, China has built more than 4 million 5G base stations, accounting for 70% of the global scale, and the number of 5G terminal connections exceeds 500 million, accounting for more than 70% of the world, and the average monthly data flow (Dataflow Of Usage, DOU) of 5G users is 1.74 times the average DOU of operator users, achieving global leadership in the scale and quality of 5G development.

1.2.6 6G

With China's comprehensive leadership in 5G development, China has accumulated rich experience in 5G network development and business expansion, and fully recognizes the capabilities and limitations of 5G networks. These provide important motivation and innovative insights for Chinese companies and universities to carry out forward-looking research on the next generation of mobile communication systems.

During the government interviews at the Two councils' meeting in 2018, Miao Wei, the Minister of MIIT, first mentioned that China had begun to initiate the research and development of 6G. In April 2019, the Wireless Communication Technology Committee (TC5) of the China Communications Standards Association (CCSA) initiated a project on "Beyond 5G System Vision and Demand Research", led jointly by Datang Mobile and China Mobile. In June 2019, the Ministry of Industry and Information Technology established the IMT-2030 (6G) study group, which includes the requirement Working Group, Wireless Technology Working Group, Network Technology Working Group, Spectrum Working Group, Standards and International Cooperation Working Group, and subsequently established the Economic and Social Working Group and the trial and Test working Group, officially launching China's 6G research process.

In September 2019, the China Mobile Research Institute held the first meeting of the "Imagining the Future" 6G series of seminars, providing an important reference for the industry to find the direction of 6G research. During the China Mobile Global Partner Conference in November 2019, China Mobile Research Institute released the "2030 + Vision and requirement Report" [6], which is the first complete and systematic 6G vision and requirement report in the world, proposing a social development vision of "digital twins, ubiquitous intelligence", hoping to reshape a new world through 6G [7]. Subsequently, Chinese companies have released 6G white papers one after another.

In November 2019, the Ministry of Science and Technology (MOST), together with the National Development and Reform Commission, the Ministry of Education, the MIIT, the Chinese Academy of Sciences, and the National Natural Science Foundation of China (NSFC), organised a 6G technology research and development work initiation meeting in Beijing, announcing the establishment of the China 6G Technology Research and Development Promotion Working Group. Prior to this, the key research and development project of the next-generation broadband communication network organized by the MOST has begun to support 6G-related technology research, mainly including wireless communication physical layer fundamental theory and technology, terahertz wireless communication technology and system, ultra-large scale antenna and RF technology, millimetre wave integrated RF front-end system key technology compatible with C band, RF front-end system technology based on third-generation compound semiconductors, etc.

In August 2020, the Future Mobile Communication Forum held a global 6G conference, and in June 2021, the IMT-2030 Promotion Group held a 6G seminar and officially released the "6G Overall Vision and Potential Key Technologies" [8] white paper, which also reveals that our country's 6G research and development has been fully launched.

Looking at the entire preliminary layout of 6G, China is basically in sync with the world, and has systematic layouts in directions such as wireless AI, visible light, terahertz, ultra-large scale antennas, integrated sensing and communication, reconfigurable intelligent surfaces, etc., and has emerged in the global industry and academia. The Chinese companies and universities are initiating systematic research and development, and carrying out more comprehensive and in-depth technology research and development, and strive for greater technological breakthroughs in 6G and make greater contributions to the global industry.

1.3 Summary

Mobile communication technology, while continuously meeting the ever-evolving communication needs of mankind, is profoundly changing the way we live and produce, stimulating new communication needs, and driving the generational change of mobile communication systems. The development of China's mobile communication industry has gone through the process of "1G blank, 2G follow, 3G breakthrough, 4G parallel, and 5G leading", with its own strength continuously increasing, and its contribution to the global industry development is growing. Looking towards 6G, Chinese industry's layout is basically in sync with the world, looking forward to making a greater contribution to the development of 6G.

References

1. Li Zhengmao, Wang Xiaoyun, Huang Yuhong, et al. TD-LTE Technology and Standards [M]. Beijing: People's Posts and Telecommunications Press, 2013.
2. IMT-2020 Promotion Group. 5G Vision and Requirements White Paper [R]. 2014.
3. Liu Guangyi, Fang Min, Guan Hao, et al. 5G Mobile Communication: Towards a Fully Connected World [M]. Beijing: People's Posts and Telecommunications Press, 2019.
4. Liu G Y, Hou X Y, Jin J, et al. 3-D-MIMO with Massive Antennas Paves the Way to 5G Enhanced Mobile Broadband: From System Design to Field Trials [J]. IEEE Journal on Selected Areas in Communications, 2017, 35 (6)
5. Wang Xiaoyun, Liu Guangyi, Ding Haiyu, et al. 5G Technology and Standards [M]. Beijing: Electronics Industry Press, 2019.
6. China Mobile Research Institute. 2030+ Vision and Requirements Report [R]. 2019.
7. Liu Guangyi, Huang Yuhong, Cui Chunfeng, et al. 6G Reshaping the World [M]. Beijing: People's Posts and Telecommunications Press, 2021.
8. IMT-2030 Promotion Group. 6G Overall Vision and Potential Key Technologies [R]. 2021.

Chapter 2
5G Development and Challenges

5G is cultivating new service and applications, with the development vision of "information at will, everything at your fingertips" being proposed. This vision includes three typical application scenarios: Enhanced Mobile Broadband (eMBB), Ultra-Reliable Low-Latency Communication (uRLLC), and Massive Machine Type of Communication (mMTC). Around these capabilities, standards such as Release 15 (Rel-15), Release 16 (Rel-16), and Release 17 (Rel-17) of 3GPP New Radio (NR) have been developed to continuously enhance and improve 5G capabilities. South Korea was the first to achieve large-scale commercial use of 5G Non-Standalone (NSA) in 2019, and China officially announced large-scale commercial use of 5G in November 2019, quickly achieving a global leading scale of network deployment, thus sparking a development craze for 5G to empower thousands of industries. This chapter first introduces 5G from the aspects of its basic capabilities, standard evolution, and frequencies used, then analyses the future and challenges of 5G development from the current status of global commercial deployment and network development.

2.1 5G Overview

With the rapid development of mobile internet services, the 5G was born to cope with the explosive growth of future mobile data traffic, the challenge of massive device connections, and the technical requirements of emerging new services. For 5G, the mobile communication industry hopes to replicate the prosperity that 4G brought to the mobile internet to all walks of life, thus proposing the development goal of the connected everything, considering not only connections between people but also between people and things, and things and things. Therefore, 5G will permeate all areas of future society, building a "user-centric" comprehensive information ecosystem, bringing users an immersive information feast, conveniently achieving intelligent interconnection between people and everything, and ultimately

G. Liu et al., *6G Migrates from Communication to XaaS*,
https://doi.org/10.1007/978-981-96-7579-1_2

Fig. 2.1 Overall vision of 5G [1]

realising the vision of "information at will, everything at your fingertips" [1], as shown in Fig. 2.1.

For this, the International Telecommunication Union (ITU) defined three typical application scenarios for 5G: enhanced mobile broadband, ultra-reliable low latency, and low power consumption large connections. To meet the needs of these three application scenarios, the 5G network will have higher performance than 4G, as shown in Fig. 2.2, including support for a user experience rate of 100 Mbps (10 times that of 4G), a connection density of 1 million per square kilometre (10 times that of 4G), millisecond-level air interface latency (1/10 of 4G), a traffic density of 10 Tbps per square kilometre, a mobility speed of over 500 km per hour, and peak rates of 20 Gbps downlink and 10 Gbps uplink. Among them, the user experience rate, connection density, and latency are the three basic performance indicators of 5G. At the same time, 5G will significantly improve the efficiency of network deployment and operation compared to 4G, with a significant increase in network spectral efficiency and an increase in energy and cost efficiency of more than 100 times [2].

Around the above 5G technical requirements defined by ITU, 3GPP began the formulation of the 5G standard—New Radio (NR) at the end of 2015, and officially released the first complete version of the 5G standard, Rel-15, in June 2018. Due to the limited research time of Rel-15, not all functions of 5G NR were completed in one version of the standard, but the basic functions of the standard were completed mainly for eMBB and some uRLLC functions.

Due to the working inertia of 3GPP and limited time, 3GPP adopted the conservative "LTE baseline" design principle, that is, the design of the LTE system is taken as the basic assumption, and modifications and new designs are introduced only when necessary. Therefore, the design of 5G NR can be said to have basically inherited

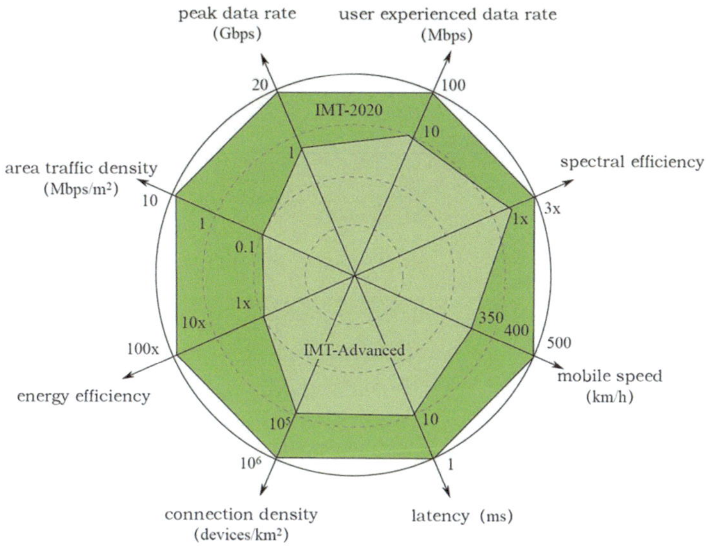

Fig. 2.2 Comparison of kPI between IMT-2020 (5G) and IMT-Advanced (4G) [3]

the mature technology and framework of LTE, while introducing some necessary designs to further expand and optimize the capabilities of 5G NR.

2.1.1 *Wireless Access Network (RAN) Architecture*

Depending on whether the 5G new radio base station can work independently and the difference in the signalling plane anchor, 3GPP defines two networking technology solutions for NR, namely Non-Standalone (NSA) and Standalone (SA). The main difference between the two networking technology solutions is whether the 5G network (5GC) can independently provide communication services to 5G users: the standalone network does not need to rely on the 4G network, and the 5G network can independently provide user access, residence and other core network capabilities, as shown in Fig. 2.3 [4]; in the non-standalone network, the LTE base station serves as the signalling plane anchor connected to the core network, and the user equipment (User Equipment, UE) needs to access the core network through the LTE base station (such as registration, authentication and mobility management, etc.), the new radio base station cannot work independently and only serves as an enhancement to the LTE data pipeline, as shown in Fig. 2.4.

The division of functions between the access network and the core network nodes is shown in Fig. 2.5, which points out the main functions contained in each network node.

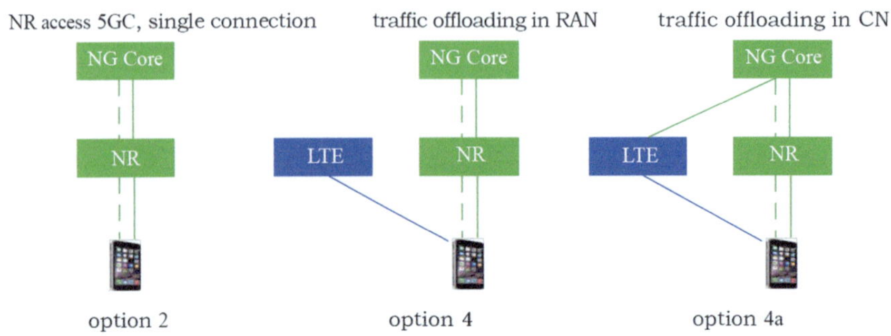

Fig. 2.3 Option 2 and Option 4 series network architecture

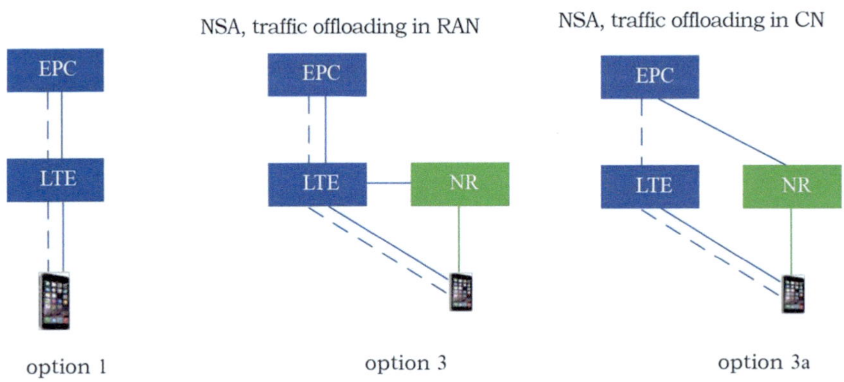

Fig. 2.4 Option 1 and Option 3 series network architecture

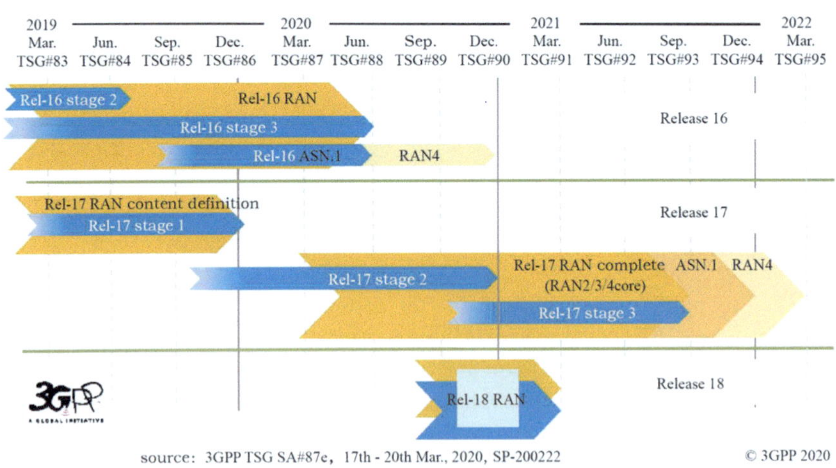

Fig. 2.5 Division of functions between access network and core network nodes [3]

The main functions included in the access network nodes, the next generation NodeB (gNB; usually refers to the 5G base station) and ng-eNB, include:

(1) Wireless resource management, including bearer control, wireless access control, mobility management when connected, dynamic resource allocation, etc.;
(2) IP header compression, encryption and integrity protection of user data;
(3) Selection of Access and Mobility management Function (AMF);
(4) User plane data routing to User Plane Function (UPF);
(5) Control plane information routing to AMF;
(6) Connection establishment and release;
(7) Scheduling and transmission of paging messages;
(8) Scheduling and transmission of broadcast information;
(9) Measurement configuration and measurement report configuration;
(10) Uplink transport layer data packet identification;
(11) Session management;
(12) Support for network slicing;
(13) Quality of Service (QoS) stream management and mapping;
(14) Support for terminal non-activated state;
(15) Non-Access Stratum (NAS) message distribution function;
(16) Wireless access network sharing;
(17) Dual connectivity;
(18) Interoperability of NR and Evolved UMTS Terrestrial Radio Access (E-UTRA).

The core network functions are mainly allocated in three nodes, including the Mobility Management Function (AMF), User Plane Function (UPF) and Session Management Function (SMF). Among them, the main functions of AMF include:

(1) NAS signalling processing;
(2) NAS security;
(3) Access Stratum (AS) security control;
(4) Inter-3GPP access network cross-core network mobility management;
(5) Terminal reachability in idle state;
(6) Registration area management;
(7) Intra-system and inter-system handover;
(8) Access authentication;
(9) Access authorisation;
(10) Idle state mobility management;
(11) Support for network slicing;
(12) SMF selection.

UPF mainly includes:

(1) Mobility anchor within Radio Access Technology (RAT) and between RATs;
(2) Protocol Data Unit (PDU) session node connected to the data network;
(3) Packet routing and distribution;

(4) Packet detection and user plane policy enforcement;
(5) Traffic usage reporting;
(6) Uplink classifier for service flow to data network routing;
(7) Branch point for multi-homed PDU sessions;
(8) User plane QoS processing;
(9) Uplink service verification (SDF to QoS flow mapping);
(10) Downlink packet buffering and downlink data notification triggering.

SMF mainly includes:

(1) Session management;
(2) Terminal IP address allocation;
(3) User plane function selection and control;
(4) Service steering configuration;
(5) Policy enforcement and QoS control part;
(6) Downlink data notification.

2.1.2 Air Interface Design

In the entire 3GPP standardization, the design and selection of 5G parameters are basically based on the design of 4G. Necessary modifications are introduced to meet new requirements or solve problems existing in the 4G standard. Therefore, 5G inherits many mature designs of 4G and only introduces necessary corrections and changes. The main parameter comparison between 4 and 5G is shown in Table 2.1.

The abbreviations in Table 2.1 are explained as follows:

OFDM, Orthogonal Frequency Division Multiplexing; SC-FDMA, Single-Carrier Frequency-Division Multiple Access; QAM, Quadrature Amplitude Modulation; BPSK, Binary Phase Shift Keying; LDPC, Low Density Parity Check; TTI, Transmission Time Interval; CRS, Cell Reference Signal; DMRS, is Demodulation Reference Signal; PDSCH, Physical Downlink Shared Channel; ACK, Acknowledgement, that is, confirmation or response; NACK, Negative Acknowledgement, that is, negative confirmation or negative response; Tx, that is, Transmit; Rx, Receive, that is, receive.

1. Changes in 5G NR compared to LTE air interface

Compared with the LTE air interface, the changes in the design of the 5G NR air interface are mainly reflected in the following aspects [3].

(1) Higher data rate

5G faces requirements such as high data rate and high capacity, and the spectral efficiency needs to be increased to 3–5 times that of 4G to meet user needs. Rel-15 enhances the peak data rate and capacity of 5G through technologies such as large bandwidth (FR1 100 MHz, FR2 400 MHz), large-scale antennas, and multi-user multiple-input multiple-output (MU-MIMO) enhancement, cancellation of CRS reference signals, channel status information feedback design, Polar/LDPC coding,

Table 2.1 Comparison of main parameters between 4 and 5G

Item	4G	5G
Wave form	Downlink: OFDM Uplink: SC-FDMA	Downlink: OFDM Uplink: SC-FDMA or OFDM
Modulation	Downlink: 256 QAM Uplink: 64 QAM	Uplink/downlink: 256 QAM as basis New π/2 BPSK on the uplink
Channel coding	Data channel: Turbo code Control channel: tail biting convolutional code	Data channel: LDPC Control channel: Polar code
Sub carrier spacing and TTI length	– 15 kHz subcarrier spacing – 1 ms TTI	– 15/30/60/120 kHz, – Shorter TTI, e.g. 0.5 ms TTI for 30 kHz
Frame structure	Fixed	Quasi static and dynamic
Carrier bandwidth /MHz	1.4/5/10/15/20	1.4/5/10/15/20/40/50/80/100
Reference signal	CRS	Cancel CRS, based on DMRS
DMRS resource	Fixed	Share with PDSCH
PBCH/SS	Broadcast beam (wide beam)	Narrow beam based on beam scanning
PDCCH	Specific weight based on CRS	DMRS based on UE specific weight
PDCCHresource	Symbol level	Symbol and frequency level
ACK/NACK delay	Minimum N + 4	Minimum N + 0
UE maximum power	Maximum 23 dBm	Maximum 26 dBm
UE capacity	1Tx/2Rx as basis, maximum 20 MHz/carrier	2Tx/4Rxas basis, maximum 100 MHz/carrier

etc. The peak data rate of 100 MHz bandwidth in the sub-6 GHz can reach 1.7 Gbps, and the peak data rate of a single user in the millimetre wave band can reach 10 Gbps.

(2) Larger bandwidth

The LTE system only supports a 20 MHz bandwidth. To meet higher data rates and system capacity scenarios, larger bandwidth can only be achieved through carrier aggregation technology. The 5G NR standard considered the need for larger bandwidth at the beginning of the design, and the data channel below 6 GHz can support a maximum of 100 MHz single carrier bandwidth. For this reason, 5G NR introduces flexible system parameter settings to support large bandwidth operations, such as below 6 GHz, 5G NR can support 15, 30 and 60 kHz subcarrier space, and the network side and terminal side bandwidth can be unequal, that is, the base station configures large bandwidth but the terminal configures small bandwidth, and the network side can flexibly configure the terminal to work in a small bandwidth mode according to needs.

(3) Larger frequency range

5G NR introduces flexible system parameter settings to support a flexible available frequency range. For example, below 6 GHz, 5G NR can support 15, 30 and 60 kHz subcarrier space, which can support the use of available frequencies from 400 MHz to 6 GHz for 5G; and above 6 GHz, it can support 120, 240 kHz and other subcarrier space, thus supporting the frequency range of 26–28 GHz, 39 GHz and even 60 GHz. At present, 3GPP has also begun to study support for frequencies above 60 GHz.

(4) Lower latency

Compared with LTE, 5G NR proposes lower latency requirements. The 5G NR control plane latency defined by 3GPP is reduced to 10 ms; the user plane latency (one-way air interface latency) is even reduced to 0.5 ms to meet the business needs of scenarios such as uRLLC. The Rel-15 standard meets low latency requirements through technologies such as flexible frame structure, short time domain scheduling unit, unscheduled transmission, mobile edge computing (end-to-end latency can be reduced to 10 ms). For the user plane, in the frame structure design of 5G, in addition to reducing the slot and OFDM symbol length through a larger subcarrier space, it also supports symbol-level data scheduling in the time domain, and introduces more uplink and downlink conversion points through a two-level design of semi-static and dynamic indication, reducing user plane latency; in the scheduling process, 5G NR has designed an uplink unscheduled transmission mode, and flexible configurable transmission can further reduce user plane latency. For the control plane, a third inactive state is introduced on the basis of the idle state and the connected state. The inactive state is an intermediate state between the idle state and the connected state, which not only saves terminal power consumption, but also quickly switches the control plane state. In addition, it can also be achieved through mobile edge computing (Mobile Edge Computing, MEC; also known as Multi-Access Edge).The method of edge computing is used to achieve local breakout and local content caching, which can further reduce the end-to-end delay of the service.

(5) Higher spectral efficiency

The spectral efficiency requirement for 5G NR is 30 bps/Hz for downlink peak spectral efficiency and 15 bps/Hz for uplink spectral efficiency. To this end, 5G NR needs to adopt a variety of technical means such as larger scale antennas, new waveforms, new coding, and reducing overhead to improve system spectral efficiency. In terms of large-scale antennas, 5G NR can support up to 32 port Channel State Information-Reference Signal (CSI-RS) designs, support a maximum of 12 stream orthogonal multi-user transmission, etc., to improve multi-user multiplexing capabilities and average spectral efficiency; in terms of new waveforms, it can reduce subcarrier interference, increase the bandwidth utilization ratio of the NR system to over 90%, and even reach a bandwidth utilization ratio of 98%; in terms of new coding, the data channel uses LDPC coding, and the control channel uses Polar codes, effectively improving coding performance; in terms of reducing overhead, 5G NR no longer supports CRS transmission for full bandwidth and every subframe period, but instead supports on-demand transmission of reference signals, reducing interference and overhead, and further improving spectral efficiency.

(6) Higher reliability

For 5G uRLLC scenarios (such as vehicle networking applications), the demand for high reliability is increasing. 5G can support 99.999% of high-reliability applications with packet sizes less than 32 Bytes by increasing coding redundancy, improving scheduling priority, reducing coding ratio, and multiple transmissions.

(7) Higher mobility

For specific scenarios such as high-speed railway, it suppresses the effects of fast channel changes, large frequency offsets, and frequent switching in 500 km/h high-speed scenarios. Rel-15 ensures high-speed mobility performance through reference signal design, random access process design, system parameter optimization, and other technologies.

(8) Wider coverage

Considering that the new frequency bands for 5G deployment are generally higher than 4G frequencies, there are certain difficulties in achieving continuous coverage based on the existing 4G site construction. Rel-15 has extended the coverage capability of the 5G network through large-scale antenna design, broadcast channel beam swapping, control channel coverage enhancement, high-power terminals, and other technologies.

2. 5G NR's optimization of LTE

5G NR has also further optimized the existing problems of LTE [3], which will be introduced in detail below.

(1) More flexible resource utilization

The LTE Time-Division Duplex (TDD) system has 7 types of uplink and downlink slot configurations, and each uplink and downlink slot corresponding special subframe configuration is also relatively limited. In the current network deployment, for the interference problem of the LTE TDD remote base station, it can only adjust the length of the Guard Period (GP) through the special subframe configuration, and the adjustable range of the GP is also limited, and it cannot flexibly and effectively solve the remote base station interference problem. In addition, LTE can only reserve part of the resources through the Multicast Broadcast Single Frequency Network (MBSFN) subframe configuration, and cannot flexibly reserve more resources for forward compatibility.

The frame structure design of 5G must focus on meeting the future different application scenarios and business needs, support flexible system parameters and frame structure configuration for multiple scenarios, have flexible forward-compatible resource configuration capabilities, to overcome the shortcomings of the above-mentioned LTE system frame structure.

(2) Lower network interference

In the LTE system, the cell reference signal (CRS) is always sent in each subframe (Always on) and occupies the full bandwidth. The CRS resource positions between adjacent cells are offset by the cell ID number modulo 6. Therefore, in the scenario of continuous LTE network coverage, the transmission of CRS can cause interference between cells, especially serious interference to users at the edge of the cell. At the same time, CRS occupies nearly 14% of the overhead of the LTE system, which may cause waste of transmission resources.

Therefore, in the 5G NR system, the transmission of CRS is cancelled, and the functions of CRS are replaced by reference signals such as CSI-RS, DMRS, and Timing Reference Signal (TRS), which can be flexibly configured in bandwidth and time. This avoids the transmission mechanism where the reference signal is sent in every subframe, realises the on-demand transmission of the reference signal, effectively reduces the interference between cells, and improves spectral efficiency.

(3) Better coverage matching of service channels and control channels

The construction of 4G TD-LTE in China is developing rapidly. By the end of 2018, the number of 4G TD-LTE base stations of China Mobile has reached 2.41 million. The 4G network has a large scale of deployed sites, and it is difficult to add new sites; while the main frequency band for 5G commercial deployment in the early stage is the mid-frequency band (such as 2.6, 3.5, 4.9 GHz, etc.), which is generally higher than the frequency band used by LTE, and the propagation loss and penetration loss of the signal are more serious. In order to ensure that the 5G commercial deployment can achieve the same coverage as the 4G network as much as possible by using the existing 4G sites, the coverage enhancement technology for the 5G NR control channel and service channel is an important content of the 5G NR design.

3. Enhancement of 5G NR Design

In the design of 5G NR, the system coverage has been enhanced in the following aspects [3].

(1) Beam swapping of broadcast channels and public control channels

In the design of the large-scale antenna system of 5G NR, a beam swapping mechanism is introduced for the transmission of broadcast channels and public control channels, using beamforming gain to counteract the propagation loss brought by the high frequency band, thereby compensating for the problem of mismatched coverage of control and service channels in LTE where only user-specific service channels can obtain beamforming gain.

(2) Higher aggregation level of control channel

Compared with LTE, 5G NR provides options for higher aggregation levels in the configuration of the public control channel search space, such as the 16 Control Channel Element (CCE) aggregation level, further ensuring the coverage range of the public control channel.

(3) Longer random access channel and uplink control channel

To enhance uplink coverage, 5G NR supports multiple repeated random access sequence formats and long format uplink control channels in the design of random access channels and uplink control channels, achieving coverage enhancement effects through time-energy accumulation.

(4) More flexible network function deployment

With the rise of new businesses such as Augmented Reality (AR)/Virtual Reality (VR), high-definition video, autonomous driving, smart factories, etc., mobile networks are facing new challenges such as real-time computing power, ultra-low latency, and ultra-large bandwidth. The existing 4G network structure can no longer meet the technical requirements of these new businesses. MEC introduced by 5G is one of the most critical technologies to meet these challenges. MEC is a new distributed computing method based on mobile communication networks, building a cloud service environment on the Radio Access Network (RAN) side, saving costs, reducing latency and round-trip time (Round Trip Time, RTT), optimising traffic, enhancing physical security and cache efficiency, etc. The value of 5G MEC is not only a new network structure and deployment method, but more importantly, it supports the underlying openness of the mobile network, thereby promoting the deep integration of mobile communication networks, the Internet and the Internet of Things. It is a technical and commercial practice under the transformation demands of operators.

4. Improvements to 5G NR high-level protocols

In terms of high-level protocols, although 5G NR is based on 4G LTE, there are significant improvements in the following areas.

(1) System messages

A method of sending system messages on demand has been introduced. Necessary system information is always broadcast periodically, while for other system information that is not necessary for all terminals, the network may only send it when the terminal has a request. In addition, a method of receiving system information based on geographical areas has been proposed from the perspective of terminal reception. The content of a system message broadcast within a specific area is the same. When the terminal moves within or re-enters this specific area, if the value tag corresponding to the relevant system information has not changed, the terminal can consider the previously stored system information to still be valid, and there is no need to read this system information again.

(2) Random access

Due to the introduction of beam management technology, random access resources (such as random access preamble codes and random access time–frequency resources) can be bound to different beams, so that the base station can select the downlink beam to send subsequent Msg2 and Msg4. In addition, due to the introduction of Supplementary Uplink (SUL), the terminal can choose one of the two uplink transmission carriers as random access based on the measurement results of the 5G downlink carrier. When the base station replies to the random access response, it also needs to consider responding separately to the two groups of terminals.

(3) Mobility management

Beam-related enhancement mechanisms have been introduced, and beam-related content has been added to messages such as measurement configuration, cell quality calculation, measurement reporting, handover request, and handover command.

(4) Terminal state transition

Due to the long delay in LTE state transition, it cannot meet the 10 ms control plane delay requirement of 5G. 5G proposes a new terminal state—inactive state. When the terminal is in the inactive state, it disconnects the wireless connection but maintains the core network connection, thereby ensuring terminal power saving while achieving fast state transition with low delay and low signalling overhead. Due to the introduction of the inactive state, terminal downlink data is sent to the base station, which triggers the paging mechanism and pages the terminal in the RAN paging area.

(5) Dual connection

5G supports dual connections between different systems with LTE, and adds different signalling plane and user plane bearers. This can effectively utilise the wide coverage of the existing 4G network, improve the reliability of signalling, and also effectively utilise the high speed, low latency characteristics of the 5G network to meet the specific needs of different technical scenarios in 5G.

(6) Business QoS guarantee

Unlike LTE's QoS management based on Service Data Flow (SDF), 5G provides a QoS management mechanism based on QoS flow, which includes two mapping mechanisms. The Non-Access Stratum (NAS) is responsible for the mapping from SDF to QoS flow, and the Access Stratum (AS) is responsible for the mapping from QoS flow to DRB. The two mappings are independent of each other. The NG-RAN can flexibly decide on the air interface treatment of the QoS flow based on the QoS profile provided by the 5G network's NAS layer.

For detailed 5G NR protocols and standard specifications, please refer to [5, 6]. Rel-15 of 5G NR defines the basic functions required for 5G eMBB and uRLLC scenarios, while the network functions for low-power, large-connection IoT scenarios are based on LTE-based Narrowband IoT (Narrow Band Internet of Things (NB-IoT) is provided, so 3GPP submits 5G NR, LTE evolution and NB-IoT as a set of 5G technologies to ITU-R, representing 3GPP's 5G technology.

2.1.3 Core Network Service and Network Slicing [4]

The traditional 4G network is a rigidly structured network with all functions included. However, for differentiated enterprise-level and vertical industry applications, the requirements for network functions vary greatly. Using this traditional, all-encompassing network construction method will inevitably lead to a huge waste of resources, and due to the rigid network structure, it is impossible to optimise latency and routing topology as needed, making it difficult to meet personalised business expansion needs.

Based on the Software Defined Network (SDN) and Network Function Virtualization (NFV) platform [7, 8], the new 5G core network is based on a Service Based Architecture (SBA), which can divide network functions into several reusable "services", with "services" communicating using lightweight interfaces. The goal of the service architecture is to achieve efficiency, softwareisation, and openness of the 5G system. The core network can be hosted on telecom cloud infrastructure, achieving IT-based operations, bringing network elasticity and agility, shortening the time to market for services and the cycle of new feature iterations. Therefore, core network equipment is usually deployed regionally in the operator's telecom cloud data centre.

Unlike the traditional 4G network's "one pipe, best effort" working mode, the 5G network introduces the concept of slicing, aiming to provide different, customised end-to-end "logical dedicated networks" based on a unified network infrastructure, optimally adapting to different business needs of different industry users. Combining the unique technical features of network slicing, "one network infrastructure, multiple logical dedicated networks", 5G can well match the core demands of industry customers for communication network services being available, secure and reliable, manageable and controllable, thereby achieving an effective balance between industry network construction costs and service experience guarantees. "No slicing, no 2B", network slicing has become one of the iconic technologies distinguishing 5G from 4G.

The 5G network supports end-to-end slicing capabilities, capability deployment as needed, etc., through disruptive designs such as function decoupling modular design, control and bearing separation, function inter-call in the form of services, and underlying cloudification, achieving network customisation, openness, and service orientation. The service-oriented architecture makes the deployment of services and functions very flexible, and the core network based on the SDN/NFV platform allows network functions to be flexibly deployed as needed, and its capacity can change elastically.

The network slicing schematic is shown in Fig. 2.6. For the 5G network, according to the deployment requirements and business requirements in different scenarios, relevant network functions can be selectively deployed, and the deployment location of network functions can be flexibly selected to best adapt to business and customer needs, while achieving the highest cost-effecincy of network investment; or based on the already deployed network, select some functions and network facilities to form a logical virtual sub-network to serve target customers, meeting personalised and

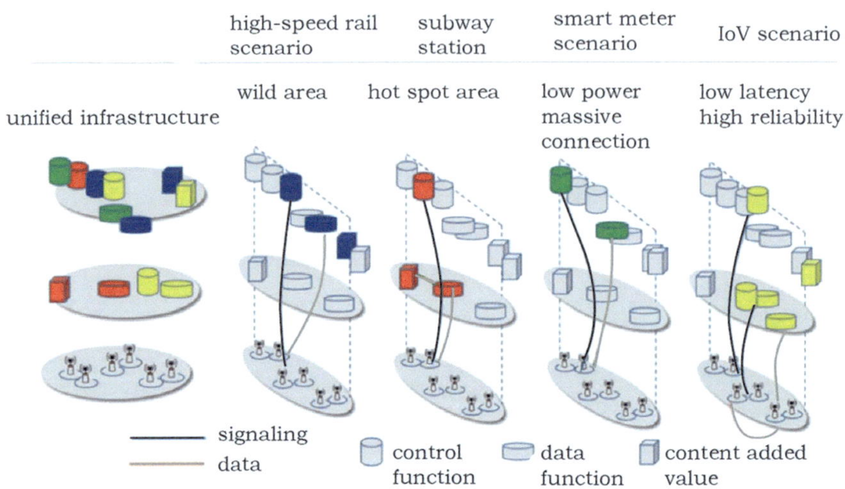

Fig. 2.6 Network slicing schematic [3, 5]

differentiated data privacy and security isolation needs. From the function selection in different scenarios in Fig. 2.6, it can be seen that the functions that need to be deployed and configured in different scenarios vary according to different requirements, and it is not necessary to deploy the entire network's function set while optimising performance, thus achieving differentiated service guarantees and saving network investment. At the same time, in the case of overlapping of multiple different application scenarios in the same physical area, the network infrastructure can also achieve dynamic sharing, meeting the needs of different businesses through dynamic generation and on-demand orchestration and deployment of slices, avoiding waste of hardware resources.

Therefore, network slicing is an effective tool for expanding differentiated and fragmented vertical industry application markets in 5G, providing agile personalised and differentiated service guarantees.

2.1.4 Mobile Edge Computing

With the continuous development of artificial intelligence and big data technology, computing power has become a very important capability in vertical industry applications. Therefore, traditional single communication connection capabilities can no longer meet the needs of many vertical industry applications. Mobile Edge Computing (MEC) [9] was born. MEC introduces functions such as computing, storage, and forwarding into the edge of the mobile network (which can be a separate network element or co-located with a wireless base station), realising intelligent user-side functions. The principle of MEC is shown in Fig. 2.7.

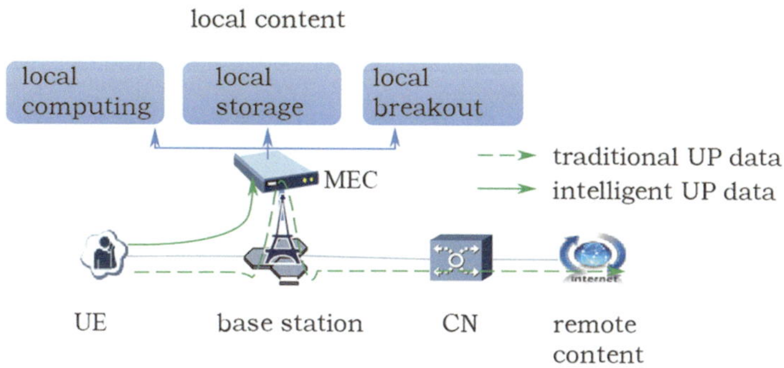

Fig. 2.7 Principle of MEC

MEC can bring the following benefits to the network:

(1) Deploy business and content as close to the user as possible to minimise the delay in accessing the service.
(2) Lower the routing function to a position as close to the user as possible, realise fast routing and local exchange of user data, shorten the delay in data interaction, and avoid sensitive data on the Internet as much as possible to protect data security and privacy.
(3) Deploy computing power close to the user, thereby transferring the user's computing to the cloud, simplifying the terminal implementation while quickly interacting with data and processing results, reducing its size, weight, power consumption, and cost. For example, for AR/VR applications, if the content processing and rendering functions are moved to MEC, it can greatly reduce the threshold for AR/VR device development, while also greatly reducing costs and weight, making the device lighter and easier to popularise.
(4) Lower the UPF of the core network to MEC, support necessary billing, security and other functions, can provide a high degree of isolation of user data, and protect the privacy of user data.
(5) Through the standard application programming interface (API), the capabilities of the wireless network can be opened up, such as location positioning, etc., to open up network capabilities to third parties, and thereby cultivate new businesses and new business models.

Edge computing can be combined with application requirements to realise the sinking of edge computing functions, flexible deployment locations, such as deployment to parks (or even base stations), cities and provincial machine rooms, forming multi-level deployment. The higher the deployment location of the edge computing device, the wider the user surface coverage, and the cost per user will also drop significantly. Edge computing devices deployed to parks (or base stations) can meet the extremely low latency requirements and data security requirements of park manufacturing, provide industrial cloud services for park manufacturing, and provide basic conditions for flexible production; edge computing devices deployed to cities can meet AR/VR business, park industrial manufacturing production platform (multi-factory interconnection requirements) and Internet of Things (IoT) data edge preprocessing requirements; edge computing devices deployed to provincial key convergence machine rooms can meet the needs of content distribution networks (CDN), cloud storage, smart city brains and other wide-area application scenarios.

2.2 Standard Evolution

After the release of the 5G Release 15 (Rel-15) standard, 3GPP did not stop. The 5G standard is constantly evolving and enhancing, as shown in Fig. 2.8.

During the formulation of Rel-15, it strives to produce a "usable" standard at the fastest speed to meet the needs of large-scale commercial deployment in 2020. On the evening of July 3, 2020, Beijing time, 3GPP Announcement of Release 16 (Rel-16) [11] standard freeze, marking the completion of the first evolution version of the 5G standard. Rel-15 is the basic version of the 5G standard, mainly focusing on the definition of the most basic functions of eMBB and uRLLC for early 5G applications; Rel-16 enhances and supplements it, making 5G NR from "usable" to "easy to use", focusing on "new capability expansion", "potential digging of existing capabilities" and "operation and maintenance cost reduction and efficiency increase", further improving and enhancing the 5G system to better serve industry applications.

Rel-16 is not only a performance enhancement based on Rel-15, but also focuses on meeting the functional requirements of multiple industry sectors, simplifying and optimising the deployment of 5G networks, and expanding the 5G ecosystem. The functions of Rel-16 in this regard mainly include uRLLC enhancement, support for vertical industries and Local Area Network (LAN) services, enhanced support for

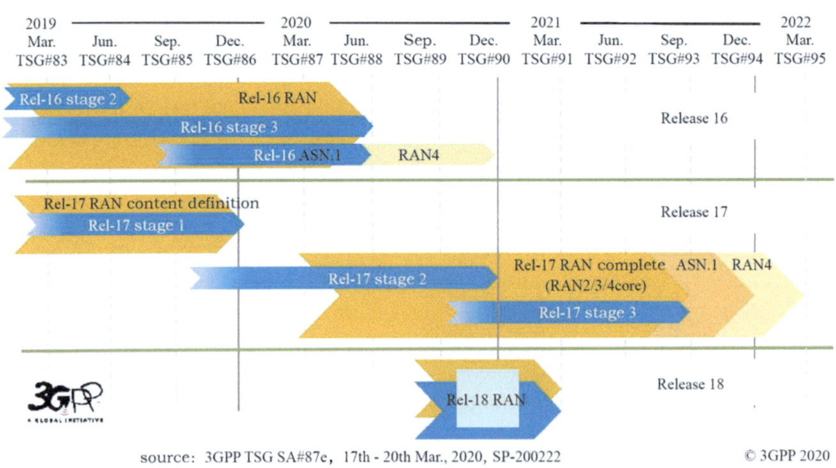

Fig. 2.8 5G standard evolution [10]

Vehicle to Everything (V2X) technology, 5G positioning and positioning services, 5G satellite access, enhanced network slicing, unauthorised frequency band access, etc. For industrial internet applications, Rel-16 supplements uRLLC and mMTC capabilities, focusing on function expansion and efficiency improvement, further enhancing the ability of 5G to serve industry applications. In Rel-15, the basic KPIs of the 5G network have been clarified, such as transmission rate, delay characteristics requirements, etc. Rel-16 introduces new technologies to support 1 μs synchronisation accuracy, 0.5–1 ms air interface delay, 99.9999% reliability and flexible terminal group management, and can achieve end-to-end delay within 5 ms and higher reliability, supporting industrial-grade time-sensitive businesses. In addition, Rel-16 introduces a variety of 5G air interface positioning technologies, with positioning accuracy improved by more than 10 times, reaching the metre level. For vehicle networking applications, it supports direct communication between vehicles (V2V) and between vehicles and roadside units (V2I), introducing various communication methods such as multicast and broadcast, as well as optimising perception, scheduling, retransmission and vehicle-to-vehicle connection quality control technologies, to realise V2X support for more rich vehicle networking application scenarios such as vehicle platooning, semi-automatic driving, extended sensors, remote driving, etc.

Overall, the technological evolution of Rel-16 helps to further improve the quality of 5G network services, enhance user experience, expand the scale of the 5G industry, and better empower the transformation and upgrade of all industries. The capability enhancements brought by Rel-16 related to vertical industry empowerment include the following:

(1) Achieve metre-level positioning (outdoor horizontal positioning accuracy is 10 m, indoor horizontal positioning accuracy is 3 m), and combine with satellite, Bluetooth, sensor and other technologies to further achieve sub-metre level positioning.
(2) By saving energy at the terminal, extending battery working time, it can save about 35% of air interface energy consumption and about 11% of overall machine energy consumption under medium business load.
(3) In terms of mobility enhancement, the "0 ms" scheme can avoid user plane data interruption during switching, ensuring consistent rate experience for UE during the switching process, while the robustness enhancement scheme based on conditional switching can improve the robustness of the control plane and increase the success rate of switching.
(4) uRLLC enhancement, while meeting the low air interface delay requirement (such as 0.5–1 ms), increases support for higher data rates, raising end-to-end reliability from 99.999 to 99.9999%.
(5) 5G V2X for vehicle networking, meeting business requirements of 3–10 ms end-to-end delay, 99.999% reliability, 10–1000 Mbps high speed, etc., achieving complementary coexistence of 5G V2X and LTE V2X vehicle networking.

3GPP launched 5G Release 17 (Rel-17) [12] in March 2020. The standardisation work was completed in June 2022 with the freezing of the Rel-17 version. Rel-17 further focuses on scenario demand expansion and network capability enhancement based on existing versions. For scenario demand expansion, lightweight terminal technology (RedCap) was introduced, targeting mid-to-high-end mMTC such as industrial wireless sensors, video surveillance, wearable devices, etc., further optimising NR design; based on Rel-16 non-terrestrial network (NTN) research, the standardisation work required for NR to support satellite communication was continued; multicast and broadcast services were introduced, supporting data reception of multicast/broadcast services under any wireless resource control (Radio Resource Control, RRC) state. For network capability enhancement, it includes NR coverage enhancement, studying coverage enhancement methods for bottleneck channels; multi-SIM (Subscriber Identity Module, user identification module) card enhancement, studying enhancement schemes when using multiple SIM cards under UE hardware constraints; small packet transmission enhancement, optimising design for small packet and non-activated packet transmission; 52.6 GHz and above NR, designing new parameters for supporting NR at higher frequencies.

In addition, Rel-17 also includes further enhancements to Rel-16 technology, such as multiple-input multiple-output (Multiple-Input Multiple-Output, MIMO) enhancement, industrial Internet of Things (IIoT) and uRLLC enhancement, integrated access and backhaul (Integrated Access and Backhaul, IAB) enhancement, self-organised network (Self-Organized Network, SON)/minimisation of drive test (Minimization of Drive Test, MDT) enhancement, non-public network (Non-Public Network, NPN) enhancement and super uplink enhancement, etc.

After three versions of evolution, the 5G NR standard has become increasingly mature. Looking to the future, 6G will have commercial capabilities around 2030. To drive a new wave of network development, 3GPP, from the perspective of market development, has defined the 5G evolution standard after Release 18 (Rel-18) as 5G-Advanced [13], and the evolution of the 5G standard is about to enter a new era. Rel-18 was initiated at the end of 2021 and is expected to be frozen at the end of 2023.

In terms of network architecture [14], the 5G-Advanced network will develop along the concepts of cloud-native, edge network and network as a service, meeting the demands for rapid deployment and on-demand iteration of network functions. In terms of network technology, the capabilities of the 5G-Advanced network will continue to be enhanced in the three aspects of "intelligence, integration and enablement". Among them, "intelligence" will focus on improving the level of network intelligence, reducing operation and maintenance costs, further promoting the application and integration of intelligent technology in telecommunications networks, conducting research on distributed intelligent architecture, and terminal and network collaborative intelligence. "Integration" will promote the integration and coordinated development of 5G networks with industry networks, home networks and integrated networks of heaven and earth. "Enablement" will continue to empower 5G networks to serve vertical industries, while improving the basic network slicing and edge computing iconic capabilities, it will support interactive communication, broadcast communication and other "more diverse" network services; in terms of end-to-end quality measurement and assurance, and scheme simplification, it will make network quality "more certain"; in terms of time synchronisation, location services, etc., it will make network capabilities "more open".

Looking forward to Rel-18, whether it is based on further enhancements of traditional technology or research and exploration in new fields and directions, it will inject fresh vitality into the evolution of 5G-Advanced, and also lay a solid foundation for the future 6G era.

2.3 5G Frequency

Frequency is the foundation of mobile communication system design and deployment, and it plays a crucial role in the development of 5G. From the historical experience of the development of mobile communications, globally unified division and planning of mobile communication frequencies help the mobile communication industry to share global industry scale, reduce equipment costs, simplify terminal implementation, and facilitate roaming between global operators.

According to ITU's forecast, the spectrum demand estimation results for different 5G scenarios are shown in Table 2.2. The low and medium frequency bands below 6GHz are mainly used to meet the continuous coverage needs of 5G macro cells, with a spectrum demand of 802–1090 MHz. The high frequency band is mainly used to meet the hotspot capacity needs of 5G indoors and outdoors.

The joint networking of low, medium, and high frequency bands is the basic spectrum strategy for future 5G wide coverage and high capacity. Among them, the low frequency band, due to its good spatial propagation characteristics, can be used to solve coverage problems. Currently, China has fewer low frequency bands below 1 GHz, which cannot meet the demand for large bandwidth and high capacity of 5G; but the low frequency band has good coverage capability, which can effectively ensure the requirements for transmission reliability. The medium frequency band below 6 GHz has good spatial propagation characteristics and larger bandwidth. If combined with the low frequency band, it can effectively balance coverage and capacity needs. Compared with low and medium frequency bands, the millimetre wave of the 5G high frequency band (> 24.25 GHz) is easier to obtain continuous ultra-large bandwidth, which is an important basis for achieving high transmission rates.

Through the coordination and three-dimensional coverage of low, medium, and high frequency bands, the 5G system can ensure a high-quality user experience in

Table 2.2 Spectrum demand estimation results

Deployment scenario	Macro cell	Outdoor hotspot	Indoor hotspot
Below 6 GHz	802–1090 MHz	–	–
24.25–43.5 GHz	–	5.3–7.58 GHz	5.3–7.58 GHz
43.5–86 GHz	–	–	9.7–12.42 GHz

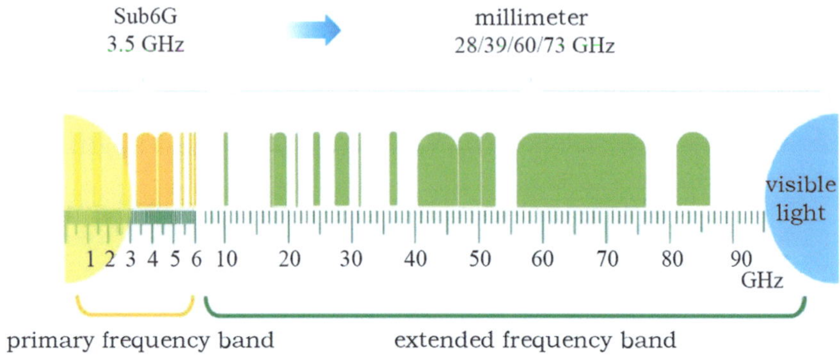

Fig. 2.9 Distribution and planning of 5G frequencies

various scenarios. In the early stage of 5G development, low and medium frequency bands (urban areas) are used for continuous coverage, ensuring deep coverage and high mobility; the medium frequency band is used as a wide capacity expansion layer, ensuring a higher network speed experience; the high frequency band is used as a solution for hotspot areas and indoor ultra-high speed and ultra-high capacity, ensuring extreme performance in single point areas. The distribution and usage planning of 5G frequencies is shown in Fig. 2.9.

In the middle and late stages of 5G development, the low frequency band is used to optimise continuous coverage, ensuring broad and deep coverage; the medium frequency band is used to enhance basic coverage, improving the average performance of the entire network; the high frequency band is used to enhance the capacity and speed of hotspot areas and indoors, ensuring extreme performance under dense networking.

According to the "5G Vision and Demand White Paper" released by the IMT-2020 Promotion Group in 2014 [1], it is estimated that China's mobile data traffic will increase by more than 300 times from 2010 to 2020, and by more than 40,000 times from 2010 to 2030. The growth rate of mobile data traffic in developed cities and hotspot areas is faster, with Shanghai's growth rate reaching 600 times from 2010 to 2020, and the growth rate in Beijing's hotspot areas reaching 1000 times. This poses a greater challenge to the demand for future spectrum resources in our country. At the end of 2018, the Ministry of Industry and Information Technology issued a 5G frequency allocation plan: China Telecom and China Unicom respectively obtained 100 MHz frequencies of 3.4–3.5 GHz and 3.5–3.6 GHz, China Mobile obtained non-continuous 100 MHz frequencies of 2.515–2.575 GHz and 2.635–2.675 GHz, as well as 4.8–4.9 GHz frequencies; China Broadcasting, China Telecom and China Unicom respectively obtained 100 MHz continuous frequencies of 3.4–3.6 GHz; China Broadcasting obtained a mobile license, its 30 MHz $\times$ 2 FDD frequency of 700 MHz (uplink 703–733 MHz, downlink 758–788 MHz) can be used for 5G, and it also obtained 4900–4960 MHz TDD spectrum. Our country's frequency allocation plan fully ensures that each operator has 100 The continuous bandwidth of MHz

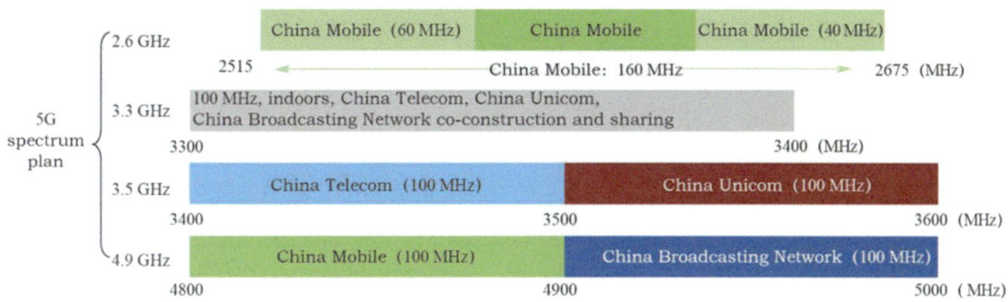

Fig. 2.10 China's new 5G frequency allocation scheme

(China Mobile's spectrum migration through the 4G network) lays the foundation for the construction of a high-quality and leading 5G network. In combination with the co-construction and sharing of China Telecom and China Unicom, as well as the co-construction and sharing of China Mobile and China Broadcasting, a good compromise has been achieved between the cost of 5G deployment and the speed of development, laying a solid foundation for the subsequent global leadership of China's 5G network development. China's new 5G frequency allocation scheme is shown in Fig. 2.10.

The 3GPP standard divides the available 5G frequencies into Frequency Range 1 (FR1) and Frequency Range 2 (FR2), where FR1 includes 450–7125 MHz; FR2 includes the millimetre wave band of 24.25–52.6 GHz. The 3GPP standard distinguishes between frequency bands, where for LTE re-farming bands, the NR band number is prefixed with n before the LTE band number, such as the band 41 used by domestic TD-LTE, which is changed to n41 in NR, but the specific working frequency range remains unchanged. The FR1 band definition in 3GPP is shown in Table 2.3, and the FR2 band definition is shown in Table 2.4.

2.4 5G Commercial Network Deployment

"Information at your will, everything in touch" paints a beautiful vision for the future after 2020, and also predicts that 5G will bring huge changes to future society. Around a series of national strategies such as network power construction, "Internet+", and manufacturing power construction, our country has also designed a grand blueprint for the development of 5G. 4G has brought unprecedented prosperity to mobile internet applications, ushering in the dawn of digital economic development. We hope even more that 5G can bring similar prosperity to all sectors of society, enabling thousands of industries and promoting the digital transformation and upgrade of the entire society. Not only China, but other major economies around the world also have high hopes for the development of 5G, hoping to seize the initiative in the new round of economic development with the help of 5G. HIS Markit predicts the contribution of 5G to driving output growth in various industries by 2035, as shown in Fig. 2.11, where key business services refer to applications that 5G will support with high

Table 2.3 3GPP FR1 band definition [15]

NR operating frequency band	Uplink operating frequency band/MHz FUL_low to FUL_high	Downlink operating frequency band/MHz FDL_low to FDL_high	Duplex mode
n1	1920–1980	2110–2170	FDD
n2	1850–1910	1930–1990	FDD
n3	1710–1785	1805–1880	FDD
n5	824–849	869–894	FDD
n7	2500–2570	2620–2690	FDD
n8	880–915	925–960	FDD
n12	699–716	729–746	FDD
n20	832–862	791–821	FDD
n25	1850–1915	1930–1995	FDD
n28	703–748	758–803	FDD
n34	2010–2025	2010–2025	TDD
n38	2570–2620	2570–2620	TDD
n39	1880–1920	1880–1920	TDD
n40	2300–2400	2300–2400	TDD
n41	2496–2690	2496–2690	TDD
n51	1427–1432	1427–1432	TDD
n66	1710–1780	2110–2200	FDD
n70	1695–1710	1995–2020	FDD
n71	663–698	617–652	FDD
n75	N/A	1432–1517	SDL
n76	N/A	1427–1432	SDL
n77	3300–4200	3300–4200	TDD
n78	3300–3800	3300–3800	TDD
n79	4400–5000	4400–5000	TDD
n80	1710–1785	N/A	SUL
n81	880–915	N/A	SUL
n82	832–862	N/A	SUL
n83	703–748	N/A	SUL
n84	1920–1980	N/A	SUL
n86	1710–1780	N/A	SUL

reliability, ultra-low latency connections, and high security and availability. It can be seen that the emergence of 5G will have a huge impact on all industries. According to HIS Markit's forecast, by 2035, 5G will drive output growth in various industries worldwide to reach $12.3 trillion, accounting for about 4.6% of the world's total actual output in 2035.

Table 2.4 3GPP FR2 band definition [16]

Operating frequency band	Uplink operating frequency band/MHz	Downlink operating frequency band/MHz	Duplex mode
	FUL_low to FUL_high	FDL_low to FDL_high	
n257	26,500–29,500	26,500–29,500	TDD
n258	24,250–27,500	24,250–27,500	TDD
n260	37,000–40,000	37,000–40,000	TDD
n261	27,500–28,350	27,500–28,350	TDD

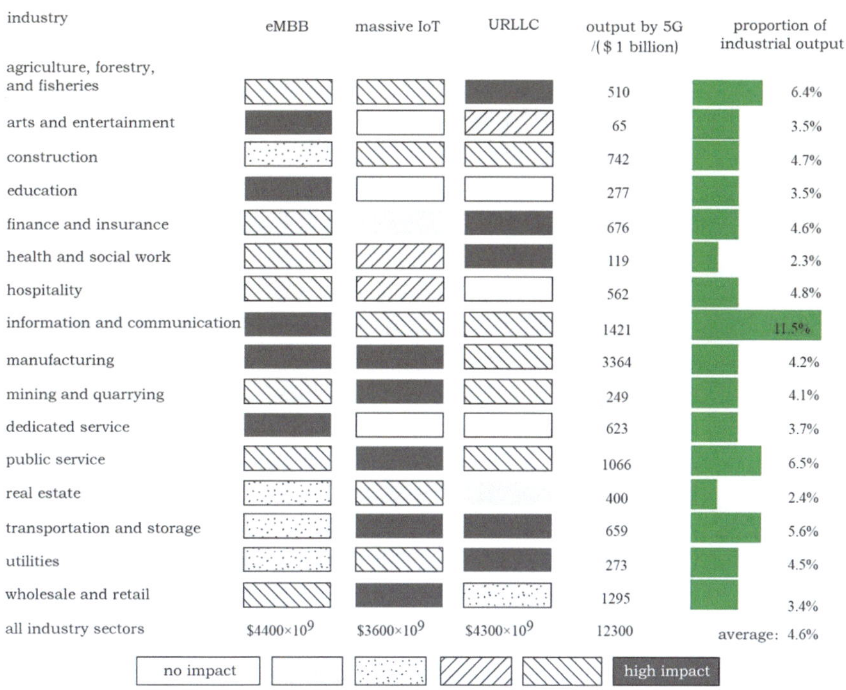

Fig. 2.11 Prediction of 5G's contribution to driving output growth of various industries in 2035 (from the consulting report of HIS Markit)

According to data from HIS Markit and the China Academy of Information and Communications Technology's "5G Economic and Social Impact White Paper", from 2020 to 2035, the overall contribution of 5G to global GDP is expected to be an annual average of $2.1 trillion, equivalent to the GDP of India, the world's seventh largest economy. By 2030, the direct and indirect contributions of 5G to GDP will be $3 trillion and $3.6 trillion respectively, driving direct and indirect employment opportunities of 8 million and 11.5 million respectively, having a huge impact on society. The above data shows the huge potential and influence of 5G, so major economies around the world have given very high expectations and attention to the development of 5G.

The large-scale construction of 4G networks has seen global operators invest a significant amount of capital. Although 4G has led to a certain degree of increase in the ARPU value of operator customers, it is not enough to quickly recover costs in the short term. Looking at the development of global operators, the financial situation of most operators is still difficult to support the large-scale construction of 5G networks. At the same time, the allocation of frequency resources in Europe and America is done through auctions, making the cost of 5G frequencies high, which has also become a heavy burden for the development of 5G by European and American operators. Therefore, it can be expected that the development of 5G will not be as fast as that of 4G. The 5G construction of most operators will adopt a progressive deployment method, gradually starting from city hotspots according to business needs, and gradually expanding to other areas. Of course, it does not rule out that regulatory authorities may mandate the scale and speed of network deployment when issuing licenses. The first to claim commercial use of 5G globally were Verizon in the United States and KT in South Korea. American operators such as Verizon claimed to achieve 5G commercial use in 2018, using a frequency of 28 GHz, and the typical application scenario is similar to China's home broadband, achieving wireless coverage through millimetre wave base stations and rooftop customer premise equipment (CPE) antennas, and then converting to Wi-Fi connections indoors. KT claimed to achieve 5G commercial use at the 2018 Winter Olympics, also using a frequency of 28 GHz. In April 2019, South Korea officially announced the commercial use of 5G, initially using a frequency of 3.5 GHz, with each operator having 80–100 MHz of frequency bandwidth, and 28 GHz becoming the second priority for deployment. China officially announced the commercial use of 5G in November 2019 at the Beijing Communication Exhibition, led by Chen Zhaoxiong, Deputy Minister of Industry and Information Technology, and the four major operators. Figure 2.11 Forecast of 5G's contribution to driving output growth in various industries by 2035 (from HIS Markit's consulting report) Japan achieved commercial use of 5G in 2020, using frequencies of 3.6–4.6 GHz and 28 GHz, with the mid-frequency band being the priority deployment band. (1) NTT DoCoMo: 3600–3700 MHz, 4500–4600 MHz and 27.4–27.8 GHz bands. (2) KDDI: 3700–3800 MHz, 4000–4100 MHz and 27.8–28.2 GHz bands. (3) SoftBank: 3900–4000 MHz and 29.1–29.5 GHz bands. (4) Rakuten Mobile: 3800–3900 MHz and 27.0–27.4 GHz bands. So far, the global 5G network deployment situation is as shown in Table 2.5. Table 2.5 Global 5G network deployment situation In the entire global development of 5G, it can be said that Asia is at the forefront, whether it is the time or scale of 5G commercial use, especially China, which is leading the world in the development of 5G. China's 5G development basically follows the development strategy of "promoting use by construction", that is, first building the network, quickly driving the maturity of terminals and chips, driving the maturity of personal business and applications, and then accelerating the application of vertical industries through mature chips and modules. Despite the impact of the new crown epidemic, with the joint efforts of the entire industry and the strong support of government policies at all levels, our country's operators have achieved a comprehensive lead in

Table 2.5 The global 5G network deployment situation

Region	Number of 4G networks	Number of 5G networks
Africa	159	8
Asia	140	45
Europe	171	101
Latin America	127	24
Middle East	47	22
Oceania	39	8
North America	19	13
Global	702	221

deployment scale and user scale at the beginning of commercial use. China Unicom and China Telecom's 5G network co-construction and sharing through the 3.5 GHz band, as well as China Mobile and China Broadcasting's 700 MHz co-construction and sharing, quickly achieved national 5G coverage, and achieved a good balance in construction speed and cost. As of the end of February 2023, the three operators in our country have 5G. The SA network has been fully commercialised, with over 2.2 million 5G base stations built, accounting for 70% of the global total. The number of 5G terminal connections exceeds 500 million, accounting for over 70% of the global total. The average data flow usage (Dataflow Of Usage, DOU) per 5G connected user per month is 1.74 times the average DOU of all users, making it the undisputed leader.

Currently, 5G commercial networks are in a growth phase, with the global architecture primarily NSA. Only 10% of operators have launched SA, with deployment frequencies mainly in the mid-band. The main commercial capabilities are currently Rel-15, with gradual support for Rel-16 to enhance and improve network capabilities.

2.5 5G Business Development

In terms of business cultivation and promotion, global operators have been focusing on the cultivation and exploration of new businesses, applications, and business models since 2016. They have established 5G joint innovation centres, and in collaboration with industry partners and vertical industry partners, have extensively carried out the cultivation and demonstration of new 5G businesses and applications. In February 2016, China Mobile was the first in the world to establish a 5G joint innovation centre, aiming to jointly promote the maturity of 5G end-to-end industries and capabilities with ICT companies and partners from all walks of life, incubate innovative applications and products, build a cross-industry integration ecosystem for mutual benefit and win–win cooperation, and launch a series of 5G business demonstrations around the shared economy demonstration platform of the National

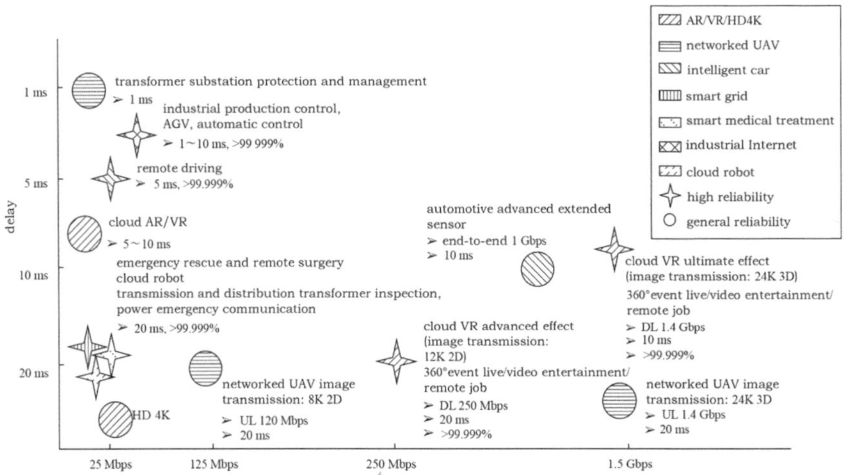

Fig. 2.12 Capability requirements of 5G characteristic services

Development and Reform Commission, including the world's first 5G + 4K sports event live broadcast, the first 5G network 4K ultra-high-definition live broadcast of the Spring Festival Gala, the first 5G train station construction in China, the first 5G remote human surgery in the country, 8K local area network cloud office video conference, etc. The Ministry of Industry and Information Technology also continuously organises the "Blossom Cup" 5G application competition to promote the integration of 5G and applications. The capability requirements of 5G characteristic services are shown in Fig. 2.12.

Drawing on the successful experience of 4G development, the development of 5G services focuses on two aspects: one is to continue to tap the market potential of personal services, enable information consumption, extend people's imagination, and make VR/AR, cloud service robots, high-definition video, naked-eye 3D, etc. possible, and enter people's lives; the other is to cultivate and incubate enterprise-level and vertical industry markets, build a smart society through the empowerment of 5G, promote industry upgrading and transformation, bring about the development and breakthrough of industries such as connected cars, networked drones, smart energy, smart healthcare, intelligent manufacturing, smart education, etc., create new opportunities for the industry, achieve the beautiful vision of promoting industry upgrading and transformation, and building a smart society.

In terms of personal services, the industry is actively investing in and cultivating characteristic services, and the development of 5G 2C applications presents features such as "new terminals, new perspectives, new ways of playing, new ways of shooting, new perspectives".

(1) New terminal—foldable large-screen mobile phone: ultra-large screen ultra-high-definition video experience, mobile office anytime, anywhere, multi-window, multi-task, dual-screen operation.

(2) New perspective—from watching videos to playing videos: ultra-high-definition, immersive, full-platform convergence, large and small screen integration, sports + performance characteristics.
(3) New way of playing—ultra-high-definition interactive cloud games: no delay, cloud operation, click and play; transition from buying equipment to renting time; massive game content; real-time interaction in different places.
(4) New way of shooting—virtual social fun shooting: star shooting, virtual idol live broadcast.
(5) New perspective—360° free viewpoint: any viewpoint + 360°, field of view + full free interaction, rich and diverse business content such as pictures, videos, 3D models, etc.

Looking at current applications, 5G has significantly alleviated the pressure of traffic growth in 2C businesses. Taking China Mobile's 5G network as an example, 5G has clearly achieved a diversion effect, accounting for nearly 30%. The traffic carried by 4G has begun to decline, and the DOU of 5G users is nearly twice that of 4G users. The advantage of 5G network traffic carrying has initially emerged. South Korean operators are pushing VR/AR businesses, and 5G traffic is growing rapidly. The data published by the Ministry of Science and ICT of South Korea in January 2020 is shown in Figs. 2.13 and 2.14.

In terms of the development of 5G 2B (toB) business, our country's integration of 5G into industries leads the world in breadth and depth. Operators are working with industries to create innovative applications, accelerating the integration of 5G into

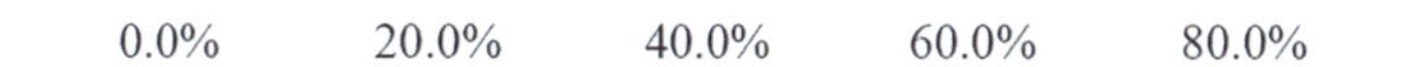

Fig. 2.13 Comparison of 4G and 5G network traffic in South Korea

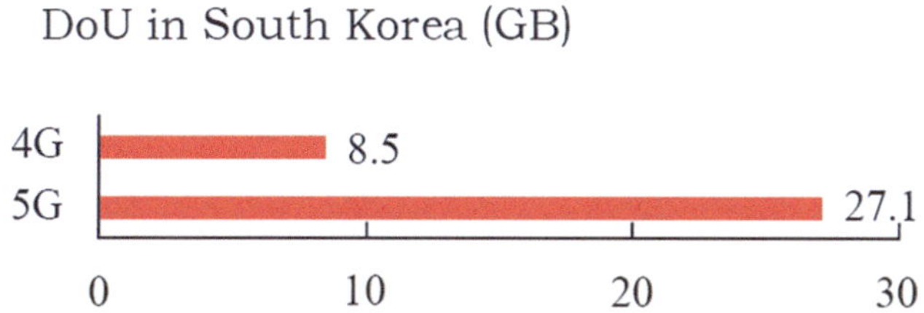

Fig. 2.14 Average monthly traffic of 4G and 5G users in South Korea

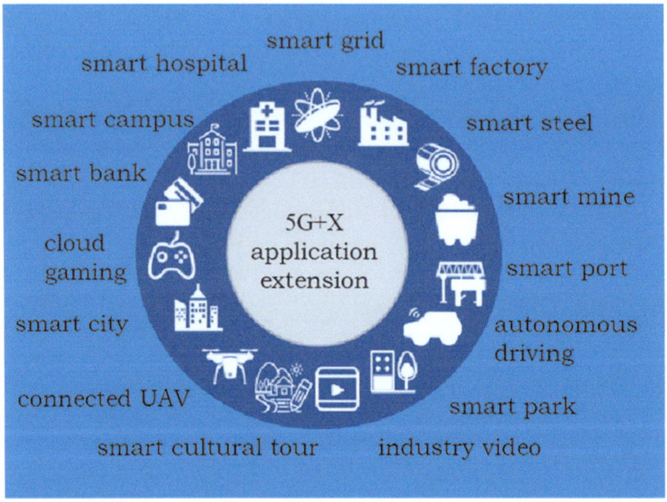

Fig. 2.15 Application fields of 5G 2B

thousands of industries, becoming an important engine for driving ICDT[1] growth. Its application fields are shown in Fig. 2.15. According to the research report of the Global System for Mobile Communications Association (GSMA) and surveys of global enterprises, the 5G 2B business market is broad. By 2030, the global vertical industry market empowered by 5G will reach 700 billion US dollars, and 2/3 of Chinese enterprises are eager to use 5G. In addition, the demand fragmentation of different vertical industry enterprises is extremely high, and there are significant differences in bandwidth, uplink speed, downlink speed, maximum number of connections, data transmission frequency, etc., requiring customised services. For this reason, the three major domestic operators are actively deploying 5G 2B applications, focusing on key vertical industries, and carrying out extensive business pilots. The pilot results show that the combination of 5G and enterprise production brings significant cost reduction and efficiency improvement effects. For example, China Mobile cooperated with a factory to carry out a pilot of intelligent flexible production based on 5G edge computing, which increased its production efficiency by 30% and reduced labour costs by 40%. Therefore, there is much to be done in the combination of 5G and vertical industries.

5G private networks are currently an important application form of 5G services in vertical industries, becoming an important part of building digital infrastructure and helping 5G integrate into enterprise production. Through 5G private networks, operators provide industry customers with customisable 5G communication services in terms of service range, network capabilities, and isolation based on authorised spectrum, which has obvious advantages over traditional vertical industry solutions.

[1] ICDT refers to the deep integration of Information Technology (IT), Communication Technology (CT), and Data Technology (DT) in the era of big data.

The advantages and disadvantages of operator 5G private networks and traditional industry private networks are compared in Table 2.6.

Therefore, overall, 5G public networks have very obvious advantages over traditional private networks: the industrial chain is more mature and robust, the response time to market demand is shorter, and it shares the network with personal business, the application scale is larger, the terminal equipment cost is lower, it uses the operator's proprietary frequency, has a professional network operation and maintenance team, network reliability and coverage are guaranteed; network development is planned, and new features are introduced faster. Therefore, the author believes that in the future development of 5G, the combination of operator public networks and vertical industry customers and enterprise-level customers will inevitably produce a win–win situation, empowering the development of vertical industries while reducing their expenditure costs, helping them achieve transformation and upgrading and intelligent development.

5G SA possesses end-to-end 5G capabilities, which can be based on network slicing and edge computing. Through virtualisation and software customisation, network capabilities can be aggregated as needed on logical or physical network elements, such as base stations or servers, to meet the specific business needs of different industries. This can effectively adapt to the differentiated and fragmented

Table 2.6 Comparison of advantages and disadvantages of operator 5G private networks and traditional industry private networks

Dimension	Item	5G private network of operator	Industry frequency private network	Wi-Fi	LTE-U
Spectrum	Bandwidth	High	Low	Medium	Medium
	Interference	Low	Low	High	High
	Coverage	High	High	Low	Low
Format	Security	High	High	Low	High
	Delay	Low	Medium	Medium	Medium
	Reliability	High	High	Low	Medium
	Mobility	High	High	Low	High
	Business concurrency capability	High	Medium	Low	Medium
Industry	Network cost	Medium	High	Low	High
	Chip cost	Low	High	Low	High
	Industry terminal richness	Low	Low	High	Low
O&M	Maintenance persistence	High	Medium	Low	Medium
	Optimization capability	High	Medium	Low	Medium

needs of vertical industry applications, as well as the need for flexible and rapid deployment. Therefore, 5G SA can provide different private network construction modes, achieving different private network services such as premium, exclusive, and VIP, as shown in Fig. 2.16 for the different service modes provided by 5G private networks.

1. Virtual Private Network (Premium Mode) [3]

A 5G virtual private network refers to a private network where users share the spectrum and wireless base station equipment with public network users. Through QoS, network slicing technology and other functional technologies and means, business priority assurance, business logic isolation, network speed, delay, reliability priority assurance needs are met, achieving business logic isolation and flexible configuration as needed. This mode is mainly aimed at most wide-area businesses and some local area businesses, and has certain requirements for network capabilities and isolation

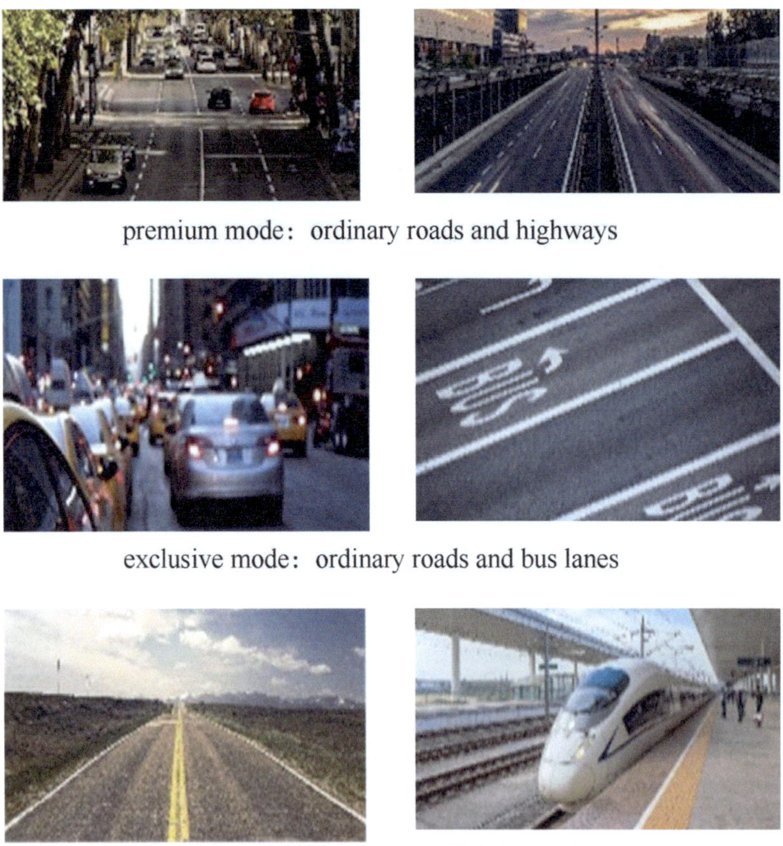

Fig. 2.16 Comparison of different service modes provided by 5G private networks

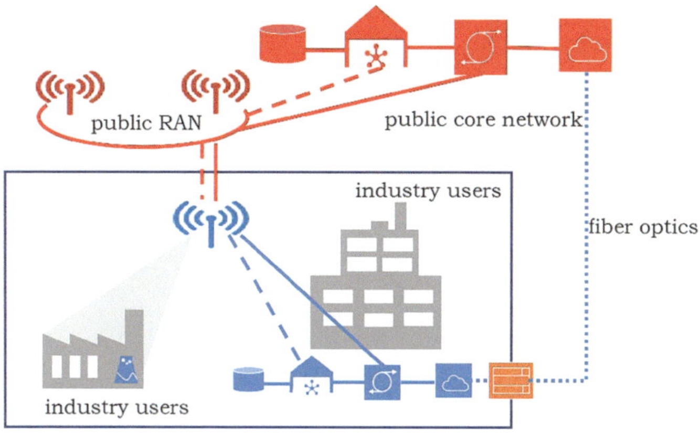

Fig. 2.17 5G Virtual private network [3]

assurance. The network deployment cost is relatively low. The 5G virtual private network is shown in Fig. 2.17.

To meet the personalised needs of vertical industries in the traditional private network way, there are problems of too small user scale and high cost. In the 5G era, this mode needs to change, which is one of the differences between 5 and 4G. Differentiated services must have flexible network support, and capabilities also need to be flexibly deployed. Only by sharing the cost of network construction and maintenance with a large group of individual users can the network deployment cost of operators be reduced and the competitiveness of operator solutions be built. Therefore, using network slicing and edge computing technologies to achieve the "privatization" of the public network should be the preferred solution for operators to serve vertical industries. In necessary scenarios, physical private networks can be supplemented to meet differentiated and fragmented market needs, so that the development of the entire business can be cost-controlled and quickly transition from quantitative change to qualitative change.

2. Quasi-Physical Private Network (Exclusive Mode) [3]

Exclusive spectrum, shared wireless base station equipment, through edge computing and other technologies that affect network architecture, combined with network slicing functions, together achieve network exclusivity, meet the needs of data not leaving the field, ultra-low latency, exclusive network, achieve data traffic offloading, local business processing effect. This mode is mainly aimed at local area businesses, and has higher requirements for network latency and isolation assurance. The network deployment cost is higher. The 5G quasi-physical private network is shown in Fig. 2.18.

3. Physical Private Network (VIP Mode) [3]

Through the exclusive construction and use of base stations, frequencies, and core network equipment, further meet the needs of ultra-high security, ultra-high isolation,

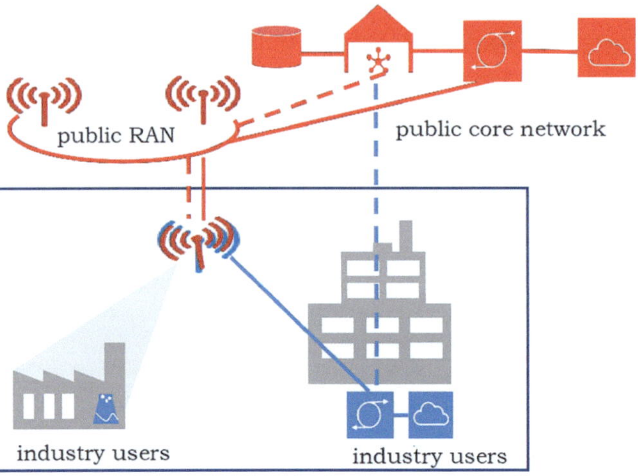

Fig. 2.18 5G quasi-physical private network [3]

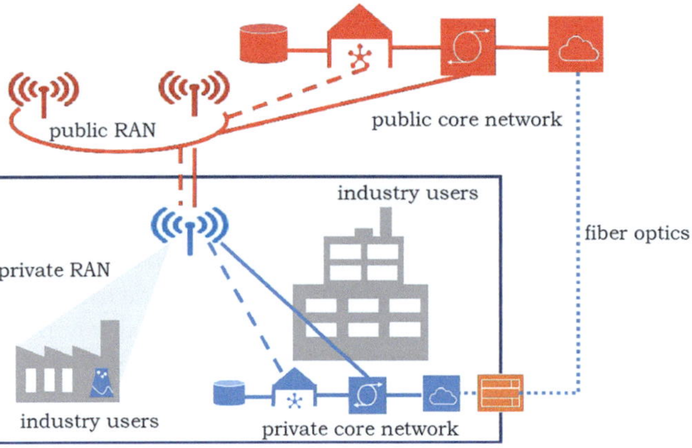

Fig. 2.19 5G physical private network [3]

and customised networks, achieving a dedicated 5G network, VIP on-site service effect.

For some industry applications with high requirements for application data and capabilities, combined with customer needs, services can also be provided to customers in the form of a private network, achieving complete isolation of data, resources, and network management from the public network, meeting the ultimate performance and security requirements. However, this network deployment method has a high deployment cost due to the customer's complete exclusive use of the entire end-to-end network and resources, and the resource utilisation efficiency usually leads to a high cost of business services. The 5G physical private network is shown in Fig. 2.19.

Through the "5G private network + AI", operators are expanding the scale of 5G applications across multiple industries, achieving development from individuality to replication, such as smart mines, smart steel, smart factories, smart power, smart ports, smart hospitals, etc.

2.6 Opportunities and Challenges of 5G Development

Looking at the history of global 4G development, since 2010, more than 5 million 4G base stations have been built globally, and the total investment is an astronomical Fig. 4G has brought growth to operators, but this growth has become saturated due to fierce market competition and intervention by regulatory agencies.

More than a decade ago, consulting agencies successfully predicted the predicament faced by mobile network operators today, as shown in Fig. 2.20. The rapid development of 4G ended the era of mobile communication market dominated by voice services and truly entered the era dominated by data. As the penetration rate of the mobile communication market has exceeded 100%, the original way for operators to increase revenue by market scale is no longer sustainable, so in the later stage of 4G development, the development of mobile operators has entered a bottleneck period. The mobile communication industry has begun to actively layout IoT applications, trying to actively explore new vertical industry markets through NB-IoT and eMTC, but it has not achieved the expected results; in the 5G era, IoT facing vertical industries is even more the focus of development.

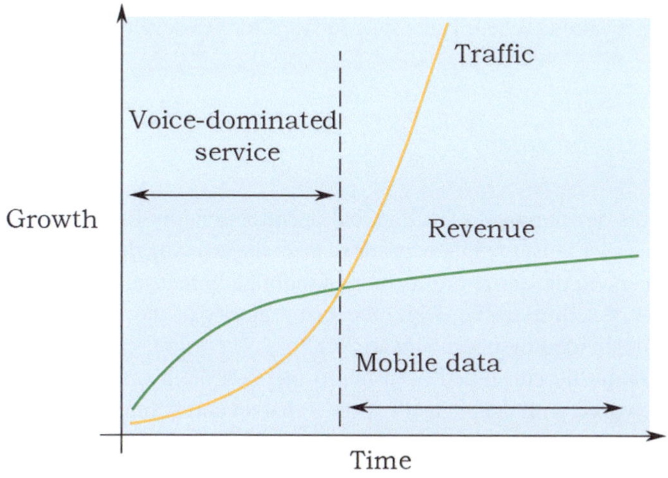

Fig. 2.20 Forecast of mobile communication market development trend

As a mobile communication infrastructure, 5G can achieve strong connections between people, people and things, and things and things, connecting all aspects of social operation; at the same time, 5G and artificial intelligence (AI), Internet of Things (IoT), cloud computing (Cloud Computing), big data (Big Data), edge computing (Edge Computing) are intertwined (AICDE), together forming a new generation of ubiquitous intelligent information infrastructure; 5G + AICDE combined with technologies such as robots, high-definition video, drones, AR/VR, will provide universal capabilities for applications in various industries, including education, medical, entertainment, transportation, etc. involved in life, industries, agriculture, energy, finance, etc. involved in production, and government affairs, security, environmental protection, etc. involved in social governance, forming 5G + X application extension, continuously introducing new products, new services, new models, new formats, accelerating the quality reform, efficiency reform, and power reform of various industries. Therefore, the industry expects 5G to replicate the unprecedented prosperity brought by 4G to the mobile Internet to all walks of life, thereby bringing a new leap to the mobile communication industry.

Looking at the current development of 5G commercial use, the development of 5G is still in the growth stage. Although our country has achieved global leadership in network scale and user numbers, there is still no killer application for 2C (toC), and individual users still find it difficult to experience the unique services and experiences of 5G. Although 2B services have some successful applications in some industries such as mines, steel, and coal mines, and thousands of 5G 2B business demonstrations have been carried out nationwide, it is difficult to replicate on a large scale. The development of 5G has raised some controversies and doubts. The high energy consumption, high cost, and difficulty in network operation and maintenance of 5G have become hot topics in the industry, and users have many complaints about service experience and operator charges. Equipment vendors hope that 5G can continue to be built on a large scale, but operators are facing the challenge of high costs and difficult monetisation. Where is the way forward for 5G development?

2.6.1 "Open Source"

Looking at the development of 4G, global operators mainly focus on the provision of personal services, using higher speeds to meet the growing demand for data traffic from customers. In the entire profit chain of mobile internet services, operators only get the value of selling traffic, while the vast majority of the value of connections and content is divided by internet manufacturers. Therefore, operators only rely on increasing the traffic consumed by users to increase the average revenue per user (ARPU). However, with the intensification of market competition between operators and the mandatory speed reduction fees imposed by regulatory agencies, unlimited packages have increasingly become the business model for the popularisation of personal services by operators. In such a large environment, operators can no longer continue the past traffic operation model and must expand new business models.

Therefore, in the later stages of 4G development, leading market operators around the world have begun to explore the Internet of Things application market, but at present, global operators have not found the secret to rapid growth in the expansion of IoT applications represented by eMTC and NB-IoT. They still rely on selling connections and traffic to make a profit, and the input–output ratio is not optimistic.

How can 5G profit and monetise? What should be the business model of 5G? Opinions vary in the industry. Some say selling traffic, 5G peak speed can exceed 1 Gbps, which is more than 10 times that of 4G, and the throughput of a single base station is dozens of times that of 4G, and there is potential for traffic supply; some say selling capabilities, because 5G can provide lower latency and higher reliability. For young people who play games, especially for those who play competitive games, the latency experience is very important and can even have a decisive impact on the results. They can provide differentiated services through acceleration packages, thereby bringing new business models; some say selling connections, facing the future society, the number of IoT terminals will be hundreds to thousands of times the number of individual users, and the market space is huge.

Although 5G can meet the needs of eMBB, uRLLC and mMTC application scenarios, many leading operators are exploring new business models for 5G, but most operators are not optimistic about the future business models of 5G in the Internet of Things and vertical industries. Therefore, most operators position the initial development of 5G on eMBB services, first considering the capacity expansion needs brought about by the development of 4G business volume.

Considering the capabilities and new scenarios of 5G as a whole, the future business development of 5G mainly focuses on the following directions [3, 4, 17, 18].

(1) Deeply cultivate the personal business market, construct a differentiated user experience, make tariff levels differentiated through different levels of experience, and improve ARPU value through the promotion of consumption levels.

In the author's daily work contact, many high-end mobile phone customers have complained: "I am your loyal user, using your number for decades, and spending hundreds to thousands of pounds on phone bills every month, but I don't experience any difference compared to those who pay 20 pounds a month? When the network is good, everyone is good, and when the network is bad, everyone is equally bad."

In this regard, mobile operators should learn from internet companies, provide differentiated business experiences, and carry out differentiated billing through different service levels, which can develop more and more flexible tariff systems on the basis of the original unlimited packages, thereby increasing the income brought by personal business. For example, considering that video business has become the main application for consuming traffic, for basic tariff packages, it can be considered to guarantee the transmission capability of standard-definition videos; for users willing to pay higher tariffs, it can guarantee the transmission capability of high-definition videos.

(2) Cultivate and introduce personal businesses with high business capability requirements to drive the upgrade of personal business experiences, such as AR/VR, which can adopt separate billing methods such as per use or duration.
(3) Expand vertical industries and enterprise markets with open and flexible product forms, and expand new business spaces.

For 5G networks, operators can utilise their high speed, low latency, high reliability, and large connection capabilities, with the help of network slicing and MEC, to layout differentiated enterprise-level and vertical industry application markets, gradually expanding the market scale of 5G applications, thereby reducing the construction and operation costs of the entire 5G network, and enhancing the profitability of 5G through new business models.

Despite the fact that there are thousands of 5G application demonstration projects and many breakthroughs in actual commercial applications, such as in the steel, coal mining, port and dock industries, the application of 5G to industries needs to fully ensure "network reliability and stability, data security isolation, simplified opening procedures, and upgraded operation and maintenance models". The "5G + industry" still faces many challenges, and there is still a long way to go for the integration of 5G and vertical industries.

① The development of each industry is chimney-like, with strict industry barriers, and 5G needs to break these barriers to integrate with other industries.
② Some industry regulatory policies are not yet perfect or complete, affecting the application of 5G, such as vehicle networking, drones, etc.
③ The 5G network, from serving the industry to integrating into production, needs to meet extremely high end-to-end reliability, and network performance and operation and maintenance levels need to be further improved.
④ The network slicing capabilities claimed by 5G in the early stages are not yet perfect and do not yet have mature commercial capabilities.
⑤ 5G 2B applications are diverse, customised, and have long processes, hindering their further scaled development.
⑥ The end-to-end cost of 5G is still not friendly enough, setting a high usage threshold, especially affecting the willingness of small and medium-sized customers to use it.
⑦ The development of 5G 2B still follows the traditional 2C product form, with single capabilities and rigid structures, unable to adapt to the differentiated and fragmented 2B market demand.

Therefore, although vertical industries and enterprise markets are the main direction of 5G's future development, there is a great deal of uncertainty about its openness based on its past development history. It requires the government to guide it through overall planning, and also requires the communications industry to introduce vertical industries into the advancement of 5G development as soon as possible, communicate needs, incubate businesses and applications, and explore and cultivate win–win business models.

(4) Cultivate a 5G application development ecosystem through network capability opening [17, 18].

The development model of traditional vertical fields is "chimney-like", each industry has its own ecosystem and industry barriers, many of which are areas that operators have not tried before. In the 5G era, due to the fragmentation of the vertical industry market and the differentiation of business needs, the traditional 2C model is difficult to replicate in vertical industries. How to calculate costs? How to price? How to meet differentiated business and deployment needs? How to expand users? How can the existing operation and management system of operators adapt to the changes in the 2B market? Due to the limitations of mechanisms and systems, traditional operators are far less innovative in new businesses and new applications than internet companies and small and micro enterprises, especially in personalised and differentiated vertical industry applications, it is difficult to gain advantages in cost and speed. These are the development challenges that operators are currently facing.

In the face of the above challenges, in addition to changes in the development, expansion, operation and management of related businesses, operators also need to focus on several aspects: firstly, to build end-to-end 5G network capabilities as soon as possible, by combining the new capabilities of 5G networks with network slicing and edge computing, to support the differentiated and fragmented demands of vertical industries; secondly, to build an open platform for 5G network capabilities, opening up various capabilities of the operator's network to third parties, encouraging more application developers to develop and create innovative services based on 5G capabilities, finding new points of revenue growth through revenue sharing and network capability leasing with third-party collaborators, and significantly enhancing the innovative vitality of 5G services and applications, accelerating the development and popularisation of 5G; thirdly, to build an application store (App Store) for vertical industry applications, and on this basis, to build an ecosystem of network operators, application developers and application users, achieving multi-party docking of network capabilities, application tools and service objects, and creating a win–win business model. Facing 5G, there are many capabilities that operator networks can open up, including high speed, low latency, high reliability, artificial intelligence, cloud computing and storage, user information, location positioning and other capabilities. Operators need to expand and cultivate this broader emerging market in conjunction with the deployment and development of 5G networks.

2.6.2 *"Throttling"*

So far, the scale of 5G networks in China has reached 5 million stations, exceeding 70% of the global scale, and the scale of the entire mobile communication base station has exceeded 10 million stations. Such a large scale of base stations brings huge construction and operation costs, especially the construction of 5G networks. Due to the adoption of large-scale antenna technology, 100 MHz bandwidth, etc., although performance and efficiency have been greatly improved, the cost of a single base station is 2–3 times that of 4G, and the energy consumption of the base station is also 3 times that of 4G. The large-scale construction of 5G networks has brought huge cost pressures to operators. In addition to construction investment, electricity bills, rent for the sky surface and machine room, and network operation and maintenance costs have also become important daily expenses. Therefore, "high cost, high energy consumption, difficult operation and maintenance" have become problems that operators urgently need to solve in the development of 5G.

In order to improve performance and efficiency, the initial base stations of 5G adopted a 192 antenna/64 channel active antenna unit (AAU) architecture. In order to meet commercial conditions, higher requirements were put forward for the architecture and chip design and process of the base station. The requirements of 5G large-scale antenna base stations for integrated circuits are shown in Fig. 2.21.

In order to meet commercial requirements, 5G base stations have undergone new design and optimisation in terms of miniaturisation and lightweighting, including the use of highly integrated integrated circuits (ICs), lightweight materials with high heat dissipation efficiency, etc., and comprehensive optimisation has also been carried out for the high complexity processing of baseband; however, its cost still reaches 2–3 times that of 4G base stations, and energy consumption is 3 times that of 4G base stations.

As shown in Fig. 2.22, from the measured data of the energy consumption distribution of 5G large-scale antenna base stations, the power consumption of the base station mainly includes baseband (BB), digital intermediate frequency and enhanced common public radio interface (eCPRI) (IF + eCPRI), radio frequency (RF) and power amplifier (PA), the power consumption of BB, IF + eCPRI, RF parts and

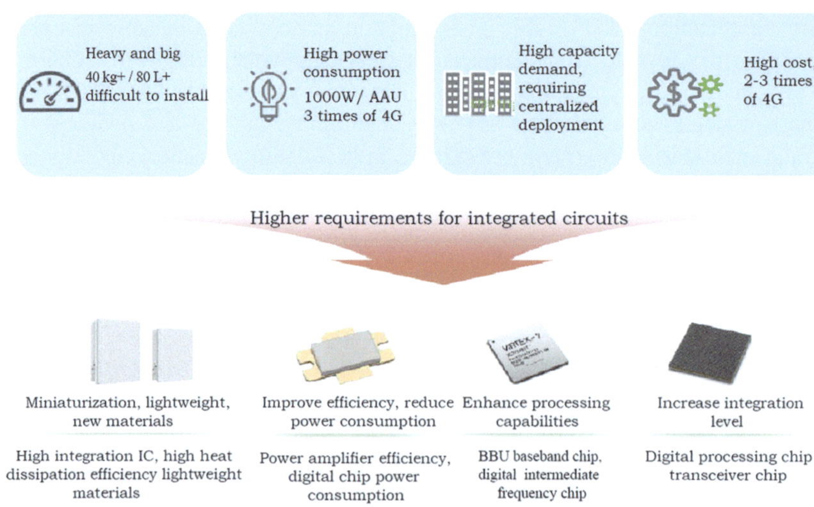

Fig. 2.21 Requirements of 5G large-scale antenna base stations for integrated circuits

the load of the base station are not linearly related, and do not change much with the increase of the base station business load, only the power consumption of PA is linearly related to the base station business load. Under full load, the power consumption of a base station (3 sectors) is as high as 3168. W, considering 1.6 million base stations, 20% of the base stations are fully loaded, 60% of the base stations are moderately loaded, and 20% of the base stations are lightly loaded, then the annual electricity consumption is as high as 392.448 billion kWh. Calculated at 0.725 yuan per kWh, the electricity cost will be as high as 284.5248 billion yuan. If you also consider the electricity costs of air conditioning in the machine room, the rent of the storefront, etc., the maintenance cost of the new 5G network brings huge cost pressure to the operators.

Therefore, operators and equipment vendors are actively promoting the energy-saving work of the 5G network, optimizing the energy consumption of 5G equipment and networks from two aspects: hardware design and optimization, and the introduction of software functions. In terms of hardware, the energy consumption of the base station is optimized from the following three aspects.

(1) Improvement of power amplifier (PA) efficiency: Compared with the first-generation product, the efficiency of the power amplifier is increased by 5%; using gallium nitride (GaN) power amplifier, compared with Lateral Diffused Metal Oxide Semiconductor (LDMOS), it will bring an 8% increase in the efficiency of the final stage power amplifier.
(2) Increase in transceiver integration: Compared with the first-generation product, the integration is doubled; the process is improved by 1–2 generations (40 nm → 28 nm → 16 nm).

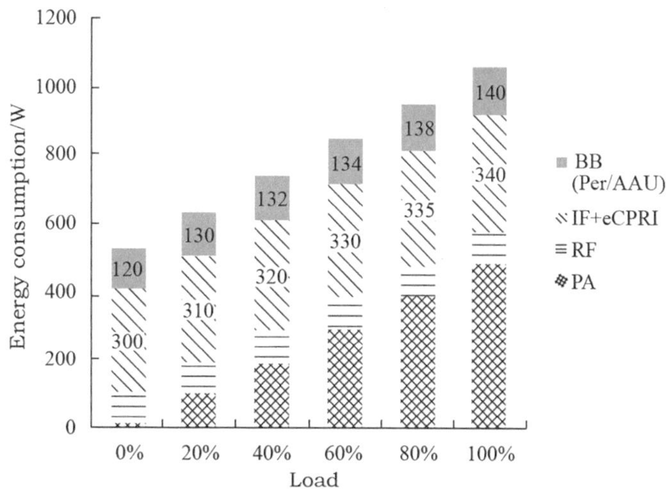

Fig. 2.22 Relationship between load and energy consumption of a single sector of a large-scale 5G antenna base station

(3) Digital chip Application Specific Integrated Circuit (ASIC) implementation: Except for some manufacturers, ASIC implementation is basically achieved; the process is improved by one generation (16 nm → 7 nm).

In terms of base station software, on the basis of traditional base station and network energy-saving functions, further introduce subframe shutdown, channel shutdown, deep sleep and other functions, which can bring significant energy consumption reduction.

(1) Subframe shutdown (see Fig. 2.23): The base station should be able to sleep some devices (at least including the power amplifier) in time according to the change in business volume, and the length of sleep time can reach the time slot level.
(2) Channel shutdown (see Fig. 2.24): The base station should be able to sleep some devices in the RF channel in time according to the change in business volume (the devices that can be slept are not limited to the power amplifier), in order to reduce energy consumption.

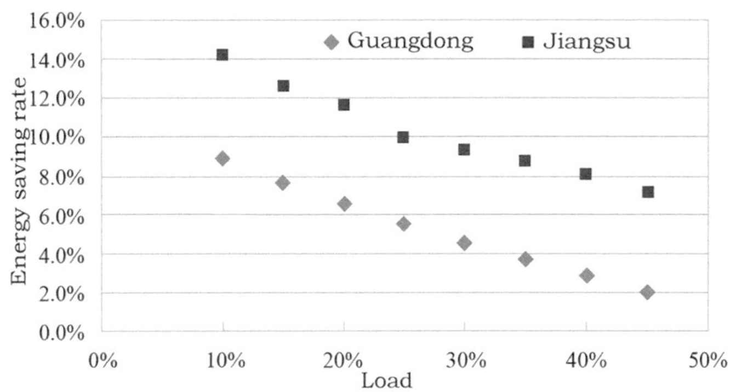

Fig. 2.23 Subframe shutdown (energy saving rate and load)

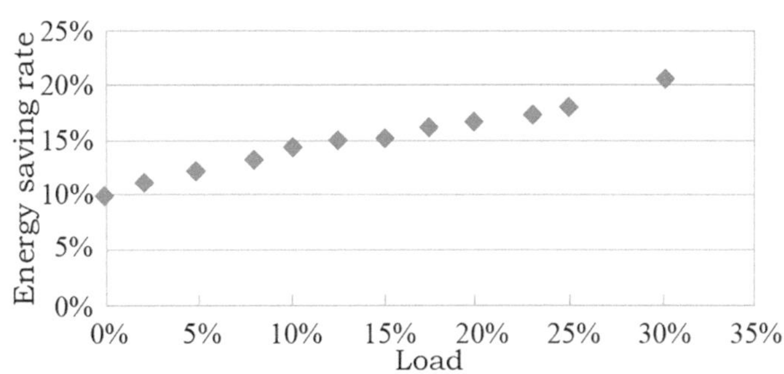

Fig. 2.24 Channel shutdown (energy saving rate and load)

(3) Deep sleep: This function means that the base station only retains the necessary function modules to work in low power mode, and can quickly return to normal state.

Judging from the application test results of the existing network, the newly introduced software functions can achieve relatively obvious energy-saving effects.

(1) Subframe shutdown: Energy saving reaches 10% under medium and low load.
(2) Channel shutdown: Energy saving is 10–20% under medium and low load.
(3) Automatic energy consumption collection: The error of energy consumption accuracy is within 5%.

In addition, in order to further reduce the rent and energy consumption of the machine room, operators' 5G construction has gradually begun to adopt the deployment method of Cloud Radio Access Network (C-RAN), as shown in Fig. 2.25. Through the centralized deployment of the central unit (CU) + distributed unit (DU) of multiple base stations, it can greatly reduce the demand for machine rooms, and also bring considerable energy-saving effects. It is expected that by 2025, more than 90% of the 5G base stations of operators will be deployed in the C-RAN mode.

The construction of 5G networks not only increases the number of base stations, but also brings more complex network optimisation and operation and maintenance issues. The design of 5G networks is more flexible than 4G, introducing more network parameters, and as new features are continuously introduced, network parameters also continue to increase. In addition, as it is difficult for 5G to provide seamless coverage in a short period of time, and the existing 2G, 3G and 4G networks are difficult to upgrade to 5G in the short term due to the demands of existing users and businesses, operators generally face the problem of coexistence of multiple networks and generations. In order to achieve the coexistence and complementary cooperation of 5G with existing networks such as 4G, the configuration and optimisation of inter-network parameters is extremely complex, which brings great difficulties and

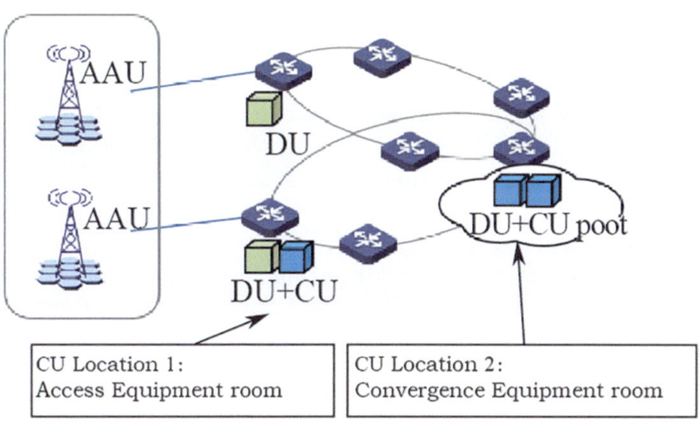

Fig. 2.25 C-RAN deployment method

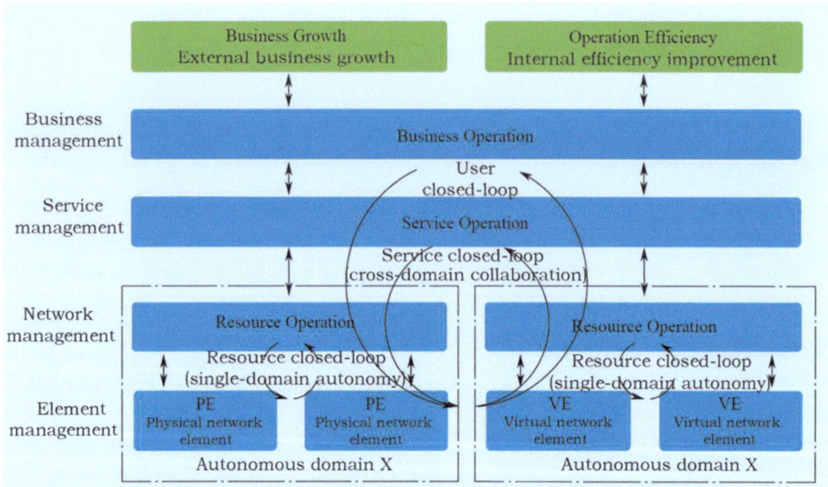

Fig. 2.26 Target framework for China Mobile's 5G network autonomous driving

challenges to the operation and maintenance of the entire network of operators, and the cost also increases significantly.

For this reason, operators have begun to use big data and AI to transform network operation and maintenance from manual to unmanned and automated, proposing the development goal of network autonomous driving [19], defining scenario-based grading standards for processes, and gradually improving the level of network operation and maintenance autonomy in steps. Global operators have established the Global Telecommunications Management Forum (TM Forum) under the European Telecommunications Standards Institute (ETSI) and initiated research on Autonomous Driving Networks (ADN) [20], aiming to provide consumers and vertical industry customers with fully automatic, zero-wait, zero-contact, zero-fault innovative network services and Information and Communication Technology (ICT) services, creating self-service, self-repairing, self-optimising communication networks, and clarifying the target architecture and implementation path for the digital transformation of communication network operation and maintenance [21].

The Autonomous Driving Network grading framework divides network autonomy capabilities into six levels: "L0 Manual Operation and Maintenance, L1 Assisted Operation and Maintenance, L2 Partial Autonomous Driving Network, L3 Conditional Autonomous Driving Network, L4 Advanced Autonomous Driving Network, L5 Fully Autonomous Driving Network". China Mobile refers to the concept of autonomous driving networks, plans for the digital transformation of network operation and maintenance, increases the construction of automation and intelligence capabilities, and sets the overall goal of reaching L4 level of network operation and maintenance autonomy by 2025. In response to the complexity of production networks and the diversity of automation capabilities, China Mobile adopts a "divide and conquer" approach to achieve a "spiral rise" in network autonomy capabilities, and refers to the TM Forum's hierarchical architecture for autonomous driving networks [20], combining production practices, proposes the internal practice target framework of "four layers and three closed loops", as shown in Fig. 2.26.

The four levels are as follows:

(1) Element management: Provides built-in automated operation and maintenance capabilities for network elements.
(2) Network management: Provides single-professional automated operation and maintenance capabilities for networks.
(3) Business management: Provides cross-professional automated operation and maintenance capabilities for networks and businesses.
(4) Business management: Provides automated service management capabilities for customers, providing unified customer touchpoints.

The three closed loops are as follows:

(1) Resource loop: Single-professional resource management, achieving single-domain autonomy.
(2) Business loop: Cross-professional end-to-end management for businesses, achieving cross-domain collaboration.
(3) User loop: User and business management, including user information, business, billing, customer service, etc.

With this as a reference, China Mobile aims to improve quality, increase efficiency, and reduce costs, building systematic capabilities at different levels to achieve network autonomy in all scenarios. Through the advancement of research on network autonomous driving, it is expected that the application of AI and network big data will be able to largely solve the difficult problem of 5G network operation and maintenance for operators.

2.7 Summary

5G, through technological innovations such as large bandwidth and massive MIMO, can bring new capabilities such as ultra-high speed, ultra-low latency, and ultra-high reliability, fully supporting the application of eMBB, uRLLC, and mMTC application scenarios. Combined with MEC and network slicing, it fully empowers the transformation and upgrade of all sectors of society. In the future, 5G will be deeply integrated with technologies such as cloud computing, big data, and artificial intelligence to accelerate the digital transformation of the entire society.

References

1. IMT-2020 (5G) Promotion Group. 5G Vision and Demand White Paper [S]. 2014.
2. ITU-R. Report ITU-R M.2410-0: Minimum requirements related to technical performance for IMT-2020 radio interface(s) [S]. 2017.
3. Liu Guangyi, Fang Min, Guan Hao, et al. 5G Mobile Communication System: Towards a Fully Connected World [M]. Beijing: People's Posts and Telecommunications Publishing House, 2019.
4. 3GPP. Study on architecture for next generation system: TSG SA meeting#70, SP-150853 [S]. 2019.
5. Wang Xiaoyun, Liu Guangyi, Huang Yuhong, et al. 5G Technology and Standards [M]. Beijing: People's Posts and Telecommunications Publishing House, 2020.
6. Shen Jia, Du Zhongda, Zhang Zhi, et al. 5G Core Technology and Enhancement: From R5 to R16 [M]. Beijing: Tsinghua University Press, 2021.
7. ETSI. Network function virtualization (NFV) management and orchestration: GS NFV-MAN 001 V0.3.14 [S]. 2014.
8. ONF. Software-defined networking: the new norm for networks [R/EB]. 2012.
9. MACH P, BECVAR Z. Mobile Edge computing: a survey on architecture and computation offloading [J]. IEEE Communications on Surveys & Tutorials, 2017, 19(3).
10. 3GPP. Release 16 Description, Summary of Rel-16 Work Items: TR 21.916 V16.1.0 [S]. 2020.
11. Chen Qin. Greater opportunities for 5G in vertical industries - Interview with Liu Guangyi, Chief Expert of Wireless and Terminal Technology Research Institute, China Mobile Research Institute [EB]. China Investment, 2020.
12. 3GPP. e-meeting document SP-200222: TSG SA#87e [S]. 2020.
13. 3GPP. Release 17 Description, Summary of Rel-17 Work Items: TR 21.917 V0.5.0 [S]. 2022.
14. China Mobile, China Telecom, China Unicom, et al. 5G-Advanced Network Technology Evolution White Paper [S]. 2021.
15. 3GPP. User Equipment (UE) radio transmission and reception; Part 1: Range 1 Standalone: TS 38.101–1, V15.2.0 [S]. 2018.
16. 3GPP. User Equipment (UE) radio transmission and reception; Part 2: Range 2 Standalone: TS 38.101–2, V15.2.0 [S]. 2018.
17. Liu Guangyi. 5G Provides Stronger Momentum for Deep Empowerment of Vertical Industries [J]. Telecommunication Engineering Technology and Standardization, 2020(6).
18. Lv Meng. China Mobile Liu Guangyi: Promoting the Development of 5G Applications, Creating the App Store of 5G [EB]. Communication World All Media, 2019-1-22.
19. China Mobile. China Mobile Network Autonomous Driving White Paper [R]. 2021.
20. TM Forum. Autonomous Networks: Empowering Digital Transformation For The Telecoms Industry [R/EB]. 2019.
21. TM Forum. Cross-Industry Autonomous Networks – Vision and Roadmap V1.0.1, IG1193[R/EB]. 2019.

Chapter 3
Why Need 6G?

With the large-scale commercial use of 5G, 6G research has become a new hot topic in the mobile communication industry. Countries and regions with advanced communication market globally have successively published 6G R&D plans. People can't help but ask why start 6G research so early when 5G is still in its growth stage and has not brought us a noticeable experience upgrade. The initial vision of 5G has not been realised yet. This chapter analyses why we should study 6G as early as possible from the perspectives of social development, sustainable development of the mobile communication industry, new scenarios and demands towards 2030, problems and challenges faced by the current network development, and the development trend of deep integration of ICDT.

3.1 The Driving Force of Social Development

3.1.1 Demand Driven by Global Sustainable Development

The United Nations' "2030 Agenda for Sustainable Development" has identified 17 sustainable development goals and 169 specific targets, showing the scale and ambition of the new global agenda, including three aspects of sustainable development—society, economy, and environment, involving poverty reduction, health, education, equality, environmental protection and other fields. This agenda is committed to eliminating poverty, protecting the planet, and ensuring human peace and prosperity through coordinated action. Massive energy consumption will drive a country's

---

With the development of concepts such as artificial intelligence, the various chip terms that will be generated from this, such as GPU, TPU, NPU, and DPU, are collectively referred to as XPU.

G. Liu et al., *6G Migrates from Communication to XaaS*,
https://doi.org/10.1007/978-981-96-7579-1_3

investment in related technologies, and the emphasis on energy-saving and emission-reducing mitigation measures and adaptation measures to adjust the energy structure are the two current responses to energy demand.

To balance economic construction and environmental protection, all countries will actively establish effective environmental management systems to develop a sustainable economic structure that can meet human basic survival needs without damaging the environment.

3.1.2 Demand Driven by China's Social Development

The intensification of population ageing will lead to the loss of human capital and the increase of social costs. Society expects better health care and emotional communication methods. The acceleration of urbanisation will bring great challenges to urban education and medical resources, road traffic, employment, housing, etc. Society expects better urban and social resource management plans. China has developed into the world's second-largest economy, insisting on technological innovation, system incentives, improving the quality of economic growth and the level of technology, is the inevitable need for China's economic development. China's economy will shift from emphasising speed to emphasising balanced development of quality.

Sustainable development and the resolution of social issues are long-term goals of socio-economic development. Achieving social equity, economic development, and environmental protection in a balanced manner is a great and complex project. China adheres to the development concept of "innovation, coordination, green, openness, and sharing", providing a good prescription for social development. In the field of information and communication, the "five development concepts" have pointed out the direction for the sustainable development of mobile communication networks and have become the focus and approach of the evolution and development from 5 to 6G [1].

Among them, "innovation" requires future networks to focus on breakthroughs in basic theory and original technology innovation, breaking through or extending traditional communication theory, building an integrated communication and computing architecture, and introducing transformative technologies such as holography, brain-like computing, and strong AI; "coordination" requires future networks to achieve cross-border integration and collaborative breakthroughs in fields such as communication, information, materials, energy, and vertical industries, requires the coordinated advancement of the innovation chain and the industrial chain, requires the international coordinated formulation of technical standards, and the global coordinated advancement of industrial development; "green" requires future networks to be low energy consumption, low emission, environmentally compatible, renewable energy supply, reconfigurable device materials, and 4D printable equipment; "open" requires future networks to have open capability interfaces, open-source hardware and software, open markets, open ecosystems, and to achieve a new situation of "integration", "interconnection", and "intelligence" in the network; "sharing" requires future networks to be universal, integrated and shared with other social infrastructures such as transportation, electricity, and urban construction, to achieve intensive infrastructure, and to improve infrastructure efficiency.

6G will add new momentum to social and economic development on the basis of 5G, and promote the implementation of the "five major development concepts" at a deeper level. In terms of "innovation", 6G provides a basic platform capability, information service capability, computing capability, and AI capability for countries, enterprises, and individuals in various aspects such as technological innovation, management innovation, business innovation, and cultural innovation; in terms of "coordination", 6G breaks through information islands between countries and industries, ensures the coordinated development of the global economy, coordinates the development of vertical industries in new business forms, and improves the national governance system; in terms of "green", 6G provides a strong environmental perception capability based on global three-dimensional coverage, forms a global cooperative environmental protection scheme, promotes the transformation and upgrading of traditional industries, reduces carbon emissions, and achieves green development; in terms of "open", 6G inherently promotes global economic openness, market openness, cultural openness, and system openness; in terms of "sharing", 6G will build a shared AI infrastructure, democratise big data and artificial intelligence, ensure the fairness of digital dividends and digital rights, further promote the upgrade of the "sharing economy", assist in "shared manufacturing" and "shared infrastructure" between countries and industries, form a new shared ecosystem, and achieve shared results.

To adhere to the "five major development concepts", 6G will reconstruct cyberspace, provide an integrated space for global economic development, and help form a global community of common development, security, and interests.

3.2 Driving Force for Sustainable Development of the Mobile Communication Industry

The mobile communication system is so far the most successful large-scale civilian system, the secret of its success lies in the continuous reduction of network infrastructure and terminal costs through global large-scale sharing and high penetration and popularity, thus better serving the whole society and becoming the most important infrastructure of the entire country. Looking back at the entire history of mobile communication development, the upgrade of the mobile communication system has never been decided by a certain person, company, or country, but is a market-oriented collective spontaneous behaviour, determined by the trend and objective laws of the entire mobile communication industry development. Therefore, whether a generation of mobile systems is successful depends on whether it can help the entire industry successfully realise commercial monetisation, and the premise of commercial monetisation is whether it can bring a better business experience to users (including individuals and enterprises), making users willing to pay more fees, not just on how many base stations have been built.

As described in Chap. 1, the upgrade of mobile communication systems has basically maintained a "decade per generation" rhythm and a "use one generation, deploy one generation, study one generation" development characteristic. The "decade per generation" upgrade rhythm covers all aspects from vision and requirement research, key technology research, standardisation, industrialisation to commercial application promotion. Each generation of mobile communication technology attempts to solve the problems faced by the industry development and application at that time, pushing the development of communication technology to new heights while better providing users with an upgraded experience. The initial 1G provided a solution for mobile voice communication, achieving a breakthrough from fixed to mobile calls, but with diversified standards, high costs, and poor experience; the subsequent 2G solved these problems well, regional standards began to unify, achieving global popularity of mobile voice applications, and successfully launched the popular short message service; 3G began to attempt data communication, although not very successful, it cultivated an application ecosystem, laying a good foundation for 4G to achieve rapid prosperity of the mobile internet; 4G truly achieved mobile broadband communication, its combination with smartphones brought unprecedented prosperity to mobile internet applications, accelerating the informatisation development of the entire society, 4G began to explore the application of the Internet of Things, but due to the split of multiple standards and the lack of breakthroughs in business models, the development of the Internet of Things business was far below expectations; the goal of 5G is to hope to replicate the prosperity that 4G brought to the mobile internet to all industries, to achieve the interconnection of all things, and to empower the digital transformation and upgrade of all sectors of society. With the rapid penetration of 5G applications, continuous breakthroughs in science and technology, and the deep integration of ICDT, 6G will fully support the digitisation of the world based on 5G, that is, to generate a digital twin world based on the physical world, and the people and things in the physical world will be able to transmit information through the digital world [1]. The digital twin world is a reflection of the physical world that can help humans further improve the quality of life, increase production efficiency, and realise the beautiful vision of "6G reshaping the world" [2]. In the entire development process of the new generation of mobile communication systems, key technology research, standardisation, and industrialisation all require huge investments from the entire industry, and the industry can only get returns in the stage of commercial application promotion. The more successful the operator's commercial realisation, the larger the scale of its network construction, and the higher the supplier's income. Therefore, during the R&D investment period of the entire new generation of mobile communication systems, the development of the industry needs the continuous income of the past generation or even multiple generations of mobile communication systems to support, in order to ensure the sustainability of the R&D of the new generation of mobile communication technology. The R&D cycle of the mobile communication system is analysed as shown in Fig. 3.1.

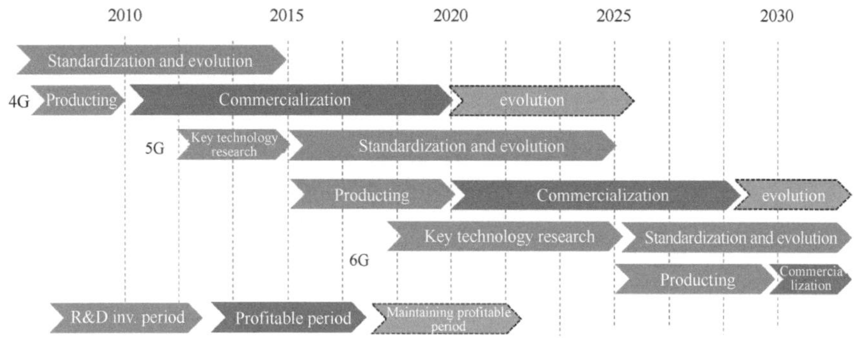

Fig. 3.1 Analysis of the R&D cycle of the mobile communication system

Recalling the development of 3G, 4G, and 5G, the mobile communication industry has the characteristic of "use one generation, deploy one generation, study one generation", and 6G is no exception. By 2019, the main use is 4G, but the industry has start to deploy and promote 5G applications on a large scale, and has also begun to pay attention to the R&D of 6G. It can be said that the design of each generation of mobile communication systems has its "time limitation", that is, the design of each generation of mobile communication systems is at a specific point in time, based on the experience and lessons of the design and application of the past generation of mobile communication systems, while facing new development needs and industrial progress, inheriting the mature technology and design of the past generation of mobile communication systems, introducing new technologies that seem to be mature enough to be widely applied and implemented at that time by considering the potential capabilitites limit of chips and devices at that time, and forming a new system design. Therefore, the design of each generation of mobile communication systems is not completely pursuing the latest and innovative technology, but as much as possible to adopt mature technologies that seem to be low-cost and mass-producible at that time, to ensure that network infrastructure and terminals can be implemented at low cost and miniaturisation, and thus to be widely applied and popularised. So, in a sense, each generation of mobile communication system design uses "old" technology. For example, the Code Division Multiple Access (CDMA) used by 3G is a technology that was already used in the military field in the 1940s and 1950s; the OFDM technology used by 4G is a technology that has been widely used in Wi-Fi, and MIMO and smart antenna technologies are technologies that appeared 20 years ago or even earlier; the LDPC and Polar codes used by 5G are technologies that appeared many years ago. At a specific point in time, some of the latest technologies and concepts may not yet have the conditions for large-scale application, such as high implementation complexity, high cost, etc., and will not be adopted by the standard; but as time progresses, the capabilities of chips and devices continue to improve, especially Moore's Law is driving the continuous improvement of computing power, some originally unachievable complex technologies and concepts become achievable after several years, these technologies or concepts become the focus of the current mobile communication system evolution or the design of the next generation of mobile communication systems. For example, massive MIMO (also known as 3D-MIMO in China), at the beginning of 4G, due to the limitations of chip technology, the prototype made at that time weighed more than 200 kg, and the power consumption and the cost was 8 times or even higher than that of the 4G base station. However, after the joint efforts of the ecosystem, after several years of iterative advancement, the chip/device process level and system optimization have improved rapidly, and massive MIMO has become feasible in the later stage of 4G. China Mobile first commercialized it in the 4G network in 2015 by modified base station, without changing the terminal, which lead to a 2 times increase in spectral efficiency in the live network; so by the time of 5G, massive MIMO has become a core technology and standard configuration of 5G network, while the supported bandwidth of massive MIMO has changed from the 60 MHz of 4G to the 160 MHz of 5G, and the cost and energy consumption of a base station have respectively dropped to 2–3 times and 3 times that of the 4G base station.

It is such continuous development and iteration that supports the mobile communication industry to continuously carry out technological innovation and breakthroughs, continuously improve the capabilities and service range of mobile communication networks, and achieve a virtuous cycle of development.

3.3 New Application Scenarios and New Business Needs

Inspired by Maslow's hierarchy of needs, China Mobile has transformed it to the level of communication needs, proposing a hierarchical communication needs model, divided into five levels: necessary communication, ubiquitous communication, information consumption, sensory extension, self-liberation [2], as shown in Fig. 3.2. In this model, communication needs and communication systems form a spiral upward cycle: the emergence of needs stimulates the development of communication technology and communication systems, and the perfection of communication systems pushes communication needs to a higher level, ultimately helping humans liberate themselves and achieve the ultimate pursuit of human intelligence [2].

According to the new Maslow's hierarchy of needs model for communication, once basic needs are met, higher-level needs will naturally emerge. 4G has brought about an unprecedented boom in mobile internet applications, while 5G will promote the interconnection of everything, extending our senses to new heights. Of course, whether from the current social development pattern or from the psychological and physiological needs of individuals, many of our desires and needs are far from being met. Issues we face today, such as traffic congestion and difficulties in accessing medical care, are not enough to truly liberate us to pursue the ultimate realisation of our own value. Combining the Maslow's hierarchy of needs model, the demand for communication will go through four ubiquitous stages: ubiquitous communication, ubiquitous information, ubiquitous perception, and ubiquitous intelligence. Ubiquitous intelligence aims to better liberate us through intelligent applications, freeing

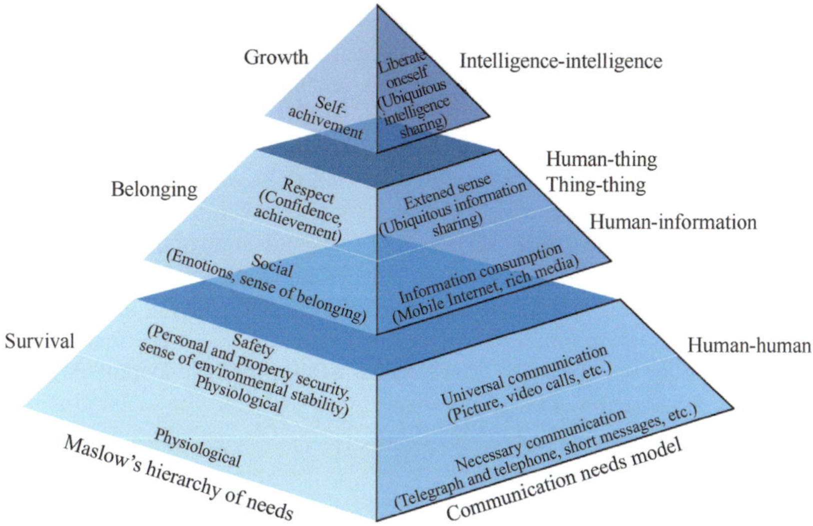

Fig. 3.2 New communication Maslow's needs model

us from the daily tedious communication and other affairs, saving us more time and energy to pursue and realise our own value. Therefore, 5G will help us achieve ubiquitous perception, and one of the goals of 6G is how to achieve ubiquitous intelligence.

"4G changes life, 5G changes society" confirms that people have never stopped pursuing higher mobile communication capabilities and a better life. The 4G era is a time of explosive growth in data services. With the popularisation of smartphones and the development of the consumer internet, from clothing, food, housing and transportation to medical care, education, and entertainment, human daily life has become greatly convenient. 5G will usher in a new era of interconnection of everything. It will achieve comprehensive interconnection between people, people and things, and things and things. 5G will penetrate all industries, bringing unprecedented vitality to the entire society. In the future, with the rapid penetration of 5G applications, new breakthroughs in science and technology, and the deep integration of new technologies and communication technologies, higher-level new needs will inevitably emerge. The combination of 5G and AI, big data, cloud computing and other technologies is accelerating the digital transformation of the entire society. So, what is the next stage of digitalisation? We believe it is digital twin.

Digital twin was first used in predictive maintenance of aircraft engines. If an aircraft engine fails during flight, the result would be disastrous, which is unbearable for both passengers and airlines. Therefore, aircraft engine manufacturers use digital twin technology to predict the entire operation of the aircraft engine, foresee potential failures, perform maintenance in advance, and ensure that the aircraft can fly freely without malfunction. We believe this concept will gradually permeate all walks of life and every corner of society, forming a world of digital twins, making the operation of the entire physical world more efficient and safe. Therefore, we believe that in the 2030s, society will not only have a real physical world but also form a virtualised digital twin world. The digital twin world is composed of digital images of the physical world. It can not only present and simulate the current operating status of the physical world but also predict the future development trend of the physical world. Based on this prediction, some preventive intervention measures can be formed, pre-verified and iteratively optimised in the digital world. After ensuring that they can achieve the expected effect, they can intervene in the operation of the physical world in advance to ensure that the operation of the physical world is in a healthy and safe state, thereby achieving "predicting the future and changing the future".

"Digital Twin, Ubiquitous Intelligence" is becoming the vision for the development of a better society in the 2030s, as shown in Fig. 3.3. "Digital Twin" and "Ubiquitous Intelligence" will greatly enhance the efficiency of social production and governance, help humans further liberate themselves, and improve the quality of life and life. The 6G mobile communication network needs to shoulder its historical mission, empower the social development vision of "Digital Twin, Ubiquitous Intelligence", and achieve the grand goal of "6G reshaping the world".

In the vision of social development towards 2030, characterised by "Digital Twin and Ubiquitous Intelligence", many new business and application scenarios can be distilled and predicted [2], such as holographic interaction, sensory interconnection (or connected senses), digital twin humans, intelligent interaction, super-transport, immersive internet (such as the metaverse), etc. These applications and scenarios will place higher demands on the capabilities of mobile communication network [3], including higher user experience rates, lower latency, higher reliability, more user connection density, and broader coverage brought about by integrated coverage of air, space, land, and sea. Among these new demands, some can be somewhat met

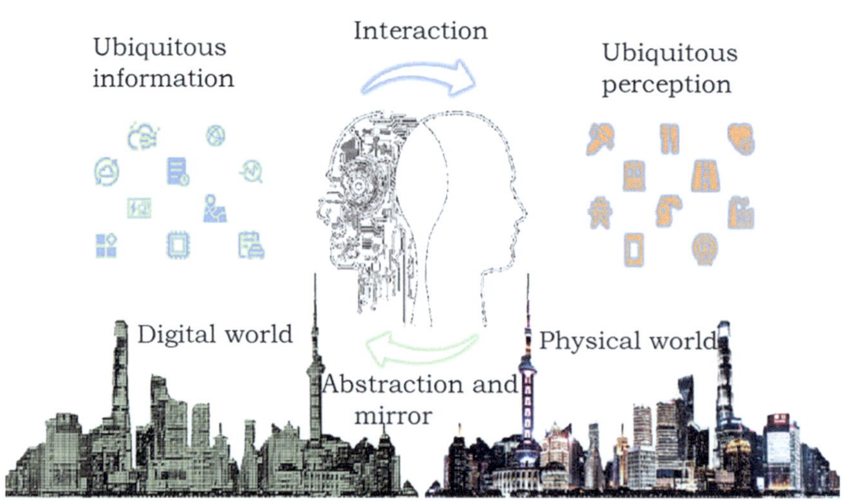

Fig. 3.3 Digital twin world [3]

by 5G and its enhancements, but will face huge challenges in terms of cost and efficiency, while most demands will exceed the capabilities of existing 5G networks and infrastructure, necessitating the redesign of the next-generation mobile communication system. Looking at the capability requirements derived from business scenarios, compared to 5G, 6G not only needs traditional communication capabilities, but also sensing, computing, AI, security and other capabilities. Therefore, from the trend of service development, the integration of communication, sensing, computing, AI and security will be an important direction for 6G development.

Below, we will further elaborate with a few typical examples. Looking towards 2030, holographic interaction will become a potential application, but it places very high demands on network transmission data rate and latency, such as user experience rates reaching up to 500Mbps, end-to-end user plane latency as low as 1–10 ms, powerful terminal computing capabilities, etc., all of which far exceed the capabilities of existing 5G networks. 5G has already begun to consider the application of the tactile internet, looking towards 2030, more sensory interconnection (e.g., touch, smell, taste, etc.) will become possible, which will bring us a more immersive interaction experience, more precise human–machine, machine-machine collaboration, and even a revolution in learning methods. Multi-sensory interconnection requires the network to have less than 1 ms end-to-end user plane latency, centimetre-level positioning accuracy, 99.999% reliability, etc. It can be seen that the requirements for end-to-end latency and positioning accuracy have already exceeded the capabilities of existing 5G networks.

From the historical evolution of the entire mobile communication system, it can be seen that the existing mobile communication systems, while meeting the capability requirements of predetermined scenarios, will also inevitably cultivate more services and applications, generate new network capability requirements, and thus give birth to a new generation of mobile communication systems. This cycle continues, driving the generation-by-generation replacement and development of mobile communication networks.

3.4 Driving by New Technology Development Trends

From the perspective of technological innovation, the continuous advancement of communication, computing, storage, and transmission technologies, the rapid development of new materials, new processes, and new devices, and the continuous integration of new technologies such as cloud computing, big data, blockchain, and artificial intelligence with communication technologies, will unlock new capabilities and architectures, promoting the generational replacement of mobile communication networks.

3.4.1 ICDT Deep Integration

The rapid development of Information Technology (IT) has accelerated the spread of the internet and the constant emergence of various applications. The advent and rapid development of cloud computing have sped up this process. Large cloud computing companies can use inexpensive commercial hardware to quickly and massively deploy IT service capabilities (e.g. computing and storage). Enterprises or individuals can rent the IT service capabilities of cloud computing companies according to business needs, store their data in cloud data centres, and call upon them as needed. These computing capabilities ensure the rapid deployment and application of internet services [4].

Communication technology is also developing and iterating at an astonishing speed. The popularity of 4G networks and smartphones has brought unprecedented prosperity to mobile internet services, changing people's daily lives. Smartphones have become an important platform for people's daily lives, meeting various needs such as travel, shopping, and entertainment. These needs generate massive amounts of data, including location, trajectory, personal preferences, entertainment, shopping habits, and other behavioural data. By collecting and analysing these user behaviour data, internet service providers can obtain user data, provide personalised services, including precise content push and convenient service acquisition, thereby promoting the rapid development of big data processing technology and applications. With the rapid rise of global mobile communication systems, Narrow Band Internet of Things (NB-IoT) [5], enhanced Machine Type of Communication (eMTC) [6] and other IoT technologies, the communication object has extended from people to things, and the number of IoT users has exceeded the number of people. For example, China Mobile has about 950 million human users, while the number of IoT users is about 2 billion. With the deployment and application of 5G networks, wireless transmission rates can reach gigabits per second (Gbps) or even 10 Gbps, wireless transmission delay can be shortened to the millisecond level, and data transmission reliability can also be increased to 99.999–99.99999%. All of these will promote the application of mobile communication technology to every corner of society, achieving the interconnection of all things.

Big data has become an important strategic resource for enterprises and society, and is a new hot topic for researchers and the industry. With the popularisation and application of 4G and 5G systems, the entire mobile communication network and its applications are generating massive amounts of data every moment. These data contain a lot of rich information from the entire society, and the rapid development of data technology (DT) such as smart shopping, smart transportation, smart medical care, smart campus, smart city will apply these data to human life and social governance, promoting the informatisation and digital development of the entire society. Digital transformation will bring new prosperity and development momentum to all walks of life, accelerate the digital process of the entire society, and achieve the goal of "5G changing society".

With the acceleration of digitalisation, every social element will generate a large amount of data, which comes from individuals, companies, infrastructure, etc. Because the ownership of these data is completely different, issues such as data storage, data management, data sharing, data security and privacy, and data transactions become difficult to handle. Therefore, there will inevitably be related big data legislation in the future to clarify data ownership and related interest distribution. At the same time, big data storage and management platforms will emerge to help everyone with data storage, management, and transactions.

Big data may be the next round of technological revolution following computers and the internet. The emerging technologies that come with it, such as data mining, machine learning (ML), artificial intelligence (AI), may change many algorithms and basic theories in the data world, achieving new breakthroughs in science and technology. Big data cannot be separated from cloud computing, which is the platform for generating big data [7, 8]. Big data and cloud computing have been closely integrated since 2013, and it is expected that their relationship will become even closer in the future.

The large-scale deployment of big data has driven the development of AI applications. AI is a branch of computer science that attempts to understand the nature of intelligence and create a new intelligent machine that can respond in a way similar to human intelligence. Since the birth of AI, related theories and technologies have become increasingly mature, and the application field continues to expand. AI can simulate the information process of human consciousness and thought. At present, the research fields of AI mainly include knowledge representation, automatic reasoning and search methods, ML, knowledge acquisition, knowledge processing systems, natural language understanding, computer vision, intelligent robots, automatic programming, etc. The applications of AI mainly include machine translation, intelligent control, expert systems, robots, language, image recognition (such as face recognition and license plate recognition), robot factories based on genetic algorithms, automatic programming, and aerospace applications. These applications profoundly affect our daily life and work.

In the design of 5G networks, the trend of ICDT integration has emerged. The core network design of 5G Service-based Architecture (SBA) fully introduces advanced IT concepts, and through Software Defined Network (SDN), Network Function Virtualization (NFV), and SBA, network slicing is achieved, which can provide flexible differentiated business QoS guarantees, providing important support for empowering vertical industries in 5G networks. 5G mobile communication networks usually consist of millions of base stations, routers, core network elements and other infrastructure equipment, as well as billions of users, which can generate a large amount of data, including the operation data of each network element, the signalling data generated during communication, event reports, and information related to user mobility in the network. If time, location and other tags are added to these data, it will bring immeasurable value to the automation and intelligence of network operation and maintenance. Therefore, based on the location information of users in the network, operators have begun to study network automation based on big data and artificial intelligence, such as large-scale Multiple Input and Multiple Output (MIMO) weight optimization, network anomaly analysis, user experience analysis and optimization [9] etc. At the same time, 3GPP has begun to study big data collection in wireless networks [10], automation and intelligence in network operation and maintenance [11] and the application of AI in wireless resource scheduling [12]. We can see that in the later stage of 5G standard evolution, the application of big data will further integrate with communication technology. Therefore, ICDT integration is becoming a new development trend, not only bringing new capabilities and characteristics to 5G, enhancing network business capabilities and user experience, but also further empowering the automation and intelligence of network operation and maintenance, reducing the operation and maintenance costs of 5G networks.

However, the existing 5G system cannot perfectly match the deep integration of ICDT, and the Network Data Analytic Function (NWDAF) is an example. NWDAF is a data analysis network element based on automatic perception and analysis of network data, participating in the entire life cycle of network planning, construction, operation and maintenance, optimization and operation, making the network easy to maintain and control, improving network resource utilization, and enhancing user experience. Although NWDAF can effectively improve network performance, this patch-based AI exposes some problems in the design of the 5G system: one is the issue of data security and the transmission overhead caused by a large amount of measurement data reporting; the other is the challenge of low latency, because all data must be uploaded to a centralized analysis unit (such as NWDAF) and processed on it, and this analysis unit may be deployed far from the data source. These problems will limit its effectiveness and efficiency, and only by evolving and transforming the existing architecture can the problems be fundamentally solved.

Therefore, we believe that the deep integration of ICDT will inevitably become an important driving force for the design of 6G networks, and cloud-native, big data, and AI will play an important role in the design of future network architectures.

3.4.2 New Hardware and Solutions

1. Moore's Law has led to a continuous increase in chip integration and computing power [13]

In 1965, Gordon Moore, while preparing a report on the trend of computer storage development, discovered an astonishing trend: each new chip generally contains twice the capacity of its predecessor, and each chip is produced within 18 to 24 months after the previous one. If this trend continues, computing power will rise exponentially relative to the time cycle. This observation by Moore is now known as Moore's Law, and the trend it describes continues to this day and remains accurate. It has also been found to apply not only to the description of memory chips but also accurately reflects the development of processor capabilities and disk drive storage capacity. This law has become the basis for many industries to make performance predictions [13].

To summarise, there are mainly three "versions" of "Moore's Law":

(1) Integrated circuit chips double the number of integrated circuits every 18 months;
(2) The performance of microprocessors doubles every 18 months, while the price drops by half;
(3) The computer performance that one dollar can buy doubles every 18 months [14].

Among the above statements, the first one is the most common, and the latter two involve price factors, but their essence is the same. Although the three statements have their own merits, they have one thing in common, that is, the "doubling" cycle is 18 months. As for whether the "doubling" is the "number of integrated circuits on the integrated circuit chip", the entire "performance of the microprocessor", or the "computer performance that one dollar can buy", it is up to the individual to decide [14].

Although theoretically the size limit of a transistor is the diameter of a silicon atom, the thresholds set by thermodynamic limits and quantum mechanical limits are larger than this diameter. Research shows that the limit size of a silicon transistor is about 1 nm (nanometre), which is the theoretical limit of a single transistor device. There will be many engineering technical problems in actual mass production. The most advanced process currently in mass production is TSMC's 5 nm transistor, and there is a good chance of achieving 3 nm mass production in two or three years. But whether 2 nm or even 1 nm transistors can be mass-produced in the future is still uncertain.

Over the past nearly 60 years, Moore's Law has been repeatedly verified in the continuous advancement of integrated circuit technology, and has driven the rapid development of the information and communication industry. We can continuously support the implementation of more advanced technologies and concepts through increasing complexity, thereby achieving better performance, so this trend will continue until the end of Moore's Law. Especially in recent years, with the continuous improvement of computing power and the continuous reduction of costs, cloud computing and artificial intelligence (AI) have begun to develop and apply in many scenarios, especially the graphics processor (Graphics Processing Unit, GPU) launched by Nvidia, has pushed the research and application of AI to a climax. Through high-complexity brute force calculations, significant improvements in performance and application effects can be achieved. This has stimulated the industry to develop dedicated computing chips for specific application scenarios, and various "XPUs[1]" can be said to be dazzling, greatly promoting the progress of information and communication technology. Moore's Law is shown in Fig. 3.4.

Especially with the advancement of integrated circuit manufacturers such as TSMC from 7 to 5 and 3 nm, under the same conditions, mobile communication systems and terminals can support more complex calculations and processing. Technologies that seemed complex and unachievable will become feasible, thereby stimulating mobile communication researchers and standard organisations to consider

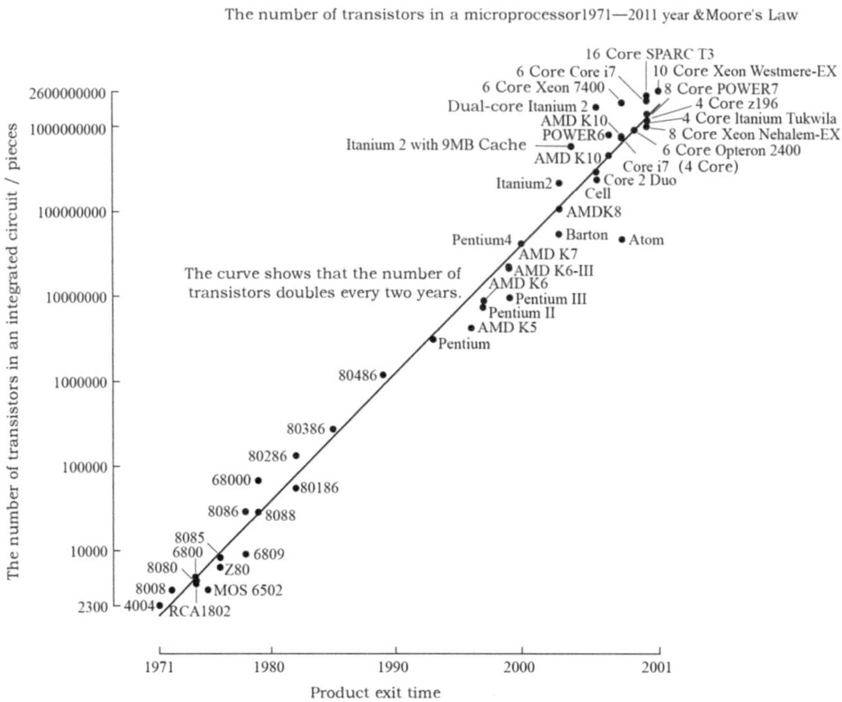

Fig. 3.4 Moore's law [13]

the research and breakthrough of more advanced technologies, which will inevitably promote the birth of a new generation of mobile communication systems.

2. Heterogeneous Computing

Heterogeneous computing is a hybrid computing system composed of CPUs, co-processors, system-on-chip (SoC), GPUs, application-specific integrated circuits (ASIC), field-programmable gate arrays (FPGA), and other computing units that use different types of instruction sets and different system architectures. Heterogeneous computing appears in the form of "CPU+", has good feasibility and versatility, and can greatly improve system performance and power efficiency. The characteristics of different heterogeneous hardware are shown in Fig. 3.5.

The acceleration of business heterogeneity needs a comprehensive software and hardware solution based on the acceleration platform to achieve stronger performance and cover more scenarios, including GPU-based, FPGA as a Service (FaaS) and Domain Specific Architecture (DSA)/ ASIC acceleration schemes. For example, NVIDIA's GPU acceleration is mainly achieved through the Compute Unified Device Architecture (CUDA) programming development framework. FaaS relies on the hardware programmability provided by FPGA, and requires users or third-party developers to complete the development of acceleration hardware and software

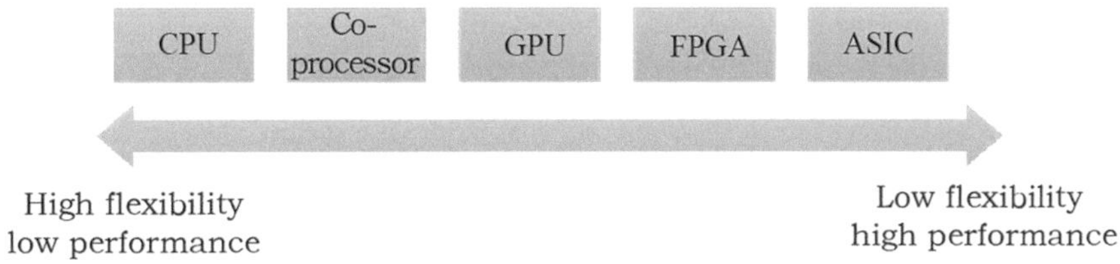

Fig. 3.5 Characteristics of different heterogeneous hardware

images for specific application scenarios; DSA provides higher flexibility for specific application scenarios on the basis of ASIC, and its efficiency is higher than that of GPU and FPGA [15].

With the development of mobile communication applications, the integration of ICDT has become a trend. Mobile communication networks need to further expand their own capabilities to better support various differentiated and fragmented application scenario requirements. The development of heterogeneous computing hardware can well meet this development demand, flexibly and efficiently support the implementation of integrated services such as communication connections, computing, AI and big data, and promote the further transformation and evolution of network architecture and technology.

3. Cloud Native

Cloud computing integrates multiple relatively low-cost computing entities into a system with powerful computing capabilities through the network, and provides software as a service (SaaS), platform as a service (PaaS), infrastructure as a service (IaaS), and management service providers through the network. Advanced business models such as Managed Service Providers (MSP) bring powerful computing capabilities into the hands of end users. A core concept of cloud computing is to continuously enhance the processing power of the "cloud", thereby reducing the processing burden on the user terminal, ultimately simplifying the user terminal into a pure input–output device, and allowing users to enjoy the powerful computing processing capabilities of the "cloud" as needed. With its large scale, virtualisation, high reliability, strong universality, high scalability and low cost, cloud computing has quickly gained widespread application among individuals and businesses. Cloud-native design has become a popular trend in IT system and information network design, bringing about agile business deployment, online and iterative operations, simplified operations and reduced costs under cloud-native.

Cloud-native is a collection of thoughts on a series of technologies, design patterns and management methods, including DevOps (a combination of Development and Operations), continuous delivery, microservices, agile infrastructure, Conway's law, and the reorganisation of company organisational structures. Gartner mentioned in its technology trend prediction report for 2022 [16] that "by 2025, 95% of digitalisation projects will adopt cloud-native infrastructure, while in 2021 this proportion was less than 40%". In the network of telecom operators, core network elements and edge computing nodes based on cloud-native technology have been widely used, and in the field of wireless access networks, base station units based on cloud-native technology have also begun to be piloted and used on a small scale.

Representative technologies of cloud-native include containers, microservices, service meshes, immutable infrastructure and declarative application programming interfaces (APIs) [17]. Container technology is based on operating system virtualisation technology, allowing different applications to run in independent sandbox environments to avoid mutual interference. The Docker container engine greatly reduces the complexity of container technology. Docker images decouple applications from the running environment, allowing applications to run consistently and reliably across different computing environments, accelerating the popularisation of container technology. Container technology has now developed into various forms such as full container, edge container, serverless container, and bare metal container. Microservices separate modules with different lifecycles through service-oriented architecture, iterate them separately, thereby accelerating the overall progress and stability. Microservices are deployed with containers, and different numbers of instances can be deployed for each microservice, achieving separate scaling, separate upgrades, making the overall deployment more economical, and improving iteration efficiency. Service-oriented architecture uses interface-oriented programming, the functions inside the service are highly cohesive, and the reuse of software is increased by extracting common function modules. Service Mesh separates the business plane and the control plane, separating service proxies, discovery and governance from the business to a dedicated Mesh infrastructure layer, and making it transparent to the business. After separation, only lightweight service proxies (Sidecars) are retained in the business process, and the service proxies are responsible for communicating with the Mesh control plane. After implementing the Mesh architecture, a large number of distributed architecture patterns (circuit breaking, rate limiting, degradation, retry, back pressure, isolation) are completed by the service mesh control plane, and a unified control plane can also ensure better security.

Serverless is an architectural concept, the core idea of which is to abstract the infrastructure that provides service resources into various services, and provide them to users on demand through APIs, truly achieving on-demand scaling and pay-per-use. This architecture eliminates the need for traditional massive continuous online server components, reduces the complexity of development and operation, reduces operating costs and shortens the delivery cycle of business systems, allowing users to focus on the development of business logic with higher value density.

In terms of cloud-native standards in the telecommunications field, ETSI NFV ISG released an enhanced NFV architecture research report for cloud-native containers and Platform as a Service (PaaS) in October 2019, and then formulated the northbound interface of the container layer and management and network orchestration (Management and Network Orchestration, MANO) manages a series of technical specifications for containers, with plans to further develop standards or establish work on container cluster management technical specifications (IFA036), container network research reports (IFA038), and container security specifications (SEC023), and to further expand the existing NFV MANO interface functions to support the life-cycle management and orchestration of containerised virtual network functions. In the open source field, CNCF has established the Telecom User Group (TUG), which introduces the needs of the telecommunications industry into the design of upstream open source projects, and has completed the white paper on "Cloud Native Thinking for Telecommunications". The Linux Foundation has established the CNTT working group to study the application of cloud native/container technology in the telecommunications industry, mainly based on open source Kubernetes to define the infrastructure architecture of cloud native networks, analyse the gap with telecommunications business needs, and provide reference implementations and testing verification frameworks for the corresponding infrastructure [18].

In the design of the 5G core network, cloud native design has been introduced, which brings advantages such as agility and elasticity, but it has not yet been introduced into the wireless access network field, so end-to-end cloud native design will be the direction of network development and the basic principle of next-generation network design.

4. Virtualisation

Virtualisation is a technology that reallocates physical resources logically, dividing "large resources" logically into "small resources with independent functions", which can maximise the use of resources and achieve user isolation based on shared resources. Virtualisation technology is the cornerstone of cloud computing and is ubiquitous in the cloud.

In the field of computing, the levels of virtualisation are as follows:

(1) Hardware abstraction layer-based virtualisation: Provides a hardware abstraction layer, including the abstraction of hardware resources such as processors, memory, I/O devices, interrupts, etc.
(2) Operating system-level virtualisation: Provides multiple isolated user-space instances, i.e., containers. Containers have independent file systems, networks, system settings, and library functions, etc.
(3) Programming language-based virtualisation, such as the Java Virtual Machine (JVM), is process-level virtualisation.

From the virtual platform perspective, virtualisation can be divided into the following categories:

(1) Full software virtualisation: No need to modify the client operating system, all operations are simulated by software, but the performance consumption is high, between 50 and 90%.
(2) Paravirtualisation: The client operating system modifies the kernel and driver program, calling the hypercall provided by the hypervisor, with performance consumption between 10 and 50%.
(3) Full hardware virtualisation: Hardware supports virtualisation, with performance consumption only between 0.1 and 1.5%.

In addition, virtualisation technology also includes CPU virtualisation, memory virtualisation, I/O device virtualisation, storage virtualisation, network virtualisation, container virtualisation, network function virtualisation, as well as 5G network slicing, etc.

For deep integration towards ICDT, different hardware needs to support different applications, such as GPUs supporting AI applications, CPUs supporting general cloud computing, structured application-specific integrated circuits supporting high real-time computing requirements, FPGAs supporting hardware acceleration, etc. Therefore, virtualisation based on heterogeneous hardware to flexibly support the integration of various applications is the trend, and will also become a new driving force for the development and evolution of mobile communication networks.

3.5 Existing Network Problems and Challenges

Since 2019, 5G networks have been deployed on a large scale worldwide. The integration of 5G networks with cloud computing, big data, and AI will inevitably give birth to many new services and applications, thereby promoting the digitisation of the entire society. With the development of 5G networks and the continuous emergence of new services and applications, 5G networks will inevitably face some new problems and challenges. Some of these problems may be resolved in the development of 5G networks, but due to the inherent limitations of 5G networks, some problems may be difficult to solve, thus becoming an important driving force and source for network evolution and transformation.

3.5.1 Layered Protocol Stack Becomes a New Bottleneck

In 5G networks, the air interface protocol adopts a layered model, including the physical layer, Medium Access Control (MAC) layer, Radio Link Control (RLC) layer, Packet Data Convergence Protocol (PDCP) layer, and Service Data Adaptation Protocol (SDAP) layer. All service data must be processed through these layers, and the processing of each layer will introduce specific delays, thus making delay a bottleneck. For example, the typical delay of eMBB data packets in the air interface is 3 ms. During the research process of 5G networks, some schemes that can reduce processing delay are introduced.

(1) Allow packet decryption to be performed before reordering at the PDCP layer to reduce packet processing volume and delay.
(2) Remove the packet cascading function of the RLC layer, decouple the automatic retransmission request and cascading/segmentation functions, and achieve more offline packet header calculations.

(3) At the MAC layer, allow MAC subheaders to be placed next to the MAC payload, addressing the potential requirement of approximately one symbol time between uplink grant and uplink transmission [19–22].

All these modifications are aimed at reducing delay by optimising the processing order. Although some of these modifications are beneficial and have been recognised in the 5G standard, these modifications are still limited to the existing layered system. If further reduction of air interface delay is needed, the most direct method is to break through the traditional layered system and create shortcuts in the data processing channel.

With the SBA thinking in IT technology, we can break down the original protocols into several services. According to the needs of the applications, these services can be dynamically orchestrated and integrated to support the required capabilities and meet personalised user needs. This can not only bring further performance optimisation space but also further enhance the agility and elasticity of the end-to-end network.

3.5.2 Continuously Evolving Network Slicing

Facing vertical industries, 5G networks need to have diverse network capabilities and deployment flexibility. To meet these needs, the 5G Core Network (CN) has introduced SBA and network slicing based on SDN/NFV. The 5G system is expected to support end-to-end network slicing. However, at the beginning of the standard design, the design and optimisation of network slicing were mainly carried out in the CN and Transport Network (TN). In the early versions of the 5G standard, RAN did not have a special design for network slicing, and related research only began in the Rel-17 version. The standardisation of network slicing involves six major industry organisations, each with their own division of work [23], and the slow progress of coordination between organisations has limited the commercialisation of end-to-end network slicing technology.

Commercial CN devices that support network slicing are relatively mature, but due to the technical difficulties of RAN slicing, there are differences in the implementation of RAN slicing among different equipment manufacturers. Overall, there are still some technical challenges in the implementation of end-to-end network slicing, which requires multi-domain coordination and connection, and the complexity is high.

In order to better exploit the value of network slicing, it is necessary to reasonably orchestrate and manage network slicing in the slicing management domain. Figure 3.6 shows a schematic diagram of the network slicing management domain, which gives the relationship between the slicing management domain and the slicing network domain. In commercial networks, these network slicing management functions (Network Slice Management Function, NSMF) are not independent, but are embedded in the operator's operation support system (Operation Support System,

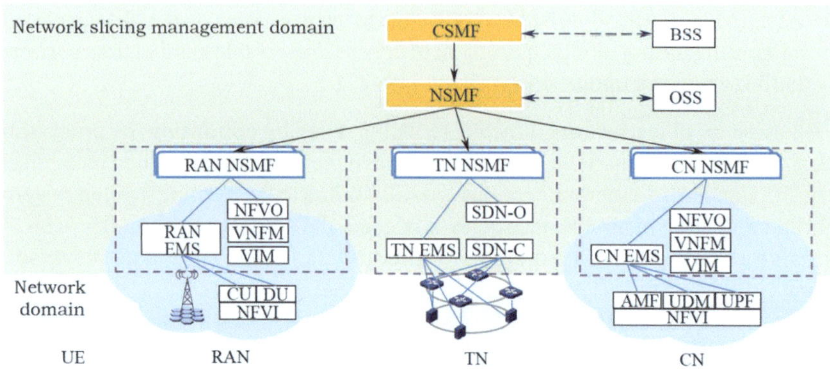

Fig. 3.6 Schematic diagram of the network slicing management domain

OSS) and business support system (Business Support System, BSS). Different operators have different network slicing management schemes, and the operator's operation and maintenance department, IT department, government and enterprise department, and network department need to participate in the integration work of the slicing management domain and BSS/OSS system. This is a challenge for operators and should be fully considered in future network design.

3.5.3 Solidified Network Architecture

For mobile networks, due to user mobility, there is a significant difference in load between base stations. Generally speaking, in China, 20% of the base stations are overloaded, while 80% of the base stations are lightly loaded. However, each base station is usually configured to full capacity, resulting in a waste of hardware processing capacity and power consumption. Since 5G base stations have larger bandwidth and massive MIMO configurations, the cost and power consumption of 5G base stations are about 3 times that of 4G base stations, further increasing network costs and power consumption. A simple way to reduce network deployment costs is to dynamically configure and share hardware and processing capabilities between different sites, and dynamically schedule and configure corresponding software and hardware resources and functions according to the services and load requirements of each cell.

The 5G system also tries to share hardware and processing capabilities. To support the flexible deployment of 5G networks, 5G NR introduces a base station CU and DU separation architecture, and discusses various possible CU and DU function splitting options [24], The function split between CU and DU is shown in Fig. 3.7. For different function splitting options, after multi-dimensional comparative analysis, the final standard only supports one high-level splitting option, namely option 2 (PDCP/RLC split).

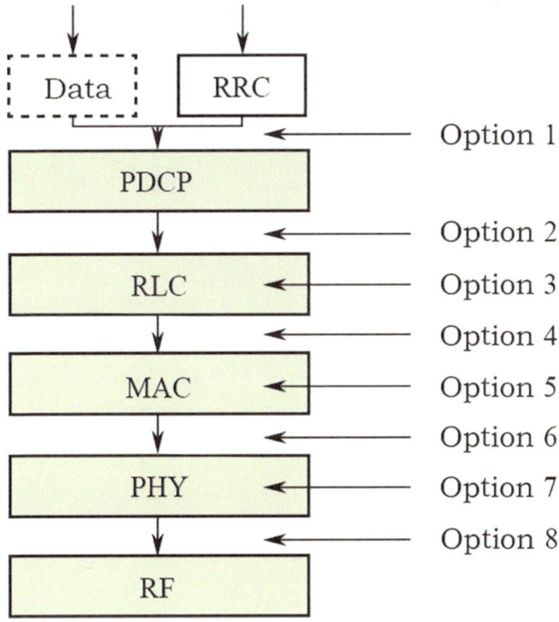

Fig. 3.7 Function split between CU and DU [24]

With the CU and DU separation architecture, the CU can be centrally deployed, sharing processing capabilities between different sites. But from the perspective of network performance, the benefits of centralized CU are not obvious, and in actual networks, CU is still deployed with DU.

On the contrary, from the perspective of saving the rent of CU and DU machine rooms, mobile operators deploy the CU and DU of multiple sites together, laying the foundation for sharing capacity and hardware processing capabilities between different sites. However, because the hardware and software of the CU and DU of the existing 5G base station are all dedicated, different sites in the pool still cannot share capacity and hardware processing capabilities. A simple and feasible way to achieve sharing is to use a cloud-native method based on SDN and NFV to design CU and DU. The schematic diagram of the CU deployment location is shown in Fig. 3.8. This method can realize the dynamic sharing of physical hardware resources and provide capacity and processing capabilities on demand. When the load of the pool is light, most of the hardware in the pool can be turned off to save power; when the load becomes heavy, the necessary hardware can be started to support site requirements.

3.5.4 5G High Power Consumption of the Network

As mentioned in Chap. 2, high power consumption is a thorny issue for 5G base stations and is the heaviest burden for mobile operators. In addition to CAPEX, the

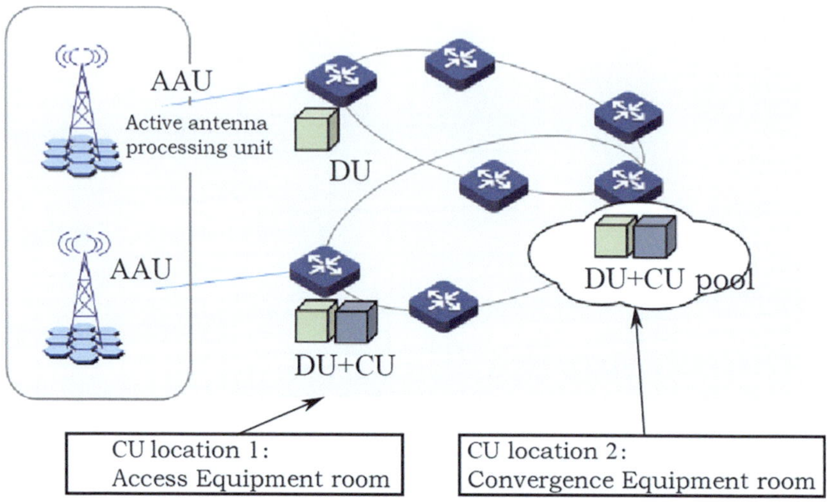

Fig. 3.8 Schematic diagram of CU deployment location

operating cost (Operating Expense, OPEX) mainly based on energy consumption is an important factor affecting the development of 5G. From 2017 to 2020, with the joint efforts of the industry, the power consumption of 5G base stations has been significantly reduced. However, the power consumption of 5G base stations is still about three times that of 4G base stations.

Currently, the power consumption of a 5G base station at full load is as high as 3800 W, while the average power consumption of a 4G base station is about 1000 W. Under normal load, the radio frequency unit consumes most of the power of the 5G base station. Due to the high requirements for baseband processing capabilities due to large bandwidth and high data rates, baseband processing consumes the remaining power.

At the same time, because 5G currently coexists with 2G, 3G and 4G, and deploy multi-frequency networks, coupled with their large scale, the power consumption of base stations has become the largest expenditure of operator OPEX. For this reason, the industry has proposed a variety of base station energy-saving solutions. The current solution is based on factors such as base station equipment type, coverage scenario, energy-saving target, shutdown time, etc., to automatically generate energy-saving strategies including cell shutdown, carrier shutdown, RF channel shutdown, and symbol shutdown through AI algorithms. Each strategy has its specific application scenarios and different degrees of impact on network quality. With the proposal of dual carbon targets, network energy saving has become an important direction for future network development, and the use of green energy is also a problem that future networks need to consider. To achieve these goals, in addition to using existing technologies, the network needs to undergo changes and evolution in the basic framework of architectural design, multi-frequency band coordination, etc.

3.5.5 *Highly Complex Operation and Maintenance Management*

Due to the relatively traditional methods of network management and maintenance, the operation and maintenance efficiency of current mobile communication networks is still low, and there has not been a revolutionary change with the multiple generational changes of the network. It has been reported that network-level shutdown incidents have occurred in China, Japan, and Europe, greatly reducing the reputation of mobile operators. For 5G networks, the complexity of operation and maintenance stems from the large-scale deployment of base stations, the interoperability of 5G with 4G/3G systems, the dynamic spectrum sharing of 5G and 4G networks, the configuration of hundreds of parameters, the core network form based on SDN/NFV, network slicing, diverse business needs of vertical industries, and other aspects [25].

As mentioned in Chap. 2 of this book, operators have begun to study the use of AI and big data to support intelligent network operations, in order to improve the efficiency of network operations and reduce the cost of network operation and maintenance. The intelligence of 5G networks mainly involves integrating AI and other technologies with the hardware, software, systems, and processes of 5G communication networks, using AI and other technologies to help the intelligent process of communication network operations, improve quality, increase efficiency, and reduce costs; the intelligence of 5G networks also needs to promote the technical and system changes of the network itself, enable agile business innovation, promote the construction of smart networks, including cloud network intelligence, network operation and maintenance intelligence, and network service intelligence.

The intelligence of 5G networks is mainly aimed at optimising communication connections. Although a service cloud has been introduced, due to the established 5G architecture, protocol functions, and processes, it can only make incremental iterations on the existing architecture scheme, introduce necessary network changes to obtain the required data and computing power through patching according to scenarios and use cases. Therefore, the intelligence of 5G networks mostly uses the mode of add-on AI, based on the characteristics of AI applications designed externally, it will face the following challenges [25].

(1) The lack of a unified standard framework leads to a lack of effective verification and safeguarding methods for AI applications. The verification of the effectiveness of AI applications is carried out post-factum, making the end-to-end overall process long and complex. The intermediate process generally requires a large amount of human intervention, and the impact on the existing network is also relatively large. This makes it difficult for AI to be truly applied to the existing network.

(2) It is difficult to achieve pre-verification, online evaluation, and fully automatic closed-loop optimisation in the add-on mode. In the add-on mode, AI model training usually requires the preparation of a large amount of data in advance, and it is difficult to collect data from the existing network centrally, and the transmission overhead is also large. This leads to problems such as a long AI

model iteration cycle, high training overhead, slow convergence, and poor model generalisation.

(3) In the add-on mode, computing power, data, models, and communication connections belong to different technical systems. For cross-technical domain collaboration, it can only be achieved through management-level communication, usually leading to second-level or even minute-level delays, and the quality of service is also difficult to effectively guarantee.

Therefore, although operation and maintenance automation and intelligence are important directions for achieving low-cost, high-efficiency operation of 5G networks, due to the lack of full consideration of big data and AI applications in network operations at the beginning of 5G network design, AI applications are difficult to get effective support under the current 5G network architecture, and it is also difficult to achieve the expected results. As a result, AI applications are also driving the transformation and evolution of network architecture design.

3.6 Summary

With the iterative upgrade of mobile communication technology and the continuous improvement of mobile communication network capabilities, its application field has expanded from traditional human services to the interconnection of all things. The mobile communication network has become the most important infrastructure for the entire society and economic development. With the large-scale application of 5G, 5G accelerates the application of cloud computing, big data, and artificial intelligence. While cultivating new businesses and applications beyond the existing network capabilities, it also brings about the deep integration of ICDT, the future capability improvement of chip and device process levels, providing a huge space for the architectural transformation, efficiency improvement, and capability expansion of mobile communication networks, thereby promoting the development of mobile communication systems towards the next generation.

References

1. LIU G Y, HUANG Y, LI N, et al. Vision, requirements and network architecture of 6G mobile network beyond 2030 [J]. China Communications, 2020, 17(9): 92–104.
2. Liu Guangyi, Huang Yuhong, Cui Chunfeng, et al. 6G reshapes the world [M]. Beijing: People's Posts and Telecommunications Publishing House, 2021.
3. China Mobile. 2030+ Vision and Demand Research Report [R]. 2019.
4. LIU G Y, LI N, DENG J, et al. The SOLIDS 6G Mobile Network Architecture: Driving Forces, Features, and Functional Topology [J].Engineering, 2022, 8(1): 42–59 .
5. XIE Y Z. The evolution of NB-IoT standard system and development of Internet of things industry [J]. Chinese Journal on Internet of Things , 2018, 2(1): 76–87.

6. ZHANG L F, LI F C, HU Z Y, et al. Study on key technology and deployment scheme of LTE-eMTC. Des Tech Posts Telecommun 2018, 7: 1–5.
7. MAYER-SCHöNBERGER V, CUKIER K. Big data: a revolution that will transform how we live, work, and think [M]. New York: Houghton Mifflin Harcourt, 2013.
8. ZHOU T, PAN Z Y, CHENG X Q. Developing tendency prediction of big data in 2019 from CCF TFBD [J]. Big data research, 2019, 5(1): 109–115.
9. ITU-T . Y.3173: Framework for evaluating intelligence levels of future networks including IMT-2020 (Study Group 13) [S]. ITU-T standard. Geneva: ITU, 2020.
10. 3GPP. TR 37.816: Study on RAN-centric data collection and utilisation for LTE and NR [S]. France: 3GPP, 2019.
11. 3GPP. TR 23.791: Study of enablers for network automation for 5G. 3GPP standard [S]. France: 3GPP, 2019.
12. LIN M T, ZHAO Y P. Artificial intelligence-empowered resource management for future wireless communications: A survey [J]. China Communications, 2020, 17 (3).
13. Moore's Law [ED/OL].[2022–10–18]. Baidu Encyclopedia.
14. Zhou Su, Wang Shuoping. Management Information Systems in the Era of Big Data [M]. Beijing: China Railway Publishing House, 2017.
15. Huang Chaobo. Hardware and Software Integration: The Path to Innovation in Ultra-Large-Scale Cloud Computing Architecture [M]. Beijing: Electronics Industry Press, 2021.
16. Gartner. Top Strategic Technology Trends for 2022 [EB]. Gartner website. 2021.
17. CNCF. Cloud Native Computing Foundation (CNCF) Cloud native definition [EB]. CNCF website. 2021.
18. China Mobile, China Telecom, China Unicom, etc. 2020 Telecom Industry Cloud Native White Paper [R]. 2020.
19. 3GPP . Way forward on U-plane modifications to LTE baseline [R]. 2016.
20. 3GPP. TS 38.321: 5G new radio-medium access control (MAC) protocol specification [S]. France: 3GPP, 2021.
21. 3GPP. TS 38.322: 5G new radio-radio link control (RLC) protocol specification [S]. France: 3GPP, 2021.
22. 3GPP. TS 38.323: 5G new radio-packet data convergence protocol (PDCP) specification [S]. France: 3GPP, 2021.
23. Wang Weibin, Lu Guanghui, Chen Xinyu. Key Technologies and Challenges for Commercial Use of 5G Core Network [J]. ZTE Communications, 2020, 26(3):9–16.
24. 3GPP. TR 38.801: Study on new radio access technology-radio access architecture and interfaces [S]. France: 3GPP, 2017.
25. China Mobile Research Institute. White Paper on 6G Inherent Intelligence Architecture and Technology [R]. 2021.

Chapter 4
Overview of 6G Research

Since 2019, with the launch of large-scale commercial use of 5G, the global mobile communication industry has begun to shift its focus to 6G, and major economies around the world have successively published research and development plans for 6G. The research layout of 6G in China is basically in sync with the international community, and the Ministry of Science and Technology and the Ministry of Industry and Information Technology have successively established the 6G Promotion Group and the IMT-2030 Promotion Group, organising domestic enterprises and universities to carry out related research on 6G. From the discussions in the industry, it is expected that 6G will begin standard research and specification around 2025, and have the capability for large-scale commercial use around 2030. This chapter will detail the current status of 6G research globally.

4.1 Overview of the 6G Research Situation

The research on 6G was first launched in the academic world [1], with the University of Oulu in Finland having a significant influence. Funded by the Finnish government, it launched its 6G flagship project, 6Genesis, and in March 2019, it organised the first global 6G wireless summit. It then released a series of technical white papers on 6G in conjunction with major universities, research institutions, and companies internationally, having a significant impact. Following this, countries and regions with advanced communication technology globally, such as the United States, the European Union, China, Japan, and South Korea, have also published corresponding 6G R&D plans. Among them, South Korea is the most aggressive, proposing to be the first to commercialize 6G globally in 2028, hoping that South Korea can become the world's number one in terms of the share of 6G core standard patents, the global market share of smartphones, and the second in the global market share of communication network equipment. Japan proposes to commercialise 6G in 2030,

G. Liu et al., *6G Migrates from Communication to XaaS*,
https://doi.org/10.1007/978-981-96-7579-1_4

hoping to build international competitiveness in integrated circuits and materials through 6G, achieving a 6G patent share of more than 10%, and a global market share of 6G infrastructure of 30%. The European Union, continuing its 5G R&D approach, is gradually introducing 6G research plans, and the 6G flagship project Hexa-X led by Nokia and Ericsson officially launched in 2020, releasing a series of research reports on 6G vision, requirements, and network architecture. Due to the constraints of the 5G deployment frequency, the progress of 5G commercialisation in the United States is relatively slow, far behind the three Asian countries. Some government officers and scholars even propose to skip 5G and develop 6G directly, so the US government is actively seeking to re-establish the leading role in the development of the 6G industry. The FCC opened the Tera Hertz (THz) frequency for 6G research and experimentation as early as 2018, and the US Department of Defense also funded research projects on communication and perception fusion. The Alliance for Telecommunications Industry Solutions (ATIS) even established the Next G Alliance at the end of 2020, hoping to reshape the dominant position of the United States in mobile communication technology standards and industry.

The R&D plan for 6G in China is basically in sync with the global community. The Ministry of Science and Technology launched a key R&D project for 6G as early as 2018, and in November 2019, it established the 6G Promotion Group, beginning to fully layout and promote 6G research. In June 2019, the Ministry of Industry and Information Technology established the 6G study Group, and in June 2021, it released the first version of the "6G Overall Vision and Potential Key Technologies" white paper. In November 2019, China Mobile took the lead in officially releasing the "2030+ Vision and Requirements Report" white paper at its global partner conference, and subsequently, China Unicom, Datang Mobile, and vivo also successively released 6G white papers.

Drawing from the entire development process of 4G and 5G, we can preliminary judge that the entire 6G development will probably be divided into two stages (see Fig. 4.1): the first stage (2018—2025), vision, requirement definition and potential key technology research, system concept design and prototype verification; the second stage (2026—2030), 6G standardization, end-to-end industrialisation, services and applications cultivation and commercial deployment. Currently, 6G research has entered the critical window period for core technology breakthroughs and is in a stage where all kinds of ideas are flourishing and competing, with the focus on vision, demand definition, and key technology research and verification.

Looking at the current global situation, there are still the following challenges in 6G development.

Firstly, breakthroughs are still needed in basic theory. Existing communication technologies are approaching the limits of Shannon's theorem and Moore's law, and 6G calls for more source technology innovation. Moreover, it is very important to do a good job in top-down and bottom-up coordinated layout and innovation in advance.

Secondly, there is a risk of fragmentation in technical standards. Every country is accelerating the development of 6G mobile communication technology, proposing different technological development paths, 6G has many candidate technologies and a wide range, the international situation is more complex, especially the uncertainty

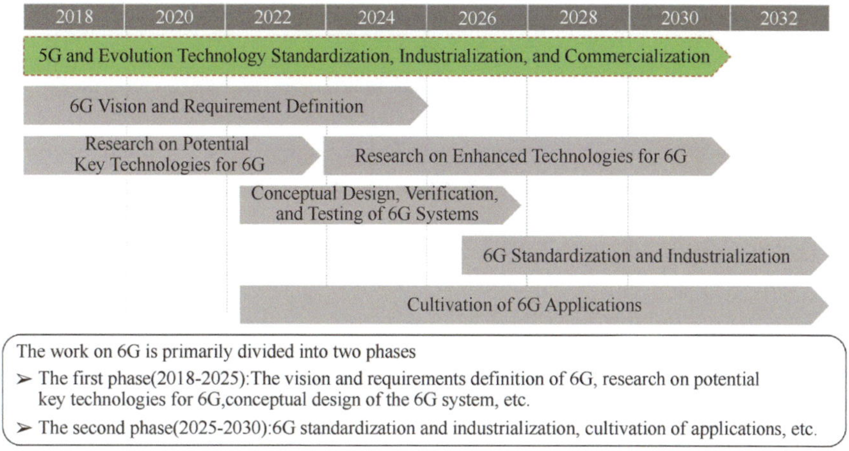

Fig. 4.1 6G R&D roadmap

of geographical politics, whether a global unified standard similar to 5G can be formed is still uncertain though the whole community is advocating and pursuing a global 6G standard.

Thirdly, there is uncertainty in the industrial model. Facing the development trend of converged ICDT, the "horizontal integration" industrial model based on "open source software + white box hardware" is developing rapidly. Compared with the traditional "vertical integration" industrial model based on "dedicated software + black box hardware", it remains to be verified which is better, and competitive industrial paths may be formed.

Fourthly, the complexity of ecosystem construction has increased. Compared with 5G, 6G will expand more scenarios, integrate more technologies, create more new fields, require new business models and more complicated industrial ecosystems, and at the same time, "dual carbon" and "safety" have become rigid demands, ensuring that it can support the low-carbon and efficient development of the economy and society with its own low carbon.

To address these challenges, the entire eco-system needs to strengthen cooperation and coordination, systematically advance 6G research and development work, strive to connect the innovation links of the 6G eco-system from theory, technology, standards, products to applications, jointly promote the formation of global unified standards and a unified ecosystem, create a cooperative and win–win industry atmosphere, and achieve the healthy and sustainable development of 6G.

4.2 Progress of 6G Research in China

In China, various universities and enterprises, drawing on the experience of 4G and 5G development, started the layout of 6G early. China's research and development progress is basically in sync with the world. Under the guidance and deployment of relevant ministries and commissions, related development plans are formulated for 6G research. Although the Ministry of Science and Technology has established a 6G promotion group in 2019, the Ministry of Industry and Information Technology has also established an IMT-2030 promotion group and organized domestic research on 6G.

4.2.1 China Mobile

As the world's largest operator, China Mobile began researching 6G technology in 2018, officially launched the 6G project in 2019, and established the Future Research Institute in May 2021. It focuses on the integration of new generation information and communication technology and pan-information science, continuously increasing investment, widely cooperating with all parties, and has achieved preliminary results in the layout of 6G technology.

1. China Mobile's exploration of 6G vision and requirements

In September 2019, at the 6G seminar held at Beijing University of Posts and Telecommunications, China Mobile first proposed the 6G social development vision of "digital twins, ubiquitous intelligence" [2–4], as shown in Fig. 4.2, it comprehensively sorted out the brand-new application scenarios of 6G after 2030, as shown in Fig. 4.3, including digital twin human, intelligent transportation, internet of senses, intelligent interaction, etc., summarising the new trends of service development such as "diversification of service needs, 3D coverage, diversification of interaction forms and content, openness and customization of service providing, integration of communication, computing, AI and security", and proposed the 6G network feature concept of on-demand services, lean networks, soft networks, native AI and native security, as shown in Fig. 4.4.

(1) "2030+ Vision and Requirements White Paper"

In November 2019, during the China Mobile Partner Conference, China Mobile Research Institute (CMRI) officially released the "2030+ Vision and Requirements White Paper" [5], systematically describing the social development vision of "digital twins, ubiquitous intelligence", focusing on intelligent life, intelligent production and intelligent society, comprehensively refining the brand-new service and application scenarios for 2030, and deriving the systematic technical requirements and its enabling technologies for different service and application scenarios, as shown in Fig. 4.5. The "2030+ Vision and Requirements White Paper" proposes that the

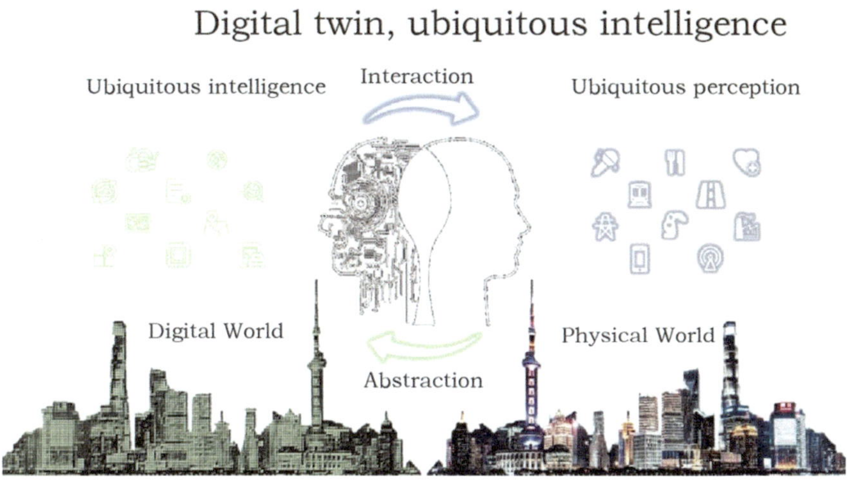

Fig. 4.2 6G vision

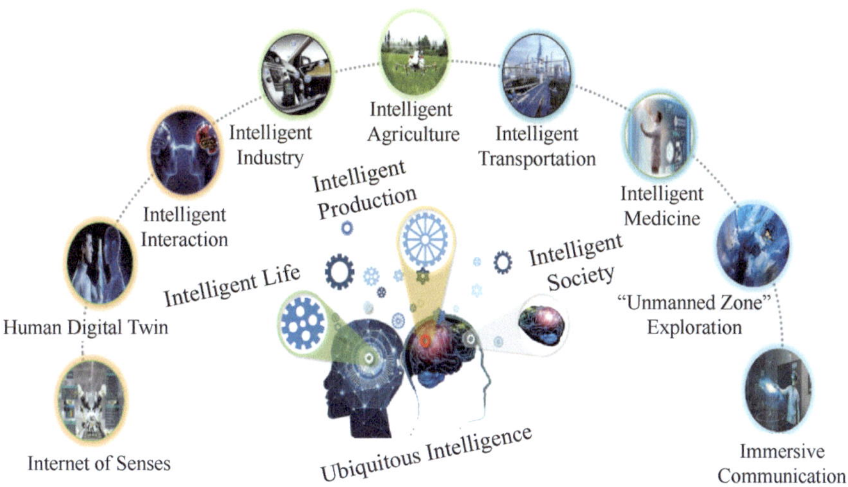

Fig. 4.3 Brand-new application scenarios of 6G beyond 2030

6G network will have multiple features such as on-demand service providing, lean network, soft, native AI, and native security.

(1) On-demand service providing: The network dynamically predicts and perceives the user's service needs and the network environment they are in, and configures the underlying resources, parameters, functions and capabilities of the network on demand to best meet the user's specific needs and provide ultimate performance services. The network based on service providing on-demand will support

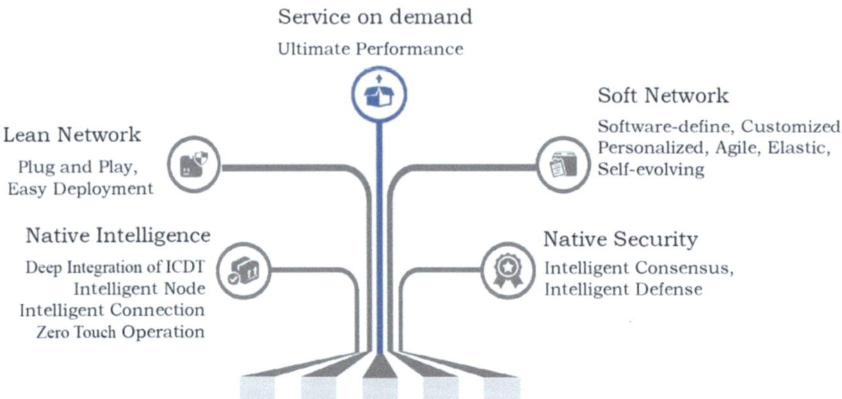

Fig. 4.4 6G network feature concept

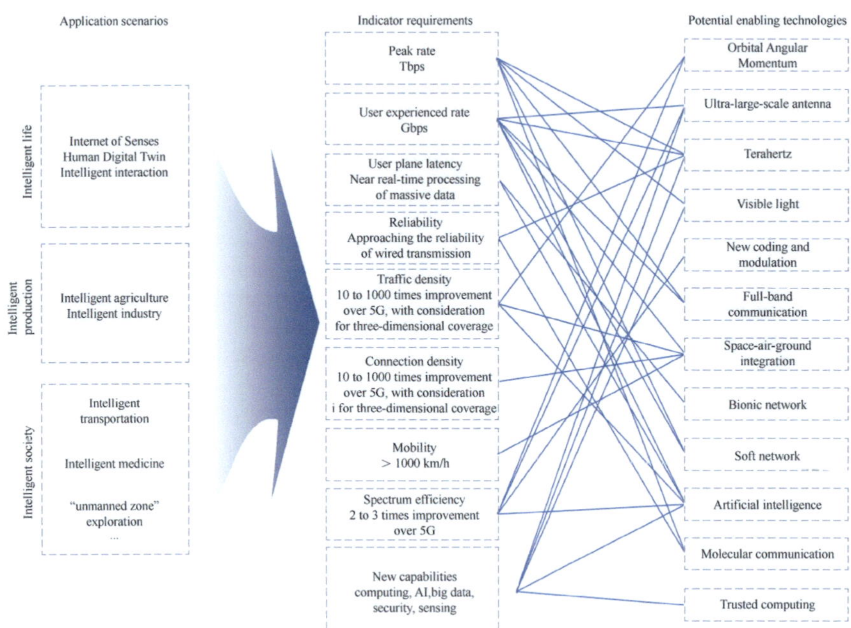

Fig. 4.5 Technical requirements and its enabling technologies for different service and application scenarios

dynamic, ultra-fine-grained service capabilities, and users can obtain corresponding service types, service levels, and combinations of different services according to their own needs. In addition, when user needs change, the on-demand service network can seamlessly switch service format and content, achieve real-time precise matching of network service capabilities and user

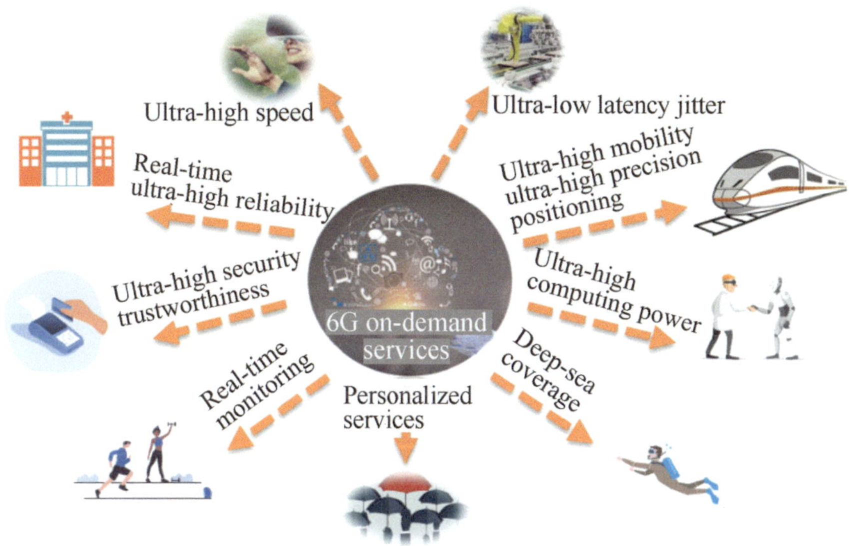

Fig. 4.6 On-demand service providing

needs, and bring users the ultimate differentiated performance experience, as shown in Fig. 4.6.

(2) Lean Network (Simplicity): With the continuous expansion of network scale and the increasing complexity, it is necessary to innovate and simplify the cellular network architecture to achieve "architecture simplification, management simplification, and transmission simplification". The volume of an ant nest in the natural world is extremely large compared to a single ant, yet they can ensure efficient division of labour and connectivity of the ant nest while continuously building and expanding the scale of the ant nest. This inspires us to think about the structural design of future 6G networks from a biomimetic perspective. Mobile communication network is inspired by cells in terms of network coverage, and perhaps future architectural evolution can draw inspiration from ant nests, further achieving network ubiquity, rapid information transmission, and natural growth and automatic evolution according to user needs, as shown in Fig. 4.7.

(3) Soft network: Enabling the network to expand as needed and achieve self-evolution of network functions, achieving "scalability on demand, autonomous evolution". Traditional networks are designed and planned according to the maximum capacity they support, while user demand and network load dynamically change due to the tidal effect brought about by user mobility. Therefore, the performance of traditional networks cannot adapt to changes in service requirements and load. The network towards 2030 needs to consider new

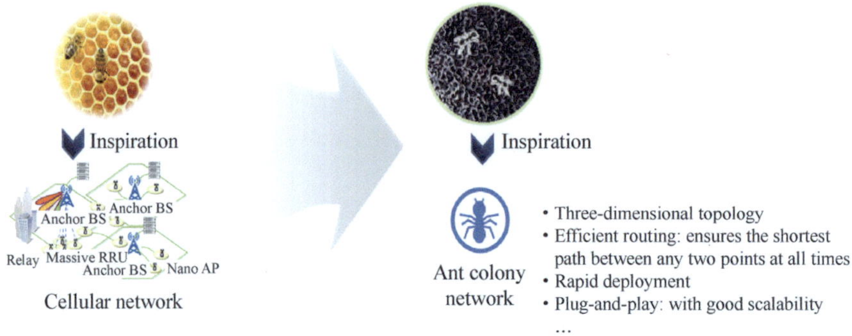

Fig. 4.7 Lean network with simplicity

design methods: fully software-defined end-to-end networks, forward compatibility design of network protocols, cell-less network structure, and user-centric network autonomy and self-evolution.

(4) Native AI: Native AI will achieve full network penetration of AI capabilities, achieving "ubiqitous network, ubiqitous computing, and pervasive intelligence". Through the infrastructure formed by the deep integration of network and computing, it provides ubiquitous computing power for AI, thereby achieving pervasive intelligence [6]. Based on the network of native AI, AI and big data capabilities empowers the high autonomy of the network, achieving zero-touch operation and maintenance. Native AI can also meet emerging new demands effectively through self-focusing, enabling intelligent decision-making in resource management, reducing costs and improving efficiency, and achieving digital transformation [7], as shown in Fig. 4.8.

(5) Native Security: The network based on native security monitors the security status in real time and predicts potential risks, combining resistance to attacks and prediction of dangers, thereby achieving intelligent security, that is, "risk prediction, proactive immunity". Intelligent consensus, through the interaction and collaboration between intelligent entities connected by the network, forms a consensus, and based on the consensus to eliminate interference, provides high-level security support for information and data. Intelligent defence, based on AI and big data technology, accurately deploys security functions and optimizes security strategies, achieving proactive in-depth security defence. Trust enhancement, using trusted computing technology, provides proactive immunity functions for network infrastructure, software, etc., enhancing the security level of the basic platform. Ubiquitous collaboration, through the ubiquitous collaboration of device, edge, and cloud, accurately perceives the entire system's security situation and agilely handles security risks. This network will achieve a comprehensive upgrade from internet security to cyberspace security, as shown in Fig. 4.9.

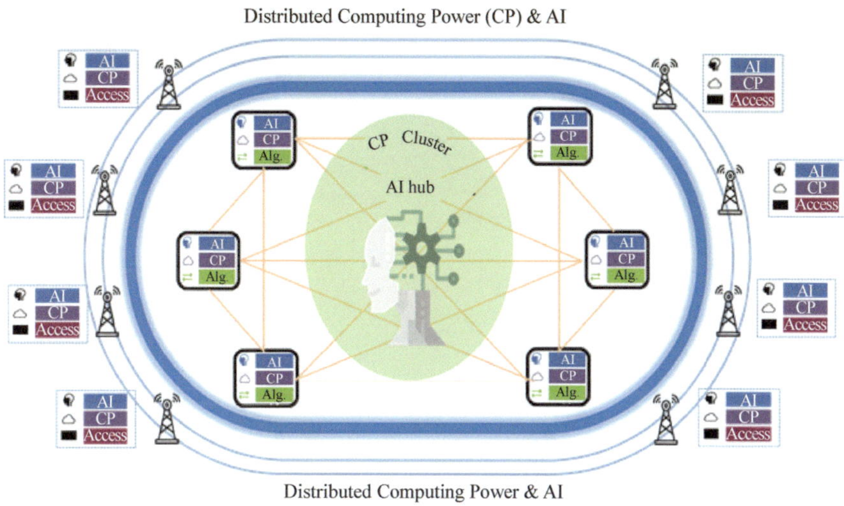

Fig. 4.8 Native AI

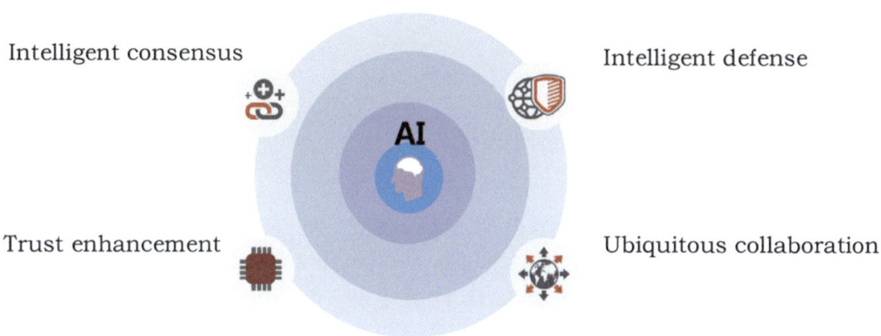

Fig. 4.9 Native security

In addition, the white paper also proposes a 6G development concept with "innovation, coordination, green, openness, sharing" as its connotation, and based on the experience of 4G and 5G development, it makes a judgement on the timetable for the entire 6G development, as shown in Fig. 4.1. The entire 6G research and development will probably be divided into two stages, the specific content is presented in Sect. 4.1.

(2) "2030+ Vision and Requirements White Paper 2.0" [8]

In November 2020, at the China Mobile Partner Conference, the CMRI released the "2030+ Vision and Requirements White Paper 2.0" [8], "2030+ Technology Trends White Paper" [9] and "2030+ Network Architecture Outlook White Paper" [10]. The "2030+ Vision and Requirements White Paper 2.0" further derives the

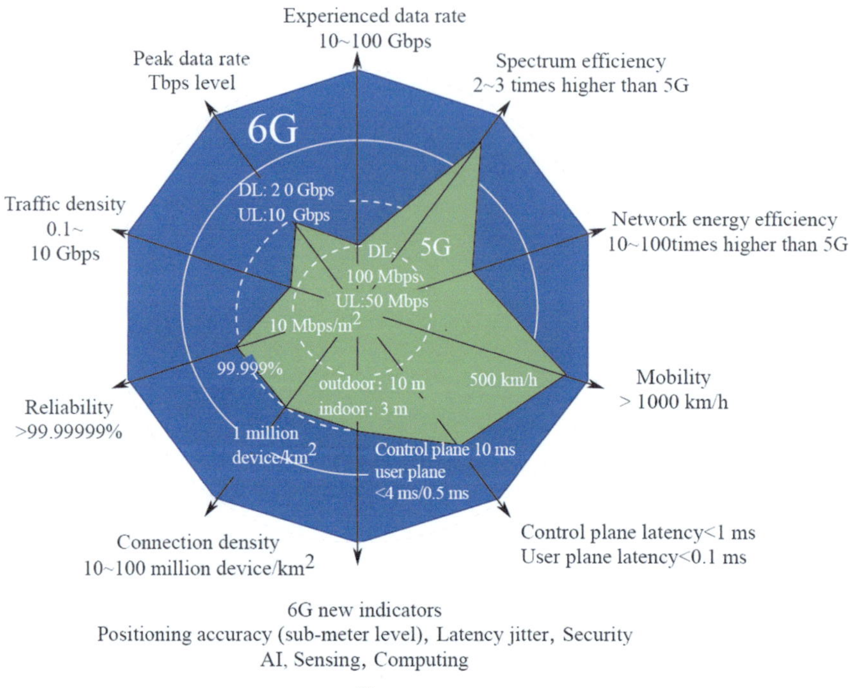

Fig. 4.10 6G technical requirements and capability

technical requirements of each typical application scenario based on the previous version, for the first time systematically proposing that 6G mobile communication networks need to provide computing, AI, sensing and security capabilities in addition to communication connections, and forming a radar chart of 6G technical requirement and capability, as shown in Fig. 4.10.

The "2030+ Technology Trends White Paper" outlines the future technology development trends of mobile communication networks, with the specific content as follows.

(1) Full-spectrum access: From the deployment of the entire 6G mobile communication network, it is necessary to comprehensively consider cost, demand and service experience, effectively use all available frequency resources in different scenarios, sub 10 GHz, millimetre wave, terahertz, visible light, etc., to achieve dynamic complementation of each frequency band, to optimize the overall service quality of the entire network and reduce network energy consumption. For example, the frequency band sub 10 GHz provides seamless network coverage, millimetre wave, terahertz and visible light frequency bands will be deployed and dynamically switched on/off in a on demand way in local and short-distance scenarios, providing higher capacity and higher data rate while minimizing network energy consumption and deployment costs.

(2) Integrated None terrestrial networks and terrestrial networks: To meet the internet connection needs of drones, planes, ships, etc., as well as the wireless connection needs of social governance areas such as instant disaster relief, unmanned area exploration, and tracking of ocean container information, it is necessary to build an integrated none terrestrial networks and terrestrial networks to expand the coverage of ground networks, forming a global "ubiquitous coverage" communication network, including the different orbit satellites, various airborne vehicles, and the traditional terrestrial networks; 6G will deeply integrate satellite communication with terrestrial networks, significantly improving air interface access capability and three-dimensional coverage capability, such as supporting on-board regeneration mode to reduce the deployment of ground cooperation stations, constructing a multi-dimensional and heterogeneous network where high and low orbit satellites coexist, achieving multi-platform function's flexible division and intelligent and reconfigurable network design, through multi-satellite multi-dimensional, inter-satellite cooperation or satellite-earth cooperation to achieve resource cooperative scheduling and seamless roaming between satellite and earth, providing users with imperceptible consistent services, ensuring network resilience and robustness as well as green.
(3) DOICT integration: 6G is a new generation of mobile information system deeply integrated with Communication Technology (CT), Information Technology (IT), Data Technology (DT), and Operation Technology (OT), showing strong cross-disciplinary and cross-field development characteristics. DOICT integration will be the core feature of the 6G end-to-end information processing and service architecture, achieving deep integration of cloud, network, edge, device, and business on the basis of large data flow, enhancing the efficiency of resource utilisation of all parties, and jointly upgrading cloud-edge computing capability, network capability, terminal capability and business capability.
(4) Communication-sensing-computation integration: An end-to-end information processing technology framework that synchronously performs information collection and computation during the information transmission process. This will break the siloed information service framework where terminals collect information, networks transmit information, and cloud edges perform computations. For the new scenarios such as unmanned drone, immersive service, and digital twin, it is a necessary means to provide highly integrated sensing, communication, and computation capabilities. Communication-sensing-computation integration is specifically divided into two levels: functional synergy and functional fusion. In the functional synergy framework, sensing information can enhance communication capabilities, communication can expand sensing dimensions and depth, computation can perform multi-dimensional data fusion and big data analysis, sensing can enhance computational model and algorithm performance, communication can bring ubiquitous computing, and computation can achieve ultra-large-scale communication. In the functional fusion framework, sensing signals and communication signals can be integrated into waveform design and detection, sharing a set of hardware equipment. Sensing and computation are fused into a computational sensing network, and computation

and network fusion achieve network end-to-end definability and microservice architecture.

The "2030+ Technology Trend White Paper" also analyses the potential key technologies for future wireless access networks from three aspects: basic transmission technology, protocol and architecture design, and network autonomy. These technologies include new waveforms, distributed large-scale MIMO, super Nyquist transmission technology, RIS, physical layer AI, service-based RAN, plug-and-play link access, adaptive air interface QoS control, light network architecture, network autonomy based on network digital twins, and a three-layer four-plane network architecture, etc.

The "2030+ Network Architecture Outlook White Paper" points out that 6G networks will undergo changes in the following aspects:

(1) Ubiquitous connections for all scenarios, evolving towards a distributed paradigm;
(2) A light network with a unified access architecture;
(3) Building a digital twin network synchronously with the physical network;
(4) An autonomous network with self-optimisation, self-growth, and self-evolution capabilities;
(5) Solving the core issue of deterministic latency;
(6) A network integrated with communication and computation;
(7) Resources on demand, services on demand.

At the same time, the "2030+ Network Architecture Outlook White Paper" proposes five propositions for consideration in the 2030+ network architecture, hoping to inspire the industry's thinking about 6G network architecture research.

(1) What is the architecture of the 6G network?
(2) How to support a new Internet of Everything with passive communication?
(3) In the era of cognitive intelligence, how can the network achieve "native intelligence"?
(4) What impact will open source technology have on the 6G network architecture?
(5) How can distributed and centralised be coordinated and unified?

2. Joint 6G research by China Mobile and universities

On May 30, 2020, the CMRI and Beijing University of Posts and Telecommunications established a joint innovation centre led by Prof. Zhang Ping. The centre conducts joint research on key technologies for 6G wireless access, wireless channel measurement and modelling, network architecture, new businesses, new antennas, and the development of 6G test verification platform prototypes.

On December 10, 2021, the Joint Innovation Centre held its first 6G R&D results showcase, releasing the "White Paper on Visible Light Communication Systems for 6G" [11] and the "White Paper on AI-based Joint Source Channel Coding" [12] and showcased the 6G prototypes of the Joint Innovation Centre, including the terahertz channel measurement platform, computing network platform, joint source channel coding platform, 6G general system-level and link-level simulation platform, and visible light communication platform. Among them, the joint source channel coding platform is an exploration platform for further obtaining system data rate and efficiency improvements; the computing network platform can empower new computing features of the network; the terahertz channel measurement platform can further explore the potential available bandwidth of 6G, spectral characteristics under higher frequency points; the 6G system-level and link-level simulation platform can efficiently support the performance and feasibility evaluation of various 6G candidate technologies.

The "White Paper on Visible Light Communication System for 6G" introduced the visible light communication technology and experimental system developed by the Innovation Centre, achieving a communication rate of 1 Gbps. The "White Paper on AI-based Joint Source Channel Coding" introduced the AI-based joint source channel coding design scheme proposed by the research team of the Innovation Centre, analysed and discussed the key technologies therein.

On 13th November 2020, the China Mobile and Tsinghua University established a Joint Research Institute with a 6G Research Centre, conducting joint research on network integrated space, air, and ground, semantic communication, reconfigurable computing architecture, autonomous network architecture, wireless and visible light networking and other technologies.

On 22nd June 2021, the CMRI and Southeast University established a Joint Innovation Centre, with the Electromagnetic Metamaterial Centre led by prof. Cui Tiejun, and the Broadband Mobile Communication Centre led by Prof. You Xiaohu and Hong Wei, conducting joint research on the application of electromagnetic metamaterials [13] in wireless communication, ultra-large-scale MIMO, high-performance millimetre-wave transceiver architecture, visible light communication and THz prototype system.

China Mobile and the team of Prof. Cui Tiejun have deep cooperation in the field of information metamaterials, jointly exploring the electromagnetic propagation characteristics of information metamaterials, jointly leading related research projects with the IMT-2030 (6G) Promotion Group and the Future Mobile Communication Forum, exploring two major application directions of intelligent metasurfaces, including intelligent reflection surfaces and metamaterial base stations. For the relatively mature technology and industry of intelligent reflection surfaces, China Mobile has proposed a three-stage development idea combined with network operation experience: the first stage is to achieve passive static reflection surfaces, which can be quickly deployed and meet the needs of extending network coverage and filling in blind spots in weak coverage scenarios; the second stage is to achieve semi-static controllable reflection surfaces, extending the coverage of metasurface beams, increasing cell capacity and data rate through device unit control to achieve beam selection; the third stage is to achieve dynamic intelligent reflection surfaces, dynamically tracking user positions and matching channel environments through coding algorithms, thereby achieving intelligent control of 6G electromagnetic wave propagation. Among them, the products of the first two stages are expected to be deployed first in some scenarios of the 5G network after the technology is mature.

In July 2021, the Southeast University-CMRI Joint Innovation Centre and Hangzhou Qiantang Information Co., Ltd. completed the industry's first field trial of intelligent metasurface (Reconfigurable Intelligence Surface, RIS) with adjustable electromagnetic unit devices and flexible control of beam direction on the 5G live network of China Mobile Nanjing.

RIS is a significant application of information metamaterials in the field of mobile communications. Its basic principle is to control the electromagnetic characteristics of metamaterials through digital programming, altering the diffuse reflection of spatial electromagnetic waves, achieving intelligent regulation and beam shaping of spatial electromagnetic waves. It also has low power consumption, low cost, and other features, and is expected to become an important infrastructure for future mobile communication networks. The technical principle of RIS is shown in Fig. 4.11, the deployment scenario of RIS is shown in Fig. 4.12, and the equipment and test site used by RIS in 5G live network testing are shown in Fig. 4.13.

Preliminary test results show that RIS can flexibly adjust the beam direction of reflected signals according to user distribution, significantly improving the signal

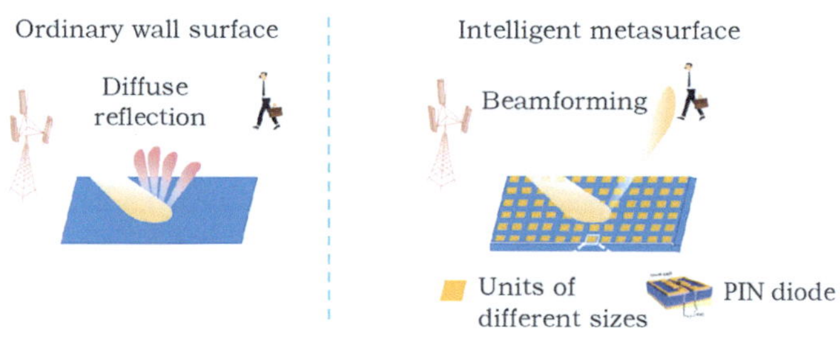

Fig. 4.11 Technical principle of RIS

Fig. 4.12 Deployment scenario of RIS

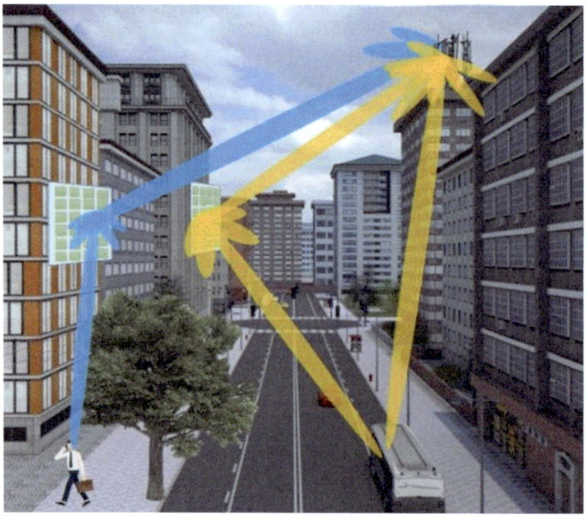

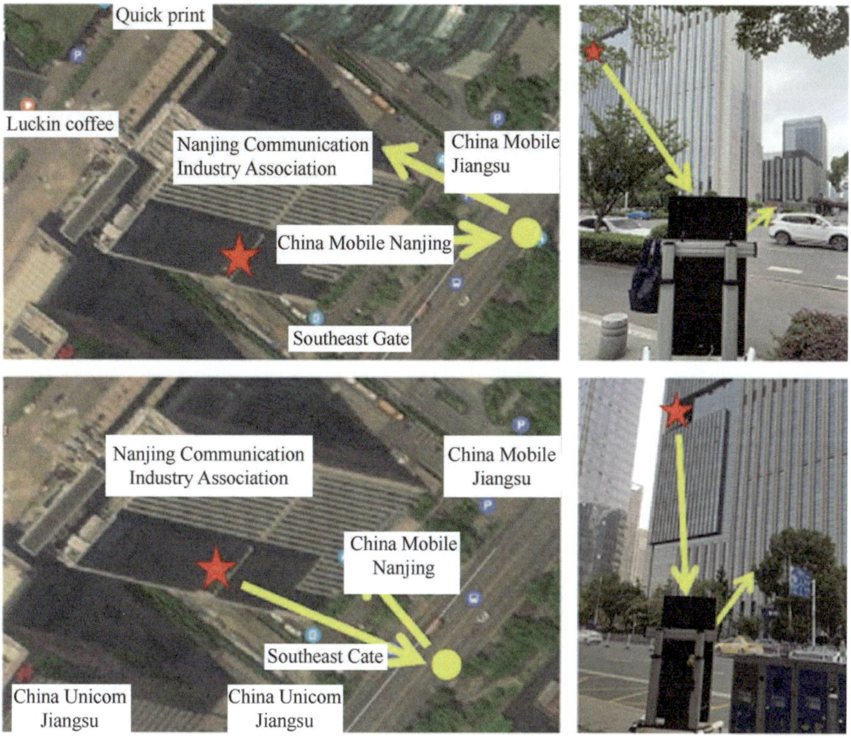

Communication

Fig. 4.13 Equipment and test site used by RIS in 5G live network testing

strength, network capacity, and user data rate in weak coverage areas of the live network. In outdoor test scenarios, the average coverage at the edge of the cell is increased by 3–4 dB, and the throughput of edge users is increased by more than 10 times; in outdoor-to-indoor coverage test scenarios, indoor coverage is increased by about 10 dB, and user throughput is increased to about 2 times.

Early field trial test results have preliminarily verified the feasibility of RIS, but there are still four challenges to standardisation and practical application.

First, the fundamental theory is not perfect, the reflection and transmission characteristics of RIS need to be clarified, the channel transmission model is not perfect, and there is a lack of modelling in the actual transmission environment.

Second, key technologies urgently need to be broken through, such as co-frequency and adjacent frequency interference characteristics, interference coordination between operators, beam shaping and channel estimation algorithms, etc., need to be researched and improved.

Third, the maturity and reliability of the device are relatively low. The current industry's RIS is a prototype, the adjustable angle is limited, the device control rate is not high, the number of unit is large and it is difficult to quickly locate and identify unit faults.

Fourth, application deployment is limited. The size and area of RIS are relatively large, and both active and wired control will limit its application scenarios. It is necessary to further optimise engineering design and enhance deployment flexibility.

In response to these challenges, China Mobile has been focusing on the fundamental theory and key technology research of RIS since 2019, proposing a series of solutions for the electromagnetic control characteristics and channel model of RIS, the joint beam shaping and channel estimation algorithm of the base station and intelligent reflection surface, and the wireless control of the intelligent reflection surface, and actively carrying out the hardware architecture breakthrough of RIS. Subsequently, relying on the Southeast University-China Mobile Research Institute Joint Innovation Centre, it will continue to improve the system design of RIS, carry out more comprehensive technical trials for richer deployment scenarios and higher frequency bands, and promote the application of RIS technology in mobile communication networks as soon as possible.

3. China Mobile 6G Joint Innovation Achievements

On 15th February 2022, China Mobile held the "Meet the Future" 6G Joint Innovation Achievement showcase, where China Mobile fully demonstrated the recent joint R&D achievements with industry partners, including a series of 6G prototypes.

(1) 2.6 GHz/26 GHz RIS prototype. China Mobile, in conjunction with Professor Cui Tiejun and Professor Chen Qiang of Southeast University, developed RIS prototypes for the 2.6 and 26 GHz bands. Figure 4.14 shows the 2.6 GHz RIS prototype (1 m $\times$ 2 m), it contains 16 columns and 32 rows, a total of 512 units. The units use PIN tubes as adjustable components, which can control the reflection phase of the unit's horizontally polarised electromagnetic waves by switching the PIN tube on and off. The phase difference of the unit under on and off states is about 180°. The 16 columns of units of this RIS prototype can be independently controlled, allowing for beam steering in the horizontal direction. Figure 4.15 shows the 26 GHz RIS prototype (35 cm $\times$ 35 cm) in actual measurement. For ease of covert deployment, its surface is decorated with a painting. It consists of 16 $\times$ 16, a total of 256 units. Due to the shorter wavelength of millimetre waves, its overall size is smaller than 2.6 GHz. The units of this prototype are fixed phase, and the development of phase-adjustable millimetre wave RIS prototypes will continue.
(2) Optically generated terahertz communication system prototype. The optically generated terahertz communication system prototype developed by Professor Zhu Min of Southeast University in collaboration with China Mobile has a carrier frequency of 101 GHz, a bandwidth of 5 GHz, a real-time rate of 3 Gbps, and a maximum communication distance of 20 m, as shown in Fig. 4.16. Currently, the team has completed the development of the next-generation optically generated terahertz real-time communication system with a carrier frequency of 370 GHz, a bandwidth of 30 GHz, and a real-time rate of 120

Fig. 4.14 2.6 GHz RIS prototype

Fig. 4.15 26 GHz RIS prototype in actual measurement

Gbps. The next step will support a real-time transmission rate of 500 Gbps, with the ultimate goal of achieving a real-time transmission rate of 1 Tbps. Currently, terahertz communication systems are divided into two different technical routes: electrical and optical. Electrical technology is limited by device bandwidth and system complexity increases with carrier frequency. The advantage of optical technology is that it achieves higher bandwidth and data rate through a simpler system architecture.

(3) 26 GHz high-performance millimetre wave transceiver. The efficient multi-channel millimetre wave transceiver component developed by Professor Hong

Fig. 4.16 Optically generated terahertz communication system prototype

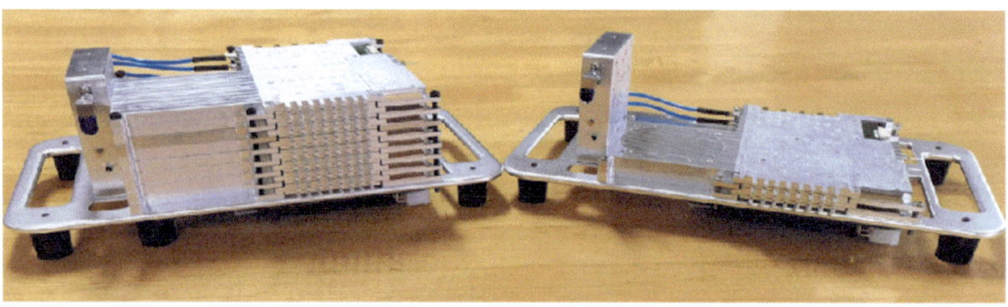

Fig. 4.17 Efficient multi-channel millimetre wave transceiver component

Wei and Zhou Jianyi of Southeast University in collaboration with China Mobile is shown in Fig. 4.17. It is a 64-channel and 16-channel component that can effectively support the implementation of more efficient millimetre wave transceivers with more channels.

(4) 1 Gbps visible light real-time communication system prototype. China Mobile, in collaboration with Professor Chi Nan and Shen Chao of Fudan University, has successfully developed a super-radiant light-emitting diode source with a single device data rate of 4.57 Gbps and a broadband visible light transceiver chip supporting a transmission rate of 1.75 Gbps. The developed visible light prototype system has a Gbps-level real-time transmission rate and is expected to achieve an industry-leading real-time transmission rate of 5 Gbps, as shown in Fig. 4.18. In the next stage, the team will challenge the development of devices with tens of Gbps rates and the development of prototype base stations. Compared with traditional frequency bands, visible light communication is low-carbon, energy-saving, free of electromagnetic pollution, rich in spectrum, and can be used without authorisation, with low deployment costs. However, visible light communication faces three challenges in its application: firstly, in terms of

Fig. 4.18 Visible light real-time communication system prototype

existing devices, the modulation bandwidth of visible light devices is small, and the peak rate that can be achieved is very limited. The chip jointly developed by China Mobile and Fudan University has a bandwidth of over 1 GHz, but to meet the more stringent requirements of 6G, further improvement is needed; secondly, in terms of networking, visible light is easily blocked and uplink is difficult. To improve the user's online experience and provide zero-interruption service, it is necessary to integrate visible light with other frequency bands for networking; finally, in terms of integration with infrastructure, the visible light communication network needs to coexist with the mature lighting industry, achieve low cost, low power consumption, high energy efficiency, and also integrate with the operator's network to support mobile services and further expand the industry scale.

(5) Sub-6 GHz, visible light, terahertz channel measurement platform. Channel measurement and modelling are the foundation and premise of wireless communication system design, evaluation, and product development. Its main purpose is to study the propagation characteristics of electromagnetic signals and characterise them using mathematical models. The reason we must understand the channel is that during the transmission process, the signal will be affected by the signal propagation environment, such as multipath and fading, which can cause distortion of the received signal. By establishing a channel model, we can optimise system design, improve the accuracy of performance evaluation, and accelerate product development based on the prior information of the channel model. China Mobile started channel measurement and modelling research in 2008, selecting dense cities and high-speed railway scenarios, and built a channel measurement platform with the team of Professor Zhang Jianhua from Beijing University of Posts and Telecommunications. After extensive field measurements, a large amount of channel data was obtained. Through complex data processing, parameter extraction, and modelling, corresponding models were formed, leading the development of ITU's IMT-Advanced and IMT-2020 channel models, 3GPP's 3D-MIMO channel model, and effectively supporting the global 4G and 5G research and standardisation. Looking towards 6G, the

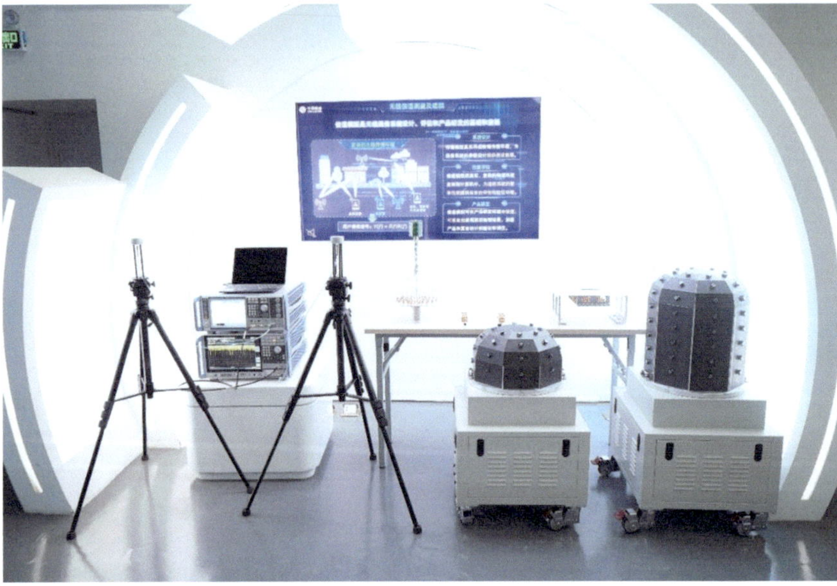

Fig. 4.19 Sub-6 GHz/mmW/THz channel measurement platform

project team has already started research on channel measurement and modelling for integrated sensing and communication, intelligent surfaces, terahertz, and visible light, with the aim of supporting the breakthrough of 6G system design and technology. The industry-leading sub-6 GHz/millimetre wave/terahertz channel measurement platform developed is shown in Fig. 4.19.

(6) 26 GHz integrated sensing and communication prototype. Based on the OFDM waveform used in existing 4G and 5G systems, the team at China Mobile Research Institute designed an integrated sensing and communication signal processing algorithm. By extracting the delay and Doppler information of the transmission channel, the distance and speed of the target object can be obtained. Simulation results show that the integrated sensing and communication signal processing algorithm based on OFDM can achieve multi-target ranging and speed measurement functions. Based on the OFDM waveform, the sensing algorithm, China Mobile and Huawei have developed an integrated sensing and communication prototype. This prototype is mainly aimed at vehicle status monitoring scenarios, with a working frequency of 26 GHz, a bandwidth of 100 MHz, and a coverage distance of 500 m. The prototype includes an active wireless unit (AAU), a baseband unit (BBU), a core network, and an interactive interface. The AAU is deployed on a crossbar above the road to complete the transmission and reception of integrated sensing and communication signals. The received signals are processed by the BBU and the core network to obtain the distance and speed information of the target vehicle, which is displayed on the interactive interface. The project team conducted field tests on Beiqing

Road in Haidian District, Beijing. Preliminary test data shows that the ranging accuracy of the prototype reached 0.5 m. When vehicles pass on the road, the position and speed information of multiple vehicles can be displayed in real time on the computer's interactive interface, effectively achieving the goal of traffic status monitoring. The field test environment and test interface are shown in Fig. 4.20.

(7) Liquid metal antenna prototype. The liquid metal antenna developed by China Mobile in collaboration with Professor Su Ming and his team from Beijing University of Posts and Telecommunications can achieve dynamically adjustable antenna radiation patterns and frequencies through antenna configuration adjustments, as shown in Fig. 4.21.

(8) Digital Twin Human Demonstration System. A digital twin human refers to the digital modelling of the human body through real-time collected human data, achieving a digital twin of a person, as well as the mapping and interaction between the physical human and the digital human. Through a person's digital twin, not only can the appearance, organs, and tissues of a person be simulated, but the health, behaviour, and emotions of a person can also be predicted in

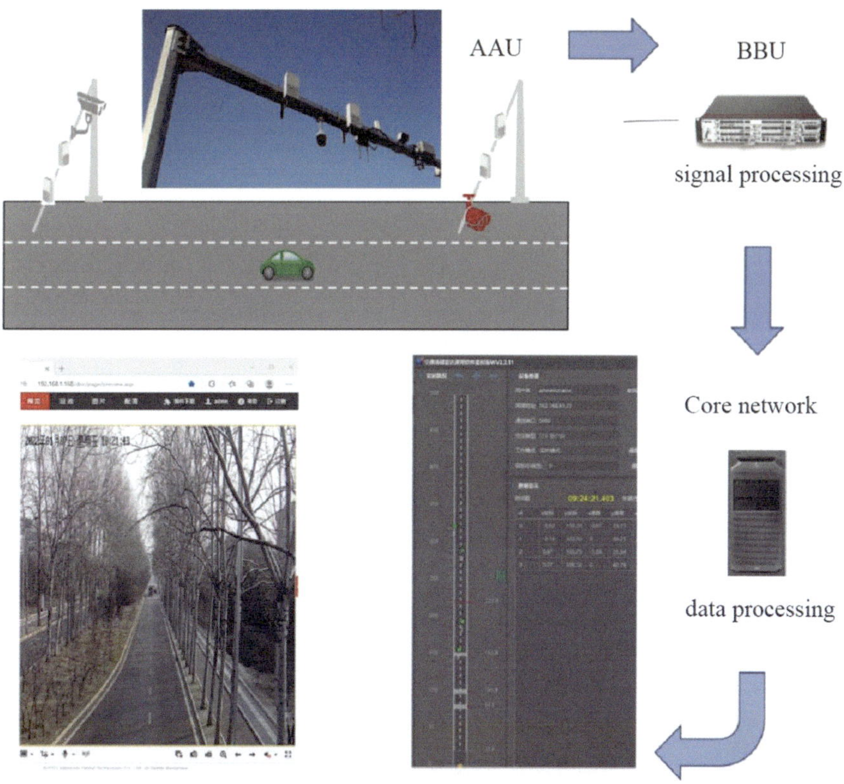

Fig. 4.20 Outdoor test environment and test interface

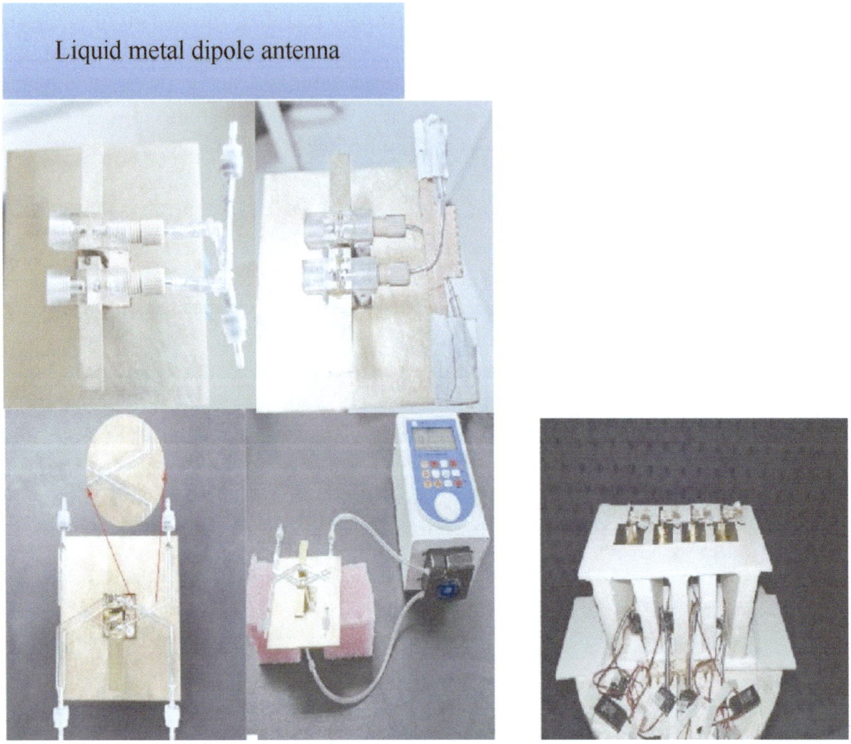

Fig. 4.21 Liquid metal antenna prototype

advance, and even the thinking of a person can be simulated, achieving spiritual immortality in a certain sense. The future business scenarios of digital twin humans are divided into three different levels: the first level is physiological twinning, which collects human information through wearable, implantable sensors, etc., combines physiological and prior knowledge, digitises the local and whole body, achieves all-round remote monitoring and precise prediction of human physiological signs and health, and achieves predictive treatment of diseases; the second level is synaesthetic twinning, which achieves modelling of human emotions and sensory perceptions; the third level is control transplantation, which achieves thought control, thought transplantation, brain-machine communication, and even brain-brain interaction. The digital twin human demonstration system developed by CMRI, as shown in Fig. 4.22, can dynamically capture human limb movements, electrocardiograms and pulses, brain waves, blood pressure and other biological characteristics, and make a comprehensive judgement of the human body state.

(9) Semantic Communication System Prototype. The semantic communication prototype developed by China Mobile in collaboration with Professor Xu Xiaodong of Beijing University of Posts and Telecommunications can achieve

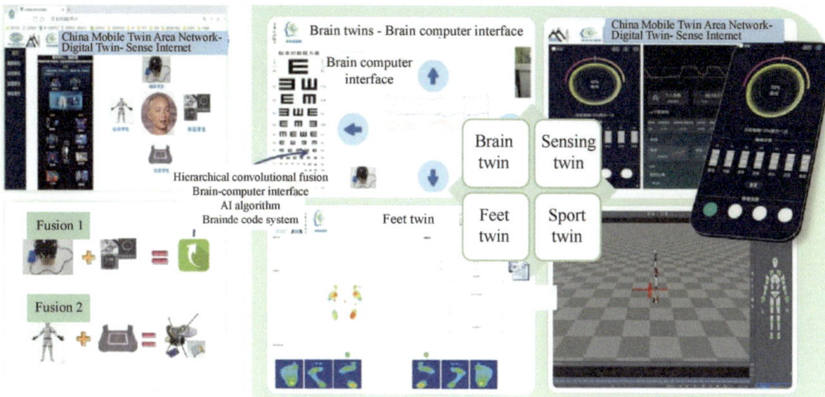

Fig. 4.22 Digital twin human demonstration system

10 times compression of transmission rate; the semantic communication prototype developed in collaboration with Professor Tao Xiaoming of Tsinghua University can achieve 90% compression of video transmission rate.

At the same time, CMRI proposes that 6G will have six main features:

Firstly, it will support scene of virtual-real interaction, supporting sensory interconnection (capable of transmitting auditory, visual, tactile, olfactory, gustatory, etc.), interactive holography (holographic objects can directly interact with each other similar to physical entities) and other new application scenarios;

Secondly, it will establish a new multi-dimensional capability system integrating communication, computing, perception, AI and big data;

Thirdly, it will be green, intelligent and secure, making efforts in energy saving and carbon reduction from the aspects of devices, equipment, networks and infrastructure, and integrating AI and security into the overall design of network architecture;

Fourthly, it will cover the network in three dimensions, with the ground mobile communication system as the main body and the satellite system as an coverage extension, creating an integrated network environment of air, space, ground and sea;

Fifthly, it will integrate interdisciplinary disciplines, technologies across fields and disciplines such as biology and bionics will accelerate integration in 6G, and new materials such as graphene will play a catalytic role in the development of 6G;

Sixthly, it will be ecologically cross-border and open, based on the model of "open source software (source code can be publicly used and modified) + white box hardware (not exclusive to a certain manufacturer, more universal)", which may become the industrial trend of 6G. At the same time, CMRI also proposed the concept of the 6G mobile information network architecture as shown in Fig. 4.23.

In addition, CMRI has also released 8 6G key technology white papers, including "6G Holographic Communication Business Development Trend White Paper (2022)" [14], "6G lean Wireless Access Network White Paper (2022)" [15], "6G SBA RAN White Paper (2022)" [16], "6G Wireless Network Autonomy White

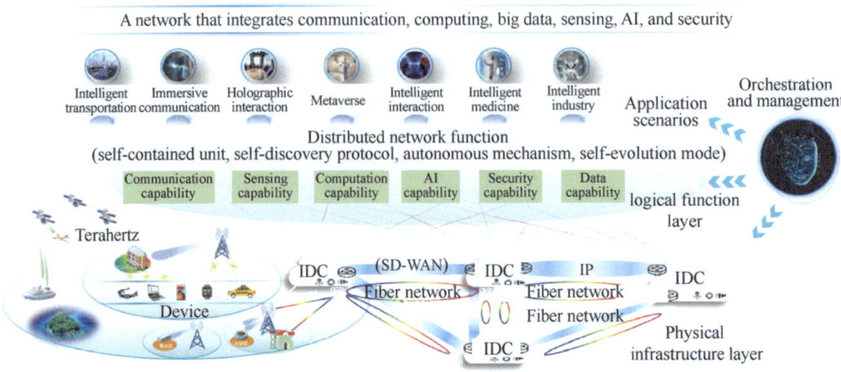

Fig. 4.23 Concept of 6G mobile information network architecture

Paper Based on Digital Twin Network (2022)" [17], "native AI Architecture for 6G Wireless network White Paper (2022)" [18], "6G Physical Layer AI Key Technology White Paper (2022)" [19], "6G Information Metamaterial Technology White Paper (2022)" [20], "6G Visible Light Communication Technology White Paper (2022)" [21]; At the same time, the "Top Ten Cross-Border Innovation Directions Affecting Future Information and Communication Development (2022)" [22] research report was also released, aiming at the pan-information communication field's 2035 vision, selecting the top ten cross-border innovation directions affecting future information and communication development, involving end, pipe, cloud, computing power, security, low carbon, paradigm, etc.

At the "Meet the Future" 6G Joint Innovation Achievement Release Conference, China Mobile officially announced the construction of a "$10 + 1 + N + 1$" 6G collaborative innovation base, as shown in Fig. 4.24, building 10 basic innovation labs, 1 end-to-end system lab, N types of new businesses and applications, and a globally leading 6G field trial network. The 6G collaborative innovation base will be open to the entire ecosystem, providing industry partners with open, scenario-based joint research and development, testing and verification of the required software and hardware tools and platform environment, as well as collaborative innovation space, connecting from basic theory, key technology to standards, industrilization and commercial applications, accelerating the breakthrough of 6G research and development, cultivating the manufracture and application ecosystem, promoting global cooperation, and forming a global unified standard.

4.2.2 Huawei

In 2021, Huawei Technologies Co., Ltd. (hereinafter referred to as Huawei) released the "6G: A New Journey of Wireless Communication" [23] white paper, summarising

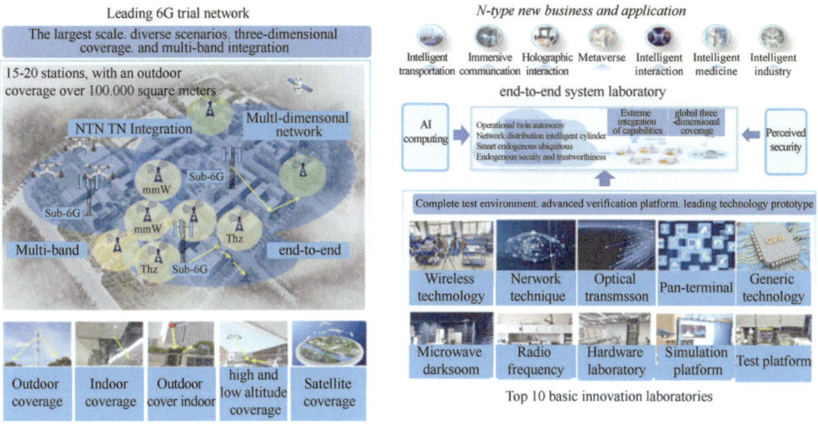

Fig. 4.24 Concept of China Mobile 6G collaborative innovation base construction

the driving forces of 6G development as new applications and businesses, inclusive intelligence, sustainable development and social responsibility, and pointed out that 6G will cross human and object connections, move towards intelligent connections of everything, make significant leaps in key performance indicators, and promote the comprehensive digital transformation of various vertical industries. 6G will be like a huge distributed neural network, integrating communication, sensing, computing and other capabilities, and the physical world, biological world and digital world will seamlessly integrate, opening a new era of interconnection of everything, intelligent connection of everything, and sensing of everything.

In this white paper, Huawei proposed six pillars of 6G, including native AI, integrated space, earth and sky, integrated sensing, ultimate connection, native trust and sustainable development, as shown in Fig. 4.25.

The "6G: A New Journey of Wireless Communication" white paper categorises the scenarios of 6G into five major categories, as shown in Fig. 4.26.

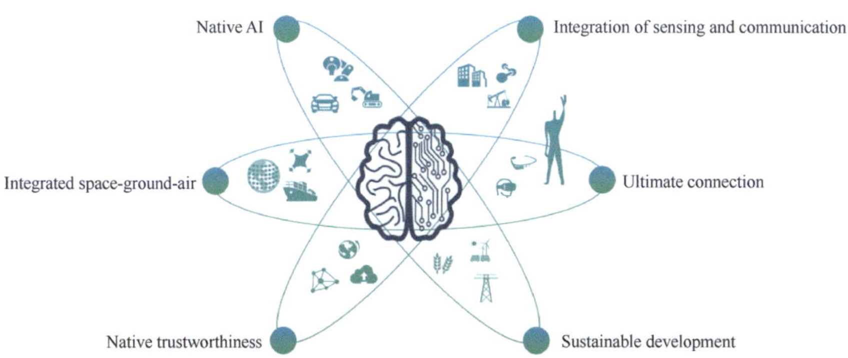

Fig. 4.25 The six pillars of 6G [23]

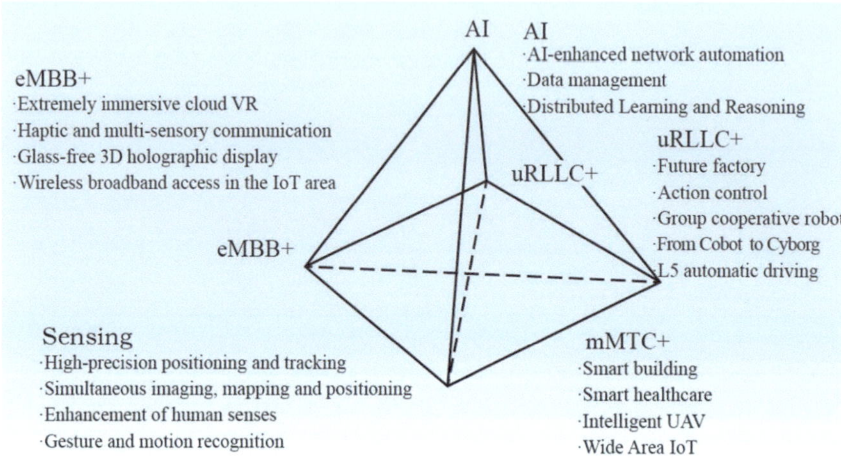

Fig. 4.26 Overview of 6G scenarios [23]

(1) eMBB+: Extreme immersive cloud VR, tactile and multi-sensory communication, glass-free 3D holographic display, wireless broadband access in the Internet of Things area.
(2) uRLLC+: Future factories, motion control, group collaborative robots, from human–machine collaborative robots (Cobot) to human–machine symbiotic cyborgs (Cyborg), L5 level autonomous driving.
(3) mMTC+: Smart buildings, smart healthcare, unmanned aerial vehicles (UAVs) enabling intelligent services, wide-area IoT.
(4) Sensing: High-precision positioning and tracking, synchronous imaging, mapping and positioning, human sensory enhancement, gesture and motion recognition.
(5) AI: AI-enhanced network automation, data management artificial intelligence as a service (AIaaS), distributed learning and inference AIaaS.

Based on the definition of the above scenarios, the white paper "6G: A New Journey in Wireless Communication" proposes the performance indicator requirements of 6G wireless networks, as shown in Fig. 4.27.

Looking to the future of pervasive intelligence, the white paper "6G: A New Journey in Wireless Communication" proposes the concept of a network for AI, hoping to make AI a native capability of 6G, providing end-to-end support for AI-related businesses and applications, achieving ubiquitous AI, as shown in Fig. 4.28.

Looking to the future design of 6G air interface and architecture, the white paper "6G: A New Journey in Wireless Communication" also discusses its paradigm shift, as shown in Figs. 4.29 and 4.30.

Finally, the white paper "6G: A New Journey in Wireless Communication" predicts the 6G timetable of various future standard organisations, as shown in

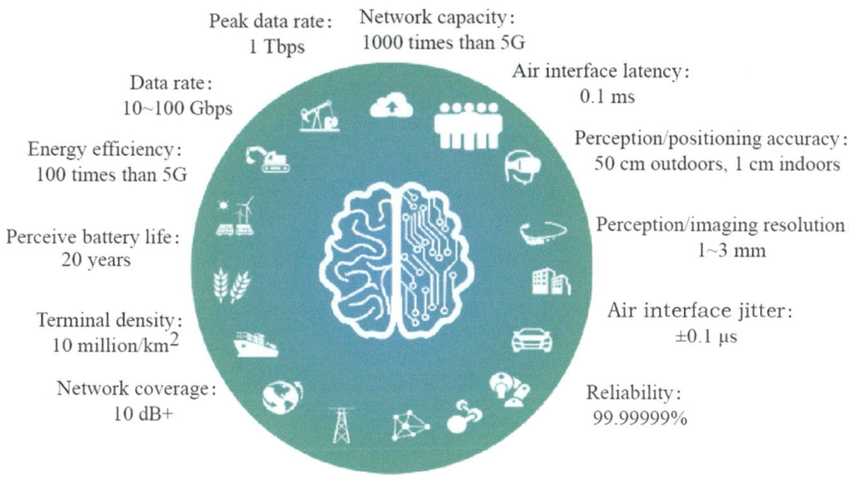

Fig. 4.27 Performance indicator requirements of 6G wireless networks [23]

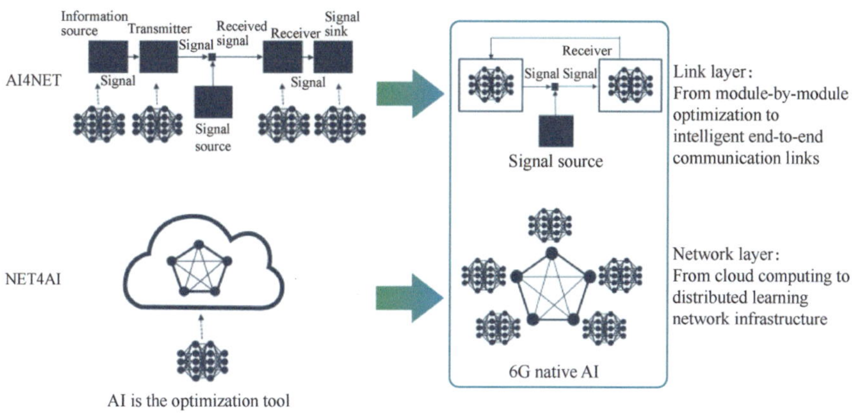

Fig. 4.28 AI4NET and NET4AI [23]

Fig. 4.31, predicting that 3GPP will release the first version of the 6G standard in 2030.

Huawei's 6G research team has built a general prototype platform for Integrated Sensing And Communication at THz band (ISAC-THz) applicable within the frequency range of 100–300 GHz [24], and has explored and validated the technical feasibility of two major challenge scenarios: high-precision sensing and imaging on the terminal side, and ultra-high-speed transmission in outdoor medium distances, gradually moving from concept to implementation. The prototype of Huawei's terahertz sensing communication integration is shown in Fig. 4.32. The objective to be tested is placed in a closed paper box, and the mechanical arm simulates a human

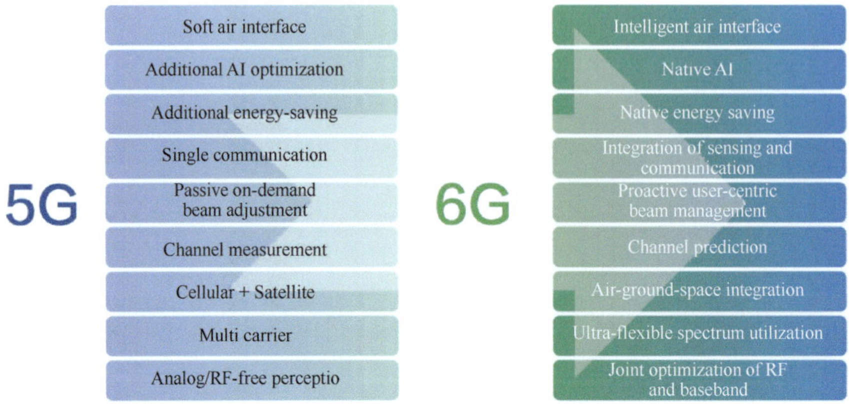

Fig. 4.29 Paradigm shift in 6G air interface design [23]

Fig. 4.30 Paradigm shift in network architecture design [23]

hand holding the terahertz terminal, scanning and imaging the object inside the paper box. The prototype uses a carrier frequency of 140 GHz, a bandwidth of 8 GHz, and a 4-transmit&16-receive MIMO array. After the terahertz wave is emitted by the terminal antenna, it penetrates the paper box, is reflected by the object to be tested, and is received by the terminal antenna. After sampling and real-time algorithm processing, an image is formed and displayed.

In order to achieve millimetre-level imaging resolution, the research team proposed the virtual aperture MIMO array technology [25] Using a small-scale array formed by a limited number of physical antenna elements on the terminal, a large-scale virtual aperture antenna array with greater degrees of freedom is formed through handheld mobile scanning. This approximates the imaging effect of a real aperture antenna with thousands of antenna elements without increasing the terminal volume. As the user's handheld scanning trajectory of the target object is usually sparse and not completely regular, the prototype uses compressed sensing, tomography, sparse

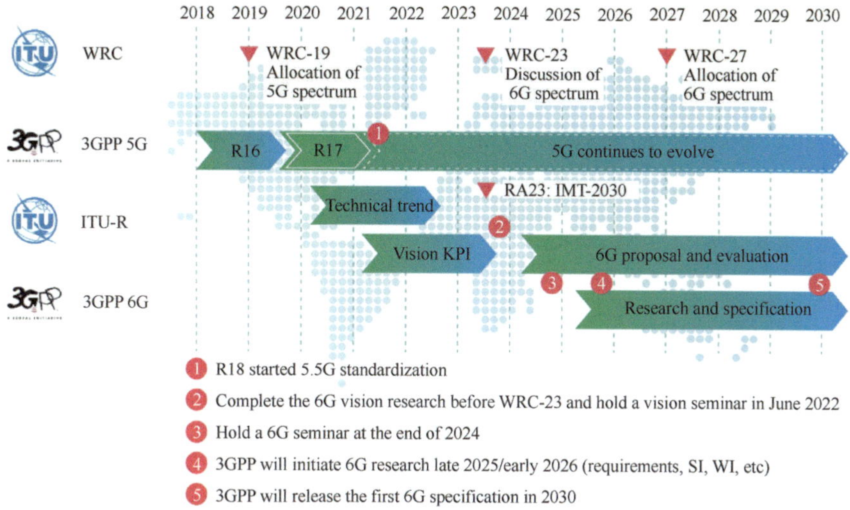

Fig. 4.31 Huawei's predicted 6G timetable for various standard organisations [23]

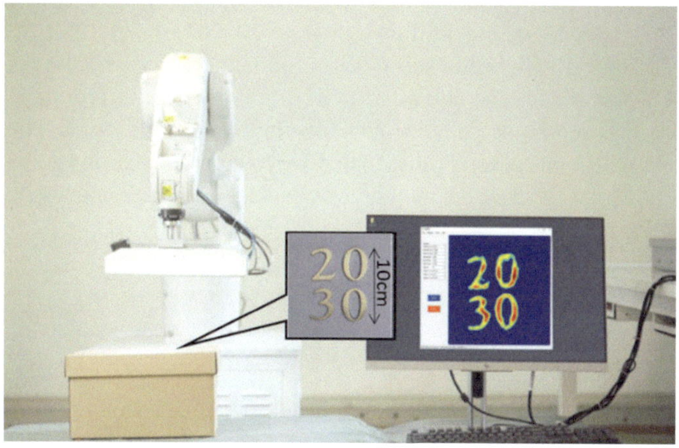

Fig. 4.32 Huawei terahertz perception communication integrated prototype [24]

aperture and other algorithms to process the signal waveform after sparse sampling, resulting in millimetre-level high-resolution images.

As shown in Fig. 4.33, the outdoor field trial verification of the terahertz communication prototype was selected in an urban scenario. The transmitter simulates a typical base station installed on the top of a building, and the receiver is set on the ground of an urban street. The distance between the rooftop and the ground is about 500 m, with a line-of-sight link. The prototype operates at a centre frequency of 220 GHz, with a bandwidth of 13.5 GHz. The system adopts a 2 × 2 polarised MIMO

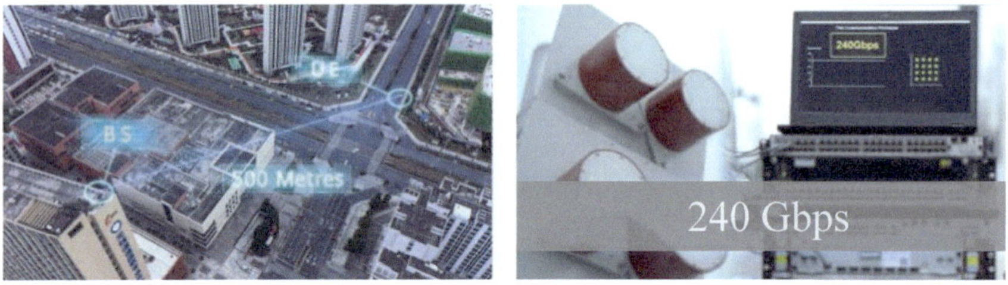

Fig. 4.33 Huawei ISAC-THz outdoor communication prototype, achieving 500 m, 240 Gbps communication rate [24]

architecture, as well as ultra-wideband and low bit-rate quantised digital baseband processing technology, to perform channel estimation and equalisation, non-linear compensation, demodulation and decoding of the baseband signal. It achieves the first outdoor medium-long distance 240 Gbps high-speed line-of-sight air interface transmission, demonstrating the technical feasibility of using the terahertz frequency band for outdoor ultra-high-speed communication.

Huawei's 6G research team has successfully demonstrated a short-distance communication prototype with ultra-low power consumption, ultra-high throughput, and ultra-low latency in the millimetre wave 70 GHz band [26], achieving a throughput of over 10 Gbps and sub-millisecond latency, and real-time demonstration of 4 K VR services. The Huawei 70 GHz millimetre wave prototype is shown in Fig. 4.34. This prototype demonstrates the ultimate experience of short-distance transmission, with a rate several times that of wired communication methods such as the Universal Serial Bus (USB), and the whole machine power consumption is less than 560 mW.

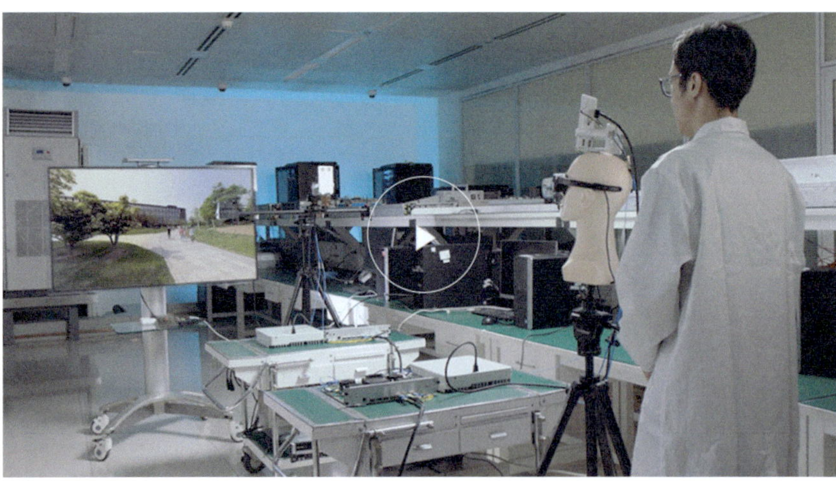

Fig. 4.34 Huawei 70 GHz millimetre wave prototype [26]

This prototype uses a variety of advanced technologies, which are introduced below:

(1) Ultra-low power Polar encoding and decoding technology for Tbps-level high throughput: Combining a minimalist algebraic decoding method, replacing medium-rate outer code, greatly simplifying the Successive Cancellation decoder (SC) process while significantly increasing the decoding throughput, reducing the chip area by 80% compared to traditional short-distance encoding schemes.
(2) Low-power single-bit ADC (Analogue to Digital Converter) technology: Using a limited number of ADC bits can significantly reduce the power consumption of the RF link; the receiving side uses oversampling zero-crossing modulation to further improve the spectral efficiency of the whole machine.
(3) High-speed short-distance phased array antenna and intelligent beam scanning technology: A newly designed dual-polarisation sub-array phased array antenna achieves high-speed transmission of dual-stream data. Combined with a fast tree search-based intelligent beam scanning algorithm, it is possible to accurately adjust the beam direction in high mobility scenarios.
(4) Efficient Silicon Germanium (SiGe) large antenna array Antenna In Package (AIP) technology: Large-scale high-gain irregular array antennas can be packaged in LTCC (Low Temperature Co-fired Ceramic) modules of mobile phone, making miniature AIPs applicable in wearable devices.

Huawei's 6G research team has achieved a breakthrough in the key technology research and integrated prototype verification of the optical wireless sensing system [27], capable of achieving an integrated architecture and capability for communication, positioning, and sensing. This Integrated Sensing And Communication with Optical Wireless (ISAC-OW) prototype aims to simulate in a medical environment, accurately sensing and positioning mobile robots through optical wireless links (visible light and infrared spectrum), thereby remotely controlling mobile robots to pick up and transport various objects. The optical link in the prototype also simultaneously carries the wireless high-speed transmission of real-time high-definition video between the mobile robot and the controller, achieving integrated sensing. In addition, this ISAC-OW prototype can monitor heart rate and respiratory status in real-time and contactlessly by sensing changes in facial colour or abdominal undulation, with detection accuracy comparable to commercial smartwatches. During the development of this prototype, the team successively broke through core technologies such as distributed optical wireless joint detection and transmission, integrated communication and sensing architecture design, and high-precision Time of Flight (ToF) modelling analysis.

(1) For integrated communication and sensing, through the design of integrated sensing waveforms, hardware architecture, and signal processing algorithms, high-speed wireless optical communication is achieved while also achieving centimetre-level indoor positioning accuracy. During the sensing process, the terminal uses an enhanced reflective surface to independently reflect the

incoming light from each base station, avoiding reflected light interfering with other base stations, allowing multiple base stations to measure phase differences in an asynchronous situation, achieving high-precision sensing effects.

(2) For contactless health monitoring, a combination of ToF modelling analysis and deep learning technology is used. By utilising the impact of heartbeats on facial blood flow, heart rate and respiratory rate are measured by accurately measuring subtle changes in the intensity of reflected light from the face (the latter can also be measured by the undulation of the abdomen during breathing), achieving detection performance equivalent to contact health monitoring devices such as smartwatches.

4.2.3 CICT Mobile

Chinese Informtion and Communication technology (CICT) Mobile began planning 6G research work in 2018, initiating a series of research work on 6G vision, requirements, capabilities, and key technologies. At the end of 2020, CICT Mobile released the "Ubiquitous Coverage·Scenario Intelligence—6G Vision and Technology Trend White Paper (V.2020)", proposing the overall vision of ubiquitous coverage and scenario intelligence, that is, 6G will be further upgraded and expanded on the basis of 5G development, achieving deep coverage in all areas, serving intelligent interconnection of scenarios, and fully supporting the social development of smart life and intelligent industries.

In December 2021, CICT Mobile further released the "Ubiquitous Coverage Scenario Intelligence—6G Scenario, Capability and Technology Engine White Paper (V.2021)" [28], further analysis and discussion were conducted on the application scenarios, key capabilities, and technology engines of 6G. Based on the three major scenarios supported by 5G, namely enhanced mobile broadband, ultra-high reliability low latency communication, and massive machine type communication, 6G will further expand to support six types of application scenarios. These include three scenarios emphasising coverage capabilities: wide area coverage, mobile broadband coverage, and hotspot coverage, as well as three scenarios emphasising connection characteristics: extreme low latency high reliability, ubiquitous massive connections, and sensing and positioning. Among them, through wide area coverage, mobile broadband coverage, and hotspot coverage, the full network coverage capability is demonstrated, representing the service capabilities of mobile broadband in space and on the ground, as well as the ultra-high capacity provided in hotspot areas; at the same time, through extreme low latency high reliability, ubiquitous massive connections, and sensing and positioning, the expansion of service capabilities for 2B business, vertical industries, and IoT connection characteristics is demonstrated.

The performance indicators of 6G wireless access networks not only include further improvements to the eight key performance indicators already covered in the 5G vision (peak rate, spectral efficiency, user experience rate, regional traffic density, network energy efficiency, mobility, connection density, and latency), but are also expected to include important new indicators for future networks, such as coverage, reliability, latency jitter, and positioning accuracy, to support a wider range of scenarios and business needs. In addition to the expansion and improvement of wireless access network performance indicators, the new capabilities of 6G systems and networks are also particularly important. These capabilities support future communication systems to become stronger, smarter, and more trustworthy, including network intelligence capabilities, trustworthy security capabilities, programmable capabilities, network-computing collaborative capabilities, three-dimensional networking capabilities, and network perception capabilities.

The innovation of 6G key technologies mainly includes three aspects: enhancement of traditional technologies, innovation of integrated technologies, and evolution of network architecture. The enhancement of traditional technologies mainly includes the evolutionary enhancement of large-scale antenna technology, high-precision positioning technology, and non-terrestrial network (NTN) technology. Among them, large-scale antenna technology will continue to play a role in increasing system capacity and coverage in 6G networks, and form super-dimensional antenna technology from the expansion of antenna dimensions, the exploration of holographic dimensions, and the enhancement of functional dimensions. CICT Mobile is currently conducting research and trial work on constructing denser antenna array forms using reconfigurable intelligent metasurface (RIS) technology. High-precision positioning technology will become one of the important technologies for 6G to enable vertical industries. As the research on 6G goes deeper, multi-source fusion positioning solutions featuring cellular network carrier phase positioning, AI positioning, and positioning and communication fusion technology are expected to become important technical solutions for 6G positioning, bringing significant improvements in positioning accuracy and resource utilisation. The space-air-ground integration technology, characterised by elastic reconfigurable network architecture and unified air interface, will enable 6G networks to become an important technology for extending network service, changing the way users connect in time and space, and comprehensively enhancing the service capabilities of 6G networks in supporting public safety, emergency communication, social governance, industrial upgrading, and other aspects. CICT Mobile is currently conducting research and testing on the integration of satellite and ground prototype development, which will lay a solid foundation for the research and verification of 6G space-air-ground integration technology.

In terms of integrating technological innovation, the main layout is along the development route of deep integration of IT, DT, and OT, including key technologies such as AI, integrated sensign and communication, programmable networks, computational networks, and trustworthy network security. Artificial intelligence will deeply integrate with communication technology to form 6G network AI technology characterised by native AI, achieving network self-optimisation and self-evolution, improving network security and reliability, significantly reducing network operation and maintenance costs, reducing network energy consumption, enhancing user experience, and achieving network automation. Integrated sensing and communication technology, under the cellular network architecture, utilises radar, imaging and other wireless sensing technologies to perceive the position, posture and environmental information of non-target terminals and other objects, obtaining multi-dimensional, multi-level perception data, extending the connection range of the mobile communication system from the target terminal to objects and the environment, truly achieving scene intelligence, and achieving the effect of mutual promotion and coordinated coexistence of communication and sensing. Programmable networks are a solid technical foundation for achieving 6G networks. By applying programmable networks in 6G networks, the flexibility and elasticity of the network can be greatly enhanced, achieving on-demand customisation and flexible deployment of the network, and achieving automated and intelligent operation and maintenance. Computating networks can achieve flexible dynamic scheduling and efficient collaboration of distributed ubiquitous computing and storage resources in 6G networks, providing basic resource guarantees for the flexible deployment and rapid collaboration of 6G network functions. 6G networks will have inherent security capabilities and, based on distributed technical means, a trustworthy 6G network security architecture has been built using cryptographic principles, solving the security and trust issues faced by the decentralised architecture of 6G networks.

In terms of network architecture evolution, in order to achieve the overall vision of 6G, the upgrade and expansion of 6G networks and services, and to meet the rich service needs of 6G networks providing intelligent ultimate connections for people, objects, environments and virtual spaces, the 5G network architecture needs to evolve and transform. Among them, the user-centric dynamic cluster-based wireless network architecture will effectively improve the consistency of business experience within the coverage area, reduce the cost and energy consumption of centralized deployment of ultra-large-scale antenna systems, and can provide different specifications/capabilities of antenna units for deployment according to the differentiated needs of deployment areas/services. The native AI network architecture characterised by "one super multi-body intelligence" supports the deployment of AI capabilities in a layered and domain-specific distributed manner, which can better play the role of AI technology in the network. In addition, the service capabilities of 6G networks will no longer be limited to being provided by the core network. The trend of lightweight core networks and the sinking of service capabilities to RAN will form a SBA RAN architecture, thereby facilitating the provision of instant, customisable, user-centric personalised and differentiated services according to the coverage range of the RAN system.

In addition to carrying out the layout and research of 6G key technologies, CICT Mobile is actively carrying out the prototype verification work of 6G key technologies, in order to identify potential key technologies for 6G through platform verification, accelerate the maturity and implementation of 6G key technologies, and support subsequent 6G standardisation and industrialisation work.

4.2.4 ZTE

ZTE Corporation (hereinafter referred to as ZTE) proposed the 6G vision of "Connected and intelligent business" in 2019, emphasising the hope that 6G will break through various business boundaries and drive mobile networks to become the core elements of business enablement, while also proposing the connotation of 6G intelligent connection. The future from 5 to 6G is a process of enhanced evolution and technological transformation. The driving forces for 6G mainly include two aspects: one is the driving forces of 5G evolution, including digital economy, energy efficiency and cost efficiency, constantly developing industry use cases, wireless sensing and network intelligence, which require 6G network enhancement to solve the capabilities and scenarios that 5G cannot support; the second is the generational driving factors of 6G, including digital and AI economy, sustainable society, developed industry use cases, perceptual internet, AI internet, robot internet, etc.

6G introduces application scenarios of connected and intelligent business such as AI internet, perceptual internet, robot internet, etc. Its typical service use cases include immersive cloud XR, holographic communication, sensory interconnection, smart interaction, communication perception, pervasive intelligence, digital twin, full coverage [29] and so on. The 6G network will not only support Tbps throughput and ubiquitous Gbps user data rates, but will also support the integration of various services and technologies such as communication, intelligence, sensing, computing, energy, space, trust, and security. Compared with the 5G network focusing on connection capabilities in three major scenarios, the 6G network will increase the network capability requirements in multiple dimensions such as intelligence, sensing, and sustainable development.

ZTE believes that the 6G network capability requirements include: peak data rate (50 times), user data rate (200 times), connection density (> 10 times), 2D business capacity (> 100 times), 3D business capacity, mobile speed (57 times), air interface delay (> 8 times), global coverage (2 times), energy efficiency (> 10 times, zero bit zero Watt), deterministic synchronisation accuracy (> 5 times), reliability (>100 times), positioning accuracy (indoor/outdoor, 10–100 times), battery/power supply life (2 times, zero consumption terminal), etc. Note that the multiples in brackets represent the multiples of 6G capabilities relative to 5G, as shown in Fig. 4.35.

Compared to the 5G network, the new network capabilities that the 6G network may introduce include AI efficiency (such as time efficiency, fidelity, performance gain, etc.), E2E network reliability (such as availability, reliability, security, integrity, maintainability or recoverability), network energy use (such as energy consumption, zero it zero Watt, zero energy consumption terminal, energy efficiency, etc.), sensing resolution (such as target position, speed, moving direction, action, image or video behaviour recognisability), semantic accuracy (such as semantic similarity, fidelity, completeness, trustworthiness), etc.

Based on the vision, scenarios and capability requirements of 6G, ZTE proposes 6G enabling technologies including network evolution technology and paradigm shift technology, as shown in Fig. 4.36. Among them, network evolution technology

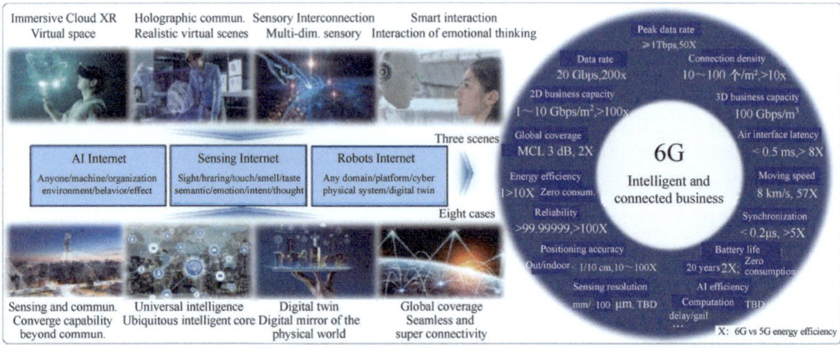

Use case reference, MIT-2030 Promotion Group, ZTE participated in drafting,"6G Overall Vision and Potential Key Technologies White Paper", June 2021.

Fig. 4.35 ZTE's vision, scenarios and capability requirements for 6G

is based on the continuous enhancement of 5G foundation, including flexible use of spectrum, utilisation of spatial resources, higher frequency band technology, and integration of air, space, land and sea. Paradigm shift technology includes native AI air interface, programmable wireless network, integration of communication, sensing, and computing and service architecture trustworthy network. Revolutionary technologies for 6G and its subsequent evolution may include service architecture trustworthy network, multimodal network fusion, intelligent air interface, programmable wireless channel, semantic communication based on end-to-end deep learning, etc., and it is believed that before 2025, 6G is in the definition and early technology research stage, and the standardisation work of the 6G will gradually begin afterwards.

Facing the development trend of cloud-native, intelligent, service-oriented, knowledge-based, and sustainable technologies for 6G networks, ZTE's 6G research team adheres to the technical innovation strategy of "consumer and industry enhanced application scenario driven, E2E horizontal and vertical technology innovation

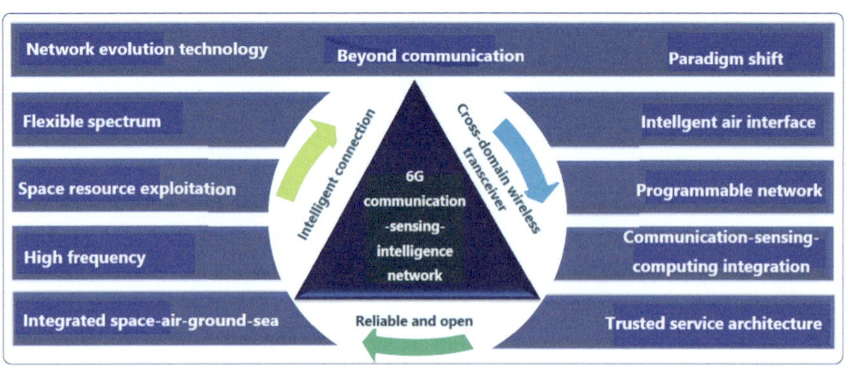

Fig. 4.36 ZTE's proposed 6G enabling technologies

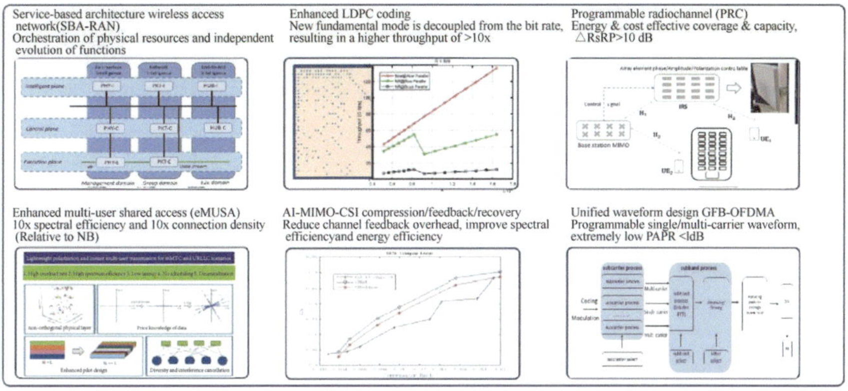

Fig. 4.37 Features and performance improvements of the six major label technologies proposed by ZTE for 6G

equally important, full frequency band technology innovation". They have completed the design of six major label technology application scenarios and technical solutions for 6G, namely SBA based wireless access network, enhanced LDPC coding, programmable wireless channel [30], enhanced multi-user shared access, AI-MIMO-CSI compression/feedback/recovery, unified waveform design GFB-OFDMA, its technical features and performance improvements are shown in Fig. 4.37.

In response to the need for more basic innovation breakthroughs in future mobile communications, ZTE is actively promoting industry-academia-research cooperation in the field of 6G, establishing 6G joint labs with several universities such as Tsinghua University and Southeast University, and conducting joint innovation research on 6G-related enabling technologies. They have achieved many fruitful results in areas such as the enhanced large-scale antenna, intelligent network, terahertz channels and devices, integration of communication, sensing and computing, and its new system architecture.

ZTE together with Chinese operators have carried out some early development of 6G enabling technologies in 5G network, such as the 6G programmable wireless channel joint test based on cellular communication instant voice (Push to talk over Cellular, PoC), including the industry's first high-frequency non-line-of-sight outdoor coverage extension, mid-frequency weak area multi-path channel rank increase, low-frequency outdoor to indoor cascade deep coverage test cases, preliminarily verifying that high, medium and low frequency super-surface artificial multi-path can reconstruct the deep coverage propagation environment of 5G network, the test results are shown in Table 4.1.

Table 4.1 Verification test results of programmable wireless channel technology based on electromagnetic surface

Frequency band	Tested cases	Received signal strength increasing/dB	Throughput increasing
mmWave	NLoS outdoor coverage extension	12.5	3 times
Intermediate frequency	Rank enhancement for multipath channels	10	Uplink 40%, downlink 2–3 times
Low frequency	Cascaded outdoor-to-indoor coverage	14	Downlink 48%, uplink 170%

4.2.5 Vivo

Vivo Mobile Communications Co., Ltd. (abbreviated as vivo) Communication Research Institute launched 6G pre-research work in 2019, proposing the 6G vision of "building a world where physical and digital are freely connected", and conducting research on 6G scenario requirements, 6G system and network architecture, 6G network and air interface intelligence, 6G new business and service system, 6G new air interface, as well as ISAC, RIS, ultra-low power communication, new waveform and other aspects.

In terms of 6G scenario requirements, vivo released the "Digital Life 2030+" [31] and "6G Vision, Requirements and Challenges" [32] white papers in 2020. In the "Digital Life 2030+" white paper, vivo Communication Research Institute analysed the development trends of 11 aspects closely related to people's lives in 2030 and beyond, and presented 29 vivid scenario cases, which include 60 creative product and service concepts, in order to give consumers and industry technicians a clear understanding of digital life in 2030 and beyond. The white paper "6G Vision, Requirements and Challenges" is based on the scenarios of people's digital life in 2030 and beyond, combined with future technology development trends, it provides preliminary views on 6G vision and requirements, and analyses the challenges of 6G for networks and terminals, contributing to the industry consensus on 6G vision and requirements. Vivo believes that, looking towards 2030 and beyond, 6G will build a ubiquitous digital world, freely connecting the physical world and the digital world, achieving mutual interaction and high integration between the two, thus providing a wealth of business applications, promoting efficient and sustainable social development, and enhancing human happiness. 6G not only provides free connection between the physical world and the digital world, but also builds a ubiquitous digital world, and achieves related interaction and high integration between the two.

In terms of 6G system and network architecture, Vivo proposes that 6G will become the basic information infrastructure for the future physical and digital integrated world. On the basis of providing stronger communication capabilities than 5G, 6G will inherently provide computing and AI services, and integrate wireless sensing and positioning, network data information to provide important basic data information services. The integration of communication, computing and information services will become the main feature of 6G, and also the basic goal of 6G system design. Based on the three basic services of 6G, Vivo has carried out technical research on mobile computing network, intelligent network architecture, ISAC network architecture, data services and data plane, mobile network privacy computing, etc.

In terms of 6G network and air interface intelligence, Vivo believes that the combination of AI and wireless will evolve gradually from single module to end-to-end. In the early stage, Vivo focuses on the research of single modules of wireless communication, and has carried out research and development in areas such as AI-based channel state information (CSI) feedback, AI-based demodulation reference signal (DMRS) demodulation, AI-based positioning, AI-based beam management, AI-based CSI prediction, AI-based power amplifier non-linear optimization, AI-based mobility management, AI-based source channel joint coding, and has developed a hardware verification platform for AI-based DMRS demodulation, preliminarily verifying the gain brought by AI.

In terms of ISAC, Vivo focuses on the scenarios and requirements of ISAC, and conducts research on the challenges faced in waveform design, signal joint detection and interference elimination, MIMO-ISAC technology, the combination of ISAC and backscatter and RIS technology, mobility management of ISAC, link adaptive technology of ISAC, channel modelling and simulation methods, the impact of hardware non-ideal factors, privacy and billing. In order to carry out performance verification of sensing and communication integration based on air interface, Vivo has developed a respiration monitoring sensing verification platform based on NR, which can accurately monitor respiration in the lab environment while achieving Gbps communication data rate.

In terms of RIS, Vivo has developed an RIS prototype in cooperation with universities, analysed and evaluated the working bandwidth and adjacent frequency characteristics of RIS devices, achieved matching of RIS working bandwidth and communication system bandwidth by optimising RIS unit structure, and conducted research on RIS beam shaping mechanism, RIS-based beam measurement method, RIS channel modelling and system-level simulation method.

In terms of ultra-low power communication, Vivo mainly focuses on low power wake-up technology, backscatter technology and energy harvesting technology. Among them, Vivo leads the research project on low power wake-up technology in 3GPP Rel-18; in terms of backscatter technology, it focuses on research in multiple access, interference deletion, space–time grouping code and beam shaping, as well as sensing and positioning based on backscatter; Vivo has developed a backscatter prototype, initially achieving 2 Mbps data rate transmission at a communication distance of 3 m, with device communication power consumption at the 100 μW level.

In terms of new waveforms, Vivo focuses on Orthogonal Time Frequency Research on Orthogonal Time Frequency Space (OTFS) technology, focusing on the design of OTFS pilot frequency, the combination of OTFS and MIMO, and the coexistence mode of OTFS and OFDM, among other areas.

4.2.6 OPPO

Guangdong OPPO Mobile Telecommunications Corp., Ltd. (OPPO) launched its future-oriented wireless communication technology research in 2019, mainly starting from the new terminal forms and new business needs of 6G, including the combination of AI and wireless communication, zero-power terminal technology, integrated sensing, space-air-ground, terahertz and other fields. On the one hand, considering that the 5G system has comprehensively covered the business needs of the Internet of People and the Internet of Things, one core development direction of 6G is to extend the network's capabilities to cover the "Internet of Intelligence" field; on the other hand, considering the power consumption and cost of 5G IoT terminals are still relatively high for a large number of minimalist applications, OPPO believes that another core development direction of 6G is to build a zero-power IoT to support the truly "ubiquitous" 6G integration of all things.

In the most core aspect of 6G Internet of Intelligence, OPPO, in its keynote speech "B5G&6G: Connecting Intelligence" at the IEEE International Conference on Communications (IEEE ICC) in May 2019, proposed the vision of "6G is a communication network shared by all intelligent entities" and the three-step plan to realise the 6G Internet of Intelligence, "For the AI, By the AI, Of the AI".

In 2021, OPPO released the white paper "AI-Cube Empowered 6G Network Architecture" [33], systematically expounding the concept of the 6G and AI integrated network, that is, when considering communication technology in the 6G era, the intelligent domain is considered as a new dimension beyond the traditional data domain and control domain. The application and evolution of intelligence require more extensive data, so faster, wider, and more ubiquitous 6G connections are needed. At the same time, future business applications and terminal types will undergo significant changes, and artificial intelligence can further enhance the efficiency of wireless connections. Finally, as the integration dimension further increases, the computing power and storage resources used for thinking and memory will further integrate with traditional communication connections and business services, serving the application and evolution of intelligence together. 5G has achieved the "mobilisation of production work", and 6G will achieve the "mobilisation of thinking and learning".

The AI emphasised in B5G/6G is the native AI in the cellular network, that is, the AI's capabilities are integrated into the network, which can be used to optimise various functions of the network and also provide AI capabilities externally. Cloud AI emphasises an application layer AI, mainly used for application business, more of a business type. The advantages of Network AI are mainly reflected in three aspects: efficiency, integration, and data privacy. The efficiency advantage is that the resource utilisation of Network AI can more efficiently optimise network functions and third-party businesses; various model training and inference functions can flexibly choose and process paths and nodes as needed, unlike Cloud AI which can only interact between terminals and application servers. The integration advantage is that Network AI is more conducive to the integration of communication resources and

AI computing resources, the network has a better perception of wireless communication conditions, and to a certain extent, it can better balance wireless resources and computing resources. The data privacy advantage is that when the network performs AI processing, it can allow each node participating in the task to use its own data to perform training or inference locally, which can better protect the data privacy of terminals and various network nodes.

In 2022, OPPO held the first 6G AI competition, focusing on the issue of dataset expansion based on AI, which received a warm response from the academic and industrial communities. Firstly, to promote the research and development of AI technology in the 6G field, OPPO established an open AI dataset and model sharing website—Wireless-Intelligence, which has so far freely released 60 sets of AI datasets and 8 AI reference models, widely used in the industry. Regarding 6G In the research of AI, OPPO first carried out a lot of research and analysis on common basic problems [34, 35], starting from the perspective of multi-module integration and systematic AI, the bottom-up multi-module intelligent cascade and the top-down integrated AI processing have formed the two basic directions of systematic AI research. Secondly, for problems such as scene intelligence adaptation under small sample limitations, a basic route of data-driven and knowledge-driven parallel wireless AI solutions has gradually been formed. Finally, considering that data is the basis of AI research, it comprehensively considered how to achieve channel modelling and virtual channel reconstruction for AI needs.

In January 2022, OPPO released the "Zero Power Consumption Communication" white paper [36], systematically elaborating on the technical vision, technical positioning, application scenarios, system principles, technical challenges and key technologies of zero power consumption IoT, as well as its combination with other 6G technologies. Zero power consumption IoT has excellent characteristics such as battery-free, small size, extremely low cost, and maintenance-free, and is expected to build the next generation of IoT communication network. In the zero power consumption IoT based on environmental energy, terminal devices collect energy from various environmental energy sources (such as radio waves, solar energy, thermal energy, kinetic energy, etc.) to drive terminal work and communication, thereby enabling terminal devices to get rid of the dependence on traditional batteries, achieve battery-free communication, which also gives zero power consumption IoT a high environmental value. Further speaking, with the use of ultra-low power consumption communication technology and extremely low complexity communication technology, smaller terminal device size and extremely low terminal cost can be achieved. Combined with sensor technology, a low-cost, maintenance-free sensor network can be achieved. Given the excellent characteristics of zero power consumption IoT such as battery-free, small size, extremely low cost, and maintenance-free, zero power consumption IoT will be widely used in various vertical industries and personal consumption fields, such as logistics, industrial IoT, intelligent transportation, smart energy, smart home, wearable and other fields. Therefore, zero power consumption IoT has great application potential and will become the next generation of IoT technology for 6G.

OPPO has developed a demonstration verification system for zero power consumption communication, and the system has verified the technology of radio frequency energy collection and reverse scattering communication.

4.2.7 IMT-2030 Promotion Group

In June 2019, the Ministry of Industry and Information Technology (MIIT) officially established the 6G study group, and renamed it as the IMT-2030 Promotion Group in 2021, organizing domestic enterprises and universities to carry out research on potential key technologies of 6G such as vision and requirements, candidate frequency, terahertz communication, visible light communication, Orbital Angular Momentum (OAM), wireless AI, new modulation and coding, and officially released the "6G Overall Vision and Potential Key Technologies" white paper in June 2021 [37]. This white paper, as a phased result of the promotion group, covers the overall vision, eight major business application scenarios, ten potential key technologies, and some thoughts on the development of 6G. The white paper points out that future 6G services will show new development trends such as immersion, intelligence, and ubiquity, forming immersive cloud extended reality (Extended Reality, XR), holographic communication, sensory interconnection, intelligent interaction, communication and sensing, native intelligence, digital twins, and ubiquitous coverage, among other major business applications. The white paper proposes ten potential key technological directions for 6G that are currently widely concerned in the industry, including new air interfaces and new network architectures with native AI, as well as enhanced wireless air interface technology, new physical dimension wireless transmission technology, new spectrum usage technology, ISAC and other new wireless technologies, distributed network architecture, computing power network, deterministic network, satellite-terrestrial integration networking, network native security and other new network technologies. The application of these and other potential key technologies in 6G will greatly enhance network performance, meet the needs of new businesses and new scenarios in future social development, serve intelligent society and life, and help realise the "Internet of Intelligence, Digital Twins" 6G vision.

In addition, the white paper proposes four key issues in the development of 6G.

(1) On the issue of the relationship between 5 and 6G development, it is proposed that 6G will leap from the Internet of Everything to the Internet of Intelligence based on 5G. The successful commercial use of 5G, especially its widespread application in vertical industries, will lay a solid foundation for the development of 6G.

(2) On the issue of 6G spectrum resources, it is proposed that 6G will extend to higher frequency bands and efficiently use low, medium and high full spectrum resources. Among them, the low-frequency spectrum will still be a strategic resource for the development of 6G, millimetre waves will play a more important

role in the 6G era, while terahertz and other higher frequency bands will mainly meet the needs of large capacity short distance in specific scenarios.

(3) On the issue of 6G intelligent evolution, it is proposed that intelligence will be endowed to everything and AI will be native, which will become an important feature of 6G. The deep integration of artificial intelligence and communication technology will trigger all-round innovation in network information technology.

(4) On the issue of the relationship between non-terrestrial communication such as satellites and cellular networks, it is proposed that the future 6G network will still be based on terrestrial cellular networks, and satellites, drones, aerial platforms and other non-terrestrial communications will play an important role in achieving seamless coverage of air, space and earth.

In September 2021, the IMT-2030 Promotion Group organised a 6G seminar and officially released a series of phased research results, including the "6G Network Architecture Vision and Key Technology Outlook" [38] white paper, "ISAC Technology Report" [39], "Ultra-Large Scale Antenna Technology Research Report" [40], "Terahertz Communication Technology Research Report" [41], "Wireless AI Technology Research Report" [42], "Intelligent Super Surface Technology Research Report" [43], "6G Network Security Vision Technology Research Report" [44] and so on.

"6G Network Architecture Vision and Key Technology Outlook" [38]: The white paper elaborates on the driving forces of 6G network architecture evolution from three aspects: business-driven, DOICT integration-driven, and IP new technology-driven. It proposes the design principles of 6G network architecture, such as "insisting on network compatibility, insisting on minimalist design" and "transition from centralisation to distribution, from increment to integration, from external to internal, from ground to ubiquity". It elaborates on the vision of a distributed autonomous 6G network architecture; introduces twelve potential key technologies of 6G network, including potential architecture technologies such as distributed network, air-ground integration networking, native intelligent, native security, digital twin network, computing power network, and potential capability technologies such as programmable network, communication and information perception fusion network, deterministic network, data service, immersive multisensory network, semantic communication.

"ISAC Technology Report" [39]: It points out that ISAC will coexist in future 6G wireless networks, endowing 6G with the ability to sense the physical world at all times and everywhere, enabling new business scenarios.

"Research Report on Ultra-Large Scale Antenna Technology" [40]: It points out that ultra-large scale antennas are the evolution and upgrade of large-scale antenna technology, which can provide higher spectral efficiency, higher energy efficiency, and higher spatial resolution for 6G networks.

"Terahertz Communication Technology Research Report" [41]: It points out that the terahertz frequency band can be used as a beneficial supplement to existing frequency bands, and can be deployed in holographic communication, micro-scale communication, ultra-large capacity data backhaul, short-distance ultra-high-speed transmission, high-precision positioning and high-resolution perception and other 6G potential business scenarios.

"Wireless AI Technology Research Report" [42]: It points out that AI has important application potential in many aspects such as complex unknown environment modelling, channel prediction, intelligent signal generation and processing, network status tracking and intelligent scheduling, network optimisation deployment, etc., and is expected to promote the evolution of future communication paradigms and the transformation of network architecture.

"Intelligent Metasurface Technology Research Report" [43]: It points out that intelligent metasurfaces have the characteristics of low cost, low energy consumption, programmability, and easy deployment. By constructing an intelligent controllable wireless environment, it is expected to bring a new communication network paradigm to 6G.

"6G Network Security Vision Technology Research Report" [44]: It proposes the "active immunity, resilient autonomy, virtual symbiosis, ubiquitous collaboration" 6G security vision, introduces key security technologies such as AI security, blockchain security, lightweight access authentication, wireless physical layer security, software-defined security, data security, enhanced cryptographic algorithms, etc., providing a reference for 6G network security research.

4.2.8 Ministry of Science and Technology 6G Promotion Group

In November 2019, the Ministry of Science and Technology, together with the National Development and Reform Commission, the Ministry of Education, the Ministry of Industry and Information Technology, the Chinese Academy of Sciences, and the National Natural Science Foundation of China, organised a 6G technology research and development work initiation meeting in Beijing, announcing the establishment of the Chinese 6G Technology Research and Development Promotion Work Group and the General Expert Group. The promotion work group is composed of relevant government departments, whose duty is to promote the implementation of 6G technology research and development work; the general expert group is composed of 37 experts from universities, research institutes, and enterprises, mainly responsible for proposing 6G technology research layout suggestions and technical demonstrations, and providing consultation and suggestions for major decisions.

Subsequently, relying on the key R&D plan for broadband mobile of the Ministry of Science and Technology, the organisation carried out the layout of 6G research work, and successively funded a series of 6G-related research projects.

4.2.9 Future Mobile Communication Forum

The Future Mobile Communication Forum [45] has always been active in the forefront of academic research on mobile communication technology, actively organising domestic and international 6G academic exchanges and cooperation, and established a 6G Special Interest Group in 2019 to carry out key technology research on 6G. During the Global 6G Conference in September 2020, the 6G Special Interest Group of the Future Mobile Communication Forum released a series of white papers, including "Multi-viewpoint Sketching 6G Blueprint" [46], "6G: Gap Analysis and Potential Enabling Technologies" [47], "Preliminary Study of Advanced Technologies towards 6G" [48], "Visible Light Communications: A Brighter Future" [49] and "ICDT Integrated 6G Network" [50].

"Multi-viewpoint Sketching 6G Blueprint"[46]: ICDT integration will comprehensively redefine the 6G technology system, with fully definable network structure and functions, supporting native intelligence and security, and achieving precise matching of demand and virtualised resources. 6G will have strong communication, computing and AI capabilities, providing "traffic", "computing power", "intelligence" services, and achieving a comprehensive upgrade from information transmission services to information processing services. 6G design should adhere to the integration of computing and communication, technology contextualisation and network native AI. At the same time, it should pay attention to new opportunities in cross-fields, seek breakthroughs from new materials, new energy, bio-information, spatial information and advanced computing, solve communication performance and cost bottlenecks, and expand communication boundaries. In addition, it should also evaluate the economic issues that 6G may face from the cross-fields of economics, law, ethics and psychology, as well as non-technical problems such as virtual space rights and interests, and the ethics of intelligent entities, and propose possible solutions from a technical perspective.

"6G: Gap Analysis and Potential Enabling Technologies" [47]: By analysing the defects and performance deficiencies of existing wireless communication systems, it reveals the development trend of wireless communication and future potential needs; by summarising the latest research progress of wireless communication, it summarises and sorts out the new wireless transmission technologies that have been vigorously developed in recent years; it outlines the key technical challenges and potential solutions related to the design of 6G communication systems, including physical layer transmission technology, network design, security methods and the development of testbeds for some key technologies.

"Preliminary Study of Advanced Technologies towards 6G" [48]: From the perspective of exploring bandwidth and information processing capabilities, this white paper re-examines two promising 6G technology drivers—terahertz and quantum information. It outlines existing terahertz technologies, including use cases and scenarios, spectrum and channel models, components/circuits/antennas, communication systems, demonstration projects, etc. With the powerful information processing capabilities of quantum information technology, it is expected to meet the huge communication and computing demands that future communication systems will face. The book discusses the research content and system structure of quantum communication.

"Visible Light Communication: A Brighter Future" [49]: Points out that visible light communication has many advantages such as large bandwidth, low energy consumption, and low cost. It is not only an effective supplement to traditional wireless communication, but also one of the important alternative technologies in the 6G era. This book systematically sorts out and analyses visible light communication in the hope of refining application scenarios, identifying main bottlenecks, and exploring corresponding solutions to promote the further practicality of this technology. In particular, it discusses the application scenarios of visible light communication, analyses the unique features and typical cases of each scenario according to different application characteristics, determines the direction of future practicality, clarifies the advantages and disadvantages and development status of different devices and equipment, and further discusses the main technical bottlenecks of devices and platforms.

"ICDT Integrated 6G Networks" [50]: ICDT integrated 6G will be an end-to-end information processing and service system, its core functions will extend from information transmission to information collection, information computing and information application, providing stronger communication, computing, sensing, intelligence and security and other multi-dimensional native capabilities, bringing a beautiful vision of "digital twins, ubiquitous intelligence" for the socio-economic development towards 2030. This white paper focuses on the network architecture and protocol stack of ICDT integration, the integration of sensing, communication and computing, the integrated architecture of air, space and earth, the native AI architecture, the intent network, the deterministic network, the digital twin network, the native security architecture, the open network architecture, the AI-enabled air interface, the multifunctional air interface, the integrated air interface of sky and earth, the new air interface, the intelligent ubiquitous terminal and other key technologies and issues, and tries to provide solutions.

During the second Global 6G Conference in March 2022, the Future Mobile Communication Forum released the "6G Overall White Paper" [51], "6G Intelligent Rail Transit White Paper" [52], "Terminal Friendly 6G Technology" [53], "Digital Twin Technology for 6G" [54], "Zero Power Consumption Communication" [55] and so on.

"6G Overall White Paper"[51]: Points out that the driving force for 6G development mainly comes from three aspects: one is the need to solve the challenges of high investment, high power consumption, and difficult operation and maintenance of 5G networks; the second is the new demand for information processing brought by future new applications and scenarios such as the "metaverse"; the third is the innovative opportunities brought by the integrated development of mobile communication technology, computer technology, artificial intelligence and big data technology. This book builds an integrated network control framework by defining key functional entities such as network brain, perception control, computing control, communication control, user control and business control, and analyses key technologies such as distributed computing, distributed perception, distributed intelligence, native security and intent management.

"6G Intelligent Rail Transit White Paper" [52]: The potential applications of future intelligent transportation networks include automatic train driving, collaborative train networks, train interconnection, ultra-high-definition (4 K/8 K) train videos, train self-organising networks, and ultra-precise (centimetre-level) train positioning. To meet the requirements of 6G intelligent transportation applications, new technological breakthroughs are needed, including but not limited to cell free large-scale MIMO network architecture, ISAC technology, artificial intelligence and security technology, new ultra-reliable low-latency technology, digital twin networks, etc.

"Terminal Friendly 6G Technology" [53]: Terminal friendly includes reducing the power consumption, cost and complexity of the terminal, supporting terminal diversity, extending access scenarios, improving uplink efficiency (energy efficiency, spectral efficiency), enhancing user experience, and even relaxing a certain point of the terminal experience to bring a better experience at other points. From the network side, terminal friendly aims to enhance network capabilities, simplifying the related processes and technical complexity of the terminal. This book details the key technologies of terminal friendly 6G, including satellite and ground integration technology, multi-band integration technology, wide-area ubiquitous access support for terminals, native networking support for flexible terminal access, communication-sensing-computation integration to extend terminal service capabilities, cell-free technology to support zero-perception mobility experience for terminals, backscatter and near-zero power receivers to support zero-power communication for terminals, new multi-address access to support unscheduled transmission and uplink asynchronous transmission for terminals, and the combination of AI and communication to enhance the terminal user experience.

"Digital Twin Technology for 6G" [54]: It points out that digital twins will be closely integrated with 6G technology and promote each other. On the one hand, 6G technology for digital twins mainly provides ultra-large capacity, ultra-low latency data and feedback information transmission for the interaction layer of digital twins, promoting better application of digital twin technology; on the other hand, digital twin technology also provides new ideas and solutions for the research of 6G key technologies. This white paper elaborates on the application scenarios of digital twins, discusses the key technologies of digital twins, the 6G network enabled by digital twins, and some solutions for vertical industries, and looks forward to the future development of digital twins.

"Zero Power Consumption Communication" [55]: Zero power consumption communication obtains energy from radio waves in space to drive terminal operation, which can be used without conventional batteries, achieving ultra-low power consumption, extremely small size and extremely low cost IoT communication terminals, which we call zero power consumption terminals, and the corresponding communication process is called zero power consumption communication. Further, the use of backscatter and low power consumption computing technology enables zero power consumption terminals to achieve extremely simple RF and baseband circuit structures, thereby greatly reducing the cost, size and circuit power consumption of the terminal. Zero power consumption technology will be widely used in applications such as industrial sensor networks for vertical industries, intelligent transportation, smart logistics, smart warehousing, smart agriculture, smart cities, energy fields, as well as applications for individual consumers such as smart wearables, smart homes and medical care. The form of zero power consumption IoT terminal is a simple tag or integrated with a sensor.

4.2.10 6GANA (6G Alliance of Network AI)

The 6G network will become the infrastructure enabling thousands of industries, realising the development vision of "digital twins, ubiquitous intelligence". This requires 6G to build a new type of open and integrated network architecture, achieving deep integration of communication, computing, AI, sensing and security, and creating communication as a service (ComaaS), Computing as a Service (CompaaS), AI as a Service (AIaaS), Data as a Service (DataaaS), Sensing as a Service (SenseaaS), and Security as a Service (SecurityaaS). Therefore, how to build a 6G network with native intelligence and realise AIaaS has become an important direction of industry research.

From 4 to 6th December 2020, during the "2020 (Third) China Information and Communication Conference" held in Chengdu, Sichuan, China Mobile, Huawei, and the Ministry of Science and Technology's 6G key R&D project group closely collaborated to jointly hold a seminar on "From Cloud AI to Network AI: Building a 6G Network Architecture". They conducted in-depth discussions and intellectual collisions on the disruptive evolution of future 6G network architecture and its key technologies, and further collaborated to jointly promote the development and innovation of AI and future 6G network technology. At this seminar, China Mobile, Huawei, and 18 other industry, academia, and research units jointly initiated the establishment of the 6G Alliance of Network AI (6GANA) [56], dedicated to promoting Network AI as a global consensus and comprehensively advancing the research and verification of 6G Network AI. The establishment of 6GANA is shown in Fig. 4.38.

On 7th April 2021, 6GANA was officially established and held a seminar at Huawei's Songshan Lake Base in Dongguan, Guangdong. The experts and invited advisors present unanimously believed that in the next 10 years, the deep integration of ICDT will bring great changes to the industrial ecosystem and application innovation. 6GANA is the first industry-academia-research exchange platform in the world focusing on this, which will promote this trend to become a basic capability of the 6G network, which is an innovation in itself in the early stage of 6G research. The mission, positioning, organisational structure, and operating mechanisms of 6GANA were discussed and determined at this establishment meeting.

Fig. 4.38 6GANA kickoff meeting

The mission and goal of 6GANA is to focus on 6G Network AI, and from a technical and ecological perspective, actively promote AI to become an inherent capability and service (AIaaS) of the 6G network, rather than just a tool. 6GANA's positioning is an open and domestic industry-academia-research platform, organising and carrying out extensive domestic and international academic and industrial exchanges, guiding the formation of industry consensus on 6G network AI, and leading future standardisation and industrialisation directions; at the same time, 6GANA will actively support and participate in the work of other global 6G industry organisations, jointly promoting the research of the entire 6G network architecture. 6GANA will focus on the research of 6G Network AI's requirements, scenarios, architecture, data, theory/algorithm, verification, operation/management, and accordingly set up 7 working groups (Technical Group, TG). The daily work of 6GANA is organised by the Technical Steering Committee. The Network AI that 6GANA focuses on AI is an native AI architecture technology that deeply integrates communication technology, information technology, big data technology, and AI technology within the communication network. It will redefine the end-to-end cloud ecosystem, support the construction of new business models for 6G networks, and enable the true arrival of the era of ubiquitous intelligence. As a spontaneous exchange platform for industry, academia, and research, current members include: China Mobile, China Telecom, China Unicom, Huawei, Tencent, Alibaba, China Information and Communication Technology Group Co., Ltd. (referred to as CICT), Unisoc group Spreadtrum, Pengcheng Laboratory, Zhejiang Laboratory, Purple Mountain Laboratory, Tsinghua University, Zhejiang University, University of Science and Technology of China, University of Electronic Science and Technology, Beijing University of Posts and Telecommunications, Harbin Institute of Technology, The Chinese University of Hong Kong (Shenzhen), Xidian University, Beijing University of Aeronautics and Astronautics, ZTE, Ericsson, Nokia, Xiaomi, Intel, Inspur and others.

In May 2022, 6GANA officially released a series of white papers, comprehensively presenting the research results of 6GANA since its establishment, including "Network AI Concept Terminology White Paper" [57], "6G Network Native AI Technology Requirements White Paper" [58], "Ten Questions on 6G Network Native AI Network Architecture" [59], "6G Data Service Concept and Requirements White Paper" [60], "B5G/6G Network Intelligent Data Analysis" [61], "Knowledge Defined Orchestration and Management White Paper" [62].

"Network AI Concept Terminology White Paper" [57]: It comprehensively explains the related concepts, terms, and definitions of Network AI, defines the basic concept terms of Network AI, and the different options for the integration of 6G network with computing, data, and AI. The white paper first clarifies the related concepts of AI4NET, NET4AI, AIaaS, Cloud AI, and Network AI, and explains their relationships. Among them, "Network AI" with its broad generality, becomes the overall overview of the white paper on the integration of AI and network. Based on the establishment of the "Network AI" concept, the white paper continues to define related fine-grained concepts, including evaluation indicators, various AI technologies, and the difference between network intelligence and native AI.

Firstly, traditional networks usually measure network performance from aspects such as bandwidth, delay, and jitter. The establishment of the Network AI concept will bring about innovations in subsequent applications, architectures, and algorithms. The existing evaluation indicators cannot fully and accurately describe the performance of Network AI. The white paper proposes the concept of AI Service Quality (Quality of AI Service, QoAIS), which measures the network performance and business experience of AI applications from the aspects of connection, computing power, algorithm, and data, and summarizes the specific indicator details, which has a high reference value for evaluating the value of innovative Network AI technologies.

Secondly, the implementation of network AI utilises a large number of AI-related technologies. AI, as a branch of computer science, has formed a system in both theory and practice. The white paper can only introduce a limited number of AI concepts to help network researchers understand the common implementation methods in network AI.

Thirdly, native AI, as a current research hotspot, has attracted widespread attention in the industry, but its essential difference from network intelligence has not yet been clearly clarified. The white paper points out that network intelligence is the use of AI and other technologies to assist in the intelligent operation of communication networks, achieving the effects of improving quality, increasing efficiency, and reducing costs. However, traditional intelligence mostly adopts the mode of external AI to supplement existing architectures, protocols, and processes, lacking a unified standard process, with a long application period, and the quality of service is also difficult to effectively guarantee. In response to these challenges, China Mobile has proposed the concept of native AI. Different from traditional network intelligence, native AI abandons the mode of external AI, arranges and controls computing power, data, and AI algorithms/models from end to end, supports the deep integration of elements such as connection, computing, data, and AI algorithms/models at the architectural level, supports the orchestration and management of AI capabilities to radio access network, transportation network, bearing network, core network, etc. as needed, provides the basic capabilities required for intelligentisation for high-level network autonomy and diversified business needs, making network intelligence more efficient and performance superior.

Finally, the white paper introduces the five stages of AI and network integration in a progressive manner from the perspective of network AI technology evolution, gradually deepening in the direction of S0-AI4NET, S1-Connection4AI, S2-(Connection + Computing)4AI, S3-(Connection + Computing + Data)4AI, S4-(Connection + Computing + Data + Algorithm)4AI to S5-AIaaS, network AI also gradually develops from the network intelligence of the external AI mode, eventually achieving native AI of 6G network. For each stage, the white paper deeply analyses different connections, integrations, and service categories, proposes potential business scenarios, service descriptions, and research directions.

"6G Network Native AI Technology Requirements White Paper" [58]: Completed by 6GANATG1, it details the current development status and challenges of network intelligence in 5G, and based on the future 6G new business forms, new industry scenario use cases, new technology and other aspects of demand and trends, systematically analyses and refines the technical requirements and impacts of native AI design, thereby strongly supporting the design and implementation of native AI-related technical solutions in the new 6G network and their final landing. Multiple AI technologies have been used in the 5G system for the application and improvement of communication problems, and the 5G system has also adapted to AI applications at the network level. However, the 5G system did not fully consider AI applications and their related capability at the beginning of its design, so in subsequent development, it used various AI technologies to self-enhance and support AI applications through add-on and patched way. In terms of self-enhancement, for identified and specific communication problems (such as mobility prediction, fault location, etc.), various AI algorithms/models are used for offline training to improve performance and reduce operation and maintenance costs.

In terms of supporting AI applications, 3GPP Rel-16 added a logical functional entity NWDAF in the 5G network architecture to implement some intelligent applications within the network and AI for internal and external use. However, the AI training method for specific problems lacks scalability and extensibility, and the lack of coordination between the upper application and the lower network pipeline also prevents NWDAF from fully utilising resources and data. These system design flaws have brought about many problems, including limited data sources, consumption of transmission bandwidth, lack of data privacy protection, lack of support for external AI services, insufficient utilisation of infrastructure, and lack of data governance/services. Therefore, at the beginning of the new 6G system design, it is fully considered how to deeply integrate with AI, aiming to provide users with ubiquitous intelligent capabilities on demand and achieve high-level autonomy and security and trust. The native AI design can better adapt to future 6G new scenarios, new use cases, bringing new business models and values. The white paper finally proposed the technical requirements for the native integration of AI and the new 6G system, and deeply elaborated from the aspects of capability requirements, service requirements, and architecture requirements.

The capability requirements of native AI include computing power, algorithms, and data. First, native AI requires the network to have the ability to perceive, measure, schedule, and trade computing power to cope with the explosive growth of data and the increase in the complexity of AI algorithms. Secondly, native AI requires AI algorithms to have the ability to quantify performance, efficient training, interaction, and evolution, to achieve native intelligence of the network and provide AI services. Finally, native AI requires the network to have the ability to collect, analyse, protect privacy, store and open data, laying the foundation for better operation of AI application.

The service requirements of native AI are still unfolded from the three aspects of computing power, algorithms, and data: first, the new 6G system with native AI needs to provide delay, energy consumption, privacy, and service experience sensitive business through multi-layer integration of computing and network; second, native AI needs to provide support for AI algorithms such as standardized description, service interface, training, testing, and inference, to achieve native AI services internally and externally; third, the new 6G network needs to support independent data plane functions, establish a knowledge graph to support native AI, to meet the end to end native AI requirements of network data collection, machine learning, intelligent services and application.

In addition, the new 6G system needs to provide a complete quality of intelligent services, that is, QoAIS, to accurately evaluate, measure, and guarantee the quality of AI applications.

"Ten Questions on the native AI Network Architecture of 6G Network" [59]: For the ten core technical issues that potentially affect the 6G network architecture, it analyses from four aspects: feature connotation, necessity, feasibility, and impact on network architecture, to study whether the technical features shown by this issue are required by the 6G native AI network architecture, and gives the views and suggestions of TG2. Following TG1's research on requirements and application scenarios, facing the diverse requirements of 6G network native AI for different industries and scenarios, the first question raised by TG2 is: How to express and import user requirements for 6G network AI services? In addition, why does the self-generation of AI use cases (forms of demand expression) become a technical feature supported by the 6G network? Further speaking, based on the service requirements, how to analyse and map to the quality requirements of AI services that the network can understand? In response to these questions, TG2 proposes the concept of QoAIS and provides a reference indicator system design. As a provider of AI services, how the network evaluates and continuously meets the above user requirements, and implements QoAIS guarantees, needs to be considered from multiple levels such as network AI management, control, and business flow. Question three proposes how to automatically implement AI from the management perspective of network AI by managing the AI during the full lifecycle. The development, deployment, and related model management of applications are implemented to ensure the Quality of AI Service

(QoAIS). The fourth question proposes, from the perspective of multi-dimensional resource integration, whether to centre on tasks, coordinating the four necessary resource elements for network AI (algorithm, computing power, data, connection) to meet the quality requirements of QoAIS services in real-time and continuously. As algorithms, computing power, and data will become new controllable resource dimensions in the network, just like traditional connections, the heterogeneity of resources and the diversity of challenges and demands make them present different technical characteristics in the design of network architecture, which need further research. The fifth question proposes from the computing power dimension, how to deeply integrate computing power and connection resources to achieve more efficient AI tasks. The sixth question, from the data dimension, proposes why it is necessary to improve the current problems of low data value density and low access efficiency through native AI, to achieve on-demand dynamic data orchestration and configuration, and to feed back to the data services of native AI. The seventh question, from the AI algorithm dimension, addresses the two major challenges currently faced in the AI field: the over-reliance of AI algorithms on big data and the credibility of AI algorithms; it proposes why the self-evolution of AI algorithms can be achieved through model-based computing, and how to provide credible AI algorithms. In terms of the deployment architecture of the above resources, 6G will present a combination of centralised and distributed characteristics. The supply of AI capabilities under centralised resources has had a lot of practice in 5G networks, so the eighth and ninth questions propose how 6G understands distributed AI architecture, what aspects it reflects, and how different nodes coordinate. Finally, from the perspective of resource opening, the tenth question proposes what specific capabilities the above resources can be packaged in the 6G network to open to third parties.

"6G Data Service Concepts and Requirements White Paper" [60]: The vision of 6G network data services is to provide trustworthy data for 6G network AI and other intelligent services. Native sensing and intelligence will be the two main new capabilities of the 6G network, the former generating massive data, the latter based on data processing. With the integration of communication and sensing as 6G Inherent capabilities provide a wealth of data for AI algorithms by sensing the network's own status, surrounding environment, and user/device behaviour. Networks, which used to serve as information transmission channels, are gradually transforming into data carrying platforms. From the perspective of data value mining, the network is both a producer and provider of data, offering reliable data services for various intelligent applications, while also being a consumer of network data, enhancing network performance and operational efficiency through data-driven intelligent applications. 6G network's native AI requires a large amount of data collaboration, sharing, and processing across regions, domains, and the entire network. The unprecedented scale and complexity of 6G networks pose challenges to data-driven intelligence in meeting the changing demands of 6G networks, necessitating a shift from data-driven to data and knowledge dual-driven intelligence. Openness and sharing of data is an important mechanism for realising data flow and value; data owners or providers must provide data in a service form to data consumers to realise value. Data monetisation is not a one-time transaction, it requires ensuring data exchange under privacy protection, achieving repeated monetisation of data. For network AI, an AI model trained on a large amount of training data is also an important intellectual property that needs protection. The overall demand for data services includes the ownership demand of data subjects, the data monetisation demand of network and service providers, the credibility and usability demand of data users, and legal and regulatory demands. Technical requirements and challenges of data services include preprocessing of multidimensional heterogeneous data, reliable storage of ubiquitous data, distributed collaboration, trustworthy data provenance, data circulation and sharing, data protection technology and privacy-enhancing computing, data and knowledge dual-drive, etc. The data carried by 6G networks includes data produced and consumed by 6G network users and terminals, network equipment and functions, infrastructure such as cloud platforms, algorithms, applications, etc. Data services are based on a data distribution and publishing framework, providing data as a service product to meet customers' real-time data needs across systems, reusable and compliant with enterprise and industrial standards, balancing data sharing and security. 6G data services aim to efficiently support end-to-end data collection, transmission, storage, and sharing, solving how to conveniently, efficiently, and securely provide data to internal network functions or external network functions, reducing the difficulty of data acquisition under the premise of complying with privacy and security laws and regulations, and improving data service efficiency and data consumption experience.

"B5G/6G Network Intelligent Data Analysis" [61]: How to embed data and intelligence to form a new technology system of network intelligence with native AI and integrated computing and communication, faces major challenges in architecture, data, and AI algorithms, among which data is an important foundation. It basically determines the performance upper limit of network intelligence, while architecture and algorithms only approach this limit. Only by solving the basic problems of how to collect, analyse, and utilise the big data of mobile network, and forming a complete data analysis system, can we accelerate the development of the new technology of B5G/6G network intelligence. The white paper mainly introduces a intelligent data analysis system of B5G/6G network, which achieves data-level information integration, information-level knowledge extraction and representation, intelligent-level knowledge computing and inference, and application-level feature customisation through the collection and storage of mobile communication network's data, translation of mobile communication principles and protocols, construction and analysis of network data and knowledge graphs, and construction and evaluation of network intelligent feature datasets. Its purpose is to create a feasible path for intelligent data analysis of network and provide a new idea and paradigm for intelligent data technology of the network.

"White Paper on Knowledge Defined Orchestration and Management" [62]: 6GANA TG5 is committed to promoting the evolution and development of the "Self-Intelligent Network", building a highly intelligent automated network, focusing on the issue of "network orchestration and management". The existing "data-driven" AI struggles to solve the pain points of 6G's native intelligence, which is an organisational form produced by the interaction of humans, machines, and environmental systems. The current autonomous systems are still in the "pseudo-freedom" stage, the reason for which lies in the limitations of their underlying technical architecture—machine learning and big data processing mechanisms. Neither behaviourist reinforcement learning nor connectionist deep learning can accurately reflect human cognition and reasoning incentives. The working group proposed the concept of "network knowledge", achieving the elastic input of objective data and subjective information and knowledge, and introducing the idea of "knowledge-driven" in the manner of software-defined networks, forming the concept of "knowledge-defined orchestration and management". The white paper systematically expounds the overall architecture of knowledge-defined orchestration and management for the first time, elaborates on the key technologies related to knowledge definition, such as network telemetry, knowledge representation, policy production, resource scheduling and policy verification, clarifies the important role of knowledge definition in promoting the evolution and development of the "Self-Intelligent Network", and in solving current network orchestration and management problems, and describes the application and practice of knowledge-defined orchestration and control in several typical scenarios from the perspective of technical requirements.

Subsequently, 6GANA will continue to carry out in-depth research and prototype verification of AI as a service and network AI, and carry out extensive domestic and international cooperation, promoting AI as a service and network AI to become the consensus of the industry, empowering the development vision of ubiquitous intelligence in the 6G era.

4.3 Progress of 6G Research Out Side of China

With the start of large-scale commercial use of 5G, academic research institutions worldwide have begun to focus on 6G research. In this regard, universities in Europe and the United States, such as the University of Oulu in Finland and New York University in the United States, are particularly active, and the academic community has suddenly set off a wave of 6G research. Leading companies in the industry and governments of various countries are also vying to introduce 6G R&D plans. The U.S. government is particularly active in this regard, claiming to want to regain the dominance of the 6G mobile communication industry. The U.S. FCC was the first to open terahertz frequencies for testing and experimentation in 2018, while the South Korean government claimed to be the first to commercialize 6G globally in 2028. Therefore, 6G quickly became a hot topic of research in academia and industry, and various journals and academic conferences have targeted 6G. This section briefly introduces the important progress and influential research projects of 6G abroad.

4.3.1 6G in Europe

In Europe, research groups and institutions such as the University of Oulu, Nokia, Ericsson, and Hexa-X/Hexa-X II have already achieved good progress in 6G research.

1. **University of Oulu**

In Europe, the first to raise the banner of 6G research was the University of Oulu in Finland. In 2018, Finland announced the 6Genesis flagship project (6G Flagship) [63] (a national 6G plan led by the University of Oulu and funded by the Finnish Academy of Sciences). The project, with a duration of eight years and a total investment of 290 million US dollars, aims to develop a complete 6G ecosystem. The research content is oriented towards the 6G vision, challenges, applications, and technical solutions for 2030, and invites universities, enterprises, and research institutions from Australia, China, Europe, the United States, etc., to participate in related research and the writing of white papers. In March 2019, with the support of IEEE, the world's first 6G Wireless Summit [64] held in Finland, it invited the industrial and academic communities to present their latest insights and innovations on 6G, discussing the theoretical and practical challenges that need to be addressed to realise the 6G vision. Based on the content of the first 6G Wireless Summit, the University of Oulu released the world's first 6G white paper in September [65], which comprehensively analysed the driving factors, research needs, challenges and issues of 6G from seven aspects: social and commercial drivers of 6G, 6G use cases and device forms, 6G spectrum and KPI targets, progress and challenges of wireless hardware, physical layer and wireless systems, 6G networks, and new service promoters. It systematically introduced the technology trends of 6G, proposed the vision of building "ubiquitous wireless intelligence", and greatly inspired the industry's 6G research.

The second 6G Wireless Summit [66] was also held from March 17 to 20, 2020. In June 2020, the University of Oulu released 12 white papers, including 6G drivers and United Nations Sustainable Development Goals, 6G business, 6G vertical industry verification and trials for 2030, 6G remote area connections, 6G networks, machine learning in 6G wireless communication networks, 6G edge intelligence, research challenges of 6G trust, security and privacy, 6G broadband connections, key and large-scale machine communication for 6G, 6G positioning and sensing, and RF-assisted 6G, etc. [67–78], these series of white papers highlighted the trend that 6G solutions need to cover multiple fields and involve multiple disciplines. Interested readers can refer to the detailed white papers.

2. **Nokia**

As one of the world's leading communication companies, Nokia actively conducts 6G research, participates in many European 6G research programs, such as 6Genesis,

European Horizon Smart Networks and Services,[1] and the EU 6G flagship project Hexa-X. Nokia also plays a leading role in 6G research in North America, such as the Next G Alliance initiative launched in 2020 to promote North American 6G research. Nokia is a founding member and holds leadership positions in the Next G Alliance Steering Group and Roadmap Working Group to promote the vision and roadmap of 6G, and as a key contributor to define the research areas, spectrum needs, use cases and application requirements of 6G.

Since 2020, Nokia has released a series of 6G technology white papers, comprehensively predicting and constructing a panoramic view of the next-generation wireless network.

The white paper "Communications in the 6G Era" [79] comprehensively expounds the new trends in future technology development and the new demands of 6G communication networks. The white paper believes that "digital twins" that truly present the physical and biological world will become the basic platform for future new digital businesses. To truly and completely reproduce every moment of the physical world in time and space, the network needs to have a huge capacity and extremely low latency. At the same time, digitisation will provide an effective way to create new virtual worlds. The digital expression of virtual objects can be combined with the digital twin world to form a multi-dimensional mixed reality, thereby shaping a brand new world beyond physics. Traditional intelligent terminals will become more sophisticated and intelligent, and devices such as subcutaneous implants, ingestible sensors, exoskeletons, and brain activity monitors will continue to emerge. With the breakthrough development of intelligent terminal equipment, human biological characteristics will be accurately and real-time mapped and integrated into the digital and virtual world, thereby generating new superhuman abilities. With the continuous progress of augmented reality and the continuous improvement of the friendliness of the human–machine interface, efficient and intuitive control of the physical world, the virtual world, and the biological world by humans is just around the corner. Therefore, the connection in the future is to achieve seamless interaction with these different worlds and create a unified experience for humans. This connection is both secure and private, can be used for preventive healthcare services, and can even be used to create a 6G network with a sixth sense, understand human intentions by intuition, accurately predict needs, make human interaction with the physical world more efficient, and thereby improve productivity.

It can be imagined that in the 6G era, humans will use "digital twins" to record and reproduce every moment of the physical world and the biological world in which we live. For this, the 6G network will be given new demands and characteristics: (1) support for new human–machine interfaces that coordinate between device clusters; (2) ubiquitous general computing that integrates local and cloud; (3) a brand new mixed reality experience based on multi-sensor data fusion; (4) the ability to implement precise sensing and feedback control of the physical world. Based on the

[1] Horizon Europe Smart Networks and Services, aims to ensure that Europe can lead the development and deployment of next-generation network technologies and services, while accelerating the digital transformation of European industry.

analysis of the future 6G network requirements, this white paper proposes five key technology directions that Nokia Bell Labs focuses on:

- AI-native air interface;
- Brand new spectrum technology;
- Intelligent network perception integrated with senses;
- Extreme connections with ultra-low latency and ultra-high reliability;
- Network architecture with cognitive abilities, more automated, professional, and inherent security and privacy policies.

In the white paper "Toward a 6G AI-Native Air Interface" [80], Nokia Bell Labs proposes that the air interface dynamically defined by AI/ML will be a key component of future 6G systems, and AI/ML in 6G. In the design of air interfaces, it will play a more fundamental and disruptive role. By deploying deep learning engines at both ends of the transceiver system, according to specific business characteristics and different data attributes, it can adaptively determine the transmission system, parameters and protocols that best match the current channel and hardware environment through end-to-end autonomous learning, such as the optimal transmission waveform, constellation distribution, optimal reception algorithm, and the best MAC access protocol. Through such a transformation of the air interface design paradigm, the intelligence and performance of the 6G system will reach new heights. In addition, the inherent air interface design of AI/ML will make the 6G network more environmentally friendly. Research from Nokia Bell Labs shows that, compared with 5G, under the same bandwidth and speed, the 6G air interface solution inherent in AI/ML can reduce the transmission power by 50%, greatly improving system energy efficiency.

The 6G network will not only provide new services, ultimate connectivity and security, wireless sensing capabilities, but the future wireless network should also provide higher capacity at a lower cost per bit and significantly reduce energy consumption. In the white paper "Extreme massive MIMO for macro cell capacity boost in 5G-Advanced and 6G" [81], Nokia Bell Labs introduces large-scale MIMO and new spectrum technologies to provide more capacity on the existing base station grid. The white paper believes that the priority spectrum for future cellular wireless is expected to choose the mid-band spectrum of 6–20 GHz for urban outdoor small cells, the low-band spectrum of 460–694 MHz for extreme coverage, and the sub THz band for providing peak data rates exceeding 100 Gbps. Compared with the 5G system deployed at 3.5 GHz, by using mid-band spectrum resources and equipping base stations with up to 1024 antenna array elements of ultra-large-scale MIMO antennas, and configuring large-scale antenna arrays on terminals, it can provide about 20 times the system capacity increase, more than 4 times the spectrum resources per cell, and more than 5 times the spectrum efficiency increase. In this way, without the need for denser site locations, it can meet the annual 35% mobile traffic growth within the next 10 years. To meet the ultimate capacity demand, Nokia Bell Labs proposes the following technical innovation research directions.

- RF front-end technology.
- Flexible MIMO processing architecture.
- Wireless system design based on narrower beams.
- Rapid evolution of SoC processing capabilities.
- Low power design.

Nokia Bell Labs foresees that 6G wireless systems will have high scalability in terms of bandwidth and antenna array size. Extremely large bandwidth spectrum resources will not be available everywhere, and it is not possible to deploy ultra-large-scale MIMO everywhere, therefore, the wireless architecture should be highly scalable and modular, so that the same basic components can be used to economically and efficiently meet various different deployment requirements such as 64–512 TRX, 100–400 MHz bandwidth.

Nokia has collaborated with AT&T to carry out the verification of distributed large-scale MIMO technology, and tested the concept verification technology in AT&T labs. Research and test results from Nokia Bell Labs show that, compared with a single-panel system with a similar configuration, it can increase the 5G uplink capacity by 60–90%. Distributed large-scale MIMO technology can significantly increase uplink capacity without sacrificing performance.

The vision of creating digital twins of the real physical world in 6G requires the adoption of new sensing technologies to construct multi-level maps of the human peripheral environment, with wireless sensing technology using mobile communication networks potentially becoming an important part of this solution. As cellular systems evolve to the millimetre wave bands in 5G and potential sub-terahertz bands in 6G, small cell deployments will begin to dominate. The large bandwidth systems deployed in small cells provide unprecedented opportunities for sensing using mobile networks. The white paper "Joint design of communication and sensing for Beyond 5G and 6G systems" by Nokia Bell Labs [82] provides a detailed introduction to the main aspects of this kind of integrated design of communication and sensing for cellular systems, including an analysis of waveform selection, which requires the ability to satisfy optimal communication and radar sensing needs simultaneously; it discusses several schemes to effectively integrate sensing capabilities into JCAS (Joint Communication and Sensing design system), some of which can be applied to the evolution of 5G based on the NR air interface, such as methods to reduce sensing costs through appropriate sensing signal design, and configuring separate transmission parameters for communication and sensing functions respectively. Research shows that in the scenario of road traffic monitoring, the use of information obtained from sensing signal detection can significantly reduce signaling overhead. The white paper also looks forward to future advanced JCAS scenarios based on distributed massive MIMO, and discusses various other research challenges that JCAS needs to solve in order to achieve the inherent 6G system design goals within JCAS.

(1) Distributed Sensing

Distributed MIMO radar systems provide the possibility for sensing in Non Line Of Sight (NLOS) situations, with technical challenges including synchronisation and clutter elimination. Since communication systems have already embedded pilots for channel estimation, these functions can potentially be used to enhance sensing performance. By additionally using mobile devices as sensors, the sensing capabilities of distributed systems can be further expanded. When considering multi-cell interference, more suitable resource allocation is needed for sensing.

(2) Multi-band Sensing

In the case of JCAS hardware with multi-band capabilities, combining low and high frequency bands can achieve ultra-fine measurement resolution, so determining the best working mode for JCAS multi-band systems and the corresponding high-resolution sensing algorithms is of significant research value.

(3) AI/ML Processing

In sensing reception processing, further research should be conducted on using AI/ML technology for simultaneously detecting and classifying multiple different types of objects. By using fine-grained channel state estimation, AI/ML provides the possibility of detecting features beyond distance and Doppler. The learning of end-to-end neural network models can be used for waveform and modulation design in JCAS.

(4) Sensor Fusion

How to fuse the information from wireless sensing with information obtained from other available sensing devices, such as various sensors on terminals and access points, is another important research area. For example, cameras installed on street light pole access points can provide additional valuable information to assist wireless sensing inference. The combination of multiple sensing modes will play a unique role in constructing the high-resolution maps needed to build powerful digital twins.

(5) Channel Modelling for JCAS

For the simulation of JCAS, the widely used communication system spatial channel models are not suitable for simulating radar perception, such as reflections and target objects are not included in these types of channel models. The combination of ray tracing models and statistical models may be a more reasonable modelling method for simultaneously evaluating data transmission and perception performance. New deep learning models, such as Generative Adversarial Networks (GAN), may also be a new method for channel modelling for JCAS.

(6) Sensing-assisted communication enhancement

Sensing and positioning can be directly applied to enhance communication itself, for example, predicting device blockage in high-reliability communication scenarios, and using geographical location information for beamforming.

(7) Waveform candidate solutions

This white paper has conducted a detailed evaluation of the candidate waveforms for JCAS, in addition to Frequency Modulated Continuous Wave (FMCW), Orthogonal Frequency Division Multiplexing (OFDM), Single Carrier (SC), FMCW +

SC, FMCW + OFDM, Discrete Fourier Transform-Spread-OFDM (DFT-S-OFDM) and Orthogonal Time Frequency Space (OTFS) also have certain potential in some scenarios.

In the white paper "Extreme communications in 6G: Vision and challenges for 'in-X' subnetworks" [83], Nokia Bell Labs introduced short-range, low-power 6G "in-X" subnetwork technology for providing wireless coverage with extreme requirements in terms of throughput, latency, and reliability. These subnetworks will be installed in robots, production modules, vehicles or human bodies, they can be part of a larger network infrastructure, and can also operate autonomously in emergency situations. Licensed spectrum can be used for static or nomadic subnetworks, while mobile subnetworks can rely on unlicensed spectrum, including operating in-X subnetworks in an underlay manner on frequencies allocated to other systems. However, to support time-sensitive businesses, new regulatory supervision may be needed. As in-X subnetworks may lead to very dense deployments, effective interference coordination must be carried out to ensure extreme performance requirements. When in-X subnetworks exceed wide-area coverage, implicit coordination schemes must complement centralized interference coordination schemes. The reliability requirements of extreme in-X subnetworks also need to be able to withstand interference attacks, impulse noise, and other wireless interference from non-cellular systems. In addition, in extreme interference situations, the requirements for communication performance may eventually need to be appropriately relaxed, provided that the operations supported by the underlying control system can also be correspondingly relaxed. These challenges are driving the research of more advanced 6G dynamic wireless resource management technologies.

In the white paper "Technology innovations for 6G system architecture" [84], Nokia Bell Labs comprehensively introduced the vision and research progress in 6G architecture, decomposing the 6G system and related architecture into functional focus areas, embedded in a broader architecture transformation theme. The 6G system includes UE, access and core network functions, covering new functions of 6G wireless access, allowing new spectrum and new use cases. It will also embed functions of previous generations (4G and 5G) as well as non-3GPP Access Technology. The Heterogeneous Cloud (Het-Cloud) provides an open, decoupled, and distributed execution environment for network functions. The most basic network structure is secure, seamless, and cross-domain, ensuring the connections required by humans, machines, industrial applications, and decoupled computing. Overall, the 6G architecture forms a heterogeneous "network within a network" from both a technical and business model perspective, covering key 6G subnets, ubiquitous wireless access networks as a service (RAN as a Service, RANaaS), and achieving "360° connectivity" through new multi-connection methods at different levels. Its services can be used seamlessly and can operate in an efficient and highly automated manner. The design principles of the 6G architecture will support AI/ML, service-oriented closed-loop orchestration automation and intent-based management, as well as adopting service-based design, decoupling and openness, platform agnosticism, and continuous integration and delivery, among other cloud-native principles.

In this white paper, Nokia Bell Labs proposes considering the evolution of the 6G network architecture from six key areas.

(1) Flexibility: The 6G architecture is expected to be more flexible in all dimensions. Firstly, the 6G architecture must be suitable for large-scale wide area network deployments as well as extremely localised local and personal area networks. Secondly, in terms of function deployment, the 6G architecture is expected to dynamically meet various latency targets and other requirements. For example, to support XR services, ultra-low latency needs to be achieved for specific users. Finally, the 6G architecture needs to have the ability to dynamically scale to meet dynamic network loads.
(2) Specialisation: Given the diversity of business use cases and deployments, the 6G architecture also needs to be able to provide customised features and functions. For example, a specific 6G subnet, a lightweight sensor network, or some extreme network use cases, may each require a set of specific network functions, while other features can be omitted as they are not needed in this type of network.
(3) Robustness and Security: Users in vertical industries expect network services to provide multiple connections in a secure, robust, resilient, and truly ubiquitous manner when in use and needed. In addition, we must leverage the best advantages from different domains, such as achieving a macro network architecture with full mobility and roaming support coexisting with architectures for short-range, non-cellular terahertz networks and non-terrestrial networks (NTNs). Another core goal of security is to meet users' strong expectations for trust and privacy.
(4) Cloud Platform: In the cloud platform, the shift to hosting network functions will continue and become more widespread. The deployment of network functions will largely shift from dedicated telecom cloud platforms to generic public, private, or hybrid clouds located locally, at the edge, or centrally. 6G will provide a unified orchestration interface for service management in the distributed cloud, supplemented by specific abstractions for the 6G RAN, allowing the use of specific functions in cloud nodes, such as hardware acceleration.
(5) Programmability: Network implementations will also reach new levels of programmability, such as adopting hardware-independent programming languages like P4, and being able to run on any cloud platform. Through open, service-based interfaces and more modular system design, it will be easier to integrate serverless services and functions from multiple vendors, allowing network deployments or network slice instantiations to only use those necessary functions and services.
(6) Simplification and Sustainability: While powerful zero-touch automation features provide a way to deal with system complexity, the introduction of 6G provides a good opportunity to reconsider architectural design, delete, redesign or merge some network functions, introduce service-based approaches for the control plane between the wireless access network and the core network, and distributed non-access layer (NAS) interfaces, simplify signalling processes, etc.

As another example, simplification can be achieved through customised protocol stacks, allowing efficient integration of customised 6G Subnets. Simplifying and providing greater flexibility and dynamism for functional deployment, which helps to achieve sustainability goals by reducing signalling and energy consumption, a clear commitment to "design for sustainability" is very important for 6G.

AI and machine learning technologies are expected to become indispensable components of the 6G architecture, helping to create a network with true cognitive capabilities to adapt to various scenarios and network deployments. AI/ML will also play a key role in the automation and optimisation of the network, as well as in enhancing the security of the system. The widespread use of AI/ML requires the network architecture to include functions and interfaces for large-scale data collection, processing and distribution, as well as model refinement training and updating inference models.

Nokia Bell Labs' innovative vision for the 6G architecture can be further summarised as follows.

From a platform perspective, 6G networks may be distributed across multiple clouds in different locations and may involve multiple stakeholders, i.e., a heterogeneous and distributed cloud environment. Cloud functions, including hardware acceleration, will be essential components for supporting deep learning processing and cloud-native implementation of lower-layer RAN functions.

At the functional level, new functional divisions need to be designed within the RAN and between the RAN and the core network, aiming to achieve optimised User-Equipment-Radio Access Network-Core (UE-RAN-Core) architecture design. This will simplify related processes and integrate similar functions. Possible solutions include simply enhancing existing interfaces to completely eliminate separation between domains. Dynamic, adaptive functional deployment needs to seek the best compromise between network performance, complexity, energy consumption, and business and consumer demands. On the same level, the ways of collecting, storing and disclosing massive data and information will change, including AI and machine learning-driven analytics functions in all layers of the mobile network. Data is created and collected in different areas and protocol layers of the network and can be provided where needed. Therefore, a common open, registration, discovery and delivery framework is needed for various types of data, such as near-real-time stream data, status data of various registration services, and accumulated data for training ML models. In addition, the convergence of networks and computing is constantly evolving. In this case, ensuring privacy and providing trustworthy solutions is crucial.

As 6G networks will support a very diverse and specialised range of uses and deployments, such as local and global service availability, enterprise and CSP deployments, best-effort and mission-critical businesses, etc. This diversity comes with specific requirements and will utilise network-specific capabilities. One such function is network slicing, which was introduced in 5G, but must be extended using dedicated software protocol stacks and dedicated hardware to meet 6G requirements, such as complete isolation of sensitive traffic. In addition, in specific 6G network environments (such as body area networks or robot networks), extreme performance requirements need to be met, such as the previously mentioned 6G subnets or in-X subnet technologies. The overall vision is to develop a modular architecture that can be easily combined according to the target deployment, use case and its specific requirements.

6G will provide unified orchestration and concurrent abstraction for all domains in the system. It is necessary to orchestrate multiple network domains to build and support end-to-end communication services that may span multiple stakeholders and management domains (such as CSPs, enterprises, and home domains). In addition, service orchestration needs to support various 6G-related businesses, including the opening of stakeholder capabilities corresponding to monetisation models, and the end-to-end orchestration of remote resources and 6G subnets. To achieve the performance goals of 6G orchestration, automation is required. Therefore, important technologies such as intent-based management, closed-loop automation, and AI/ML will continue to be used in architectural design.

In the white paper "Security and Trust in the 6G era" [85], Nokia Bell Labs deeply analysed the security threats faced by 6G, including the 6G decomposition architecture, open interfaces, an environment with a large number of stakeholders, and the need to ensure reliable trust established between devices, subnets, hybrid clouds, and applications. This white paper comprehensively elaborates Nokia's views and progress on 6G security enabling technologies, including automated software creation, automated closed-loop security operations, privacy protection technologies, hardware and cloud embedded anchor trust, quantum security, interference protection, physical layer security, and distributed ledger technology. The white paper believes that AI/ML will play a broad and critical role in the entire 6G security technology system and architecture.

3. **Ericsson**

As a global equipment supplier, Ericsson started its research on 6G early and is the technical coordinator of the European 6G flagship project Hexa-X/Hexa-X II. It has a strong influence in 6G research, and its many views are worth referencing and learning from.

In 2021, Ericsson released the "Cognitive Networks" [86] research report. Firstly, it proposed the future development vision of cognitive networks, where artificial operators will interact with the network using high-level demands in the form of intentions. Then, the network needs intelligence ("brain") to understand these intentions and translate them into action plans and settings to meet all the requirements given by the artificial operator. Decisions need to be made based on existing knowledge, the decision-making process improved, and new knowledge acquired, which requires data from operations. Finally, all actions and configurations should be applied in the infrastructure, which is evolving into a software-based system deployed in a distributed cloud-native platform.

Cognitive networks mainly have two aspects of content.

(1) Zero-touch deployment and operation. In the future network vision, there are many different services, which are dynamically deployed and adjusted; zero-touch means that once a service is defined, all stages of its lifecycle will be automatically managed.
(2) Continuous real-time performance improvement. Some services have extremely high performance requirements and need careful tuning to meet these requirements. We also need to ensure that the overall network performance is optimised and improved based on performance measurements.

With these requirements, it is no longer possible to manually configure services and infrastructure, manually carry out all detailed configurations and fine-tuning. Instead, the network needs cognitive abilities: to learn from data, make wise choices, and apply these choices to the infrastructure. This applies to all parts of the system, from antenna sites and air interface configurations to end-to-end service parameters.

In 2021, Ericsson CTO ERIK EKUDDEN released a report on the five major network trends for the 6G era [87]: digital representation of network reality, Adaptive Infinite Connectivity, Integrity of Trustworthy Systems, Joint Cognitive Networks, and Unified Network Computing Structure.

The convergence of the internet, telecommunications, media, and information technology will lead to the creation of a unified global system composed of interconnected components. The network computing structure facilitates unification across ecosystems, application management, execution environments, and the opening of network and computing capabilities. 6G will act as the controller of the network, covering a variety of applications from simple devices to advanced network physical systems. At the same time, 6G will integrate storage, computation, and communication to create a distributed unified network computing architecture. This structure will enable service providers to access tools and services, not just connections that can be provided to the open market.

Future use cases (such as IoS and connected intelligent machines) will be highly distributed and require guaranteed low deterministic latency, high throughput, and high reliability. The network will supplement connection products by hosting and managing these types of use cases in an embedded unified execution environment. Hardware and software acceleration, time-sensitive communication technologies (such as time-sensitive networks and ultra-reliable low-latency communication), and time-aware functions will ensure end-to-end timing and reliability.

Considering the data streams of billions of sensors and actuators, in-network stream processing will be provided by deterministic computing capabilities offered by interoperable application programming interfaces. Preparation of information such as data sources, metadata extraction, and annotation will provide lightweight universal data pipelines and computing structures. In addition, the network will provide application deployment, scaling, and resource allocation in a serverless manner to support the execution of all possible sensor and actuator functions in the future.

This computing structure will evolve around a unified ecosystem involving participants and users of air interfaces, the internet, cloud services, and devices. The scale of the ecosystem requires extensive use of bilateral agreements between ecosystem participants. The computing structure will provide a decentralised and agentless exchange for virtual services, including identity and relationship processing. Smart contract technologies such as distributed ledgers will automate contract negotiations between parties, supporting service delivery models such as automated sales, delivery, and billing.

In February 2022, Ericsson released the white paper "6G—Connecting a Cyber-Physical World" [88]. The white paper believes that the four main driving factors for 6G development are: the reliability of core social systems; achieving sustainability through the efficiency of mobile technology; accelerating automation and digitisation to simplify and improve people's lives; meeting the growing demand for unlimited connections, communicating anytime, anywhere, for anything.

Increasingly high expectations have set a clear goal for 6G—it should contribute to an efficient, human-centric, sustainable society through ever-present intelligent communication. Therefore, the vision of 6G is to transform society through technology, as shown in Fig. 4.39, 6G makes it possible to move within a network-physical continuum that connects the senses, actions and experiences of the physical world with its programmable digital representation. The 6G network provides intelligent, unlimited connections and complete synchronisation of the physical and digital worlds. A large number of sensors embedded in the physical world send data to update the digital representation in real time. Actuators in the real world execute commands from intelligent agents in the digital world. Events can be traced and analysed, observed and acted upon in real time, and future actions can be simulated, predicted and planned. Compared to the virtual world (interaction with avatars in virtual reality/augmented reality), the network-physical continuum provides a close connection with reality, where digital objects are projected onto digitally represented physical objects, enabling them to coexist seamlessly with fused reality and enhance the real world.

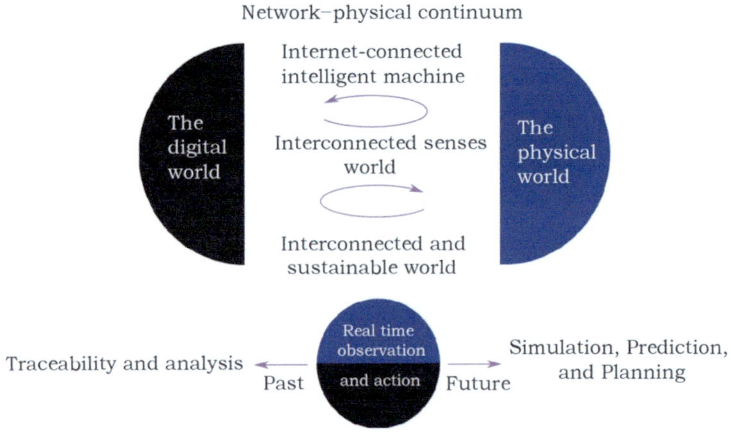

Fig. 4.39 Vision of 6G transforming society through technology [88]

Against such a backdrop, the white paper expands the future 6G use cases into six categories: immersive communication, ubiquitous broadband, space–time business, emergency business, computing-AI business, ubiquitous IoT, as shown in Fig. 4.40, and derives the performance indicators of the 6G system, as shown in Fig. 4.41.

The white paper suggests that compared to today's networks, the capabilities of future wireless access networks need to be enhanced and expanded in all aspects. In addition to traditional functions such as achievable data rates, latency, and system

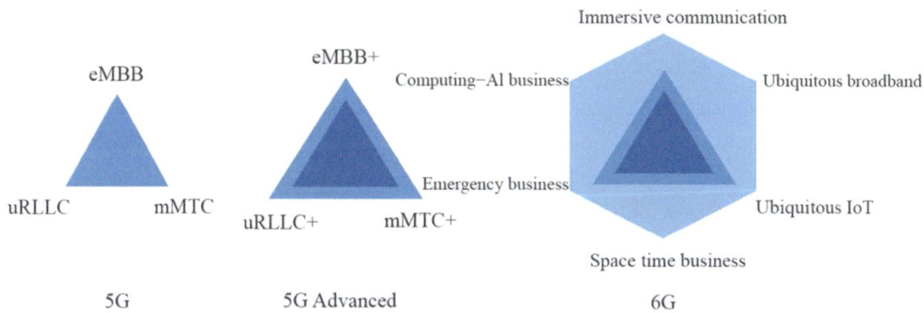

Fig. 4.40 Ericsson's proposed 6G typical scenarios [88]

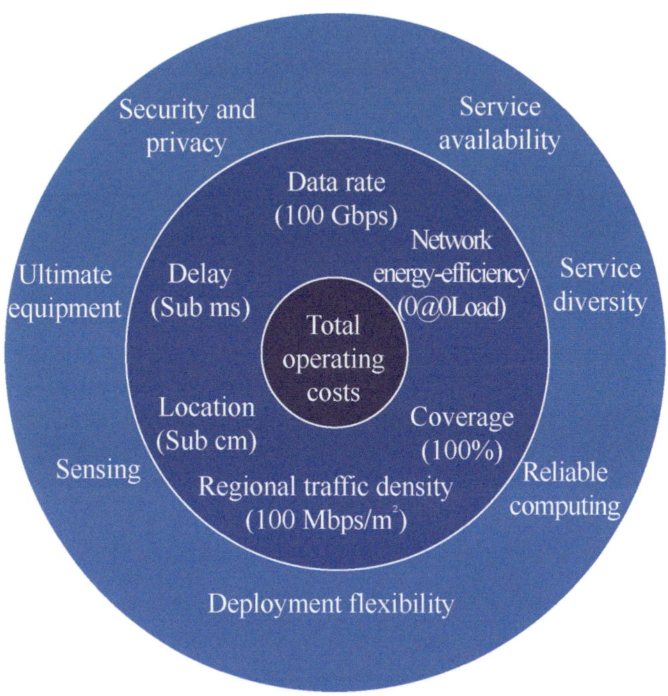

Fig. 4.41 Performance indicators for the 6G system proposed in the white paper [88]

capacity, it also includes new features, some of which may be more deterministic in nature. It is worth noting that the capabilities of future wireless networks should not only match the use cases currently envisaged but also support future services that have not yet been envisaged.

In terms of traditional functions, 6G should achieve higher achievable data rates and lower latency in all relevant scenarios, including the possibility of providing hundreds of Gbps per second and end-to-end sub-millisecond latency in specific scenarios, providing high-speed connections with predictable low latency and low jitter.

In terms of cost, future wireless access networks should be able to meet the exponentially growing traffic demand in a cost-effective manner. Higher spectrum efficiency is an important aspect, and additional spectrum is naturally another important aspect, as well as being able to economically and efficiently deploy very dense networks.

In terms of coverage, it is necessary for 6G to continue to expand to global coverage—narrowing the digital divide in remote areas and ensuring that the overall costs of users and service providers are at a sustainable level.

In terms of energy consumption, a significant increase in traffic volume will not lead to a corresponding increase in energy consumption. When there is no traffic within the node, the energy consumption should be close to zero. As wireless networks become an increasingly important part of society, resilience and security capabilities are crucial.

In terms of trust, the network should be able to utilise new privacy computing technologies, improve service availability, and provide enhanced secure identities and protocols with end-to-end guarantees.

In addition, 6G also needs reliable computing and AI integration capabilities, infrastructure capable of quickly developing and deploying distributed applications and network functions, as well as data and computing acceleration services, and performance guarantees throughout the network. Finally, to drive the comprehensive digitisation and automation of society, the network needs high-precision positioning and detailed perception capabilities from the surrounding environment.

The white paper believes that the 6G network for 2030 should broadly consider promising technologies. Numerous 6G elements will form a seamless system with all the necessary capabilities to realise the vision of intelligent communication that is always present in the connected network physical world. With the foundation of a trustworthy system and an efficient computing structure with built-in cognitive capabilities, the future network will provide unlimited connections for brand new applications and services, making 6G a broad platform for innovation and the information pillar of society.

The foundations of the 6G network listed in the white paper are shown in Fig. 4.42. The white paper lists the technical elements related to 6G, as shown in Table 4.2.

Looking towards 2030, a strong demand for communication and frontier technologies is imminent, societal expectations are constantly increasing, enabling technology progress is accelerating, the key elements of 6G are the ultimate performance of

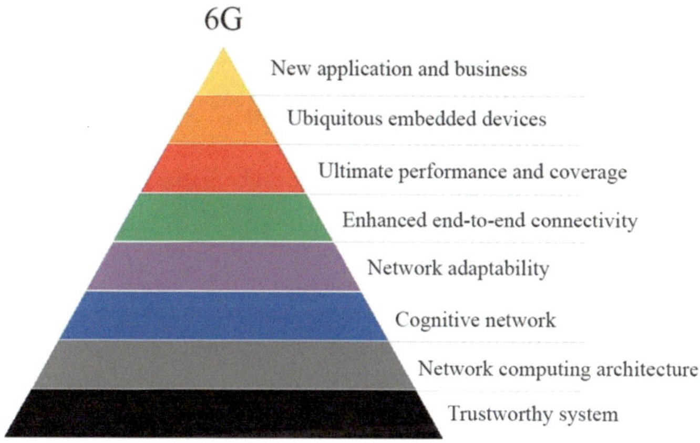

Fig. 4.42 The foundations of the 6G network listed in the white paper [88]

Table 4.2 6G related technical elements

Technical elements	Detailed characters
Network adaptability	Dynamic network deployment
	Network simplification and cross-RAN/CN optimization
	Programmable equipment and network
Enhanced end-to-end connection	Network collaboration
	Elasticity
	Protocal evolution
	Predictable latency
Ultimate performance and coverage	Frequency
	Non-terrestrial access
	Multi-connectivity and distributed MIMO
Ubiquitous embedded devices	Zero-power devices
	Immersive interactive devices
Network autonomy	Intent-driven management
	Autonomy system
	Explainable and trustworthy AI
	Datra-driven architecture
Network computing architecture	Reliable computing
	Enabler of ecosystems
	Unified fluid computing
Reliable system	Privacy computing
	Service availability
	Security assurance

wireless access, network adaptability and global coverage, 6G should become a trustworthy platform for intelligence, computing and spatial data, encourage innovation and act as the information pillar of society.

4. **Hexa-X**

In 2020, the EU launched the "European Horizon Plan (2021–2027)", which includes research on six key technologies including the next-generation network, and officially launched the 6G flagship project Hexa-X in December 2020 [89]. Nokia acts as project coordination and Ericsson acts as technical coordination. The team brings together 25 businesses and research institutions, including Orange, Atos, B-COM Technology Research Institute, French Alternative Energies and Atomic Energy Commission (CEA), Siemens AG, Telecom Italia, University of Pisa, Telefónica, Nokia, Oulu, Ericsson, and Intel, etc. At the same time, the project has invited industry leaders from China Mobile, NTT DoCoMo, Vodafone, Qualcomm, MTK, etc., to serve as project consultants. Hexa-X is an important step in bringing together key industry stakeholders in the EU to jointly advance 6G. The project aims to create unique 6G use cases and scenarios, develop 6G foundational technologies and define new intelligent network architectures. The project's vision is to connect people, the physical world, and the digital world through networks built with 6G technology.

The 6G vision defined by Hexa-X [89] as shown in Fig. 4.43, 6G will serve three intertwined worlds through seven enabling technologies: (1) a human world full of wisdom and values; (2) the digital world of information; (3) the physical world. Real-time interaction is crucial for integrating the world and meeting future challenges. Future 6G networks should be able to tightly synchronise and integrate these worlds, even seamlessly move between them, thereby bringing many new scenarios, applications, and services, benefiting people at all levels, including consumers, businesses, or society.

The interaction between the physical world and the digital world will realise the digital twin of the world, where rich sensor information can be used for deep data mining and analysis. Intelligent agents can act on digital twins and trigger actions in the physical world through actuators. Such actions will improve the efficiency and resilience of operations in the physical world through better planning and control and preventive measures (such as maintenance before problems arise). This could lead to the widespread use of digital twins, resulting in a large amount of communication demand.

The interaction between the human world and the physical world will realise effective control and feedback between worlds, such as based on efficient human–machine interfaces. The interaction between the human world and the digital world enables AI to help improve our lives and enhance fully immersive communication between people. Knowledge about the human body is obtained through electronic health applications, collecting information about the human body's state and synchronising it from the human world to the digital world, thereby achieving preventive healthcare, etc.

This vision centres around three core values, with the three worlds unfolding around these three core values (see Fig. 4.43). They are: the trustworthiness of 6G

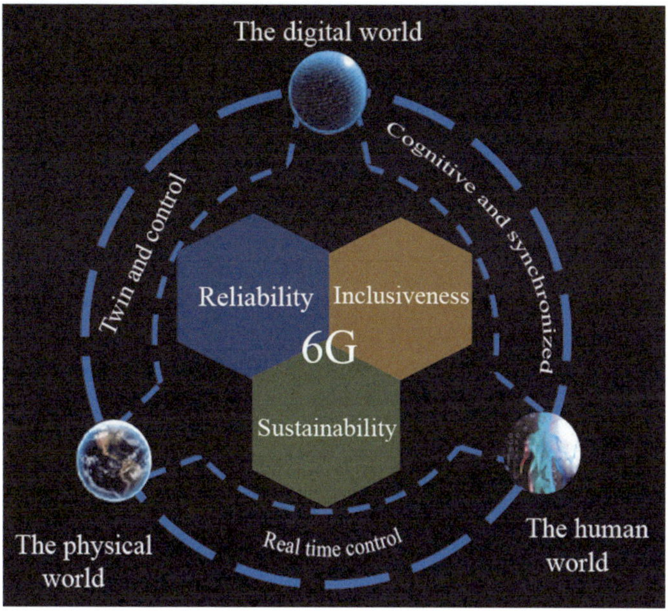

Fig. 4.43 Hexa-X's 6G vision [89]

as a pillar of society; the inclusiveness of 6G, making it accessible to everyone, everywhere; the sustainability of 6G in making the maximum contribution to global development in environmental, social, and economic aspects. These three core values influence the target capabilities and requirements of 6G and guide the project together with the interaction with the three worlds. In summary, to realise this vision, Hexa-X needs to address a series of research challenges to lay the technical foundation for the wireless systems of the 6G era.

(1) Intelligent Connectivity: 6G will deploy intelligence on a large scale in broader society, taking on key roles and responsibilities. 6G should provide a framework to support (such as through advanced resource management), enhance (such as by supplementing data, capabilities, insights, etc.), and ultimately realise real-time trustworthy control—transforming AI/ML technologies into important and trustworthy tools that can significantly improve efficiency and service experience, and integrate human factors ("human–machine loop").
(2) Network of Networks: 6G will aggregate various types of resources, including communication, data, and AI processing. These resources will be optimally connected at different scales, ranging from within the body, within machines, indoors, data centres to wide area networks. Their integration forms a massive digital ecosystem that becomes increasingly capable, intelligent, complex, and heterogeneous, ultimately creating a unified network of networks. It will serve various needs, support different nodes and connection methods, and handle large-scale deployment and operation to meet various needs with the greatest

(cost) efficiency and flexibility, promoting business and economic growth and addressing major societal challenges such as sustainable development, health, safety, and the digital divide.

(3) Sustainability: 6G will transform the network into a high-energy-efficient digital infrastructure and will deeply modify the complete resource chain of wireless networks to reduce the global ICT's consumption of resources. Its digital structure will also create the ability to perceive and understand the physical world's state in real-time, thereby promoting sustainability from environmental, economic, and social perspectives—providing effective and sustainable digitalisation tools for global industries, societies, and policies, turning the United Nations Sustainable Development Goals into reality, and assisting in the implementation/operation of the European Green Deal, especially after the novel coronavirus pandemic, to achieve the sustainability of economic and world development.

(4) Global Service Coverage: 6G should take digital inclusivity as a primary task, including efficient and affordable solutions for global service coverage, connecting remote areas such as rural areas, oceans, or large land transportation, supporting new services and businesses will promote economic growth, narrow the digital divide, and improve the safety and operational efficiency of currently uncovered areas.

(5) Ultimate Experience: 6G will provide extremely high data rates (access levels from hundreds of Gbps to several Tbps), extremely low (hard to perceive) latency, nearly infinite capacity, and precise positioning and sensing, driving a leap in network performance to become possible—releasing the commercial value of new technologies within the GHz to THz range, supporting ultimate service experiences such as fully immersive communication or large-scale remote control, and accelerating the pace of digitalisation.

(6) Trustworthiness: 6G should ensure the privacy and integrity of end-to-end communication and guarantee data privacy, operational resilience, and security, establishing trust in wireless networks and their enabling applications between consumers and businesses—supporting and promoting European values of security, trust, and privacy protection as well as EU sovereignty objectives, to nurture an open, trustworthy, and deeper democratic Europe in the digital age.

To achieve the above vision, the basic network design paradigm needs to be extended from performance-oriented to performance and value-oriented. Values need to be intangible but important, such as sustainability, trustworthiness, and inclusivity, thereby generating a new class of evaluation criteria, namely Key Value Indicators (KVI), which must be understood and used in the network design towards 6G. Hexa-X believes that the development towards 6G requires broad support and global efforts, and it will strive for openness and cooperation between European and global research communities, standardisation bodies, and policy-makers through organising public workshops, writing joint white papers, and actively participating in major events. It will develop an open, modular, and flexible framework—the X-enabler structure—as a foundation, integrating global technological innovations addressing the above six research challenges.

For this, Hexa-X has set up five major technical goals [90]: the foundation of the 6G end-to-end system architecture, oriented towards 6G, the wireless performance, connecting intelligence to 6G, network evolution and expansion towards 6G, and influence creation for 6G. The project carries out relevant research through seven task packages, and the five main objectives and their relationships of Hexa-X are shown in Fig. 4.44. The realization diagram of goals 2–4 is shown in Figs. 4.45, 4.46 and 4.47.

Currently, Hexa-X has output a series of research reports, covering vision, use cases and requirements, key technologies and network architecture, etc., the specific content can be referred to references [91–98]. The 6G use cases released by Hexa-X are shown in Fig. 4.48.

The latest requirement indicators defined by Hexa-X are shown in Table 4.3.

For the detailed technical indicator values of each use case, please refer to reference [91]. For end-to-end architecture design, Hexa-X has proposed the following 8 design principles.

Principle 1: Capability openness. The architectural solution should expose new and existing network capabilities to E2E applications and management (such as predictive orchestration).

Principle 2: Designed for (closed-loop) automation. The architecture should support full automation, allowing the network to be managed and optimised without human intervention.

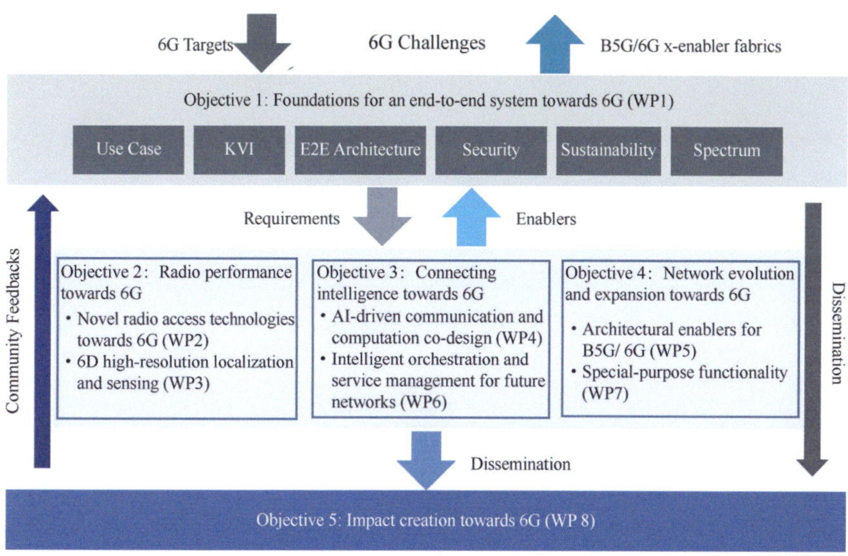

Fig. 4.44 Five main objectives and their relationships of Hexa-X [90]

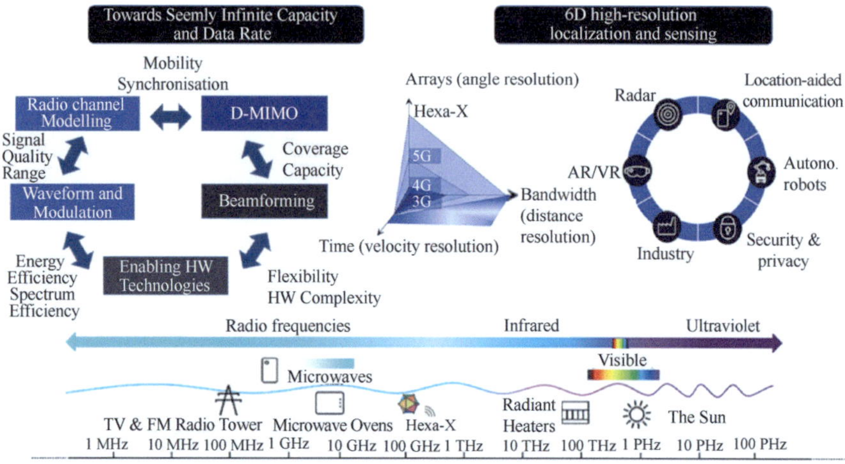

Fig. 4.45 Schematic diagram of realization of objective 2 [90]

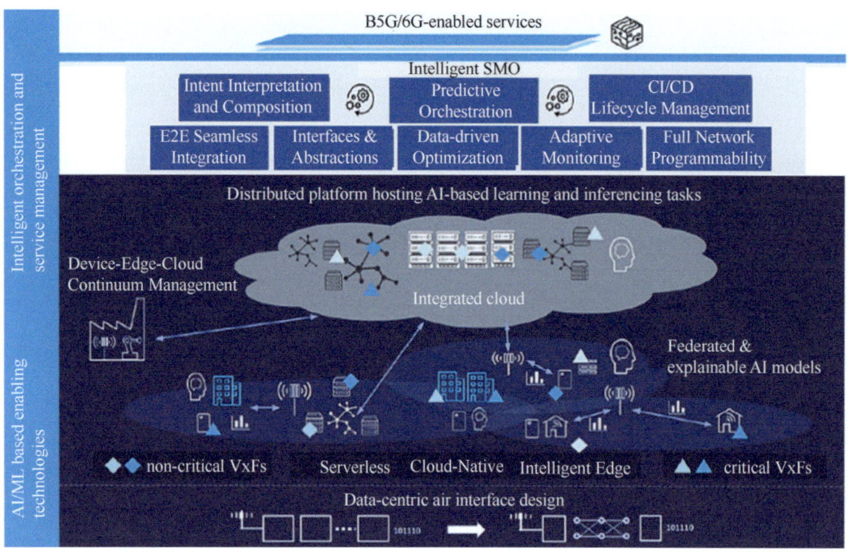

Fig. 4.46 Schematic diagram of realization of objective 3 [90]

Principle 3: Flexibility of different topologies. The network should be able to adapt to various scenarios without loss of performance, while still being easy to deploy.

Principle 4: Scalability. The architecture needs to be scalable in supporting deployments from very small to very large scales, expanding and reducing network resources as needed.

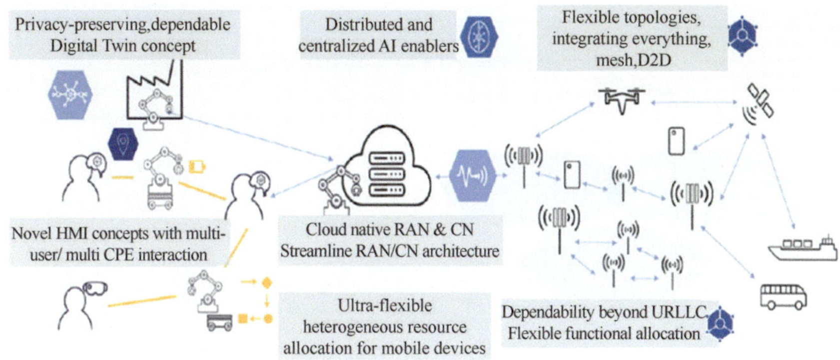

Fig. 4.47 Schematic diagram of realization of objective 4 [90]

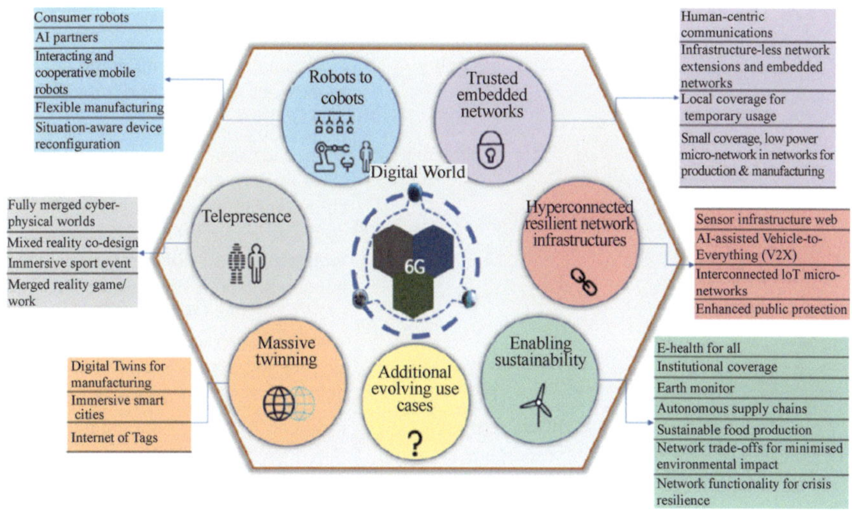

Fig. 4.48 Latest 6G use cases released by Hexa-X [91]

Principle 5: Elasticity and availability. The architecture should be resilient in terms of service and infrastructure supply, using methods such as multi-connection and separation of control plane (CP) and user plane (UP) to eliminate single point failures.

Principle 6: Capability exposure interfaces are service-based. Network interfaces should be designed to be cloud-native, utilising the most advanced cloud platforms and IT tools in a coherent and consistent manner.

Principle 7: Separation of network function concerns. Network functions have bounded contexts, all dependencies between services are realised through their Application Programming Interface (API), with minimal dependency on other

Table 4.3 Hexa-X latest requirement indicator definitions [91]

Communication	Dependent property	Availability/%	The percentage of time during operation that QoS targets are met and services are provided. This should include additional information regarding application resilience, such as uptime as an acceptable service downtime(ms)
		Reliability/%	Percentage of the number of sent packets transmitted within the QoS constraint
		Security	Refer to applicable standards and regulations in the field; functional requirements, if any
		Impartiality	Refer to applicable standards and regulations in the field; functional requirements, if any
		Maintainability	Functional requirements and regulations (e.g., recovery time, auditability)
	QoS property	Service delay/ms	From a use case perspective, the E2E (End-to-End) latency for communication services between two application endpoints, with allowable variability/jitter and/or expected upper bounds, should distinguish between uplink (UL) and downlink (DL) and detail the expected packet sizes and traffic characteristics (if any) in the deployment specifications
		Data rate (Expected minimum, expected value, maximum)/ Mbps	Refers to a single User Equipment (UE). The minimum expected data rate ensures correct operation of the use case (with potential degradation in Quality of Experience, QoE, or limited functionality). The required data rate ensures the desired QoE and full functionality of the use case. If an upper limit can be specified, it represents the maximum data rate that may occur within the use case
		Resource limit	Please refer to the deployment instructions (e.g., frequency, energy consumption, …)
		Scalability	Please refer to the deployment instructions (e.g., number of users, mobility, …)

(continued)

Table 4.3 (continued)

AI and computing	Dependent property	Agent availability/%	The percentage of time that an AI agent can receive and respond to inference requests (i.e., the agent can be used for decision-making) while meeting agreed-upon QoS (Quality of Service) objectives
		Agent reliability/%	Percentage of requests completed within agreed-upon QoS targets
		Security	Requirements or regulations related to the use of artificial intelligence in application fields
		Impartiality	Requirements or regulations related to ensuring that AI/computation operates as expected
		Maintainability	Functional requirements or specifications regarding how the system responds to defective/ erroneous operations (for example, the ability and speed of an AI agent to recover from attacks)
	QoS property	AI service RTT/ms	Maximum tolerable latency between the time a request is issued from an application and the time an AI service makes a response available to the application
		Inference accuracy/%	Domain-specific metrics used for estimating AI accuracy (if available) (i.e., quality functions)
		Interpretability	Qualitative indicators of interpretability for AI-based models and decisions, specific to the use case
		Training/model transfer delay/ms	The tolerable time for training new/ evolved models relevant to the use case or transferring models from one AI agent to another
		Resource limit	Please refer to the deployment instructions (e.g., frequency, energy consumption, …)
		Scalability	Please refer to the deployment instructions (e.g., number of users, mobility, …)
Localization sensing	Dependent property	Service availability/ %	Percentage of time to respond to localization or sensing service requests and meet given QoS objectives

(continued)

Table 4.3 (continued)

		Service reliability/%	Percentage of requests completed within the agreed QoS target
		Security	Localization or sensing requirements or regulations related to safety in various use cases
		Impartiality	Requirements or regulations for positioning or sensing integrity, such as robustness against potential interference or attack
		Maintainability	When it comes to the use of services, functional requirements or regulations in the use case domain
	QoS property	Localization accuracy/m	Estimated position accuracy, reported in terms of horizontal and vertical position precision
		Localization/sensing service RTT/ms	Maximum tolerable service delay from the time a request is issued by a consumer (application, service) to the time a location/sensing response is provided. This should not be confused with RTT (Round-Trip Time) measurements, which are a method of measuring the distance between a user and a base station
		Localization accuracy/(°)	Accuracy of UE estimated direction: roll angle, pitch angle, and yaw angle
		Refresh rate/(1/s)	The rate at which an application needs to obtain new location estimates
		Minimum and maximum resolvable range/m	Minimum and maximum distinguishable distance between two objects required by the application/use case
		Angular resolution (°)	Minimum distinguishable angle (direction) required between two objects
		Minimum and maximum speed/(m/s)	The range of speeds (minimum and maximum velocities) of the objects that the service needs to measure
		Speed resolution/(m/s)	Minimum measurable change in object speed
		Resource limit	Please refer to the deployment instructions (e.g., frequency, energy consumption, …)
		Scalability	Please refer to the deployment instructions (e.g., number of users, mobility, …)

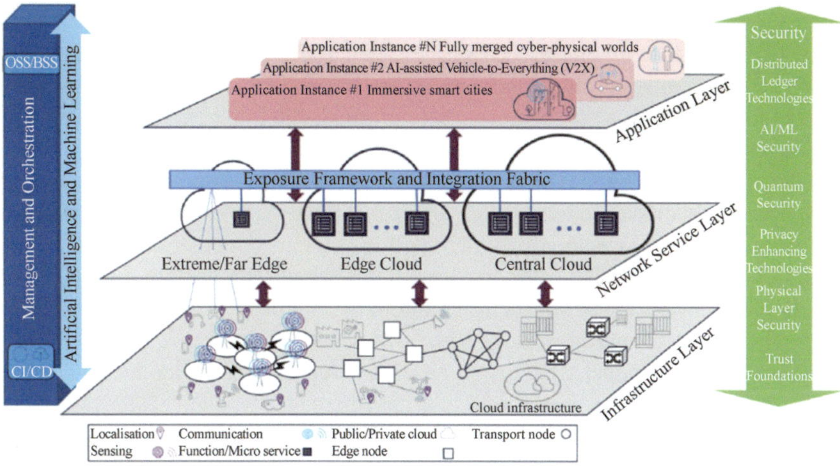

Fig. 4.49 Preliminary design of Hexa-X's end-to-end 6G network architecture [91]

network functions, thus network functions can be developed, deployed, and replaced independently.

Principle 8: Network architecture is simplified compared to previous generations. Simplify network architecture, utilise cloud-native upper layer RAN and CN functions to reduce complexity, configure fewer (well-motivated) parameters and fewer external interfaces.

Based on these principles, the preliminary design of Hexa-X's end-to-end 6G network architecture is shown in Fig. 4.49. For detailed end-to-end architecture content, please refer to reference [91].

Currently, Hexa-X is continued as Hexa-X II, its related research work and will update and publish its latest research results. It can be said that Hexa-X's research is the most systematic globally, and during the research process, the project is constantly absorbing the latest breakthroughs and progress from external research, and the related design of the project is constantly iterating, so Hexa-X's research results well represent the latest progress in the industry and will have a profound impact on global 6G research and subsequent standardisation.

4.3.2 6G in South Korea

The South Korean government announced a "6G R&D Implementation Plan" on 23 June 2021, which will invest 220 billion won (approximately RMB 1.25 billion) over the next five years to develop 6G key technologies and strive to become the first country in the world to achieve 6G commercialisation by 2028 through cooperation with countries such as China and the United States. This plan was proposed by the

South Korean Ministry of Science and ICT at a 6G strategy meeting on the same day, which specifically includes securing next-generation core original technologies, taking the lead in international standards and patents, and building a research industry foundation. In terms of next-generation core original technologies, the South Korean government will focus on promoting six key areas such as low-orbit communication satellites and ultra-precision network technology, to ensure a commanding height in 6G communication core technology. In terms of international standards, the plan states that the South Korean government will lead the development of a 6G vision, working with the Patent Office to provide comprehensive support for the R&D and patenting of 6G core technologies. In addition, the South Korean government aims to operate 6G R&D centres at three universities within the year, cultivating high-level talent through industry-academia-research cooperation, laying the foundation for related technology R&D and industrial development.

The Minister of Science and ICT of South Korea, Lim Hyesook, once stated in a declaration: "The next-generation mobile network is the foundation of digital innovation, we should exert creativity and lead the international market in the 6G era based on our experience and technology in the network field".

As early as July 2020, Samsung released the white paper "The Next Hyper—Connected Experience for All" [99], it proposed a 6G vision to bring the next generation of super-connected experiences to every corner of human life, as well as new services that 6G can provide, such as truly immersive XR, high-fidelity mobile holography, digital twins, etc. It also elaborated on the key performance indicators of 6G, the overall architecture and new requirements for reliability. In addition, the white paper pointed out some possible 6G enabling technologies, such as terahertz technology, new antenna technology, duplex evolution technology, spectrum sharing technology, native AI technology, distributed computing technology and high-precision network technology, etc. Finally, the white paper also provided Samsung's views on the timeline for 6G research, standardisation and deployment, that is, the standardisation of 6G will be completed in 2028, with initial deployment, and then large-scale commercial use of 6G will be realised in 2030.

In May 2022, Samsung released another white paper on 6G spectrum, "6G Spectrum—Expanding the Frontier" [100], defining the spectrum of 6G, as shown in Fig. 4.50.

- Low frequency band: Spectrum less than 1 GHz, used to achieve wide area coverage or deep indoor network coverage.
- Mid-frequency band: 1–24 GHz band. Traditionally, the mid-frequency band refers to 1–7 GHz, in the 6G era, it needs to consider expanding it to 24 GHz,

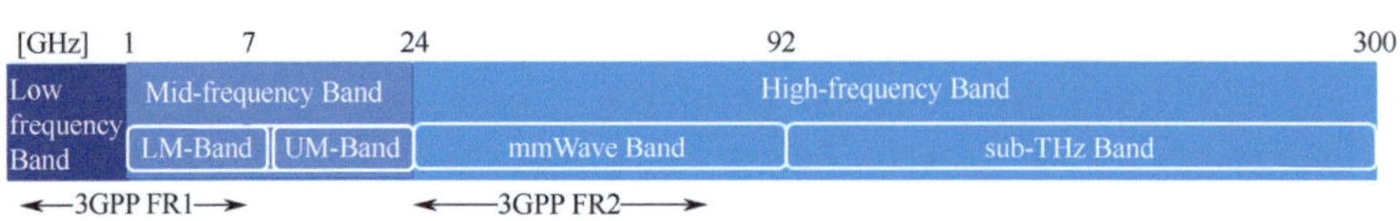

Fig. 4.50 Definition of 6G spectrum

and ITU-R is currently studying possible candidate bands for cellular mobile communication.

- High frequency band: Compared to 5G, the upper limit of the high frequency band needs to be further expanded, so the high frequency band is defined as 24–300 GHz. Among them, 24–92 GHz is the millimetre wave band, this band will not only be used in 5G, but also in 6G, and the sub-terahertz band (92–300 GHz) will become a new band considered in 6G.

The white paper elaborates on methods to ensure 6G spectrum such as opening up new bands, spectrum refaarming, and also points out that the sub-terahertz band and the upper part of the mid-frequency band (7–24 GHz) will become the pioneer bands for 6G mobile communication research.

In the keynote speech at the IEEE ICC 2022 conference held in May 2022, Samsung further updated its main technical views on 6G, as shown in Table 4.4.

At the same time, Samsung has also released a series of prototypes of key technologies, which are briefly introduced below.

Table 4.4 Samsung's main technical views on 6G

Frequency band	
THz/mmWave	Low frequency
Upper intermediate frequency	Frequency band sharing
Lower intermediate frequency	Unlicensed frequency band
Wireless access technology	
Super MIMO (X-MIMO)	LoS MIMO (e.g., Orbital angular momentum)
THz communication	Integration of communication and sensing
The 2nd generation mmWave technology	Near-zero energy communication
Advanced duplex technology	Waveform, coding, and modulation
Intelligent metasurface technology	Energy-saving for base stations and terminals
Network architecture	
Integration of communication and computing	Distributed cloud system
RAN-Core network architecture	
Trustworthy system	
Trustworthy Authentication and Protocols	Recoverability
Reliability	Quantum Security
Native AI system	
Native AI air interface	Split/federated learning
Distributed intelligence	
Others	
Green network	Low cost and high efficiency
Evolution from 5 to 6G	

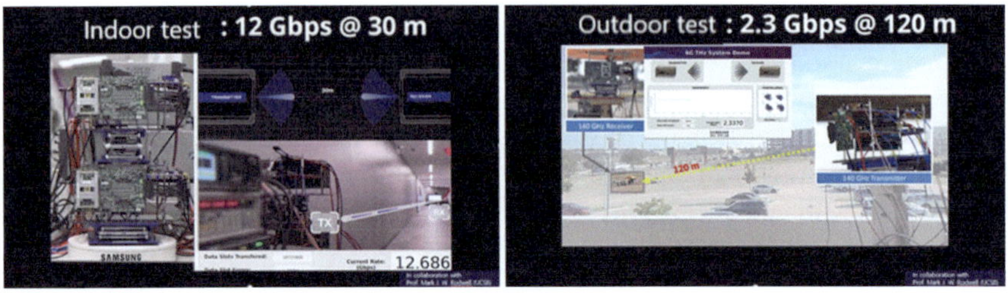

Fig. 4.51 Terahertz high-speed transmission prototype and indoor and outdoor test results

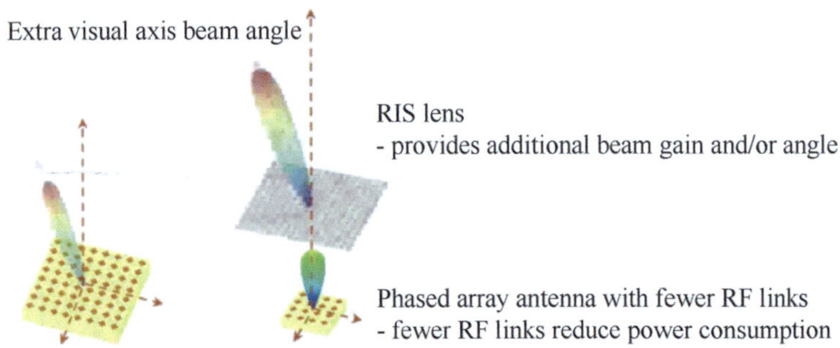

Fig. 4.52 Intelligent super surface lens prototype

(1) Terahertz high-speed transmission prototype: As shown in Fig. 4.51, a transmission rate of 12 Gbps is achieved at an indoor transmission distance of 30 m; a transmission rate of 2.3 Gbps is achieved at an outdoor transmission distance of 120 m.
(2) Intelligent Super Surface (RIS) lens prototype: As shown in Fig. 4.52, it can improve beam gain and control beam range.
(3) Cross Duplex (XDD) prototype: As shown in Fig. 4.53, it verifies the feasibility of eliminating interference between different frequency bands in adjacent uplink and downlink transmission.
(4) Millimetre wave full-duplex prototype: As shown in Fig. 4.54, compared to traditional TDD systems, the throughput has achieved twice.
(5) AI-based non-linear equalisation technology prototype: As shown in Fig. 4.55, it has been verified that this technology can nearly double the coverage distance.

4.3.3 6G in Japan

In April 2020, the Japanese Ministry of Internal Affairs and Communications released Japan's "6G Comprehensive Strategy" and key technology strategic goals for 2025, aiming to compile a comprehensive strategy to promote the establishment of key

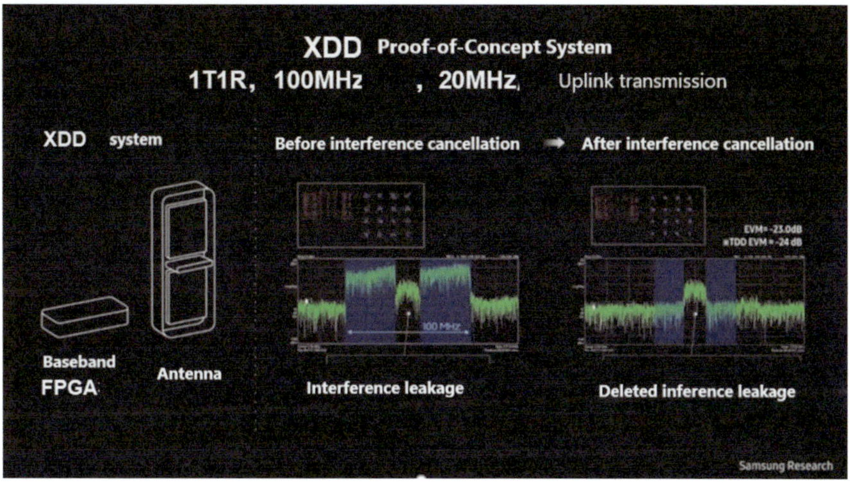

Fig. 4.53 Cross-duplex prototype

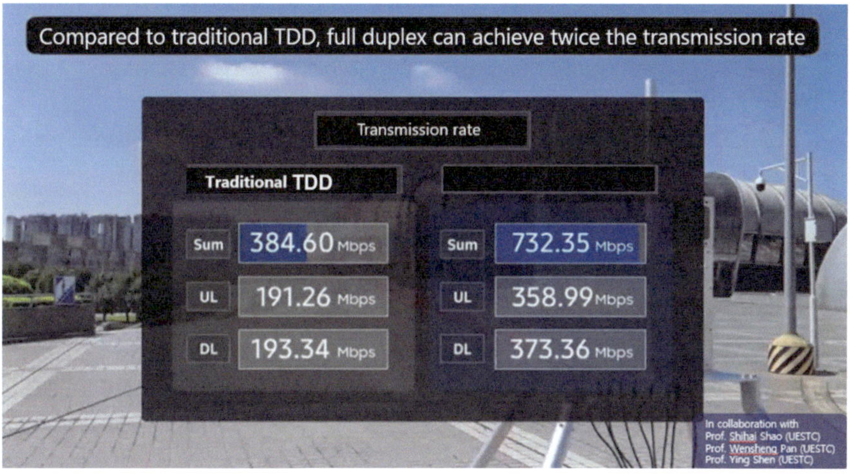

Fig. 4.54 Millimetre wave full-duplex prototype and outdoor test results

technologies in 2025 through fiscal support and tax incentives, ensuring international competitiveness. The goal is to increase the global market share of 6G from the current 2–30%, and the global patent share of Japanese companies in 6G-related fields from 5.5% to more than 10%.

In June 2020, the Japanese Ministry of Internal Affairs and Communications launched the “Beyond 5G Promotion Strategy”, initiating support for enterprises. Within 2020, the centre set up by the Ministry will subsidise enterprises for travel expenses to attend standardisation-related international conferences, and will also

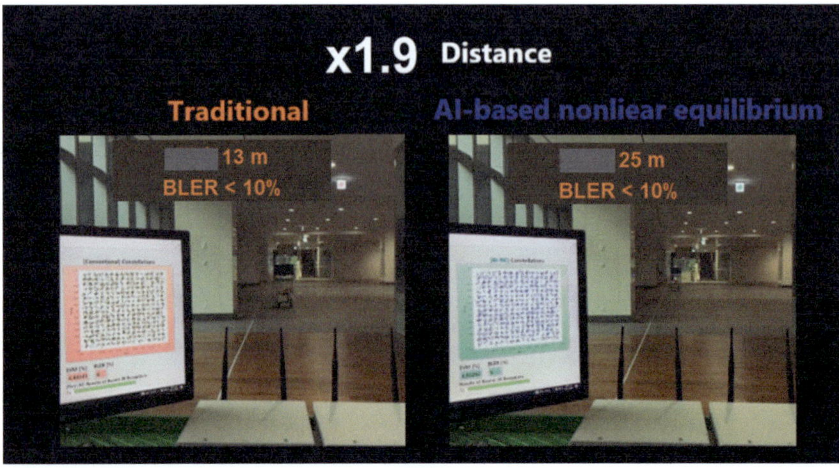

Fig. 4.55 AI-based non-linear equalisation technology prototype and indoor test results

establish a reserve of talents with experience in international conference negotiations. Through the national talent pool, it hopes to obtain talents who can achieve results in actual negotiations. At the same time, it will discuss the direction of key technology research and development in standardisation with enterprises, government, and universities.

The Japanese Ministry of Internal Affairs and Communications and the National Institute of Information and Communications Technology (NICT) collaborated at the end of 2020 to establish the "Beyond 5G New Business Strategy Centre", aiming to start and carry out 6G-related research work as soon as possible. The Beyond 5G New Business Strategy Centre will work together with industry, academia, and government to jointly commit to the acquisition of intellectual property rights and standardisation work. Future main activities include: studying specific action guidelines according to the latest global trends; providing a place for Beyond 5G participants to establish partnerships; establishing a database of experts related to standardisation and intellectual property rights, and sending consultants; holding human resource development seminars, etc.

In Japan, the most active in conducting 6G research is the Japanese operator NTT DoCoMo, which released the white paper "5G Evolution and 6G" (*5G Evolution and 6G*) [101] report in January 2020, including thoughts on 5G evolution, world view of the 6G era, 6G needs and use cases, 6G technology research areas.

As shown in Fig. 4.56, NTT DoCoMo proposed a vision similar to digital twins, and will pay great attention to ageing issues, communication between people and things, expansion of communication environment, complexity of information-physical world integration in future social development, and based on the scenarios defined by 5G, it has further expanded and formed the demand for 6G, as shown in Fig. 4.57.

The typical scenarios and their performance indicators of 6G are as follows.

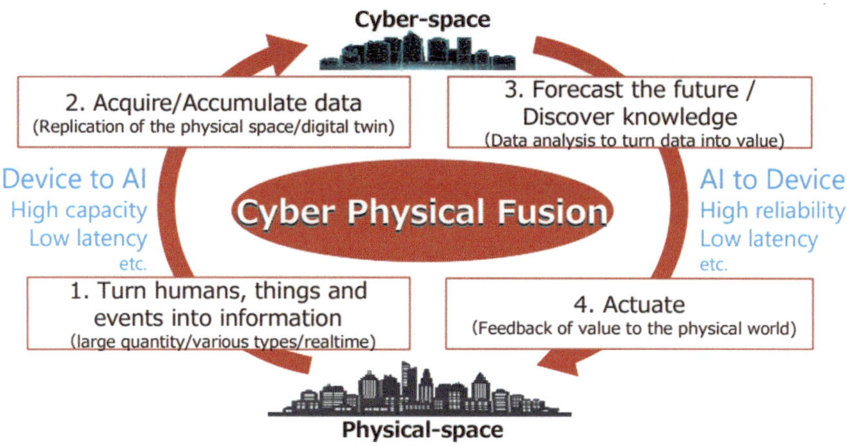

Fig. 4.56 Vision of digital twins [101]

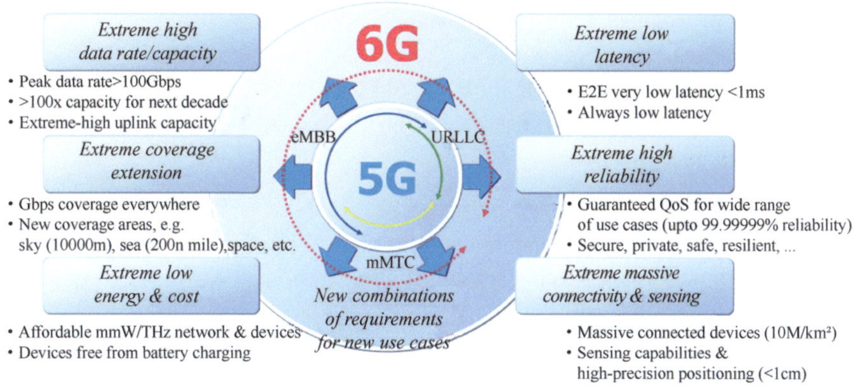

Fig. 4.57 6G scenarios and requirements proposed by NTT DoCoMo [101]

(1) Ultra-high speed/capacity communication, using new frequency bands to achieve peak data rates greater than 100 Gbps.
(2) Extreme coverage expansion, ubiquitous Gbps coverage, new coverage areas, such as the sky (10,000 m), ocean (200 n miles, 1n mile = 1.852 km), space, etc.
(3) Ultra-low energy consumption and cost, reasonably priced millimetre wave/terahertz devices, devices that do not require battery charging.
(4) Ultra-low latency, end-to-end (End to End, E2E) extremely low latency less than 1 ms, always maintaining low latency.
(5) Ultra-high reliability, providing QoS guarantees for a wide range of use cases (reliability up to 99.99999%), ensuring privacy, security, resilience, etc.

(6) Ultra-large scale connections, 10^6 connected devices per square kilometre, transmission capability and high precision positioning (cm level).

In terms of technical research, NTT DoCoMo focuses on the following aspects.

(1) New network topology.
(2) Coverage extension, including non-terrestrial networks.
(3) Frequency extension and improved spectrum utilisation.
(4) Further development of wireless transmission technologies, such as Faster-Than-Nyquist signalling (FTN) and virtualised massive MIMO.
(5) Enhanced uRLLC and industrial IoT networks.
(6) Expansion of the integration of various wireless technologies.
(7) Ubiquitous multifunctionality and AI in mobile networks.

In 2021, NTT DoCoMo released the white paper "5G Evolution and 6G V4.0" [102], updating its understanding of 6G technology.

- In terms of new wireless topology, further consideration is given to distributed antenna deployment, RIS, cooperative transmission and reception between terminals, and distributed antenna deployment including sensing and low power communication.
- Coverage extension considers non-terrestrial communication systems.
- Further extension of frequency range and technologies to improve spectrum usage efficiency.
- Massive MIMO wireless transmission.
- Expansion of uRLLC and industrial internet.
- Multifunctional wireless communication systems.
- Integration of multiple wireless access technologies.

In terms of network architecture, 6G should have the following features.

- Flat network topology.
- Flexible network function deployment.
- Simple network.
- Advanced operation and maintenance.
- Integrated operation technology of multiple access technologies.
- Core network and switching control technology supporting ultra-low latency.
- Digital twins supported by wide-area time synchronisation and generalised deterministic communication.
- Mobility control based on location for ultra-large coverage.
- Advanced security.
- Distributed computing resources.

In December 2020, Japan established the Beyond 5G Promotion Alliance [103], with the aim of preparing for the realisation of 6G around 2030 and promoting the healthy development of telecommunications applications. The alliance achieves its goals through the following activities.

(1) Engaging in 6G-related research and development and standardisation studies.

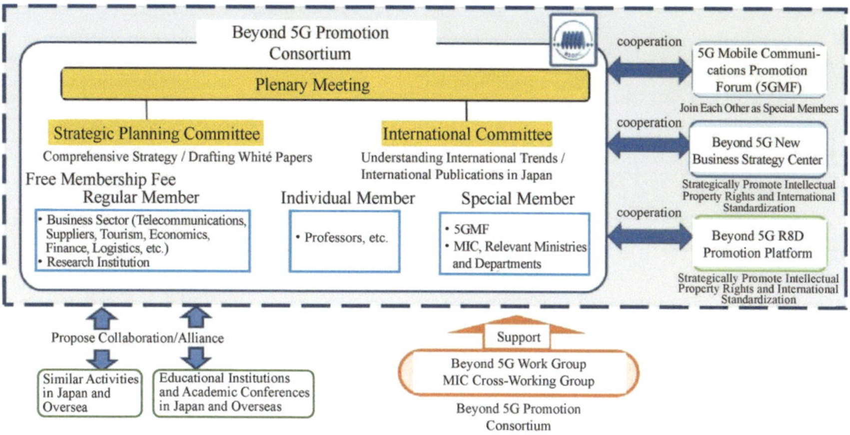

Fig. 4.58 Organisational structure of the beyond 5G promotion alliance [103]

(2) Collecting 6G information and exchanging it with other organisations.
(3) Providing liaison and coordination work for 6G-related organisations.
(4) Strengthening education and understanding of 6G.
(5) Other activities necessary to achieve the goals of this consortium.

The organisational structure of the Beyond 5G Promotion Alliance is shown in Fig. 4.58.

The Beyond 5G Promotion Alliance believes that Beyond 5G is the next generation mobile communication system after 5G, with characteristics that help create sustainable and new value, such as ultra-low power consumption, ultra-security and reliability, autonomy and scalability. In addition to further enhancing the high-speed, large-capacity, low-latency, multi-connection features of 5G, it is the next generation mobile communication system after 5G. The positioning of Beyond 5G by the Beyond 5G Promotion Alliance and Japan's advantages are shown in Fig. 4.59.

In March 2022, the Beyond 5G Promotion Alliance published the white paper "Post-5G White Paper ~ Message to the 2030s", which discussed business trends, market trends in the telecommunications industry and trends in other industries, identified unique use cases in each industry and summarised the capabilities and key performance indicators (KPIs) required by Beyond 5G, as shown in Table 4.5. It also analysed the technology trends towards Beyond 5G (B5G), and the value and roles these technologies are expected to play.

Of particular note, the white paper also defined the scope of application for the listed demand indicators, as shown in Fig. 4.60.

The white paper comprehensively explained future technology development trends from the aspects of system platforms and applications, trustworthiness (security, privacy and attainability), network energy efficiency improvement, coverage expansion based on non-terrestrial networks (NTN), network architecture, wireless and light. Especially for wireless and optical communication, it further discussed

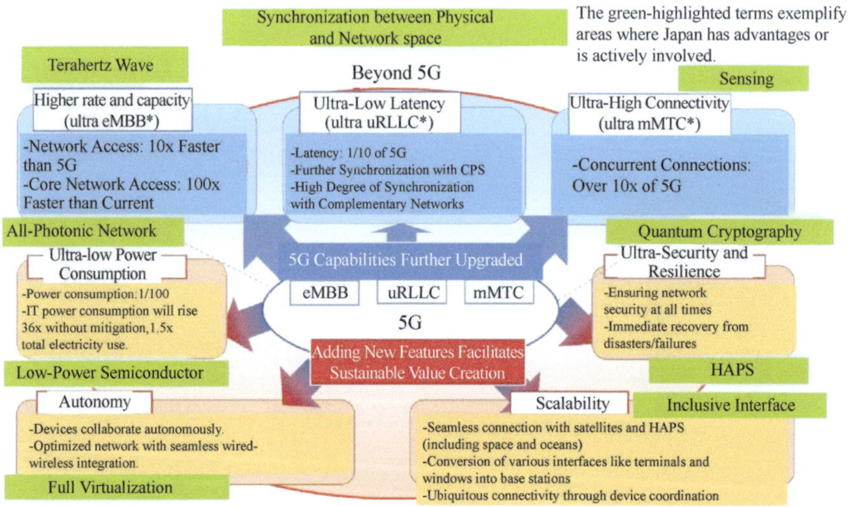

Fig. 4.59 Positioning of beyond 5G by the beyond 5G promotion alliance and Japan's advantages [103]

new wireless network topologies, technologies supporting wider bandwidth and higher spectral efficiency, air interface enhancements, technologies supporting extremely low latency and ultra-high reliability, high energy efficiency and low power consumption technologies, integrated sensing and communication, network management, communication based on native AI, optical communication technology, radio-over-fibre (RoF), optical wireless and chaotic communication.

Furthermore, the white paper also proposes the future network architecture as shown in Fig. 4.61, which has the following characteristics.

- Network architecture: Virtualised RAN and core network, allocation of computing resources using virtualised RAN (vRAN), IP connections between service entities, network AI architecture.
- User/application-centric communication architecture: User-centric architecture, application-aware network optimisation.
- Autonomous network operation.

4.3.4 6G in United States: Next G Alliance

In October 2020, the Alliance for Telecommunications Industry Solutions (ATIS) in the United States established the Next G Alliance (Next G Alliance), with the aim of establishing North America's leadership in 6G and future mobile communication technologies over the next decade [104]. The Next G Alliance will focus on

Table 4.5 Capability requirements for B5G [103]

Classification	Requirements	Performance requirements across industries
Quantitative requirements	Ultra-high speeds and ultra-large capacities	10–100 Gbps (uncompressed transmission for holographic communication) 50 Gbps (remote monitoring and remote control) 10–100 Gbps (intelligent logistics) Several tens Gbps (intelligent medicine) 48–200 Gbps (volumetric video) Several tens Gbps (low and medium orbit communication) 10 Mbps (natural disaster warning)
	Ultra-low delay	Millisecond level (fully automated logistics operations) Several millisecond (emergency stop of ultra-high-speed trains) 100 ms (immersive remote control system) 1 ms (remote monitoring and remote control) 100 microsecond (local communication) 1 ms (remote control of robots) MTP (Motion-to-photon) delay 10 ms, TTP (Time-to-present) delay 70 ms
	Precision of time synchronization	Tiime synchronization compatible with the Precision Time Protocol (PTP) to ensure the accuracy of internal clocks, including radio segments at the microsecond level (fully automated operations of logistics facilities (warehousing and logistics))
	Ultra-high security and resilience	10^{-6} (remote monitoring and remote control (automotive)) 10^{-7} (telesurgery (medical)) (unit: block error rate)
	Localization sensing	Localization accuracy of 1–2 cm (civil engineering (construction and real estate)) Centimeter-level sensing accuracy (vehicles driving alone or at night in rural areas (Automotive))
	Ultra-large-scale connectivity	10^{-6} to 10^{-7} devices per square kilometer (implantable devices (medical))
	Covrage	Supersonic aircraft flying at an altitude 10 km higher than current passenger aircraft Coverage of outer space areas beyond 100 km (Aircraft) 100% land coverage (telecommunications and IT) Outer space and lunar coverage (Space) High-Altitude Platform Station (HAPS) covering an airspace with a radius of tens to hundreds of kilometers and at an altitude of several kilometers above the ground (HAPS)

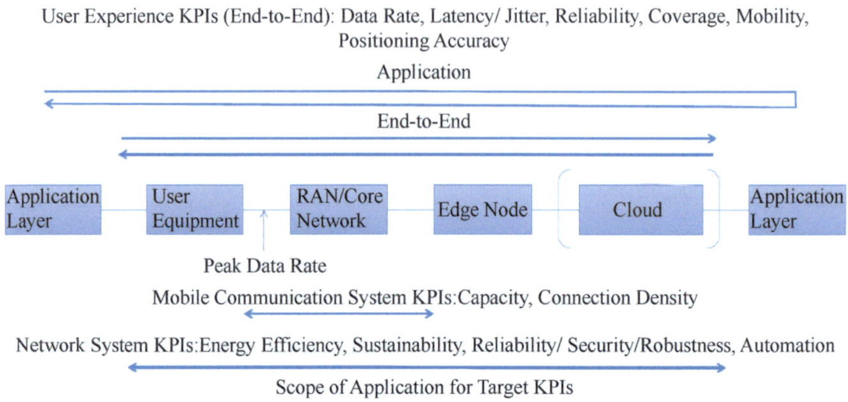

Fig. 4.60 Definition of the scope of application for demand indicators [103]

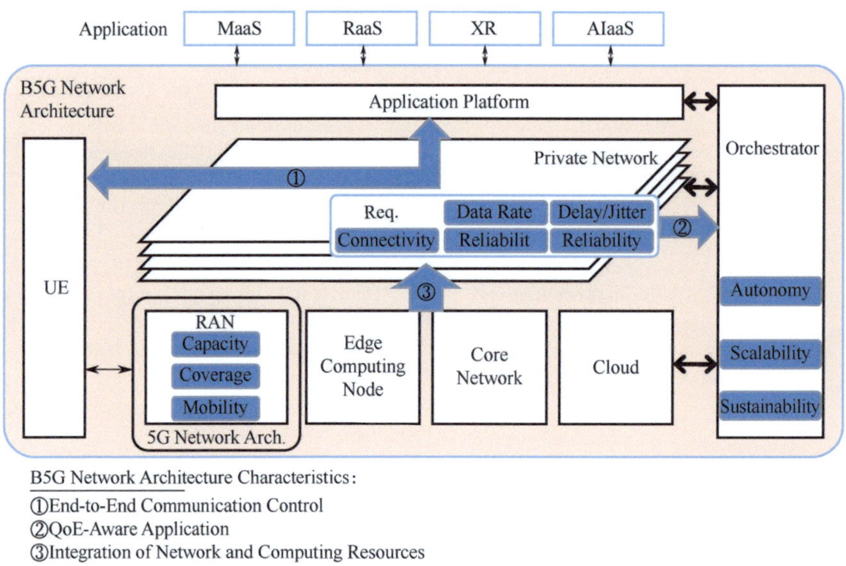

Fig. 4.61 Future network architecture [103]

research and development, standardisation, and commercialisation, actively developing a national 6G roadmap, establishing core 6G technologies, influencing government policies and investments, and promoting the commercialisation of Next G technologies. Currently, the Next G Alliance has 42 members, mainly North American companies, but also includes well-known technology and telecommunications companies from Europe, Japan, Korea and other countries/regions.

The organisational structure of the Next G Alliance, as shown in Fig. 4.62, aims to ensure North America's lead in 6G technology in key consumer and industrial

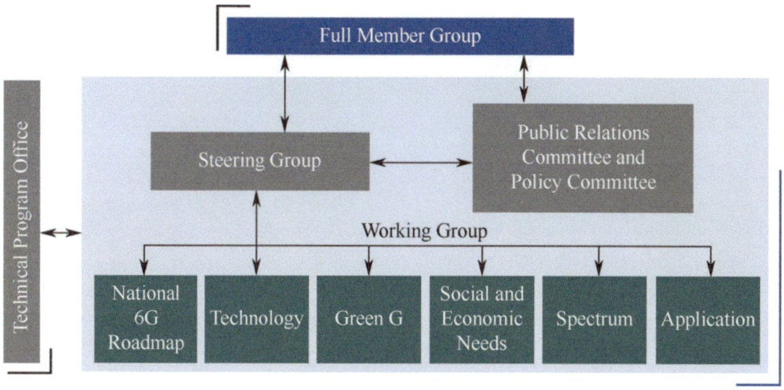

Fig. 4.62 Organisational structure of the Next G Alliance [104]

sectors, promoting and strengthening North America's economic interests locally and globally. To achieve success, the Next G Alliance believes it is necessary to rethink the existing North American model, with the industry leading North American 6G research, development and commercialisation, while the government should play a significant role in promoting and supporting these efforts through policy and partnerships. To enhance North America's leadership in the 6G field, the government should support at least three key areas.

(1) Government decision-makers and the industry need to work together to build a vision for North American 6G leadership, construct a policy framework to support industry innovation, including coordination with other partner countries with similar goals, and achieve policy priority consistency.
(2) To ensure that these North American policy priorities are designed into the 6G framework, the government must actively support domestic research and development, including substantial funding, government guidance, and tax incentives.
(3) The government needs to start implementing process incentives to promote public and private investment in 6G commercialisation and deployment. Although 6G may not be available until a decade from now, to ensure that North America is in a leading position in this regard, the government needs to take action now, such as planning sufficient and appropriate spectrum.

The Next G Alliance in its report "Road to 6G" [105] proposed its future vision, as shown in Fig. 4.63, which includes six aspects.

(1) Enhance the trust, security, and resilience of the network, so that future networks can be fully trusted by individuals, businesses, and governments under any circumstances.
(2) Enhanced digital world experiences, including multisensory experiences that transform human collaboration, human–machine collaboration, and machine-to-machine interaction, will change work, education, and entertainment, improve quality of life, and create great economic value.

Fig. 4.63 Vision of the Next G Alliance for 6G [105]

(3) Economically efficient solutions should cover all aspects of network architecture, including terminal devices, wireless access, cell site backhaul, and energy consumption. These solutions must be improved and be able to provide services in various environments, including cities, rural areas, and suburbs, while also supporting higher data rates and services expected in future networks.
(4) Distributed clouds and communication systems built on virtualisation will increase flexibility, critical performance, and resilience, especially in certain scenarios and use cases, such as mixed reality, uRLLC, interactive gaming, and multisensory applications.
(5) An AI-Native wireless network is needed to increase the robustness, performance, and efficiency of wireless and cloud technologies to cope with

more diverse business types, ultra-dense deployment topologies, and more challenging spectrum conditions.

(6) Sustainability related to energy efficiency and the environment must be prioritised throughout the 6G lifecycle to achieve carbon neutrality in the mobile communications industry by 2040.

In addition, the report also provides enabling technologies for different objectives, as shown in Tables 4.6, 4.7, 4.8, 4.9, 4.10 and 4.11.

In addition, the Next G Alliance agrees with IMT-2030's goal of completing ITU-R standards by 2030 and has submitted a 6G timeline proposal to ITU-R WP5D [104], as shown in Fig. 4.64.

In terms of technical research, the Next G Alliance will focus on the evolution technologies of existing 5G and revolutionary technologies, including new air interfaces, network architectures, spectrum access, X-haul, trust/privacy/security platforms, distributed communications and cloud, and sensing technologies. As there are many technology areas related to the 6G system, the key technology elements can be divided into five categories:

- Device technologies;
- Wireless technologies;
- System and Network Architecture (SNA);
- Operations, Administration, Maintenance (OAM) and Service Support (SE);
- Trustworthy.

In terms of specific wireless technologies, the Next G Alliance focuses on the following directions:

- Wireless technologies for spectrum expansion and efficiency improvement, such as THz/sub THz, millimetre wave enhancement, spectrum sharing, advanced MIMO technology, advanced duplex schemes, waveforms, coding, modulation and multi-addressing;
- Wireless technologies for artificial intelligence and distributed cloud, such as AI-Native air interface, cross-device and network distributed computing and intelligent air interface enabling;
- Green communication wireless technologies, such as green networks, device energy saving, zero power consumption communication, ultra-low resolution communication systems;
- Wireless technologies for advanced topology and networks;
- Communication-Sensing Integration (JCS).

In terms of system and network architecture, the Next G Alliance focuses on network topology, network adaptability, distributed cloud and computing, artificial intelligence in networks and devices (including AI-based PHY/MAC, AI-assisted mobility management, AI-optimised resource scheduling, AI-based orchestration, AI-based security).

Network operation, management and service support networks are also important research contents, specifically including:

Table 4.6 Technical considerations for trust, security, and resilience

Technical challenge	Detailed description	Technical consideration
Service availability	End-to-end observability is crucial for establishing KPIs, which can collcctively form a comprehensive measure of trustworthiness	Data pipelines for enhancing observability of RAN, cloud, core, and service metrics
	Data pipelines for enhancing observability of RAN, cloud, core, and service metrics	Provide efficient and robust resource allocation for critical services
	Through service tracking, control data, and service exposure interfaces	Enhancement of service quality
Elasticity	Using data-driven techniques to assess the network's capacity, constraints, and operational status for each service; automated and dynamic methods to predict and modify network behavior in order to meet quality of experience	Network performance optimization
	Applying machine learning techniques to data-driven risk identification and mitigation methods in order to overcome vulnerabilities, failures, and interference	Threat detection and response
	Disaster relief and fault tolerance during crises	Local survivability
	Utilizing AI/ML for predictable and verifiable global supply chains, automated security operations, automated software generation and verification	Automation in 6G
Trustworthy artificial intelligence/machine learning	Explainable and ethical behavior from AI models hosted on the network, along with service exposure audits of AI functionalities; privacy protection, and robust checks and balances for the use of machine learning for legitimate and ethical purposes	Explainable Artificial Intelligence
	Data protection for third-party functionalities, including confidential handling of datasets and model parameters used in machine learning	Security and confidentiality of AI/ML

(continued)

Table 4.6 (continued)

Technical challenge	Detailed description	Technical consideration
Security and privacy	Chain of trust for all hardware and software components; technologies such as blind signatures, zero-knowledge proofs, and group signatures	Identity based on a root of trust
	Protect data in transit, at rest, and in use. Techniques such as homomorphic encryption, multi-party computation, and federated learning	Privacy protection technologies and protocol stacks
	Interference detection and mitigation	Physical layer security
	Confidential computing for critical mission network slices	Privacy computing
	Utilizing trusted execution environments to mitigate the risk of potential breaches using quantum computers against traditional cryptographic methods	Post-quantum cryptography techniques
Observability and resource optimization	Continuously monitor observable and measurable indicators that can be analyzed for compliance with key performance and security metrics	Machine learning and data analysis
	Real-time prioritization of resources to achieve high levels of performance and resilience in peer-to-peer and device-to-network communications	Solutions covering all information and telecommunications services

- Automated service management/orchestration, data management and AI/ML based intelligent network controllers;
- Public safety in emergencies and disaster scenarios;
- Technology drivers for business service integration;
- Green energy-saving networks.

In terms of network trustworthiness, the focus is on communication security (specially designed beamforming and MIMO pre-coding methods, encryption key generation and user verification programs based on the inherent randomness of wireless channels, symmetric key generation based on fingerprints, quantum key distribution and quantum encryption), system reliability, data and privacy protection.

Table 4.7 Technical considerations for the Digital World Experience (DWE)

Technical challenge	Detailed description	Technical consideration
Innovations in cyber-physical technologies	This involves blending the virtual/digital and physical worlds through innovative sensing and feedback methods that go beyond vision and sound	Point technology innovations (e.g., holographic communication, haptic interfaces). An approach exemplified by linking localization/sensing and XR/haptics through cross-technology fusion
Development of knowledge system technologies	Its purpose is to accelerate intelligent data processing and task automation based on new AI/ML technologies that absorb and integrate data from cyber-physical subsystems	Design 6G systems with endogenous AI/ML enablers (such as semantics) and tools (such as predictive models) to manage the resources involved in delivering Distributed World Experience (DWE)
Enable extreme automation technologies	Extreme automation applies to processing activities within the technology stack and service delivery chain, aiming to provide intuitive and seamless digital experiences. For example, distributed computing and communication technologies enable network operators to automate the dynamic orchestration of network resources. Other examples apply to subscription and service activation tasks, concealing complexity from end-users, and enforcing finer-grained privacy and security management policies	Promote enabling technologies and spectrum resources for communication, computing, devices, interfaces, and services
Provide support tools and processes	APIs and SDKs enable developers to leverage new 6G capabilities to create innovative services. Consumers can rely on consumer protection controls and tools to trust and manage their digital world experiences	Enhance the availability and accessibility of tools for different user categories, such as 6G system operators, developers, and end-consumer users

(continued)

Table 4.7 (continued)

Technical challenge	Detailed description	Technical consideration
Evolution of computing and communication platform capabilities	Horizontal driving factors common to multiple use case scenarios that can be applied to operational environments with multiple users and service providers	Development and deployment of personalization technologies (e.g., identity management, processing of personal data)
Market creation based on exhibition-enhanced applications and 6G capabilities	Applications showcasing the value of DWE to a broader market, incorporating various aspects of the 6G innovation portfolio	Service-oriented robotics, combining ultra-high-speed data communication, synchronized localization and mapping, and high degrees of automation. Real-time personalization. Converged reality telepresence. Immersive communication using XR and wearable devices

In terms of frequency, the Next G Alliance will research and evaluate the applicability of the spectrum to 6G requirements, and evaluate the spectrum management and access mechanisms needed to maximise North American 6G opportunities, and develop a regulatory policy roadmap, as shown in Fig. 4.65.

In addition, in "The Road to 6G" [105], the Next G Alliance introduces four categories of future-oriented application scenarios.

- Networked robots and autonomous systems: online collaborative operation of group service robots, on-site robots suitable for hazardous environments.
- Multi-sensory extended reality: ultra-realistic interactive sports—drone racing, immersive gaming/entertainment, mixed reality collaborative design, mixed reality remote presentation, 6G immersive education, high-speed wireless connections in aircraft for entertainment services.
- Distributed sensing and communication: remote data collection, wireless wearable devices and implants, bridging the North American digital divide, public safety applications, synchronous data channels, healthcare—in-body networks.
- Personalised user experiences: personalised hotel experiences, personalised shopping experiences.

Based on the above application scenarios, "The Road to 6G" [105] also derives the types and ranges of performance requirements that 6G should have, as shown in Table 4.12.

Table 4.8 Technical considerations for economically efficient solutions

Technical challenges	Specific description	Enabling technology
Capacity	The growing traffic demand will affect the cities and suburbs and consume capacity. The cost per bit must be continuously reduced by improving the spectral efficiency per unit area. Cell splitting and technology improvement help improve efficiency	Cell densification (see coverage for details) Wireless technology: » Advanced large-scale MIMO » New waveform, coding, modulation and multiple access schemes » AI in air interface » Ultra low resolution data converter Spectrum efficiency: » Spectrum sharing: licensed, unlicensed, local, coexistence with 5G or non cooperative systems » Carrier aggregation, ultra wideband carrier » Higher spectrum
Coverage	The spectral efficiency needs to be improved by increasing the coverage reliability in a large number of new environments to relieve the pressure on frequency resources	Cell densification: » Low latency/cost backhaul » Open and virtualized RAN » Integrated Access Backhaul (IAB) » Intelligent repeater » Reconfigurable intelligent surface (RIS) » Improved AI based planning and self optimization

(continued)

Table 4.8 (continued)

Technical challenges	Specific description	Enabling technology
Lack of supply side competition	Reduce deployment costs and increase competition through innovative network architecture and standard interfaces	Business model innovation network architecture: » Tighter integration of building owned networks » Off ground networks, including collaboration between ground and off ground networks » Distributed cloud platform » Network decomposition » RAN core splitting » Wireless grid network (Mesh) and direct connection technology (Sidelink) » Collaborative communication » Lightweight protocol stack » Embedded subnet connection
Economics of low user density	Develop deployment architecture to maximize coverage of unconnected, sparsely populated rural areas	Business model innovation
Distribution cost structure	Providing capacity for wide areas will require cost-effective remote backhaul to support small communities in sparsely populated areas. Rural communities should use turnkey projects to allow the deployment of solutions that rarely customize community planning	» The project reference design helps to easily deploy a set of typical rural deployment scenarios » Provide "interstate" information of local distribution points for rural communities and major transportation arteries

(continued)

Table 4.8 (continued)

Technical challenges	Specific description	Enabling technology
Penetration loss of outdoor to indoor coverage	Capacity must be connected from outdoor space to indoor space. Once inside, it must be effectively allocated to adapt to the indoor environment structure of people's life and work	Cooperative use of authorized and unauthorized spectrum
Physical and organizational partitions in the space	An indoor space distribution system that is deployed by the site owner, industrial partner or management company in partnership to provide commercial services	Tighter integration of building owned networks (interoperability, business model)

Table 4.9 Technical considerations for distributed cloud and communication systems

Technical challenges	Specific description	Technical considerations
Deployment of large-scale network computing architecture	Realize large-scale consolidation of distributed and interconnected mobile networks and cloud systems, and achieve ubiquitous access	» Joint and coordination » Homomorphic computing environment » Autonomous and autonomous networks » ML for network computing structure optimization » Seamless computing allocation and offload
Innovation, integration, and/or interoperability of edge devices	Unified computing expansion across devices, network computing resources, and data centers	Support the continuum from device to edge and cloud
Use autonomous decision-making technology	Apply a large amount of data in networks at all levels to improve decision-making speed, privacy, security and reliability	Distributed learning, federated learning, and split AI/ML
Provide trust root	Support zero trust and cloud native architecture	» Data trust structure » Security and signature telemetry » Distributed ledger and blockchain
Enable secure communication between entities	Support zero trust for cloud native architecture, communication between cloud and physical network elements	» Application and session based endpoint security » Network slices for service and customer isolation
Provide policy control with corresponding monitoring capability	Capabilities required to support orchestration functions and monitor compliance, as well as security performance	» Closed loop control loop with integrated AI/ML and analysis » Adaptive policies and contextual software policies » Unified telemetry of network and platform resources

Table 4.10 AI-native technical considerations for wireless networks

Technical challenges	Specific description	Technical considerations
Compatibility and interoperability	When end-to-end links are dynamically adjusted through learning, a key challenge is to be able to utilize spectral efficiency gains while ensuring broad and global interoperability	» Diversified data collection » Distributed and supervised end-to-end learning » Migration and joint learning » Endogenous AI interface » Cooperative reasoning and learning » AI based privacy and security
Minimum performance guarantee	In essence, AI/ML algorithms cannot be strictly defined in the same way as traditional algorithms. This requires a more comprehensive performance evaluation. Resilience under adverse conditions should be a key consideration. Failsafe backup mechanism may be required	» Performance monitoring of deployed AI/ML algorithms » Failsafe backup mechanism » Dynamic model adaptation
Datasets and AI/ML validation	Due to the key role of data in AI/ML, the creation of data sets, the depth and breadth of data coverage, and the methods used for interoperability, verification, and test cases of AI driven methods need to be carefully considered in the context of data and simulators	» Diversified data collection » Proof of concept » Extensive verification framework
Computation complexity	ML inference engine can have very high computing requirements, but it needs to be evaluated according to the expected rapid growth of ML hardware accelerator capability compared with the complexity of existing methods	» Dedicated hardware accelerator » Distributed and supervised end-to-end learning » Computational complexity assessment » Delay reduction

(continued)

Table 4.10 (continued)

Technical challenges	Specific description	Technical considerations
Overhead management	In some aspects of AI enabled systems, overhead may be increased (for example, sensor data), and overhead may also be reduced (for example, the reference channel in the Over the Air (OTA)). There are tradeoffs between stimulus response delay, overhead, and performance gains, which must be studied	» Built in AI air interface » Distributed and supervised end-to-end learning » Data driven modeling » Migration and joint learning
Provide policy control with corresponding monitoring capability	Ability to support orchestration functions and monitor compliance and security performance	» Closed loop control loop with integrated AI/ML and analysis » Adaptive policies and contextual software policies » Unified telemetry of network and platform resources

The Next G Alliance, in collaboration with most companies from North America, Europe, and Japan and South Korea, has participated in relevant 6G work, demonstrating strong capabilities, and is expected to have a significant impact on future 6G research, standardization, and industrialization.

4.4 Summary

This chapter comprehensively summarizes the 6G global research progress of enterprises, important organizations, and landmark projects. In the overall layout of 6G, all countries are fully laying out and proposing development goals, especially Japan and South Korea. The study on the 6G scenarios and use cases has been triggered in 3GPP in September 2024. The work on 6G is transfering from research to standardization. Although the descriptions vary, the definition of the 6G vision is basically around "digital twins, ubiquitous intelligence". From the current layout of key technology research, there has not been a breakthrough technology in the physical layer of air interface, more is the optimisation and improvement for specific application scenarios, while a large number of new concepts and paradigms have appeared in the network architecture, and it is expected that there will be a major change in the 6G network architecture.

Table 4.11 Technical considerations for sustainability

Technical challenges	Specific description	Technical considerations
Energy conservation	Wireless technologies and components designed to improve energy efficiency, including data converters, system on chip (SOC), and power amplifiers	New waveform, coding, modulation and multiple access schemes, predistortion, ultra-low resolution data converters, SOC with higher integration and large-scale MIMO with higher dimensions
	Energy efficient infrastructure with optimized computing, network, storage, and data center facilities	Dynamic optimization of infrastructure components to meet the needs of the 6G network, but no idle resources
	Use AI/ML to dynamically optimize the network, so as to reduce the energy consumption of base stations, equipment and core networks	Dynamic optimization of 6G network to ensure optimal service availability and no idle resources
	Adjust the connection protocol to allow idle connections and significantly reduce device and radio power consumption, saving battery and RF power	RF protocol and spectrum power optimization
	Device energy collection with a truly ultra-low power interface reduces battery consumption or minimizes the need to replace batteries during their service life	Regeneration/zero power equipment
	End to end observability to establish KPIs that can form a comprehensive measure of component energy consumption and efficiency	Data pipeline used to enhance the observability of RAN, cloud, core and service indicators
Reduce environmental impact	Improve water consumption for liquid cooling or monitor water consumption. Indicators such as water efficiency introduced by the Green Grid can be used to track water-saving progress	Classification of "Eco Rating" of life cycle assessment of products and services based on KPI reports
	Reduce direct and indirect impacts on land. Improve the mining process of rare materials needed to manufacture electronic products. Track and reduce the impact of materials and rare materials extraction on the environment. Develop into bioplastics and more sustainable materials. Improve recyclability, reduce waste, and minimize the impact of waste on the environment	

(continued)

Table 4.11 (continued)

Technical challenges	Specific description	Technical considerations
	Consumers should be more responsible for the environment, protect their health and reduce energy consumption. Provide consumers with tools to monitor their consumption and take action to reduce their impact on the environment	Including that the network slice used by the application conforms to the specific green certificate, providing consumers with the choice of green communication channels
Technical empowerment	Use green certificates and indicators supporting communication infrastructure to measure and select the most effective resources	Real time supply chain management and KPI reporting

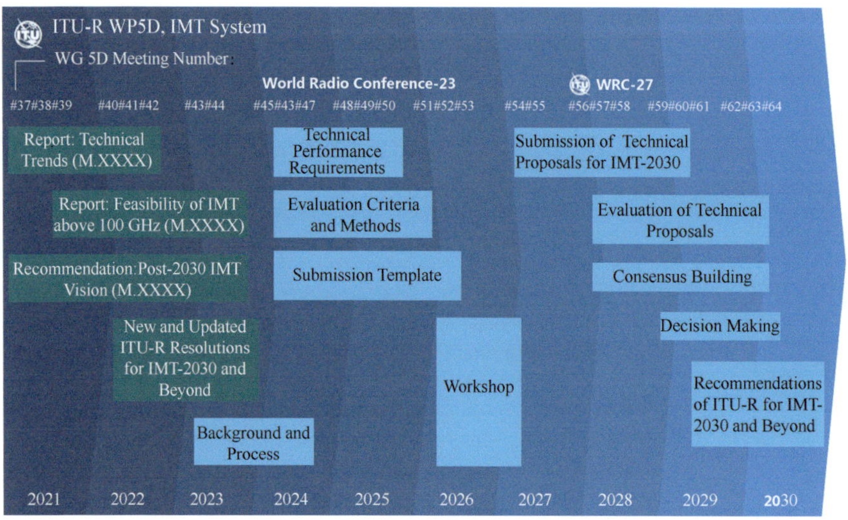

Fig. 4.64 Proposed ITU-R IMT-2030 timeline by the Next G Alliance [104]

Fig. 4.65 Next G Alliance's spectrum work approach [104]

Table 4.12 Next G Alliance's suggested 6G performance requirement indicators and their range

Category	Index	Suggestions on the scope of demand indicators
Performance	Downlink (DL) user experience data rate	Very low: < 100 kbps Low: < 10 Mbps Medium: 10–50 Mbps High: 51–100 Mbps Very high: 101–1000 Mbps Superelevation: > 1000 Mbps
	Uplink (UL) user experience data rate	Very low: < 100 kbps Low: < 10 Mbps Medium: 10–50 Mbps High: 51–100 Mbps Very high data rate: 101–1000Mbps Ultra high data rate: > 1000 Mbps
	Synchronization accuracy	• Low: > 1 ms • Medium: > 100 us • High: > 10 us • Very high: > 1 us • Superelevation: < 1 us • N/A: Not applicable
	Service continuity	• Not required • Required
	Mobility (type/speed)	• Fixed (including nomadic) • Pedestrians: 0–10 km/h • Vehicle: 10–120 km/h • High speed vehicles: 120–350 km/h • Super high speed: > 350–500 km/h
	DL end-to-end packet delay	• Very low: < 1 ms • Low: < 10 ms • Medium: < 50 ms • High: < 100 ms • Very high: < 500 ms • Best effort
	UL end-to-end packet delay	• Very low: < 1 ms • Low: < 10 ms • Medium: 10–100 ms • High: 100–500 ms • Best effort
	End-to-end packet jitter	High: very sensitive (microseconds) Medium: sensitive (ms) Insensitive

(continued)

Table 4.12 (continued)

Category	Index	Suggestions on the scope of demand indicators
	Availability	• Low: < 90% • Medium: 90–95% • Medium high 95–99.9% • High: > 99.9% • Very high: > 99.999% • Best effort
	DL end-to-end packet reliability	• Very low: < 10% • Low: < 1% • Medium: < 10–3 • High: < 10–6 • Very high: < 10–8 • Best effort
	UL end-to-end packet reliability	• Very low: < 10% • Low: < 1% • Medium: < 10–3 • High: < 10–6 • Very high: < 10–8 • Best effort
Positioning and perception	Positioning accuracy	• Fixed: no positioning required • Loose: 3–30 m • Medium: 1–3 m • Strict: 0.1–1 m • Very strict: 1 mm–10 cm
	Range resolution	Loose: 50–200 m Medium: 10–50 m Strict: 0.1–10 m Very strict: 1 mm–10 cm
	Object induction accuracy	Loose: < 10% Strict: < 1%
Connectivity	Survival time	Very low: < 0.1 ms Low: 0.1–1 ms Medium: 1–50 ms High: 50–100 ms Very high: > 100 ms
	Connection type	BAN, PAN, LAN, WAN, Multiple
	Side chain connection	Necessary, possibly beneficial, not applicable
	Criticality	Non critical, safety critical, mission critical
	Priority Services (NS/EP)	Yes/No

(continued)

Table 4.12 (continued)

Category	Index	Suggestions on the scope of demand indicators
	Connection density (pcs/km2)	Low: < one thousand Middle: 1000–10,000 High: ≥ ten thousand Very high: ≥ 1 M Superelevation: ≥ 10 M Variable
Communication	Communication direction	One way, two way
	Common communication mode	Unicast, multicast, broadcast
	Data reporting method	Time driven, query driven, event driven, hybrid driven
Service	Edge computing service	Required, not applicable
	Artificial intelligence/machine learning services	Required, not applicable
	Adaptability	Network based, device based, both, no API required
Terminal/equipment	Equipment life	Short: 2–4 years Medium: 4–8 years Long: more than 8 years
	Device power limit	Unlimited Restricted: < 2 W Restricted: < 500 mW Very limited: < 50 mW
	Number of subscriptions supported by the device	1 or > 1
	Equipment configuration and customization	No need, need to deploy before, need to deploy after

References

1. Liu Guangyi, Wang Yingying, Wang Ailing. Progress and Future Prospects of 6G [J]. Radio Communication Technology, 2021, 47(6): 665-678.
2. Liu Guangyi. 6G Vision and Demand Imagination [R]. 2019.
3. LIU G Y, HUANG Y H, LI N, et al. Vision, requirements and network architecture of 6G mobile network beyond 2030 [J]. China Communications, 2020, 17(9): 92-104.
4. Liu Guangyi, Huang Yuhong, Cui Chunfeng, et al. 6G Reshaping the World [M]. Beijing: People's Posts and Telecommunications Publishing House, 2021.
5. China Mobile Research Institute. Vision and Requirements White Paper for 2030+ [R]. 2019.
6. ELSAYED M, EROLKANTARCI M. AI-Enabled Future Wireless Networks: Challenges, Opportunities, and Open Issues [J]. IEEE Vehicular Technology Magazine, 2019, 14(3): 70-77.
7. ZHANG Z, XIAO Y, MA Z, et al. 6G Wireless Networks: Vision, Requirements, Architecture, and Key Technologies [J]. IEEE Vehicular Technology Magazine, 2019, 14: 28-41.
8. China Mobile Research Institute. Vision and Requirements White Paper 2.0 for 2030+ [R]. 2020.
9. China Mobile Research Institute. Technology Trend White Paper for 2030+ [R]. 2020.

10. China Mobile Research Institute. 2030+ Network Architecture Outlook White Paper [R]. 2020.
11. Beijing University of Posts and Telecommunications - China Mobile Research Institute Joint Innovation Centre. Visible Light Communication System for 6G White Paper [R]. 2021.
12. Beijing University of Posts and Telecommunications - China Mobile Research Institute Joint Innovation Centre. AI-based Joint Source Channel Coding White Paper [R]. 2021.
13. BASHARAT S, HASSAN S A, PERVAIZ H, et al. Reconfigurable Intelligent Surfaces: Potentials, Applications, and Challenges for 6G Wireless Networks [J].IEEE Wireless Communications, 2021, 28 (6): 184 - 191.
14. China Mobile Research Institute. 6G Holographic Communication Business Development Trend White Paper (2022) [R]. 2022.
15. China Mobile Research Institute. 6G intelligent and lean Wireless Access Network White Paper (2022) [R]. 2022.
16. China Mobile Research Institute. 6G Service-based RAN White Paper (2022) [R]. 2022.
17. China Mobile Research Institute. 6G Wireless Network Autonomy Based on Digital Twin Network White Paper (2022) [R]. 2022.
18. China Mobile Research Institute. 6G Wireless native AI Architecture and Technology White Paper (2022) [R]. 2022.
19. China Mobile Research Institute. 6G Physical Layer AI Key Technology White Paper (2022) [R]. 2022.
20. China Mobile Research Institute. 6G Information Metamaterial Technology White Paper (2022) [R]. 2022.
21. China Mobile Research Institute. 6G Visible Light Communication Technology White Paper (2022) [R]. 2022.
22. China Mobile Research Institute. Ten Cross-Border Innovation Directions Affecting Future Information Communication Development (2022) [R]. 2022.
23. Huawei Technologies Co., Ltd. 6G: A New Journey in Wireless Communication [R]. 2022.
24. Huawei Technologies Co., Ltd. 6G Terahertz Communication Sensing Integration, Opening New Possibilities for Wireless [EB/OL]. (2022–03). https://www.huawei.com/cn/technology-insights/future-technologies/6g-isac-thz.
25. LI O, HE J, ZENG K, et al. Integrated Sensing and Communication in 6G A Prototype of High Resolution THz Sensing on Portable Device [C]// 2021 Joint European Conference on Networks and Communications & 6G Summit (EuCNC/6G Summit). 2021: 544–549.
26. Huawei Technologies Co., Ltd. Ultra-Low Power High-Speed Short-Range Wireless, Enabling 6G Immersive Experience [EB/OL]. (2022–03). https://www.huawei.com/cn/technology-insights/future-technologies/6g-short-range-communications.
27. Huawei Technologies Co., Ltd. 6G Optical Wireless Communication Sensing Integration, Expanding New Boundaries of Wireless Communication Spectrum [EB/OL]. (2022–03). https://www.huawei.com/cn/technology-insights/future-technologies/6g-isac-ow.
28. CITIC Mobile. Full Coverage Scene Intelligence - 6G Scene, Capability and Technology Engine White Paper (V.2021) [R]. 2021.
29. Fang Min, Duan Xiangyang, Hu Liujun. 6G Technology Challenges, Innovations and Prospects [J]. ZTE Communications, 2020, 26(3): 61–70.
30. Xu Jun, Peng Focai, Xu Jin. 5G NR Channel Coding Research [J]. Post and Telecommunication Design Techniques, 2019, 3: 16–21.
31. VIVO. Digital Life 2030+ [R]. 2020.
32. VIVO. 6G Vision Requirements and Challenges [R]. 2020.
33. OPPO. AI-Cube Empowered 6G Network Architecture [R]. 2021.
34. XIAO H, WANG Z, TIAN W, et al. AI enlightens wireless communication: Analyses, solutions and opportunities on CSI feedback [J]. China Communications, 2021, 18(11): 104–116.
35. LIU W, TIAN W, XIAO H, et al. EVCsiNet: Eigenvector-Based CSI Feedback Under 3GPP Link-Level Channels [J]. IEEE Wireless Communications Letters, 2021, 10(12): 2688-2692.
36. OPPO. Zero Power Consumption Communication [R]. 2022.
37. IMT-2030 Promotion Group. 6G Overall Vision and Potential Key Technologies [R]. 2021.

38. IMT-2030 Promotion Group. 6G Network Architecture Vision and Key Technology Outlook [R]. 2021.
39. IMT-2030 Promotion Group. Communication Perception Integrated Technology Report [R]. 2021.
40. IMT-2030 Promotion Group. Ultra-Large Scale Antenna Technology Research Report [R]. 2021.
41. IMT-2030 Promotion Group. Terahertz Communication Technology Research Report [R]. 2021.
42. IMT-2030 Promotion Group. Wireless AI Technology Research Report [R]. 2021.
43. IMT-2030 Promotion Group. Intelligent Hyper Surface Technology Research Report [R]. 2021.
44. IMT-2030 Promotion Group. 6G Network Security Vision Technology Research Report [R]. 2021.
45. Future Mobile Communication Forum. Homepage [EB/OL]. [2023-1-30]. Future Mobile Communication Forum Official Website (future-forum).
46. Future Mobile Communication Forum. Multi-Perspective Depiction of 6G Blueprint [R]. 2019.
47. Future Mobile Communication Forum. 6G: Gap Analysis and Candidate Enabling Technologies [R]. 2019.
48. Future Mobile Communication Forum. Preliminary Study of Advanced Technologies towards 6G [R]. 2019.
49. Future Mobile Communication Forum. Visible Light Communications A Brighter Future [R]. 2019.
50. Future Mobile Communication Forum. ICDT Integrated 6G Network [R]. 2019.
51. Future Mobile Communication Forum. 6G Overall White Paper [R]. 2019.
52. Future Mobile Communication Forum. 6G Intelligent Rail Transit White Paper [R]. 2019.
53. Future Mobile Communication Forum. Terminal Friendly 6G Technology [R]. 2019.
54. Future Mobile Communication Forum. Digital Twin Technology for 6G [R]. 2019.
55. Future Mobile Communication Forum. Zero Power Consumption Communication [R]. 2019.
56. 6GANA. Home [EB/OL] [2023-1-30]. https://www.6g-ana.com.
57. 6GANA. Network AI Concept Terminology White Paper [R]. 2022.
58. 6GANA. 6G Network Native AI Technology Requirement White Paper [R]. 2022.
59. 6GANA. Ten Questions on 6G Network Inherent AI Network Architecture [R]. 2022.
60. 6GANA. 6G Data Service Concept and Requirement White Paper [R]. 2022.
61. 6GANA. B5G/6G Intelligent Data Analysis [R]. 2022.
62. 6GANA. White Paper on Knowledge Definition, Arrangement and Control [R]. 2022.
63. University of Oulu. 6Genesis Flagship Program[EB/OL].(2019-1-17) [2023-1-30]. 6G Flagship official website.
64. 6G Mobile Communications Event. A first for Finland and the World [EB/OL].(2019-3-24).[2023-1-30]. 6G Summit official website.
65. 6G Flagship. Key drivers and research challenges for 6G ubiquitous wireless intelligence [R]. 2019.
66. 6G Mobile Communications Event. Home [EB/OL].(2020-3-17).[2023-1-30]. 6G Summit official website.
67. 6G Flagship. White Paper on 6G Drivers and the UN SDGs [R]. 2020.
68. 6G Flagship. White paper on business of 6G [R]. 2020.
69. 6G Flagship. 6G white paper on validation and trials for verticals towards 2030's [R]. 2020.
70. 6G Flagship. 6G white paper on connectivity for remote areas [R]. 2020.
71. 6G Flagship. White paper on 6G networking [R]. 2020.
72. 6G Flagship. White paper on machine learning in 6G wireless communication networks [R]. 2020.
73. 6G Flagship. 6G white paper on edge intelligence [R]. 2020.
74. 6G Flagship. 6G white paper: research challenges for trust, security and privacy [R]. 2020.
75. 6G Flagship. White paper on broadband connectivity in 6G [R]. 2020.

76. 6G Flagship. White paper on critical and massive machine type communication towards 6G [R]. 2020.
77. 6G Flagship. 6G white paper on localization and sensing [R]. 2020.
78. 6G Flagship. White paper on RF enabling 6G: opportunities and challenges from technology to spectrum [R]. 2021.
79. NOKIA Bell Labs. Communications in 6G Era [R]. 2020.
80. NOKIA Bell Labs. Toward a 6G AI-Native Air Interface [R]. 2021.
81. NOKIA Bell Labs. Extreme massive MIMO for macro cell capacity boost in 5G-Advanced and 6G [R]. 2021.
82. NOKIA Bell Labs. Joint design of communication and sensing for Beyond 5G and 6G systems [R]. 2021.
83. NOKIA Bell Labs. Extreme communications in 6G: Vision and challenges for 'in-X' subnetworks [R]. 2021.
84. NOKIA Bell Labs. Technology innovations for 6G system architecture [R]. 2022.
85. NOKIA Bell Labs. Security and Trust in the 6G era [R]. 2021.
86. Ericsson. Cognitive Networks [R]. 2021.
87. Ericsson. 5 network trends towards the 6G era [R]. 2021.
88. Ericsson. 6G-Connecting a cyber-physical world [R]. 2022.
89. Hexa-X. Vision [EB/OL]. [2023-1-30]. Hexa-X official website.
90. Hexa-X. Objectives [EB/OL]. [2023-1-30]. Hexa-X official website.
91. Hexa-X. Targets and requirements for 6G - initial E2E architecture [R]. 2022.
92. Hexa-X. AI-driven communication & computation co-design: Gap analysis and blueprint [R]. 2021.
93. Hexa-X. Special-purpose functionalities: intermediate solutions [R]. 2022.
94. Hexa-X. Design of service management and orchestration functionalities [R]. 2022.
95. Hexa-X. Gaps, features and enablers for B5G/6G service management and orchestration [R]. 2021.
96. Hexa-X. Towards Tbps Communications in 6G: Use cases and Gap Analysis [R]. 2021.
97. Hexa-X. Expanded 6G vision, use cases and societal values – including aspects of sustainability, security and spectrum [R]. 2022.
98. Hexa-X. 6G Vision, use cases and key societal values [R]. 2021.
99. Samsung Research. The Next Hyper – Connected Experience for All [R]. 2020.
100. Samsung Research. 6G Spectrum – Expanding the Frontier [R]. 2022.
101. NTT DOCOMO. 5G Evolution and 6G [R]. 2020.
102. NTT DOCOMO. 5G Evolution and 6G V4.0 [R]. 2022.
103. B5GPC. Beyond 5G White Paper–Message to the 2030s [R]. 2022.
104. Next G Alliance. 6G applications and use cases [R]. 2022.
105. Next G Alliance. Road to 6G [R]. 2022.

Chapter 5
How to Define 6G

As 6G research gradually becomes a hot topic, the industry's views on 6G are diverse, such as some people believe that 6G is terahertz, 6G is satellite communication, 6G is AI, 6G is semantic communication, etc. This chapter revolves around how to define 6G, introducing 6G from the perspectives of vision, requirements, frequency, wireless key technology, network architecture, etc., hoping to give readers a systematic and comprehensive understanding.

5.1 Overview of Industry Views on 6G

Mobile communication has experienced evolution and development from 1 to 5G. In different eras, mobile communication technology and system design have achieved different breakthroughs, from the earliest mobile voice communication to the Internet of Things in the 5G era. So, facing 6G, what should be the iconic features of mobile communication? The academic and industrial sectors have put forward many distinctive views. These views are reasonable from different angles, but they all have a strong one-sidedness, which has caused great confusion to the industry and even led to misguidance. Let's take a look at these typical views.

5.1.1 First Representative View—6G is Terahertz

The first representative view is "6G is terahertz", and the representatives who hold this view are mostly experts and professors from universities. Looking at the development history of the mobile communication system, the transmission rate is a very iconic indicator. Each generation of mobile communication systems regards the improvement of the transmission data rate as a very important indicator. The iconic peak

G. Liu et al., *6G Migrates from Communication to XaaS*,
https://doi.org/10.1007/978-981-96-7579-1_5

data rate of 4G is 100 Mbps, while 5G is 10 Gbps (millimetre wave band), which has increased by 2 orders of magnitude, so the peak data rate of 6G should naturally increase by 2 orders of magnitude compared to 5G to reach Tbps level.

So how to achieve a transmission data rate at the Tbps level? Looking at the system design and configuration of 5G, the increase in peak data rate mainly comes from larger bandwidth, more antenna ports, and higher modulation orders. 5G has adopted more antenna ports on the base station and terminal side, such as 192 antennas and 64 digital channels of the sub-6 GHz base station, and 2 transmit antennas and 4 reception antennas of the terminal; although the millimetre wave has adopted a larger antenna scale, such as thousands of antenna units on the millimetre wave base station side and dozens of antenna units on the terminal side, the number of parallel data streams that a single user can support is very limited, usually 1–2.

In terms of bandwidth, 5G can support a maximum single carrier bandwidth of 100 MHz in the sub-6 GHz, and supports a maximum of 400 MHz single carrier bandwidth in the millimetre wave. In terms of modulation order, 4G supports 64 QAM (Quadrature Amplitude Modulation), while 5G has risen to 1024 QAM.

So how can 6G further increase the peak data rate? Looking at the several aspects analysed above, due to the limitations of the cellular communication system in terms of base station and terminal size, cost, power consumption, etc., especially the limitations of terminal implementation, it is difficult to make a big breakthrough in the number of antennas and digital channels, and the modulation order is also limited by the interference of the communication environment and the impact of the device on the Error Vector Magnitude (EVM), so the greatest expectation is larger bandwidth. 5G is gradually expanding the frequency range that can be supported from hundreds of megahertz to 100 GHz in the standard, and the frequency for larger continuous bandwidth can only be found in higher frequency bands. The frequency band above 100 GHz has become the direction to find larger continuous bandwidth, so the terahertz frequency band has become the potential frequency band. The terahertz frequency band is generally considered to be above 300 GHz to 10 THz, or even relaxed to above 100 GHz. The terahertz frequency band contains a large amount of undeveloped spectrum, which can bring potential ultra-large capacity and ultra-high speed, so terahertz technology has become a 6G technology direction that is highly anticipated by the academic community.

However, as shown in Fig. 5.1, the terahertz frequency band is far higher than the existing mobile communication frequency band. Due to the greater attenuation caused by electromagnetic propagation distance and rain and fog, its propagation distance is relatively limited, similar to the millimetre wave of 5G, and it needs to adopt specific designs to compensate for the shortcomings of coverage, such as larger antenna array scales. In addition, the frequency of the terahertz frequency band is higher, and the process and materials of its signal generator devices are between electronics and optics. The process and level of its devices are still very immature, especially from the current level, its effective transmission power, device noise level, etc. are all worse than millimetre waves, and the power efficiency is even lower. Although from the perspective of the existing terahertz technology demonstration system, some researchers are considering using the terahertz frequency band

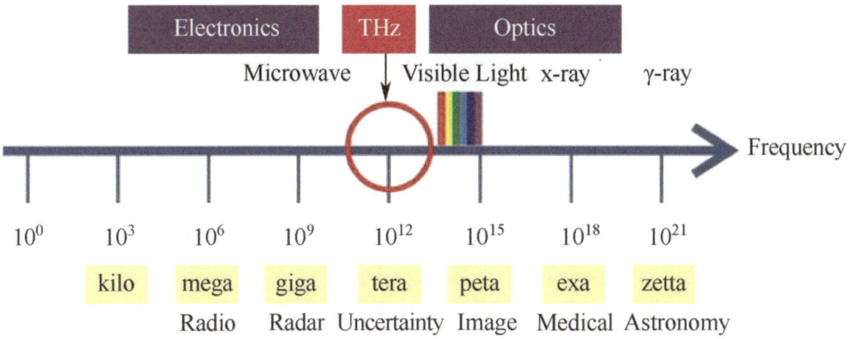

Fig. 5.1 Terahertz frequency band distribution

for wide-area transmission, but considering the cost and power consumption of the system, the terahertz frequency band is most suitable for short-distance line-of-sight transmission scenarios, providing ultra-high capacity and ultra-high speed. In addition, because the terahertz frequency band has high frequency, short wavelength, and wide bandwidth, its application in sensing and detection can bring higher accuracy, making it a high-quality candidate frequency band for the integration of sensing and communication. In summary, the terahertz frequency band is mainly suitable for short-distance, high-capacity and high-speed line-of-sight transmission scenarios, and it is difficult to use for continuous coverage of wide-area systems.

5.1.2 *The Second Representative View—6G Is Satellite Communication*

Many scholars believe that 5G and 4G have already provided very good terristrial network coverage, and there are no services that require higher transmission data rates, so the transmission data rates of 4G and 5G are already sufficient. 6G only needs to solve a larger range of coverage on the existing 5G basis by using satellite communication, to achieve truely seamless coverage. This view has a very heavy influence in the academic world, especially after the launch of Musk's Starlink plan, low-orbit satellite communication is maturing, and the costs of its satellite design and manufacturing, and launching are decreasing year by year, and high-throughput satellite technology has also appeared.

Therefore, satellite communication and networking have once again become the focus of attention of governments and scholars around the world, especially from the political and military perspectives of a country. Judging from the current application of the Starlink plan, it still needs ground stations to convert satellite signals into Wi-Fi signals, and it cannot directly serve mobile phone-like terminals. Compared with the existing ground mobile communication methods, its application cost is still high, and the conditions for ground coverage and deployment are not flexible enough.

In addition, the coverage method of the satellite communication network still faces great challenges, such as the limited uplink budget, which makes it impossible to directly send mobile phone signals to the satellites, and it cannot directly provide indoor coverage. The antenna must be placed in a position outside the building that can directly receive satellite signals, so the satellite communication system must cooperate with the ground cellular system to truly guarantee the seamless coverage experience of users.

In addition, the relative closure of the satellite ecosystem, the long industrialisation cycle, the chaos of the system, the limited orbit and the long application cycle are also the main reasons for the slow development of satellite communication industry. In summary, facing large-scale civilian communication scenarios, the typical characteristics of satellite communication determine that satellite communication can only be an important supplement to the terristrial system, and cannot become the protagonist.

5.1.3 The Third Representative Viewpoint—6G Is AI

In recent years, the development of Moore's Law has led to a rapid increase in computing power. High complexity calculations can be realised with the support of dedicated computing chips such as Graphics Processing Units (GPUs), enabling AI to move towards large-scale applications. Companies researching AI algorithms and applications are springing up like mushrooms after the rain, with thousands of AI companies in Beijing alone, and AI is being pursued by many businesses. On the other hand, due to the bottleneck encountered in the research and development of traditional mobile communication technology, it is difficult to make significant breakthroughs in the short term. Cross-border integration has become the direction of transformation, and many scholars who originally studied mobile communication technology have turned to AI and big data research, trying to find a new way in the integration of communication and AI technology, hoping to apply existing AI algorithms and models to solve some problems faced by communication networks. The communication community has also quickly set off a wave of AI research, with AI being used for wireless resource management, channel prediction and feedback, network planning and optimisation, network management and maintenance, etc., becoming a hot topic in the industry. Some scholars have even proposed an ideal goal: based on the propagation environment characteristics of the transmitter and receiver, a neural network can be trained to replace all existing communication processing procedures, to achieve the best combination of modulation, coding and MIMO processing procedures, as well as the most efficient communication transmission.

However, in actual communication systems, there are very few AI algorithms and models that can be maturely applied, and the vast majority of application cases belong to the inference and prediction of non-real-time complex problems, such as fault cause analysis in network operation and maintenance automation, business volume prediction, etc. The widespread application of AI in communication systems still needs further research and verification in combination with the characteristics of the scene, the scale of data, the sensitivity of power consumption and complexity, processing delay, etc.

Looking towards 6G, the potential value of AI has been recognised by the industry. In terms of the architectural design and capability expansion of 6G networks, AI as a Service has basically become a consensus in the industry, that is, through "Native AI" design, AI is made into a basic capability and service of 6G networks, not only serving internal application scenarios of the network (AI for Network, AI4Net), but also as a capability open to third parties or directly serving third-party customers (Network for AI, Net4AI). It is expected that AI capabilities will further expand the boundaries of 6G networks together with big data, computing, sensing and security capabilities, better serving a richer and more colourful new application scenarios, supporting the realisation of the social development vision of "digital twins, ubiquitous intelligence". In summary, AI will be an important capability and feature of 6G, its introduction will bring new thoughts and inspirations to the design of the entire 6G system, solve many problems, but may also bring out new problems, all of which need further research by the academic and industrial communities.

5.1.4 The Fourth Representative Viewpoint—6G Is Semantic Communication

Past communication systems have faithfully served the transmission of data bits. With the advancement of AI and big data technology, it is possible to extract and compress the semantics of the data to be transmitted, only needing to transmit a small amount of semantics. The receiving end, based on the same background and context information and the same model, can recover the received semantic information, achieving the same information transmission effect. This transmission method can significantly compress the amount of data to be transmitted, resulting in a several-fold increase in transmission efficiency. Therefore, semantic communication has become one of the hot topics of research in the academic community, and its basic principle diagram is shown in Fig. 5.2.

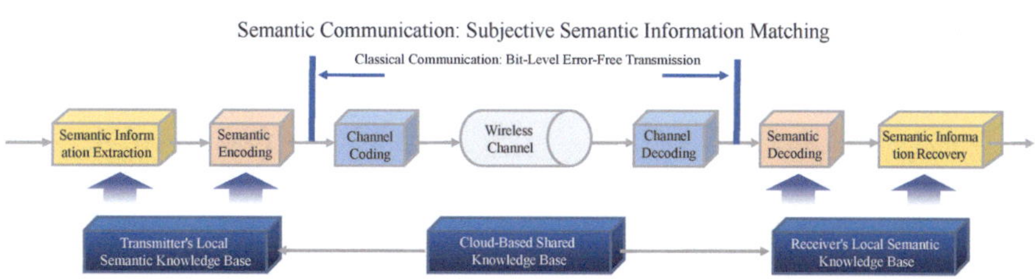

Fig. 5.2 Principle diagram of semantic communication [1]

At the sending side, the data generated by the source is first sent to the semantic information extractor for semantic encoding, i.e., generating a semantic representation sequence, then sent to the semantic encoder for encoding and compression of semantic features, and then sent to the channel encoder to generate a channel encoding sequence, which is sent into the transmission channel. At the receiving side, the opposite operation is performed, the channel output signal is first sent to the channel decoder, the output decoding sequence is then sent to the semantic decoder, the resulting semantic representation sequence is sent to the semantic information restorer, and finally, the source data is obtained. At present, semantic communication has made good progress in video transmission, especially for video conference scenarios. Considering that the conference environment is basically static, it is only necessary to recognise and encode some dynamic changes of the face to ensure the effect of the video, so semantic communication can achieve good data rate compression. From some reported simulation and demonstration prototype results, it can reduce the data transmission rate by 90%.

Therefore, semantic communication is a brand new communication architecture. By integrating user's demand for information and semantics into the communication process, it is expected to significantly improve communication efficiency and improve the quality of user experience (Quality of Experience, QoE). Although semantic communication has great potential, its development still faces the following challenges: difficulty in sharing knowledge, difficulty in perceiving and recognising context, difficulty in meeting resource requirements, and contradictions between model collaboration and privacy security [2]. The current research on semantic information theory is not perfect, and the application scenarios of semantic communication need to be further explored. Its large-scale application in mobile communication systems is expected to have a long way to go.

The above views all touch on some aspect of 6G, but they cannot yet depict the full picture of 6G. The basic features of 6G are systematically introduced below.

5.2 6G Vision and Requirements

With the large-scale deployment of 5G networks, 5G is accelerating the application and development of AI, big data, cloud computing, etc., jointly promoting the digital transformation and upgrade of society. It can be foreseen that with the deep development of 5G applications, the world after 2030 will inevitably be a digital world. So, what is the next stage of digitalisation? The author believes it is digital twin, and the vision of the digital twin world is shown in Fig. 5.3.

Digital twins can not only play a role in the industrial field, but also have great potential in communication, smart city operation, home life, monitoring and management of human body functions and organ activities, and so on, thereby realising the digital twin of the entire world. Through the ubiquitous sensing ability of the mobile communication network, real-time data of each physical entity can be obtained, and a digital image of each physical entity can be formed through big data modelling. All

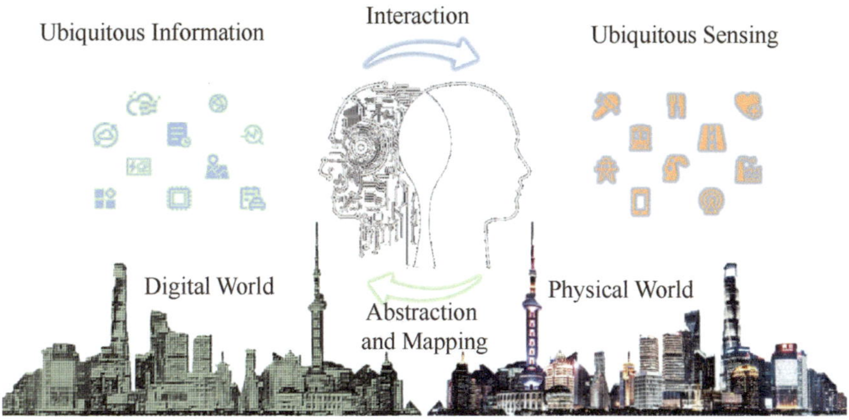

Fig. 5.3 Vision of the digital twin world

these digital images constitute a virtual digital world. After establishing necessary interaction mechanisms and means between the digital world and the physical world, a digital twin world can be formed [3–5]. Based on real-time and historical information from each object in the physical world, relevant background knowledge, etc., the digital world can not only perceive real-time changes in the physical world, but also predict future development trends and directions of the physical world, generate intervention measures and mechanisms for certain events and trends in advance, and carry out feasibility verification and tuning in the digital world in advance. After ensuring that it can achieve the expected effect, it is applied to the physical world, so that the trend of the physical world can meet expectations, avoid accidents and failures, maintain the healthy and sustainable development and operation of the entire physical world, and thus achieve "predict the future, change the future".

In the context of the digital twin world, data is everywhere, and the application of AI will become more convenient and ubiquitous, so another feature of society after 2030 is "ubiquitous intelligence".

In summary, looking towards 2030, the vision for the development of the entire society is "Digital Twins, Ubiquitous Intelligence", which will also give birth to many new application scenarios, such as holographic interaction, sensory interconnection, digital twin humans, intelligent interaction, super transportation, immersive internet (metaverse), smart industry, smart agriculture, etc. As shown in Fig. 5.4, the application of AI will fully empower smart life, smart production and smart society. The design of 6G networks needs to fully consider the needs of these new application scenarios and provide the necessary capabilities to support them. The following will detail these application scenarios.

The first new application scenario—Holographic Interaction. Holography will completely change our forms of communication and interaction, upgrade immersive experiences, and bring about a revolution in ways of life, entertainment, and work. Applications of holography can be seen in many technology demonstrations and

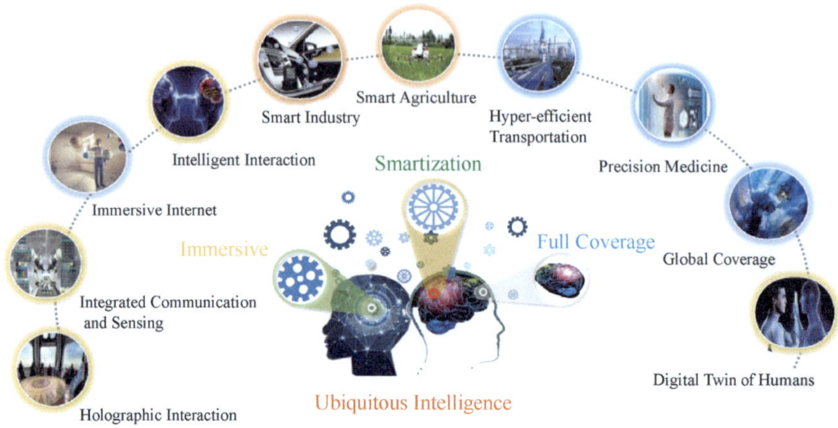

Fig. 5.4 6G typical application scenarios [3]

science fiction films. Perhaps in future meetings, the speaker will be a holographic projection, and the audience may also appear in the form of holographic projections. Looking towards 2030, holographic display technology will be integrated into many application scenarios, such as communication, meetings, remote medical care, office design, military and entertainment games, etc., completely changing people's living habits and bringing high-quality life experiences and working environments. In addition, holographic interaction can also achieve interaction between the projected content and the user. Just like the scene depicted in the movie "Iron Man", you only need to swipe your hand in the air to switch the display content, and you can also design and assemble digital mechanical devices and test their performance. Holographic interaction makes the way of information dissemination no longer a fixed mode, it greatly narrows the distance between the dissemination of information and the user, and establishes an information dissemination method centred on satisfying human experience. The demand for 6G networks for holographic interaction to achieve the ideal effect is very demanding, its demand indicators (see Fig. 5.5): peak rate reaches Tbps level, user experience rate is greater than 50 Gbps, user plane delay is less than 10 ms, in addition, it also needs strong computing power to support real-time scene rendering and processing.

The second application scenario—Sensory Interconnection. For individual consumers, the design goal of the existing communication system is to achieve the transmission and interaction of vision and hearing, as well as the interaction of data bits. With the advancement of technology, the transmission of more senses becomes possible, such as touch, smell, taste, and even emotions and consciousness, to achieve the interconnection of the five senses, we call it sensory interconnection. With sensory interconnection, the collaboration between humans and machines, machines and machines, and virtual socialising are no longer a dream, we can transmit a hug, we can improve the efficiency and effect of skill learning through the sensory interconnection between coaches and athletes, and even bring about a revolution in

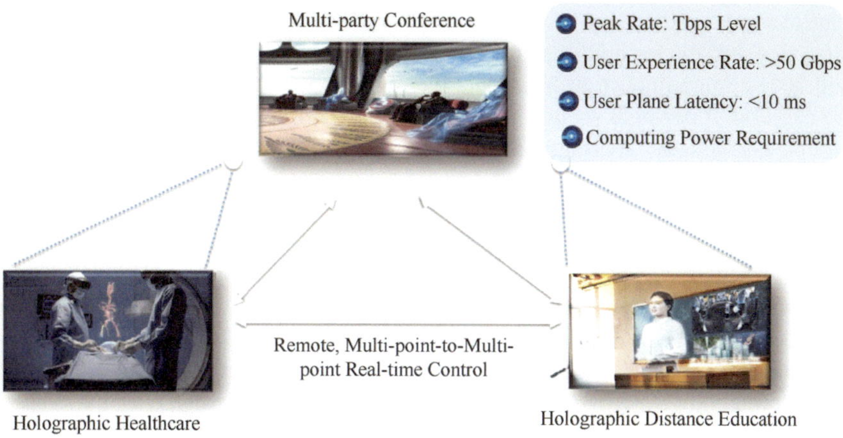

Fig. 5.5 Holographic interaction and its demand indicators

learning. The specific demand indicators of sensory interconnection for 6G networks (see Fig. 5.6): user plane delay is less than 1 ms, reliability is greater than 99.99999%, positioning accuracy reaches cm level.

The third new application scenario—Digital Twin Humans. China is a populous country where the lack of per capita medical resources and uneven distribution of medical resources have led to the social pain point of difficulty in accessing medical care. Particularly in recent years, due to high social pressure and a fast-paced lifestyle, the number of people in sub-health conditions is constantly increasing. The onset of

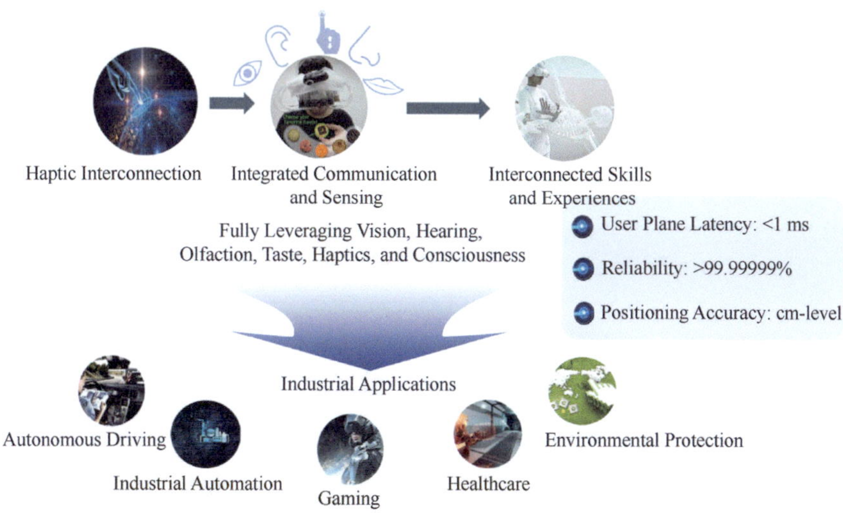

Fig. 5.6 Specific demand indicators of sensory interconnection for 6G networks

cardiovascular and cerebrovascular diseases (such as cerebral infarction, myocardial infarction, etc.) is becoming increasingly younger. Once these diseases occur, they will have a huge impact on individual quality of life, family happiness index, and the exertion of personal social value, and will also bring huge losses to enterprises, groups, and society. Therefore, the prevention and treatment of cardiovascular and cerebrovascular diseases are of great significance to human society. Through wearable or implantable sensing devices, various information about the human body can be comprehensively collected. Based on existing medical knowledge and experience, digital modelling can be carried out on local organs or even the entire function and operation of the human body, constructing a digital human body, also known as a digital twin. Through digital twin technology, the digital twin can simulate and emulate local organs of the human body, the body's circulatory system, etc., predict possible pathological changes in the human body, locate the cause of the disease through the digital twin, form means of early intervention (treatment measures), and based on the digital twin, verify and optimise the intervention measures to ensure that they can achieve the expected effect, then apply them to the physical body, thereby avoiding the occurrence of diseases (such as cerebral infarction, myocardial infarction, etc.), and achieving the effect of "treating diseases before they occur". We call this application a digital twin human. Of course, digital twin humans have many application scenarios, such as organ research, digital life, virus mechanisms, and surgical assistance, etc. It will play a very important auxiliary role in improving the quality of human life and effectively solve the difficulty of accessing medical care faced by today's society. The specific demand indicators of digital twin humans for 6G networks (see Fig. 5.7): user plane delay less than 0.1ms, reliability higher than 99.99999%, positioning accuracy reaching cm level, connection density greater than 10 per/m^2.

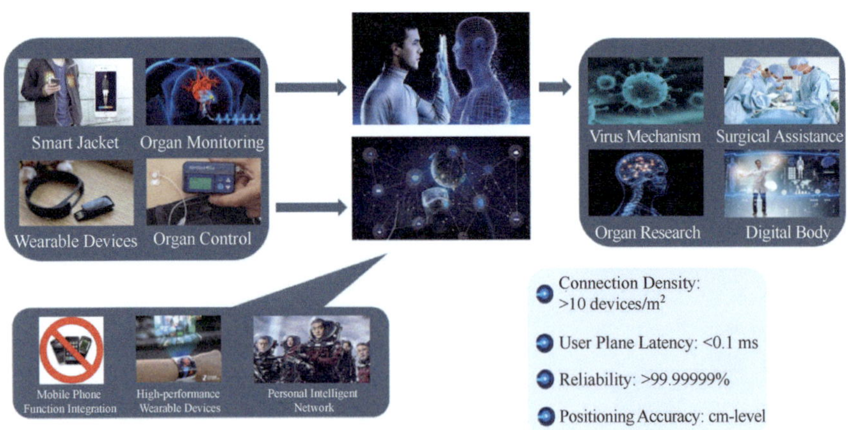

Fig. 5.7 Specific demand indicators of digital twin humans for 6G networks

Fig. 5.8 Intelligent interaction technology

The fourth application scenario—Intelligent Interaction. As shown in Fig. 5.8, intelligent interaction technology will become more intelligent, especially human–machine interaction will become more contextual and personalised, bringing a deeper level of humanistic care experience, especially in emotional care for the disabled, mentally challenged, patients, children and the elderly. On the other hand, intelligence as the refinement of skills and experience can directly interact between people, people and machines, and machines and machines, greatly improving the efficiency of learning and collaboration, and even bringing about a revolution in learning. The specific demand indicators for 6G networks: connection density greater than 10 per/m^2, user plane delay less than 0.1ms, reliability higher than 99.99999%, positioning accuracy less than 1 m.

China is a populous country, and as the process of urbanisation continues to advance, traffic congestion has become a common problem and social issue in large cities. People spend an average of 2–3 h a day on transportation, greatly reducing work efficiency and quality of life. One important reason for traffic congestion is the single mode of transportation. Future transportation could be an integrated air-space-ground transportation system. In addition to traditional ground and underground transportation tools, there are also automotive cars, flying cars, delivery drones, smart balance cars, smart bicycles, water surface and underwater transportation vehicles, etc. People can customise their travel methods according to their location, traffic conditions, destination, etc., to maximise the satisfaction of personalised and efficient travel needs. For this, mobile communication networks must continuously upgrade their capabilities to support the realisation of this scenario, including three-dimensional coverage, precise positioning, high reliability and security, high business connection density, etc. The specific demand indicators of super transportation for 6G networks (see Fig. 5.9): mobility greater than 1000 km/h, reliability higher than 99.99999%, positioning accuracy less than 1 m, capacity density greater than 1 Gbps/m^2.

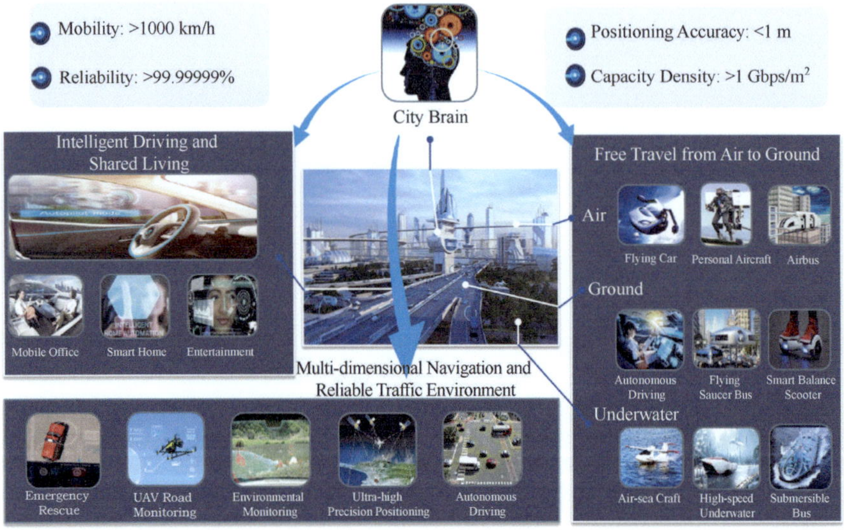

Fig. 5.9 The fifth application scenario—super energy transportation

The sixth new application scenario—immersive internet, also known as the metaverse. The development of 5G will inevitably accelerate the development of immersive services, bringing an upgrade to the user entertainment experience. This experience will gradually become the basic form of our daily life, work and entertainment, thereby accelerating the evolution of the traditional internet to the immersive internet. Since 2021, with Facebook announcing its metaverse development vision, the metaverse has quickly become a hot topic in the industry, receiving a lot of attention from capital and research institutions, and also attracting widespread attention from government departments at all levels. However, since the metaverse does not have a precise definition and the imagination space is huge, everyone has their own metaverse in their minds. Some are oriented towards entertainment and games, some are oriented towards work and life, but they all have immersive business experiences in common, so we can collectively refer to them as the immersive internet. The specific demand indicators of the immersive internet for 6G networks (see Fig. 5.10): traffic density greater than 10 Gbps/m^2, powerful computing power, trustworthy network, and precise positioning, etc., depending on the specific scenario. Figure 5.9 Super transportation demand indicators for 6G networks.

Looking at the above new application scenarios, the future business and applications will undergo significant changes, presenting many new features: firstly, the demand will become more diversified and fragmented, and the dynamic range of network capability requirements will be much larger, including data rate, delay, reliability, etc.; secondly, the coverage will be three-dimensional, the network will not only consider terrestrial coverage, but also need to consider three-dimensional space coverage, and need to consider satellites as a supplementary coverage means; thirdly, the form and content of interaction will become more diverse, no longer just a simple

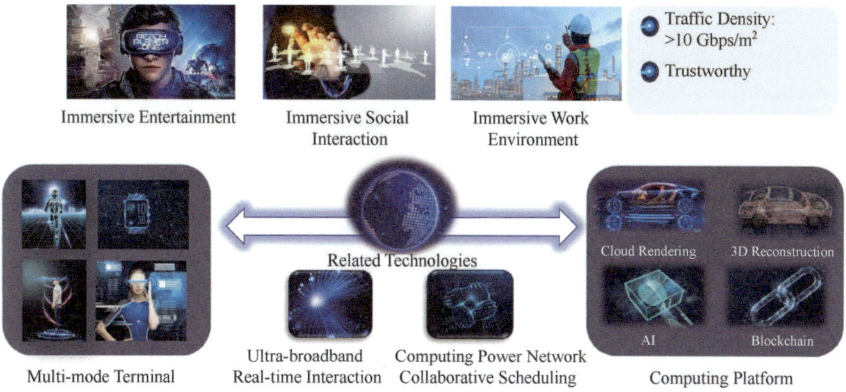

Fig. 5.10 Specific performance indicators of the immersive internet for 6G networks

human–machine interface or simple communication content; fourthly, the business will be open and customised, the pursuit of personalisation will be an important manifestation of human self-liberation and self-realisation of value, personalised business customisation will bring more business models to the industry, open business can make everyone a business provider, not just passively accept the services of operators and internet service providers; fifthly, the deep integration of communication, sensing, computing, AI, big data and security, etc., the future network will not only provide communication connection capability, but also include computing capability, AI capability, big data capability, security capability and perception capability, etc., to build a more comprehensive capability system, thereby expanding a broader 6G application space.

Based on the detailed analysis of the above application scenarios, the demand for network capabilities can be deduced, forming the overall KPI requirements for the 6G network, as shown in Fig. 5.11. Of course, similar to the scenario and requirements indicator definition of 5G, the final 6G network key technology indicator requirements need to be defined separately for several typical scenarios. Although the industry has not yet reached a consensus on this, higher technical indicator requirements are inevitable. The industry needs to make breakthroughs in aspects such as spectrum, wireless transmission technology, network architecture, network functions, security, and technology platforms to achieve a qualitative leap in the 6G network technology system and truly empower the social development vision of "digital twins, ubiquitous intelligence".

From Fig. 5.11, it can be seen that in order to further enhance communication capability and efficiency, traditional communication indicators need to be further improved than 5G, such as the 6G vision requires network sensing capability to reach the level of human tactile perception, which puts forward higher requirements for the reliability and air interface delay of mobile communication networks. In order to support the sustainable development of mobile communication networks and society,

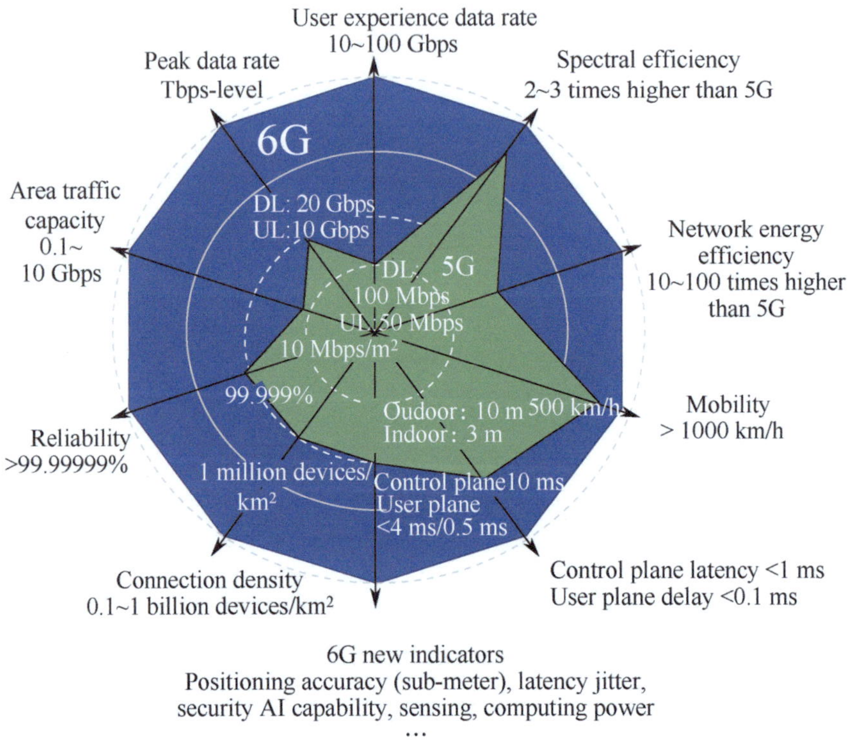

Fig. 5.11 Overall KPI requirements of the 6G network [4]

spectrum efficiency, energy efficiency, and cost efficiency are also indicators that 6G needs to focus on.

In addition, the scenario-based KPI definition of 6G will be an important feature of the requirements indicator definition, but the values of the network technology indicator system under each 6G scenario have not yet reached a consensus in the industry, and need to be further enriched and perfected in combination with the development of applications and technologies.

Furthermore, facing the development vision of "digital twins, ubiquitous intelligence", many new application scenarios have put forward higher and more requirements for network capabilities, as shown in Fig. 5.11, in addition to traditional communication functions, more precise positioning, posture perception, computing capability, AI capability, security capability, etc. are also needed, and how to define these new capabilities is a problem that the industry urgently needs to solve.

At present, China Mobile has taken the lead in completing the research report on the 6G vision and driving force [6] and the research report on 6G use cases [7]. ITU-R WP5D has also published the 6G vision report, and will start the research and formulation of the 6G technology demand report in 2024, which is expected to be completed in 2026.

5.3 6G Frequency

Frequency is the basic resource of mobile communication and is also the first factor to consider in the design of a new generation of mobile communication systems. With the large-scale application and popularisation of 5G, while bringing a further upgrade of personal business experience, it will also cultivate new businesses and applications, promoting the continuous growth of mobile communication business volume; in addition, with the digital transformation and upgrade of the whole society, the number and types of IoT connections will continue to rise, and the demand for IoT communication will also continue to upgrade, constantly expanding the connotation of connections. Therefore, towards 2030, the trend of continuous growth in mobile communication business volume is very obvious. Although the development of 5G is still in the growth stage and the characteristic business has not yet matured, the exact numbers are still difficult to predict, but combining the development experience of past mobile communication systems, a 100-fold increase in business volume is obvious. In addition, from the demand of future business and application scenarios, 6G needs to provide higher quality coverage, higher security and reliability, as well as higher energy efficiency ratio and lower bit cost, high-quality spectrum resources are particularly important, especially in the low-frequency band. Therefore, the development of the 6G network requires the support of high-quality frequencies, including the quantity and quality of new frequencies.

In terms of frequency quality, considering that 6G has higher requirements for the sustainable development indicators of the network's data rate, coverage, cost, and energy saving, high, medium, and low frequency bands are all needed to meet the needs of differentiated scenarios and the upgrade requirements of business experience. The low frequency band provides basic wide-area seamless coverage and indoor deep coverage, including the provision of basic services, the medium frequency band is mainly used to provide continuous coverage in urban areas, and the high frequency band is deployed on demand according to the scenario. In addition, considering that the allocation of medium and low frequencies is relatively scattered and fragmented, it brings great challenges to the network deployment of operators, including the cost efficiency of the sky surface and radio frequency units, the scale and robustness of the industry, and the complexity and cost of terminals.

Multi-frequency band deployment also makes the setting of network interoperability parameters complex, increasing the difficulty and complexity of comprehensive network operation and maintenance, and also affecting the user experience. Therefore, the medium and low frequencies of 6G should be as much as possible a globally unified planning of continuous large bandwidth spectrum, to ensure the global scale effect and low cost of 6G deployment. At the same time, regulatory agencies also need to study how to reduce the fragmentation of frequency allocation, and the reintegration and large bandwidth allocation of expired frequencies, to improve the cost efficiency and value of frequency use, the use efficiency of towers and machine rooms, reduce operator expenses, and help the entire industry to achieve

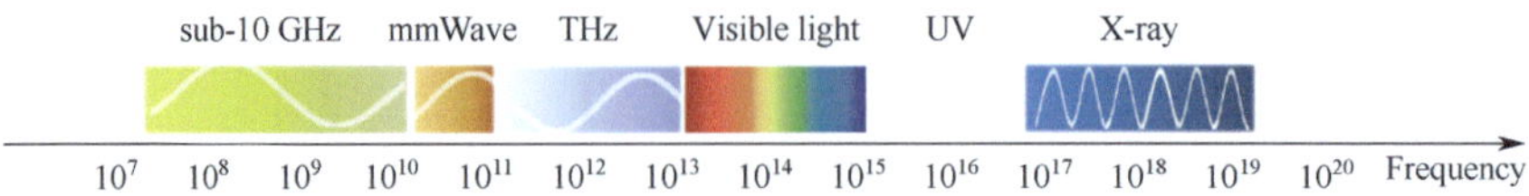

Fig. 5.12 Schematic of frequency bands considered by 6G

user experience upgrades in a simpler and more efficient way. The frequency bands considered by 6G are shown in Fig. 5.12.

In terms of quantity, considering that the improvement of spectrum efficiency becomes more and more difficult in the 6G era, the increase of site will also lead to linear growth of network cost and power consumption, so the 100-fold growth of business volume will mainly rely on the increase of frequency quantity to meet. According to ITU-R's prediction of 5G frequency demand [8], 802–1090 MHz is needed below 6 GHz, and 15–20 GHz frequency is needed in the millimeter wave band, but there is still a big gap from the 5G frequency planned by various countries at present. At the same time, according to the joint research report of GSMA and the authoritative analysis agency Coleago Consulting in the spectrum field [9], in order to meet the future business needs during 2025–2030, each country needs to add 2 GHz medium frequency spectrum, among which 3.5, 4.9, 6 GHz are the most critical frequency bands, as shown in Table 5.1. If the above medium frequency bands are lacking, it will consume 2–3 times the energy, and the network construction cost will also increase by about 4 times.

From Table 5.1, it can be seen that the United States, Japan, and South Korea will mainly focus on the C-band 3.1–4.2 GHz band in the future, intending to connect the high and low sections of the entire C-band, obtaining 1100 MHz continuous spectrum. But for China and Europe, due to satellite occupation, partial issuance, etc., the total amount of spectrum that can be obtained in the C-band is far less than that of the US, Japan, and South Korea, so it needs to rely on the 700 MHz provided by the 6 GHz band (6425–7125 MHz).

Therefore, looking forward to 2030–2040, regulatory agencies in various countries still need to plan enough medium and low frequency resources for the development of 6G in advance, especially the continuous large bandwidth above 500 MHz, to achieve a quantum leap in user experience compared to 5G systems, while considering the cost of initial network deployment. In addition, in order to achieve Tbps-level transmission rates, continuous large bandwidth above tens of GHz in the medium frequency band will also be indispensable, so millimeter waves, terahertz, and visible light bands are also important candidate bands.

Based on the above analysis, the frequency deployment strategy for future 6G can be basically predicted as follows: As an initial deployment of 6G, each 6G network needs at least a continuous 500 MHz of large bandwidth below 10 GHz, through the increase in bandwidth and the number of antennas to achieve a basic communication capability relative to 5G networks can have an order of magnitude increase, can achieve continuous deployment in urban areas; while millimetre wave, terahertz and visible light frequencies, are according to the development of business to achieve

Table 5.1 2025–2030 Reserve of medium and low frequency resources in various regions/countries (italic ones represent potential unissued)

Country	Total (macro network)	Released (macro network)	Sub 2 GHz	2.3 GHz	2.6 GHz	3.1–3.4 GHz	3.4–3.8 GHz	3.8–4.2 GHz	4.5–5.0 GHz	U6 GHz
China	**1685**	**985**	465	90 (indoors)	160	100 (indoors)	200	0	160	*700*
Europe	**1840**	**1140**	550	0	190	0	400	400 (Local)	0	*700*
Japan	**1740**	**1450**	490	100	160	待定	400	300	*490 (Local 300)*	*100*
Korea	**1620**	**1000**	450	80	190	*300*	300 *(20)*	*300*	*100 (Local100)*	*0*
United States	**1770**	**1250**	500	30	190	*300*	400 (Local 50)	180 + *220*	0	0

on-demand deployment and dynamic switching, providing high capacity and ultra-high data rate, while taking into account the network's low energy consumption needs. Especially in China, considering that the 6 GHz (5925–7125 MHz) band has a potential 1.2 GHz frequency available for mobile communication services, it can be assumed as the first frequency of 6G network deployment, considering the network sharing of operators, each network can guarantee about 500 MHz of continuous large bandwidth, which will be very beneficial to the user experience upgrade and rapid deployment of 6G.

5.4 6G Network Architecture Outlook

Combining the driving forces of 6G network evolution in Chap. 3 of this book, it can be seen that the requirements for future 6G network architecture comes from three aspects: one is to solve the problems and challenges faced by existing 5G networks, such as high energy consumption, high cost and complicated operation and maintenance, rigid network architecture and equipment formfactor, long duration for network function update, etc.; the second is to further expand the capability of mobile network, support more diverse capability dimensions, such as communication, sensing, computing, AI, big data and security, to meet the more diverse and varied needs of 6G application scenarios; the third is to further enhance the end-to-end business adaptability of the mobile network, to meet the more fragmented and differentiated deployment needs from new application scenario.

Therefore, taking into account the experience and lessons of 5G development, and future development needs, 6G network should support the following typical characteristics: on-demand service providing, lean (simplicity), flexibility, native AI, native security and digital twin, as shown in Fig. 5.13.

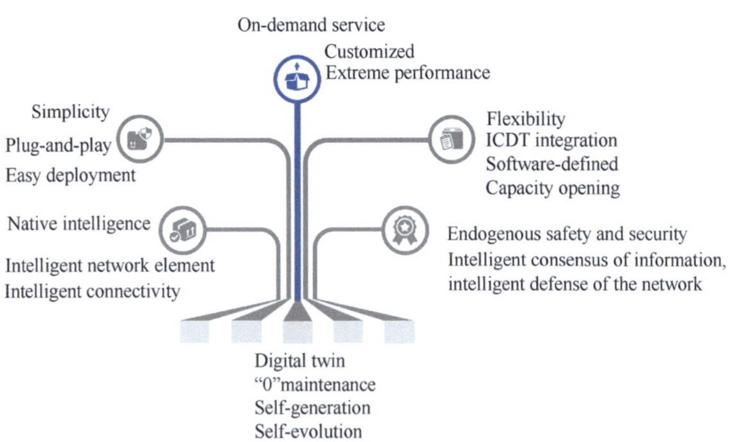

Fig. 5.13 Typical features of 6G network

The following focuses on the typical features of the 6G network.

5.4.1 On-Demand Service

The digital twin world of 2030 will give birth to a large number of new applications and scenarios. Therefore, the 6G network must have better perception capabilities of user behaviour and user intent, and be able to deploy functions, configure parameters and allocate resources according to user needs, as shown in Fig. 5.14. In addition, the 6G network should provide services with smaller granularity, and users can freely combine service types and service levels according to their own needs.

The first prerequisite for achieving on-demand service providing is real-time perception of application and business needs. This can be achieved by collaborating among the application and service layer, data plane, and AI plane to predict user's business and application needs in advance. When user needs change, the network provides seamless switching of service modes and service content on demand for users. The second prerequisite is on-demand configuration of resources. The distributed resource layer will serve as a unified infrastructure platform for the 6G network, providing physical resources such as wireless, computing, and storage. Taking into the application and business needs with collected network status and resource status, the 6G network can achieve on-demand resource allocation. The third prerequisite is on-demand orchestration of network functions, which targets for on-demand supply of multi-dimensional capabilities and end-to-end QoS assurance.

In the future, the customers will require more diverse and personalised requirements for RAN, Transport Network (TN) and Core Network (CN). The network needs to percept the user's need, interpret as user needs of RAN, TN and CN, orchestrate the most appropriate function and resource combination and launch the instance of the network deployment.

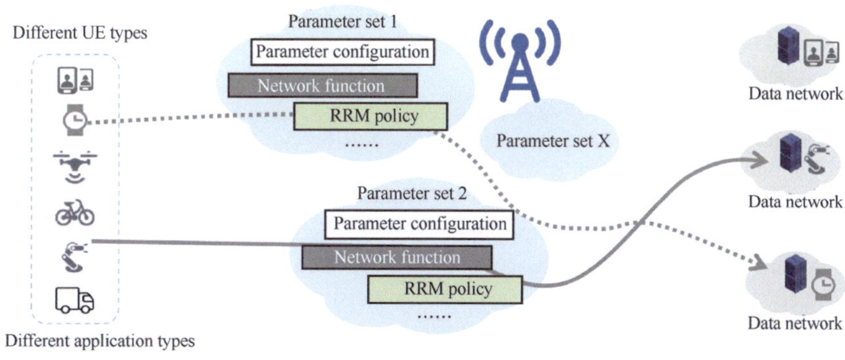

Fig. 5.14 Example of on-demand service providing network [10]

To support on-demand service providing, 6G networks need to provide ultimate network capabilities. In addition to traditional communication, more comprehensive network capability is needed. The design of the network architecture needs to support the integrated communication, sensing, computing, AI, big data, and security, etc., to meet the more differentiated and fragmented requirements in the future. The following content of this section will focus on the integration of communication and sensing, computing, and big data, while the integration of communication and AI, security will be detailed in subsequent chapters.

1. ***Integration of Communication and Sensing***

From the typical application scenarios of 6G, it can be seen that sensing capability has become an important requirement for many application scenarios, such as robot collaboration, vehicle positioning, drone navigation and supervision, recognition of human gestures and postures, etc., all require high-precision sensing capabilities. In addition, for many applications in the human body domain, various sensing capabilities are also needed, such as blood pressure, heartbeat, pulse, etc.

For traditional applications, the sensing of location and targets is mostly achieved through external systems or components, such as the current mobile phone navigation is achieved through the integration of GPS or Beidou receiving modules. Vehicle anti-collision or autonomous driving uses millimetre-wave radar, laser radar or even video recognition for sensing, most of these methods require independent sensing modules, and have high requirements for device size, power consumption, cost, etc., greatly limiting the application of sensing capability in many scenarios.

For this reason, existing mobile communication systems have begun to consider the integration of communication and sensing (Sensing as a Service). 5G system has begun to design mechanisms for precise positioning based on base station reference information, such as Positioning Reference Signal (PRS), etc., which can already achieve metre-level positioning accuracy (indoor and outdoor distinction accuracy). However, for the more complex sensing capabilities required in 6G application scenarios, such as shape sensing, as well as imaging, obstacles, etc., the current communication system does not yet have, mainly relying on dedicated millimetre-wave and visible light radar to support, can achieve short-distance precise positioning and situation sensing.

Therefore, if we can directly use the hardware and signals of the communication system to provide richer sensing capabilities, and replace the dedicated radar with the communication system, it will certainly empower more industrial and civilian fields related to location and sensing capabilities, and thus expand the value and market space of the mobile communication industry. In this regard, the academic community has started research on the integration of communication and perception since 2011 [11], hoping to use the echo of the transmission signal of the base station or terminal to achieve radar sensing capabilities, and call it integrated sensing and communication (ISAC). Therefore, the future base station or terminal will no longer be a simple base station or terminal, but can be a radar.

The ISAC aims to achieve the synergy of sensing and communication functions through integrated design (spectrum resource sharing, integrated air interface, integrated hardware, etc.), multi-point cooperation and intelligent information interaction, effectively improving system spectrum efficiency and hardware resource utilization, and has great application value and implementation significance [12, 13] Early communication and radar systems were independently studied due to different business needs. Driven by various new application requirements and technological developments, the wireless communication spectrum is evolving towards higher frequency bands such as millimetre waves, terahertz and even visible light that support much larger bandwidths. The boundary between the two is gradually blurring, and more system-level similarities are gradually emerging. In terms of working frequency bands, high frequency bands and large bandwidths can support higher resolution and faster sensing capabilities. The working frequency bands of communication and radar are constantly expanding, gradually overlapping, and achieving communication and sensing functions in the same spectrum, which enhances spectrum utilisation, is the preferred path for technological and industrial development. In terms of system architecture, communication and radar systems have similarities in baseband signal design and RF parts, and are expected to achieve integrated design of shared baseband signals and shared RF. Research on integrated design of communication and sensing will be detailed in subsequent chapters.

ISAC for 6G will be a kind of ubiquitous collaborative sensing [14]. Building wide-area high-precision sensing capabilities based on mobile communication networks can deeply empower existing life, production, and social development, thereby promoting the rapid development of the digital economy, and has great social and commercial value. ISAC has rich application scenarios, as shown in Fig. 5.15.

2. ***Integration of communication and computing***

There is a complex relationship between communication and computing. On the one hand, communication and computing are two independent disciplines, each

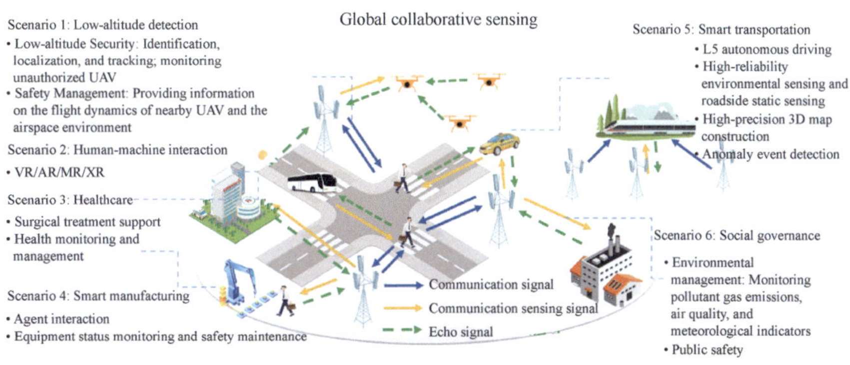

Fig. 5.15 Application scenarios of communication and perception integration

focusing on completely different content. Communication focuses on how to efficiently transmit data and information, while computing focuses on how to perform fast and efficient information processing. On the other hand, the process of communication greatly depends on computing, because the process of communication itself is the processing of signals and information, especially as computing power continues to increase, promoting the application and popularisation of more advanced communication technologies, and in turn promoting the rapid development of computing.

With the advent of the Internet, computing is not only limited to traditional scientific research, but has begun to enter ordinary households. Through computer software, many complex calculations can also be completed; at this time, the role of wired communication networks is to connect different computers and serve as transparent pipes for information transmission between each other. With the development of smartphones and mobile internet, numerous internet applications have generated massive amounts of data, driving the development of cloud computing and AI. People have begun to perform complex big data processing through the cloud, and use AI applications to mine the value of data and serve their own business operations and customer services, etc., and have achieved remarkable results. For example, the customer big data mining in Alibaba and JD's e-commerce platforms provides a powerful tool for precise marketing for merchants and platforms. Therefore, for a time, cloud computing has become a hot frontier technology, and Amazon, Google, Alibaba, Tencent, and Huawei have all launched their own cloud computing platforms, leasing them to government agencies and small and medium-sized enterprises. For considerations of data security and privacy, IT personnel of large enterprises have built their own private cloud platforms, for private data storage and processing, supporting internal applications.

In the face of the vigorous development of cloud computing and big data, traditional telecom operators have also seen business opportunities, hoping to get a share of the pie, and hope to bring more user-friendly cloud computing and cloud storage services with their own network connection advantages. Therefore, starting from the 4G era, ETSI began to formulate edge computing (Mobile Edge Computing, MEC, later renamed as Multi-access Edge). The technical standard of computing aims to provide operators with a tool to increase business revenue. As shown in Fig. 5.16, by moving cloud computing from outside the mobile communication network to inside, bringing it closer to users, there is the possibility of providing better services to users.

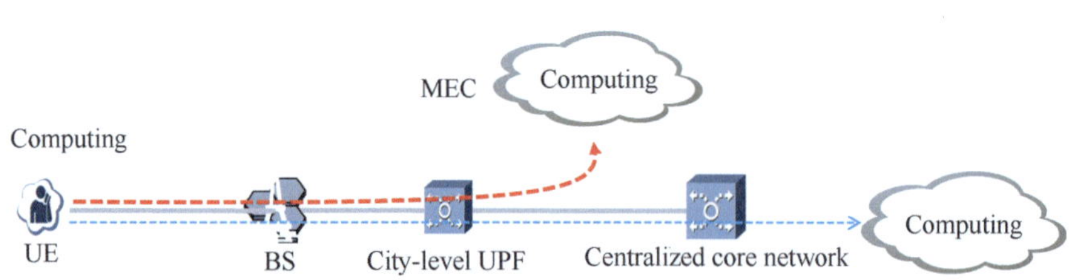

Fig. 5.16 From external computing to 5G's add-on mobile edge computing

(1) Deploy business and content as close to the user as possible to minimise the latency of business access.
(2) Lower the routing function to as close to the user as possible, achieve fast routing and local exchange of user data, shorten data interaction latency, while avoiding exposing sensitive data on the Internet, and protect data security and user privacy as much as possible.
(3) Deploy computing capabilities close to the user, thereby transferring the user's computing to the cloud, achieving fast interaction of data and processing results, and at the same time simplifying the implementation of the terminal, reducing its size, weight, power consumption and cost. For example, for AR/VR applications, if the content processing and rendering functions are moved to mobile edge computing, it can greatly reduce the threshold for AR/VR device development, while also greatly reducing costs and improving quality, making the device lighter and easier to popularise.
(4) Lower the user plane function (User Plane Function, UPF) of the core network to MEC, support necessary billing, security and other functions, can provide high isolation of user data, and protect user data privacy.
(5) Through standard APIs, the capabilities of wireless networks can be opened up, such as location positioning, etc., exposing network capabilities to third parties, and thereby cultivating new services and new business models.

Whether it is from the traditional cloud computing deployment form or MEC, computing is just an add-on to the communication network, and communication and computing are still independently designed and optimised, with communication still playing the role of a dumn pipeline. In practice, MEC has been doing business demonstrations and demonstrations since the 4G era, but due to the lack of effective business models, and no obvious advantages compared to public cloud, it has not been widely used. Large companies use their own private clouds, small companies use public clouds, and operators' MECs are hard to attract attention.

With the large-scale deployment of 5G, operators have begun to deploy 5G base stations with centralized RAN (C-RAN) in large scale to save costs. The large-scale concentration of base stations lays the foundation for further cloudification of base stations. Through base station cloudification, the different base stations can share unified cloud virtual resources, achieve dynamic scheduling of software and hardware resources, better cope with the tidal effect of traffic brought by user mobility, and help operators further reduce network deployment, operation and maintenance costs. At the same time, with cloud computing resources, base stations can also provide computing services, making computing another basic service and capability of the network, that is, computing as a service, achieving the integration of communication and computing. As shown in Fig. 5.17, for different computing applications, the type of hardware chosen may be different, so the future base station access cloud will be build on various heterogeneous hardware, such as CPU, GPU, ASIC, field programmable gate arrays (FPGA), etc., which can be dynamically orchestrated and scheduled according to different services and applications.

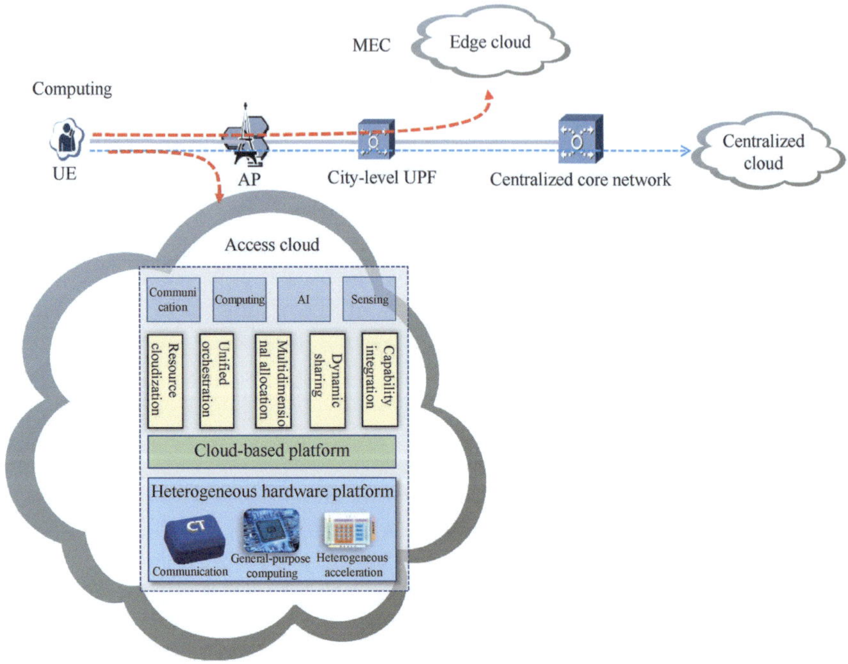

Fig. 5.17 From edge cloud to access cloud, further integration of communication and computing

Of course, simply sharing physical processing resources together cannot be completely called integration. It is also necessary to extend the traditional communication protocol and QoS dimensions, create an end-to-end computing service protocol system similar to the communication protocol, as shown in Fig. 5.18, to ensure that the end-to-end QoS of computing services is effectively guaranteed like the QoS of communication services.

3. ***Integration of Communication and Big Data***

5G is accelerating the realisation of connected everything, driving the digital transformation and upgrade of the entire society. The application of big data and AI has begun to play an increasingly important role in various internet applications, production management, and social governance. By 2030, human society will enter a digital twin world where the virtual digital world and the physical world are bridged. Data will be ubiquitous and will become a new asset and wealth in the virtual digital world. People can trade data and use data provided by others through services. The generation, management, and use of data present trends such as "ubiquitous data, collaborative data, knowledgeable data, and assetised data" [15]. Data is considered the "new oil" of the digital society and a new driving force for social progress. However, how to "mine" and "refine" and exploit and realise its value is a challenge faced by 6G networks. Especially in the current context of enhanced awareness of data security and privacy protection, and increasingly perfect laws and regulations,

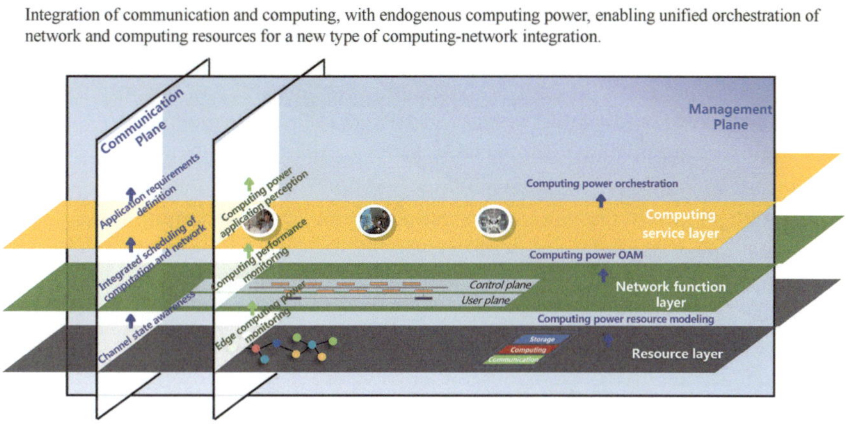

Fig. 5.18 Protocol design for integrated communication and computing

how to achieve trustworthy data management under controllable conditions requires corresponding architectural design in line with the development trend of data itself. A direct question is, can data be designed and supported as a service (Data as a Service, DaaS) in the network? The answer is yes.

In addition, mobile communication networks are both producers and providers of data, providing trustworthy data services for various intelligent applications, and at the same time, they are consumers of network data, improving network performance and operational efficiency with data-driven intelligent applications. Within the mobile communication network, millions of base stations and other network equipment and hundreds of millions of users also generate massive amounts of data in their daily life. Compared with 5G network data mainly based on communication network operation data and user contract data, the scope and type of 6G data will be more abundant as 6G services extend from communication to sensing, computing, and AI services. Especially with the further integration of 6G communication and sensing, data on the network's own status, surrounding environment, and user/device behaviour will add more precise location information, which can not only better serve AI-based experience enhancement, network management, maintenance and optimisation, but also serve a wider range of social governance applications. Therefore, 6G data shows more massive, polymorphic, temporal, and related characteristics than 5G data.

Combining the above two aspects of analysis, for the design of the 6G network architecture, we will break the limitation of the network as an information transmission pipeline, making 6G gradually transform into a data carrying platform, turning data into a capability and service of the network, provided for internal and external users of the network.

Data service is an abstract function between data providers and data consumers, decoupling data consumers and physical data providers. Especially for multiple data providers or multiple data consumers, data services help maintain data integrity and improve data service efficiency through reusability. Data service is based on a data distribution and publishing framework, providing data as a service product to meet customers' real-time data needs across systems, reusable and compliant with enterprise and industrial standards, balancing data sharing and security. 6G data services aim to efficiently support end-to-end data collection, transmission, storage, processing and application, solving how to conveniently, efficiently, and securely provide data to internal or external network functions, reducing the difficulty of data acquisition and improving data service efficiency and data consumption experience under the premise of complying with privacy and security laws and regulations.

By implementing data sharing applications through data service, ensuring "data out of one hole", improving data consistency; by building platform capabilities, providing different forms of data services (data consumers do not need to pay attention to technical details), meeting flexible and diverse service needs; by implementing automated development and testing of data services, real-time deployment capabilities through the data service centre, improving the agile response capability of data services.

Based on the "data plane" architecture that unifies data logic management and operations and separates physical deployment, it is based on the stratification and domain of data, and uses the improvement of data quality as a means to provide data for cloud-edge AI model training through distributed data collaboration at the cloud edge, flexibly and efficiently. It enhances the cognitive ability of data and creates new value through AI algorithms, while providing secure and controllable data capabilities to the outside world with the help of key technologies such as privacy computing, to ensure the self-growth of data services from bottom to top. 6G data services will be characterised by "self-growth", forming efficient and secure data capabilities through distributed collaboration, trustworthy openness and other technologies.

To meet the new requirements of 6G network architecture and service scenarios, we strive to change the current situation in the network architecture where there is only point technology and lack of systematic data services, innovatively design a unified 6G data service framework, to comply with the trend of 6G data management and utilisation, and fully meet the demands for data service compliance and data value realisation. The design of the 6G data service framework should follow the following principles.

(1) *Transformation from point technology to unified system architecture*

Existing communication networks, as the "pipeline" for data transmission, implement data processing, services and privacy protection through point technology. In the 6G era, to systematically solve data service challenges, meet new businesses, and legal and regulatory requirements, it will be necessary to introduce an independent data plane, build a unified and trustworthy data service framework at the architecture level, solve the problem of data islands, provide trustworthy data services while meeting the regulatory requirements of data laws, realise cross-domain and cross-manufacturer data sharing, improve operational efficiency for operators, and realise the value of data through various intelligent applications.

(2) *Transformation from centralised trust to decentralised trust*

The network architecture of the mobile communication system is moving from the traditional layered structure to a flat structure, and the control method is evolving from centralised to distributed, while adapting to the natural distributed attributes of data, the distributed deployment of computing power, and the decentralised trust model becomes inevitable. The development of blockchain technology provides technical support for building trustworthy authentication, authorisation and data access control. The decentralised trust system transforms the anchor of trust from traditional authoritative institutions or third-party endorsements to multilateral consensus, and relies on underlying cryptographic technology to provide secure and trustworthy services for distributed data processing, storage, access control, traceability, etc.

(3) *Transformation from data-driven to data and knowledge dual-driven*

The 6G data service framework needs to adapt to the diversity of terminals, support heterogeneous multi-source data access, collection, processing and storage of massive data. The high value of data is highly expected by enterprises, and the development of technologies such as AI and knowledge graphs provides support for knowledge extraction from massive data. By innovatively mining the associations between data through knowledge graphs combined with AI and other means, new opportunities and values are discovered from diverse and inherently associated data, transforming data into knowledge to realise cognition-based intelligence, enabling the intelligence and diversity of applications. On the one hand, by associating business scenarios with data through knowledge graphs, network AI, etc., it demands standards for data, improving data quality; on the other hand, by gradually precipitating the formed business association knowledge into knowledge graphs, network AI, etc., it further releases new value.

Based on the above considerations, at the protocol design level of the 6G network architecture, we systematically propose to add a data plane on the basis of the traditional communication protocol plane, as shown in Fig. 5.19. The data plane systematically manages and stores data inside and outside the network, while providing QoS guarantees for end-to-end data services, supporting the implementation of DaaS.

On the data plane [10], the involved functional modules are shown in Fig. 5.20. It requires both global data collection, storage and processing, and local data collection and processing. Because in many application scenarios, for the sake of data privacy, customers do not want their data to be uploaded to the public network, the network needs to provide the ability to process and store these data locally, therefore, a data management architecture that combines distributed and centralized is needed. Training data management provides the training samples required for model training, and customizes and preprocesses the training data according to the performance requirements of the model.

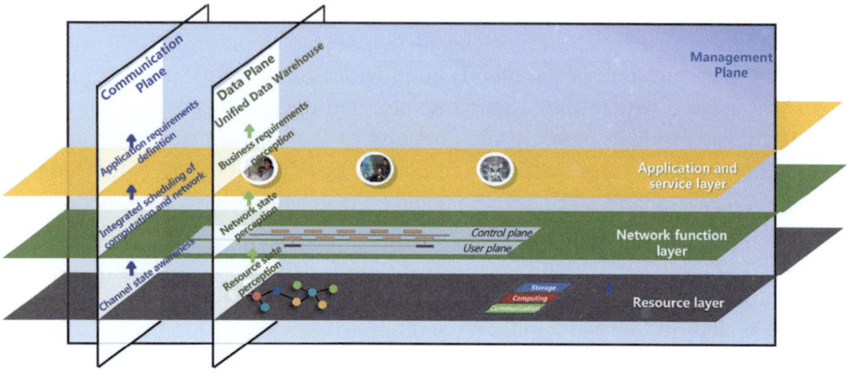

Fig. 5.19 Addition of data plane to 6G network architecture protocol

5.4.2 *Lean Network*

From the experience of large-scale deployment of 5G, overly complex core networks, a large number of base stations, high costs caused by a single base station formfactor, complex operation and maintenance, etc., are all problems that the new generation of mobile communication networks need to solve. Therefore, from the perspective of network deployment and operation, operators hope that 6G networks can be extremely simple, and can achieve simplicity in deployment, architecture, and management.

First, the 6G network will be deployed in a cloud-based manner, achieving a lightweight core network. Looking at the deployment development trend of 5G, the deployment method of C-RAN is becoming mainstream, so the 6G network will also be deployed in this way, while considering the centralized baseband resources to further carry out cloud-native design, to support the agility and elasticity of the entire wireless network, achieve dynamic resource sharing, on-demand function configuration and activation, integration of communication and computing capabilities, forward compatibility, rapid iteration and evolution, on-demand interface data collection and other new features; through the lightweight core network design, further decoupling and separation of the control plane and user plane can be achieved, making functions and services closer to users, better adapting to differentiated and fragmented application scenarios and business needs, as shown in Fig. 5.21.

Second, the deployment of base stations will support plug-and-play and on-demand deployment [16]. With the vigorous development of emerging businesses, the demand for networks continues to increase. In some areas where business is sudden, network coverage that can be expanded on demand is provided through plug-and-play. Plug-and-play can achieve network edge construction and use, demand-driven deployment, and reduce network costs from the perspective of network architecture design.

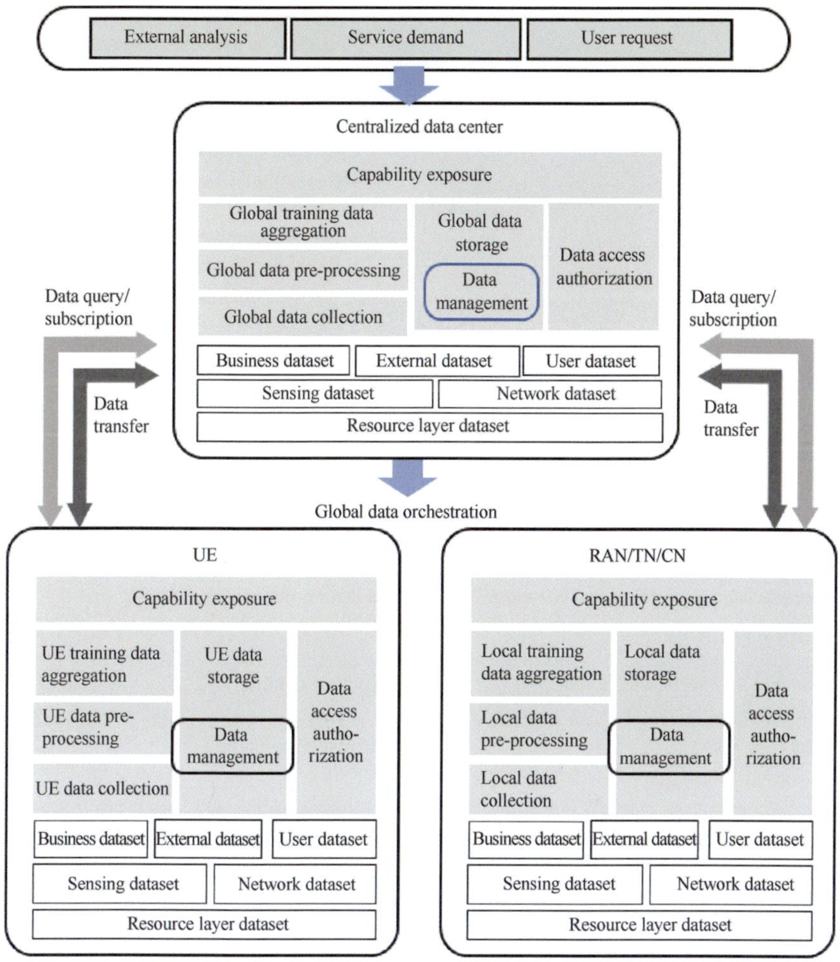

Fig. 5.20 Functional modules of the data plane

To support the above characteristics, wireless network functions can be divided according to the dependence on air interface technology, as shown in Table 5.2.

Combining the characteristics of different network functions, network functions are divided into wireless resource control, high-level data packet processing, air interface adaptation, and physical layer processing. The simple wireless access network architecture adopts a cloud-edge-end collaborative form, and through orchestration management, dynamically allocates wireless, computing, storage and other resources as needed to generate network functions. High-level data packet processing and wireless resource control can be deployed in the cloud, fully utilising the advantages of DOICT convergence, combining intelligent means, analysing network status and

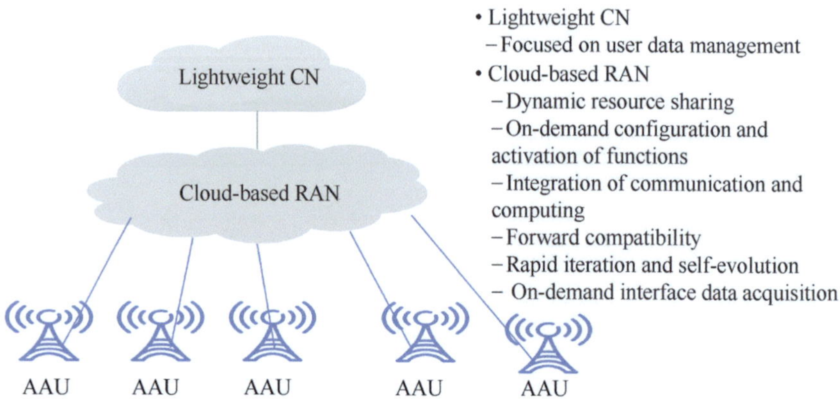

Fig. 5.21 Cloud-native wireless network

Table 5.2 Dependence of wireless network functions on air interface technology

Air interface technology dependence	Network functions
Closely related to air interface technology	Physical layer processing, scheduling and resource allocation in Medium Access Control (MAC)
Related to air interface technology	Radio resource control
Relatively independent from air interface technology	High level packet processing, such as encryption, integrity protection, sorting, etc.

business models, possessing business packet type awareness, and achieving self-optimisation of network control. Based on the analysis results of the data packet type, combined with the needs of the business, choose the appropriate data transmission link. Through air interface adaptation, the network selects the appropriate physical layer processing according to the requirements of high-level data packet processing, achieving network-business adaptation and unification of multiple air interface accesses. Air interface adaptation also includes network management of edge device functions, and self-configuration of edge devices through the cloud, achieving plug-and-play of edge devices. By detecting the use of network resources, network status, and user and business situations, combined with technologies such as AI, network orchestration management predicts network status and possible performance degradation, and updates network orchestration management strategies in advance.

Third, 6G supports the integrated core network design and unified access mechanism design of air, space, ground and sea [17]. For 6G networks, it is necessary to consider three-dimensional coverage of air, space, land, and sea. In traditional methods, different networks need to be designed to support different scenarios, such as satellites, cellular, cable, and underwater, etc. To ensure business continuity, interoperability has been introduced between different networks, but this interoperability

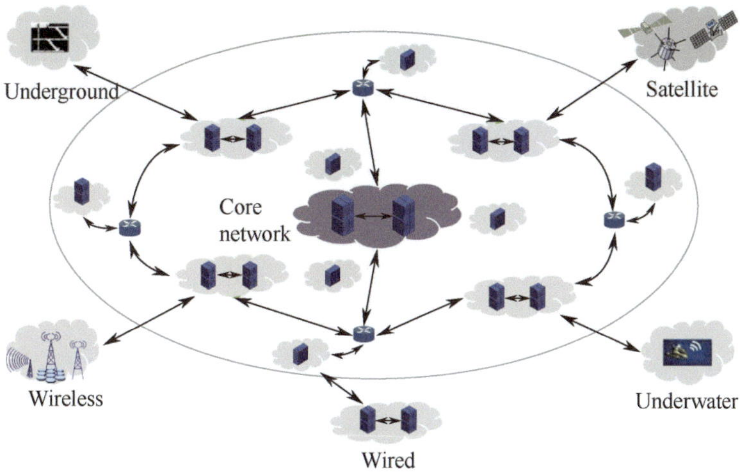

Fig. 5.22 Unified access core network

is very limited. Due to differences in protocol design and access mechanisms between different networks, traditional networks cannot support business continuity with QoS guarantees. When a user leaves the target network, his service may be dropped or degraded. In future 6G networks [17], homogeneous or heterogeneous networks need to be managed uniformly to provide users with a consistent user experience. The unified access core network shown in Fig. 5.22 is an example of future network architecture, where a unified CN is designed to simplify the network architecture. By integrating wireless access technology (RAT) and unified access mechanisms, different RATs can connect to the same CN, achieving seamless switching between different RATs for users.

Fourthly, it supports further decoupling of control and data, achieving wide-area coverage of control base stations and dynamic on-demand access of data base stations [16]. As shown in Fig. 5.23, future networks will be divided into two layers: one is the wide-area control layer, where control base stations can work in low-frequency bands such as 700 MHz, achieving wide-area coverage of control signalling; the other is the data access layer, achieving dynamic on-demand access of data base stations. The data access layer can work in high-frequency bands with large bandwidth, such as 7 GHz, millimetre wave, terahertz, and visible light. On-demand activation of data base stations can significantly reduce the number of concurrent live base stations, thereby greatly reducing network costs and power consumption. Through unified signalling and dynamic data access, reliable mobility management and fast business access, consistent user business experience, low access latency, and low inter-cell interference can be ensured.

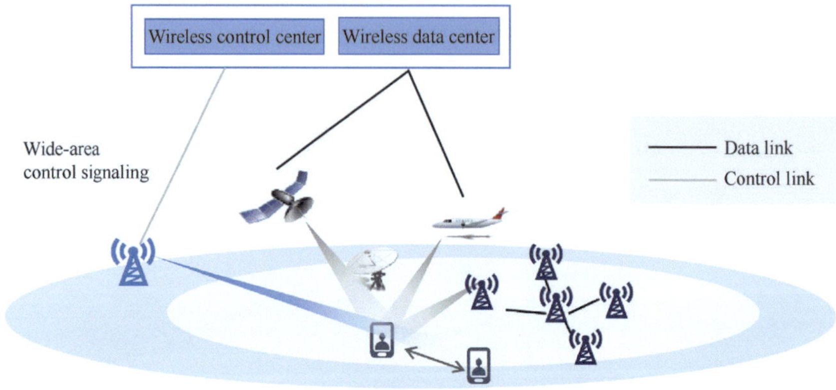

Fig. 5.23 Wide area control and dynamic data access

5.4.3 Soft Network (Flexibility)

Faced with the uncertain new business and new scenario requirements of the future [4, 5], we should focus more on enhancing the network's adaptability to all scenarios. Therefore, the future network will be an end-to-end software-definable and cloud-native network, which will help to achieve rapid business deployment, rapid iteration of functional software versions, dynamic sharing of resources (such as wireless spectrum, computing, and storage), and network automation and intelligence, greatly enhancing the adaptability of 6G networks to more differentiated and fragmented new services and scenarios.

The flexibility of 6G networks is reflected in the following three aspects: firstly, at the network function level, necessary network functions are flexibly combined as needed to provide customised network service capabilities; secondly, at the infrastructure and resource level, appropriate network resources (including computing, storage, spectrum, power, deployment location and other resources) are allocated as needed to maximise network efficiency; thirdly, at the application and service level, on the one hand, service needs are accurately perceived, on the other hand, multi-dimensional intelligent orchestration and management of network functions and network resources are realised to fully adapt to various scenarios. We believe that an end-to-end service-oriented architecture based on cloud-native technology is a necessary technical means to create network adaptability to all scenarios. In order to maximise the adaptability of the network, the research of service-based RAN is of paramount importance in the design of future network architecture [18, 19].

Currently, research on service-based architecture mainly focuses on the control plane of the core network. The research on service-based RAN is still in its early stages. For a long time, base stations have been developed in an integrated monolithic manner to ensure the ultimate performance of the "last mile". Due to performance concerns, academia and industry have a conservative attitude towards service-based RAN. However, we believe that performance concerns should not be a stumbling block on our path to exploring service-based RAN. On the one hand, the performance indicators defined by the current 3GPP/ITU are all for the air interface, but in some scenarios, high reliability and low latency of the air interface are not the most important, and end-to-end performance indicators are what we need to care more about; on the other hand, there is a "game" relationship between system performance indicators, although it seems that service-based RAN may reduce some performance, but its performance loss can be compensated by improving system stability, availability and other capabilities, just like the trade-off relationship between block error rate and rate; in addition, the further development of cloud-native technology and chip capabilities will also reduce these performance losses.

In order to agilely respond to future more diversified service function requirements, Quality of Service (QoS) requirements, management strategy requirements, deployment requirements, and open requirements, and to make the network have stronger forward compatibility, the next-generation wireless access network needs to start from the perspective of service capabilities, and reconstruct the wireless network into a service-based RAN with finer-grained functions [18], to better play the platform advantages of C-RAN, the service-based 6G network is shown in Fig. 5.24.

(1) Through the redefinition of RAN function services, the upgrade of base station function versions will be realised faster, and business function requirements will be met in a timely manner.

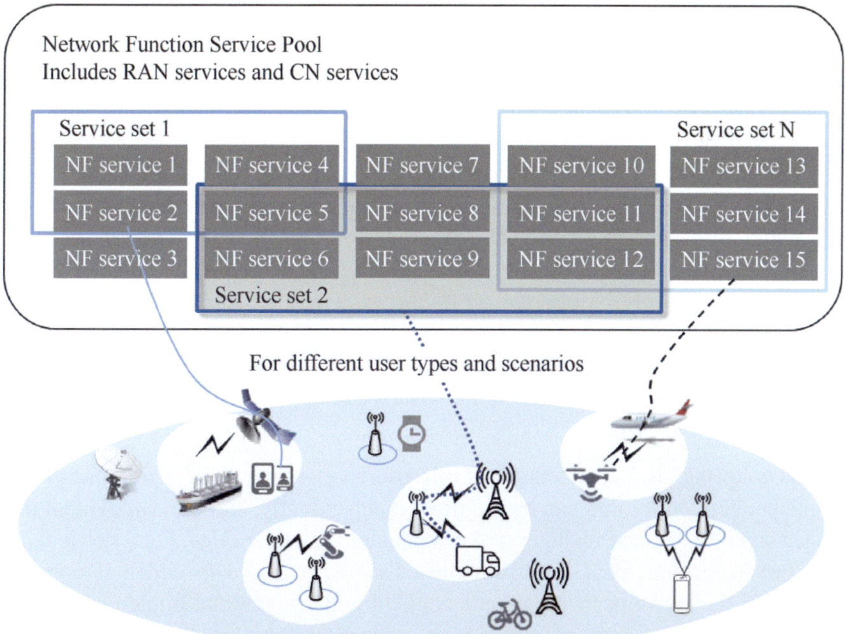

Fig. 5.24 Service-oriented 6G network

(2) Through the definition of service-based interfaces between RAN services and CN services, new interaction methods will be brought to the end-to-end network process, while reducing the impact of the introduction of cross-domain new functions on existing services.
(3) Through the definition of service-based interfaces between RAN services and third-party services, more timely and multi-dimensional opening of wireless network capabilities will be realised.
(4) Through the cloud-native infrastructure platform, hardware resource pooling and sharing will be realised, reducing network deployment costs and overall network power consumption, and agilely responding to business deployment requirements.
(5) Through the integrated orchestration management of RAN services and CN services, the complexity of network-wide operation and maintenance management will be reduced, and the network's adaptability to new businesses will be improved.

Before reconstructing the next-generation wireless network architecture through the microservice method, it is necessary to first determine the basic principles of service-based RAN design to avoid the thorny problem of "distributed monolith" caused by unreasonable service division. The design principles of service-based RAN mainly include the following five aspects [18].

(1) The services provided by RAN need to be defined according to external demands. The network is used to handle business requirements, so the first step in defining the design of service-based RAN is to refine external requirements into key requests. For the access network, requirements can come from core network functions, access network nodes, third-party applications, or user equipment (UE).
(2) The definition of RAN services needs to meet the characteristics of "loose coupling" and "high cohesion" [5, 6]. Robert Martin has a discussion on the single responsibility principle: "Aggregate things that change for the same reason, and separate things that change for different reasons." That is to say, there should only be one reason to change a service, and each responsibility carried by the service is a potential reason for its modification. Microservices should be designed as small, cohesive, and single-responsibility as possible, which will reduce the "size" of the service and improve its stability, but "small" is not the main goal of microservices.

"Loose coupling" is a core feature of microservice architecture. A microservice is an independent entity that can be modified independently, and the deployment of a service should not cause changes to the consumers of that service. For a service, it is necessary to consider what should be exposed and what should be hidden. If too much is exposed, the consumer will become coupled with the internal implementation of the service, which will create additional coordination work between the service and the consumer, thereby reducing the autonomy of the service.

(3) RAN services should strive to maintain the independence of their own data. Ensuring the private attributes of data is one of the prerequisites for achieving loose coupling. If services need to maintain data synchronisation, then changes by the service provider will greatly affect the service consumer.
(4) The goal of service-based RAN is to define RAN services and the relationships between them. Currently, even in the information technology (IT) field where microservices are very mature, there is still no specific algorithm to assist in service splitting or service definition. Based on previous research, we believe that the design of a service-based RAN architecture can be achieved through the following three steps: first, refining external demands into various key requests, i.e., system operations, there are two system operation modes in 5GC, namely request-response and subscribe-notify; second, determining the decomposition of RAN services; third, appropriately allocating system operations to the decomposed RAN services. Therefore, the definition of RAN services and each RAN service API is the core of service-based RAN.
(5) The control plane functions and user plane functions are deeply decoupled to meet the requirements of lightweight, low-cost and flexible deployment of the user plane.

5.4.4 Native AI

In recent years, with the rapid increase in computing power, a variety of AI-specific computing chips have emerged, driving continuous significant progress in AI research and application, especially in machine translation, intelligent control, expert systems, robotics, language and image understanding, genetic programming, robot factories, automatic program design, aerospace applications, massive information processing, storage and management, and the execution of tasks that are either too complex or large-scale for living organisms. This has sparked a new wave of AI enthusiasm worldwide. Looking towards 2030, as the digitisation process of the entire physical world continues to accelerate, big data will be ubiquitous, better supporting the widespread application of AI, and promoting the realisation of the beautiful development vision of "ubiquitous intelligence" in the entire society. Therefore, the design of future 6G network architecture needs to fully consider support for AI applications, that is, to make AI a basic capability and service of the network, available to customers at any time as needed, that is, AI as a Service, to achieve "Network for AI".

Furthermore, as a tool, AI also has the potential to be applied in various fields of mobile communication. AI can effectively help operators improve the operational efficiency of the network, as well as the efficiency and capability of services. Mobile communication networks are typically composed of millions of base stations, routers, core network elements and other infrastructure equipment, as well as hundreds of millions of users. Massive amounts of data are generated in the network, including operational data of various network elements, signalling data generated during communication, event reports, and information related to user mobility within the network. If these data are tagged, it will bring immeasurable value to the automation and intelligence of network operations. Starting from the 5G era, operators have begun to research the intelligence of 5G networks based on big data and AI.

These studies can be divided into application and demand research (including 5G + AI scenarios and demands, 5G application scenarios and demands, intelligent mapping of applications and network resources), intelligent wireless access technology (such as AI-based massive MIMO design, wireless resource scheduling, positioning technology, mobility management, signal detection, channel estimation, etc.), and network intelligence (such as MEC architecture based on federated learning, intelligent network slicing, multi-RAT cooperation, intelligent network planning, intelligent energy saving, network anomaly analysis, user experience analysis and optimisation, etc.) [20].

3GPP began studying big data collection in wireless networks [21], automation and intelligence of network operations [22], and the application of AI in wireless resource scheduling [23] from 2017. The research on AI function architecture in 3GPP mainly focuses on NWDAF introduced at the network function level and MDAF (Management Data Analytics Function) introduced at the management level. NWDAF is a data analysis network element that automatically perceives and analyses the network based on network data, participates in the entire lifecycle of network planning, construction, operation, network optimisation and operation, making the network easy to maintain and operate. 3GPP defined the basic framework of NWDAF in the Release 16 stage, and defined the decomposition of NWDAF functions, data collection optimisation, coordination between NWDAF instances during multi-instance deployment in Release 17, but did not standardize AI datasets or training models. Although the network performance has been effectively improved through NWDAF, this patchwork integration of AI functions has exposed some problems: one is the data security problem and excessive signalling overhead caused by massive measurement reporting; the other is the challenge of low latency, because all data must be uploaded to the central analysis processing unit for processing, i.e., NWDAF, which may be deployed far from the data source. There are similar problems in the design of MDAF at the management level. MDAF analyses management domain data, supports data analysis within the RAN or CN domain or cross-domain data analysis, and supports interfaces and interactions with NWDAF, but its essence is still a central analysis processing unit.

The European Telecommunications Standards Institute (ETSI) defined the ENI (Experiential Network Intelligence) system in 2017 as an independent AI engine providing intelligent services for network operations, network assurance, device management, service orchestration and management, etc. [24]. The ENI system is a system that integrates data analysis functions into the closed-loop management and orchestration process, focusing on defining functions related to network autonomy, internal interfaces of the ENI engine, and external interfaces facing other network functions. For example, it cleans and classifies features of raw data through data processing and normalisation, translates strategies or instructions generated by the ENI system, and outputs in a language that the service object can understand.

The Telecom Management Forum (TM Forum) has also begun research related to AI. In June 2019, the ITU-T SG13 research group of the International Telecommunication Union Telecommunication Standardisation Bureau started research on machine learning use cases [25].

In the same month, the Global Mobile Communications Systems Association (GSMA) began work on the white paper for intelligent autonomous network cases [26].

In summary, current research on the integration of mobile communication systems and AI mainly focuses on using AI technology to enable network automation, enhance the performance of existing communication systems, and improve the user experience of communication services, which we can collectively refer to as "AI for Network". Most of the work in standard organisations is focused on discussing intelligent application scenario cases, the data required for collection, the interfaces between functional entities, and the content of data analysis output.

There is less discussion on the functional architecture of the AI lifecycle workflow in the network and the integration and orchestration with network functions. At present, the intelligent work of mobile communication networks is still in its initial stage, with a lot of work focused on the exploration of requirements and solutions, while mature network intelligent applications are relatively few. At the same time, due to the fact that the design of the 5G network architecture did not consider support for AI, the existing scenario-driven network intelligence is facing many challenges.

Firstly, data acquisition is very difficult and the quality of data is hard to guarantee. Because the previous network architecture and protocol design did not predefine the interface for data collection, the current data collection servers/devices based on implementation, such as deep packet inspection or data probing, cannot provide sufficient data in a timely manner. Data collection based on network management also has problems such as fewer types of data, longer collection cycles (15 min), inconsistent data formats, naming and calculation methods from different manufacturers, and difficulties in opening up southbound network management data. At the same time, due to the instability of data collection within the device, loss of transmission links, limited storage space of network management devices, and difficulty in obtaining labels, the data often has quality problems such as missing, serial, unlabeled or label errors. Before training the AI model, a lot of time and manpower costs need to be spent on data preprocessing.

Secondly, the application effect of the AI model lacks effective verification and guarantee methods. The training and iterative optimisation of the current intelligent model are all completed offline, and the effect of the intelligent model after going online lacks direct verification methods. Most of them indirectly judge through the network performance evaluation indicators related to intelligent applications, which are poor in real-time and lack direct correlation. When the network performance indicator is lower than expected after the model goes online, it can only avoid the negative impact of the AI model through the "fallback" mechanism, which is lagging.

Thirdly, the application of AI in 5G networks is handled on a case-by-case basis, where data collection, algorithm optimisation and processing are patched to the corresponding network elements or external processing units as new network elements added to the network. For different AI use cases, different modifications may need to be made to the network, which brings difficulties to network management and operation.

Finally, the research and development and innovation of AI algorithms for mobile communication network features are still in their infancy, and AI technology and communication technology are loosely coupled in a superimposed fusion mode.

All these factors determine that the performance and efficiency of AI are far below expectations. Therefore, considering the combination of "Network for AI" and "AI for Network", in the design of the 6G network architecture, we need to consider how to make AI capabilities ubiquitous and omnipotent in the network, and like the human brain and neural network, provide AI capabilities as needed in a distributed or centralised manner to serve network management and optimisation. Through the intelligent platform, the 6G network can introduce external AI capabilities into the network, provide new services and capabilities, introduce external data into the network, and further improve data processing efficiency. The analysis data and AI capabilities within the network can also be opened to third parties through intelligent bodies to provide them with various intelligent services they need.

The way to achieve the above goals is the architectural design of native AI. By analysing the user's AI application requirements, the necessary elements such as algorithms/models, computing power and data are formed, and the above resources are obtained through unified orchestration and scheduling. Through the full lifecycle AI QoS management and evaluation, the user's AI application requirements are met, including the internal needs of the network and the needs of external network customers.

Literature [4] proposes a functional deployment architecture for native AI in 6G networks, as shown in Fig. 5.25. Intelligent terminals and intelligent network elements provide edge intelligence and terminal-side intelligence, meeting the ultra-low latency requirements of user AI service requirements. In addition to automatically orchestrating AI services across the network, the intelligent platform also opens the network's AI capabilities to third parties. It can also crowdsource external AI algorithms, models, and computing power for internal use or for third-party services. The cloud network super brain provides training for large-scale, complex AI models and processing and storage of large-scale data, collaborating with intelligent network elements and terminals on the edge to provide on-demand AI services.

In terms of protocol design, in order to make AI a basic capability and service of 6G networks, we also need to introduce corresponding protocol design to achieve end-to-end management and resource orchestration of AI service QoS, ensuring the QoS of the entire lifecycle of AI services. The protocol architecture of native AI in 6G networks is shown in Fig. 5.26. On the basis of the communication, computation, and data planes, we further introduce the intelligence plane. Detailed design still needs further research and improvement.

The native AI system includes functional entities such as AI model management, training data management, AI model evaluation, AI model optimisation, data analysis, and knowledge base management. AI model management includes the selection, generation, storage, update, transmission, and deletion of models. The models

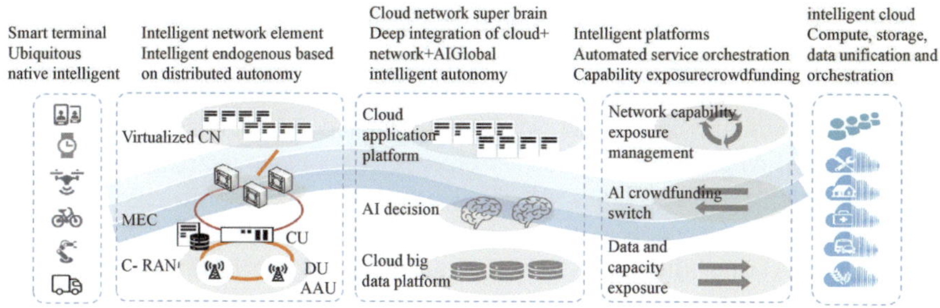

Fig. 5.25 Functional deployment architecture of endogenous AI in 6G networks

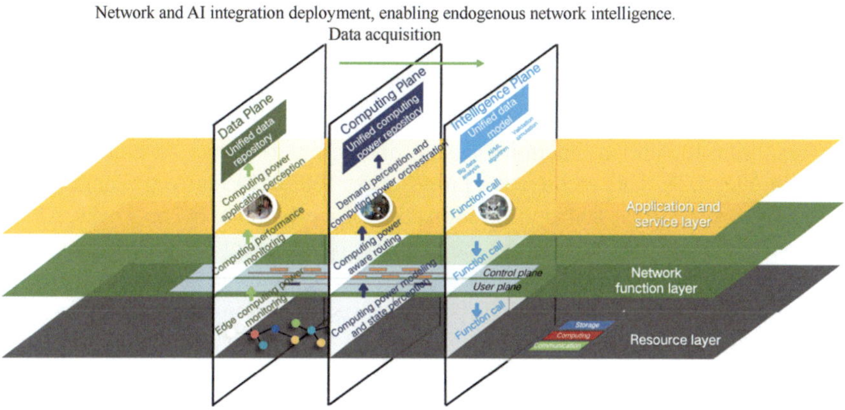

Fig. 5.26 Protocol architecture of native AI in 6G networks

it manages include various models at the resource layer, network function layer, and application and service layer, which are used to optimise the related entities on these layers.

Training data management provides training samples needed for model training and customises and preprocesses training data according to model performance requirements. AI model evaluation is used to evaluate the performance of AI models. After optimising each entity using the inference results of the AI model, the generated performance indicators are collected, the evaluation results are calculated, and then passed to the AI engine. The AI model is then optimised accordingly (such as changing the model structure, changing the algorithm, retraining, etc.).

Data analysis is a main function, including AI model training and inference based on real-time data. The required data is obtained through the data collection interface. Knowledge base management summarises and extracts the application results of AI models, extracts rules or relationship patterns, and imports expert experience from outside. AI models, training data, and knowledge can be reused and shared by third parties. At the same time, the six functions/services of the native AI system can also be shared with the network function layer and other parties, providing the required applications and services.

The RAN domain, TN domain, CN domain, UE domain, and other domains also need to combine centralised AI with distributed AI capabilities. The globally centralised AI platform can globally process external and internal data models, orchestrate intelligent capabilities according to the requirements of specific use cases, and then distribute the results to the AI platform of the specific subdomain for execution. For local AI capabilities, they need to be deployed as close to the user as possible, providing real-time AI capability support, including models and algorithms, etc.

As one of the three pillars of AI, computing power is crucial. With the integration of communication, computation, and storage, computing power is also something we need to manage. From the perspective of network deployment, its cost-effectiveness must be considered, comprehensively considering how to utilise this distributed, centralised computing power. We also need to consider methods for sharing the computing power load between terminals and networks.

The intelligent layer function module is shown in Fig. 5.27.

Combining the above functional design, the complete native AI architecture is shown in Fig. 5.28. This architecture proposes the following innovative concepts.

Firstly, to support the self-optimisation and self-evolution of 6G networks, AI use cases will no longer be proposed by the network management or optimisation team, but by the network through the central AI use case generator or imported externally through APIs. The network provides performance evaluation and assurance for AI services, which is achieved through QoAIS (Quality of AI Service). The performance requirements of AI use cases are derived by the interpreter and translated into specific

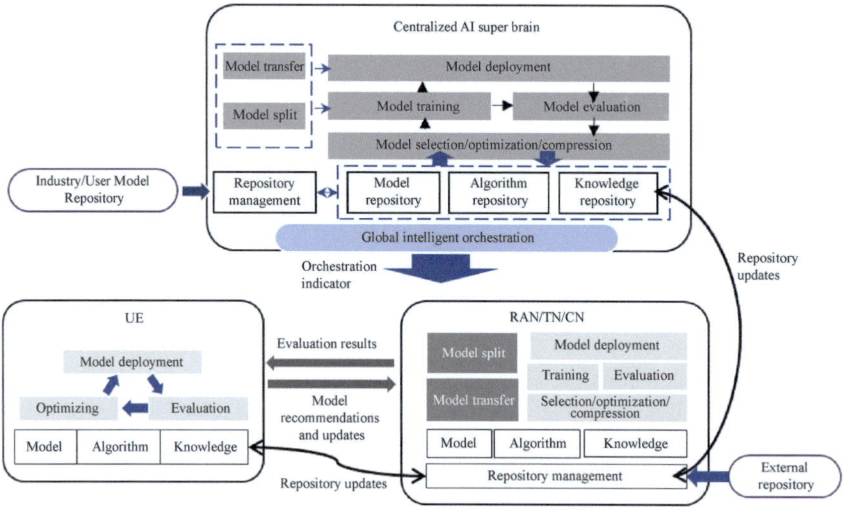

Fig. 5.27 Intelligent layer function module

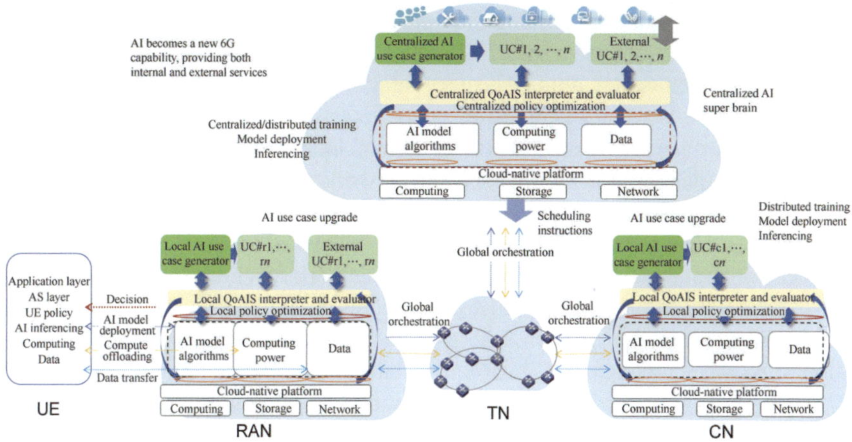

Fig. 5.28 Complete endogenous AI architecture

requirements for models, algorithms, computing power, data, and work modules in the AI workflow.

Then, the orchestrator aggregates network capabilities, supported by hardware resources, to meet the above requirements of AI use cases. Of course, the central AI super-brain can also deliver some orchestration results that require global coordination to each sub-domain for execution, ensuring that AI applications receive global support. This architecture provides a similar framework for local use cases to support real-time use cases. As it is an end-to-end architecture, the AI capabilities of the terminal are also considered during orchestration, supporting intelligent applications of the terminal.

Finally, AI models, data, and knowledge can be reused and shared with third parties. At the same time, the various functions and services in the native AI architecture can also be opened to third parties, providing the required intelligence.

In addition to the design of the above functional architecture, we have summarised and refined three technical features of native AI in 6G networks to address the technical challenges brought by the AI full lifecycle workflow and cloud network AIaaS.

The first technical feature is AI full lifecycle service orchestration based on QoAIS. This feature is already reflected in the functional architecture. To meet the different demands of the network itself and various industries for intelligent service quality, 6G networks need to build a quality evaluation and guarantee system for AI services, and based on the target performance of AI services, orchestrate the functional services, related data and models, and required resources within its lifecycle. To support this technical feature, further research is needed on AI use case self-generation technology, QoAIS parsing and evaluation technology, API design of data/model/resources, automatic model search and construction technology (such as AutoML), online learning data processing, model training inference and self-updating process orchestration, etc.

The second technical feature is the deep integration of native AI and communication. This feature is what distinguishes operators from cloud service providers and any other AI service providers. This integration includes both the application of AI to enhance network operation intelligence and network performance "AI for Network" scenarios, and changes made to the network architecture, protocols, and functions to guarantee the performance of AI services, i.e., "Network for AI" scenarios. It also includes joint optimisation of AI functions and network functions to improve overall network resource efficiency and reduce costs. In this direction, many key technologies are currently under research, such as distributed model training based on data segmentation or model segmentation [25, 26], distributed real-time collaborative inference [27], optimisation of air interface transmission and resource scheduling based on federated learning [28], joint coding of AI model source and channel, second-order model training algorithm to reduce air interface training overhead [29], etc.

The third technical feature is the integration of native AI and network digital twin. One of the features of the 6G network architecture is the digital twin network [30]. Society after 2030 will be a digital twin society, and digital twin technology can also be applied to 6G networks to achieve full digitalisation. Through digital twin, each network entity and user service can be digitised through real-time information collection. Real-time status monitoring, trajectory prediction, and predictive intervention for possible faults and service dropouts will become possible, thereby improving the operating efficiency of the entire network and service efficiency; it can also verify the effect of deploying new network features in advance, accelerate the improvement and optimisation of new features, and achieve rapid automatic introduction of new functions, thereby achieving network self-evolution [31]. The native AI architecture can utilise the simulation capabilities provided by the digital twin network to generate sample data required for model training, improving the completeness of the input sample space. It can also pre-validate and optimise the effects of AI workflows or AI models in the digital twin network, avoiding performance damage to the real network caused by the activation of AI functions. Conversely, the digital twin network can use the native AI architecture to solve its own construction problems, such as training an AI model that can simulate the function by collecting relevant data of the network function, acting as its twin body, and constructing the twin body with data generated by GAN, thereby reducing the consumption of network resources by data transmission.

Therefore, the native AI design of the 6G network architecture will completely change the plug-in and grafting AI application modes, provide ubiquitous AI capabilities, and provide the resources required for AI such as algorithms/models, computing power, and data as needed, efficiently and quickly meeting the needs of various AI use cases. This AI capability can be distributed or centralised, achieving the best AI services through edge-cloud collaboration. At the same time, 6G introduces external AI functions into the network through the AI platform, provides new businesses and new capacities, and also introduces external data into the network, further improving data processing efficiency. In addition, the analysis data and AI capabilities in the network can also be opened to third parties to provide them with services and the support they need.

5.4.5 Native Security

Facing the social development vision of "digital twins, ubiquitous intelligence" after 2030 [5], the digital twin of the world is an inevitable trend for future social development. Every object in the physical world will be digitised, have a digital replica, and digital twins of operator networks, people, factories, enterprises, government infrastructure, etc. will exist in the digital space, achieving interaction between the physical world and the digital world through digital twins, achieving the effect of predicting and changing the future. Therefore, major changes will occur in many aspects of society such as production, life, and social governance. Changes and operations of digital twins in virtual space will directly affect physical entities, thus bringing more security risks. If the security of virtual space is threatened, it will directly affect the safety of the physical world. Some people predict that future wars will not be attacks on the physical world, but attacks on the digital twin world, which may directly lead to the destruction of the physical world. Therefore, security will be more important for future society.

The new business, scenario and technology development trends pose new challenges to the security architecture design of the 6G network. Industries related to high-tech technologies such as industrial intelligent control, autonomous driving, and virtual reality will increasingly rely on the 6G network. While data is exploding, data confidentiality and identity privacy should be fully protected and reliably transmitted.

At the same time, to continuously improve user experience, the 6G network will tend to be more distributed and provide unique services centred on users. The change in network architecture will prompt the 6G security architecture to make corresponding changes. From the Internet of Everything to the Internet of intelligence, the business application value that the 6G network will carry in the future will also be greatly enhanced, which undoubtedly has great appeal to interest-driven attackers. 6G needs to provide stronger security protection capabilities for new business application scenarios and new technologies; at the same time, according to the characteristics of network flexibility and dynamic expansion, it requires security capabilities to be deployed on demand and elastically at a fine granularity, dynamically adapting to the network, terminal environment and business scenarios, ensuring network and business security.

In the 6G era, while supporting traditional security capabilities, it should extend to support trustworthiness, encompassing security (Security), physical security (Safety), privacy, resilience and reliability, achieving the generalised 6G network security trust [32].

Privacy leaks, man-in-the-middle attacks, distributed denial of service attacks, etc., have always been security challenges faced by mobile communication networks. To solve these problems, the industry often uses patches, plug-in security services, etc. These passive protection methods cannot avoid security damage in time, and it is difficult to adapt to the development of the network [32], there is still room for improvement in terms of security and efficiency.

In order to solve the existing security architecture problems and face future new challenges, we should rethink the design of security mechanisms. The rapid development and progress of cloud computing, big data, and AI technology provide new means and support for the security design of 6G networks. Drawing on the principles of the human immune system, we should integrate the security mechanism and communication system, and build the 6G security system into an inherent immune system in the network, so that security is ubiquitous and everywhere, can provide passive defence, and can also provide active defence, and can also gain immunity from attack events. With the development of AI, blockchain, quantum communication and other technologies, it is possible to embed security processes in communication networks. By embedding basic security capabilities in network elements, providing collection, control, isolation and other capabilities support, and redesigning the 6G network security architecture based on distributed technology means, we can build an native security trust mechanism in the 6G network and solve the security trust issues faced by the current centralized network architecture from the root. AI will empower 6G network security in all aspects, and AI technology can help enhance security analysis and decision-making capabilities, thereby improving the overall security capabilities of the 6G network. The native security mechanism based on AI makes the 6G network have intelligent consensus, intelligent defence, self-immunity and self-evolution, ubiquitous collaboration capabilities, and realizes the transition from "network security" to "secure network". The characteristics of native security are shown in Fig. 5.29.

Literature [32] proposes that the native security of 6G networks should have the following characteristics: first is active immunity, based on trustworthy technology, to provide active defence functions for network infrastructure, software, etc.; the second is resilient autonomy, according to the security needs of users and industry applications, to achieve dynamic orchestration and flexible deployment of security

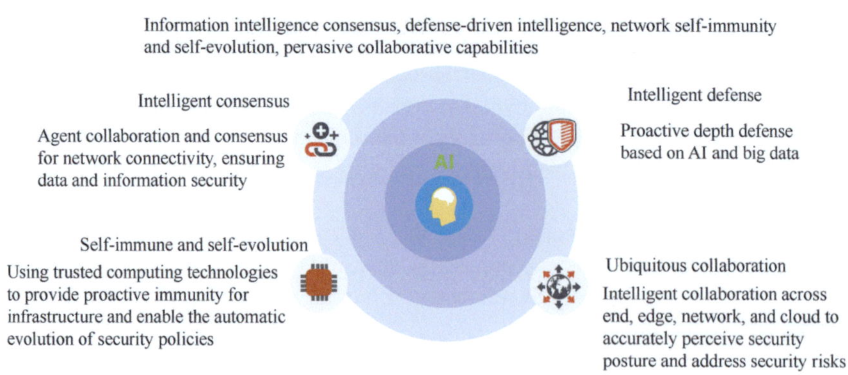

Fig. 5.29 Characteristics of native security

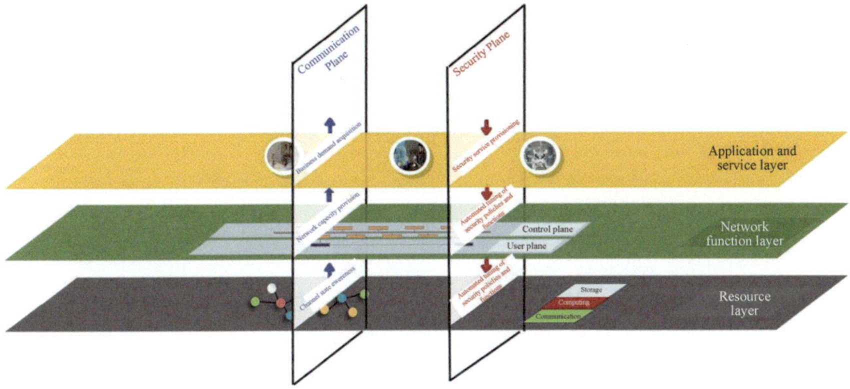

Fig. 5.30 Integrated communication security, creating security as a service

capabilities, to enhance network resilience; the third is virtual symbiosis, using digital twin technology to achieve the unity and evolution of physical network and virtual digital twin network security; the fourth is ubiquitous collaboration, through the intelligent collaboration of end, edge, network, and cloud, to accurately perceive the entire network's security situation and agilely handle security risks.

At the same time, considering the future trend of data and application localization and regionalization, every person, every intelligent body, every terminal, every network node may become a provider of data, business and capabilities, so we also need to make security a basic capability and service of the network, provided on demand to objects that need security protection. For this reason, when designing the overall network protocol architecture, we need to consider introducing a new security plane to achieve end-to-end orchestration, management and QoS assurance of the entire security service, as shown in Fig. 5.30.

The functional design of the security plane is shown in Fig. 5.31 [10], the network will monitor its security status in real time, predicting potential risks. By combining attack defence with risk prediction, it achieves intelligent security such as risk prediction and proactive immunity. Through the interaction and cooperation between intelligent entities in the network, an intelligent consensus is formed, eliminating interference, and providing a high level of security for information and data. Based on AI and big data technology, it accurately deploys security functions, optimises security strategies, and achieves proactive, in-depth security defence. It provides proactive immunity for network infrastructure, software, etc., and uses trusted computing technology to enhance the security level of the basic platform. Through the ubiquitous collaboration of terminal, edge, network, and cloud, it accurately perceives the security status of the entire system, properly handles security risks, maximises the security of the 6G network, and achieves a comprehensive upgrade from network security to cyberspace security.

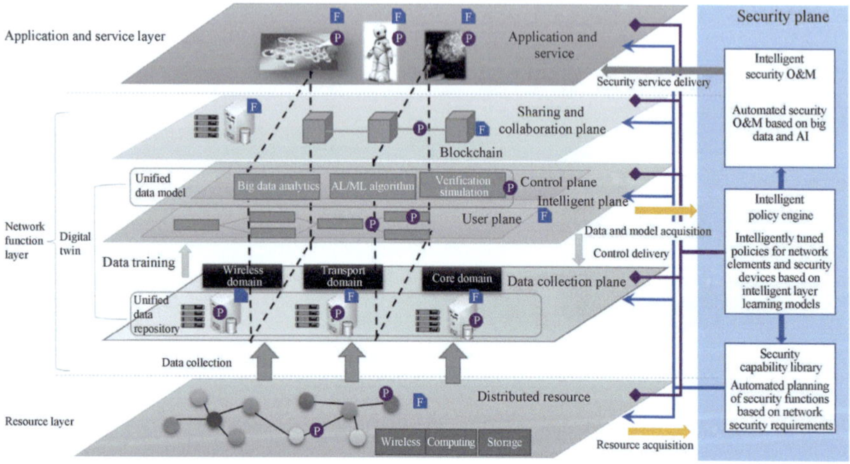

Fig. 5.31 Functional design of the security aspect

5.4.6 *Digital Twin*

With the large-scale deployment of 5G, the scale of operators' mobile communication networks is getting larger and larger. Taking mobile operators as an example, there are millions of base stations, and there are 2G, 3G, 4G and 5G, even Wi-Fi and other IoT systems, each network has multiple frequency bands, the interoperability parameters between networks are complex, and the planning, construction, operation and maintenance, optimisation processes are completely separated. The operation and maintenance system of the 5G network in the early stage is shown in Fig. 5.32, which brings great challenges to the operation and optimisation of the network and becomes an important part of the daily expenses of operators.

For this reason, the community has begun to practice network intelligence, hoping to gradually automate the original manual operations through the use of big data and AI. For this reason, ITU has also defined five levels of network intelligence [33], as shown in Table 5.3.

Although operators have been exploring, researching and deploying network intelligence for many years, they still cannot effectively solve problems such as high network energy consumption, complicated multi-system interoperability, high operation and maintenance costs, and low efficiency. At the same time, as the network evolves towards programmable, software-driven, and service-based architectures, the complexity and scale of network operation and maintenance will reach unprecedented heights. The introduction of new services and technologies also puts more stringent requirements on the sensitivity of network operations. Operators urgently need a more comprehensive, intelligent, scalable and cost-effective network automated operation and maintenance system.

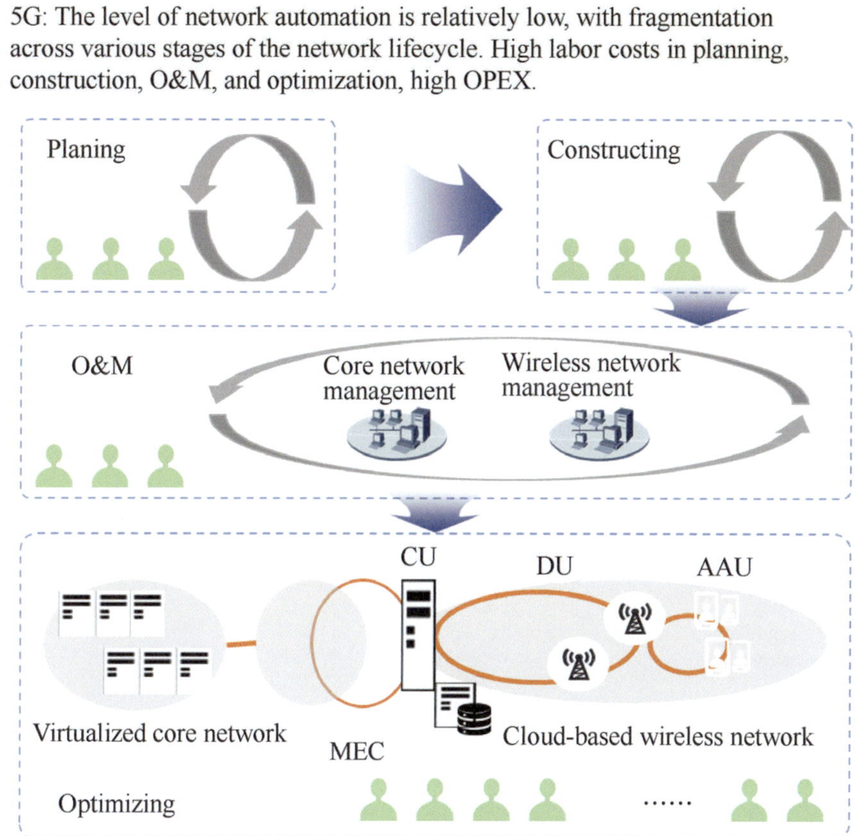

Fig. 5.32 Operation and maintenance system of 5G network in the early stage

The automation of network operation and maintenance has different granularities. It can be the automation of tasks, functions or processes, or it can be the automation of network and service lifecycle management. The current level of 5G network operation and maintenance automation is low, most of which rely on solidified expert rules and automatic scheduling to achieve, and some scenarios still rely on manual operations. The network operation and maintenance automation achieved by intelligent means is still "patch" and "plug-in". "Patch" refers to achieving a higher degree of automation for some specific functions through use case-driven methods, reducing manual intervention, such as self-organising network (SON), base station self-start, automated neighbour relationship, PCI self-optimisation, mobile robust optimisation (MRO) etc.; "plug-in" refers to the collection and aggregation of relevant data to the network management or related platforms for training, and the model is distributed to the corresponding network elements to generate the intelligence required for operation and maintenance.

Table 5.3 The 5 levels of network intelligence

Grade/name		Hierarchical evaluation dimension				
		Execution	Data acquisition	Analysis	Decision making	Demand mapping
L0	Manual operation network	Manual	Manual	Manual	Manual	Manual
L1	Auxiliary operation network	**Manual and system**	**Manual and system**	Manual	Manual	Manual
L2	Primary intelligent network	*System*	**Manual and system**	**Manual and system**	Manual	Manual
L3	Intermediate intelligent network	*System*	*System*	**Manual and system**	**Manual and system**	Manual
L4	Advanced intelligent network	*System*	*System*	*System*	*System*	**Manual and system**
L5	Fully intelligent network	*System*	*System*	*System*	*System*	*System*

Remarks:

1. All levels of decision-making and execution support manual intervention, and manual review conclusions and execution instructions have the highest authority
2. In the implementation of intelligent grade evaluation, each dimension can be evaluated separately

This "chimney-style" automation system and R&D model can improve the level of network management automation to a certain extent under the existing network architecture, but due to the limitations of the existing network structure, and the difficulty in ensuring the validity and real-time nature of the data, it is difficult for data from different manufacturers to communicate and share, resulting in low efficiency of network automation and difficulty in achieving the expected results.

Digital twin technology provides a new approach and solution for the realisation of 6G network autonomy, that is, by digitally twinning the network itself to build a twinned digital network [34]. As shown in Fig. 5.33, the digital twin network is a network system composed of a physical network entity and its twinned digital network, and the physical network and the twinned digital network can interact in real time. The twinned digital network element corresponding to the physical network element can be constructed through various data collection and simulation methods, thereby forming the digital twin of the network element and the network in the digital domain. The digital domain generates perception and cognitive intelligence through

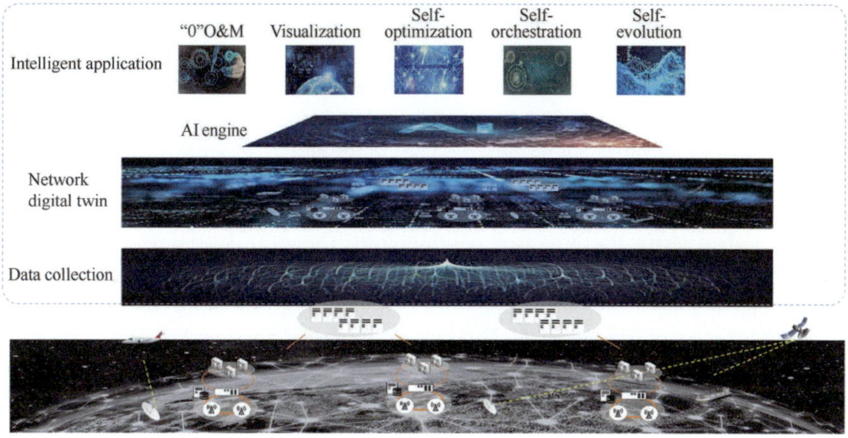

Fig. 5.33 Network autonomy based on digital twin network

rich historical and real-time data and advanced algorithm models, continuously optimises and simulates the optimal state of the physical network, and issues corresponding operation and maintenance operations in advance, automatically corrects the physical network, and predictively solves possible faults of the network element or network in advance to achieve the effect of "curing before illness"; then evaluates the effect of operation and maintenance by collecting corrected data, forming a closed loop. Through this closed-loop interaction between the digital and physical domains, cognitive intelligence and automated operation and maintenance operations, the network can quickly recognise and adapt to complex and changing dynamic environments, efficiently analyse, diagnose, simulate and control the physical network based on data and models, and achieve "autonomy" throughout the network lifecycle of planning, construction, maintenance, optimization and healing.

The digital twin of the network, as a digital mirror of the physical network infrastructure, has the same network elements, topology, and data as the physical network, and can realise the fine "replication" of the entire process of the network and equipment, providing a digital verification environment close to the real network for network operation and maintenance optimisation operations and strategy adjustments. Therefore, compared with traditional simulation platforms, the AI models and pre-verification results trained based on the digital twin of the network have higher reliability. In addition, the digital twin network also records and manages the behaviour of the network's digital twin, supports its traceability and playback, and can complete pre-verification without affecting network operations, greatly reducing the cost of trial and error. In addition, the digital twin network has the ability to build and expand autonomously, and can be combined with AI technology, enabling it to explore new business requirements that have not yet been deployed to the live network and verify the effects in the twinned digital network, thus realising the self-evolution of the network.

Digital twin technology can also be applied to 6G networks. Through digital twinning, each network entity and user service can be digitised through real-time information collection and modelling. The simulation and emulation of the digital network make real-time network status monitoring, trajectory prediction, and early intervention for possible faults and service degradation possible, achieving high levels of network automation, "zero-touch maintenance" and "predictive maintenance" [16], not only can it improve the efficiency of network operation and service, but it can also verify the effect of new network function deployment in advance through digital twins, accelerate the improvement and optimisation of new functions, realise the rapid and automatic introduction of new functions, and thus achieve the self-evolution of the network. The digital twin of the network will greatly help operators achieve a high degree of network autonomy and structurally solve the problem of difficult operation and maintenance of mobile communication networks.

For the digital twin of the network, it is crucial to obtain data from all network domains [10]. As shown in Fig. 5.34, network autonomy based on digital twins can be achieved by data processing and parametric modelling of various network entities and functions, resulting in a digital model of the entire network in virtual space. A user's digital twin may include data such as trajectory, speed, service type, service model, data rate, and channel status. The data of cloud-native RAN may include wireless resource model, hardware resource model, air interface protocol model, wireless channel model, network service status, as well as KPIs and statistical indicators, etc. Similar modelling processes and digital twins can also occur in TN and CN. The AI engine predicts network status using AI algorithms based on the obtained digital twin data. The network planning entity will continuously seek the optimal state of the physical network, verify it through simulation, and then execute corresponding operations in the management domain to map it to the network in the real world.

Any network resource object can generate a corresponding digital twin in the digital domain, including underlying physical resources, network functions, and various applications and services at the upper level. The generation of digital twins depends on various data collection, data processing and storage, as well as digital twin modelling techniques. To optimise resource objects, the digital twin network establishes its optimisation model based on the collected data and information, and generates its digital planning entity at future time points [16]. Then, by calling the configuration function, data planning can be implemented for each resource object. The aforementioned digital twin network functions, as well as the connection management, synchronisation, optimisation, etc., between the digital domain and the physical domain are all network functions. The data of the digital twin network includes digital twins, digital planning entity, and intelligent models. The digital twin network provides various applications and services upwards and calls and optimises various resource objects downwards through these functions and data. At the same time, the digital twin network can share the aforementioned functions and data with third parties through various sharing technologies. In addition, it can prevent attacks and tampering through various security technologies.

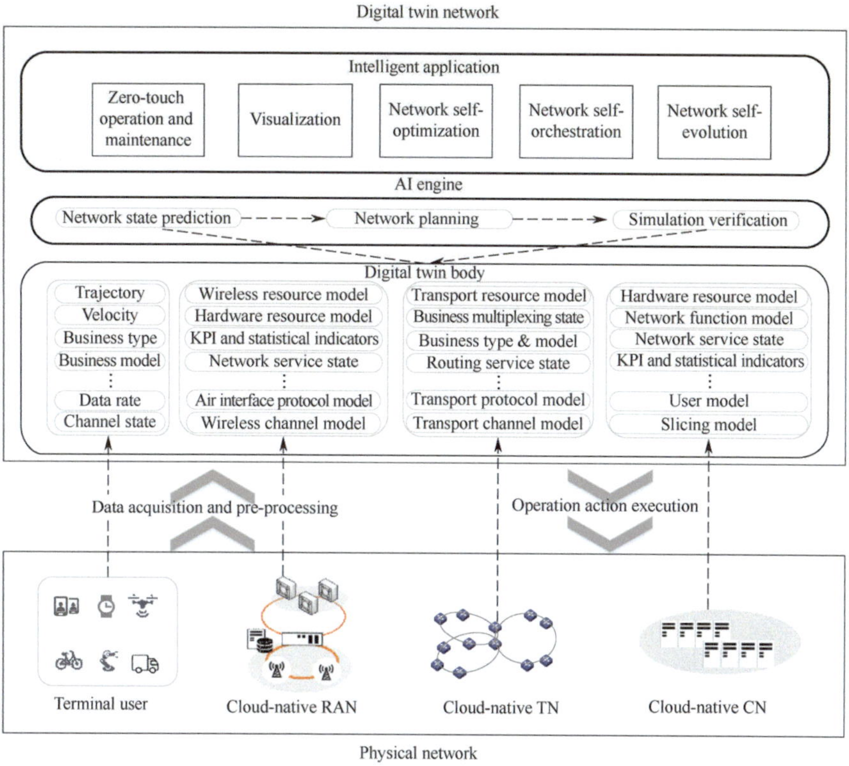

Fig. 5.34 Network autonomy based on digital twins [10]

5.4.7 6G Network Architecture Overview

Combining the above analysis and design, the 6G network will be a new generation of mobile information network that integrates communication, computing, big data, perception, AI, and security, as shown in Fig. 5.35. Its underlying layer is the physical infrastructure layer, which includes base station antennas and active antenna units (AAU)/radio frequency (RF) in different frequency bands (possibly including sub-10 GHz, millimetre wave, terahertz, visible light and satellite, as well as aerial and ground satellite infrastructure), virtualised and cloud-based computing and storage facilities. Above the physical infrastructure layer is the logical function layer, which forms communication, computing, big data, sensing, AI, security and other capabilities by calling the basic resources of the physical infrastructure layer through unified management and orchestration, serving various application scenarios at the upper level. The three-layer logical structure decouples the underlying hardware and upper-level functions to support end-to-end cloud-native design, bringing agility and elasticity to the end-to-end network, and enhancing the adaptability of the end-to-end network to new businesses and scenarios.

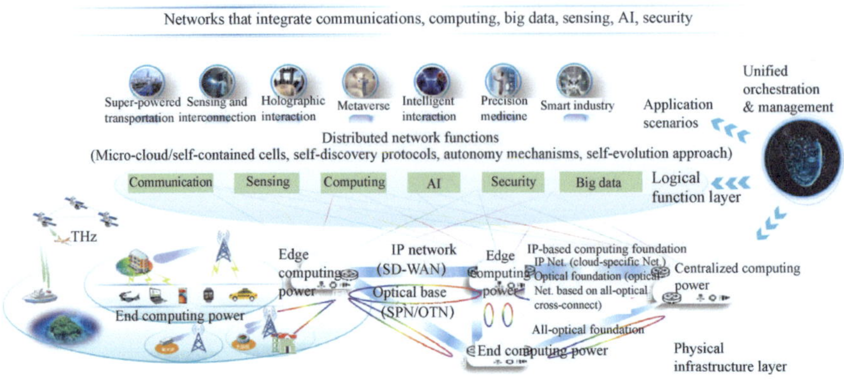

Fig. 5.35 6G network

In addition, in order to further expand the capability dimensions of the 6G network, communication, big data, computing, AI, and security are all built into the basic capabilities and services of the network, achieving their integrated design, thereby ensuring their end-to-end multi-dimensional QoS experience. For this reason, the end-to-end logical protocol of the 6G network should be expanded as shown in Fig. 5.36, introducing new data plane, computing plane, intelligent plane, and security plane in addition to the traditional communication plane (which can be further divided into control plane and user plane in specific standards), and managing and orchestrating each protocol plane through the management plane.

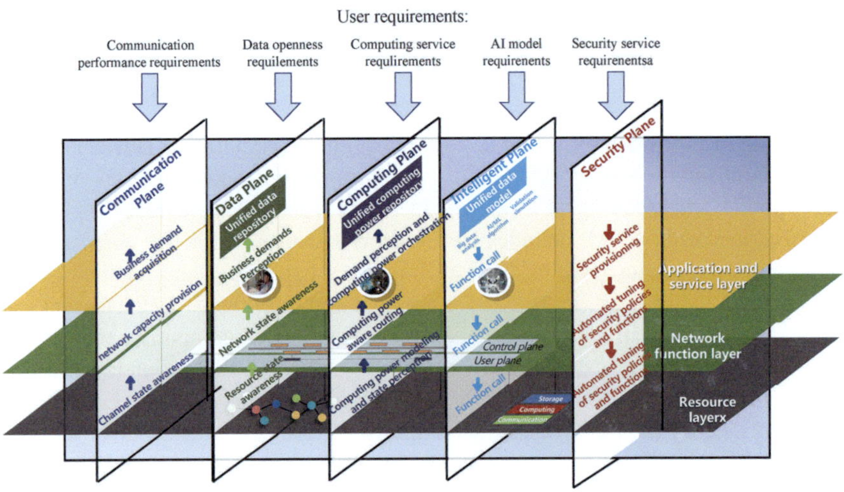

Fig. 5.36 Protocol integrating communication, big data, computing, AI, and security

5.5 Air Interface Technology

The air interface technology of mobile communication networks has become quite mature after decades of research, especially the transmission efficiency of the physical link has approached the Shannon limit. Although the industry has explored many research directions for 6G, such as OAM [35], beyond Nyquist transmission [36], holographic radio [37], full duplex [38], AI [39], etc., there has not been a significant breakthrough that can be matured enough and affect the entire 6G air interface transmission design in the short term.

Therefore, the focus of air interface research for 6G is on four aspects: firstly, how to further improve various technical indicators of the air interface, such as peak rate, spectral efficiency, reliability, delay, etc.; secondly, how to adapt to more differentiated and fragmented service and scenario requirements; thirdly, the fusion design with newly introduced capability dimensions, such as AI, sensing, computing, etc., and how to adapt to the future end-to-end full cloud evolution; fourthly, to solve problems found in the development of 5G networks, such as large gaps in uplink and downlink link performance, large gaps in user experience between cell edges and cell centre, etc. Therefore, the focus of 6G air interface design is to carry out scenario-based air interface optimisation design on the basis of inheriting mature technologies and designs that have been fully verified in 5G (such as large-scale antennas, Polar/LDPC coding, OFDM modulation, etc.), in order to efficiently meet the key air interface technical indicator requirements under different service and application scenarios, and to study cloud-native air interface protocol architecture and protocol systems, etc.

To support the future new use cases flexibly, a unified system design is necessary. As shown in Fig. 5.37, the minimal subset of common functions are extracted and set as the system kernel, while the different use cases can be supported by the common functions of the system kernel combined with some additional functions necessarily. By this way, the new use cases can be introduced in a incremental way.

In the entire 6G air interface technology, there are several key directions worth paying attention to and researching, such as distributed massive MIMO [40], intelligent air interface [41], etc., which will be detailed in subsequent chapters.

The ultimate dream of intelligent air interface is to train a neural network through the joint optimisation design of the transmitter and receiver [42, 43], to replace all existing modulation, coding and antenna processing, and achieve the optimal transmission error rate, as shown in Fig. 5.38. Using deep learning technology, with the goal of minimising information transmission error probability, etc., the transmitter and receiver modules are jointly optimised, and end-to-end communication can break the modular barriers of traditional communication systems, and carry out joint optimisation of the transmitter and receiver, thereby achieving the globally optimal bit error rate performance of the communication link.

(a) Classic wireless communication system architecture
(b) New architecture of wireless communication system based on deep learning.

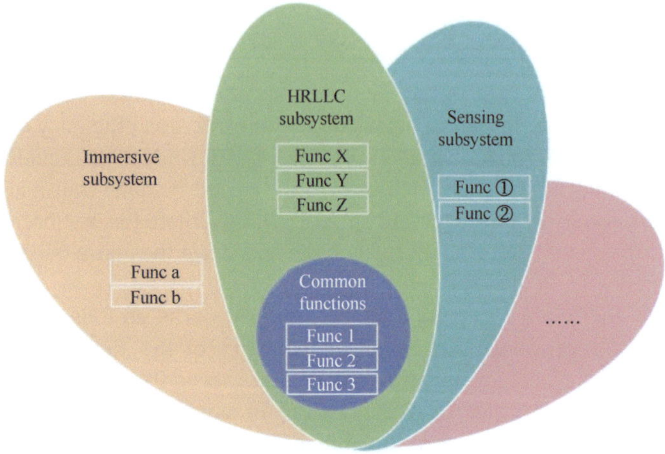

Fig. 5.37 Unified design for diverse scenarios

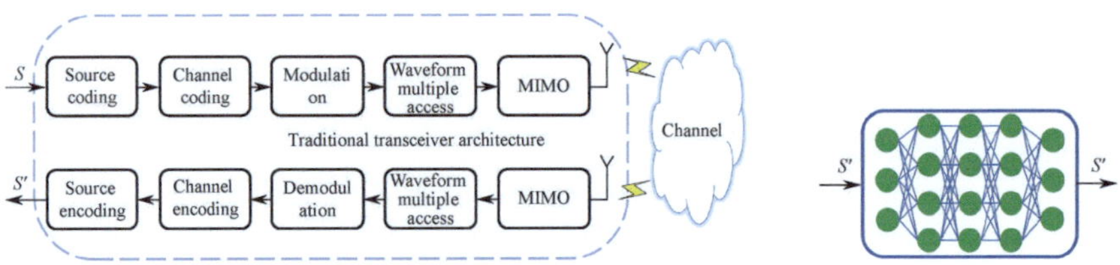

Fig. 5.38 Comparison of classic wireless communication system architecture and new architecture of wireless communication system based on deep learning

However, in actual communication systems, due to the dynamic changes of channel fading, interference, resource scheduling, etc. caused by complex environments, and the delay and cost of obtaining the required system context information by the transmitter and receiver, it is almost impossible to achieve the above ultimate goal. At present, the industry's application of AI in the air interface mainly focuses on enhancing traditional physical layer function modules, such as channel coding and decoding, modulation and demodulation, waveform multiple access, multi-antenna MIMO, channel estimation, channel information feedback and prediction, receiver algorithm, etc.

The main issues currently affecting the application of AI in air interfaces include adaptability to scenarios, computational complexity and energy consumption, interpretability, etc. The industry has not yet reached a consensus on some key issues in intelligent air interface research, such as dataset establishment, generalisation measurement, model calibration, evaluation criteria and evaluation indicators. Evaluation criteria and indicators are the benchmarks for judging the quality of wireless AI algorithms and are important bases for future research and standardisation of wireless AI algorithms; a good dataset is the foundation of wireless AI algorithm research, and different datasets directly affect the design and performance of wireless AI algorithms; wireless AI algorithms designed based on a certain dataset often have insufficient generalisation in cross-scenario applications, and there is a need to vigorously research generalisation enhancement techniques. 3GPP has already begun to study intelligent air interface in Release 18, but the existing methods for evaluating wireless solution performance cannot meet the needs for systematic evaluation of wireless AI algorithms. Therefore, the industry needs to work together to promote the formation of relevant consensus and accelerate the research of air interface AI.

Considering the high deployment cost of massive MIMO in 5G networks, and the large gaps in the capabilities of uplink and downlink in existing 5G networks, and the large gaps in user experience between cell centres and cell edges, distributed massive MIMO with coordination is expected to play an important role in solving these problems. Supporting cell-free (Cell-Free or No Cell) distributed massive MIMO has become one of the important directions of industry research, for detailed content please refer to references [44–46].

5.6 Summary

This chapter starts from the representative views of the industry on what 6G is, analyses the industry's preliminary understanding and definition of 6G, and then defines 6G from the perspectives of 6G's vision and requirements, potential frequencies and their use and deployment strategies, integrated coverage expansion of air, space, land and sea, integrated capabilities of communication, computing, big data, sensing, AI, security, high autonomy based on network digital twins, network architecture and air interface design, and proposes that 6G is a new generation of mobile information network integrating communication, computing, big data, sensing, AI, security, etc., which will have new features such as on-demand service providing, simplicity, flexibility, native AI, native security and network digital twins.

References

1. Niu Kai, Dai Jinsheng, Zhang Ping, et al. Semantic Communication for 6G [J]. Mobile Communications, 2021, 45(4):6.
2. Shi Guangming, Xiao Yong, Li Yingyu, et al. Semantic Communication Network for Internet of Things [J]. Journal of Internet of Things, 2021.
3. Liu Guangyi, Huang Yuhong, Cui Chunfeng, et al. 6G Reshaping the World [M]. Beijing: People's Posts and Telecommunications Press, 2021.
4. LIU G Y, YUANG Y H, LI N, et al. Vision, requirements and network architecture of 6G mobile network beyond 2030 [J]. China Communications, 2020, 17(9).
5. China Mobile. 2030+ Vision and Requirements White Paper [R]. 2nd Edition. 2020.
6. NGMN. 6G vision and drivers white paper [R]. 2021.
7. NGMN. 6G use cases white paper [R]. 2022.

8. Fang Jian, Li Jingchun, Huang Biao, et al. Current Status and Prospects of 5G Spectrum Research [J]. Telecommunications Science, 2015, 31(12):8.
9. GSMA. GSMA Calls for 2 GHz of Mid-Band Spectrum to Meet UN Targets [R/OL]. 2021–7.GSMA official website.
10. LIU G Y, LI N, DENG J, et al. The SOLIDS 6G mobile network architecture: Driving forces, Features and functional topology [J]. Engineering, 2022, 8.
11. STURM C, WIESBECK W. Waveform Design and Signal Processing Aspects for Fusion of Wireless Communications and Radar Sensing [J]. Proceedings of the IEEE, 2011, 99: 1236-1259.
12. TAN D K P, HE J, LI Y, et al. Integrated Sensing and Communication in 6G: Motivations, Use Cases, Requirements, Challenges and Future Directions [J]. 2021 1st IEEE International Online Symposium on Joint Communications & Sensing (JC&S), 2021: 1–6.
13. AHANG J A, RAHMAN M L, WU K, et al. Enabling Joint Communication and Radar Sensing in Mobile Networks—A Survey [J]. IEEE Communications Surveys & Tutorials, 2022, 24(1): 306-345.
14. Liu Guangyi, Lou Mengting, Wang Qixing, et al. Research on Integrated Communication and Sensing Architecture and Key Technologies for 6G [J]. Mobile Communications, 2022(6).
15. 6GANA. Concept and Demand of Data Services White Paper [R]. 2022.
16. China Mobile Research Institute. 6G Ultra-Simple Wireless Access Network White Paper [R]. 2022.
17. Liu Chao, Lu Lu, Wang Shuo, et al. Outlook on the Fusion 6G Network Architecture for Integrated Air, Space, and Ground Multi-Access [J]. Mobile Communications, 2020, 6.
18. China Mobile Research Institute. 6G Service-Oriented RAN White Paper [R]. 2022.
19. LI N, LIU G Y, ZHANG H M, et al. Micro-service-based radio access network [J]. China Communications, 2022, 19(3).
20. ITU-T. Framework for evaluating intelligence levels of future networks including IMT-2020 (Study Group 13): Y.3173 [S].2020.
21. 3GPP. Study on RAN-centric data collection and utilization for LTE and NR: TS 37.816 Version 1.0.0 [S]. 2019.
22. 3GPP. Study of Enablers for Network Automation for 5G: TR 23.791 [R]. 2018.
23. 3GPP. New WID Self-Organizing Networks (SON) for 5G networks: TSG-SA SP-190785 [S]. 2019.
24. ETSI. Improved operator experience through Experiential Networked Intelligence (ENI) [S]. ETSI WhitePaper, 2017, 22.
25. ITU-T. Architectural framework for machine learning in future networks including IMT-2020 (Study Group 13): Y.3172 [S]. 2019.
26. GSMA. AI in Network use cases in China [R]. 2019.
27. AMIRI M M, GUNDUZ D. Federated Learning over Wireless Fading Channels [J]. IEEE Transactions on Wireless Communications, 2020, 19(5): 3546-3557.
28. HUANG Y, CHENG Y, BAPNA A, et al. GPipe: Efficient Training of Giant Neural Networks using Pipeline Parallelism [J]. Google, 2018.
29. TEERAPITTAYANON S, MCDANEL B, KUNG H T. Distributed Deep Neural Networks Over the Cloud, the Edge and End Devices [C]// 2017 IEEE 37th International Conference on Distributed Computing Systems (ICDCS). IEEE, 2017.
30. WANG S, TUOR T, SALONIDIS T, et al. Adaptive Federated Learning in Resource Constrained Edge Computing Systems [J]. IEEE Journal on Selected Areas in Communications, 2019:1–1.
31. DÜNNER C, LUCCHI A, GARGIANI M, et al. A Distributed Second-Order Algorithm You Can Trust [C]// International Conference on Machine Learning. 2018.
32. IMT-2030 Promotion Group. 6G Network Security Vision White Paper [R]. 2021.
33. ITU-T. Grading method for intelligent capability of mobile networks: ML5G-I-151 [S]. 2018.
34. China Mobile Research Institute. Autonomous Wireless Network Based on Digital Twin Network White Paper [R]. 2022.

35. MOHAMMADI S M, DALDORFF L K S, BERGMAN J E S, et al. Orbital Angular Momentum in Radio-A System Study[J]. IEEE Transactions on Antennas and Propagation, 2010, 58 (2): 565-572.
36. WANG Q, CHANG Y, YANG D. Deliberately Designed Asynchronous Transmission Scheme for MIMO Systems [J]. IEEE Signal Processing Letters, 2007, 14(12), 920-923.
37. PAN Shilong, ZONG Baiqing, TANG Zhenzhu, et al. Intelligent Holographic Radio for 6G [J]. Radio Communication Technology, 2022, 048(001) : 1-15.
38. WANG Jun, ZHAO Hongzhi, QING Zhaojin, et al. Adaptive Interference Cancellation in RF Domain in Simultaneous Same Frequency Full Duplex Scenarios [J]. Journal of Electronics & Information Technology, 2014, 36(6):6.
39. IMT-2030 Promotion Group. Research Report on Key Technologies of Wireless Artificial Intelligence (AI) [R]. 2021.
40. JIA X, XIE M, MENG Z, et al. Achievable Uplink Rate Analysis for Distributed Massive MIMO Systems with Interference from Adjacent Cells [J]. China Communications (English Edition), 2017, 14(5):12.
41. China Mobile Research Institute. White Paper on Key AI Technologies for 6G Physical Layer [R]. 2022.
42. SHEA T O', HOYDIS J. An introduction to deep learning for the physical layer [J]. IEEE Trans Cogn Commun Netw, 2017, 3(4): 563–575.
43. JIANG H, DAI L. End-to-end learning of communication system without known channel [J]. Proc. IEEE Int. Conf. Commun. (ICC), 2021.
44. SHUAI F, CHEN J, ZHANG Y, et al. Wireless Powered IoE for 6G:Massive Access Meets Scalable Cell-Free Massive MIMO [J]. China Communications (English Edition), 2020, 17(12):18.
45. ZHANG Y, ZHOU M, ZHAO H, et al. Spectral Efficiency of Superimposed Pilots in Cell-Free Massive MIMO Systems with Hardware Impairments [J]. China Communications (English Edition), 2021, 18(6):16.
46. UNNIKRISHNAN K G, EMIL B, ERIK G L. Clustering-Based Activity Detection Algorithms for Grant-Free Random Access in Cell-Free Massive MIMO [J]. IEEE Transactions on Communications, 2021, 69(11).

Chapter 6
Service-Based Architecture for 6G RAN

The 5G's commercialization shows that the current 5G network faces the challenges from the customizing and personalizing deployment and service providing, which leads to high cost and management efficiency, low resource efficiency and long time to market. Although 5G core network revolutionary adopts a service-based architecture (SBA), which targeting for agility and elasticity of the network to fulfill the differentiated requirements from vertical and enterprise applications, the radio access network (RAN) keeps the traditional integrated and rigid architecture and suffers the difficulties from customizing and personalizing the functions and capabilities. To solve this problem, open RAN (ORAN) is proposed to introduce the cloudification, openness and intelligence to RAN by an implementation manner. Due to the 5G RAN specifications are not designed for cloudification, the ORAN achieves limited gain from the cloudification. Towards 2030, the use cases and corresponding requirements will be more diverse and demands the enhanced network capability to serve the personalized and customized services. Based on the experiences and lessons from the past research, a fully service-based design is proposed for the RAN architecture design of the 6G in this chapter, which introduces the service-based transformation of user plane function, RAN, and various new capabilities, and further explores the mechanisms to improve service decoupling and service invocation efficiency. Starting from the driving forces behind the service-based RAN in 6G, this chapter analyzes the experience and insights from 5G SBA, and conducts a systematic study on the service-based RAN, including the definition of RAN services, interface protocol stacks, service orchestration mechanisms, impact analysis on the air interface, NAS evolution, radio capability exposure, and joint optimization design between RAN and core network.

G. Liu et al., *6G Migrates from Communication to XaaS*,
https://doi.org/10.1007/978-981-96-7579-1_6

6.1 Introduction

5G's [1, 2] commercialization has made good progress in past 5 years, and many vertical and enterprise uses cases succeed in replacing the wired network by wireless network. However the past practices also shows that the current 5G network faces the challenges from the customizing and personalizing deployment and service providing for diverse vertical industrial use cases, which leads to high cost and low management efficiency, low resource efficiency and long time to market. Although 5G core network revolutionary adopts a service-based architecture (SBA), which targeting for agility and elasticity of the network to fulfill the differentiated requirements from vertical and enterprise applications by enabling the independent scaling, tailoring, independent evolution, and on-demand deployment of network functions, the radio access network (RAN) keeps the traditional integrated and rigid architecture and suffers the difficulties from customizing and personalizing the functions and capabilities, which put a constraint on the penetration of 5G network in vertical and enterprise use cases.

The open RAN (ORAN) [3] have tried to solve this problem by introducing the cloudification, openness and intelligence to RAN in an implementation manner based on the 5G specifications released by 3rd generation partner project (3GPP). Typically, shown as Fig. 6.1, a base station is split as several modules, e.g. O-RU, O-DU, O-CU, and orchestration and automation, and the open interfaces are defined for inter-working among the different modules. To make the O-CU being able to deploy in the open hardware, e.g. COTS platform, the virtualization of O-CU is supported as Fig. 6.1. A new function module termed as RAN intelligent controller (RIC) is introduced to embed the intelligence in the RAN. RIC is further split as near real time (near RT) part and none real time (non-RT) part based on the processing latency requirements of different functions. By this reference O-RAN architecture and its supporting technologies above, O-RAN claims to bring cloud scale economics, intelligence and agility to the RAN. However, due to the constraint of the current 5G RAN specifications and the capability of COTS server, the benefits claimed above are far behind what expected in practical deployment. The tightly coupled protocol stacks of 5G NR is not friendly for virtualization and cloudification, which lead that the agility of O-RAN only makes sense for the operator featured operation and maintenance, but not the reality for the mobile network to adapt to the diverse and customized use cases.

In order to meet the requirements of diverse applications for network agility, flexibility, and scalability, the transformation of networks towards virtualization and cloud-native is an inevitable trend, and the cloud-native RAN have also gained widespread favor from global operators. Containerization, microservices architecture, and Kubernetes orchestration are the core characteristics of cloud-native RANs. Among them, containerization and Kubernetes orchestration are more about the implementation methods, while the microservices architecture is reflected in the reconstruction of functions. Based on the microservices architecture, RAN is constructed as a collection of loosely coupled services. Each service performs a

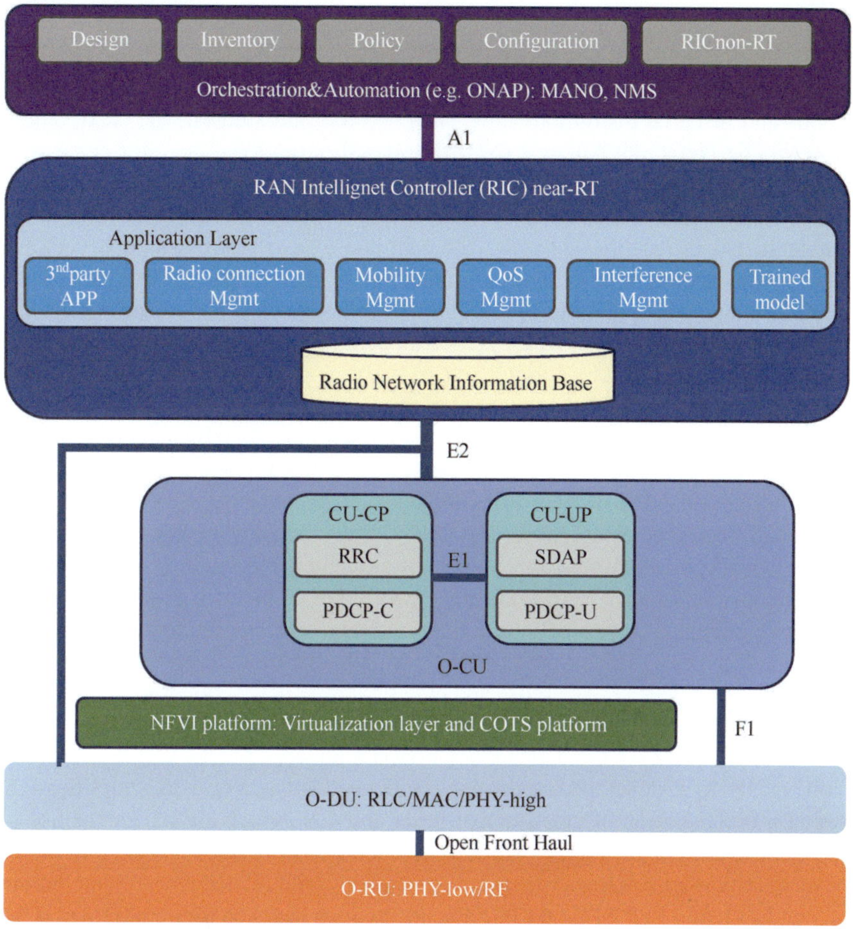

Fig. 6.1 O-RAN's reference architecture

specific function, and can be independently developed, deployed, and scaled on any infrastructure and container platform. However, the microservices architecture is not the only architectural pattern that can be adopted in the evolution of RAN towards cloud-native RAN. Other architectural patterns that can be adopted include Service-Oriented Architecture (SOA), Service-Based Architecture (SBA), etc. Therefore, RAN that is functionally reconstructed based on cloud-native design principles is referred to as a service-based RAN.

The design of a service-based RAN is key to achieve a cloud-native RAN. Only by refactoring the RAN software and fully considering cloud-native design principles (such as microservices architecture, separation of storage and state, etc.), that is, by implementing cloud-native RAN design, can the technical advantages brought by cloud-native infrastructure, such as independent deployment, on-demand scaling, observability, and high availability, be fully utilized, thereby optimizing network resource utilization, cost efficiency, and stability.

Centered around cloud-native RAN/service-based RAN, major industry organizations and research institutions have launched relevant studies. The Next Generation Mobile Networks Alliancc (NGMN), based on a global operator consensus, published the "6G Position Statement" [4], proposing important principles that future 6G network construction needs to follow. It is clear that the overall positioning of 6G is that the network will continue to evolve towards a decoupled, open, and interoperable cloud-native architecture, fostering new services by natively supporting network application programming interfaces (APIs). At the same time, the "CLOUD NATIVE MANIFESTO—An operator view by NGMN Alliance" [5] was released, stating that by adopting cloud-native approaches, telecommunications operators can fully unleash the potential of open and interoperable systems of infrastructure, container platforms, and functions/applications, thereby achieving greater flexibility and scalability, improving operational efficiency and cost-effectiveness, and offering innovative services to customers.

Hexa-X identifies intelligence, flexibility, and efficiency as key directions for future network architectures [6]. The goal of efficient networks is to streamline and redesign the cloud-native RAN and core network (CN). The three major design principles of efficient networks include: a. Service-based open interfaces: Network interfaces should be designed for cloud-native use, with attention to the rational division of services to achieve service reuse and ease of adding new services to the network. b. Network function separation, which means achieving interaction between services through APIs, reducing the dependency between network functions (NFs), thereby allowing NFs to be developed and replaced independently. c. Network simplification, based on orchestration of cloud-native RAN and CN, which allows for the use of fewer configuration parameters and external interfaces. SBA is considered one of the most important solutions for efficient networks, with related research including: interactions between RAN and CN NFs, interactions between user equipment (UE) and NFs, and independently, non-nested signaling processes.

O-RAN alliance identifies and analyzes various architectural principles related to cloud-friendly future 6G RAN architectures [7], including principles for future requirements, future standardization, and future deployment and management of the 6G RAN architecture. The report's recommendations include ensuring good separation between interfaces and layers of different vendors. Relevant solutions include minimizing the information exposed by NFs, allowing for significant changes to NFs without affecting other NFs. It also includes assigning functions to NFs to avoid splitting a specific function across different NFs, which would create unnecessary dependencies and signaling, affecting complexity and latency.

3GPP has discussed the service-based N2 interface [8]. The main objective of this research project was to introduce a new service-based N2 interface between NG-RAN and CN while retaining the traditional N2. Messages related to UE mobility and non-access stratum (NAS) signaling are transmitted through the traditional N2 interface, while other messages are transmitted through the new service-based N2 interface.

Since the study item was discussed after 5G had been widely commercialized, to avoid impacting deployed networks, the study item did not successfully proceed.

In order to fully leverage the advantages of the SBA between the RAN and the core network, and to enable the RAN to communicate with various types of NFs in a one-to-many (1:N) manner through the SBI interface, a distributed NAS model design has been proposed in [9]. For a single UE, multiple NAS connections are terminated at multiple NFs, with each NAS connection having its own security context, as well as its own set of keys and integrity and encryption algorithms. This new NAS model is more secure, as the compromise of one NAS connection does not imply the compromise of other NAS connections for the same UE.

6G-Cloud is a project funded by the European Union's Horizon Europe research and innovation program [10], aimed at building an AI-native and cloud-friendly system architecture on top of the cloud continuum through disruptive and innovative methods and solutions. It will integrate cloud resources provided by multiple stakeholders and allow for flexible, dynamic composition of network functions from different 6G network domains in a hybrid cloud environment according to service needs. The 6G-Cloud project will address the following key issues: designing a comprehensive, AI-based service-oriented 6G system architecture, developing AI-driven cloud continuum and management frameworks as well as business interfaces for multi-stakeholder environments, defining AI/machine learning (ML) frameworks to support native AI in 6G systems, and using service-oriented network design to support the 6G "network of networks" concept. The 6G-Cloud project will also incorporate scalability, sustainability, resilience, and security requirements into the system design.

Reference [11] believes that in the era of 6G, service-based architectures will extend to the RAN and be deployed in heterogeneous cloud environments of private, public, and hybrid clouds. Furthermore, as the core network becomes more distributed and the higher layers of the RAN become more centralized, costs may be reduced through the integration of RAN-CN functions.

The academic community has also conducted extensive research on the design of 6G service-based RAN/cloud-native RAN. Wang et al. [12] proposes a new, fully service-based architecture (H-SBA) for space-air-ground integrated networks (SAGIN), extending SBA from the control plane to the user plane, and from the core network to the access network. Based on H-SBA, a protocol design for the next generation of networks is proposed, and it is suggested to use the Internet Engineering Task Force (IETF) QUIC (Quick UDP Internet Connections) and SRv6 to replace 5G's HTTP/2.0 and GTP-U (GPRS Tunneling Protocol—User Plane). Song et al. [13] proposes a prototype of a microservices-based virtualized and distributed unit (vDU), redesigning the user plane applications as microservices with reasonable granularity. It also verifies that the proposed vDU prototype can bring capacity enhancement benefits due to its multi-server pooling characteristics. The vDU proof of concept is an internal implementation of the gNB (gNodeB) in a cloud environment, with no impact on the 3GPP 5G New Radio (NR) specifications. However, at present, this research only considers the reconstruction of the user plane using the microservices approach, and the design of the control plane still needs to be considered in the future. Jinho et al. [14] proposes a new service-based control plane architecture for the integration of RAN and CN, where all interfaces in the network are SBI interfaces. In this architecture, each NF can be freely deployed at any position in the network according to the characteristics of the services it provides, without being limited by the positions of other NFs. Simulation verification shows that the proposed architecture can reduce the average completion time of control procedures by 27% and the number of signaling messages by 45%.

Although the academia and industry has conducted relatively extensive research on service-based RAN, a systematic and comprehensive understanding of the technology has not yet been formed. Based on the experiences and lessons from the past research, a fully service-based design is proposed for the 6G RAN architecture design in this chapter, which introduces the service-based transformation of user plane function, RAN, and various new capabilities, and further explores the mechanisms to improve service decoupling and service invocation efficiency.

6.2 Drivers for the Design of Service-Based RAN

The drivers for the service-based RAN comes from two aspects: empowering the new business and new value, and simplifying 6G networks.

6.2.1 Empowering New Business and New Values

1. ***Capability exposure and sharing***

The direct and prompt capability exposure of RAN is expected to better empower industry and enterprise applications. Direct exposing through SBI can avoid excessive point-to-point (P2P) interface definitions and unnecessary access and mobility management function (AMF) transparent forwarding, thus enhancing communication efficiency, and is therefore considered one of the core values of service-based RAN. This specifically includes directly exposing RAN information to internal network data analysis functions (NWDAF)/location management functions (LMF)/peer RAN nodes, or directly to third-party applications. Although direct interaction between RAN and these network functions can also be achieved through defining P2P interfaces, communication based on SBI can avoid the problem of defining too many P2P interfaces when adding new functions and services, especially in future networks where communication, computing, sensing, data, AI are deeply integrated, and related network functions such as communication, computing, sensing, data, AI may all require similar RAN information, SBI will help improve the efficiency of RAN capability exposure.

2. ***Enhancing Customer Experience***

Industry and enterprise applications are a key potential market for 6G, which demands 6G networks to have the capability to flexibly adapt to the customized the needs of diverse and differentiated scenarios. 5G has proposed corresponding solutions for three major scenarios: enhanced mobile broadband (eMBB), ultra-reliable low-latency communication (uRLLC), and massive machine-type communication (mMTC). However, there are still many challenges in practical implementation. The main challenges lie in the fact that different vertical industry customers require extreme performance and customized capabilities in different dimensions, while also demanding controllable costs, agile deployment, ease of use, on-demand expansion of computing power, and customization for secondary development. The traditional 5G gNB products are designed for individual consumers (2C) with common requirements and cannot fulfil these needs simultaneously. If the approach of 5G is still followed, the pain points of commercial practice in industry applications will inevitably not be resolved. Service-based RAN, with its capabilities of elastic scaling, on-demand composition, and agile development, is taken as a primary solution to address this issue.

3. ***Supporting New Business Models***

The service-based architecture will enable third-party functions to be flexibly embedded into the RAN processing, achieving cross-layer optimization, which may in turn trigger new business models. With the deep integration of communication and computing, RAN can provide customized computing services to third parties, allowing them to embed necessary computing functions into the data processing chain rather than introducing this capability through multi-access edge computing (MEC) as an add-on, thereby optimizing user experience. For instance, for typical 6G services such as cloud gaming, cloud extended reality (XR), and holographic communication, third-party video frame rate adjustment services can be embedded in the RAN processing procedure to allow the quicker adaption of the business characteristics to the radio link quality.

6.2.2 Optimizing Network Design

1. ***Integrated Management***

Service-based RAN will achieve integrated fusion management with service-based CN and information technology (IT) support systems, reducing the management complexity of network operation and maintenance. In the telecommunications industry, the cloud-native concept and its technologies have gained widespread attention for their flexibility, agility, and convenience, and have been preliminary introduced into the 5G core network, gradually expanding to other subdomains. Additionally, the IT support systems for the telecommunications networks' operations and maintenance management are also gradually introducing cloud-native design concepts (such as microservices, containers, DevOps, etc.) to adapt to the rapidly developing needs of telecommunications business, the need for efficient and reliable platforms, and the need for agile and fast operational and maintenance requirements. Integrated management of telecommunications networks based on cloud-native will greatly reduce management complexity.

2. ***Lean Network Design***

The basic design principles of service-based RAN, such as "high cohesive", "loose coupling" and "stateless design", will help achieve lightweight design of network protocols. For example, for a large number of low-cost and low-complexity terminals (such as zero-power terminals), stateless management or weak state management will be more conducive to efficient small data transmission. The design of highly cohesive and functional services can reduce standardization overhead and avoid protocol stack function stacking and inefficiency caused by introducing similar features at different protocol layers.

3. ***Accelerating Innovation***

The service-based RAN will facilitate the rapid introduction of new RAN functions and the quick iteration and evolution of existing functions to meet market demands. Based on traditional layered and stacked design of the air interface protocols, a new feature to be introduced to RAN need to follow the long procedures of standardization (study item first, followed by a works item), production, trial and commercial launch, which typically takes few years. The long time to market prevents the network from monetizing the value and capability in many emerging industrial use cases. Although the processes of the standardization and industrialization can be shortened by reducing the Release cycle, this may sacrifice the technological maturity and stability. With a service-oriented design, new features will appear in the form of highly cohesive and loosely coupled services, each can be released and updated independently and the impact on existing functional services can be avoided and network stability can be ensured. This facilitates the stable, agile development of functional services, their rapid introduction, and iterative evolution on demand.

4. ***Network Cost Optimization***

RAN is the largest capital expenditure (CAPEX) and operational expenditure (OPEX) item for mobile network operators, e.g., its share in the total cost of ownership (TCO) rises from 45–50% to 65% for 5G. It is also the resource with the lowest utilization ratio, with most base stations typically operating with low traffic load [15]. Migrating RAN's baseband units to the cloud can take advantage of numerous benefits offered by cloud computing, such as improving utilization ratios through shared cloud infrastructure, thereby helping operators to minimize both CAPEX and OPEX. The current cloudification of RAN is more about implementing, simplifying network management processes through containerization and Kubernetes orchestration, enhancing operational and maintenance efficiency, and reducing site power consumption. This simple cloud migration approach (i.e., the Re-platform mode) does not reconstruct the RAN protocol and software architecture; it merely modularizes the design of the centralized unit (CU) and distributed unit (DU), packages and deploys them in containers, changing only the platform on which the software runs and the technical system for operations and maintenance. Therefore, issues that need to be addressed in distributed scenarios, such as data synchronization between components or services, stability, and low resource utilization rates, still need to be resolved separately [16, 17]. Only by refactoring the RAN software (i.e., the Re-build mode) can the full advantages of cloud-native infrastructure be leveraged, such as independent deployment, on-demand scaling, observability, and high availability, thereby optimizing network resource utilization, cost efficiency, and stability.

Summarily, the design of 6G RAN needs to transit from the Re-platform mode to the Re-build mode, to achieve a truly cloud-native RAN and service-based RAN. The careful study will be presented in Sect. 6.4.

6.3 Insights from the Development of 5G SBA

To support network slicing and explore the diverse vertical use cases, 5GC has adopted SBA. The experiences and lessons from the design and commercialization of 5GC is a good basis for further study on service based RAN.

6.3.1 The Design of 5G SBA Evolves Gradually with the Advancement of IT Technology

The SBA has evolved iteratively with the progress of Microservices Architecture (MSA). The SBA of 5GC Rel-15 is primarily designed with the second generation of MSA as a Ref. [18], and it mainly includes three aspects of improvement [19]:

1. ***Functional Softwareization***

Traditional CN elements with tightly coupled functions are restructured into highly cohesive and loosely coupled Network Functions (NFs). This facilitates the changes of requirements to be met by only modifying the affected network functions and services.

2. ***Lightweight Network Interfaces***

Interfaces between different network services use lightweight Restful/HTTP protocols, which facilitates rapid interface development and quick upgrades of NFs. These interfaces can also expose network capabilities to business applications.

3. ***Unified Service Framework***

By introducing the Network Repository Function (NRF) and corresponding mechanisms for service registration, discovery, and invocation, a basic communication

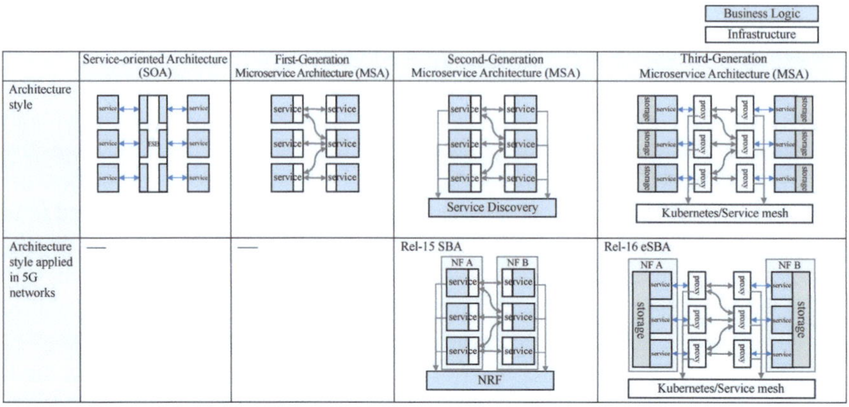

Fig. 6.2 The evolution of 5GC SBA with MSA

framework is established between network functions and network services, enabling automated service management. Service providers automatically register with the NRF when they come online, and can then be invoked by users without the need to update network routing configuration data.

5GC SBA is enhanced in Release 16 by drawing on the third generation of MSA, introducing an indirect communication method based on the Service Communication Proxy (SCP). Consequently, each Network Function (NF) only needs to focus on its own business logic, while delegating its original functions of service discovery and routing to the SCP, making the NF more lightweight [20]. Additionally, a stateless management mechanism is introduced to enhance the reliability of the SBA. Drawing on the concept of the AMF Set, the notions of NF Set and NF Service Set are introduced across all NFs. Similar NFs and services form a Set, sharing user context within the Set. Instances of NFs/NF services within a Set have identical functions, share context data, or achieve context migration, thereby ensuring the network's resilience and scalability characteristics [16, 17].

Rel-17/Rel-18 is primarily designed for the exposure of User Plane Function (UPF) events [21] and Network Exposure Function (NEF) [22], enabling the 5G network to better empower vertical industry applications. It can be anticipated that the 6G SBA will further evolve with the development of IT technologies, for example, by introducing a higher-performance, lightweight interface protocol, and a distributed service orchestration framework [20], etc. (Fig. 6.2).

6.3.2 *Insights from 5G SBA Design*

1. ***Architectural Issues***

To meet the demands of 5G service and applications, the 5GC architecture has fully embraced the principles of cloud-native design from the outset. The network architecture employs SBA, modularizing the network functions and abstracting them as independent NFs. Each NF can be scaled, evolved, and deployed on demand independently; each NF encompasses several network function services that are loosely coupled, with the characteristics of independent release, deployment, upgrade, and maintenance. A complete system process can be accomplished through the invocation of a series of services.

Under the SBA architectural pattern, multiple services jointly access and maintain the same monolithic database. For instance, services such as Namf_Communication, Namf_Location, and Namf_MT of the AMF all access and maintain the same UDSF database. This data association increases the coupling between services, making them still reliant on network elements, which to some extent limits the flexible upgrade and free combination of services.

Revelation: There are various architectures for service-based design, including Service-Based Architecture (SBA), Microservices Architecture (MSA), and Service-Oriented Architecture (SOA), each with its own characteristics and applicable scenarios [23]. Compared to SBA, MSA places greater emphasis on data isolation at the service granularity and the interaction between services, which is more conducive to the flexible combination of services oriented towards specific scenarios. In the process of future architecture design and standardization, it is necessary to fully explore the technical advantages of MSA on the basis of the SBA architectural pattern. For example, designing interface protocols that support efficient interaction between services and designing truly independent functional services (that is, not only loosely coupled in logical functions but also independent in the corresponding context data).

2. ***Service Granularity Issues***

Since there were no precedents to refer to in the division of 5GC services, the overall approach was based on an understanding of 4G processes, starting with the interaction messages of network elements to dismantle services, which resulted in a large granularity of NF services. The number of services in each NF does not exceed 10, for example, there are only 4 services in the AMF, with most functions implemented by the Namf_communication service, and the functions were not effectively separated [19]. The problem this brings is that when new services are launched, all related NFs need to be upgraded, and it is not possible to achieve independent deployment and loading of services oriented towards specific services. In addition, in some scenarios, there are too many service operation definitions, leading to lengthy signaling in the service invocation process, long process execution times, and low efficiency.

Revelation: Service granularity is key to influencing the performance of service-oriented architectures and requires continuous optimization. "Small" and "single responsibility" are not immutable principles for service division; only the most suitable service granularity can better enhance the network's agility, testability, and deployability.

3. ***Synchronous Communication Issues***

Unlike IT services, telecommunications services are serially executed and stateful. Therefore, when using synchronous remote procedure call protocols (such as REST) for service invocation, there is a need to wait for conditions to be met before responses can be given, leading to nested waiting relationships. This results in deep process call chains, and when a call fails, the rollback process is also lengthy. Additionally, the threshold for introducing new features is relatively high; introducing a new feature often affects a series of subsequent services that are called.

Revelation: The introduction of SBA in 5G has achieved a breakthrough in architecture, but the processes still follow the 4G design. This is the root cause of the aforementioned problem. Looking ahead, optimizations can be made by leveraging the characteristics of the SBA. For example, taking advantage of the direct invocation between NFs, parallelized sub-procedure can be introduced to reduce the latency of the entire procedure. Alternatively, for certain scenarios, considering the introduction of asynchronous messaging-based communication can help avoid the issues caused by message waiting and the resulting problems.

6.4 Systematic Design of Service-Based RAN

6.4.1 Overall Architecture of Service-Based RAN

The 6G service-based RAN aims to decouple traditional integrated monolithic base station functions into network functions and services based on cloud-native thinking. It achieves interaction and capability exposure between functional services through service-based interfaces, providing more flexible and streamlined network service capabilities in a composable manner. This approach helps to enhance the network's adaptability to various industries, especially meeting the needs for custom application functions, controllable costs, agile deployment, on-demand scaling, and rapid iteration [24–27].

Overall, the service-based RAN mainly includes the following four technical characteristics [26]:

- ***Functional Decoupling***, Only by building services that are highly cohesive and loosely coupled can we maximize functional reuse and leverage the advantages of cloud-native platforms. Therefore, functional decoupling is a necessary step to achieve a service-based RAN.
- ***On-Demand Composition***, which is the core value of the service-based RAN. By composing the necessary functional services on demand, it meets the customized, cost-controlled, flexible adaptation, and agile response needs of operators and customers.
- ***Capability Exposure***, which is a fundamental capability provided by the service-based RAN.
- ***Service-Based Interfaces***, through which specific services provided by network functions are accessed via APIs.

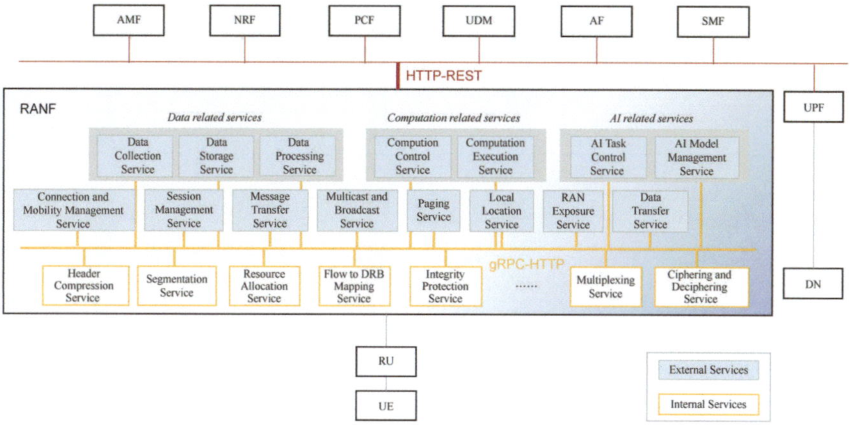

Fig. 6.3 Service-based RAN

Unlike the 5G Core (5GC) which only uses the SBA, the service-based RAN will consider both SBA and MSA to fully leverage the advantages of each pattern. When interacting with CN NFs, the RAN can act as an NF and call the services of CN NFs through standardized HTTP-REST Service-Based Interfaces (SBIs). Internally, the RAN adopts the MSA architectural pattern, where services can call each other using more efficient interface protocols, such as gRPC inline proposed in [27].

Under this architecture, the definition and implementation of RAN services will be somewhat different from those of the 5GC, with the main difference being that RAN services will be further divided into internal and external services:

- ***External services are similar to 5GC NF services***, primarily used to provide services to other NFs within the SBA; unlike 5GC NF services, RAN external services need to support at least two sets of SBI protocols. One set is for APIs that enable interoperability between NFs from different vendors or core network NFs (such as HTTP-REST as shown in Fig. 6.3), and the other set is for more efficient APIs that facilitate interaction between services within the MSA (such as Google's gRPC-HTTP remote procedure call as shown in Fig. 6.3). For example, when an external service receives a request from a core network NF, it needs to call internal services to jointly satisfy the request.
- ***Internal services are only used within the MSA*** and can be called by services within the MSA. They do not provide services to external NFs.

6.4.2 SBI Protocol

The NFs within the 5G core network are connected via service-based interfaces, utilizing a protocol stack that includes the Transmission Control Protocol (TCP), Hypertext Transfer Protocol version 2 (HTTP/2), RESTful services, JavaScript

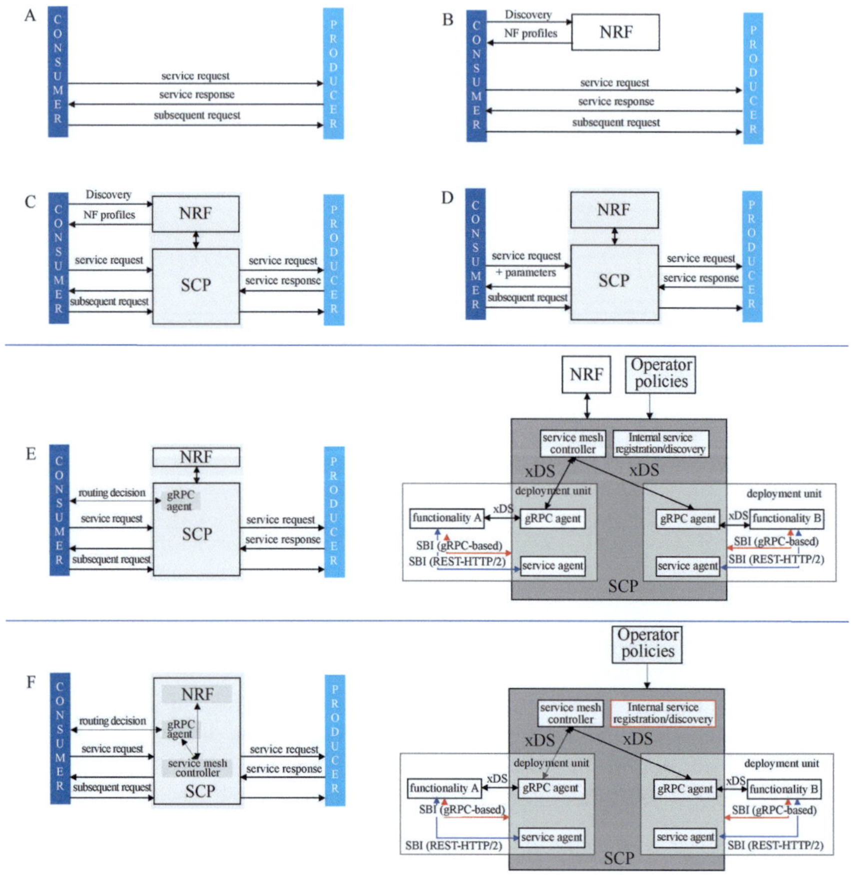

Fig. 6.4 **a** The four communication model defined by 3GPP; **b** The proposed Model E and Model F

Object Notation (JSON), and the OpenAPI 3.0 protocol suite. With the rapid advancement of IT technologies, protocols such as QUIC, gRPC, and Remote Direct Memory Access (RDMA) are maturing, and service interfaces will continue to evolve towards greater flexibility and openness, as well as higher efficiency and reliability.

- ***QUIC is a low-latency internet transport protocol based on the User Datagram Protocol (UDP)***. It ensures network reliability while reducing the number of handshakes required to establish a connection, effectively lowering the time to establish a connection. At the same time, multi-path parallel transmission effectively improves transmission efficiency and reduces TCP congestion. 3GPP Rel-15 also discussed the HTTP-over-QUIC protocol (i.e., HTTP/3), but QUIC was not mature enough at the time to be adopted by the standard. However, with the mature application of HTTP/3 in the IT field, it is expected to become the interface protocol for NF communication in the 6G phase.

- ***gRPC is a high-performance open-source RPC software framework carried by the HTTP 2.0 transport layer protocol,*** providing a method for designing API interfaces for managing and configuring network devices. The Istio gRPC-based inline service mesh proposed by Intel and SKT can reduce the latency of communication between microservices by 70% and the CPU usage by 33% [27].
- ***RDMA can solve the problem of long synchronization latency*** for remote data by bypassing the remote CPU and directly accessing the memory of the remote machine. It offloads the protocol stack to the network card, providing lower data acquisition latency.

In the SBA, HTTP/3 based on QUIC can be used as the interface protocol to achieve a smooth upgrade from HTTP-REST. In the MSA, gRPC inline or even more efficient technologies can be used to facilitate calls between services. When an external service receives a request through the HTTP/3 interface and requires assistance from an internal service, it can be encapsulated into a gRPC message and sent to the corresponding internal service.

6.4.3 Communication Model

As of now, 3GPP has defined four communication models between services [19]:

(1) Model A, Direct communication without service discovery;
(2) Model B, Direct communication with service discovery;
(3) Model C, Indirect communication without proxy discovery;
(4) Model D, Indirect communication with proxy discovery.

Overall, direct communication between services is achieved through Models A and B. In these two models, the service consumer performs service selection based on locally configured information or service producer information (NF profile) obtained from the NRF. The NF profile includes the NF type, the fully qualified domain name (FQDN) or IP address of the NF, and the names of supported services. In the Release 16 eSBA phase, indirect communication based on the Service Communication Proxy (SCP) was introduced, including Models C and D. Consequently, each NF only needs to focus on its own business logic, while delegating its original functions of service discovery and routing to the SCP, making the NF functionality more lightweight. In these two models, the SCP acts as a proxy function for the service consumer, performing the selection of service producers based on the NRF or operator policies.

To support the new gRPC-based SBI protocol stack proposed in Sect. 6.4.2, or to achieve effective intercommunication between external and internal services, the existing communication patterns need certain enhancements. For instance, during the registration process, each NF or NF service needs to register the supported interface protocols with the NRF to assist in the service discovery and selection process, selecting service providers that support the corresponding interface protocols. In addition, new communication patterns may be introduced, such as Models E and

F. Model E is similar to the enhanced Model C, with the difference that Model E lacks a service discovery process, replaced instead by a routing decision interaction process. Specifically, the SCP generates routing decisions and directly sends them to the consumer via the xDS (eXtensible Discovery Service) interface, with the consumer selecting the producer based on the routing decisions. Model F is similar to the enhanced Model D, with the main difference being that Model F also lacks a service discovery process and adds a routing decision interaction process. The difference between Model E and Model F lies in the relationship between the NRF and the SCP; in Model F, the NRF is a function of the SCP (such as Service Mesh), so when using Model F for service communication, there is no need to apply the current standardized service registration process, as the SCP will actively monitor NFs and complete the registration on their behalf. In Models E and F, interactions with external services can be achieved through REST-HTTP SBI, while interactions with internal services can be achieved through gRPC-based SBI.

6.4.4 Service Definition

6.4.4.1 Principles of Service Definition

Whether applied to services within the SBA or services within the MSA, the basic principles of service definition are consistent [28, 29], including:

- ***Functional self-containment***. The execution of a service should not inherently depend on other services.
- ***Data independence***. If services need to maintain data synchronization, then modifications to the service provider will greatly affect the service consumer, leading to service coupling. Under the principle of independence, services can undergo independent lifecycle management, meaning different services can be deployed, upgraded, and migrated independently.
- ***Reusability***. A service that is not invoked by more than one service user should not stand alone as a service.

6.4.4.2 Definition of External Communication Services

The two primary service decomposition models commonly used in the IT field are sub-domain decomposition based on Domain-Driven Design (DDD) and decomposition by business capability [28]. The overall approach involves analyzing what the application does, with each business capability potentially being considered a separate service. For the RAN, its business capabilities are primarily reflected in its interface interactions. The control plane external interaction interfaces include NG-C and XN-C, with the corresponding service entities being AMF and peer gNB. The user plane external interaction interfaces include NG-U and XN-U, with the corresponding service entities being UPF and peer gNB. Referring to the main functions and processes supported by NGAP/XnAP [30, 31] and the PDU Session user plane protocol/NR user plane protocol, the RAN can generally be broken down into the following services:

(1) The Session Management Service (PMS) is primarily responsible for the creation, updating, release, and status notification of sessions, as well as the creation, updating, release, and status notification of contexts. For example, the AMF can establish, modify, or release the necessary UE contexts in the access network through Create/Update/UpdateTrigger/Release requests, including: Security Key, Mobility Restriction List, UE Radio Capability, and UE Security Capabilities, etc.
(2) The Connection Mobility Management Service (CMS) overall includes UE context management, mobility management, and multi-connectivity management. UE context handling also includes context creation, suspension/resumption, release/release trigger, modification, push (which can be triggered by OAM to place a UE context with another Connection Mobility Management Service), transmission, information verification, and information query services.
(3) The Paging Service (PS) is used to ensure UE reachability or to send light control signaling/data to the UE through paging.
(4) The Signaling Transmission Service (STS) provides the transmission of UE NAS messages and non-UE messages, such as the AMF can reroute the Initial UE Message to another AMF through the NASMessageReroute request.
(5) The Multicast Broadcast Service (MBS) is used to provide multicast broadcast services. For example, consumers such as AMF/LMF can use Transfer/Cancel requests to perform or cancel multicast broadcast data transmission, which can carry multicast broadcast resource requirements, QoS requirements, transmission methods, etc.; they can also use StatusNotify to learn about the multicast broadcast status, such as whether the multicast broadcast message has been successfully received by the UE.
(6) The Data Collection Service (DCS) is used to obtain access network measurement data. For example, consumers can activate/deactivate a trace session for a UE through ActivateTrace/DeactivateTrace requests to DCS.
(7) The Local Location Service (LLS) is used to provide UE location information, such as positioning information, geographical location, used positioning methods, and failure reasons.
(8) The RAN Exposure Service (RES) is used to subscribe/unsubscribe to related events from the access network.
(9) The Bearer Management Service (BMS) provides the management of bearers, including the establishment, modification, and release of bearers.
(10) The Data Transfer Service (DTS) mainly realizes the data transfer of the user plane or data plane, equivalent to an API gateway service, calling internal user plane functions based on related data/business QoS requirements, providing reliable, fast, and large-volume data transfer services (Fig. 6.5).

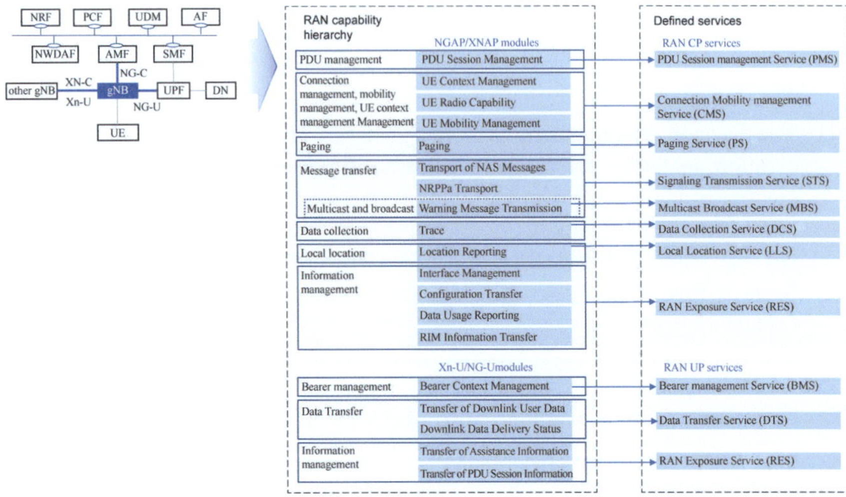

Fig. 6.5 Services provided by RAN

6.4.4.3 Definition of Internal Communication Services

Regarding the definition of internal services, we can gain some insights from existing 3GPP standards [32–36]. Taking the Packet Data Convergence Protocol (PDCP) as an example, TS38.323 not only specifies the functions of PDCP but also provides the "services" it offers to the upper layer (i.e., transmission control plane data service, transmission user plane data service, header compression service, uplink data compression service, encryption, integrity protection), as well as the "services" it expects from the lower layer (i.e., Radio Link Control (RLC) acknowledged data transfer service, unacknowledged data transfer service). Some functions can also be considered as its services, such as header compression and encryption, while others cannot be categorized as services, which we believe is because they do not conform to the principles of service definition. For example, functions such as RLC's resegmentation and Automatic Repeat reQuest (ARQ) error detection are only applicable to AM data transmission and are closely coupled, which does not meet the "self-contained" principle of service definition. Based on the above considerations, we believe that internal services can include: configuration and reporting, header compression service, encryption, integrity protection, QoS flow to Radio Bearer (RB) mapping, RLC transparent mode (TM) data transfer, RLC unconfirmed (UM) data transfer, RLC acknowledged (AM) data transfer, data multiplexing, hybrid automatic repeat reQuest (HARQ), and radio resource allocation (Table 6.1).

Table 6.1 Services and functions defined by 3GPP specifications

Layer	Services provided to upper layers	Functions
RRC	– Broadcast of common control information; – Notification of UEs in RRC_IDLE, e.g. about a mobile terminating call; – Notification of UEs about ETWS and/or CMAS; – Transfer of dedicated signalling; – Broadcast of positioning assistance data; – Transfer of application layer measurement configuration and reporting	– Broadcast of system information: – RRC connection control: – Inter-RAT mobility including e.g. AS security activation, transfer of RRC context information; – Measurement configuration and reporting: – Configuration of BAP entity and BH RLC channels for the support of IAB-node – Configuration of SRAP entity and Uu/PC5 Relay RLC channels for the support of L2 U2N relay – Other functions including e.g. generic protocol error handling, transfer of dedicated NAS information, transfer of UE radio access capability information – Support of self-configuration and self-optimisation – Support of measurement logging and reporting for network performance optimisation; – Support of transfer of application layer measurement configuration and reporting
SDAP	– Transfer of user plane data	– Transfer of user plane data; – Mapping between a QoS flow and a DRB for both DL and UL; – Mapping between an MBS QoS flow and an MRB for DL; – Mapping between a PC5 QoS flow and a SL-DRB for NR sidelink communication; – Marking QoS flow ID in both DL and UL packets; – Marking PC5 QoS flow ID in unicast of NR sidelink communication packets; – Reflective QoS flow to DRB mapping for the UL SDAP data PDUs
PDCP	The following services are provided by PDCP to upper layers(RRC or SDAP layers): – Transfer of user plane data; – Transfer of control plane data; – Header compression; – Ciphering; – Integrity protection	– Transfer of data (user plane or control plane); – Maintenance of PDCP SNs; – Header compression and decompression using the ROHC protocol; – Header compression and decompression using the EHC protocol; – Ciphering and deciphering; – Integrity protection and integrity verification; – Timer based SDU discard; – For split bearers and DAPS bearer, routing; – Duplication; – Reordering and in-order delivery; – Out-of-order delivery; – Duplicate discarding

(continued)

Table 6.1 (continued)

Layer	Services provided to upper layers	Functions
RLC	– TM data transfer; – UM data transfer; – AM data transfer, including indication of successful delivery of upper layers PDUs	– Transfer of upper layer PDUs; – Error correction through ARQ (only for AM data transfer); – Segmentation and reassembly of RLC SDUs (only for UM and AM data transfer); – Re-segmentation of RLC SDU segments (only for AM data transfer); – Duplicate detection (only for AM data transfer); – RLC SDU discard (only for UM and AM data transfer); – RLC re-establishment; – Protocol error detection (only for AM data transfer)
MAC	– Data transfer; – Radio resource allocation	– Mapping between logical channels and transport channels; – Multiplexing of MAC SDUs from one or different logical channels onto transport blocks (TB) to be delivered to the physical layer on transport channels; – Demultiplexing of MAC SDUs to one or different logical channels from transport blocks (TB) delivered from the physical layer on transport channels; – Scheduling information reporting; – Error correction through HARQ; – Logical channel prioritisation; – Priority handling between overlapping resources of one UE; – Radio resource selection

6.4.4.4 Definition of New Capability Services

6G will achieve a deep integration of communication, sensing, computation, and intelligence, and new capabilities will also be introduced as services.

1. ***AI-related services.***

The AI Task Control Service can invoke connection, computation, and data services to execute relevant AI tasks. For example, by interacting with the Data Collection Service through the AI task control function to obtain UE-related data, and interacting with the connection function to establish a connection between the UE and the AI Task Control Service, enabling the UE to serve as an execution node for AI tasks, providing services such as AI model training/inference. The AI Model Management Service offers functions such as model registration, model selection, and model performance monitoring.

2. ***Data-related services***.

Existing networks only serve as data transmission pipelines, which cannot meet the future ubiquitous AI application requirements for the entire process of data collection, transmission, processing, and storage. Introducing a Data Collection Service on the

RAN side can effectively meet the large volume and real-time requirements of 6G AI big data, while avoiding performance impacts on other functional services and UEs due to data collection. The Data Storage Service can evolve the Unified Data Storage Repository (UDR)/UDSF to achieve more reliable and efficient data storage and read/write operations. The Data Processing Service provides functions such as privacy removal processing and data refinement.

3. ***Computation-related services***.

The Computation Control Service provides functions such as computation resource perception and allocation, computation node selection, and computation deployment/suspension/recovery/termination. The Computation Execution Service provides functions for AI model training or AI model inference.

6.4.5 Service Orchestration

6.4.5.1 Service Orchestration Methods

The Data Transfer Service is the primary service provided externally to users, which is implemented by calling a series of internal services. Overall, service orchestration can be achieved through two methods: centralized orchestration and decentralized collaboration.

Orchestration Model 1: Decentralized Collaboration.

The decentralized collaboration method is similar to the chain pattern in IT, where there is no unified orchestration service, and the services in the workflow share the coordination responsibilities. The service orchestration strategy needs to be pre-distributed to each internal service, so that the internal services can directly select the next internal service based on specific identifiers.

Orchestration Model 2: Centralized Orchestration.

The centralized orchestration method is similar to the aggregate pattern, where the aggregate service (i.e., the Data Transfer Service) selects internal services one by one based on the orchestration strategy to achieve customized data processing (Fig. 6.6).

Flexible service orchestration can bring the following two advantages:

Advantage 1: Enable parallel processing of multiple services.

The processing order not only includes serial chaining but also parallel processing. For example, the existing PDCP layer would first perform integrity protection, then

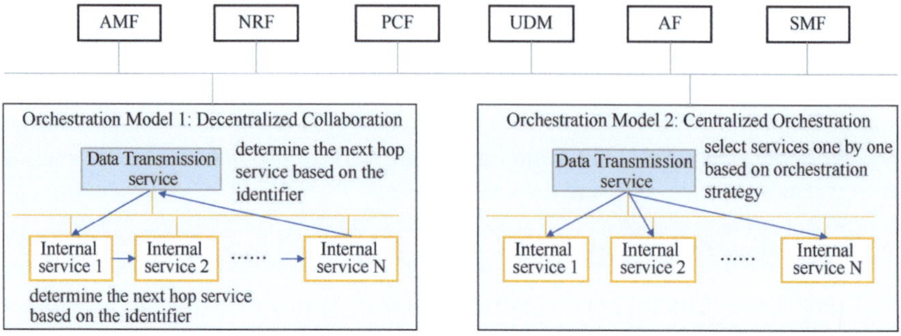

Fig. 6.6 Two models of services orchestration

encryption, and finally add the PDCP header. Considering that the main objects of encryption are the Internet Protocol (IP) and the packet headers, data fields, and MAC-I above IP, but the necessity of encrypting MAC-I is not significant, services such as encryption and integrity protection can be executed in parallel. This can greatly enhance the efficiency of data processing (Fig. 6.7).

Advantage 2: Allows for dynamic scaling at the service level.

Tidal phenomena are common dynamic changes in networks, where the number of users accessing the network and the volume of data transmitted both vary with time across different spaces. A service-based RAN can monitor these tidal variations in the network, dynamically deploying or deleting service instances to achieve dynamic scaling at the service level. For example, when the number of access users increases, only the services related to user management in the control plane need to be expanded. When the volume of downlink data increases, only additional instances of the downlink service are required. This is in contrast to the existing 5G data wireless bearers, which can only achieve scaling at the bearer level. As a result, the utilization rate of computing resources can be further optimized.

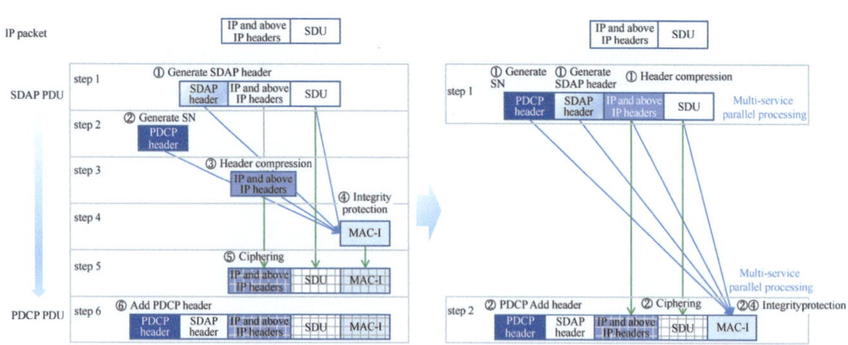

Fig. 6.7 Multi-services processed in paralleled

6.4.5.2 Analysis of the Impact on the Air Interface

1. ***Impact on the CP (RRC) Protocol***

With the service-based transformation of the RAN control plane, the functions of RRC have essentially been decomposed into services within the control plane. However, to avoid affecting the existing protocol system, control signaling over the air interface can still be sent via Radio Resource Control (RRC) messages. Similar to how the DU sends RRC messages to the CU via the F1AP, control plane services (such as the Connection and Mobility Management Service) can send connection and mobility-related RRC messages to the Signaling Transmission Service via a gRPC interface. The Signaling Transmission Service then generates the complete RRC messages directed at the UE and completes the transmission of the messages over the air interface. Of course, as UEs are gradually upgraded and evolved, UEs could also completely replace RRC with HTTP, which could reduce the RRC protocol encapsulation process and accelerate the deployment of new capabilities on the UE.

2. ***Impact on the UP Protocol***

As mentioned earlier, the control plane can reuse the existing RRC protocol to carry service-based configurations, which have a minimal impact on the UE. However, the user plane is different. The service-based approach of the RAN user plane breaks through the limitations of the traditional layered protocol stack, allowing functions to be flexibly invoked and combined on demand, without being restricted to the upper and lower layers. But this flexibility brings the challenge of redesigning the data processing order and header formats. This means that the UE needs to have the capability to support the on-demand combination of protocol functions.

In the traditional layered protocol model, data packets are processed layer by layer, and correspondingly, headers are added layer by layer, with the format of each layer's header being essentially fixed. Unlike the traditional fixed-format headers, the header format in the service-based RAN architecture is related to the services on the processing path [38]. Therefore, the header may include two parts: a common part and a service-specific part. In the service-specific part, only that service can modify the corresponding section of the header. As shown in Fig. 6.8, only Service 1 can modify the middle blue part of the header, Service 2 can modify the light blue part, and Service 3 can modify the dark blue part. To avoid data security issues between services from different vendors, for the blue parts of the header, Service 1 is allowed to modify, Service 2 is only allowed to access, and Service 3 is neither allowed to modify nor access. Therefore, a key for data access will be assigned to Service 2, and a key for modifying data will be assigned to Service 1.

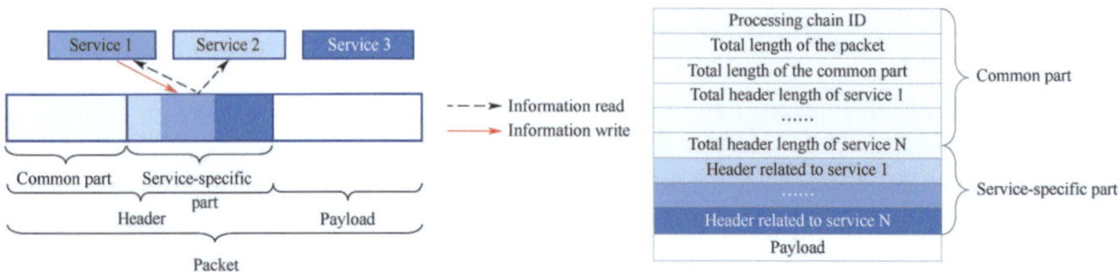

Fig. 6.8 Service-specific header design

6.4.6 The Evolution of NAS

The evolution of NAS unfolds along two dimensions. First, there is the change in the NAS model, which is adapted to the characteristics of the service-based architecture through a distributed transmission approach [9]. Second, there is the change in the NAS messages, where the content of NAS messages is carried by HTTP to accelerate the introduction of new capabilities and rapid iterative evolution.

6.4.6.1 Distributed NAS Model

Although the service-based architecture allows direct interaction between the RAN and CN NFs, messages such as the LTE Positioning Protocol (LPP) between the UE and the Location Management Function (LMF) are still required to be forwarded to the AMF due to the limitations of the NAS model. This is because messages like LPP are carried by NAS-MM, and the anchor point for NAS-MM is the AMF. To avoid unnecessary AMF forwarding of messages such as LPP and NAS-SM, the NAS model needs to be adaptively modified. The new protocol model will break through the dependencies between NAS protocols, allowing different NAS messages to be directly forwarded to the corresponding CN NF without going through the AMF. To support this change, the RRC layer needs to be enhanced to have the capability to recognize different NAS messages.

6.4.6.2 HTTPization of NAS Messages

The previous subsection discussed the improvement of the NAS model, while this section focuses on the improvement of NAS messages, specifically defining traditional NAS messages as HTTP messages. Consequently, the UE can interact with NFs and NFs can interact with each other using the same protocol model without the need for protocol translation. As shown in Fig. 6.9, the improvement of NAS messages can be divided into two specific options.

Option 1: Define only the messages related to the new capabilities of 6G as HTTP messages. Following the traditional design approach, new capability messages (such

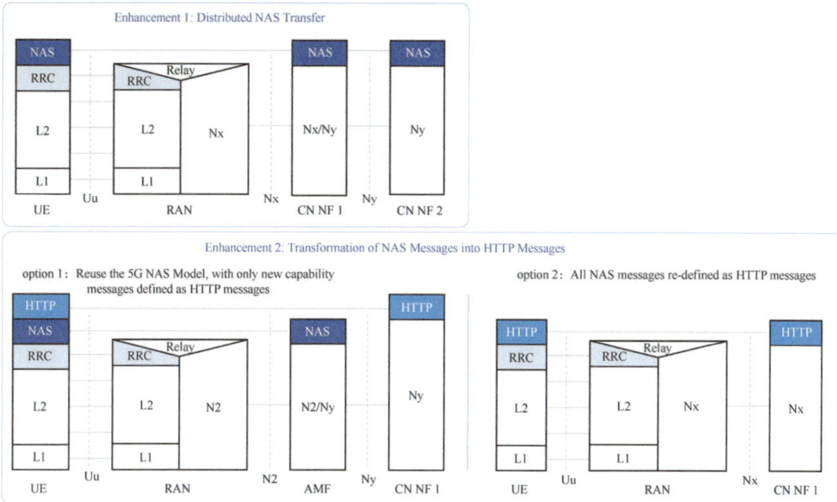

Fig. 6.9 Comparison between 5G NAS model and new NAS model

as 6G computation, data, etc.) that need to be exchanged between the UE and CN would be defined as NAS or other messages carried by NAS. However, in a service-based architecture, these signals can be defined as HTTP messages. When transmitted over the air interface, these HTTP messages can be encapsulated in the existing NAS-MM and sent to the UE. This can reduce the standardization workload brought about by NAS design.

Option 2: In addition to the new capability signals of 6G being designed as HTTP messages, traditional NAS-SM, UE policy, LPP, and other messages can also be converted into HTTP message formats. This means that the traditional NAS protocol will no longer exist, and when transmitted over the air interface, the RRC protocol will directly carry HTTP messages. This change will naturally support the distributed transmission mechanism mentioned in Sect. 6.4.6.1 and is the ideal state for the development of future cloud-native networks.

The distributed NAS model and the HTTPization of NAS messages represent two distinct dimensions in the evolution of NAS. Both may be concurrently enhanced during the standardization process for 6G.

6.4.7 Radio Capability Exposure

Starting from Release 15, 3GPP has been researching the opening of network capabilities, providing relevant specifications for the Network Exposure Function (NEF), and designing 36 standard capability exposure services for Application Functions (AF) and third-party applications. These include services such as QoS capabilities, event monitoring, parameter configuration, device triggering, Packet Flow Description (PFD) management, traffic steering, background traffic, and policy charging, etc.

With the development of technologies such as big data and AI in network applications, the RAN also needs to abstract various open capability services to provide to the business application layer, empowering industry applications in a timely and efficient manner. Broadly speaking, the open capability services that the RAN can provide include the following two categories:

① ***Event Exposure Service***: Allows internal or external AFs to subscribe to various RAN/UE-related events, such as changes in the location of the UE.
② ***Parameter Configuration Service***: Allows AFs or third parties to provide RAN/UE configurations or policies. For example, allocating specified bandwidth and computing resources for certain types of applications.

To avoid the fragmentation of API capability exposure standards, 3GPP TS 23.222 [22] defines the Common API Framework (CAPIF) capability exposure framework standard as a universal standard. According to this framework, capability exposure can be structurally divided into the following four logical functions. When building actual network architectures, these four functional modules can be flexibly combined and deployed either centrally or in a distributed manner:

(a) ***API Exposing Function (AEF)***: Provides different capability exposure API services based on its own capabilities, and works with the CAPIF Core Function to authenticate and authorize API callers.
(b) ***API Publishing Function (APF)***: Publishes API information provided by the API Provider to the CAPIF Core, so that API invokers can discover the API services.
(c) ***API Management Function (AMF)***: Provides functions for API Providers to manage APIs.
(d) ***CAPIF Core Function***: Defines functions such as API authentication, authorization, logging, and billing, which are common requirements for various capability exposure functional entities.

The CAPIF Core Function and API Exposing Function provide services externally through the Cccf and Caef service-based interfaces.

In actual deployment, the NEF can implement the complete CAPIF architecture or only partial CAPIF functions. When introducing RAN capability exposure, the relationship between the RAN Exposure Function (REF), NEF, and CAPIF needs further consideration. Here, the REF can be the RANF or a new NF added by the RAN to achieve external capability exposure. For example, when the NEF implements the complete CAPIF architecture, i.e., option 1, the REF needs to leverage the NEF's functions to expose RAN capabilities. The REF and NEF can have a hierarchical relationship or can be independently deployed. In a hierarchical relationship, the NEF API exposure function can act as the API gateway for REF1. In an independent deployment relationship, the REF publishes services to the NEF and provides them externally to API invokers for discovery through the NEF. However, unlike the

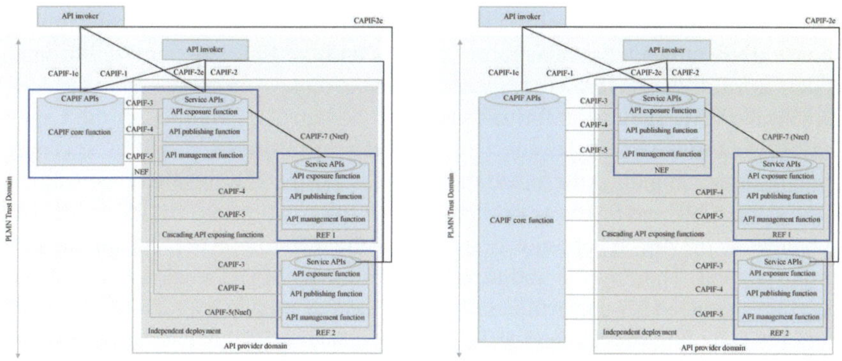

Option 1: NEF implements the CAPIF architecture

Option 2: NEF implements the service specific aspect compliant with CAPIF architecture

Fig. 6.10 Relationship between REF and NEF

hierarchical model, the REF can directly provide services to API invokers through CAPIF-2 or CAPIF-2e. When the NEF is mapped only to the AEF, APF, and AMF of CAPIF, and the CAPIF core function is implemented by the network capability operation platform, i.e., option 2, the REF and NEF can also have a hierarchical relationship or be independently deployed. In this scenario, the REF and NEF functions are similar, both requiring the independent provision of open services through the network capability operation platform (Fig. 6.10).

6.4.8 *Joint Design of RAN and CN*

6G will achieve holistic service-based architecture, connecting the RAN and CN. This will also facilitate the joint design of RAN and CN, including process optimization, functional integration, flexible deployment, and more.

6.4.8.1 Procedure: Parallel Optimization

With the distributed NAS model, the RAN can communicate directly with CN NFs. In addition to reducing unnecessary AMF forwarding steps, another advantage is that the end-to-end process can be further broken down into multiple sub-processes for parallel execution. For example, the registration process and the PDU session establishment process can be executed in parallel, enabling the rapid establishment of service channels.

As shown in Fig. 6.8, when a new base station determines to quickly establish a service channel through this mechanism, it can obtain the AUSF/UDM addresses based on the RAN/CN UE context, and then obtain subscription data related to the selection of the AMF/session management function (SMF) from the AUSF/

UDM, achieving the selection of the AMF/SMF. Based on the selection results, the new gNB initiates the registration process and the PDU session establishment process towards the AMF/SMF respectively. In the request message sent to the AMF, the selected SMF address information should be carried; similarly, in the request message sent to the SMF, the selected AMF address information should be carried. Thereby, after completing the registration process and the PDU session establishment process, the AMF and SMF can exchange status information and UE identifiers with each other, achieving synchronization between the registration context and the PDU session context. Thereby, the traditional registration and PDU session establishment processes that need to be executed serially can be executed in parallel, greatly shortening the signaling process and avoiding the end-to-end delay caused by response waiting (Fig. 6.11).

6.4.8.2 Functionality: Integrated Design

From an end-to-end network perspective, there are many similar functional services in the RAN control plane and the CN control plane, including connection and mobility management functions, and bearer and session management functions. Considering the trend of 6G core network decentralization and the centralization of higher RAN layers, similar functions of RAN and CN can be designed integrally to enhance the flexibility of NF deployment and reduce signaling overhead. For example, the connection and mobility management functions of RAN and CN can be integrated into a uAMF (unified AMF), and the session and bearer management functions of RAN and CN can be integrated into a uSMF (unified SMF). This can lead to more streamlined processes for service establishment, handover, QoS management, and more.

6.4.8.3 Deployment: On-Demand Composition

The value of service-based architecture lies in the ability to compose network function services according to the needs of business applications. On-demand composition can be divided into two levels: services are composed on-demand into NFs, and NFs are composed on-demand into subnets.

Composition of services into NFs. Unlike the service-based 5GC, the service-based RAN breaks through the binding relationship between 5G NFs and services, allowing services to be flexibly defined as NFs for specific scenarios. This can reduce the cost of base station equipment by streamlining NF functions. At the same time, interactions between NFs that were traditionally realized through standardized SBI interfaces can be transformed into interactions through internal interfaces (such as gRPC inline), thereby shortening the latency of service invocation. This process can be achieved based on the service registration process, which can be specific to: NRF, through a service discovery request, learns about the multiple services that need to

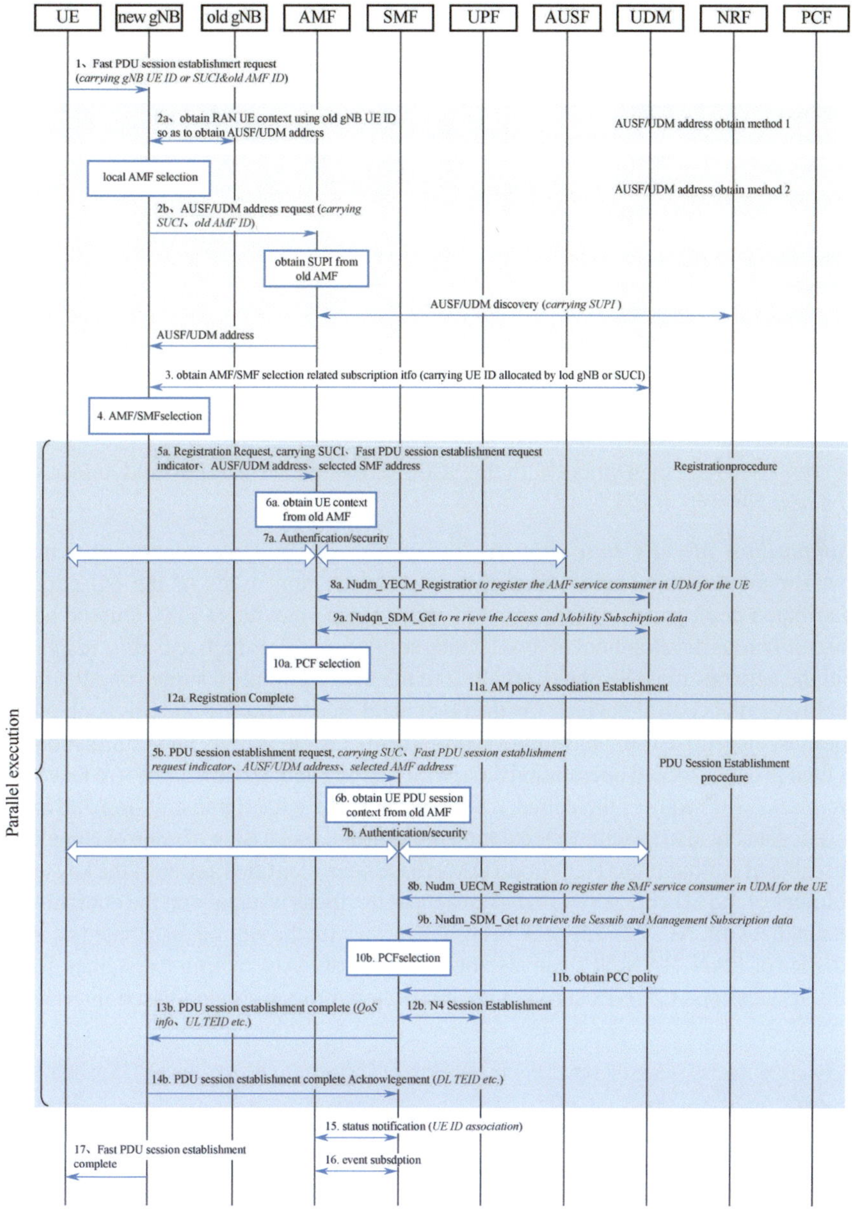

Fig. 6.11 A fast a PDU session establishment procedure: parallel execution of registration and PDU session establishment procedures

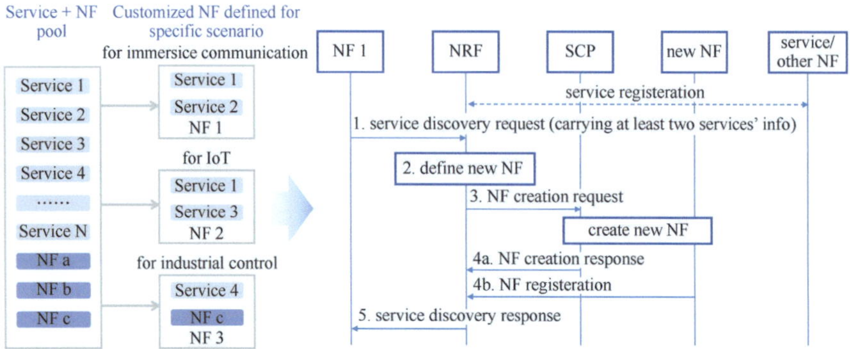

Fig. 6.12 New NF defined for specific scenario

be composed, and then works with the SCP to complete the creation and registration of a new NF (Fig. 6.12).

Composition of NFs into subnets. To address the massive connections, ultra-low latency communication, and diverse customization needs of the 6G era, the distributed deployment of 6G networks will become inevitable [38]. On one hand, the continuous development of low-latency services requires high real-time response, and the network must be deployed close to the user equipment, supporting the local deployment of control devices. On the other hand, due to security and privacy considerations, enterprise users have high requirements for the privacy and confidentiality of their production and operational data, with a strong demand for data not to leave the premises. This requires the network not only to achieve local transmission and analysis of data but also to achieve local transmission and processing of control signaling. Distributed Autonomous Networks (DAN) have been identified as one of the key technologies of the 6G core network, but currently, the focus is mainly on the constitution of core network NFs into subnets. In combination with the natural distributed characteristics of RAN, RAN NFs can also become part of DAN. Within the subnet, RAN NFs/services and CN NFs/services are connected through service-based interfaces.

6.4.9 Service-Based UE

As the capabilities of UEs continue to grow, UEs can be more than just consumers of network services; they can also be producers of services. With the help of service-based interfaces, UEs can communicate directly with relevant network functions, reducing unnecessary transmission paths and lowering end-to-end transmission latency. When UEs provide or consume services to/from the CN, traditional NAS messages need to be replaced with HTTP messages (as described in Sect. 6.4.6.2); when UEs provide or consume services to/from the RAN, traditional RRC messages need to be replaced with HTTP messages (as described in Sect. 6.4.5.2).

Services that UEs can provide include: data collection services, computing services, and open services. UEs can offer information such as wireless measurement reporting, location measurement, and logged measurements through data collection services to CN NFs or RAN NFs. They can also assist the network in completing AI training, inference, and other tasks through computing services. Additionally, they can expose information related to UE applications through open services, such as notifications of changes in UE location and UE capability information.

6.5 Performance Verification

This section conducts testing and verification of the RAN user plane service-based solution. As shown in Fig. 6.13, the Data Transfer Service defined in Sect. 6.4.4.3 is further divided into Uplink Data Transfer Service (ULT) and Downlink Data Transfer Service (DLT). Based on these two services, as well as the User Plane Control Service (UPC) and Data Traffic Distribution Service (DTD), a service scaling mechanism is designed to expand or reduce the resources corresponding to the RAN user plane services according to the changes in traffic volume. This is compared with a non-service-based RAN solution to verify the performance of compute resource utilization. The functions of the user plane services are as follows:

- ***UPC Service***: Includes the E1AP module and the UPC module, and simultaneously receives control messages from the CU-CP. It specifically involves: establishing an E1 connection with the CP, receiving bearer context maintenance messages sent by E1AP, and maintaining the UPC state machine.
- ***DLT Service***: Equipped with SDAP/PDCP, NG-U, F1-U entities, and downlink processing functions, it is mainly responsible for receiving configuration

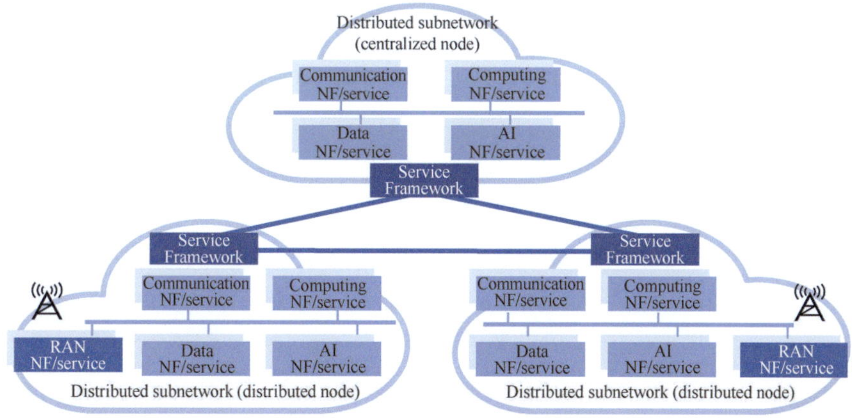

Fig. 6.13 Customized sub-net (NF set) defined for specific scenario

messages from UPC and completing the processing of downlink data packets. The interaction between NUPC and DL is primarily the maintenance of bearer contexts.

- ***ULT Service***: Equipped with SDAP/PDCP, NG-U, F1-U entities, and uplink processing functions, it is mainly responsible for receiving configuration messages from UPC and completing the processing of uplink data packets.
- ***DTD Service***: Responsible for distributing the UE's uplink and downlink traffic to the corresponding DL or UL instances. Upon UE access, UPC first allocates DL instances and UL instances to the UE. After the UE successfully accesses the network, UPC synchronizes the UE-DL instance status relationship information and the UE-UL instance status relationship information to DTD. Upon receiving the status relationships, DTD saves them in memory (Fig. 6.14).

This demonstration uses the number of CPU cores to represent computing power, and calculates the average number of CPU cores $x = \frac{\sum_t \tau_t}{T}$ occupied by the user plane in real time during the test period using the Grafana tool, where τ_t is the total number of CPU cores occupied by the user plane at time t, and T is the total length of the test time interval. Under the service-based architecture, when there are no UEs connected, 4 CPU cores are occupied by the user plane, with UPC, DLT, ULT, and DTD each occupying one CPU core. As the number of UEs increases, DLT scales instances based on the DL load $z = y/(\alpha * n)$, where y represents the current number of UEs connected for the test case, n is the actual number of DL instances launched for the current test, and α is the benchmark for the number of UEs per single DL instance, with $\alpha = 5$ in this test. When z > 0.8, the DL instance expands; when z < 0.5, the DL instance contracts.

Simulate terminal access for 14 UEs, observe the number of DL instances and the total number of CPU cores occupied by the user plane in Grafana after the user plane is scaled out. Then log out 13 UEs, retain 1 UE, and observe the number of DL instances and the total number of CPU cores occupied by the user plane in Grafana after the user plane is scaled in. During the test period, the average improvement in

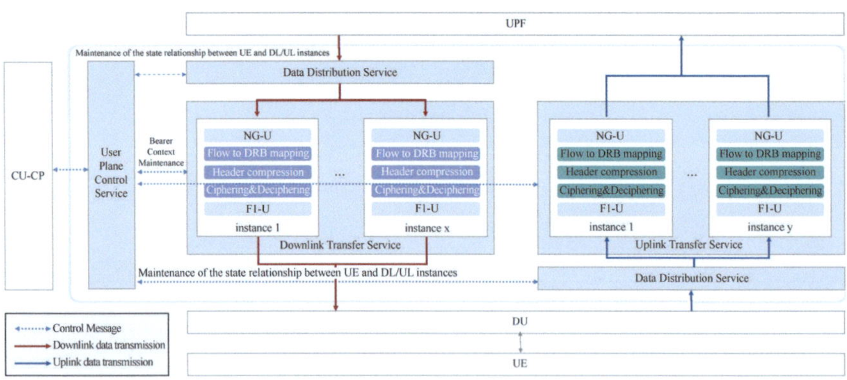

Fig. 6.14 Illustration of RAN user plane services

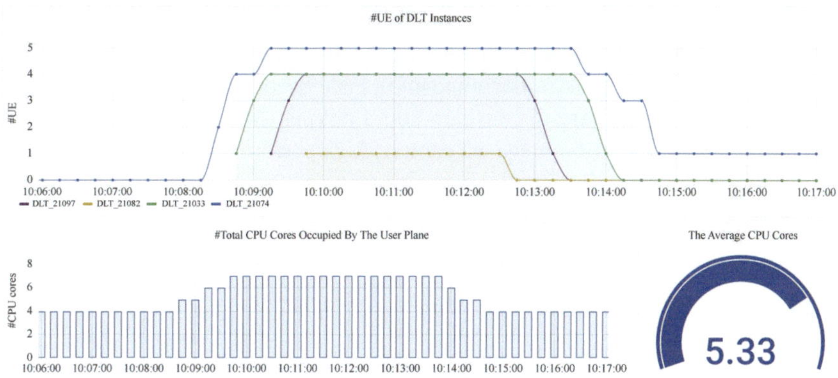

Fig. 6.15 Performance of computational resource utilization

computing resource utilization is calculated using the formula $\frac{\psi - x}{\psi} \times 100\%$, where ψ is the number of CPU cores occupied by the user plane at the peak traffic time.

As shown in the figure, after connecting 14 UEs, the DLT expands, and the total number of CPU cores occupied by the user plane is 7; after logging out 13 UEs, the DLT contracts to 1 instance, and the total number of CPU cores occupied by the user plane at this time is 4. During the time period from 10:06:00 to 10:17:00, the user plane underwent scaling out and scaling in, with an average occupation of CPU cores by the user plane, $x = 5.33$. Compared to the 5G planning deployment method (which is based on the resource requirements at peak traffic for fixed deployment $\psi = 7$), the utilization rate of computing resources has increased by $\frac{\psi - x}{\psi} \times 1\ 00\%$ $= 23.9\%$ (Fig. 6.15).

6.6 Summary

The RAN has been developed as an integrated monolithic entity for decades, with tightly coupled functions and strongly related context data, making it difficult to separate independent logical functional services. With the rapid development of the integration of ICDT technologies, the evolution of RAN towards cloud-native and service-oriented architectures has become a trend. This chapter deeply analyzes the driving forces behind the service-based RAN, based on the experience and lessons learned from the service-based 5G core network, and provides a systematic design for the service-based RAN, as well as an analysis of its impact on the CN and UE. For the service-based RAN, it will inevitably cause a certain degree of traditional network performance loss, but it enhances the system's agility, customization, adaptability, stability, and other dimensions of network capabilities, making it more adaptive for the diverse requirements of future scenarios. The service orientation implies hardware-software decoupling, functional decomposition, and interface opening, which will inevitably lead to performance losses. Compared to the core network, the RAN has even more stringent performance requirements. The necessity and rationality of RAN service decomposition have become a focus of attention in the industry and require careful consideration. Looking towards future scenarios, it is necessary to re-examine the trade-offs among high performance, diversity, and low cost, rather than simply defining high performance as the optimization goal of the system.

References

1. Liu, G., Huang, Y., Wang, F., Liu, J., & Wang, Q. (2018). 5G features from operation perspective and fundamental performance validation by field trial. China Communications, 015(011), 33-50.
2. Liu, G., Hou, X., Huang, Y., Shao, H., & Wang, Q. (2019). Coverage enhancement and fundamental performance of 5g: analysis and field trial. IEEE Communications Magazine, 57(6), 126-131.
3. C. Coletti et al., "O-RAN: Towards an open and smart RAN," O-RAN Alliance White Paper, Oct. 2018.
4. 6G Position Statement. NGMN. 2023.
5. CLOUD NATIVE MANIFESTO-An operator view, NGMN, 2023.
6. Deliverable D5.3 Final 6G architectural enablers and technological solutions. Hexa-X. 2023.
7. Architecture principles for a cloud-friendly future 6G RAN architecture, Report, O-RAN next Generation Research Group (nGRG), 2024.
8. New WID Study on Service-based interface between NG-RAN and 5GC [presentation]. In: 3GPP TSG-SA2 Meeting 148e; 2021 Nov 15–19; online; 2021.
9. 6G Technologies white paper, Next G Alliance, 2022.
10. Service-oriented 6G network architecture for distributed, intelligent, and sustainable cloud-native communication systems. 6G-CLOUD. 2024.9.5
11. 6G explained, Nokia Communications, 2024 Sep 5. Available from: https://www.nokia.com/about-us/newsroom/articles/6g-explained/.
12. Wang X, Sun T, Duan X, et al. Holistic Service-Based Architecture for Space-Air-Ground Integrated Network for 5G-Advanced and Beyond. China Commun 2022;19(1):14–28.
13. Song P, Peng H, Zhang X. A Micro-Service Approach to Cloud Native RAN for 5G and Beyond. IEEE Access 2023;11:130257 - 130271.
14. Jinho C, Neha S, Shiva S, et al. RAN-CN Converged Control-Plane for 6G Cellular Networks. In: Proceedings of 2022 IEEE Global Communications Conference Workshops; 2022 Dec 4–8; Rio de Janeiro, Brazil: IEEE; p. 1253–1258.
15. RAN-in-the-Cloud: Bringing Cloud Economics to 5G RAN, NVIDIA Corporation Enterprise Solutions, 2023.
16. Alibaba Group; Alibaba Cloud Intelligent Business Group; Cloud Native Application Platform. Alibaba Cloud Cloud Native Architecture Practice. China Machine Press, 2021.
17. Neil Ford, et al. Software Architecture: The Hard Parts. China Machine Press.
18. Cloud Native Architecture White Paper, Alibaba Cloud Computing Co., Ltd., 2022.
19. TS 23.501: System architecture for the 5G System (5GS) specification. 3GPP standard. France: 3GPP; 2024.
20. TS 23.742: Study on Enhancements to the Service-Based Architecture. 3GPP standard. France: 3GPP; 2018.
21. New WID New SID on Study on UPF enhancement for Exposure and SBA [presentation]. In: 3GPP TSG-SA Meeting 94e; 2021 Dec 14–20; online; 2021.

22. TS 23.222: Functional architecture and information flows to support Common API Framework for 3GPP Northbound APIs. 3GPP standard. France: 3GPP; 2024.
23. Mark Richards, Neal Ford. Fundamentals of Software Architecture. 2021.
24. Liu G, Li N, Degn J, et al. The SOLIDS 6G Mobile Network Architecture: Driving Forces, Features, and Functional Topology. Engineering 2022;8(2022):42–59.
25. Li N, Liu G, Zhang H, et al. Service-based RAN: The next phase of cloud RAN. In: Proceedings of 2022 IEEE Global Communications Conference Workshops; 2022 Dec 4–8; Rio de Janeiro, Brazil: IEEE; p. 1206–1211.
26. 6G service-based RAN white paper, China Mobile Research Institute, 2022.
27. Toward 6G Core Architecture Using an Inline Service Mesh white paper, Intel + SK Telecom, 2023.
28. S.Newman. Building Microservices. Beijing, Post&Telecom Press, 2021.
29. C. Richardson. Microservices pattern. Beijing: China Machine Press, 2020.
30. TS 38.413: NG-RAN; NG Application Protocol (NGAP). 3GPP standard. France: 3GPP; 2024.
31. TS 38.423: NG-RAN; Xn application protocol (XnAP). 3GPP standard. France: 3GPP; 2024.
32. TS 38.321: 5G new radio-medium access control (MAC) protocol specification. 3GPP standard. France: 3GPP; 2024.
33. TS 38.322: 5G new radio-radio link control (RLC) protocol specification. 3GPP standard. France: 3GPP; 2024.
34. TS 38.323: 5G new radio-packet data convergence protocol (PDCP) specification. 3GPP standard. France: 3GPP; 2024.
35. TS 37.324: E-UTRA and NR-Service Data Adaptation Protocol (SDAP) specification. 3GPP standard. France: 3GPP; 2024.
36. TS 38.331: 5G new radio-Radio Resource Control (RRC) protocol specification. 3GPP standard. France: 3GPP; 2024.
37. Robert B, Ted F, Mark H. From Protocol Stack to Protocol Heap-Role-Based Architecture. ACM SIGCOMM Computer Communication Review; 2003, 33(1), p.17–22.
38. Duan X, Wang X, Lu L, et al. 6G Architecture Design from Overall, Logical and Networking Perspective. IEEE communications magazine. 2023; 61(7): 158-164.

Chapter 7
Distributed Massive MIMO Systems

This chapter starts with the technical background and evolution trend of MIMO, and provides a detailed introduction to the architecture and principles of future distributed massive MIMO technology and Cell-Free MIMO systems, covering technical features, system models, application scenarios, network deployment and key technologies, etc.

7.1 Background of MIMO Technology

Over the past forty years, wireless technology has evolved from enabling simple voice calls to supporting high-speed data transmission. Compared to 4G LTE technology, the 5G network being deployed globally can continue to support a variety of typical services and their enhancements, including Enhanced Mobile Broadband (eMBB), Ultra-Reliable Low Latency Communications (uRLLC), and Massive Machine Type Communications (mMTC), etc., to provide faster data rates and a better user experience than current mobile broadband. To further increase data transmission rates or cell throughput, strategies such as using larger bandwidth, increasing cell density, and improving cell spectral efficiency can be implemented. The technology to improve spectral efficiency by increasing the number of transmit and receive antennas on the base station and user side is the well-known Multiple Input Multiple Output (MIMO) technology [1–3]. The schematic diagram of single antenna and multiple antenna systems is shown in Fig. 7.1.

G. Liu et al., *6G Migrates from Communication to XaaS*,
https://doi.org/10.1007/978-981-96-7579-1_7

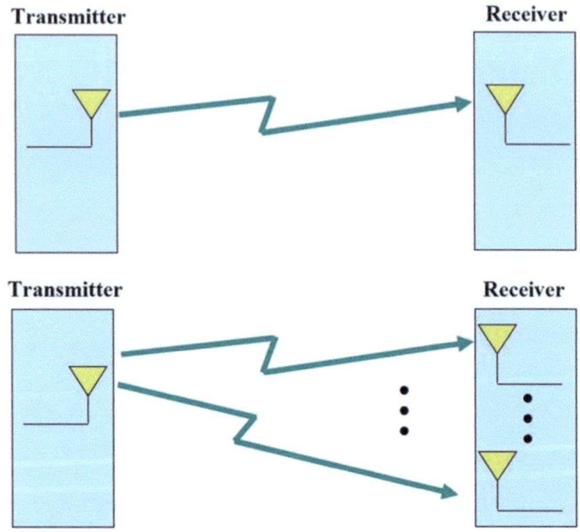

Fig. 7.1 Schematic diagram of single antenna and multiple antenna systems

7.1.1 *MIMO Technology and Massive MIMO Technology*

With the further development of signal processing technology, MIMO systems can achieve higher beamforming gains and spatial multiplexing gains to improve spectral efficiency, and enhance link robustness and reliability [4]. Although the concept of MIMO has been around for a long time, LTE is the first technical standard to fully implement the concept of multi-user MIMO. 4G LTE base stations can be equipped with up to 8 antenna ports, covering a 120° horizontal sector. Therefore, a total of 24 antenna ports are set up within a single cell, providing significant MIMO performance gains. However, compared with LTE, 5G eMBB services require higher system performance gains, which requires the deployment of more transmit and receive antennas at the base station or terminal. Massive MIMO technology can serve more users under the same time–frequency resources, significantly improving beamforming gain, spatial multiplexing gain, and spatial diversity gain, thus greatly improving spectral efficiency. Therefore, massive MIMO technology will be one of the key technologies for future wireless network communication. Massive MIMO technology can be divided into centralized and distributed forms in terms of system implementation, and Sects. 7.1.2 and 7.1.3 of this book will provide a brief introduction to these two systems.

7.1.2 *Centralised Massive MIMO System*

In the gradual evolution from point-to-point MIMO to multi-user MIMO, the centralized massive MIMO system based on cellular networks (see Fig. 7.2) can better solve

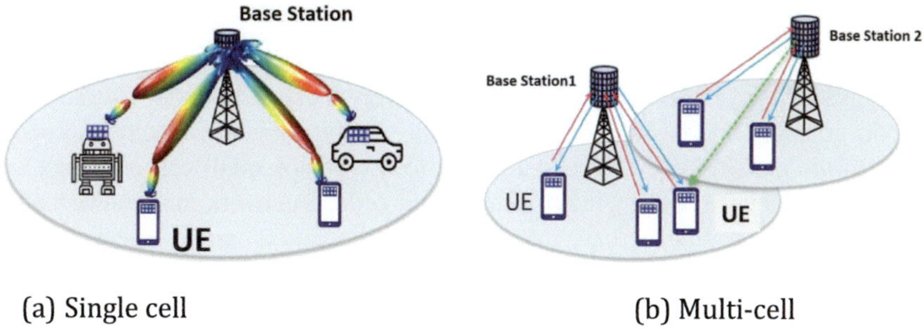

Fig. 7.2 Centralised massive MIMO systems

user interference and has good scalability, achieving higher system throughput in both low frequency (below 6 GHz) and high frequency (such as millimetre wave and above) applications. Centralised massive MIMO systems have the following key features.

The method of deploying a large number of base station antennas in a concentrated manner can achieve higher array gains, spatial reuse gains, and spatial diversity gains. In addition, when the number of base station antennas increases to a certain scale, two typical characteristics will be exhibited: one is the favourable propagation characteristic, which effectively eliminates multi-user interference to improve system performance, and can achieve data rates close to the system capacity limit through simple linear signal processing schemes; the second is channel hardening, which reduces the random fluctuations of the user's equivalent channel gain and improves the reliability of transmission. Time-Division Duplex (TDD) mode is conducive to the implementation of massive MIMO technology, mainly due to its uplink and downlink channels having reciprocity characteristics. The base station can obtain the downlink channel characteristics by measuring the uplink channel and use it for downlink data processing. Its channel estimation resources are only related to the number of users and will not change with the number of antennas, which has good scalability. However, in Frequency-Division Duplex (FDD) mode, base stations equipped with large-scale antennas need more reference signal resources to obtain downlink channel characteristics, and consume a lot of uplink resources to support channel state information feedback. These resource overheads pose a great challenge to the implementation of massive MIMO technology. Centralised massive MIMO technology is one of the key driving forces for the standardisation of 5G New Radio (NR). Ericsson took the lead in launching the world's first commercial 5G NR base station AIR 6468 based on massive MIMO technology in 2017, which uses 64 transmit and receive antennas and operates below 6 GHz. Subsequently, other companies such as Huawei, Nokia, ZTE and Facebook respectively demonstrated massive MIMO base stations equipped with 128 antennas, achieving a huge increase in spectral efficiency.

7.1.3 *Distributed Massive MIMO System*

Centralised massive MIMO systems can effectively increase network capacity and support gradually increasing network demands. As the network capacity is proportional to the density of the base station, various types of small cell networks are emerging. Although reducing the coverage of the cellular network can achieve higher system capacity, it will increase inter-cell interference to a certain extent, leading to a serious decline in system performance. In addition, the number of base station antennas is also limited by the size of the antenna panel and power consumption, especially in cellular systems operating in low-frequency bands, the number of antennas that can be centrally deployed is very limited, so the Cell-Free MIMO system based on distributed massive MIMO technology was born. In the distributed massive MIMO system, a large number of antenna units or antenna panels serve as access points (APs) widely distributed in the service area. These access points cooperate with each other through the backhaul network to jointly provide data transmission services for users, and use the advantages of ultra-large-scale antennas to obtain strong macro diversity gains, solve inter-cell interference problems, reduce channel shadowing effects and improve signal-to-noise ratio and coverage. Therefore, to a certain extent, distributed massive MIMO systems can provide uniform high data rate transmission services for all users.

7.2 Evolution and Technical Features of Distributed MIMO Systems

7.2.1 *Evolution of Distributed MIMO Systems*

As early as around 2000, the academic community proposed cooperative multi-antenna transceiver technology and began extensive research [10]. The corresponding distributed MIMO systems are mainly divided into the following evolutionary forms according to different technical characteristics [11].

(1) Network MIMO: As early as 2006, Bell Labs proposed the Network MIMO technology [12, 13], different cells serve users within the joint coverage range of these cells in the form of cell clusters through cooperation. In this scenario, base stations form non-overlapping cooperative clusters, each of which is served by a central processing unit (CPU). The central processing unit implements centralized signal processing, is responsible for cooperation between base stations, and obtains channel information for receiving and transmitting signal processing from the base station through uplink and downlink signalling. Network MIMO transforms the interference channel into a multipoint access channel, thus effectively suppressing inter-cell interference.

(2) Cooperative multi-cell MIMO: Since traditional network MIMO relies on centralized signal processing and needs to exchange instantaneous channel information between base stations, the IEEE 802.16 m standard gradually proposed cooperative multi-cell MIMO technology in 2008 [14, 15]. This technology can support distributed signal processing, does not require the exchange of channel information, and can greatly reduce the computational complexity and the requirements for the capacity of the fronthaul link [16].
(3) Cloud Radio Access Network (C-RAN): Around 2010, research related to the C-RAN based on cloud computing architecture was in full swing [17], where the baseband processing of C-RAN is completed by the CPU connected to multiple remote radio units (RRU). The RRU can be seen as a signal forwarding device, with all complex signal processing concentrated in the central processing unit.
(4) Coordinated MultiPoint (CoMP) transmission technology: This is the commercial name used by t LTE-Advanced standard for the distributed MIMO in around 2013. The relevant standards have adopted different cooperative MIMO technologies, but the actual commercial deployment of related products has not achieved the expected performance.

7.2.2 Cell-Free MIMO System and Technical Features of Distributed Massive MIMO

The Cell-Free MIMO system based on distributed massive MIMO combines the advantages of high-density, centralized massive MIMO technology and Coordinated MultiPoint (CoMP) transmission technology. Unlike many antennas centrally deployed at one base station, the Cell-Free MIMO system deploys a large number of antennas separately within a certain area, these geographically separated antennas and related systems form access points (AP). These access points are connected to the central processor through the fronthaul network to achieve synchronization and coherently serve users under the same time–frequency resources, as shown in Fig. 7.3. In addition to the excellent beamforming gain and spatial reuse gain of the massive MIMO system, this system also has the following performance characteristics.

(1) **Macro Diversity gain**: Different access points separated geographically can obtain higher spatial diversity gain, and each user in the network can be served by several adjacent access points simultaneously. This technology can reduce the path loss and shadow effect brought by channel propagation while enhancing the reliability of the link. Literature [18] shows that the distributed massive MIMO system can achieve a channel gain of 5–20 dB higher than the centralized network, as shown in Fig. 7.4.
(2) **Anti-inter-cell interference**: Through cooperative processing between different access points, this system can effectively transform interference signals into useful signals, achieving a network architecture without cells and boundaries within the distributed system area. Therefore, compared to small cell

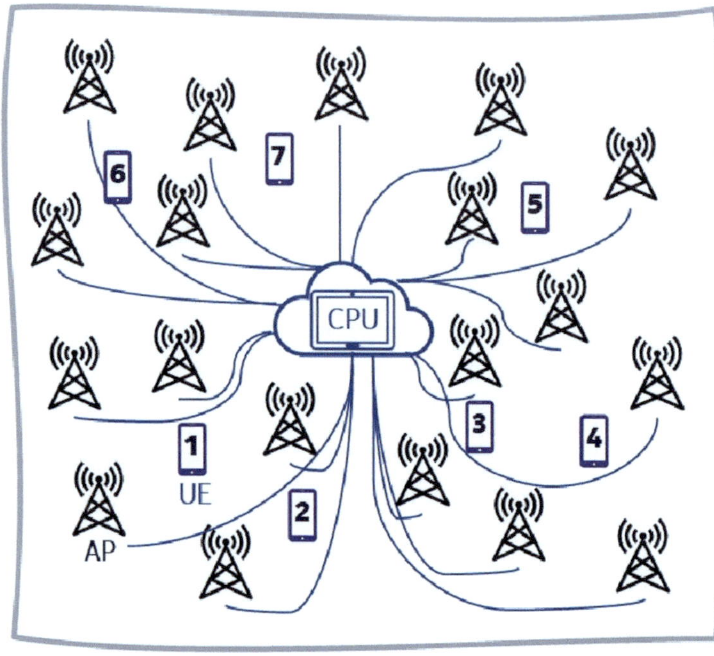

Fig. 7.3 Cell-free MIMO system

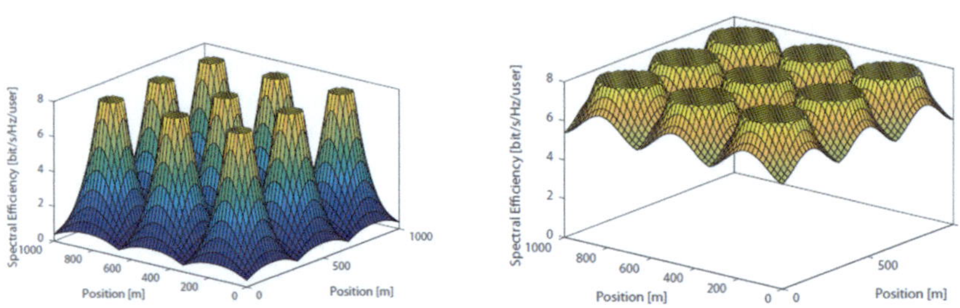

(a) Data coverage under cellular network (b) Data coverage under distributed network

Fig. 7.4 Spectrum efficiency distribution of user terminals at different locations [18]. *Note* 8 bps/Hz as the maximum spectral efficiency corresponds to the unencoded 256 QAM modulation mode

networks, the Cell-Free MIMO network has solved the problem of inter-cell interference, achieving a significant performance improvement [19, 20].

(3) **User-centric processing**: The network-centric architecture provides services to users inside non-overlapping cooperative clusters composed of distributed access points. The problem with this network deployment mode is that users at the edge of the cluster will be interfered by other cooperative clusters, resulting in performance loss. In the user-centric network deployment, users can access

the network through nearby access points based on signal strength. The Cell-Free MIMO system achieves a user-centric dynamic cooperative cluster through optimised power control strategy, ensuring fair and uniform service quality for all users.

7.3 Introduction to Cell-Free MIMO System

This section introduces three different modes of Cell-Free MIMO systems based on distributed massive MIMO technology [21].

7.3.1 TDD Mode

Assume that there are M access points (APs) equipped with multiple antennas serving K multi-antenna users in the Cell-Free MIMO system. All access points are connected to a central processing unit (CPU) through a fronthaul network. In the TDD working mode, each access point obtains the channel information between the user and itself through uplink channel estimation, such as the Minimum Mean Square Error (MMSE) method. In uplink signal processing, the access point achieves data combination through uplink combining reception, such as Maximum Ratio Combining (MRC) or Localised MMSE, and then the central processing unit performs centralised data demodulation. In downlink signal processing, the measured uplink channel information can be used for the pre-coding design for downlink transmission, thereby performing transmit beamforming and pre-coding processing for data, achieving spatial multiplexing of users and data streams.

7.3.2 FDD Mode

Due to the different working frequency bands of uplink and downlink in FDD mode, it does not have the reciprocity of TDD mode channels, and the measured uplink channel information cannot be directly used for downlink signal processing. The Cell-Free MIMO system under FDD mode requires feedback of downlink channel information. Each access point needs channel feedback from all users, so the cost of channel acquisition and feedback is very large, which is also one of the difficulties in network deployment under this mode. However, there is partial reciprocity of uplink and downlink channels in FDD systems, such as angle, delay and other large-scale information reciprocity [21–23]. Therefore, these partial reciprocity characteristics can be used to reduce the cost of channel acquisition and feedback to a certain extent [24].

7.3.3 Scalable Mode

In traditional Cell-Free MIMO systems, the related signal processing, such as channel estimation, pre-coding/receiving merging, and power control, etc., as well as the computational complexity and the required fronthaul overhead will increase proportionally with the increase in the number of users. Therefore, the scalability of distributed networks is very important. The user-centric dynamic cooperation cluster (Dynamic Cooperation clustering) approach can achieve scalable Cell-Free MIMO systems [25].

Figure 7.5 shows a dynamic cooperation clustering based Cell-Free MIMO architecture with four users as an example. The dashed line area corresponds to the area where a group of multiple access points form a cooperation cluster to serve a user. The key to dynamic cooperation clustering is to allow partial overlap of clusters, which poses a great challenge to uplink and downlink scalable signal processing.

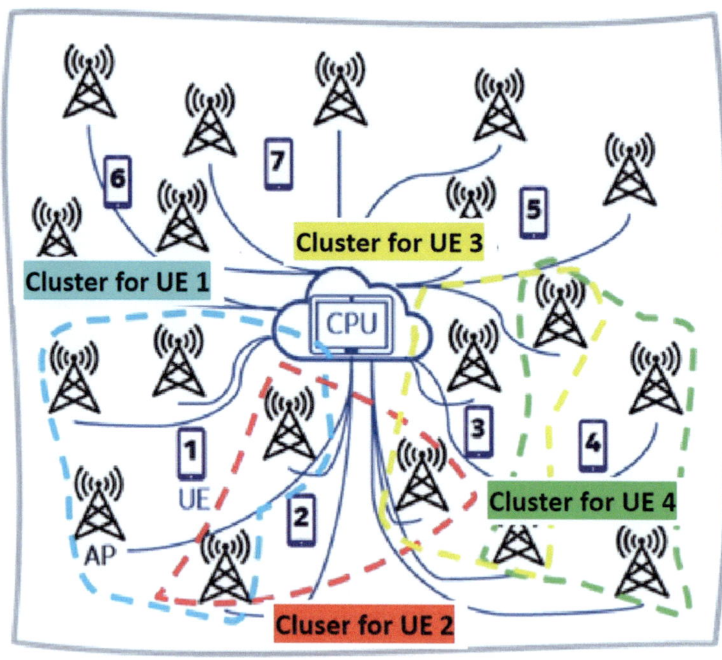

Fig. 7.5 Dynamic cooperation clustering based Cell-Free MIMO architecture

7.4 Cell-Free MIMO System and Key Technology of Distributed Massive MIMO

7.4.1 User Association to Access Point, Pilot Allocation and Channel Estimation

1. ***User association to access point***

In traditional centralised mobile communication systems, users rank the surrounding service access points according to various criteria (such as strongest received signal power, highest signal-to-noise ratio, load balancing, etc.) and connect to the best one. The connection between users and access points is many-to-one, that is, a user can only connect to one access point while an access point can connect multiple users.

In order to shorten the communication distance between users and access points and improve the service experience of the worst users, Cell-Free MIMO systems can allow multiple access points to serve a user at the same time, and multiple users can connect to the same access point, that is, the connection between users and access points is many-to-many.

For simplicity, each user can be connected to all access points [19]. However, this connection method has many disadvantages, such as, access points far from the user cannot provide reliable service due to the limited transmission power. At the same time, the data sent by this access point will also interfere with other users. In addition, the disadvantages of large backhaul link overhead, large delay and poor network scalability make this connection method inapplicable in actual systems. Therefore, the cluster-based connection method has more research and practical value, that is, users select some access points from all access point sets according to certain criteria for connection. For single-antenna access points and users, reference [26] collects the channel estimation values of all users at each access point, arranges them in descending order, and selects the strongest N users for connection. Reference [27] uses a method similar to reference [26], but uses the Frobenius norm as the measurement standard. Reference [28] provides two methods for pairing users and access points in the scenario of multi-antenna access points and users. In the first method, each access point calculates the average Frobenius norm of all users, that is $\overline{G}_m = \frac{1}{K}\sum_{k=1}^{K} ||\hat{G}_{k,m}||_F \overline{G}_m = \frac{1}{K}\sum_{k=1}^{K} ||\hat{G}_{k,m}||_F$, each user is only connected to access points that exceed this average value. The second method is the same as the method in reference [27]. Reference [29] provides two strategies for improving energy efficiency in access point selection. In the method based on received power selection, user k chooses service access point m to satisfy: the power received by user k from access point m accounts for more than δ % of its total received power. Due to the high computational complexity of the method based on received power selection, a method based on large-scale fading is also proposed in the paper. This method requires that a set of service access points for user k meet the following criteria: the sum of the large-scale channel fading between this set of service access points and user k accounts for more than δ % of the sum of the large-scale channel fading between the user and all sets of service access points. Reference [30] uses an iterative elimination algorithm to eliminate each user's invalid service access points, and removes access points with a signal-to-noise ratio below the threshold while ensuring that power consumption changes within a certain range. Reference [31] proposes a service access point selection strategy under multiple CPUs. This strategy has good network scalability and is suitable for deployment in actual networks. All access points connected by each CPU form a cluster centred on the cell, and all access points selected by each user form a cluster centred on the user. The cluster centred on the user may include access points belonging to different clusters centred on the cell. When selecting an access point, all access points connected by each CPU first estimate the channel. When there is an access point whose channel conditions are sufficient to support user communication, the access point reports to the CPU, and the CPU informs all connected access points to serve this user and sends the user's data to the access point for pre-coding and other operations.

2. ***Pilot Allocation***

In cell-free MIMO systems, uplink channel estimation is crucial. To perform uplink channel estimation, pilots must first be allocated to all users in the service area. Since the length of the pilot sequence is limited by the channel coherence time, and the number of service users is generally more than the number of orthogonal pilots, it is impossible to allocate orthogonal pilots to all users in the cell-free MIMO system. Therefore, pilot reuse must be carried out, which will inevitably cause pilot contamination. How to allocate pilots to suppress pilot contamination is one of the important issues in cell-free MIMO systems. Existing literature mainly uses methods of user grouping before pilot allocation, or designs appropriate pilot frequency division algorithms for pilot allocation. The method based on user grouping groups users with similar channel characteristics or short geographical distances into one group, and suppresses pilot contamination by using orthogonal pilots within the group and reusing pilots between groups; the algorithm design for pilot allocation mainly studies the trade-off between algorithm complexity and performance, and the protocol design for measurement.

(1) *Pilot allocation method based on user grouping*

Firstly, consider the pilot allocation method based on user grouping. This section assumes there are K users, M access points, and τ_{p} orthogonal pilots within the region. Both [32] and [35] use the K-means method for user grouping. Attarifar et al. [32] divides users into several disjoint subsets, each containing users with the largest possible minimum distance using the same pilot, and different subsets use different orthogonal pilots. Firstly, multiple geographical locations are randomly selected as centroids, each centroid is a class, and each user is assigned to the centroid closest to them. Then the position of the centroid is updated, that is, the arithmetic mean of the geographical locations of the users belonging to each class is calculated as the position of the new centroid. Finally, all users are reassigned to their classes based on the updated centroid positions. Repeat the above iterative process to obtain the final clustering result. Attarifar et al. [32] only considers the geographical location information of users. To consider the impact of access points, Chen et al. [33] proposed a grouping method using the distance between users and access points, called Interference-Based K-means (IB-KM) pilot allocation scheme. This scheme randomly generates $\left\lceil \frac{K}{\tau_p} \right\rceil$ centroids in the region, calculates the distance d between each user and the access point as well as the distance μ between the centroid and the access point, selects a centroid for each user based on the values of d and μ and forms clusters, recalculates and updates the centroids based on the cluster results, and iterates until a threshold is reached. Each user calculates the distance to the access point and forms clusters in combination with the distance to the centroid. In addition to user grouping, Chen et al. [33] also proposed a pilot allocation scheme based on user grouping, which assigns users with the fewest common access point sets to a group and assigns the same pilot to this group. Specifically, the CPU collects the access point selection set of each user and constructs the access point-user selection relationship matrix $\boldsymbol{S}$, calculates the number of common access points each user has with other users based on matrix $\boldsymbol{S}$ to form matrix $\boldsymbol{T}$ as a measure of interference between users, and decide which users can be grouped based on matrix $\boldsymbol{T}$. Hao et al. [34] uses the K-means method to cluster users served by the same access point into the same group and assigns orthogonal pilots to them. The specific process is to select $\frac{K}{\tau_p}$ users served by different access points. A user is selected as the initial centroid. If there are no users with completely different access point services, the user with the fewest identical access point services is chosen as the centroid. Other users are grouped according to the Euclidean distance to the centroid, the CPU allocates pilot frequencies and calculates the achievable uplink sum rate. The centroid is updated based on the positions of the users in each group. Then the user grouping, pilot allocation, and calculation of the achievable uplink sum rate are repeated. The results are compared with the previous achievable rate, and the update is stopped when the results converge. As described in reference [35], under the condition that the CPU knows the large-scale gain, a vector composed of the Frobenius norm with each access point is generated for each user $\xi_k = [\xi_{1k} \cdots \xi_{Mk}]$ as a measure of similarity, calculate $\xi_C = \frac{1}{K} \sum_{k=1}^{K} \xi_k$ as the hypothetical centroid, calculate the similarity of each user to the centroid:

$$f_{\mathrm{D}}(\boldsymbol{\xi}_k, \boldsymbol{\xi}_{\mathrm{C}}) = \frac{\boldsymbol{\xi}_k^{\mathrm{T}} \boldsymbol{\xi}_{\mathrm{C}}}{||\boldsymbol{\xi}_k||_2 ||\boldsymbol{\xi}_{\mathrm{C}}||_2}$$

Based on similarity, they are arranged in descending order, and the users with a gap of τ_p after sorting are divided into a group, that is, the grouping result is $\kappa_t = O(t : \tau_{\mathrm{p}} : K) = \{o_t, o_{t+\tau_{\mathrm{p}}}, o_{t+2\tau_{\mathrm{p}}}, \cdots\}, \forall t \in \{1, \cdots, \tau_{\mathrm{p}}\}\kappa_t = O(t : \tau_{\mathrm{p}} : K) = \{o_t, o_{t+\tau_{\mathrm{p}}}, o_{t+2\tau_{\mathrm{p}}}, \cdots\}, \forall t \in \{1, \cdots, \tau_{\mathrm{p}}\}$.

(2) *Straightfoward algorithms for pilot allocation*

In addition to the pilot allocation method based on user grouping, straightforward algorithms can also be designed for pilot allocation. Literature [19] considers that the random pilot allocation method may allocate the same pilot to users in close geographical locations, causing serious pilot contamination. Therefore, a greedy pilot allocation method is proposed, which first randomly allocates pilot sequences to all users, and each time updates the pilot sequence for the user with the lowest downlink rate to minimise the effect of pilot contamination. After a certain number of iterations, the algorithm will converge, and the pilot sequence allocated to each user is finally obtained. Considering that the greedy pilot allocation method is prone to fall into local optimal solutions, literature [36] proposes a pilot allocation method based on taboo search. Firstly, initialise a pilot allocation result and calculate the throughput under this result. Secondly, obtain the K neighbourhoods of the initial pilot allocation result, each neighbourhood only differs from the initial result in one pilot allocation result. Then, calculate the throughput of all neighbourhoods and sort them in descending order, and take the result with the maximum throughput as the candidate solution. Finally, compare the current optimal solution with the candidate solution. When the candidate solution has a larger throughput, update the result and the current optimal throughput. Otherwise, add the candidate solution to the taboo list, and stop searching when the number of iterations is K. Literature [19] and [36] did not consider the impact of user mobility. In response to this issue, literature [37] proposes an orthogonal pilot allocation scheme based on the utility function, ensuring that orthogonal pilots are allocated to high-mobility users to cope with the problem of outdated Channel State Information (CSI). Literature [38] uses the Hungarian algorithm to solve the pilot allocation problem, taking large-scale fading information as a measure of the distance between users, and allocating pilots from two perspectives of maximising system throughput and maximising user fairness. Compared with many other literatures, the algorithm proposed in literature [38] has significant performance advantages. The pilot allocation problem can also be solved using graph theory. Literature [39] constructs an interference graph based on the access point selection result, representing the interference relationship between users, and uses the graph colouring algorithm for pilot allocation. This scheme is a low-complexity solution that can well balance throughput and complexity. Literature [40] uses a Weighted Graph Framework (WGF), which introduces a new metric to represent the severity of pilot contamination. By constructing the WGF, the pilot contamination problem can be mapped to the maximum k-Cut problem for solution. The above literature assumes that the channel follows a Rayleigh distribution, but in the actual deployment environment, the channel may also follow a Rician distribution. Literature [41] analysed the phase shift caused by user mobility and phase noise when the channel follows a Rician distribution. Since the phase shift changes frequently and is difficult to estimate, when the pilot length is short, the performance loss is severe. Therefore, in systems with high mobility or low hardware complexity, it is necessary to adjust the pilot length. The above literature is dedicated to assigning pilot sequences to all users in the region. When a new user accesses the network, literature [25] proposes that the user should first send a periodic broadcast synchronisation signal to all nearby access points, detect the access point with the maximum channel gain as the main access point, and the main access point calculates and obtains the pilot with the minimum pilot contamination effect τ assigned to the user. For nearby access points, when the access point does not have a user served by the pilot τ or the new user's channel is better than the user previously served by the pilot τ, the new user can be served by the access point.

3. ***Channel Estimation***

Traditional mobile communication systems often use methods such as least squares or minimum mean square error for channel estimation. These methods require the number of transmit and receive beamforming vectors to be larger than the number of transmit and receive antennas. However, in millimetre wave communication systems, in order to compensate for larger path losses to meet coverage requirements, large-scale antenna arrays need to be deployed at the transmitting and receiving ends. At the same time, limited by the channel coherence time, the training time for channel estimation should also be shortened as much as possible. Therefore, traditional channel estimation methods are not applicable to millimetre wave communication systems. Fortunately, millimetre wave channels have sparse characteristics, so the industry is studying channel estimation methods from some new perspectives.

Subspace matching methods can leverage the characteristics of signal subspace and noise subspace for channel estimation [42, 43]. Literature [42] uses the beam domain two-dimensional multiple signal classification (MUSIC) method to estimate the departure angle and arrival angle of each multipath in the channel, and uses the least squares method to estimate the path gain of each multipath. It is worth noting that the estimation accuracy of the beam domain MUSIC method is related to the designed beamforming vector. Inaccurate beamforming vectors will lead to estimation ambiguity and increase estimation errors. Literature [43] analysed the discrete Fourier transform based beamforming vector for a uniform linear array with half-wavelength spacing, and found that it can avoid estimation ambiguity and obtain the maximum number of distinguishable path directions.

Compressive sensing is another type of method that uses channel sparsity for channel estimation [44, 45]. Literature [44] uses orthogonal matching pursuit (OMP) and de-gridding methods to estimate channel by using compressed measurement results. In order to maximise the joint likelihood function of all parameters in each multipath, and to achieve fine estimation of de-gridding, the residual function of each multipath is used in each iteration of the OMP algorithm.

7.4.2 Uplink and Downlink Signal Processing Techniques

1. ***Uplink Combination Techniques in Distributed Massive MIMO***

During uplink transmission in MIMO systems, multiple users send multiple data streams to the base station using the same time–frequency resources. In traditional MIMO systems, multiple antennas are centrally deployed at the base station, which simultaneously receives overlapping signals from multiple users. How can the data of a specific user be recovered from these overlapping uplink signals? To solve this problem, uplink combination technology needs to be used.

The specific processing procedure of uplink combination includes the base station designing appropriate combining vectors based on the estimated channel state information (CSI) of each user, combining the received signals and the separation of multiple user signals. In traditional MIMO systems, three schemes are generally used: Maximum Ratio Combining (MRC) [46], Zero-Forcing (ZF) combining and Minimum Mean Square Error (MMSE) [47] combining. MRC is achieved by adjusting the phase of the signals from multiple antennas at the receiving end and adding them together with appropriate weighting coefficients. The weighting coefficient corresponding to each antenna signal is proportional to the signal-to-noise ratio corresponding to that antenna. The larger the signal-to-noise ratio, the larger the weighting coefficient, and the greater the contribution to the merged signal. Specifically, the weighting coefficient vector is the conjugate transpose of the channel matrix. The MRC scheme is simple to process and easy to implement, but because it directly treats the interference from other users as noise when processing overlapping multi-user signals without doing any interference elimination processing, so the performance of separating multi-user signals is poor. In order to eliminate inter-user interference and improve combining performance, ZF combining designs the combining vector as the pseudoinverse of the channel matrix. After multiplying the channel matrix, it becomes a diagonal matrix, equivalent to forming multiple interference-free links, which means the signals between each user can be separated without interference. Compared with MRC, ZF combining effectively eliminates inter-user interference at the cost of higher processing complexity because there is a matrix inversion operation in the processing process. In addition, ZF combining has the problem of noise enhancement. To solve this problem, MMSE combining technology can be used. Its design idea is to minimize the mean square error of the combined signal and the original signal, so that the combined signal is close to the original signal. This scheme is optimal in performance because it eliminates both noise and interference. But it also requires matrix inversion operations, so the complexity is also high.

The distributed massive MIMO system has a dual-layer network architecture of AP and CPU, where the processing capacity of the AP is limited and the CPU has higher processing capacity. However, all the tasks of uplink combining processing cannot be handed over to the CPU. This is not conducive to the system scalability on the one hand, and on the other hand, it will bring a huge load to the fronthaul link. Therefore, in this system, in addition to considering the uplink combining methods of traditional MIMO systems, the degree of cooperation between APs and the load of the fronthaul link also need to be taken into account.

Literature [48] classifies the uplink combining schemes of distributed massive MIMO into various schemes according to different degrees of cooperation. Literature [20] introduces four combining schemes, in which the APs have different levels of cooperation. Uplink channel estimation and uplink combining are executed inside the CPU in some schemes while in other schemes, they are executed at each AP. Therefore, there is a big difference in the amount of information exchanged between the AP and the CPU, and the system performance that can be obtained, among different schemes. They will be introduced in the following paragraph in order of their level of cooperation, from highest to lowest.

The first uplink combining scheme [20] is a completely centralised processing method, inheriting from traditional MIMO. The implementation process includes each AP forwarding the received uplink pilot signal and data signal to the CPU via the fronthaul link. The CPU obtains the CSI of all users based on the uplink pilot and merges the uplink data according to the CSI. Because the CPU understands the global CSI, if ZF or MMSE combining is used here, it can effectively suppress interference between all network users. It can be seen that this combining scheme not only perfectly inherits the advantages of traditional MIMO but also utilises the macro diversity gain of distributed massive antennas. Therefore, it is the optimal combining scheme in terms of performance under ideal conditions. However, the processing capacity of the CPU is limited in reality. As the number of users in the network continues to increase, the CPU will be overloaded. So this scheme is not conducive to system expansion. In addition, the pilot signal, unprocessed data signal and large-scale fading information of the network all need to be sent to the CPU via the fronthaul link. When the capacity of the fronthaul link is limited, the system performance will be greatly affected.

To solve the problem of excessive fronthaul load and CPU processing pressure in the first combining scheme, the second uplink combining scheme [20] uses a two-stage processing approach with AP based local processing plus CPU processing for large-scale fading decoding (Large Scale Fading Decoding, LSFD). LSFD is a method of weighted data combining using large-scale fading information. It can design weighting coefficients based on the large-scale fading information in the network to optimise the performance after combining. The basic process is that the AP performs channel estimation locally to obtain the CSI of local users, and then uses the local CSI to perform the first level of combining for the user data it serves. The AP then sends the first-level combined signal to the CPU, and the CPU performs LSFD on the data based on the large-scale fading information. LSFD is used here because channel estimation is performed at the AP, so the AP can only combine the received data based on the local CSI. However, a user may be served by multiple APs, so these APs send the user data they serve after combining to the CPU. The CPU needs to use large-scale fading information for secondary combining to finally recover this user's data. This uplink combining scheme assigns the tasks of channel estimation and part of the signal combining to the AP, significantly reducing the CPU's processing pressure. In addition, the pilot signal does not need to be transmitted to the CPU, which somewhat reduces the load on the fronthaul link. However, its performance is worse than the first combining scheme. On the one hand, due to the limited processing capacity of the AP, it is difficult for the AP to perform better but more complex combining operations. Even if the more complex MMSE combining is used, it can only eliminate local interference. On the other hand, the amount of large-scale fading information data required for CPU decoding is quite large, and it needs to be transmitted to the CPU via the fronthaul link, so the fronthaul load is still high.

To further reduce the fronthaul and CPU load, the third uplink combining scheme [20] is simplified based on the first uplink combining scheme. It only directly adds the first-level combined signals of the same user sent from multiple APs at the CPU to complete the second-level combining. It does not require large-scale fading information, thereby further reducing the fronthaul load. However, this simple centralised decoding without using any information further reduces system performance compared with the second uplink combining scheme. In related study, its performance may be worse than the fourth uplink combining scheme mentioned later, so it has poor practicality.

The fourth uplink combining scheme [20] is a completely distributed processing scheme, with each AP independently completing its signal processing tasks. The basic process is that the AP performs local channel estimation through the pilot signal sent by the UE, obtains the CSI of local users, and then uses the local CSI to combine the data and upload the data to the CPU. The CPU does not do any second-level processing. Since there is no second-level combining, each user can only be served by one AP in this scheme, which is theoretically similar to a small cell network (Small Cell). The only difference is that users can choose the best AP to serve them during the access stage. In this scheme, the CPU only receives uplink data and does not perform any processing. As the number of users increases, the processing pressure inside CPU will not be increased, making it a completely scalable system.

Overall, the first uplink combining scheme performs channel estimation and signal combining at the CPU, which can utilise the CSI of all network users for global interference elimination and combining processing. Therefore, its performance is the best, but all processing is done at the CPU, resulting in a large CPU load and fronthaul load. The second uplink combining scheme performs channel estimation at the AP, which uses the local user's CSI to perform the first level of signal combining. The CPU uses large-scale fading information to perform the second level of signal combining (LSFD). This results in reduced performance, CPU load and fronthaul load. The third uplink combining scheme also performs channel estimation at the AP, which uses the local CSI to perform the first level of signal combining. The CPU directly adds received signal linearly for simple central decoding, further reducing the CPU load and fronthaul load, but the performance is further degraded. The fourth uplink combining scheme performs channel estimation and signal combining at the AP, with the CPU having no processing during the transmission. The system overhead does not increase with the increase in the number of users, making it a fully scalable solution. These four combining schemes reflect the degree of cooperation among distributed massive MIMO antennas, achieving a trade-off between system performance and system overhead.

2. ***Downlink precoding of distributed massive MIMO***

In wireless systems, channels have random and complex characteristics. There is severe interference in the process of signal transmission in wireless channels, especially interference between users, which leads to a decline in system performance. Due to the limited processing power at the user end, it is difficult to take effective measures to eliminate channel interference and user interference at the user end. Therefore, precoding processing is generally performed on the transmitted signal at the transmitting end. Precoding refers to the method of using CSI to calculate the precoding vector and pre-processing the transmitted signal to improve system capacity and reduce the bit error rate. Traditional MIMO precoding calculates the precoding vector at the base station using global CSI. In distributed massive MIMO system, a large number of APs are deployed throughout the network. Their precoding vectors can be either calculated at the AP using only local CSI or be calculated at the CPU. According to the location of the calculation, precoding schemes can be classified into centralised precoding and distributed precoding in distributed massive MIMO systems. Centralised precoding refers to each AP sending the obtained CSI to the CPU through its own fronthaul link. The CPU uses global CSI to calculate the precoding vector and then sends the precoding vector back to the corresponding AP through the fronthaul link. This can effectively reduce the interference caused by other APs, but it requires a large fronthaul link overhead and has a high computational complexity and slow calculation speed. Distributed precoding is generally calculated at the AP using only local CSI, which reduces the amount of information exchanged between the AP and CPU, lower the computational complexity, and speed up the calculation. However it cannot eliminate the interference caused by other APs.

Similar to the uplink combining scheme, the commonly used downlink precoding schemes in distributed massive MIMO systems include Conjugate Beamforming (CB), ZF precoding, and MMSE precoding. Similar to uplink MRC technology, CB precoding uses the conjugate of the estimated channel coefficient as the precoding vector. In this way, the data signal after precoding can maximise the channel gain of the target user [19]. The CB precoding scheme has low computational complexity and only needs to know the local CSI of the AP to realise distributed precoding calculation. Therefore, it is widely used in distributed massive MIMO systems, but its disadvantage is that it does not consider interference between users. Similar to uplink ZF/MMSE combining technology, downlink ZF/MMSE precoding technology can eliminate interference between users through matrix inversion operations. It can be completed by each AP for simple distributed precoding, or it can be implemented at the CPU for complex centralised precoding. Although it is superior to CB precoding in downlink transmission performance, the operation of matrix inversion in downlink ZF/MMSE precoding will increase with the increase in the number of users, and the corresponding hardware implementation difficulty will also increase.

3. ***Power allocation in distributed massive MIMO systems***

Power allocation is an important technology in wireless communication systems to combat fading and co-channel interference. Power allocation technology calculates the transmission power corresponding to different channel conditions according to certain criteria under the premise of limited system resources, thereby improving resource utilisation, and enhancing the capacity and performance of the system. In the downlink data transmission of distributed massive MIMO systems, a large number of distributed APs serve users simultaneously. The distances between each AP and the users are different, and the channel conditions also vary greatly. Therefore, the significance of power allocation in distributed massive MIMO systems is more prominent, and the calculation of power allocation coefficients is more complex. By adjusting the transmission power on the AP side in distributed massive MIMO systems, co-channel interference can be reduced, and the utility function within the system coverage can be maximised, which optimizes the network performance and achieves the efficient resources utilization.

Before power allocation, the utility function must first be determined. Different utility functions have different design objectives, corresponding to different types of optimisation problems and different solution methods. In distributed massive MIMO systems, to achieve the goal of consistent user rates within the region, user fairness should be considered as the optimisation target first. In addition, in traditional MIMO systems, spectral efficiency and energy efficiency are usually used as optimisation targets, which are also applicable in distributed massive MIMO systems. Based on these optimisation targets, the main utility functions in the current research on power allocation in distributed massive MIMO systems are maximising the minimum user rate (Maximum Minimum Rate, max–min Rate), maximising spectral efficiency (Maximum Spectral Efficiency, max-SE), and maximising energy efficiency (Maximum Energy Efficiency, max-EE).

In distributed massive MIMO systems, in order to provide all users with uniform service quality, most of the existing research adopts max–min Rate power allocation. Max–min Rate power allocation aims to achieve fairness among users in the region by maximising the minimum user rate in the system. This problem is a quasi-convex optimisation problem in the field of convex optimisation, and can be solved using convex optimisation tools such as CVX to obtain the optimal power allocation coefficient. Literature [19] provides a detailed introduction to this scheme and its solution. Compared with full power transmission, the concentration of user spectral efficiency distribution under max–min Rate power allocation significantly increases, which means the fairness of the system under max–min Rate power allocation is greatly improved.

However, the disadvantage of max–min Rate power allocation is that it needs to allocate a large amount of power to a small number of users with poor channel conditions. These users will drag down the performance of the entire system. Based on this observation, some research has considered the problem of maximising the spectral efficiency of the entire network. Literature [50] proposes a max-SE power allocation scheme. This problem is a non-convex optimisation problem, which can be converted into a Weighted Minimum Mean Square Error (WMMSE) problem, and a local optimal solution can be obtained using the Block Coordinate Descent (BCD) method. By adopting max-SE power allocation, the user's rate is significantly higher than the full power transmission. Compared with max–min Rate power allocation, only a small number of poor users have lower spectral efficiency under max-SE power allocation while the total spectral efficiency is higher.

The focus of the above two criteria is on spectral efficiency and does not comprehensively consider the energy consumption of the system. By 2026, the number of 5G users is expected to reach 3.5 billion. With the increase in wireless connected devices, carbon dioxide emissions and network operator's electricity consumption are increasing sharply. In the next generation of mobile systems, to fit the development concept of green communication, the increase in data rate should not be at the expense of a significant increase in transmission power. Therefore, some research has considered the power allocation problem from the perspective of energy saving. In order to reliably transmit a certain amount of information with less energy consumption, literature [51] proposes to use the ratio of data rate to total energy consumption as the utility function, which is the max-EE power allocation scheme. This problem is a non-convex optimisation problem, which can be solved by the Sequential Convex Approximation (SCA) method. Compared with traditional MIMO systems, the energy efficiency of distributed massive MIMO systems is greatly improved by adopting max-EE power allocation.

In addition to the three utility functions mentioned above, other different utility functions can be considered according to different expected effects, such as the trade-off between user fairness and spectral efficiency, power allocation that changes priority according to different user needs, and so on. Furthermore, power allocation schemes under more practical conditions, such as fronthaul link constraints and system hardware impairment, can also be considered.

The power allocation problem is generally solved by optimising the performance within the system coverage area to achieve the expected effect. However, the computational overhead will inevitably be increased sharply with the increase in the number of users or APs. In practical applications, such schemes are difficult to apply to large networks due to poor scalability. In large networks, distributed and heuristic power allocation schemes can be considered [49], which can achieve acceptable performance with a small amount of information exchange, i.e., scalable power allocation schemes. In addition to traditional optimisation methods or heuristic methods, machine learning is also a possible solution. One of the advantages of machine learning is that it can significantly reduce online computational complexity at the cost of offline training, without affecting performance. In the application of machine learning, a smart power allocation model is first obtained through a large amount of offline training. In subsequent online calculations, only the corresponding parameters need to be input, and the model can quickly calculate the power allocation coefficient. Some research has used machine learning methods to reduce the complexity of the max–min Rate power allocation problem [52]. In addition, machine learning methods can escape from local optimal and seek global optimal solutions. The WMMSE method for the max-SE power allocation problem may fall into a local optimal solution, and some research has found a better solution through machine learning [53].

Based on the above situation, there are still some unresolved issues in the power allocation problem in distributed massive MIMO systems. For example, how to consider other utility functions or constraint conditions for power allocation according to actual needs; whether a joint optimisation scheme can be designed to achieve a trade-off between complexity and performance, so that power allocation can play a greater role in distributed massive MIMO systems.

7.4.3 Network Synchronisation and Calibration

1. ***Network Synchronisation***

Network synchronisation is key to wireless network performance. Although the basic synchronisation requirements have not changed in the transition from 4 to 5G, the widespread use of TDD radio technology and the growing demand for new network architectures to support 5G use cases have made the need for time synchronisation more prominent in 5G. Industrial automation is an example that requires precise timing [54] and may generate additional synchronisation requirements in the near future. To further promote IoT applications and reduce transmission latency, future 6G networks will adopt higher carrier frequencies and distributed massive MIMO technology. Therefore, achieving synchronisation and calibration of 6G networks is crucial and will inevitably face more challenges.

In simple terms, radio access network synchronisation is to prevent co-channel interference (UE to UE, AP to AP) and ensure complete signal reception. Synchronisation requirements can be divided into two categories: TDD cell frame synchronisation and symbol-level synchronisation based on multiple AP cooperation. The former is the basis for the normal operation of wireless cellular networks, such as traditional centralised antenna systems; the latter is based on the former and futher take into account symbol-level synchronisation for multiple AP cooperation, which is an inherent requirement of distributed massive MIMO systems.

(1) *Cell frame synchronisation*

TDD cells operating at the same frequency (or adjacent frequencies) in overlapping coverage areas need time-domain isolation to prevent base station to base station and UE to UE RF interference, ensuring the synchronisation between the cellular network's transceivers. Time-domain isolation has two requirements:

(1) The same TDD configuration must be used in the cells;
(2) The frame start timing deviation between cells must be lower than the maximum value of the cell phase synchronisation accuracy specified in 3GPP.

The key to solving TDD cell synchronisation and interference problems is to set the interval time (called the protection interval) between transmission and reception, as shown in Fig. 7.6. The protection interval can achieve isolation of uplink and downlink switching, and its total configurable protection time is an integer number of symbols. The duration of the protection interval needs to consider four influencing factors:

(1) Propagation time in the air (T_{prop});
(2) Sufficient transient time to meet the transmitter's change between the specified on/off power levels ($T_{\text{On}\rightarrow\text{Off}}$, $T_{\text{Off}\rightarrow\text{On}}$);
(3) UE and base station have enough time to switch between transmitting and receiving modes ($T_{\text{RX}\rightarrow\text{TX}}$, $T_{\text{TX}\rightarrow\text{RX}}$);
(4) Cell frame synchronisation error (T_{Sync}).

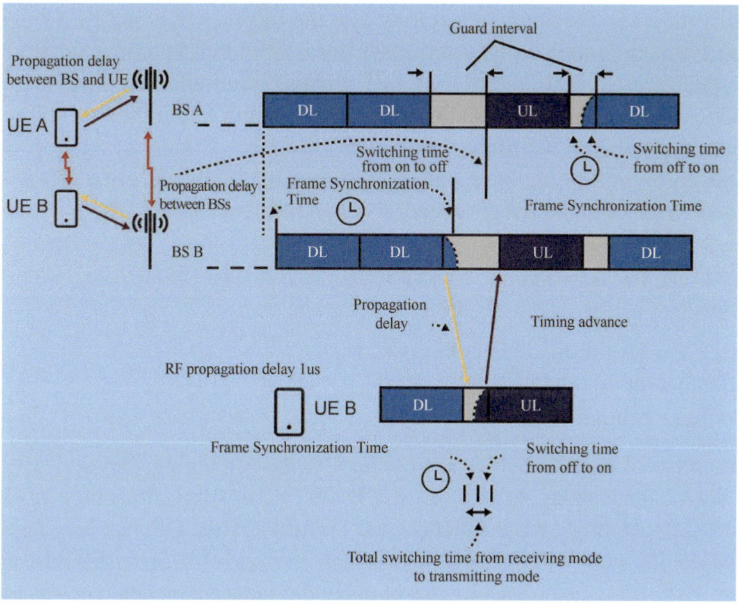

Fig. 7.6 TDD cell phase synchronisation [55]

Since there is no data communication during protection intervals, they reduce the spectral resource utilisation in time. The focus of reducing resource overhead is to shorten the protection interval as much as possible while still achieving network synchronisation. The protection interval is composed of the above four influencing factors, and the budget allocation for each part is the result of a trade-off between cost (product and deployment), availability, TDD cycle, and resource overhead.

In 3GPP NR, the cell synchronisation accuracy is specified as 3 μs [55], the same as LTE. This is because the reduced transient time in NR makes it possible to maintain the same synchronisation accuracy as LTE at a low cost.

(2) Symbol-level synchronisation based on multiple AP cooperation

Over the years, various functions that benefit from coordinated transmission or reception from multiple APs have been standardised, each with different usage and characteristics. Some involve combining spectral assets, allowing for higher aggregated bandwidth and throughput (carrier aggregation, dual connectivity, etc.). Others involve improving link performance at the cell edge (such as variants of Coordinated Multi-Point operation). Some involve specific services, such as multimedia broadcast multicast services on single-frequency networks. Coordination may also occur between NR and LTE. In addition, multi-access point cooperative transmission can better solve the edge coverage in high-frequency 6G networks.

These functions and related timing requirements apply to a single operator network, so it is sufficient to control the relative time error between different APs using this function, focusing on ensuring that the receiver (UE end or AP end) can fully receive data symbols from different service APs within the time window allowed by the specification, that is, different APs achieve symbol-level timing alignment. Therefore, 3GPP defines the Maximum Receive Timing Difference (MRTD) as the maximum relative receive timing difference that the UE must be able to handle. MRTD is composed of the base station relative time alignment error (Time Alignment Error, TAE) and the RF propagation delay difference (ΔT_{prop}). That is, MRTD $= \text{TAE} + \Delta T_{\text{prop}}$.

Depending on the level of the required MRTD, three main categories can be determined:

(1) MRTD as part of the Cyclic Prefix (CP);
(2) MRTD with no relation to CP;
(3) No time requirements.

In the case of MRTD being part of the CP, the remaining duration of the CP allows for channel delay spread. In 3GPP TS 38.104 [56], the range of TAE is 65–260 ns, depending on the function and duration of the CP, but only applies to intra-site/deployment within the same address: $\Delta T_{\text{prop}} \approx 0$. Strict MRTD, and the resulting TAE, have the characteristics of using the same frequency (such as MIMO) or adjacent spectrum (such as continuous carrier aggregation), where common or shared functions in the RF chain may cause strict timing dependencies.

Although such characteristics exist in distributed APs (usually $\Delta T_{\text{prop}} \neq 0$), that is, not limited to multi-AP deployment scenarios of the same address/different sites, it is difficult to specify and authorised a fixed TAE. Instead, the required TAE will depend on the actual deployment, i.e., RF characteristics and the relative position of the UE between the cooperating APs, as shown in Fig. 7.7.

As shown at the bottom of Fig. 7.7, $\Delta T_{\text{prop\_UE1}}$ (User 1 propagation delay difference) and $\Delta T_{\text{prop\_UE4}}$ (User 4 propagation delay difference) are less than $\Delta T_{\text{prop\_UE2}}$ (User 2 propagation delay difference), which is less than $\Delta T_{\text{prop\_UE3}}$ (User 3 propagation delay difference). For the same delay spread, UE_1 and UE_4 can tolerate a larger TAE than UE_2 and UE_3. For the same positions of D and E, the TAE of D-E is usually less than that of A-B.

In contrast, MRTD multi-AP transmission in heterogeneous networks with no relation to CP allows for a relatively large ΔT_{prop}, enabling more flexible deployment. For example, synchronous dual connectivity allows for a maximum of 33 μs MRTD, of which 30 μs can be allocated to ΔT_{prop}, with the remaining 3 μs used for TAE.

Asynchronous dual connectivity with no TAE requirements and no ΔT_{prop} restrictions is an example of no time requirements. The category of no time requirements also includes cooperative scheduling, cooperative zero-beamforming, and dynamic AP transmission selection in multi-AP cooperation.

In conclusion, correct network synchronisation is a prerequisite for excellent wireless network performance. Some of the most eye-catching use cases in 5G, including industrial automation, rely on more precise timing and may generate additional synchronisation requirements in the near future. Although the level of required

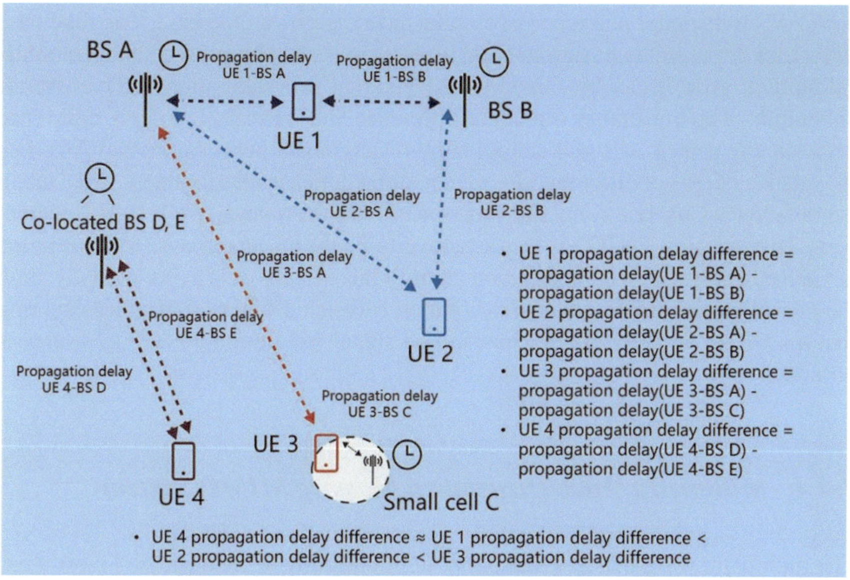

Fig. 7.7 Synchronisation for coordinated transmission/reception

synchronisation accuracy depends on several factors, treating the strictest synchronisation requirements as general 5G requirements would be a mistake, as doing so would make the cost of 5G and the future development of mobile technology unsustainable. The strictest requirements can only be addressed in real-time and on-site. Future 6G will also face the same requirements.

2. ***Channel Calibration***

The physical radio channels in TDD systems are reciprocal, but the RF front-end uses different circuits when transmitting and receiving, so the downlink and uplink channels do not match the channel reciprocity. Considering the mutual coupling effect of the antenna array, the asymmetry of the channel is more complex.

Reducing the user feedback of the downlink channel and utilising the reciprocal characteristics of the TDD system's physical radio channel requires channel calibration. The modelling of the non-reciprocity of TDD MIMO channel and corresponding calibration methods to minimise mismatch is channel calibration.

There are generally two types of methods to calibrate different RF links in the same antenna array:

(1) The transmitter and receiver self-calibration through internal circuits;
(2) Over-the-air calibration through radio loop.

To utilise multiple APs to serve UE jointly, the network infrastructure needs to be synchronised. The network can have an absolute time (phase) reference, but the phases between different APs are not synchronised. This means that in practice, each AP's transmitter and receiver circuits have their own time base. The time base difference between the transmitter and receiver in each AP represents the reciprocal calibration error. It can be solved by AP internal self-calibration or Over-the-air calibration. The transmitter time base difference and transmission delay difference between any pair of APs will cause phase misalignment between the two APs, i.e., the relative phases of different APs are not aligned. Phase misalignment will cause a degradation in Coherent Joint Transmission (CJT) performance, while Non-Coherent Joint Transmission (NCJT) is less sensitive to phase misalignment. To limit phase deviation, a correlation compensation algorithm is usually used to periodically align the phases between APs (relative calibration between different APs), thus ensuring that the UE receives the jointly transmitted signal from multiple APs as strong as possible.

7.4.4 Millimetre Wave Distributed Massive MIMO System

The millimetre wave massive MIMO systems currently being discussed mostly use a centralised array composed of a large number of antenna units, which significantly increases system throughput through its higher array gain and spatial multiplexing gain [57]. There are a series of problems such as high power consumption, high complexity, and high cost in the ultra-large-scale antenna array in the millimetre wave band, so the hybrid antenna array architecture and hybrid beamforming are more often adopted. However, the coverage problem of millimetre wave communication, such as large path loss, shadow fading, blockage, etc., greatly limits its performance and application scenarios. Section 7.1.3 of this book discusses a key feature of the distributed massive MIMO system–macro diversity gain, which is particularly prominent in high-frequency bands such as millimetre waves and terahertz. Therefore, the combination of distributed massive MIMO and high-frequency bands has great application prospects. The hardware complexity and power consumption of massive MIMO can be reduced by splitting the very large centralised array into smaller arrays and installing them at multiple distributed access points. Unlike the distributed massive MIMO system in the low-frequency band, a key problem to be solved in the high-frequency band such as the millimetre wave distributed massive MIMO system is beamforming, especially the implementation of analogue beamforming or beam alignment. In the high-frequency band, each user will also be equipped with multiple antennas or even hybrid antenna array architecture, which poses a huge challenge to beamforming.

Literature [57, 58] propose the analogue beam alignment technology of joint collaboration between access points and users in the millimetre wave distributed massive MIMO system. As shown in Fig. 7.8, analogue beamforming and digital precoding are only implemented at each access point in a distributed manner. Data and related scheduling control are completed by the central processor. The collaborative analogue beamforming between access points and users is based on the principle of maximising the total effective beamforming gain in the region, and quickly achieves

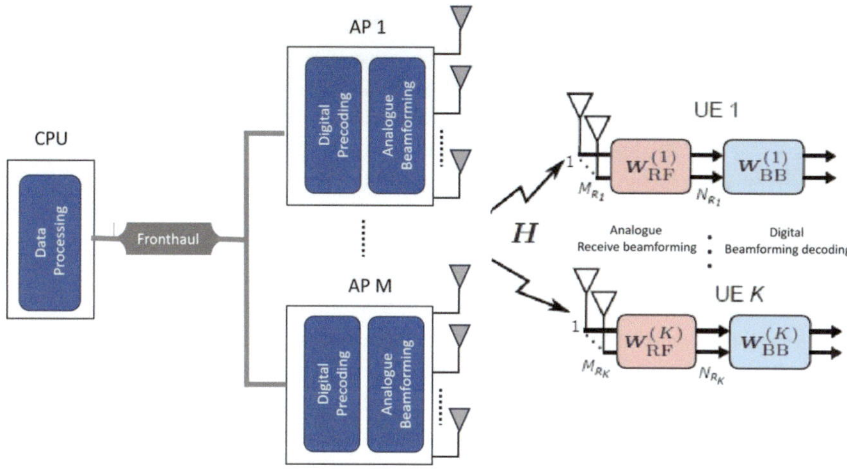

Fig. 7.8 Schematic diagram of the mmW distributed massive MIMO system

effective analogue beam alignment through iterative calculations between access points and users.

The system is based on hybrid beamforming, which includes M access points serving K users equipped with hybrid antenna architecture at the same time.

7.5 Network Deployment of Cell-Free MIMO Systems

Cell-Free MIMO systems have different deployment options. This is because the core of the Cell-Free MIMO system lies in the high degree of coordination between APs, and there is a lot of room for choice in the specific implementation of this coordination. For example, depending on the different ways in which the physical layer functions are split between the CPU and AP, or the different architectures of the fronthaul network, there are different classification methods for system deployment options, corresponding to different trade-offs between performance, cost, and complexity.

If classified roughly according to the split of physical layer processing, Cell-Free MIMO systems can be divided into centralised and distributed types. The following qualitatively compares small cell networks (Small Cell) with these two types of Cell-Free MIMO systems in terms of capacity, fronthaul/backhaul impact, and scalability. The physical layer (L1) processing of small cell networks is completed at the location close to antenna inside small cell. The connection between the small cell and the central unit (CU) belongs to the "backhaul" network. The capacity need of the backhaul is only equivalent to the actual traffic load. Therefore although the throughput of the small cell network system is far less than the potential of the Cell-Free MIMO system,, its scalability is better from the perspective of deployment. The other extreme opposite to the small cell network is the completely centralised computing Cell-Free MIMO system. Theoretically This situation can achieve the highest system capacity without considering cost. Because all data processing and control are completed at the CPU, in principle, it can achieve the optimal solution of all AP physical layer joint processing. In this scenario, the transmission network between the AP and the CPU belongs to the "fronthaul". Its requirements for bandwidth and delay are much higher than the backhaul network under the same traffic load conditions. In the centralised processing Cell-Free MIMO system, the computing power required by its CPU and the extremely high demand for fronthaul bandwidth will become key factors restricting the system scalability in practical deployment.

Taking the capacity demand of the fronthaul network as an example, the following shows the growth of transmission rate demand with the increase in the number of antennas or RF bandwidth in the ideal fully centralised Cell-Free MIMO system deployment. The relationship between the total rate of the Common Public Radio Interface (CPRI) fronthaul interface and t the number of antennas/bandwidth is shown in Table 7.1. The values in the table are calculated according to the CPRI protocol [59]. It can be seen that when the number of antennas in the system increases to 256, a Cell-Free MIMO system with 100 MHz RF bandwidth requires a fronthaul network throughput of more than 1 Tbps. If only 20 MHz RF bandwidth is supported, the total transport bandwidth will also reach to about 1 Tbps when the number of antennas is increased to 1024. Such high transport capacity is generally considered to be unacceptable from cost point of view.

In fact, small cell networks and fully centralised Cell-Free MIMO systems are two extreme choices. Therefore, in actual systems, the design of Cell-Free MIMO systems can appropriately balance performance and transport network costs, as shown in Fig. 7.9. The core of distributed processing is to offload part of the computing load to each AP, and only retain a part of the global processing function necessary for joint transmission at the CPU. For example, APs that need to transmit information in coordination can independently obtain channel state information from mobile terminals (TDD mode) and perform part of the physical layer processing including MIMO precoding. This can reduce the CPU processing load and also reduce the fronthaul load. However, in downlink transmission, all data transmitted to the terminal must be sent from the CPU to all relevant APs through the fronthaul network.

Table 7.1 Relationship between the total rate of the CPRI fronthaul interface and the number of antennas/ bandwidth

Antenna number	10 MHz	20 MHz	100 MHz
1	0.49 Gbps	0.98 Gbps	4.9 Gbps
4	1.96 Gbps	3.92 Gbps	19.6 Gbps
16	7.84 Gbps	15.68 Gbps	78.4 Gbps
64	31.36 Gbps	62.72 Gbps	313.6 Gbps
256	125.44 Gbps	250.88 Gbps	1.254 Tbps
1024	501.76 Gbps	1.004 Tbps	5.018 Tbps

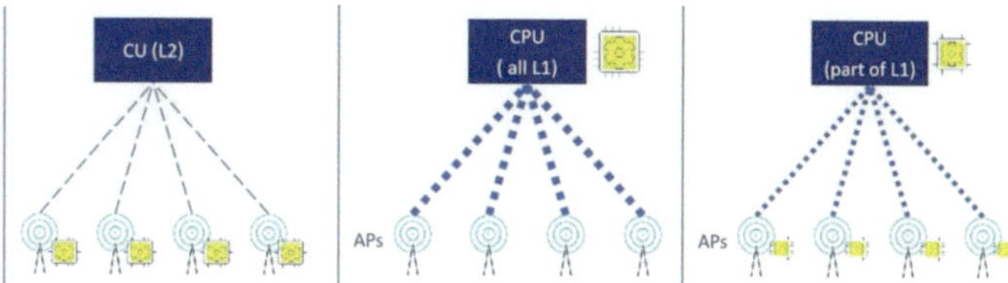

Deployment mode	Small Cells	Cell-free MIMO (centralized processing)	Cell-free MIMO (distributed processing)
System capacity	Low	High	Medium
Fronthaul/backhaul bandwidth requirement	(backhaul) Low	(fronthaul) High	(fronthaul) Medium
Scalability	High	Low	Medium

Fig. 7.9 Comparison of different network architecture modes, their performance and transport network requirements

In addition, how to split the computing functions of AP and CPU also has a certain adjustable space, which in turn affects the final transport network deployment cost and the overall system scalability. Figure 7.10 shows the main network function splitting methods defined by 3GPP, where the higher the sequence number (integer position) of the split point option, the simpler the function of the RF end network element. Note that for Cell-Free MIMO systems, in principle, only split points above option 6 (fronthaul) can be used to implement AP coordinated transmission (the split point option corresponding to the traditional small cell is 2, which cannot support physical layer cross-AP coordination). In principle, the higher the split point, the more conducive to centralised processing to obtain the best coordinated transmission performance, such as option 8 (CPRI). But the cost is greater fronthaul capacity requirement and higher energy consumption for signal transmission.

Figure 7.11 shows the front/backhaul network bandwidth requirements of mainstream 3GPP function split options. Under the conditions of 64 QAM modulation format and 20 MHz transmission bandwidth full load, the front/backhaul bandwidth required for the RF signal corresponding to different physical layer function splits (relative values) [60]. For options 8 and 7-1, the bandwidth required for the front haul is fixed, regardless of whether it is fully loaded. The front haul bandwidth required by other options is closely related to the user's load, so it represents its possible peak value. From the perspective of front haul capacity requirements, the main reason why option 8 is significantly higher than other options is that its transmission content is 15-bit quantized time domain sampling point information, while other options are frequency domain symbol information.

It should be noted that the calculation of the relative bandwidth value of the fronthaul in Fig. 7.11 does not reflect the impact of the Cell-Free MIMO system itself on the resource scheduling mechanism. It simply statically calculates the full load fronthaul bandwidth requirement on the data channel corresponding to a single RF channel. The distributed MIMO processing mechanism itself also has more detailed

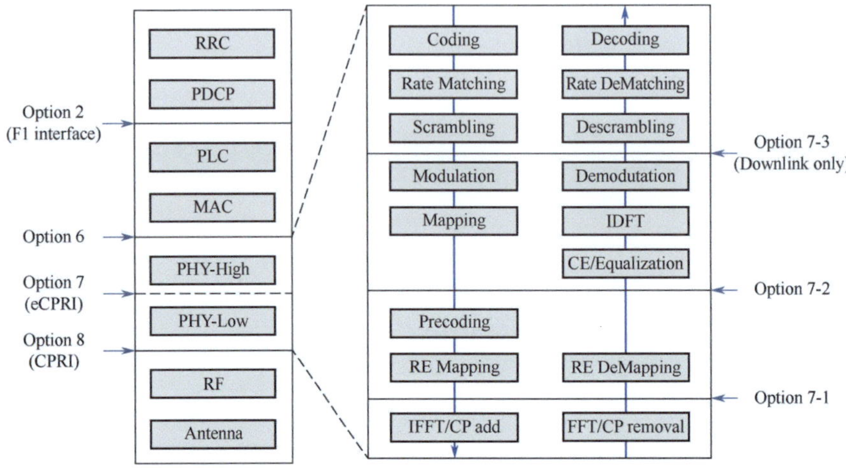

Fig. 7.10 Main network function split methods defined by 3GPP

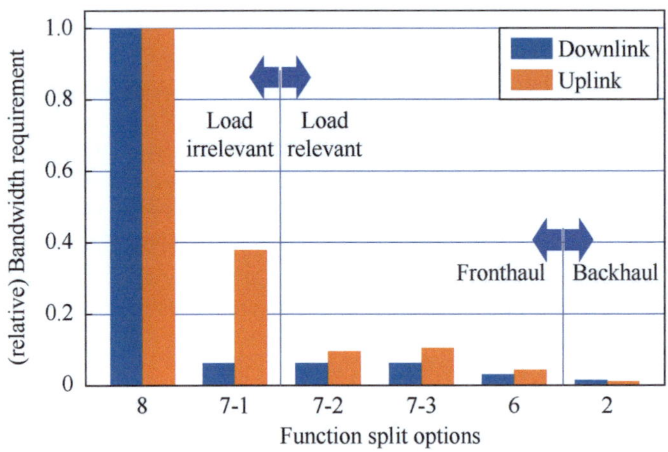

Fig. 7.11 Front/backhaul network bandwidth requirement for 3GPP mainstream function split options

implementation options, such as TDD, FDD and dynamic AP grouping methods, etc., which will have a great impact on the actual throughput requirements to the fronthaul network.

In addition, different scheduling mechanisms and function split methods will also cause many differences in the interaction of control information. But generally speaking, the bandwidth consumed by control information is much smaller than that of the data channel, so the impact on the overall transport network capacity is small.

From a macro perspective, by designing different distributed MIMO processing mechanisms and using corresponding fronthaul split points, in principle, it can achieve "continuously adjustable" from small cell networks to various modes of centralized Cell-Free MIMO systems. The actual system design and deployment should comprehensively consider performance requirements, the number of APs required, algorithm complexity, transport network costs and other factors to get the most reasonable configuration (see Sect. 7.4.2). However, how to conduct specific and quantitative analysis of these factors involves a huge optimization space and requires interdisciplinary research and engineering verification. This is a challenging problem that is currently in the research and exploration stage.

In addition to the above-mentioned options for distributed or centralized computing, another cost-related consideration when deploying a Cell-Free MIMO system is how to optimize the physical topology of the fronthaul network. Logically, each AP has direct communication with the CPU. If the number of physical ports on the CPU corresponds one-to-one with the number of APs, it may not be practically feasible. Therefore, an appropriate physical port multiplexing mechanism is also an important factor in enhancing the feasibility of deploying a Cell-Free MIMO system. Below are two typical ways of building physical topology, namely chain (or ring) topology and tree topology.

The chain (or ring) topology connects all APs in series to a simple optical fiber chain, where each AP needs to have at least the ability to add/drop signals. When the AP has computing capabilities, it may also be responsible for "online" data aggregation processing, thereby further reducing the overall load on the fronthaul network. This type of deployment is also referred to as the Radio Stripe mode in some real cases.

The tree topology has certain advantages in reducing deployment costs, and its principle mainly lies in utilising the time domain or wavelength domain multiplexing characteristics provided by the optical transport network itself. The physical topology of the real network deployment is shown in Fig. 7.12. Only one optical port is needed at the CPU to connect to multiple APs via connecting to one or more cascaded multiplexers/demultiplexers Wavelength multiplexer/demultiplexer corresponds to the Wavelength Division Multiplexing (WDM) scheme, and passive splitter corresponds to Time Division Multiplexing (TDM) scheme. They can also be further combined. In practical deployment, these schemes generally directly utilise mature, low-cost Passive Optical Network (PON) equipment to carry wireless fronthaul traffic [61].

For Cell-Free MIMO systems, there is another direction worth paying attention to in the use of the low-cost optical transport network, which is Analog Radio-over-Fiber (ARoF) technology [62, 63]. In principle, the analog radio-over-fiber link between the CPU and AP is to directly extend the signal on the RF link in the analog signal domain, which is a slightly higher split than option 8. This type of networking method logically implements a purely centralised Cell-Free MIMO system, but it can effectively avoid the large fronthaul bandwidth requirements brought about by digital quantisation. Since the transmission of analog signals in the optical domain will introduce additional noise, such schemes will also affect the performance of wireless physical layer processing to a certain extent. Overall, the Cell-Free MIMO system based on analog radio-over-fiber can reduce system complexity and fronthaul network power

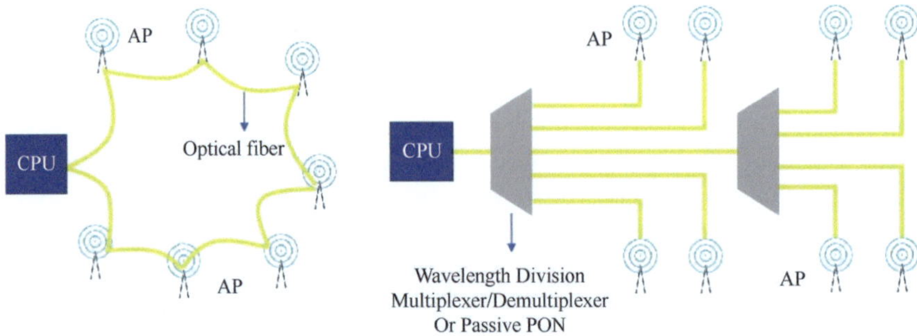

Fig. 7.12 Physical topology of real network deployment

consumption. It has a good practical prospect, but further interdisciplinary research and verification are still needed.

7.6 Summary

Ubiquitous coverage, large capacity and reliable high data rate transmission are key requirements for all types of indoor, outdoor and hotspot scenarios. The excellent macro diversity gain and anti-interference ability of the Cell-Free MIMO system based on distributed ultra-large scale antennas make it effectively applicable in scenarios such as stadiums, airports, smart factories and city centre squares, as shown in Fig. 7.13.

Distributed massive MIMO technology and Cell-Free network architecture are promising technologies for future 6G. However, its related technologies have not yet matured and there are many open problems, such as user access point association and pilot allocation, channel estimation, uplink and downlink signal processing based on beamforming, power allocation, calibration and synchronisation, combination with millimetre waves, fronthaul and backhaul design, and network architecture, etc., which need further research. We expect that future MIMO technology will appear in the scenarios as shown in Fig. 7.14, in the form of Cell-Free MIMO system architecture with distributed ultra-large scale antennas.

Fig. 7.13 Example of application scenarios for distributed massive MIMO

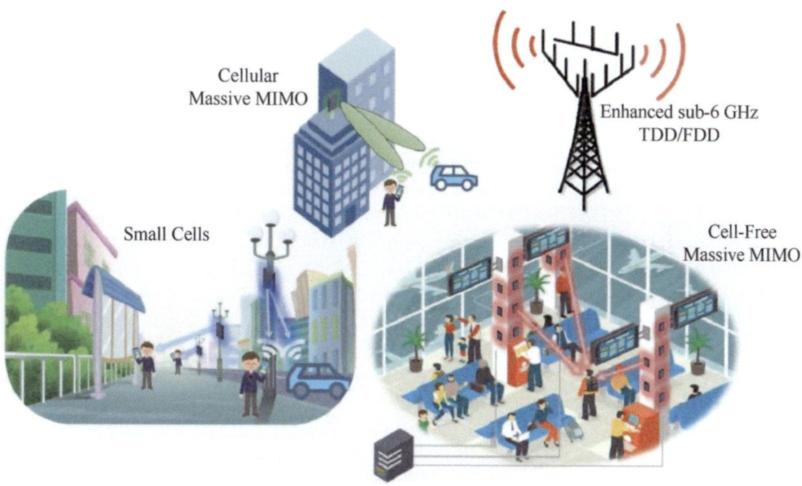

Fig. 7.14 Example of future diverse MIMO system scenarios

References

1. TELATAR E. Capacity of multi-antenna Gaussian channels [J]. European transactions on telecommunications, 1999, 10(6): 585–595.
2. TSE D, VISWANATH P. Fundamentals of wireless communication [M]. Cambridge University Press, 2005.
3. ROBERT W, HEATH JR. Foundations of MIMO communication [M]. Cambridge University Press, 2018.
4. OESTGES C, CLERCKX B. MIMO wireless communications from real-world propagation to space-time code design [M]. Academic Press, 2010.
5. MARZETTA T L, LARSSON E G, YANG H, et al. Fundamentals of massive MIMO [M]. Cambridge University Press, 2016.
6. NGO H Q, LARSSON E G, MARZETTA T L. Aspects of favorable propagation in massive MIMO [J]. 2014 22nd European Signal Processing Conference (EUSIPCO), 2014, 76–80.
7. WU X, BEAULIEU N C, LIU D. On favorable propagation in massive MIMO systems and different antenna configurations [J]. IEEE Access, 2017, 5: 5578–5593.
8. GUNNARSSON S, FLORDELIS J, PERRE L V D, et al. Channel hardening in massive MIMO: Model parameters and experimental assessment [J]. IEEE Open Journal of the Communications Society, 2020, 1: 501–512.
9. ERICSSON. Ericsson first to deliver 5G NR radio [R/OL]. 2016.
10. SHAMAI S, ZAIDEL B M. Enhancing the cellular downlink capacity via co-processing at the transmitting end [J]. IEEE VTS 53rd Vehicular Technology Conference, 2001, 3: 1745–1749.
11. GANESAN K, UNNIKRISHNAN. Distributed Massive MIMO: Random Access, Extreme Multiplexing and Synchronisation. Diss [M]. Linköping University Electronic Press, 2022.
12. KARAKAYALI M K, FOSCHINI G J, VALENZUELA R A. Network coordination for spectrally efficient communications in cellular systems [J]. IEEE Wireless Communications, 2006, 13(4): 56–61.
13. VENKATESAN S, LOZANO A, VALENZUELA R. Network MIMO: Overcoming inter cell interference in indoor wireless systems [G]. Proceedings of the Forty-First Asilomar Conference on Signals, Systems and Computers. 2007, 83–87.
14. IEEE 802.16M BROADBAND WIRELESS ACCESS WORKING GROUP. Uplink Multi-cell MIMO for Inter-cell interference mitigation: C80216m-Link-08_094 [J]. 2008.
15. IEEE. Downlink Multi-BS MIMO: C802.16m-09/1099 [J]. IEEE 802.16m Broadband Wireless Access Working Group, 2009.
16. GESBERT D, HANLY S, HUANG H, et al. Multi-cell MIMO cooperative networks: A new look at interference [J]. IEEE journal on selected areas in communications, 2010, 28(9): 1380–1408.
17. CHINA MOBILE RESEARCH INSTITUTE (2011). C-RAN: The Road Toward Green RAN (PDF) [R]. Archived from the original (PDF) on 2013–12–31, 2013.
18. INTERDONATO G, BJÖRNSON E, QUOC NGO H, et al. Larsson, Ubiquitous Cell-Free massive MIMO communications [J]. EURASIP Journal on Wireless Communications and Networking, 2019(1): 1–13.
19. NGO H Q, ASHIKHMIN A, YANG H, et al. Cell-Free massive MIMO versus small cells [J]. IEEE Transactions on Wireless Communications, 2017, 16(3): 1834–1850.
20. BJÖRNSON E, SANGUINETTI L. Making Cell-Free massive MIMO competitive with MMSE processing and centralised implementation [J]. IEEE Transactions on Wireless Communications, 2020, 19(1): 77–90.
21. HE H, YU X, ZHANG J, et al. Cell-Free Massive MIMO for 6G Wireless Communication Networks [G]. Journal of Communications and Information Networks, 2010, 6(4): 321–335.

22. HUGL K, KALLIOLA K, LAURILA J. Spatial reciprocity of uplink and downlink radio channels in FDD systems [R]. Proc. COST, 2002, 273(2).
23. IMTIAZ S, DAHMAN G S, RUSEK F, et al. On the directional reciprocity of uplink and downlink channels in frequency division duplex systems [J]. 2014 IEEE 25th Annual International Symposium on Personal, Indoor, and Mobile Radio Communication (PIMRC). IEEE, 2014.
24. 3GPP . Rel-17 work scope on NR MIMO and sub-3GHz FDD enhancements: RP-191762 [S]. Huawei, HiSilicon, 3GPP TSG RAN Meeting #85, 2019.
25. BJÖRNSON E, SANGUINETTI L. Scalable Cell-Free massive MIMO systems [J]. IEEE Trans. Commun., 2020, 68(7): 4247–4261.
26. BUZZI S, D'ANDREA C. Cell-Free Massive MIMO: User-Centric Approach [J]. IEEE Wireless Communications Letters, 2017,6(6): 706–709.
27. BUZZI S, D'ANDREA C, D'ELIA C. User-Centric Cell-Free Massive MIMO with Interference Cancellation and Local ZF Downlink Precoding [J]. 2018 15th International Symposium on Wireless Communication Systems (ISWCS), 2018: 1–5.
28. BUZZI S, D'ANDREA C. User-Centric Communications versus Cell-Free Massive MIMO for 5G Cellular Networks [J].WSA 2017; 21st International ITG Workshop on Smart Antennas, 2017: 1–6.
29. NGO H Q, TRAN L, DUONG T Q et al.On the Total Energy Efficiency of Cell-Free Massive MIMO [J]. In IEEE Transactions on Green Communications and Networking, 2018, 2(1): 25–39.
30. NGUYEN T H, NGUYEN T K, HAN H D, et al. Optimal Power Control and Load Balancing for Uplink Cell-Free Multi-User Massive MIMO [J]. IEEE Access, 2018, 6: 14462–14473.
31. INTERDONATO G, FRENGER AND P, LARSSON E G. Scalability Aspects of Cell-Free Massive MIMO [J]. ICC 2019 - 2019 IEEE International Conference on Communications (ICC), 2019: 1–6.
32. ATTARIFAR M, ABBASFAR A, LOZANO A. Random vs Structured Pilot Assignment in Cell-Free Massive MIMO Wireless Networks [J]. 2018 IEEE International Conference on Communications Workshops (ICC Workshops), 2018: 1–6.
33. CHEN S, ZHANG J, BJÖRNSON E, et al. Structured Massive Access for Scalable Cell-Free Massive MIMO Systems [J]. In IEEE Journal on Selected Areas in Communications, 2021, 39(4): 1086–1100.
34. HAO Y, XIN J, TAO W, et al. Pilot Allocation Algorithm Based on K-means Clustering in Cell-Free Massive MIMO Systems [J]. 2020 IEEE 6th International Conference on Computer and Communications (ICCC), 2020: 608–611.
35. FEMENIAS G, RIERA-PALOU F. Cell-Free Millimeter-Wave Massive MIMO Systems With Limited Fronthaul Capacity [J]. In IEEE Access, 2019, 7:44596–44612.
36. LIU H, ZHANG J, ZHANG X, et al. Tabu-Search-Based Pilot Assignment for Cell-Free Massive MIMO Systems [J]. In IEEE Transactions on Vehicular Technology, 2020, 69(2): 2286–2290.
37. INTERDONATO G, FRENGER P, LARSSON E G. Utility-based Downlink Pilot Assignment in Cell-Free Massive MIMO [R]. WSA 2018; 22nd International ITG Workshop on Smart Antennas, 2018: 1–8.
38. BUZZI S, D'ANDREA C, FRESIA M, et al. Pilot Assignment in Cell-Free Massive MIMO Based on the Hungarian Algorithm [J]. In IEEE Wireless Communications Letters, 2021, 10(1): 34–37.
39. LIU H, ZHANG J, JIN S, et al. Graph Colouring Based Pilot Assignment for Cell-Free Massive MIMO Systems [J]. IEEE Transactions on Vehicular Technology, 2020, 69(8): 9180–9184.
40. ZENG W, HE Y, LI B, et al. Pilot Assignment for Cell-Free Massive MIMO Systems Using a Weighted Graphic Framework [J]. In IEEE Transactions on Vehicular Technology, 2021.

41. ÖZDOGAN Ö, BJÖRNSON E, ZHANG J. Performance of Cell-Free Massive MIMO With Rician Fading and Phase Shifts [J]. IEEE Transactions on Wireless Communications, 2019, 18(11): 5299–5315.
42. GUO Z, WANG X, HENG W. Millimetre-Wave Channel Estimation Based on 2-D Beamspace MUSIC Method [J]. In IEEE Transactions on Wireless Communications, 2017, 16(8): 5384–5394.
43. ZHANG D, WANG Y, SU Z, et al. Millimetre Wave Channel Estimation Based on Subspace Fitting [J]. In IEEE Access, 2018, 6: 76126–76139.
44. WEILAND L, STÖCKLE C, WÜRTH M, et al. OMP with Grid-Less Refinement Steps for Compressive mmWave MIMO Channel Estimation [J]. 2018 IEEE 10th Sensor Array and Multichannel Signal Processing Workshop (SAM), 2018: 543–547.
45. ANJINAPPA C K, GÜRBÜZ A. C, YAPICI Y. Off-Grid Aware Channel and Covariance Estimation in mmWave Networks [J]. In IEEE Transactions on Communications, 2020, 68(6): 3908–3921.
46. MARZETTA T L. Noncooperative Cellular Wireless with Unlimited Numbers of Base Station Antennas [J]. IEEE Transactions on Wireless Communications, 2010, 9(11): 3590–3600.
47. BJÖRNSON E, HOYDIS J, SANGUINETTI L. Massive MIMO Has Unlimited Capacity [J]. IEEE Transactions on Wireless Communications, 2018, 17(1): 574–590.
48. DEMIR Ö T, BJÖRNSON E, SANGUINETTI L. Foundations of User-Centric Cell-Free Massive MIMO [M]. Now Foundations and Trends, 2021.
49. NAYEBI E, ASHIKHMIN A, MARZETTA T L, et al. Precoding and power optimisation in cell-free massive MIMO systems [J]. IEEE Transactions on Wireless Communications, 2017, 16(7): 4445–4459.
50. CHAKRABORTY S, DEMIR Ö T, BJÖRNSON E, et al. Efficient Downlink Power Allocation Algorithms for Cell-Free Massive MIMO Systems [J]. IEEE Open Journal of the Communications Society, 2020, 2: 168–186.
51. NGO H Q, TRAN L N, DUONG T Q, et al. On the total energy efficiency of cell-free massive MIMO [J]. IEEE Transactions on Green Communications and Networking, 2017, 2(1): 25–39.
52. ZHAO Y, NIEMEGEERS I G, GROOT S H D. Power Allocation in Cell-Free Massive MIMO: A Deep Learning Method [J]. In IEEE Access, 2020, 8: 87185–87200.
53. LIANG F, SHEN C, YU W, et al. Towards Optimal Power Control via Ensembling Deep Neural Networks [J]. IEEE Transactions on Communications, 2020, 68(3): 1760–1776.
54. ALRIKSSON, F, BOSTRÖM L, SACHS J, et al. Critical IoT connectivity: Ideal for time-critical communications [R]. Ericsson Technology Review, 2020.
55. 3GPP. NR; Requirements for support of radio resource management: Technical specification TS 38.133 [S].2017.
56. 3GPP. NR; Base Station (BS) radio transmission and reception: technical specification TS 38.104 [S].2018.
57. XIAO M, MUMTAZ S, HUANG Y, et al. Millimetre wave communications for future mobile networks [J]. IEEE Journal on Selected Areas in Communications, 2017, 35(9): 1909–1935.
58. SONG N, YANG T. Distributed Hybrid Beamforming for MmWave Cell-Free Massive MIMO [J]. IEEE International Conference on Acoustics, Speech and Signal Processing (ICASSP), 2022: 5373–5377.
59. COMMON PUBLIC RADIO INTERFACE (CPRI). Interface Specification 7.0 [S]. CPRI official website, 2022.
60. AGIWAL M, ROY A, SAXENA N. Next Generation 5G Wireless Networks: A Comprehensive Survey [J]. IEEE Communications Surveys & Tutorials, 2016,18(3).
61. NAKAYAMA Y, DAISUKE D. Wavelength and bandwidth allocation for mobile fronthaul in TWDM-PON [J]. IEEE Transactions on Communications 2019,67(11) : 7642–7655 .

62. HABIB U, AIGHOBAHI A E, , QUINLAN Q, et al. Analogue Radio-Over-Fiber Supported Increased RAU Spacing for 60 GHz Distributed MIMO Employing Spatial Diversity and Multiplexing [J]. Journal of Lightwave Technology: A Joint IEEE/OSA Publication , 2018,36(19): 4354–4360.
63. KIM J, SUNG M, CHO S H, et al. MIMO-Supporting Radio-Over-Fiber System and its Application in mmWave-Based Indoor 5G Mobile Network [J]. In Journal of Lightwave Technology, 2020, 38(1): 101–111.

Chapter 8
Reconfigurable Intelligent Surfaces

Reconfigurable Intelligent Surfaces (RIS) have attracted widespread attention from academia and industry due to their ability to flexibly manipulate electromagnetic characteristics in the channel environment. RIS typically consists of numerous carefully designed electromagnetic units. By applying control signals to the adjustable components on these units, their electromagnetic properties can be dynamically controlled. This allows for active and intelligent control of spatial electromagnetic waves in a programmable manner, forming an electromagnetic field with controllable amplitude, phase, polarization, and frequency characteristics. RIS inherently offers low cost, low power consumption, and easy deployment, making it a promising solution for addressing the new demands and challenges of future wireless networks. RIS has the potential to transform the wireless propagation environment from passive adaptation to active control, and may be implemented in 5G networks ahead of schedule. It is also likely to enable intelligent wireless environments in future 6G networks [1], bringing a new network paradigm [2]. This chapter provides a comprehensive introduction to RIS technology, covering its fundamentals, hardware materials and electromagnetic regulation, channel models, RIS-assisted wireless communication, and network deployment.

8.1 Overview

RIS is a programmable artificial two-dimensional material with sub-wavelength dimensions, developed from meta-material technology. It is typically composed of metal, dielectric, and adjustable components and can be modeled as an RLC circuit. RIS generates the electromagnetic behavior required for each unit by controlling varactor diodes, PIN diodes, Radio Frequency Micro Electro Mechanical Systems

G. Liu et al., *6G Migrates from Communication to XaaS*,
https://doi.org/10.1007/978-981-96-7579-1_8

(RF-MEMS), liquid crystals, or graphene [3]. As a reconfigurable spatial electromagnetic wave modulator, it adjusts the signal propagation direction in three-dimensional space, suppresses interference, and enhances signals, thereby intelligently reconstructing the wireless propagation environment between transceivers. RIS can overcome the uncontrollability of traditional wireless channels, actively control the wireless propagation environment, and build a new paradigm of smart radio environments. Compared with traditional technologies, RIS offers low cost, low power consumption, and easy deployment. It has broad application prospects in non-line-of-sight (NLOS) transmission, edge user enhancement, secure communication, passive Internet of Things, and high-precision positioning. Therefore, it is expected to enhance 5G networks and become one of the potential key technologies for 6G networks [4, 5].

In short, RIS plays a significant role in solving non-line-of-sight transmission problems, expanding coverage, increasing transmission degrees of freedom, reducing electromagnetic pollution, enabling ultra-large-scale terminal access, enhancing environmental perception and positioning, and supporting future perception communication integration. It also helps in obtaining effective three-dimensional environmental data.

As a key candidate technology for future wireless networks, RIS offers several advantages over existing communication technologies. The four most typical advantages are:

(1) It is almost passive, RIS does not require a dedicated power supply to regulate electromagnetic waves, does not introduce noise, has low power consumption, and meets green communication requirements.
(2) It is low cost, as it does not need mixers, digital/analog converters, power amplifiers, and other high-cost components.
(3) It is simple and easy to deploy. Its scalable and lightweight design makes it easy to install and disassemble.
(4) It provides excellent control of electromagnetic waves, allowing precise manipulation of their propagation characteristics, such as phase, amplitude, polarization mode, and time delay. Additionally, it can achieve negative refraction and reflection of electromagnetic waves.

In addition, RIS supports full-duplex transmission, is generally lightweight, can flexibly form various surface shapes, and either does not require a return link or only needs a low-rate one. It is convenient for widespread deployment and offers significant advantages for engineering applications. Unlike traditional technologies such as relay, backscatter communication, and massive MIMO, RIS represents a completely new technical field. Relay communication involves active reception and transmission, requiring numerous RF links and incurring high hardware costs and energy consumption. Backscatter communication, while passive, does not require RF links and functions as a low-rate transmitter. Massive MIMO uses an active array structure, also needing many RF links and high costs. RIS, however, employs a passive array structure, avoiding the need for costly, high-power devices and enabling low-energy green communications. Thus, RIS offers unique advantages over traditional relay, backscatter communication, and massive MIMO technologies. RIS facilitates the creation of an intelligent, controllable wireless propagation environment, enhancing transmission reliability, capacity, and spectral efficiency, ensuring its broad application prospects.

8.2 Hardware Materials and Electromagnetic Control

Conceptually, RIS is a two-dimensional surface made of reconfigurable novel metamaterials that can manipulate incident electromagnetic waves in a variety of ways. It contains a large number of independent, low-cost, sub-wavelength resonant units (i.e. array elements), which can fully control the amplitude, phase, polarization state, and the reflection, refraction, and scattering of radio waves by adjusting their parameters and spatial distribution.

8.2.1 RIS Hardware Structure

Based on RIS's response to electromagnetic waves, it can be divided into reflective, refractive, and simultaneous reflective and refractive types. This section discusses the reflective RIS as an example.

Reflective RIS generally consists of sub-wavelength units made of ultra-thin metal or non-metal materials arranged periodically. By regulating the frequency response characteristics of each reflection unit, the output radiation signal's beam shape, direction, and focus position can be varied to achieve the system performance optimization goal of joint deployment of the base station and the super-surface panel. As shown in Fig. 8.1, the hardware structure of the reflective RIS usually includes several components such as reflection units, control switches, metal back-plates (such as copper backplates), control circuit boards, controllers, and their control interfaces. Among these components, the unit structure integrates some adjustable components to obtain the reconfigurability of electromagnetic waves, such as PIN diodes, varactor diodes, liquid crystals, RF-MEMS switches, etc. Based on the specific control ability of the unit structure on electromagnetic waves, the reconfigurable super-surface can be roughly divided into amplitude reconfigurable, phase reconfigurable, polarization reconfigurable, etc. Through the frequency response control of the reflection unit, the reflected signal of the yellow incident beam shown in Fig. 8.1 can be a red reflection beam or a blue reflection beam in a direction different from that of the red reflection beam.

Reflective RIS supports super-normal reflection of surfaces in accordance with the generalized Snell's reflection law [6]. The greater the spatial degrees of freedom of the reflected signal, the stronger the ability to flexibly shape and focus the three-dimensional beam. That is, the direction, shape, and intensity distribution of the reflected signal's wave vector can all be artificially controlled according to the

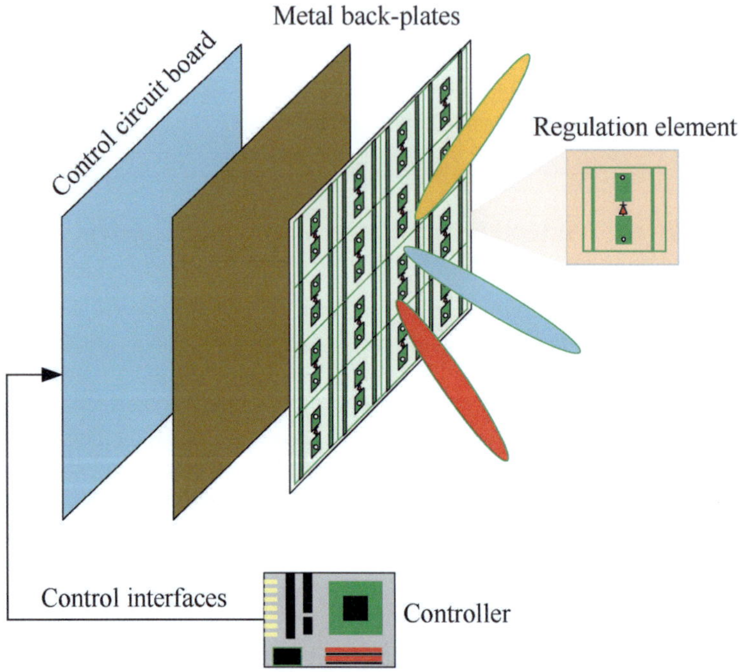

Fig. 8.1 Hardware structure of reflective RIS

network deployment environment. Traditional reflectors are high-gain antennas with normal incidence and normal emission; the reflection of electromagnetic waves by ordinary metal plates is a reflection that follows Snell's law, meaning the reflected wave is a mirror reflection of the incident wave, and the direction, shape, and intensity distribution of the signal reflection and scattering cannot be artificially controlled.

From the perspective of RIS deployment, RIS can be divided into two categories based on the need for real-time user tracking: one is a semi-static panel that does not need to support real-time user tracking, meaning it does not require frequent use of controllers and control interfaces, thus reducing power supply requirements; the other is a dynamic panel where all reflection unit frequency response characteristics can be flexibly adjusted by the controller, requiring frequent use of controllers and control interfaces, and a more stable power supply.

The main feature requirements of the semi-static panel are energy-saving, easy to deploy, low cost, and easy to achieve high gain. Its key technology is the semi-static panel design and algorithm implementation of the joint optimization deployment of the base station and the intelligent reflective panel. The main feature of the dynamic panel is that the base station can adjust the RIS panel frequency response characteristics as needed based on the deployment environment and its random changes. Its key technologies include dynamic panel design and algorithm implementation, low-cost low-power control circuit and interface implementation, etc.

8.2.2 RIS Regulation Mechanism

RIS, as a digitally programmable electromagnetic meta-surface, can flexibly control electromagnetic waves by designing and arranging electromagnetic unit structures, bringing new physical phenomena and applications. As mentioned earlier, the RIS panel consists of a series of units integrated with adjustable components; these adjustable units aim to achieve a certain phase response, amplitude response, and polarization response. Based on the different mechanisms of the control unit, RIS can be divided into:

(1) Electronically reconfigurable, using RF-MEMS switches, PIN diodes, varactor diodes, and field-effect tubes.
(2) Optoelectronically reconfigurable, using photoconductive switches.
(3) Novel reconfigurable materials, using liquid crystals, graphene, ferrite materials, etc.
(4) Physically reconfigurable, changing the direction, shape, or placement of the constituent units through a motor.

1. Electronically Reconfigurable

Electronic reconfiguration can utilize RF-MEMS switches or PIN diodes to redistribute the surface current of the unit in a discrete manner. Varactor diodes can be used to continuously control the surface current of the unit. The primary advantage of this scheme lies in the ease of integrating electronic components into the meta-surface.

PIN diodes can act as variable resistors at RF and microwave frequencies. They are current-controlled, implying that different input currents will produce lower resistance (corresponding to ON or "connected" state) or higher resistance (corresponding to OFF or "disconnected" state). PIN diodes are currently the most commercially available.

Field-effect transistors (FETs) utilize an electric field to control the shape of the conductive channel, thereby controlling the conductivity of a certain type of carrier in the semiconductor material. Field-effect transistors are voltage-controlled components, with very small direct current under on or off states, making them superior to PIN diodes in terms of lower direct current power consumption. However, PIN diodes perform better than field-effect transistors in terms of radio frequency insertion loss. Field-effect transistors have excellent isolation characteristics in the low-frequency range, but due to the presence of drain-source capacitance, the isolation characteristics will deteriorate at high frequencies.

RF-MEMS switches have good direct current and radio frequency (DC-RF) isolation, low loss, high linearity, extremely low direct current power consumption, and higher power capacity. However, compared with PIN diodes and varactor diodes with response times in nanoseconds (ns), RF-MEMS switches respond slower, with response times in microseconds (μs). RF-MEMS switches are easy to integrate with antennas using monolithic integration technology [7, 8].

A varactor diode is a semiconductor PN junction component that can provide variable junction capacitance through controlled bias voltage. Although these electronic components have many advantages, they also have some disadvantages. For instance, RF-MEMS switches and varactor diodes show resistance and capacitance non-linearity in addition to high operating voltage. The direct current bias lines required for these electronic components can deteriorate radiation patterns and may increase additional losses. Bypass capacitors and bias inductors make the meta-surface unit structure more complex. In addition to losses, especially at high incident radio frequency power, these electronic components may generate harmonics and inter-modulation distortion. The reconstruction based on varactor diodes is easily affected by nonlinear effects.

The key characteristics of electronic components such as varactor diodes, PIN diodes, and RF-MEMS switches are compared as shown in Table 8.1 [9].

2. *Optoelectronic Reconfigurable*

Reconfigurable meta-surfaces based on optoelectronic components, such as photo-conductive switches, work by triggering photosensitive semiconductor materials with incident lasers, causing a sudden change in their conductivity. Although these photo-conductive switches are inherently lossy and require lasers to work, they have the potential advantages of linearity, no harmonics, and no inter-modulation distortion. Additionally, they do not require bias lines, which eliminates performance degradation and complexity in meta-surface unit design. The main problem with these switches is activating the semiconductor, i.e., how to illuminate them with lasers. Although photo-conductive switches have linear advantages and avoid bias lines, they are lossy and require complex activation mechanisms. In addition, each switch activation mechanism has significant power consumption. In order to be integrated into the meta-surface, they must be low-loss, extremely compact, and highly power efficient.

Table 8.1 Comparison of key characteristics of tunable components

Tunable components	Voltage	DC power consumption	Switching speed	Power capacity	RF insertion loss	Control method
PIN diode	⚡	🔋	Ns level	Low	🟡 −	Digital
Field-effect transistor (FET)	⚡	🔋	ns level	Low	🟡 +	Digital
RF-MEMS	⚡⚡⚡	🔋	Ns level	Medium	🟢	Digital
Varactor diode	⚡⚡	🔋	Ns level	Medium	🟡	Analog

3. New Reconfigurable Materials

New reconfigurable materials, such as ferrites, graphene, and liquid crystals, can have their electrical properties changed. Ferrites can change their magnetic permeability in response to an applied static magnetic field and their dielectric constant in response to an applied static electric field. Graphene can change its equivalent conductivity by regulating its chemical potential energy. Liquid crystals can change their dielectric constant by reorienting their molecules with a change in applied voltage.

Although liquid crystal antennas have been reported in the millimeter wave low-frequency band, their losses are still relatively large compared to electronic components [10]. They are expected to have greater advantages in millimeter wave high-frequency bands and terahertz frequency bands. In addition, due to the slow switching speed of liquid crystals (ms level), the reconfiguration refresh time is longer compared to meta-surface based on PIN diodes and RF-MEMS.

Ferrite materials are often used for single antennas due to their large volume, high loss, and magnetic control characteristics, so they are not commonly seen in large arrays. Graphene may be a promising candidate material in the future, but this technology is still in the development stage and is not yet mature. The advantages of graphene have been demonstrated in the terahertz frequency band, but it experiences higher loss in the lower millimeter wave frequency band.

4. Physically Reconfigurable Structures

Physical structures can be reconfigured by mechanisms such as motors that change the direction, shape, or arrangement of the constituent units. In this method, the physical size or direction of the component is changed by mechanical drives to achieve phase tuning. For reflective arrays, linear (translation) and rotational motion can be used to achieve phase tuning.

Literature [11] has implemented a new phased array prototype with 756 micro-motor-controlled units operating at a center frequency of 8.3 GHz. The control circuit consists of a microprocessor, a field-programmable gate array (FPGA), and a micro-motor driver. Its working principle is as follows: the computer passes the beam pointing information to the microprocessor; the microprocessor calculates the unit rotation angle using the ray tracing method and assigns it to 30 FPGAs. Each FPGA controls the corresponding micro-motor driver, rotating the corresponding antenna unit to obtain different radiation phases, thereby achieving beam scanning. This control circuit adopts a parallel mechanism, meaning all units can rotate simultaneously. It should be noted that unit rotation is only applicable to circular polarization.

In summary, when choosing a tunable component, there is no single best answer. The best choice is the one that best suits the constraints of the application under consideration. We need to consider the maturity of the technology, performance, availability, power consumption, switching time, components linearity, and applications for specific frequency bands and working environments.

Table 8.2 Beam gain loss under different phase quantization

Phase quantization	Directional loss (MATLAB simulation)	Directional loss (Full wave simulation)
Continuous	0 dB	–
3 bit	0.22 dB	–
2 bit	0.99 dB	0.92 dB
1.5 bit	1.66 dB	1.12 dB
1 bit	3.92 dB	2.92 dB

8.2.3 *Phase Discretization*

The control ability of RIS over the beam is reflected in the scanning resolution and pattern control, which usually depends on the unit's phase control range and available phase states, i.e., the phase quantization effect in the meta-surface. The use of continuous phase control or a higher number of phase quantization states in the meta-surface is usually not feasible for engineering implementation. Due to the large number of components in the meta-surface, the reconstruction design of each unit needs to be greatly simplified. Table 8.2 lists the beam gain loss under different phase quantization compared to continuous phase shift [12].

In Table 8.2, the beam gain loss of 3-bit phase quantization can be ignored. A phase quantization accuracy of 2 bits or more can result in a smaller beam gain loss. Additionally, 2 bits provide a larger gain benefit than 1 bit. The lower the phase quantization accuracy, the wider the bandwidth, due to the greater tolerance of phase quantization.

8.2.4 *RIS Research Hardware Prototype*

At present, most of the research work in the literature is focused on theoretical derivation. The design and field measurement work of RIS hardware prototypes have just begun, but there are also some inspiring designs. The RIS prototype designs are shown in Table 8.3.

8.3 RIS Channel Modeling

Channel modelling is the basis for subsequent RIS deployment planning, coverage/capacity, network performance analysis and model-based intelligent control [18]. The general RIS includes passive RIS and active RIS, reflective surfaces and transmissive or a simultaneously transmitting and reflecting RIS (STAR-RIS). This section targets passive reflective RIS with capability of spatial modulation, assuming that

Table 8.3 RIS prototype design

Features	Solutions
RF switch	Reference [13] describes a structure consisting of 3720 antenna elements arranged on a 6 m^2 plane. The structure operates in a nearly passive mode, since the plane itself does not emit new radio waves. However, it can be adaptively configured via a low-power control circuit to focus incident radio waves and perform beamforming towards specific directions and positions
Piezoelectric ceramic mechanical tuning	Reference [14] presents the design of an intelligent glass meta-surface. The meta-surface is an artificially designed thin layer placed on a two-dimensional surface having a glass substrate, where numerous sub-wavelength elements are periodically arranged. A transparent glass substrate is superimposed on another transparent glass substrate, and the distance between the laminated substrates is fine-tuned using piezoelectric ceramics. By slightly moving the glass substrates, the system can dynamically control three modes: transmission of incident waves, partial transmission and partial reflection of incident waves, and reflection of all incident waves
Varactor diode	Reference [15] introduces a 5.8 GHz tunable RIS based on varactor diodes, consisting of 55 $\times$ 20 1-bit elements, measuring 80.08 cm $\times$ 31.30 cm. The control method utilizes an FPGA and shift registers, with column elements grouped into sets of five for control, resulting in a total of 220 bias signals. In indoor tests, where the transmitter and receiver were separated by a 30 cm thick concrete wall, the RIS prototype provided a 26 dB gain compared to the baseline case where the RIS was replaced with a copper plate. During short-range outdoor measurements, a 27 dB gain was observed, while successful long-range outdoor testing transmitted a 32 Mbps data stream over a distance exceeding 500 m
Photodiode + Varactor diode	Reference [16] details the design of an optically driven RIS, utilizing a series-connected photodiode array. This array receives different optical signals and converts them into corresponding voltage signals, which in turn drive varactor diodes to tune the meta-surface elements
PIN diode	Reference [17] reports the design and testing of a 2-bit RIS with 256 elements operating at 2.3 GHz and 28.5 GHz. The RIS achieved an antenna gain of 21.7dBi at 2.3 GHz and 19.1dBi at 28.5 GHz

RIS does not amplify the fed signal by power amplifier and mainly focuses on spatial modulation by RIS codebook, discussing the methods of wireless channel modelling when RIS is deployed in complex wireless scenarios.

The wireless channel model describes the statistical characteristics of wireless signal propagation from the base station to the terminal and vice versa. In the performance evaluation of 4G/5G wireless communication systems, the Geometry Based Stochastic channel Models (GBSM) [18–20] and the Map-based Hybrid Channel Model (MHCM) are adopted by the 3rd Generation Partnership Project (3GPP) and the International Telecommunication Union (ITU) [18–20], for 5G scheme selection and performance evaluation. The relevant models provide the large-scale and small-scale (3 Dimension, 3D) channel characteristics between the Base Station

(BS) and User Equipment (UE), supporting modelling of such features as 0.5–100 GHz frequency band, large bandwidth, large-scale MIMO, spatial consistency, blockage, oxygen attenuation, time-varying Doppler shift, absolute time of arrival, dual mobility, etc. [20]. Among them, MHCM balances the accuracy of the channel model and the computational load, divided into deterministic and statistical parts, i.e., deterministic calculation through ray tracing technology, reflecting the deterministic component of the channel; the statistical part supplements the impact brought by map errors, configuration simplification and rough surface scattering, reflecting the statistical component of the channel.

8.3.1 Model Assumptions and Design Principles

The deployment of RIS changes the original connection relationship between BS and UE, adding propagation paths between BS and RIS, RIS and UE, as well as between RISs:

(1) BS-RIS channel;
(2) RIS-UE channel;
(3) RIS-RIS channel.

The RIS panel itself is a periodic or non-periodic structure composed of sub-wavelength units. The geometric structure, size, direction or deployment pattern of the element determines the electromagnetic characteristics of the material, which can absorb, enhance or change the direction of electromagnetic waves. The amplitude, phase, frequency, polarization, beam direction and other characteristics of the reflected wave are determined by the structure or codebook of the RIS panel. There are many physical implementation schemes for reconfigurable RIS, including RF-MEMS switches, PIN diodes, varactor tubes, Field Effect Tubes (FET), liquid crystals, ferroelectric materials and graphene, etc. Some change the permittivity of the substrate, some change the reflection and transmission coefficients by changing the impedance of the equivalent circuit, and some use micro-stepping motors to control the direction of the unit to change the radiation phase, etc. The main considerations for implementation are power consumption, insertion loss, response time, control accuracy, process maturity and cost [18].

The following factors need to be considered for the RIS channel model.

Firstly, the deployment of RIS in the network. The deployment of RIS is not arbitrary, but varies according to different objectives such as coverage supplementation or channel rank increase. It can be mainly based on single reflection, or it can achieve the goal through multiple reflections by cascading multiple RIS panels.

Secondly, the abstraction of the RIS physical model. Different RIS physical implementation schemes have different characteristics, and the abstraction of the RIS physical model in the channel model needs to reflect the commonality of various physical models, such as the phase response, amplitude response, polarization response, anisotropy, loss, dispersion, reciprocity of the RIS unit, etc.

Thirdly, the modelling of the cascaded channel from BS to RIS and then to UE. It is necessary to consider the impact of RIS deployment on the original channel model based on the current statistical channel model such as GBSM or deterministic channel model such as MHCM. For example, the characteristics of the reflected waves of multiple different directions or polarized waves fed into the RIS under the far-field condition on the same RIS codebook are different. One can split the feed excitation of a polarized wave in one direction into two virtual RIS base stations with different polarization reflection characteristics. In addition, for the sake of simplification of calculation, "keeping the strong and removing the weak" must also be considered in the process of channel modelling.

The radiation of the RIS panel is related to the following factors:

(1) Far/near field;
(2) The radiation characteristics of the element as well as the codebook of the RIS panel;
(3) The position of the RIS element on the panel, different positions have different radiation characteristics;
(4) The phase response and insertion loss characteristics of the RIS element are related to the element structure, carrier frequency, direction of the incident wave vector, and polarization.

In addition, based on the relative positions of BS, RIS and UE, and assuming d_1 is the distance from BS to RIS, d_2 is the distance from RIS to UE, the following scenarios can be identified with respect to the RIS panel.

Scenario 1: d_1 is far field, d_2 is far field.

Scenario 2: d_1 is far field, d_2 is near field.

Scenario 3: d_1 is near field, d_2 is far field.

Scenario 4: d_1 is near field, d_2 is near field.

Considering the simplification and usability of the channel model, in the above scenarios, from the perspective of the RIS sub-wavelength element, whether BS-RIS or RIS-UE both satisfy the radiation near field or radiation far field assumption (the distance between the RIS and the target field point is greater than 10 wavelengths).

Since the deployment, algorithm, and performance evaluation of RIS are carried out in a complex scenario with multiple cells and multiple terminals, therefore, the RIS channel modeling needs to balance between complexity, accuracy, usability, and simplification. A feasible modelling approach is to decouple the basic model and the extended model of RIS separately.

1. Basic Model Assumptions

(1) A consistent, position-independent radiation model of RIS element;
(2) No polarization leakage in RIS element;
(3) Perfect Electric conductor with simplified insertion loss model (frequency independent);
(4) Frequency independent ideal phase.

2. Extended Model Assumptions (Optional)

(1) Frequency-dependent phase model (dispersion model);
(2) Frequency-dependent insertion loss model (dispersion model);
(3) RIS polarization leakage model;
(4) RIS polarization twist model;
(5) Other non-ideal factor models.

8.3.2 RIS Physical Model

The abstraction of the RIS physical model is key to RIS channel modelling. Based on the principles of the basic model assumptions, it is necessary to construct a simplified physical model that is decoupled from the RIS implementation scheme but can reflect the key characteristics of RIS. Abstracting the RIS physical model based on Huygens' principle or antenna theory, and abstracting the RIS element as a electromagnetic surface is a research direction in academia. Based on antenna theory, it can be assumed that the radiation pattern of the RIS element is Exponential-Lambertian [21] or directional model [22].

Exponential-Lambertian model: $f(\theta) = (\cos\theta)^{\alpha}, \theta \in [0, \pi/2]$, where, θ is the angle between the scattering direction and the normal of the RIS element, α is an adjustable coefficient ($\alpha \leq 0.57$).

Directional model: $f(\theta) = ((1 + \cos\theta)/2)^{2}, \theta \in [0, \pi]$, where, θ is the angle of the scattering direction relative to the surface of the RIS.

This method has a low computational load, but considering the coupling between elements, the spacing between RIS elements cannot be too small [23].

When the RIS panel is large, it can be abstracted as equivalent current or equivalent magnetic flow based on geometric optics, Huygens' principle, and physical optics, and then the field strength at the target point can be obtained by numerical integration. This method has clear physical meaning and is applicable to scenarios of radiation near field or radiation far field, but not suitable for induction near field, and the computational load is slightly larger [23].

Based on electromagnetic scattering theory [24], the radiation pattern of the RIS element will be used as the basis for the abstraction of the basic RIS element model. Based on Shooting and Bouncing Ray method (SBR) [25, 26], geometric optics is used to obtain the reflected field strength, and then the far-field integration method of physical optics is used to solve the scattering field of the target point. The principle is a high-frequency approximation based on the Stratton-Chu equation [27].

8.3.3 MHCM RIS Channel Model

Since the deployment of RIS needs to be combined with the environmental characteristics of the actual scenarios, the RIS deployed can effectively enhance coverage and capacity and reduce unnecessary interference, this article will carry out relevant research based on the map-based hybrid channel model (MHCM), making the deployment of RIS more physically meaningful. At the same time, the idea of this article is also effective for channel modelling based on GBSM. On the basis of MHCM, when the base station serves as the signal radiation source Tx, the terminal serves as the receiving device Rx, and RIS serves as the passive controllable reflection node, the channel model between Tx and Rx, in addition to considering the existing model wireless channel link between Tx and Rx, should also consider the link between Tx-RIS-Rx. The key lies in: (1) Tx-RIS channel model; (2) The response model of RIS to the polarized incident signal induced from different wave incident based on the codebook control strategy; (3) RIS-RIS channel model; (4) RIS-Rx channel model.

As shown in Fig. 8.2, the signal received on the receiving side (Rx side) comes from the following logical links:

$$
\begin{aligned}
&T_x \rightarrow R_x;\\
&T_x \rightarrow RIS_i \rightarrow R_x;\\
&\vdots\\
&T_x \rightarrow RIS_i \rightarrow \ldots RIS_j \rightarrow R_x
\end{aligned}
$$

where, $i, j \in \{1, \ldots, N\}$, N is the optional number of RIS panels.

Considering the calculation load and channel accuracy requirements, it is not necessary to include all possible logical link combinations. It can be specifically determined based on the principles of RIS deployment and control objectives. This is because only when the signal received by RIS is strong enough, RIS has a significant impact on the final channel. For example, when n = 2, the deployment and control objective of RIS is to enhance the coverage of Tx NLOS domain through controllable single reflection by RIS, only need to consider the following links: $T_x \rightarrow R_x$, $T_x \rightarrow RIS_1 \rightarrow R_x$, $T_x \rightarrow RIS_2 \rightarrow R_x$. For $T_x \rightarrow R_x$, 3GPP or ITU's MHCM can be adopted; for $T_x \rightarrow RIS_i \rightarrow R_x$ or $T_x \rightarrow RIS_i \rightarrow \ldots \rightarrow RIS_j \rightarrow R_x$, the first sub-logical link

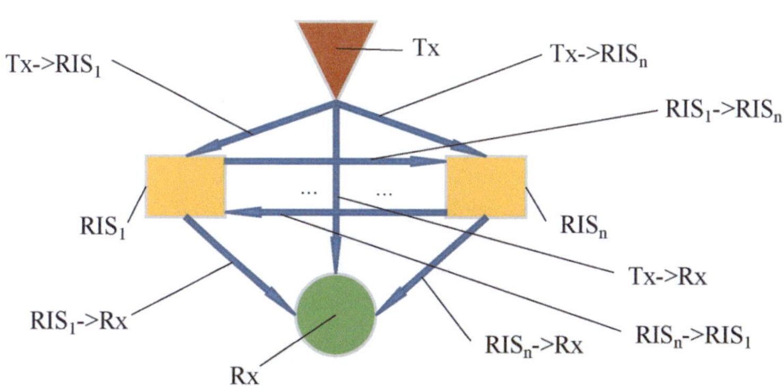

Fig. 8.2 Directed graph of transmit and receive channel logical links

$T_x \rightarrow RIS_i$ can adopt 3GPP or ITU's MHCM, while the middle and last sub-logical links need to be determined based on Channel Impulse Response (CIR), direction of arrival, and other factors in combination with the configuration parameters of the RIS. Under the far-field assumption, the direction diagrams of the RIS reflection signals of wireless signals from different directions of arrival are different. Considering the balance between computational load and accuracy, the principle of "keeping the strong and removing the weak" needs to be considered in the selection of related logical links. Figure 8.3 shows the RIS channel modelling process.

Each pair of T_x-R_x can establish a logical link directed graph, as shown in Fig. 8.2. Based on the relative position/the target of RIS deployment, select the logical links that have a significant impact on the T_x-R_x wireless channel. For each selected logical link between T_x and R_x, perform sub-logical link decomposition as follows.

Input (logical link): $T_x \rightarrow RIS_{s_1} \rightarrow \ldots \rightarrow RIS_{s_N} \rightarrow R_x$.

Output (sub-logical link): $T_x \rightarrow RIS_{s_1}$; $RIS_{s_1} \rightarrow RIS_{s_2}$; …; $RIS_{s_N} \rightarrow R_x$.

Where, s_j represents the logical number of the RIS at the j level of the directed graph, $j \in \{1, \ldots, N\}$.

The reflection phase of each element in RIS panel can be independently controlled based on the control strategy. The reflection direction coefficient of the RIS control element induced by impinging electromagnetic waves with different incident is different, which can be calculated based on the physical model of RIS. Its normalized directivity coefficient is as follows:

$$\begin{bmatrix} f_{\mathrm{hh},k,l}(\vartheta_i, \phi_i, \theta_r, \varphi_r) \; f_{\mathrm{vh},k,l}(\vartheta_i, \phi_i, \theta_r, \varphi_r) \\ f_{\mathrm{hv},k,l}(\vartheta_i, \phi_i, \theta_r, \varphi_r) \; f_{\mathrm{vv},k,l}(\vartheta_i, \phi_i, \theta_r, \varphi_r) \end{bmatrix} \tag{8.1}$$

where, k and l represent the row number k and column number l of RIS element in RIS panel, respectively; ϑ_i and ϕ_i represent the angles of the inverse incident wave vector relative to the global coordinate system's Z axis and X axis respectively; θ_i and φ_i represent the angles of the reflected wave vector relative to the global coordinate system's Z axis and X axis respectively; f represents the directional amplitude coefficient of the reflection by the RIS control element; *hh*, *hv*, *vh*, and *vv* represent different combinations of polarization of the incident and reflected waves, where v represents vertical polarization, and h represents horizontal polarization.

The RIS panel is composed of RIS control elements, each of which can independently control such parameters as phase based on the RIS control strategy. This article mainly considers the phase controllability of RIS, but the method's idea can be applied to the control of amplitude/polarization as well.

The following introduces the configuration parameters of the RIS panel.

RIS panel size (assumed to be rectangular): K × L, where K is the number of control elements in the length direction, L is the number of control elements in the width direction.

RIS control unit size: a · λ × b · λ, λ is the wavelength of the carrier frequency.

RIS control unit phase control granularity: $\log_2$B bit represents the granularity of the control unit adjustment is $\frac{2\pi}{B}$, B is the number of phase control segments.

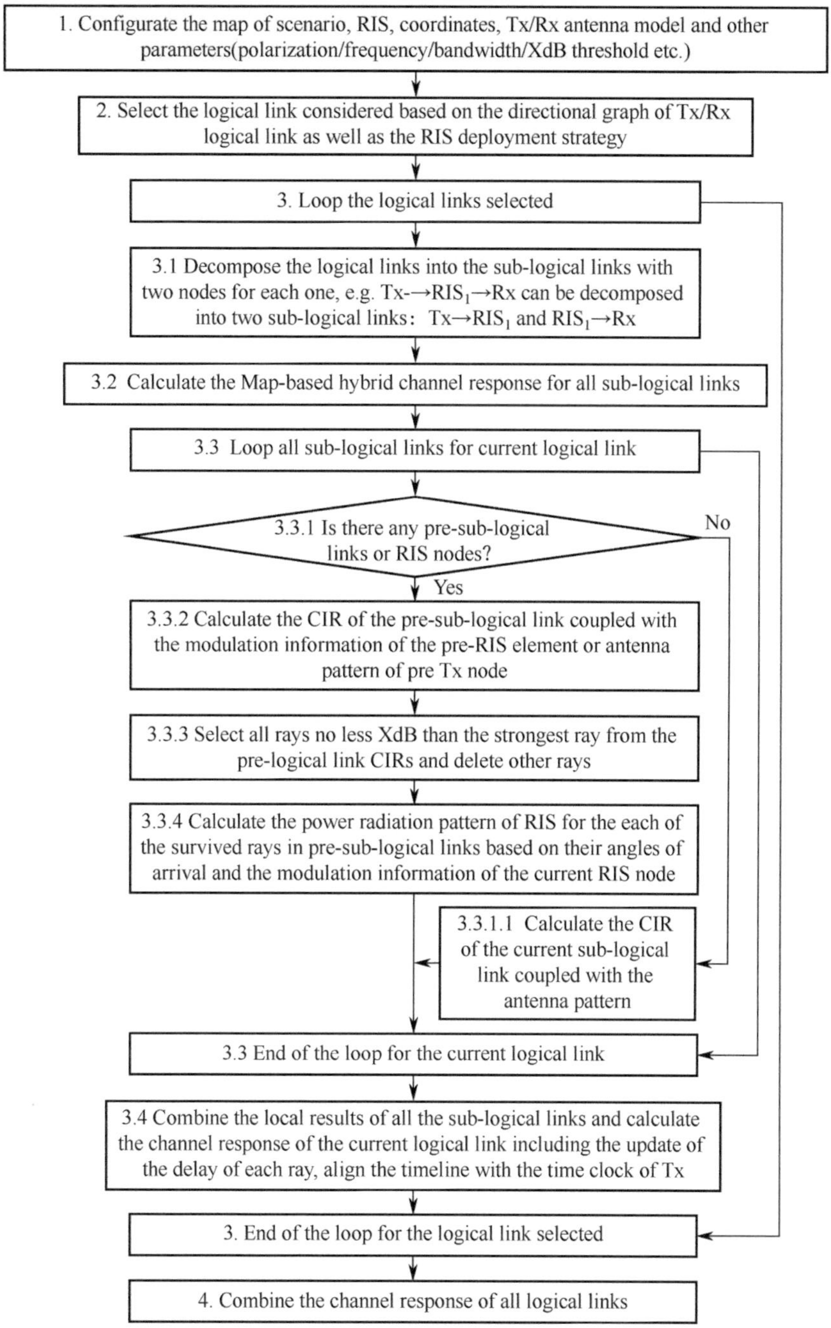

Fig. 8.3 RIS channel modelling process

The reflection field pattern of the RIS panel is composed of the field pattern of RIS elements that make up the panel and array factor pattern determined by control phases of each element. The control phase of each reflection element $\psi_{\mathrm{pq,k,l}}$ is determined by the RIS control centre based on the control strategy, an optional configuration is:

$$\psi_{pq,k,l} \in \left[0, \frac{2\pi}{B}, \frac{2 \cdot 2\pi}{B}, \cdots, \frac{(B-1) \cdot 2\pi}{B}\right], \text{and}\, p, q \in \{h, v\} \tag{8.2}$$

In fact, for the control phase of each RIS control element $\psi_{\mathrm{pq,k,l}}$ the optional set is not limited to formula (8.3), nor does it need to be exactly the same, it can be determined based on the control strategy.

The reflection directionality coefficient of the RIS panel is as follows:

$$\begin{bmatrix} \Gamma_{\mathrm{vv}}(\vartheta_i, \phi_i, \theta_r, \varphi_r) & \Gamma_{\mathrm{hv}}(\vartheta_i, \phi_i, \theta_r, \varphi_r) \\ \Gamma_{\mathrm{vh}}(\vartheta_i, \phi_i, \theta_r, \varphi_r) & \Gamma_{\mathrm{hh}}(\vartheta_i, \phi_i, \theta_r, \varphi_r) \end{bmatrix} \tag{8.3}$$

The specific calculation is as follows:

$$\Gamma_{pq}(\vartheta_i, \phi_i, \theta_r, \varphi_r) = \sqrt{4\pi KLab} \sum_{k,l=1}^{k=K,l=L} f_{pq,k,l}(\vartheta_i, \phi_i, \theta_r, \varphi_r) \cdot \exp\big(j\psi_{pq,k,l}\big) \cdot \exp\left(\frac{j2\pi\left(\hat{\mathbf{r}}_{\mathrm{in}}^T \cdot \overline{\mathbf{d}}_{\mathrm{ices},k,l}\right)}{\lambda}\right) \cdot \sqrt{\eta} \tag{8.4}$$

where, Γ represents the directivity of RIS panel; η is the reflection efficiency of the RIS; $\overline{\mathrm{d}}_{\mathrm{ices,k,l}}$ is the position vector of the center of the RIS element in the kth row and the lth column 1920; ϑ_{i}, φ_{i} represent the angles of the inverse vector of the incident wave relative to the Z-axis and X-axis of the global coordinates, respectively; θ_{r}, ϕ_{r} represent the angles of the reflected wave vector relative to the Z-axis and X-axis of the global coordinates; f represents the directional amplitude coefficient of the reflection of RIS element; hh, hv, vh and vv represent different combinations of polarization of the incident wave and reflected wave, where v represents vertical polarization, h represents horizontal polarization, and p, q $\in$ {h, v}, that is,

$$\hat{\boldsymbol{r}}_{in} = \begin{bmatrix} \sin\vartheta_i \cos\varphi_i \\ \sin\vartheta_i \sin\varphi_i \\ \cos\vartheta_i \end{bmatrix} \tag{8.5}$$

The MHCM 1819 takes into account the effects of ultra-wideband, oxygen attenuation, and blockage. Most of these models can be directly applied, but for ultra-wideband scenarios, considering the possible frequency dependence of metamaterials or electromagnetic surface electrical parameters, it is necessary to update Eqs. (8.1)–(8.4) based on frequency segmentation (bin).

When the sub-logical link calculates the channel coefficient according to the MHCM, it needs to be classified based on the type of transmitting and receiving nodes.

When the Tx node is an RIS panel, it is necessary to split the Tx node based on all the preceding sub-logical links of the current RIS, each path corresponds to a wave direction, and also corresponds to the two Tx branch nodes of the current sub-logical link (corresponding to horizontally polarized wave and vertically polarized wave respectively). All these Tx branch nodes are in the same position, but the directional coefficients of the transmitting antennas are independent, and the results calculated based on the wave direction are shown in formula (8.4). For the y − th sub-logical link of the x − th logical link, the amplitude directional coefficient of the transmitting antenna of the Tx branch node corresponding to the n − th cluster and the m − th path is as follows:

$$\begin{bmatrix} F_{tx,s,\theta}^{x,y,v}\left(\theta_{n,m,ZoD}, \varphi_{n,m,AoD}\right) \\ F_{tx,s,\varphi}^{x,y,v}\left(\theta_{n,m,ZoD}, \varphi_{n,m,AoD}\right) \end{bmatrix} = \begin{bmatrix} \Gamma_{vv}\left(\vartheta_{i,x,y}, \varphi_{i,x,y}, \theta_{n,m,ZoD}, \phi_{n,m,AoD}\right) \\ \Gamma_{vh}\left(\vartheta_{i,x,y}, \varphi_{i,x,y}, \theta_{n,m,ZoD}, \phi_{n,m,AoD}\right) \end{bmatrix} \tag{8.6}$$

$$\begin{bmatrix} F_{tx,s,\theta}^{x,y,h}\left(\theta_{n,m,ZoD}, \varphi_{n,m,AoD}\right) \\ F_{tx,s,\varphi}^{x,y,h}\left(\theta_{n,m,ZoD}, \varphi_{n,m,AoD}\right) \end{bmatrix} = \begin{bmatrix} \Gamma_{hv}\left(\vartheta_{i,x,y}, \varphi_{i,x,y}, \theta_{n,m,ZoD}, \phi_{n,m,AoD}\right) \\ \Gamma_{hh}\left(\vartheta_{i,x,y}, \varphi_{i,x,y}, \theta_{n,m,ZoD}, \phi_{n,m,AoD}\right) \end{bmatrix} \tag{8.7}$$

where, $\vartheta_{i,x,y}$, $\phi_{i,x,y}$ correspond to the zenith angle (Zenith Of Arrival, ZOA) and azimuth angle (Azimuth Of Arrival, AOA) of the incoming wave of the pre-logic link, respectively, and tx and rx refer to the transmitting and receiving nodes of the sub-logic link.

When the Tx or Rx node is a BS or UE, the field pattern of the actual transmitting antenna of the BS or UE is used, i.e., $F_{tx,s,**}^{x,1,*}$ or $F_{rx,u,**}^{x,y_{last},*}$ [19, 20, 23], where $*$ represents h or v, $**$ represents θ or φ, ylast represents the number of the last sub-logical link of the current logical link.

When the Rx node is an RIS panel, the field pattern of the receiving antenna is $F_{rx,u,\theta}^{x,y,*} = F_{rx,u,\phi}^{x,*} = \sqrt{\cos\Theta}$, where, Θ is the angle between the normal vector of the RIS panel and the inverse vector of the incident wave $\hat{r}_{in}$, see Eq. (8.5). Since the RIS is a single-sided reflection, when $\Theta \geq \frac{\pi}{2}$, $F_{rx,u,\theta}^{x,y,*} = F_{rx,u,\varphi}^{x,y,*} = 0$.

For the channel coefficients and delay in NLOS and LOS (line of sight) scenarios, please refer to Reference [18].

The NLOS path channel coefficient is $H_{u,s,n,m}^{x,y,*}(t)$, $\tau_{u,s,n,m}$;

The LOS path channel coefficient is $H_{u,s,n=1}^{x,y,*}(t)$, $\tau_{u,s,n=1}$.

As shown in Fig. 8.4, Tx is vertically polarized (v-pol), based on the principle of "**keep the strong and remove the weak**", the solid line represents the effective path that meets the screening criteria, the dashed line represents the path that does not meet the screening criteria. Next, based on the results of the sub-logical channel, the channel parameters and delay of the logical channel are merged. The merging of channel parameters is a multiplication relationship, and the delay merging is an addition relationship [18], that is

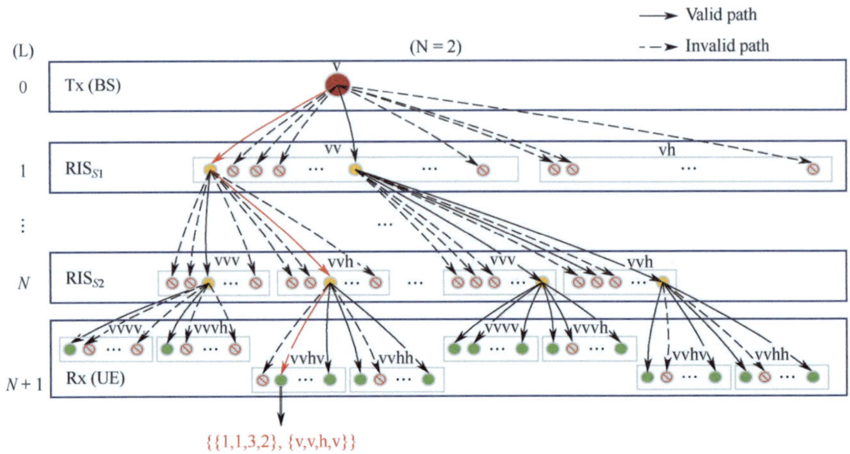

Fig. 8.4 Multi-path tracing of sub-logical links

$$H^{\Lambda}_{u,s,\{\{k_0,k_1,\cdots,k_{N+1}\},\{p_0,p_1,\cdots,p_{N+1}\}\},1}(t) = \prod_{L=1}^{N+1} H^{\{\{k_0,k_1,\cdots,k_{L-1}\},\{p_0,p_1,\cdots,p_{L-1}\}\},\Lambda}_{u_L,s_L,\{\{k_0,k_1,\cdots,k_L\},\{p_0,p_1,\cdots,p_L\}\},1}(t) \quad (8.8)$$

$$\tau^{\Lambda}_{u,s,\{\{k_0,k_1,\cdots,k_{N+1}\},\{p_0,p_1,\cdots,p_{N+1}\}\}} = \sum_{L=1}^{N+1} \tau^{\Lambda}_{u_L,s_L,\{\{k_0,k_1,\cdots,k_L\},\{p_0,p_1,\cdots,p_L\}\}} \quad (8.9)$$

For the absolute delay model of the statistical path, please refer to reference [19]. The effective logical link between Tx and Rx is as follows:

No. 0: Tx → Rx.

No. 1: Tx → RISi → Rx.

…

No. Λ: Tx → RISi → … → Rx.

…

No. K: Tx → RISi → … → RISj → Rx.

The wireless channel between Tx and Rx is the collection of all effective sub-paths for the above logical links:

$$\begin{Bmatrix} H^{0}_{u,s,\{\{k_0\},\{p_0\}\},1}(t),\ \tau^{0}_{u,s,\{\{k_0\},\{p_0\}\}}; \\ H^{1}_{u,s,\{\{k_0,k_1\},\{p_0,p_1\}\},1}(t),\ \tau^{1}_{u,s,\{\{k_0,k_1\},\{p_0,p_1\}\}}; \\ \cdots \\ H^{\Lambda}_{u,s,\{\{k_0,k_1,p_1,\ldots,p_{N_\Lambda+1}\}\},1}(t),\ \tau^{\Lambda}_{u,s,\{\{k_0,k_1,\ldots,k_{N_\Lambda+1}\},\{p_0,p_1,\ldots,p_{N_\Lambda+1}\}\}} \\ \cdots \\ H^{K}_{u,s,\{\{k_0,k_1,\ldots,k_{N_K+1}\},\{p_0,p_1,\ldots,p_{N_K+1}\}\}}(t)\tau^{K}_{u,s,\{\{k_0,k_1,\ldots,k_{N_K+1}\},\{p_0,p_1,\ldots,p_{N_K+1}\}\}} \end{Bmatrix} \quad (8.10)$$

The above combinations of {{k*, ...}, {p*, ...}} are the effective path combinations for each logical link.

8.4 RIS-Assisted Wireless Communication

8.4.1 Performance Evaluation Analysis

Many studies aim to reveal the fundamental performance limitations of RIS and the impact of various engineering imperfections on its achievable performance. Against this backdrop, to quantify the advantages and limitations of RIS in different network scenarios, researchers have provided various precise, approximate, and asymptotic analysis frameworks. The typical performance indicators suggested for RIS-assisted wireless communication systems include the following aspects:

(1) Coverage Probability or Interruption Probability: The probability that the received SINR is higher or lower than the target threshold.
(2) Bit Error Rate: The probability that the decoded information bits differ from the transmitted information bits.
(3) Ergodic Capacity: The expected value of the channel capacity measured by Shannon's formula.
(4) Transmission Capacity: The aggregate data rate that can reliably communicate throughout the system.

Table 8.4 lists various results of the performance study of the RIS-assisted wireless communication system. In addition to the performance analysis of the RIS-assisted wireless communication system itself, reference [28] also compares RIS with full-duplex relay, analyzing the spatial throughput of a single-cell multi-user system assisted by multiple RIS randomly deployed in the cell, showing that as long as the number of RIS elements exceeds a certain value, the RIS-assisted system can outperform the full-duplex relay in terms of spatial throughput. Some studies have compared the performance of distributed and centralized RIS, and reference [29] derived the capacity region of the distributed deployment scenario in closed form, providing the outer boundary of the capacity region for the centralized deployment. These research results show that even considering the limited precision of phase quantization and some nonideal factors of hardware, the RIS-assisted wireless communication system has very attractive performance improvement potential.

8.4.2 Precoding and Beamforming

RIS needs to reflect as much of the incident electromagnetic signal as possible to the target area to achieve higher received gain and a larger coverage range, and to reduce

Table 8.4 Performance study results of RIS-assisted wireless communication system

Achievable data rate	Reference [30] analyzed the asymptotic achievable rate of RIS-assisted downlink systems
	Reference [31] examined the achievable data rate for mobile users, based on optimal phase shifts of RIS, with limited user feedback
	Reference [32] provided an approximate expression for the achievable data rate in RIS-assisted communication systems and derived the minimum number of phase shifts required to guarantee rate degradation constraints
	Reference [28] investigated the spatial throughput of a single-cell multi-user downlink system assisted by multiple randomly distributed RISs, averaged over the random positions of all users and RISs
	Reference [33] considered the non-ideal hardware of RIS reflection units, modeling the relevant structures based on the distances between reflection units, and analyzed the achievable rate degradation using a simplified receiver structure
	Reference [34] addressed capacity degradation in RIS-assisted systems due to potential hardware imperfections by proposing a hardware impairment model and deriving closed-form expressions for effective noise density and capacity loss
Increasing rank	Reference [35] analyzed the application of RIS in LOS scenarios to increase the rank of the channel matrix, resulting in substantial capacity gains. By optimizing the phase matrix of RIS, the rank of the channel matrix is enhanced, enabling point-to-point MIMO system capacity gains; deployment and phase matrix optimization potentially exceed 100% capacity performance gain
Outage probability	Reference [36] studied the outage probability in single-user multi-RIS link communication systems, finding that the outage probability decreases with the number of RISs and the number of reflection units per RIS
	Reference [37] presents an asymptotic expression for the distribution of uplink sum rate in RIS systems and proposes a framework for analyzing outage probability. Under large-scale assumptions, the asymptotic analysis is highly consistent with the results of numerical simulations
	Reference [38] calculated the outage probability, average symbol error probability, and achievable rate of RIS-assisted systems, demonstrating that the accuracy of the derived analytical expressions improves with an increasing number of RIS units. In high signal-to-noise ratio conditions, the achievable diversity order was quantified using polynomial approximations
Coverage probability	Reference [39] considered a generalized millimeter-wave downlink cellular network with random obstacles and reflectors, proposing a stochastic geometry approach to analyze the downlink coverage probability

system energy consumption and the bit error rate. At the same time, because RIS reflection units have reconfigurable capabilities, they can generate artificial multi-path, thereby assisting the MIMO system in achieving better spatial resource utilization and degrees of freedom, thereby increasing channel capacity. To achieve the above goals, beamforming technology needs to be used in the RIS-assisted wireless communication system. The principle of RIS beamforming is to change the phase response of the reflection units to the incident electromagnetic wave, so that all the reflection units on the panel add up in phase in a specific direction. In this way, the reflected energy is concentrated in a narrower solid angle. Beamforming can not only provide higher directional gain but also reduce the interference of the reflected signal with the neighboring area. Through the joint optimization of precoding beamforming on the Base Station (BS) side and the RIS side, effective and secure wireless communication can be achieved.

The RIS-assisted wireless communication system can be modeled as follows.

$$\mathrm{y} = (\mathrm{H}_2\Phi H_1 + \mathrm{H})\mathrm{Ws} + \mathrm{n} \tag{8.11}$$

The received signal y_k of the $k-th$ user and the joint received signal y of K users can be modeled as

$$y_k = \left(h_{2,k}\Phi H_1 + h_k\right)x + n_k, x = \sum_{k=1}^{K} \sqrt{p_k} w_k s_k, \tag{8.12}$$

For non-LOS scenarios, the LOS channel components in the brackets h_k and H can be removed. Φ is the RIS weight adjustment matrix, which is in diagonal matrix form. Depending on the panel capabilities, the regulation matrix can include amplitude regulation and phase regulation parts, or only contain the phase regulation part. The BS precoding and RIS weights can be optimized independently or jointly, but generally, joint optimization can achieve higher channel capacity [40]· The parameters for the RIS transmission system analysis are shown in Table 8.5.

For the scenario where RIS is used to enhance coverage, there is usually no direct link between the BS and the terminal due to obstructions. It is necessary to rely on RIS to generate artificial reflection paths to form virtual direct links. If there are only direct paths between the BS and the RIS, and between the RIS and the terminal, or if the multipaths other than the direct paths can be ignored, then the beamforming design is straightforward. However, actual application scenarios are generally more complex, which increases the complexity of the joint optimization of the precoding beamforming on the BS side and the RIS side. For example, there may only be NLOS paths or both LOS and NLOS paths between the BS and the RIS, and between the RIS and the terminal. This makes the channel complex. For instance, a single RIS may serve multiple UEs, multiple RISs may serve a single UE, multiple RISs may serve multiple UEs, or even multiple BSs and multiple RISs may jointly serve multiple UEs. Additionally, multiple RISs can cooperate to achieve multi-hop transmission. For these complex scenarios, considering factors like interference suppression, channel capacity, energy efficiency, and spectral efficiency, it is necessary to jointly design the beamforming of the BS and the RIS. The problem to be solved at this time is no longer the simple RIS reflection beam control problem, but the performance optimization problem of the RIS-assisted communication system under certain constraints.

1. Based on the Alternating Optimization Algorithm

It is often difficult to obtain a closed-form solution for joint beamforming design due to multiple constraints, and it can only be solved by numerical methods. The

Table 8.5 Analysis parameters of RIS transmission system

Parameter	Definition	Description
$\boldsymbol{h}_k$	LOS channel from the k − th UE to the AP	The dimension is $[n_r \times n_t]$
$h_{2,k}$	Channel between the k − th UE and the RIS	The dimension is $[n_r \times N]$
H_1	Channel between the AP and the RIS	The dimension is $[N \times n_t]$
$\boldsymbol{\Phi}$	Phase adjustment matrix on the RIS panel	$\boldsymbol{\Phi} = \text{diag}([e^{j\theta_1}, e^{j\theta_2}, ..., e^{j\theta_N}])$
η_k	Noise signal during transmission from the $k - th$ UE	Gaussian white noise
y_k	Received signal from the k − th UE	The dimension is $[n_r \times 1]$
$\boldsymbol{y}$	Joint received vector of all K UEs	$\boldsymbol{y} = \left[y_1 \; y_2 \; ... \; y_K \right]^T$
H_2	Joint channel between K UEs and the RIS	The dimension is $[n_r K \times N]$
$\mathbf{W}$	Joint precoding matrix for K UEs	The dimension is $[n_t \times r]$
H	Joint channel between K users and the AP	The dimension is $[n_r K \times n_t]$
r	Total number of transmission layers for K UEs	Scalar, less than or equal to n_r
p_k	Transmission power of the k − th UE's signal	Total power after summing the transmission powers of K UEs
s_k	Original signal transmitted by the k − th UE	Complex symbol
w_k	Precoding of the k − th UE at the AP	The dimension is $[n_t \times 1]$

Note n_r is the number of antennas for the UE; n_t is the number of antennas for the BS; K is the number of UEs; N is the number of units in the RIS

non-convex nature of its optimization model makes it difficult to solve by conventional methods, and it usually needs to be solved iteratively by alternating optimization. Table 8.6 lists the beamforming of the BS side and the RIS side based on the alternating optimization algorithm. Although they are all based on alternating optimization, the constraints, optimization objectives, or research scenarios they use are different.

The main problem with the alternating optimization algorithm is that there is a processing delay, and the complexity is relatively high. Improving the convergence efficiency of iterative solutions and avoiding falling into local optima are critical

Table 8.6 Beamforming of BS Side and RIS Side based on alternating optimization algorithm

Research scenarios	Constraints	Optimization objective	Methodology	References
RIS-assisted multi-user downlink links	Maximum power constraint	Spectral efficiency and energy efficiency maximization	Alternating optimization of base station beamforming and RIS phase shifts is performed using gradient search methods and sequential optimization	[41, 42]
RIS-assisted single-user links	Maximum power constraint	Achievable rate	RIS phase shifts are related to the amplitude of reflection coefficients. Alternating iterative optimization methods are employed to optimize base station beamforming and RIS phase shifts	[43]
RIS-assisted multi-cell system	Maximum power constraint of a single base station	Maximization of the minimum signal-to-noise ratio for users	The beamforming subproblem is formulated as a second-order cone programming problem, with semidefinite relaxation and successive convex approximation used to address the optimization of RIS phase shifts	[44]
RIS-assisted single-user MISO system	Nonlinear ratio constraint	Spectral efficiency maximization	Alternating optimization of base station transmission power and RIS reflection phase shifts is considered, addressing complexity issues	[45]
RIS-assisted multi-group multicast system	Base station maximum power constraint	Sum rate maximization	The corresponding subproblems are second-order cone programming problems, solvable using convex optimization theory. Sequential optimization methods are used to construct approximate problems with closed-form solutions	[46]
RIS-assisted broadband OFDM system	Base station maximum power constraint	ubcarrier average efficiency maximization; Mean square error minimization	Modeling is based on the relationship between sum rate maximization and mean square error minimization, with solutions derived through alternating optimization	[47]

(continued)

Table 8.6 (continued)

Research scenarios	Constraints	Optimization objective	Methodology	References
RIS-assisted multi-user network downlink physical layer broadcasting	Minimum signal-to-noise ratio constraint	Base station power consumption minimization	Alternating optimization methods are used to design base station transmission beamforming and RIS phase shifts, deriving lower bounds for the optimal solution	[48, 49]
RIS-assisted multi-user MISO system	Outage probability constraint	Minimum transmission power	A robust beamforming method is designed for cascade channels with incomplete channel state information	[50]
RIS-assisted multi-user MISO system	Base station maximum power constraint	Minimum symbol error probability	Alternating optimization methods are applied to ensure stable solutions for the non-convex symbol error probability minimization problem	[51]
RIS-based wireless power transfer system	Maximum power constraint	Energy efficiency maximization	The Dinkelbach method is used to convert the fractional problem of the joint optimization model into an equivalent subtractive form, which is then solved iteratively	[52]
Single RIS-assisted multi-user system	User association constraints among multiple base stations	Rate maximization	User association constraints among multiple base stations are considered, and alternating optimization methods are used to achieve robust and high-rate multi-user communications	[53]

issues in practical applications, especially in ultra-low latency transmission or ultra-high-speed mobile scenarios. In addition, it is very difficult to obtain accurate high-dimensional cascaded channel information, and the performance of the algorithm is highly dependent on this.

2. Based on Other Optimization Algorithms

In the scenario of a single RIS serving multiple users, literature [49] analyzed the joint beamforming optimization to minimize transmission power, using a semi-definite relaxation method for solving. Simulation results show that in multi-user scenarios, joint beamforming optimization between the BS and RIS to suppress interference between users can significantly improve system performance. To address the increased complexity of solving joint beamforming optimization, literature [54] used fractional programming to solve the joint beamforming problem, decomposing the original problem into three independent subproblems, thus using a low-complexity solution algorithm. Literature [55] considered a wireless network with a single BS, multiple RISs, and multiple users, aiming to maximize the weighted sum rate of all cell edge users. This work used the Lagrange method to find the optimal beamforming of the BS, and the Riemannian manifold conjugate-gradient method to find the RIS phase shift. Literature [56] studied the problem of passive beamforming and information transmission in RIS-assisted MIMO systems. The RIS is used to enhance the main communication while transmitting a second data stream by adjusting the state of the reflective elements. By describing the RIS phase shift problem as a two-step stochastic program, and using the stochastic average approximation algorithm to solve it, the RIS phase shift and capacity maximization problem were optimized. Literature [57] studied the design of a single-user MISO wireless communication system assisted by RIS, jointly optimizing the beamforming of the BS and the phase shift of RIS. The resulting non-convex optimization problem was solved using a branch-and-bound algorithm, which can handle modulus constraints that need to be executed on the RIS phase shift, at the cost of an exponential increase in the complexity of the entire optimization algorithm.

3. Considering Non-Ideal CSI and Discontinuous Phase Shift

Existing RIS beamforming research generally uses ideal channel state information (CSI) and ideal hardware conditions to simplify the problem. However, actual wireless applications need to handle a large number of non-ideal factors. For example, the reflection link between the RIS and the terminal can be affected by the mobility of the terminal and the passive nature of the RIS, making channel measurement challenging. Literature [58] considered the design of active and passive beamforming under the condition of RIS reflection channel errors. Simulation results show that this model can achieve high robustness transmission under channel measurement errors. In contrast, the interruption probability of traditional beamforming schemes increases with the increase of channel measurement errors. Literature [59] extended the restriction of single-antenna terminal users and studied joint beamforming with multiple antennas on the user side. It also analyzed the impact of random phase errors uniformly distributed on the RIS element on the received signal-to-noise ratio (SNR).

Literature [60] studied a downlink multi-user multi-antenna system based on finite resolution phase shift RIS, assuming knowledge of the actual situation of large-scale fading gain only. It proposed a hybrid beamforming scheme to achieve sum rate maximization, where the BS uses digital beamforming and the RIS uses analog beamforming. The resulting sum rate maximization problem is handled by alternating optimization. Literature [61] considered solving the sum rate maximization problem by optimizing the BS's digital beamforming and RIS's discrete phase shift. The resulting resource allocation problem is handled by an iterative algorithm based on alternating optimization. In literature [62], the authors studied a RIS-assisted multi-user communication system, where the RIS has discrete phase shifts, considered the dual-objective maximization of system ergodic capacity and delay-limited capacity, and characterized the corresponding Pareto boundary. The study shows that by using a dynamic alternating transmission strategy to change the RIS phase shift, ergodic capacity can be achieved. To achieve delay-limited capacity, it was proposed that the RIS phase shift matrix needs to be fixed at a specific value.

The current research on RIS beamforming design has covered various application scenarios and proposed corresponding solutions, but there are still some problems. For example, some studies have adopted simplified models or ideal CSI, which limits their applicability. Furthermore, considering various constraints and non-ideal factors, the model becomes extremely complex, and the space and time complexity to solve the problem is high. The introduction of new technologies such as AI is expected to improve the efficiency of beamforming design. For instance, literature [63] trains a Deep Neural Network (DNN) in an unsupervised manner offline and then deploys it to the RIS-assisted communication system for rapid beamforming on the RIS side. Literature [64] further applies unsupervised learning to the joint beamforming prediction of the BS and RIS to provide near-real-time performance. However, the cold start problem of the AI model and the robustness issue in different environments still need further verification.

8.4.3 Acquisition of Cascaded Channel State Information

In RIS-assisted wireless communication systems, in addition to the direct channel between the BS and the terminal, there is also a cascaded channel regulated by RIS. This cascaded channel consists of two sub-channels, including the channel between BS and RIS and the channel between RIS and UE. The acquisition of CSI of the cascaded channel is one of the basic problems to be solved in the quasi-passive RIS system. To achieve RIS performance enhancement and intelligent control of the electromagnetic environment, obtaining accurate CSI is crucial.

The acquisition of CSI includes channel estimation, codebook feedback, beam training, and other technologies. Channel estimation is mainly for the BS to restore the complete CSI based on the received uplink pilot information. Codebook feedback refers to the receiver measuring the pilot and feeding it back to the receiver through the codebook. Beam training is to approximate the feature vectors of the uplink and downlink channels based on beam measurement and feedback. The main difficulties in obtaining RIS CSI are shown in Fig. 8.5.

Next, we will analyze the main schemes for obtaining RIS CSI.

1. Timing Design and Feature Utilization

The timing design and feature utilization of RIS CSI acquisition play a positive role in reducing channel estimation overhead. Literature [65] analyzes the dual-time-scale channel estimation, utilizing the slow-changing property of the BS-RIS semi-static channel, estimating the high-dimensional BS-RIS channel over a longer time, and performing multiple low-dimensional RIS-UE channel estimations under one BS-RIS channel estimation, achieving accurate channel estimation with low pilot overhead. Literature [66] analyzes the single structure sparsity in the mixed space and angular domain, feeds back user-independent CSI and user-specific CSI

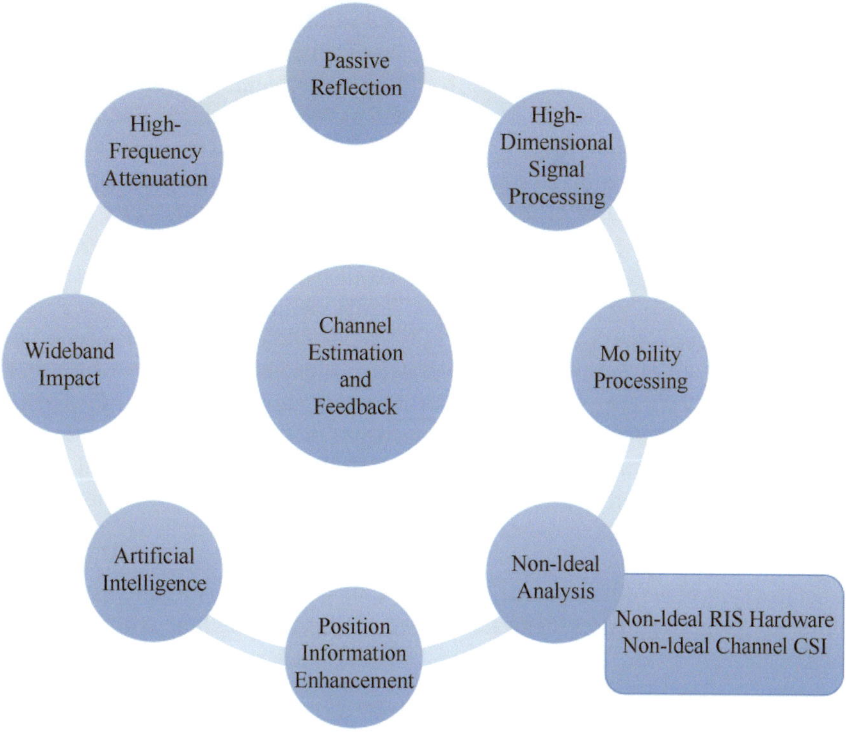

Fig. 8.5 Main difficulties in obtaining RIS CSI

through dimension reduction methods, and designs dynamic codebooks through channel angle information to accurately estimate the dynamic BS-RIS-UE cascaded channel. On this basis, literature [67] analyzes the dual-structured sparsity in the angular domain of the cascaded channel. It utilizes the characteristics of the shared BS-RIS channel and the shared part of the environmental scatterer, reveals that the angular domain cascaded channels related to different users have completely identical non-zero rows and partially identical non-zero columns, and implements channel estimation through the Double Structured Orthogonal Matching Pursuit (DS-OMP) algorithm. Moreover, Literature [68] designs a compressed sensing based channel estimation scheme for RIS-aided terahertz MIMO systems and developes a multiuser downlink strategy where the limited scattering nature of terahertz channel is exploited. A region-separation aided user grouping algorithm is proposed.

2. Low Overhead Codebook Feedback Design

Reducing feedback overhead is an important issue in RIS-assisted communication research as it plays a significant role in reducing system overhead costs and improving communication efficiency. For the RIS-assisted MIMO system, the frequency division duplex downlink channel feedback is a huge challenge. This is due to the expansion of cascaded channel dimensions. To solve this problem, some research studies are based on transmission schemes with limited parameter feedback. Reference [69] analyzed the specific triple structure sparsity of the beam space cascaded channel. It returned all the common parameters shared by all users, i.e., the path angle, offset value, and amplitude ratio on the BS side, through some active users, while the remaining user-specific information is compressed and quantized for effective feedback. Reference [70] adopted a two-stage channel estimation strategy and designed a super-resolution atomic norm algorithm. It first estimates some channel parameters and then uses these estimates in subsequent training to significantly reduce training overhead. In addition, codebook-based design can effectively reduce feedback overhead. Reference [71] studied the beam training codebook for RIS-assisted uplink communication, using semi-passive RIS to help the RIS controller achieve the minimum overhead feedback. Reference [72] studied a hierarchical analog codebook-based design, using limited feedback message overhead to design analog phase shifters at the RIS to indicate the optimal beam index. Reference [73] achieved adaptive bit splitting by flexibly using different feedback bits to design codeword structures to meet the different distributions of LOS and NLOS path gains. It designed a limited channel feedback scheme that splits feedback bits into four parts to generate sub-codebooks.

3. Design for Passive Reflection Characteristics

RIS has passive reflection characteristics and does not require a large number of RF chains. However, it has a large number of units, which makes the acquisition of CSI difficult. If RIS elements are connected to the baseband through all-digital or hybrid analog/digital architectures, the high hardware complexity and power loss will severely affect the power cost advantage of RIS. To solve this problem, some research adopted RIS hardware structure design with a small number of active units with sensing and signal processing capabilities. Reference [74] studied a new RIS architecture based on sparse channel sensors, deploying a small number of active components connected to the baseband on the passive RIS surface. It performed RIS channel estimation and reconstruction with significantly reduced pilot numbers through compressed sensing tools, and used deep learning to study reflection matrix design, achieving performance close to the rate upper limit. Reference [75] analyzed the CSI acquisition under the hardware design of uniformly distributed active sensors, using the rank-deficient structure of the RIS channel to design a residual neural network. Reference [76] studied the channel estimation performance in the fully passive RIS architecture and the hybrid active sensor RIS architecture. For pure passive RIS, the angle of departure (AOD) from the BS and the angle of arrival (AOA) to the UE are estimated in the first stage, and then other channel components are estimated. For hybrid active sensor RIS, assuming alternating uplink and downlink training to estimate segmented channels, channel estimation can be achieved with lower channel path losses.

4. Design for High Frequency and Terahertz Bands

RIS has broad application prospects in the millimeter wave band. A major feature of high-frequency scenarios is that the attenuation faced by the transmitted signal is more severe. Multiple reflection paths are usually ignored, and only the direct path and a small number of once-reflected paths can reach the receiver. A common assumption is that high-frequency channels usually have sparse characteristics, which significantly simplifies RIS channel estimation. Reference [77] considered RIS-assisted millimeter wave communication for the BS-RIS-UE path, noting that the direct path from BS to UE can be achieved by turning off the RIS. This reference adopted a two-stage channel estimation for parameter decoupling, using Estimation Signal Parameters via Rotational Invariance Techniques (ESPRIT) or Compressive Sensing (CS) techniques to implement low-cost channel estimation. Literature [78] adopts the inherent sparse quasi-static BS-RIS channel assumption and utilizes the long-term statistical information of slow-varying channel components. It jointly carries out BS-RIS channel calibration and RIS-UE channel estimation, infers the cascaded channel through Bayesian posterior estimation, and uses the Approximate Message Passing (AMP) algorithm to reduce complexity in the estimation. Literature [79] proposes a multi-objective optimization problem based on the joint received signal atomic norm and channel gain norm objective functions. It transforms the cascaded channel estimation problem into a sparse matrix recovery problem and achieves high-precision channel estimation using a method based on alternating direction methods.

5. Design for Broadband Effects

One of the major advantages of high-frequency transmission is that the available bandwidth is large. However, in ultra-wideband scenarios such as millimeter wave and low-frequency terahertz bands, spatial broadband effects may occur. These effects can affect the accuracy of channel estimation and cause dispersion problems in beam control. In addition, the large deployment area of RIS will further exacerbate the delay between RIS elements, which may cause different RIS elements to receive different signals. Correspondingly, when transmitting signals, different subcarriers may direct their signals in different directions, leading to dispersion problems. In fact, the cross-correlation function between the spatial steering vector and the cascaded channel reflected by RIS has two peaks in the angular domain: the frequency-related real angle and the frequency-independent error angle. The existence of the error angle will seriously affect the angle estimation performance 80. To solve this problem, literature [80] designs a Two-Stage Orthogonal Matching Pursuit (TS-OMP) algorithm. Considering the impact of broadband on the transmission model, literature [81] transforms broadband channel estimation into a recovery problem for parameters such as angle, delay, and gain. It then detects these channel parameters using the Newton OMP algorithm to achieve channel estimation with low pilot overhead.

6. Dimension Reduction Design for Ultra-High Dimensional Channels

Channel decomposition is an effective means to solve the high-dimensional channel estimation of RIS. The high-dimensional channel can be equivalently split into multiple low-dimensional channels by certain methods, which can greatly reduce the complexity of channel estimation. Literature [82] models the channel passing through each RIS unit, simplifies estimation by utilizing channel rank characteristics, and performs separate cascaded channel estimation for RIS-assisted communication using eigenvalue decomposition. This approach significantly reduces the time cost of estimation. Literature [83] decomposes the BS-RIS-UE cascaded channel into a combination of keyhole channels of individual RIS units, designs a sub-channel estimation scheme based on multi-round pilot training, appropriately configures reflection phase shifts in each round of training, and uses sub-channel CSI to design a joint RIS-transmitter precoding model. Literature [84] models the channel estimation problem of RIS-assisted communication as a third-order tensor using the Parallel Factor model and performs channel estimation through cascaded channel decoupling.

7. Design for Terminal Mobility

The mobility of the terminal poses challenges to channel estimation, and the limited signal processing capability of the RIS passive unit highlights the need to study time-varying transmission channels in mobile scenarios. If mobility is not well managed, channel estimation deviations are inevitable, leading to channel matching bias whether in link adaptation or beamforming, and creating a gap between achievable capacity and ideal channel capacity. Literature [85] models the cascaded channel as a mobile state space model, utilizes time correlation and prior information, and designs a Kalman filter-based scheme to track the time-varying channel, thereby improving estimation accuracy. Literature [86] addresses the problem of time-varying cascaded channel estimation in RIS-assisted communication and designs a low-pilot overhead channel extrapolation scheme using deep learning methods. Literature [87] addresses the Doppler effect and proposes a broadband channel estimation scheme based on Doppler frequency shift adjustment for both multipath and single-path propagation environments. For multipath scenarios, the study proposes a quasi-static channel estimation mechanism that adjusts Doppler distortion through joint RIS reflection matrix design and time–frequency domain conversion. For single-path scenarios, it uses the constant modulus constraint of the RIS shift matrix to design mobile scene channel estimation. Literature [88] presented a RIS channel estimation solution employing a cascaded channel decoupling strategy, which splits the RIS regulation processing into virtual receiving response and virtual regulation transmission. It further explored practical implementations in single-user, multi-user, and multiple RIS scenarios.

8. Design based on Location Information

By utilizing the fixed characteristics of the BS and RIS locations, location information can be effectively used to design low-complexity channel estimation schemes that obtain key information such as the AOA of the signal. The incorporation of location information also enhances other channel estimation schemes, providing more channel information, improving accuracy, and reducing complexity. Literature [89] uses information from an external positioning system to design directional training beams based on rough location data. Channel angles and other parameters are then extracted through atomic norm optimization methods, achieving performance not constrained by the predetermined beam codebook and accelerating the beam alignment and parameter estimation process. Literature [90] examines a RIS-assisted single-input single-output (SISO) multicarrier system, leveraging the large-scale, multi-element, and low-cost characteristics of RIS for joint three-dimensional positioning. A low-dimensional parameter space search algorithm is designed to reduce the four-dimensional parameter estimation problem, performing one-dimensional searches for time delay and two-dimensional searches for angles to achieve sub-meter positioning and synchronization accuracy. Literature [91] explores channel estimation and beamforming with imperfect user location information, using this imperfect data to estimate LOS path angles and for BS passive beamforming and RIS reflection matrix design. It also provides a closed-form expression for the achievable rate, analyzing how positioning accuracy, RIS element number, channel assumptions, and other conditions affect this rate.

9. Design Using Artificial Intelligence Tools

To fully utilize information from collected data or to address channel estimation issues under unknown channel models, applying new Artificial Intelligence (AI) methods presents a potential solution. The rapid development of AI in recent years has introduced a new processing paradigm for traditional wireless communication, offering innovative solutions for channel estimation in RIS-assisted wireless communication. Reference [92] examines an approximate optimal MMSE channel estimation scheme, proposing a data-driven non-linear solution based on deep learning. This approach employs Convolutional Neural Networks (CNN) to denoise and approximate the global optimal MMSE. Reference [93] adopts a strategy of reducing the number of active users to minimize pilot overhead and designs a two-stage scheme for overall channel estimation. It creates a spatio-temporal spectrum framework based on deep CNN to estimate the channels of non-active users by exploiting the spatio-temporal correlation between neighboring users, thus achieving channel mapping for both active and non-active users. Reference [94] highlights the inherent features of real data in AI, processing signals in a data-driven manner, which is applicable to non-ideal RIS scenarios such as model mismatch, resource shortage, hardware damage, and dynamic transmission. This study specifically analyzes how to handle estimation errors in sub-sampled channels by incorporating denoising networks or designing original extrapolation networks.

10. Design Considering Non-ideal Factors

The aforementioned analysis of RIS channel estimation and feedback mainly focuses on the design of ideal RIS assumption schemes, but errors are likely to occur in practical applications. For example, the ideal RIS assumption is that each reflective element has a constant amplitude, variable phase shift, and the same response to different frequencies across RIS elements. Different types and magnitudes of errors may impact channel estimation schemes in various ways. Reference [95] considers the discrete phase shift RIS model, analyzing the relationship between the actual response amplitude, phase shift, and frequency under practical RIS hardware. This non-ideality can introduce errors in existing channel estimation schemes. Reference [96] examines RIS-assisted communication with hardware damage, designing a scheme using linear minimum mean square error estimation, and analyzes the relationship between channel estimation performance and factors such as damage level, the number of reflective elements, and pilot power, confirming that hardware damage limits channel estimation performance at high SNRs. Reference [97] explores the use of a maximum ratio combining receiver under non-ideal CSI with added phase noise randomness, providing a closed-form expression for the achievable uplink spectral efficiency that depends only on large-scale statistics for BS precoding and reflective beamforming matrix design. Achieving high-precision channel estimation requires a substantial amount of pilot resources, and applications need to balance channel estimation accuracy, cost, and efficiency. Evaluating the impact of CSI errors on beamforming performance is crucial in RIS-assisted communication system design. Reference [98] studies robust beamforming based on non-ideal cascaded channels, considering bounded CSI error models and statistical CSI error models, and jointly designing the BS precoder matrix and RIS reflective beamforming matrix. This approach achieves better performance in terms of the minimum transmit power, convergence speed, and complexity. Reference [99] addresses the impact of estimation errors on RIS channels, where the interference channel is a spatially correlated Rayleigh fading channel and there is hardware damage. It analyzes the channel hardening effect, derives the terminal uplink ergodic rate and RIS system performance boundaries, and proves that hardware damage, noise, and NLOS interference can be ignored when there are enough antennas.

8.5 RIS Network Deployment and Performance Evaluation

RIS, with its low cost, low power consumption, and easy deployment characteristics, has the opportunity to be ubiquitously deployed in networks, intelligently controlling the electromagnetic propagation environment and thereby introducing a new network paradigm. However, as a newly introduced network element, the unique technical characteristics and application scenarios of RIS pose significant challenges when RIS is ubiquitously deployed in networks. Given the diversity of application scenarios, it becomes imperative to adopt flexible deployment approaches. Different application scenarios may require different RIS deployment strategies, and Deployment designs must be customized to meet actual needs, such as improving coverage, reducing electromagnetic interference, and enhancing positioning accuracy. Designing RIS networks needs to consider the implementation scheme of the network architecture in traditional cellular networks and also explore the implementation scheme in emerging network architectures, such as network implementation based on the Cell-Free architecture. In this section, we will identify typical RIS deployment scenarios, study the possible problems and challenges for the identified communication scenarios, and propose corresponding candidate deployment schemes.

8.5.1 RIS Network Deployment

1. Deployment Scenarios

From the perspective of communication environment complexity and RIS deployment and control complexity, deployment scenarios can be divided into small-scale, controllable restricted areas and large-scale, complex environments. These two types of scenarios have significant differences in RIS network deployment principles and requirements [100].

Small-scale controllable restricted areas have the opportunity to deploy a sufficient density of RIS and achieve precise intelligent control of the electromagnetic environment, such as typical indoor hotspot coverage areas. In such areas, the wireless propagation environment is relatively independent, the number of main scatterers is limited, and it is convenient to deploy RIS panels on the corresponding surfaces. These are generally hotspot areas for business needs, where a large number of business demands are relatively concentrated and stably distributed in this geographical area. For such areas, a sufficient number and larger size of RIS can replace the main scatterer surfaces in the original natural environment, and additional RIS can be deployed at appropriate positions to increase the scattering surface. This means increasing the deployment of RIS at appropriate positions, artificially introducing more scattering paths, and achieving the purpose of propagation path reselection and channel reconstruction through flexible RIS selection and scattering control. In a limited geographical area, a sufficient density of RIS can jointly optimize scheduling and control, precisely control the wireless propagation environment as needed, and build a wireless intelligent environment that can be almost precisely described and controlled. The topology of this scenario can not only suppress large-scale fading but also achieve dynamic tracking and control of phase/amplitude, Doppler frequency shift, etc., of small-scale multi-path channels through precise control of RIS, thereby suppressing multi-path fading effects. From the perspective of RIS form requirements, dynamically adjustable RIS is needed to achieve precise control. Therefore, such RIS will be higher in terms of structural and control complexity, cost, and power consumption. However, such scenarios are generally hotspot areas, insensitive to cost, and geographically limited, making RIS deployment and optimization relatively simple.

In a wide and complex environment, where business distribution is relatively sparse, it is neither convenient nor necessary to achieve precise control of the wireless propagation environment. For such environments, the focus can be on controlling the large-scale characteristics of the wireless propagation channel, including shadow fading, free-space path loss, and other large-scale characteristics. For scenarios where the natural propagation channel is severely blocked by shadows, an RIS can be deployed at an appropriate location to control the scattering angle, constructing a new BS-RIS-UE propagation path to overcome the shadow blocking problem. As shown in Fig. 8.6a, when there is a high building blocking between the BS and UE, an RIS can be deployed on the surface of the adjacent building to scatter and control the signal, constructing a new propagation path. In addition, by deploying a large-scale array RIS to achieve a higher beamforming gain, free-space path loss can be overcome to some extent. As shown in Fig. 8.6b, when the signal strength of the UE at the edge of the cell is limited, the RIS can be deployed near the edge of the cell. The large antenna aperture of the RIS provides a significant antenna gain; the downlink can improve the received signal strength of the UE at the edge of the cell, and the uplink can enhance the beamforming gain of the UE uplink transmission signal. For such scenarios of large-scale channel characteristic control, since the channel characteristics change slowly or basically do not change, the dynamic requirements of RIS control are relatively low, and an RIS with a low response rate or even fixed weights can be considered. It can be seen that for a wide and complex environment, the RIS will mainly control the existing or newly introduced main propagation paths and main scatterers, achieving semi-dynamic or static control of the large-scale characteristics of the wireless channel. The required RIS form is simple and easy to deploy, and the cost is relatively low.

Different from the above classification method, literature [101] classified the typical deployment scenarios of RIS from four aspects, namely network deployment mode, coexistence and sharing, rank increase and coverage enhancement, and coverage area. From the perspective of network deployment mode, RIS deployment scenarios can include independent mode and network control mode. These two modes have differences in control link requirements, measurement/control signaling interaction, and network deployment complexity, and each has its own advantages and disadvantages. From the perspective of coexistence and sharing, RIS deployment scenarios can include multi-operator network coexistence, single-user access and multi-user access, multi-RISs deployment, and spectrum properties (such as licensed spectrum and unlicensed spectrum). From the perspective of rank increase and coverage enhancement, RIS deployment scenarios can include deployment near the base station, at the edge of the cell, in the middle of the cell, or ubiquitous deployment everywhere. Among them, ubiquitous deployment may truly bring about a revolution in wireless network architecture. From the perspective of coverage area, RIS deployment scenarios can include remote areas, urban indoor/outdoor, Non-Terrestrial Networks (NTN), etc.

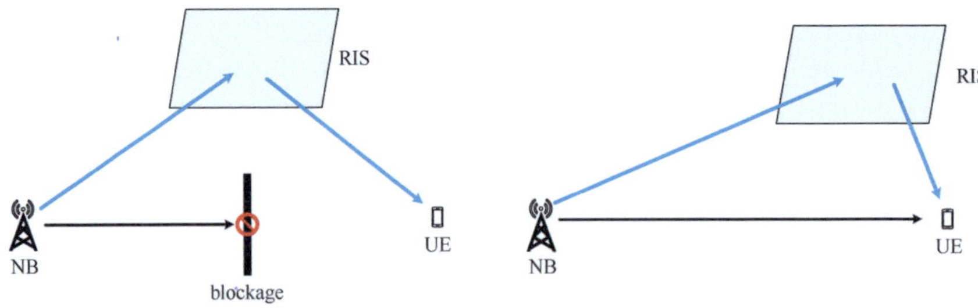

(a) The new transmission path of NB-RIS-UE overcomes shadow blockage.

(b) Enhancement of the coverage at the edge of the cell.

Fig. 8.6 Typical scenarios of RIS deployment

Among them, the RIS deployment mode is classified based on whether it is controlled by the network. The deployment mode of RIS controlled by the network is called "network control mode (NCM)", while the self-controlled deployment mode of RIS is called "standalone mode (SAM)". Table 8.7 compares the advantages and challenges of the network control mode and the independent mode.

By comparing the advantages of the above two modes and combining the characteristics of licensed and unlicensed spectrum, we can conclude that: (1) the NCM is suitable for complex networks and licensed spectrum scenarios with high network coexistence requirements (cellular networks); (2) the SAM is suitable for simple networks and local coverage of unlicensed spectrum technology scenarios (such as Wi-Fi).

2. Support for multi-bands

To meet the higher throughput requirements of future networks, it is necessary for the network to support multiple bands from 6 GHz to terahertz frequency bands. Traditional solutions require the deployment of multiple sets of distributed radio frequency units supporting these various bands to simultaneously support both high and low frequency bands, which poses significant challenges in terms of cost and complexity. RIS has the potential to support cross-band operations, with the opportunity to deploy as few RISs as possible to cover the full band, or even a single set of RIS to meet the full band requirements. That is to say, rather than deploying different base stations to support high and low frequency bands separately, RIS only requires one set to significantly reduce deployment costs and complexity. It should be noted that the cost of RIS supporting the full band is high, especially for RIS supporting high frequency bands. However, only hotspot areas generally need high frequency band coverage to support large bandwidth services, so RIS supporting high frequency bands can be deployed only in these areas. Additionally, different regions

Table 8.7 Comparison of network control mode and independent mode

Types	Advantages	Challenges
Network control mode (NCM)	1. Supports multi-network cooperation 2. Supports multi-user access 3. Better meets the coexistence requirements for networks operating in licensed spectrum	1. Network deployment is relatively complex 2. Requires the deployment of network control links 3. Necessitates the design of interaction protocols for control and measurement signaling
Standalone mode (SAM)	1. No need for network control links 2. The network is simple and easy to deploy 3. Suitable for unlicensed spectrum with low coexistence requirements	1. Difficult to overcome coexistence interference between multiple networks 2. May result in significant inter-cell interference 3. Does not support multi-user access effectively

and operators use different bands, so RIS supporting various band combinations can be deployed according to the coverage needs of different regions and operators, thus balancing the relationship between RIS cost and deployment complexity.

3. Basic Principles and Process of RIS Deployment and Optimization in Typical Communication Scenarios

The network deployment in traditional communication scenarios can include indoor coverage, outdoor coverage, and indoor coverage from outdoors, etc. RIS can be used to support these scenarios by supplementing blind spots, strengthening weak areas, and increasing channel degrees of freedom. The basic principles of network deployment include: (1) ensuring that the signal strength in the coverage area is above the expected threshold to meet the minimum transmission rate; (2) ensuring that the signal strength or SINR distribution in the target coverage area is stable, avoiding unexpected fluctuations. For the latter, reasonable RIS deployment and regulation can be used to maintain high signal strength in the coverage area. Alternatively, RIS dynamic beam tracking can be scheduled to ensure the signal strength for users with service demands, while considering the access needs of idle users (Idle State) at any time and place. That is, the basic coverage signal strength must meet the initial open-loop access requirements. After the user equipment (UE) accesses and enters closed-loop control, RIS can dynamically control the beam to track the UE, providing stronger signals for connected state (Connected State) UEs and achieving higher transmission rates. It should be noted that to improve coverage capacity, especially in outdoor scenarios, due to limited RIS deployment, RIS may be far from the base station, resulting in weak signal strength at the RIS location. Therefore, even with RIS antenna gain, the coverage distance is relatively short. At this point, higher complexity and cost are required to increase RIS gain (such as deploying RIS with a larger antenna aperture) to improve coverage extension capability as much as possible. Figure 8.7 shows the basic process of RIS deployment and optimization in typical communication scenarios. Firstly, under constraints of complexity and cost, the initial RIS deployment topology is designed based on the natural wireless channel and business demand distribution in typical scenarios. Then, based on the adaptive wireless transmission control performance of RIS, the RIS deployment topology is further updated and optimized to build an intelligently controllable wireless environment. The goal of RIS deployment and optimization design is to balance complexity, cost, and performance, outputting the RIS topology, including deployment location, density, RIS form, control/cooperation relationships, and other parameters.

Figure 8.7 shows the basic process of RIS deployment and optimization in typical communication scenarios. Firstly, under constraints of complexity and cost, the initial RIS deployment topology is designed based on the natural wireless channel and business demand distribution in typical scenarios. Then, based on the adaptive wireless transmission control performance of RIS, the RIS deployment topology is further updated and optimized to build an intelligently controllable wireless environment. The goal of RIS deployment and optimization design is to balance complexity, cost, and performance, producing the RIS topology, including deployment location, density, RIS form, control/cooperation relationships, and other parameters.

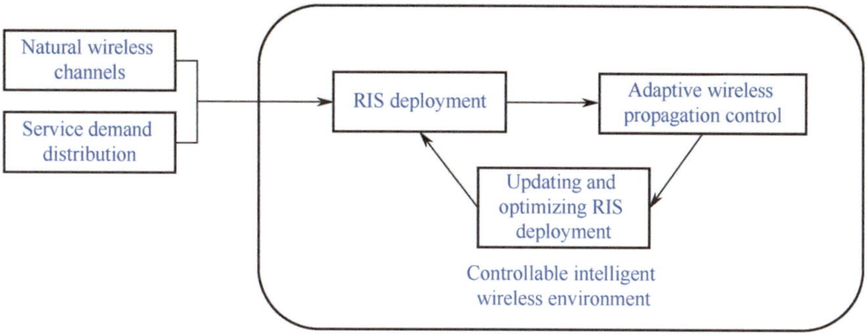

Fig. 8.7 Basic process of RIS deployment and optimization in typical communication scenarios

4. RIS Constructing Near-field Wireless Propagation Environment [102]

The near-field propagation characteristics offer increased potential for the upcoming 6G network; however, constructing a near-field propagation environment using traditional active phased array antennas poses several challenges. Firstly, ultra-large active phased array antennas (APAAs) have significantly increased hardware costs, complexity, power consumption, weight, and volume, making it challenging to implement dense deployments and limiting the near-field coverage area. Secondly, the near-field distance peaks near the axis of the APAA array and diminishes as the angle increases, further restricting the near-field's coverage range. Thirdly, while communication services typically benefit from NLOS multipath environments, the ideal propagation environment for sensing positioning and wireless power transmission services is the near-field LOS channel. By exclusively deploying centralized APAAs, the target is likely to be in a NLOS multipath channel. Numerous studies have explored Cell-free/Coordinated Multi-Point (CoMP) [103, 104] techniques using traditional active phased array antennas, yet these distributed antenna technologies, while mitigating the challenges of centralized APAA deployments to an extent, remain constrained by the inherent technical features of active phased array antennas, making it difficult to achieve dense and ubiquitous deployments.

The unique technical features of RIS can be used as an effective means to solve the challenges faced by traditional active phased array antennas. First, RIS, as a programmable two-dimensional electromagnetic meta-surface, can abnormally control electromagnetic waves in a passive manner, with the advantages of low cost and low power consumption, and can be easily made into a large antenna aperture, thus realizing dense deployment at a low cost; second, RIS types are diverse and can flexibly adapt to complex and diverse deployment environments. From a functional perspective, RIS types can include channel control type (e.g., reflective RIS, transmissive RIS, and semi-transparent semi-reflective RIS), information modulation type RIS (e.g., RIS-based new base station, RIS-based backscatter transmitter, and RIS-based companion communication), and RIS-based new phased array antennas, etc. RIS can be easily made into different sizes, shapes, and surface forms to meet different deployment requirements; finally, the simple and easy-to-deploy features of

RIS can also easily build a near-field LOS environment, thereby better supporting the needs of perception positioning and wireless power transmission. In addition, since RIS is passive control, it naturally has a low level of electromagnetic radiation, which can still meet the human electromagnetic radiation safety index Specific Absorption Rate (SAR) in the ubiquitous near-field environment.

In summary, compared with traditional active phased array antennas, RIS has the characteristics of passive control, low cost, and easy deployment, and can be densely and widely deployed, thus having the opportunity to build a ubiquitous near-field channel environment for future 6G networks.

5. An Adaptive Network Capacity and Coverage Adjustment Mechanism

In real-world scenarios, the demand for network capacity and coverage is unevenly distributed across geographical spaces, and this distribution is dynamic or semi-dynamic. For example, during large-scale events, morning and evening peak times, and high-speed rail communications, the demand manifests as semi-static changes in different geographical areas (capacity demand migrates semi-statically across these spaces). In such scenarios, the network topology needs to be able to adaptively implement the migration of network capacity distribution across geographical dimensions. Therefore, network topology and regulation design must adjust coverage and network capacity resources semi-dynamically, i.e., maximize energy efficiency while minimizing cost overhead by using as few base stations and cells as possible to address coverage hole issues and meet capacity demand. In response to these demand characteristics, we provide an example of adaptive network capacity and coverage adjustment based on RIS under cost constraints.

The first step is to adaptively adjust network coverage semi-statically in response to the unevenness and semi-static changes in coverage and capacity demand across geographical areas. Specific implementations include three optional methods: (1) Using movable drones or high-altitude platforms to carry base stations (BS), adjusting capacity and coverage over a large physical range in the air; (2) Using unmanned or manned vehicles to carry base stations (BS) or relays and other ground platforms, adjusting capacity and coverage over a large geographical area; (3) Using higher-altitude platforms, such as satellites, to carry BSs, adjusting coverage over an even larger geographical range. The first two methods can also use RIS instead of base stations on drones, high-altitude platforms, or unmanned vehicles, controlling coverage through RIS signal regulation. In specific engineering implementations, some fixed candidate positions can be optimally selected, quantifying geographical space to a limited extent, thereby reducing the complexity of engineering implementation.

The second step is to use the CoMP/Cell-Free mechanism to achieve small-scale coverage and capacity adjustment. CoMP/Cell-Free can adaptively adjust the set of cooperating access points (APs) and the logical cellular coverage area. Generally, the set of APs cooperating with CoMP is small, while Cell-Free can achieve cooperation among a larger set of APs. In a network introducing RIS, using a concept similar to CoMP/Cell-Free, it is possible to control the cooperation set of multiple RISs within a single cell and manage the sharing of RISs between multiple cells to achieve small-scale coverage and capacity control.

The third step involves the adaptive adjustment of the cell's coverage area. The cell achieves dynamic adjustment of the coverage range through adaptive power control or by using RIS regulation. The coverage and interference control are managed through adaptive switching (cell ON/OFF) of base stations or by scattering/absorption via RIS.

The fourth step is to use RIS to finely control both the large-scale and small-scale characteristics of the local area wireless channel. For example, RIS can be used to semi-statically control large-scale characteristics to overcome coverage holes and fill in blind spots. RIS can also dynamically control the phase, amplitude, and other parameters of multipath signals to suppress multi-path effects.

Through the above RIS-based adaptive network capacity and coverage adjustment mechanism, there is an opportunity to achieve an adaptive and intelligent flexible wireless network topology, thereby constructing a new paradigm of wireless network topology.

To implement the RIS-based adaptive network capacity and coverage adjustment mechanism, the following issues need to be specially studied:

(1) The impact of coverage and capacity migration on spectrum allocation and sharing.
(2) The need for adaptive adjustment of the network topology structure during cell migration.
(3) Adaptive planning of Cell IDs.
(4) Adaptive modulation of other network resources, such as computing resources.
(5) Mobility requirements of base stations (BS) and RIS for the return link, including the potential use of wireless return links.
(6) Power requirements for BS and RIS mobility include deploying drone stations at appropriate locations and densities to serve as charging stations, thereby providing charging support. Alternatively, sufficient temporary docking platforms (mobile platform stations) can be provided at fixed coverage locations. These platforms can accommodate drones, offer charging functions, and provide recycling and parking capabilities.

6. Coexistence analysis of RIS networks

Existing RIS research mainly focuses on the new challenges posed by classic communication problems since the introduction of RIS, such as channel estimation and beamforming. These studies primarily examine these issues under the assumption of a single network system model. In actual wireless mobile communication networks, the coexistence of multiple networks is a traditional problem, and the introduction of RIS may introduce new network coexistence challenges. Reference [100] preliminarily analyzed the coexistence problem of RIS networks and proposed possible solutions. Building on Reference [100], Reference [105] further conducted in-depth analysis and modeling of RIS network coexistence, evaluating two new types of RIS structures with out-of-band filters and RIS block mechanisms. In practical networks, the wireless signals incident on the RIS panel include both "target signals," which are optimized and regulated by RIS, and "non-target signals." RIS regulates both types of signals simultaneously. It can enhance the "target signal" by adjusting the amplitude, phase, polarization mode, etc., of electromagnetic waves, but it may also unexpectedly regulate the "non-target signal." In uncontrolled situations, RIS may improperly regulate "non-target signals" from other networks, leading to significant network coexistence issues. The document proposes two solutions: a multi-layer RIS structure with a bandpass filter layer and a RIS block mechanism. It provides theoretical analysis and numerical simulations of these two newly proposed RIS structures, demonstrating their effectiveness in addressing RIS network coexistence problems. Additionally, this issue highlights the need for network-controlled large-scale deployment of RIS to prevent its arbitrary and unexpected regulation of "non-target signals," thereby avoiding serious network performance degradation.

7. A Typical Use Case: Deployment of RIS in High-Speed Rail Communication Scenarios

The communication needs of intelligent high-speed rail can be divided into four major scenarios: continuous wide-area coverage of the main railway line, hotspot areas such as railway yards and hubs, ground infrastructure monitoring along the railway, and broadband applications for intelligent trains. These communication needs exhibit distinct characteristics, primarily manifesting as the extremely uneven distribution of capacity and coverage requirements in geographical space. This unevenness follows a specific pattern: business needs are confined to the railway corridor, and with the high-speed operation of trains, there is a collective migration of capacity and coverage needs. These needs are linearly distributed along the railway line.

For the typical deployment methods of high-speed rail communication based on RIS, considering the environmental and channel characteristics of high-speed rail and combining the technical features of RIS, Reference [106] proposed three deployment modes. Reference [107] further summarizes four typical network deployment modes: deploying RIS along the railway, placing RIS on top of the carriage for high-speed rail mobile relay or customer premise equipment (CPE) enhancement, applying transparent enhanced RIS on window glass, and installing RIS on the inner walls of the carriage. Among these, the second mode is a new type.

High-speed rail communication services exhibit the characteristic of migrating with the movement of the train, meaning that only the cells through which the high-speed rail passes require business connections. After the train passes and before the next one arrives, the cell does not need to support high-speed rail communication. A practical solution is for adjacent base stations along the railway to deploy and share RIS. By adopting a shared RIS scheme, only low-bandwidth control signaling needs to be switched between base stations, and the delay requirement for RIS control signaling can be appropriately reduced to balance the dynamism of RIS control with the real-time nature of shared switching. In contrast, when adjacent base stations in a traditional high-speed rail network share a Remote Radio Unit (RRU) or distributed antenna, it requires switching large-bandwidth business data and control signaling with low latency, which is more complex to implement.

The backhaul link of the base station controlling the RIS has different design constraints depending on the deployment method, leading to different implementation options. For the RIS deployment mode along the railway, the backhaul link between the base station and the RIS can use wired or wireless communication methods. Wireless backhaul is flexible in deployment but requires spectrum resources to transmit control signaling between BS and RIS, resulting in some spectrum resource overhead. However, the control signaling information rate of this backhaul link is low, so the spectrum occupancy cost is not high. For the two modes of RIS deployment on the train window glass and the interior of the carriage, the backhaul link between RIS and BS can only use wireless communication methods. The RIS deployed on the roof to enhance the mobile relay or CPE antenna is controlled by the mobile relay or CPE, and the backhaul link is connected with the mobile relay or CPE, usually via wired methods.

These deployment modes all need to ensure the air interface synchronization relationship between base stations and between base stations and RIS, to ensure the synchronization relationship between RIS amplitude phase control and channel/signal. This is especially critical when multiple RIS beamforming serves a UE simultaneously, similar to the joint transmission of traditional CoMP, requiring precise time synchronization and phase alignment. Reference [108] provided a detailed analysis of this issue and proposed a solution for multiple RIS collaborative regulation mechanisms based on network control.

Tunnel coverage can use RIS to enhance the traditional antenna form of the existing Distributed Antenna System (DAS), which is smaller in size and easy to deploy on the side of the tunnel wall without obvious protrusion. Considering the low-cost feature of RIS, more passive reflection RIS can be deployed on the side of the tunnel wall, enhancing the signal coverage in the tunnel with high-density RIS.

8.5.2 Performance Simulation and Testing

1. RIS Physical Model Abstract Simulation

The parameters of the RIS physical model abstract simulation are shown in Table 8.8.

The simulation of the RIS physical model is shown in Table 8.9.

The radiation direction diagram of a single RIS unit is shown in Fig. 8.8, and the coverage result of the RIS panel in the target area is shown in Fig. 8.9.

2. Complex Scenario RIS Deployment Simulation

Consider deploying BS in a dense urban block and improving NLOS area coverage through a RIS panel. The simulation area is 160,000 m^2, and the base station BS is deployed on the roof of a building at a high position of 43 m, as shown in Fig. 8.10, and the simulation parameter configuration is shown in Table 8.10.

The logical link decomposition is shown in Fig. 8.11. There are 2 logical links from BS to UE, and the decomposition results are as follows:

Table 8.8 RIS physical model abstract simulation parameters

Number	Parameters	Values	Unit
1	Base station position	$[40 \times \cos(10°),\ -40 \times \sin(10°),\ 1]^T$	m
2	RIS position	$[0, 0, 1]^T$	m
3	RIS normal horizontal angle (AZ)	0	°
4	RIS normal down tilt angle (EL)	0	°
5	RIS normal rotation angle (SL)	0	°
6	High-gain beam pointing to the UE position	$[0, 10, 1]^T$	m
7	Four-point coordinates of the RIS beam projection rectangular area	$[10, -5, -2]^T$ $[10, -5, 2]^T$ $[2, 5, 10]^T$ $[10, 5, -2]^T$ The grid granularity is 0.1 m	m
8	RIS codebook modulation bit number	1	–
9	Incident wave vector polarization angle	0	° (0-V,90-H)
10	RIS polarization type	Single polarization, Dual polarization (45)	–
11	RIS polarization angle (relative to the RIS local coordinate x-axis)	0, 20, 45, 70, 90	°
12	RIS orthogonal polarization phase offset	0	°
13	BS EIRP	36.7515	dBm
14	Carrier frequency	26.9	GHz
15	Number of RIS elements ($Nx \times Ny$)①	64×64	–
16	RIS horizontal element spacing	0.45	λ (Wavelength)
17	RIS vertical element spacing	0.45	λ (Wavelength)
18	RIS reflection efficiency	0.6	–

Note ① Nx and Ny respectively represent the number of antennas in the x and y directions

Table 8.9 RIS physical model simulation example

Number	RIS polarization type	RIS polarization angle (relative to the RIS local coordinate x-axis)
S-0	Single polarization	0
S-20	Single polarization	20
S-45	Single polarization	45
S-70	Single polarization	70
S-90	Single polarization	90
D-45	Dual polarization	45

L1: BS → UE; Decomposition: L1-1: BS → UE;

L2: BS → RIS1 → UE; Decomposition: L2-1: BS → RIS1, L2-2: RIS1 → UE.

The coverage results of BS + RIS are shown in Fig. 8.12.

The meanings of the related identifiers are introduced below.

Ia: Power coverage of logical link L2 (v-pol + h-pol).

Iv: Power coverage of logical link L2 (v-pol).

Ih: Power coverage of logical link L2 (h-pol).

Ga: Power coverage of logical link L1 (v-pol + h-pol).

Gv: Power coverage of logical link L1 (v-pol).

Gh: Power coverage of logical link L1 (h-pol).

Ca: Power coverage of logical link L1 + L2 (v-pol + h-pol).

Cv: Reception coverage of logical link L1 + L2 (v-pol).

Ch: Reception coverage of logical link L1 + L2 (h-pol).

From the simulation results, it can be seen that when RIS is not deployed, the coverage of BS in the street where the target UE is located is very weak (see Ga in Fig. 8.12). After deploying RIS, due to the controllable beam of RIS pointing to the street where the target UE is located (see Ia in Fig. 8.12), the coverage of the related street is significantly enhanced (see Ca in Fig. 8.12).

The power delay spectrum of the RIS beam pointing to the terminal is shown in Fig. 8.13. The strongest path is the path that arrives from the BS → RIS → UE logical link. From the above simulation results, it can be seen that RIS is very effective in improving the coverage of the NLOS area.

8.6 Outlook on Development Trend of RIS

RIS, as a newly emerging dynamic electromagnetic parameter control technology, has initially demonstrated its powerful performance in multiple fields. However, many problems and challenges remain before its large-scale commercial use [2, 109]. The challenges and trends of RIS technology mainly involve theoretical models, application technologies, and engineering research.

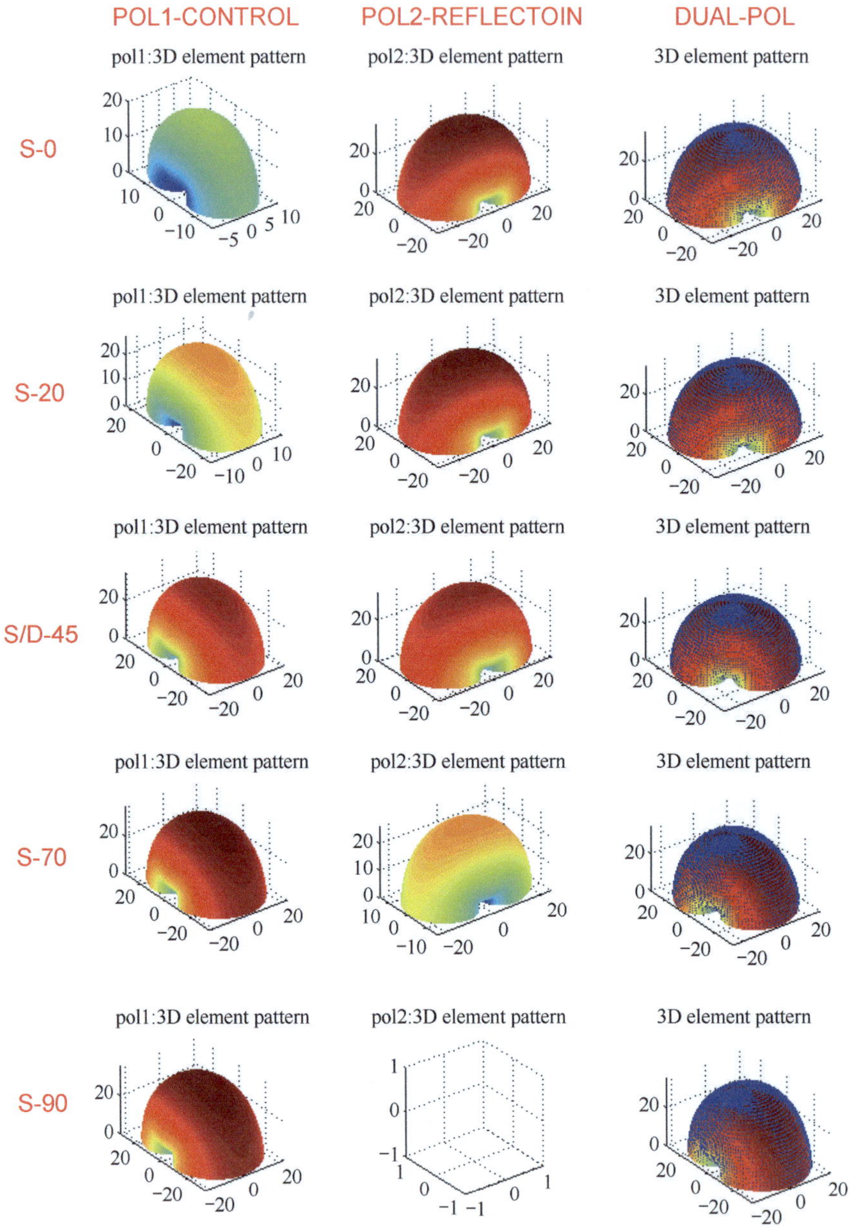

Fig. 8.8 Radiation direction diagram of RIS unit

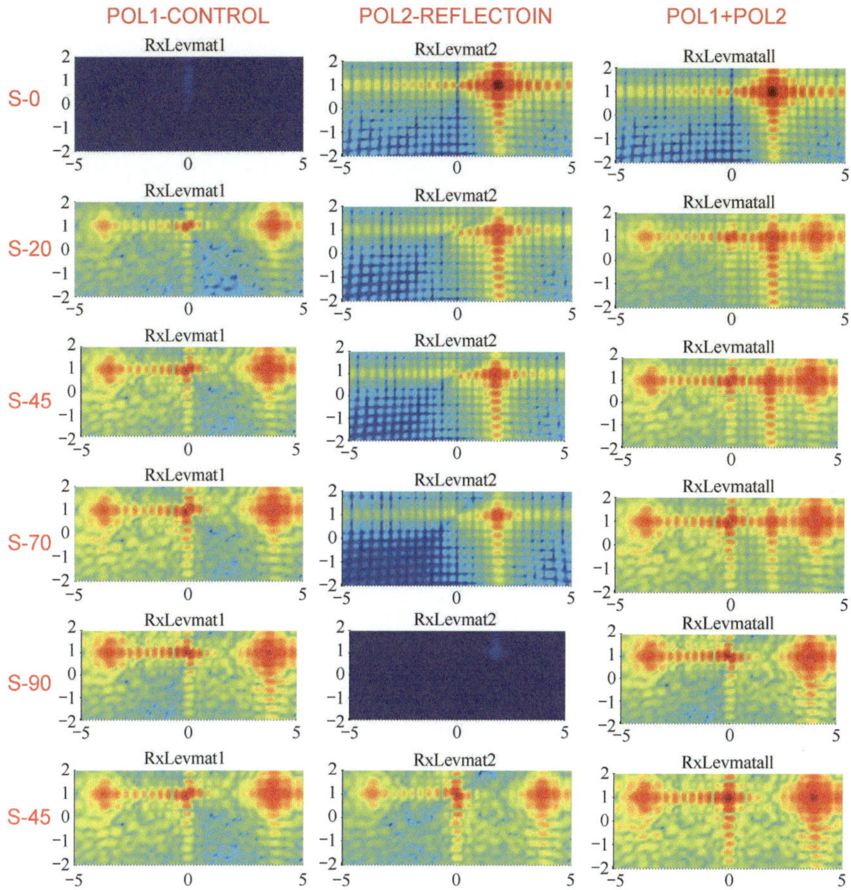

Fig. 8.9 Coverage result of RIS panel in target area

Regarding the theoretical models that characterize RIS, although some progress has been made, it is still far from establishing a complete theoretical system. Further in-depth exploration is needed in the physical mechanisms of electromagnetic control related to RIS technology, electromagnetic informatics, channel models, etc., to quickly build a complete theoretical system. Additionally, RIS technology requires interdisciplinary integration of material science (mainly meta-materials), electromagnetics, information and electronics, and communication engineering, necessitating coordinated advancement across multiple disciplines.

In the field of applied technology research, existing achievements are mainly focused on solving classic problems in traditional wireless communication, such as channel estimation, beamforming, and information modulation. However, there is insufficient research investment in novel application areas such as RIS-based Integrated Communication And Sensing (ISAC), AI-enabled RIS, RIS-based Secure Communication, and RIS-based Air Computing, with few reports on related research

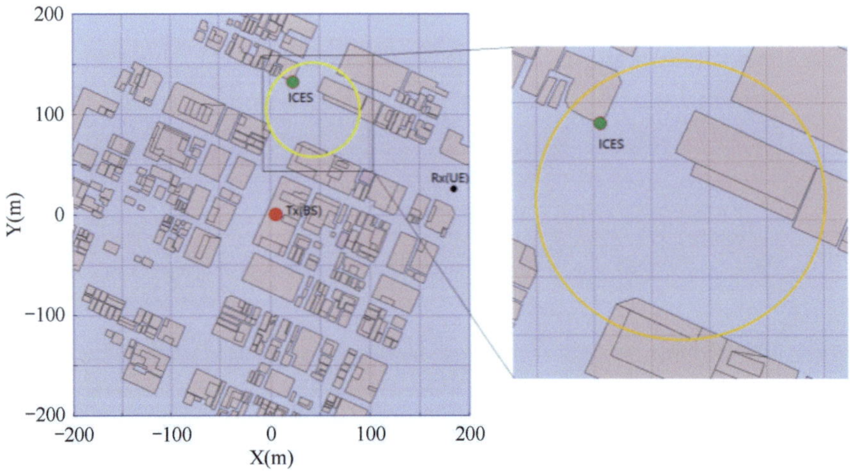

Fig. 8.10 3D map of the simulation scenario

Table 8.10 Simulation parameter configuration

Parameters	Values
BS coordinates (m)	[0, 0, 43]
BS main lobe mechanical azimuth angle Az (°)	120
BS main lobe down tilt angle (DT) (°)	10
ICES coordinates (m)	[21.67, 133.2, 36.2]
RIS normal vector azimuth angle (Az) (°)	−60
RIS normal vector down tilt angle (Dt) (°)	0
Carrier frequency f_1 (GHz)	2.6
BS transmit antenna polarization	Vertical polarization (v-pol)
Equivalent isotopically radiated power (EIRP) (dBm)	43
BS half power beam width (HPBW) (°)	14
BS beamforming direction	From BS pointing to the center of the RIS panel
RIS panel dimensions	14 × 14 (RIS elements)
RIS unit dimensions	$\frac{\lambda}{3} \times \frac{\lambda}{3}$
RIS reflection efficiency	0.8
RIS phase control granularity	2-bit
Adjustable phases of RIS control units	$[0, \pi/2, \pi, 3\pi/2]$
RIS pointing at target terminal UE coordinates (m)	[184, 26, 1.5]
Height of the simulated coverage area from the ground (m)	1.5

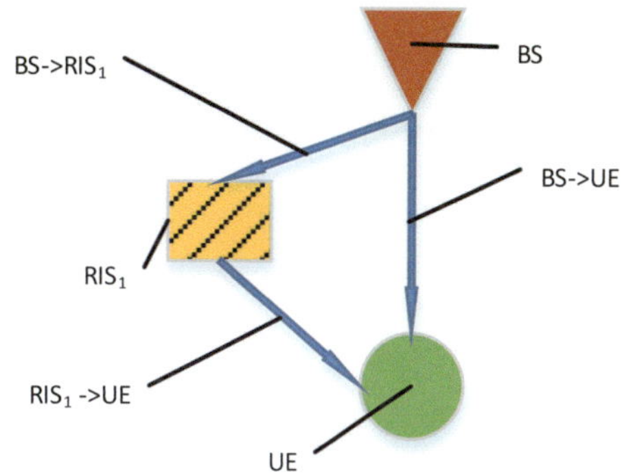

Fig. 8.11 Logical link decomposition

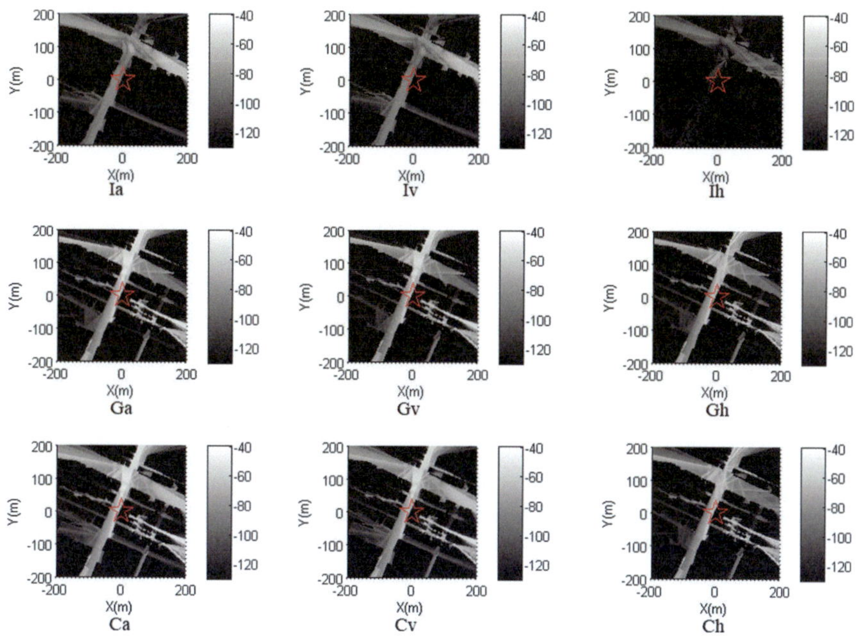

Fig. 8.12 Coverage results of BS + RIS

results. Moreover, most of the existing research is based on simple system models, and the provided mechanisms are generally applicable only to ideal scenarios. Therefore, future research needs to invest more in new application areas of RIS and further consider more complex model assumptions.

In terms of engineering research, although there have been some preliminary tests of RIS prototypes and some performance gains demonstrated in specific scenarios, there is still a considerable gap from real engineering applications. This gap exists particularly in hardware design and regulation, RIS standardization, new wireless network architecture design based on RIS, and RIS network deployment.

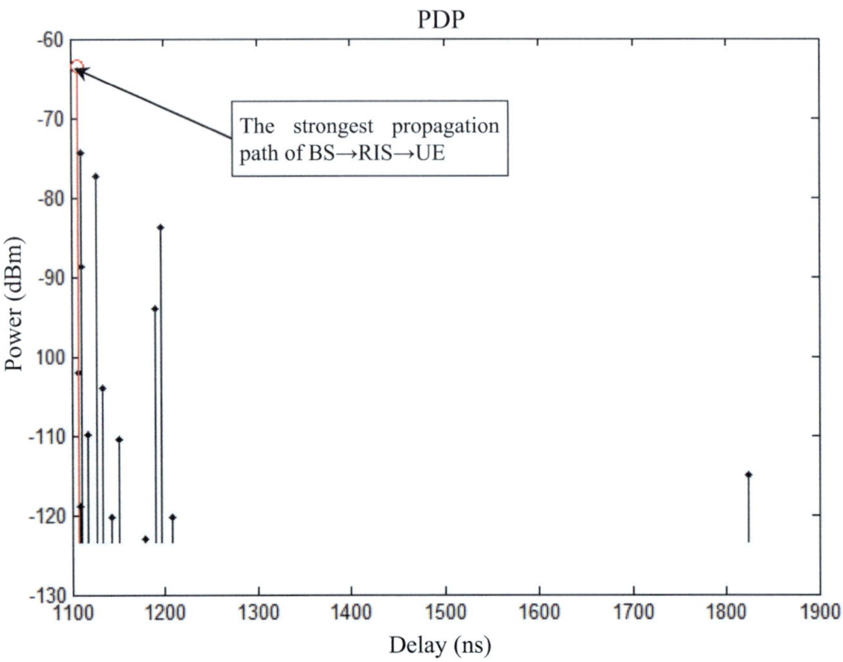

Fig. 8.13 Power delay spectrum of the RIS beam pointing to the terminal

(1) **Hardware design and regulation**: From the perspective of RIS hardware, there are problems such as the low maturity of materials and components, a significant gap between cost and power consumption and ideal targets, and scaling difficulties. Future research on RIS hardware structure and regulation algorithms needs to address the expansion of unit components capable of independently controlling the characteristics of electromagnetic signals, such as high-speed control of electromagnetic parameters like amplitude, phase, spectrum, or polarization within a large bandwidth. It also involves designing RIS arrays with diverse functions, optimizing the spatial arrangement of RIS elements, integrating heterogeneous multi-type atomic arrays, and developing multi-layer, multi-functional RIS modules. Additionally, efforts should be made to improve the effective working bandwidth of RIS arrays, enhance energy conversion efficiency, and develop basic and expanded control algorithms to flexibly extend the function set of RIS arrays and explore new application scenarios.

(2) **RIS standardization**: 3GPP Release 18's 5G-Advanced (5G-A) has completed the standardization of Smart Repeater (SR) technology characteristics, which are very similar to those of RIS. This work has laid a foundation for the future standardization of RIS. In the 6G stage, the standardization of RIS will differ from 5G-A as the 6G standard is a new protocol. We won't need to consider compatibility with traditional systems. By then, RIS technology will be more

mature, allowing us to standardize more complete functional features of RIS. The design of RIS standards and protocols involves:

① Terminal access process, including measurement, discovery, random access, transmission, and target signal recognition.
② Wireless resource management and scheduling processes mainly focus on managing and scheduling RIS resources. This includes the selection of RIS channels and BS-UE direct channels, as well as the scheduling and selection of RIS in scenarios involving multiple RIS.
③ The impact of RIS on the handover process (mobility management), such as handover between different RIS or frequency bands.
④ Design of control information exchange between the network and RIS, including beamforming information, synchronization information, and channel status information.

(3) **New Wireless Network Architecture Design Based on RIS**. From the perspective of future wireless network architecture, RIS-based wireless networks are not yet clearly defined. As a new type of network node, there is currently a lack of in-depth research on RIS interface protocols, network element functions, and topology. It is necessary to explore the definition of RIS network element functions in various transmission scenarios, control methods for RIS and wireless networks, and corresponding interface protocols; study the network topology of RIS within homogeneous or heterogeneous wireless networks; and examine the scalability, security, robustness, and delay characteristics of the new wireless network architecture integrating RIS. Promoting the standardization process related to this new architecture is also essential.
(4) **RIS Network Deployment**. Regarding actual networking methods, there is limited research on RIS supporting multiple bandwidths, systems, and complex networks. RIS deployment must address the coexistence of same and different frequencies between multiple communication systems [100, 105]. Most existing RIS research focuses on optimizing performance for a single communication system. Although some literature has explored coexistence issues in RIS for multiple communication systems, further research on engineering implementation details is needed [100, 105]. Coexistence problems related to RIS should be systematically simulated and evaluated once the RIS channel model matures. Based on these evaluations, RIS structure design, equipment specifications, spectrum planning, and network deployment should be adjusted to address coexistence issues. For RIS engineering application implementation, a three-stage deployment approach can be adopted: first, deploying a small number of non-standard static RIS panels in the 5G network to address coverage gaps, particularly for high-frequency millimeter waves; second, deploying semi-dynamic RIS based on the 5G-A standard to optimize network coverage; and third, implementing intelligent and flexible RIS in future wireless networks to introduce a new communication network paradigm for 6G.

8.7 Summary

As a highly promising key technology for 5G-A and 6G, RIS has rapidly advanced in academic research in recent years. RIS will introduce a new communication network paradigm to future 6G by creating an intelligent and controllable wireless environment to meet future mobile communication needs. A simplified version of RIS may see initial commercial deployment and standardization during the 5G-A stage, particularly to address the coverage issues of 5G millimeter waves. However, RIS-enabled future 6G networks still face many technical problems, deployment challenges, and standardization hurdles. Therefore, in-depth research and comprehensive evaluation of RIS key technologies and schemes are necessary. Due to the scope and depth of this research, this article only provides a preliminary analysis of the challenges faced by RIS engineering technology research and applications, proposes potential solutions, and primarily engages in qualitative analysis discussions. These related issues still require further theoretical study, simulation, and evaluation to verify the feasibility and performance limits of the proposed schemes and to lay the foundation for the eventual industrial deployment of RIS.

References

1. RENZO D M, ZAPPONE A, DEBBAH M, et al. Smart Radio Environments Empowered by Reconfigurable Intelligent Surfaces: How It Works, State of Research, and The Road Ahead[J]. IEEE Journal on Selected Areas in Communications, 2020, 38(11): 2450–2525.
2. Cui Tiejun, Jin Shi, Zhang Jiayi, et al., Research Report on Intelligent Hyper Surface Technology[R]. IMT-2030 (6G) Promotion Group, 2021.
3. Jian M, Alexandropoulos G C, Basar E, et al. Reconfigurable intelligent surfaces for wireless communications: Overview of hardware designs, channel models, and estimation techniques[J]. Intelligent and Converged Networks, 2022, 3(1): 1–32.
4. YUAN Y F, ZHAO Y J, ZONG B Q, et al. Potentional key technologies for 6G mobile communications[J]. Science China: Information Sciences, 2020, 63(8):1–19.
5. ZHAO Y J, JIAN M N. Applications and challenges of Reconfigurable Intelligent Surface for 6G networks[J]. arXiv preprint arXiv:2108.13164, 2021.
6. Nanfang Yu et al. ,Light Propagation with Phase Discontinuities: Generalized Laws of Reflection and Refraction. Science334, 333–337(2011). https://doi.org/10.1126/science.1210713
7. KURMENDRA, KURMENDRA R. A review on RF micro-electro-mechanical-systems (MEMS) switch for radio frequency applications[J]. Microsystem Technology, 2020.
8. CAO T T, Hu TT, ZHAO Y L. Research Status and Development Trend of MEMS Switches: A Review[J]. Micromachines, 2020, 11(7): 694.
9. HUM SV, PERRUISSEAU-CARRIER J. Reconfigurable Reflectarrays and Array Lenses for Dynamic Antenna Beam Control: A Review[J]. IEEE Transactions on Antennas and Propagation, 2014, 62(1):183–198.
10. KARABEY O H, BILDIK S, FRITZSCH C, et al. Liquid crystal based reconfigurable antenna arrays[C]. Proceedings of the 32nd ESA Antenna workshop, 2010.
11. YANG FAN, XU SHENHENG, LIU XIAO, et al. Novel phased array antennas based on surface electromagnetics[J]. Chinese Journal of Radio Science, 2018, 33(3): 256–265.

12. AHMAD G, BROWN T W C, UNDERWOOD C I, et al. How coarse is too coarse in electrically large reflectarray smart antennas?[C]. 2017 International Workshop on Electromagnetics: Applications and Student Innovation Competition, 2017: 135–137.
13. ARUN V, BALAKRISHNAN H. RFocus: Beamforming using thousands of passive antenna[C]. 17th USENIX symposium on networked systems design and implementation (NSDI 20). 2020: 1047–1061.
14. NTT DOCOMO. DOCOMO conducts world's first successful trial of transparent dynamic metasurface[R]. 2020.
15. PEI X, YIN H, TAN L, et al. RIS-Aided Wireless Communications: Prototyping, Adaptive Beamforming, and Indoor/Outdoor Field Trials[J]. IEEE Transactions on Communications, 2021, 69(12): 8627–8640.
16. ZHANG X G, JIANG W X, JIANG H L, et al. An optically driven digital metasurface for programming electromagnetic functions[J]. Nature Electronics, 2020, 3(3): 165–171.
17. DAI L, WANG B, WANG M, et al. Reconfigurable Intelligent Surface-Based Wireless Communications: Antenna Design, Prototyping, and Experimental Results[J].IEEE Access, 2020, 8:45913–45923.
18. DOU J W, CHEN Y J, ZHANG N, et al. On the channel modeling of intelligent controllable electro-magnetic-surface[J]. Chinese Journal of Radio Science, 2021, 36(3): 368–377.
19. 3GPP. Study on channel model for frequencies from 0.5 to 100 GHz: TR38.901 V16.1.0[S]. 2019.
20. ITU-R. Guidelines for evaluation of radio interface technologies for IMT-2020: M.2042[S]. 2017.
21. TANG W, CHEN M Z, CHEN X Y, et al. Wireless communications with reconfigurable intelligent surface: Path loss modeling and experimental measurement [J]. IEEE Transactions on Wireless Communications, 2021, 20(1): 421–439.
22. KILDAL P S. Foundations of Antenna Engineering: A Unified Approach for Line-of-Sight and Multipath[M]. Gothenburg:Kildal Antenn AB,2015.
23. DEGLI-ESPOSTI V, VITUCCI E M, RENZO M D.Reradiation and Scattering from a Reconfigurable Intelligent Surface: a General Macroscopic Model[J]. IEEE Transactions on Antennas and Propagation, 2022,70(10): 8691–8706.
24. He G Y,Lu C C,Hong J C,et al.Calculation and measurement of electromagnetic scattering[M].Beijing:Beihang University Press,2006:163–167(in Chinese)..
25. Ling F, Jin J M. Hybridization of SBR and MoM for scattering by large bodies with inhomogeneous protrusions – summary. [J]. Journal of Electromagnetic Waves and Applications, 1997, 11:1249–1255.
26. Guo, L., Zhang, M., & Wu, Z. Fundamental Theories and Methods of Electromagnetic Scattering by Random Rough Surfaces and Targets [M]. Beijing: Science Press, 2015.
27. STRATTON J A, CHU L J. Diffraction theory and electromagnetic waves[J]. Physical Review, 1939, 56:99–107.
28. LYU J, ZHANG R. Spatial throughput characterization for intelligent reflecting surface aided multi-user system[J]. IEEE Wireless Communications Letters, 2020, 9(6): 834–838.
29. ZHANG S, ZHANG R. Intelligent reflecting surface aided multiple access: Capacity region and deployment strategy[C]. 2020 IEEE 21st International Workshop on Signal Processing Advances in Wireless Communications (SPAWC), 2020.
30. JUNG M, SAAD W, DEBBAH M, et al. Asymptotic optimality of reconfigurable intelligent surfaces: Passive beamforming and achievable rate[C]. IEEE International Conference on Communications (ICC), 2020.
31. HE J,WYMEERSCH H,SANGUANPUAK T, et al. Adaptive beamforming design for mmWave RIS-aided joint localization and communication[C]. IEEE Wireless Communications and Networking Conference (WCNC), 2017.
32. ZHANG H L, DI B Y, DONG L Y, et al. Reconfigurable intelligent surfaces assisted communications with limited phase shifts: How many phase shifts are enough?[J]. IEEE Transactions on Vehicular Technology, 2020, 69(4): 4498 - 4502.

33. ALEGRÍA J V, RUSEK F. Achievable rate with correlated hardware impairments in large intelligent surfaces[C]. 2019 IEEE 8th International Workshop on Computational Advances in Multi-Sensor Adaptive Processing (CAMSAP) , 2019:559–563.
34. HU S, RUSEK F, EDFORS O. Capacity degradation with modeling hardware impairment in large intelligent surface[C]. IEEE Conference Global Communications, 2018:1–6.
35. ÖZDOGAN Ö, BJÖRNSON E, LARSSON E G. Using intelligent reflecting surfaces for rank improvement in MIMO communications[C]. 2020 IEEE International Conference on Acoustics, Speech and Signal Processing (ICASSP) , 2020.
36. ZHANG Z J, CUI Y, YANG F, et al. Analysis and optimization of outage probability in multi-intelligent reflecting surface-assisted systems[J].arXiv, 2019.
37. JUNG M, SAAD W, JANG Y, et al. Reliability analysis of large intelligent surfaces (liss): Rate distribution and outage probability[J].IEEE Wireless Communications Letters, 2019, 8(6):1662–1666.
38. D. Kudathanthirige, D. Gunasinghe and G. Amarasuriya, "Performance Analysis of Intelligent Reflective Surfaces for Wireless Communication," ICC 2020 - 2020 IEEE International Conference on Communications (ICC), Dublin, Ireland, 2020, pp. 1–6, https://doi.org/10.1109/ICC40277.2020.9148760..
39. ZHANG H, DI B, SONG L, et al. Reconfigurable intelligent surfaces assisted communications with limited phase shifts: How many phase shifts are enough?[J]. IEEE Transactions on Vehicular Technology, 2020:1–1.
40. PEROVIĆ N S, RENZO M D, FLANAGAN M F. Channel Capacity Optimization Using Reconfigurable Intelligent Surfaces in Indoor mmWave Environments[C]. 2020 IEEE International Conference on Communications (ICC), 2020: 1–7.
41. HUANG C, ZAPPONE A, DEBBAH M, et al. Achievable rate maximization by passive intelligent mirrors[J]. IEEE International Conference Acoustics, Speech and Signal Processing, 2018: 3714–3718.
42. HUANG C, ZAPPONE A, ALEXANDROPOULOS G C, et al. Reconfigurable intelligent surfaces for energy efficiency in wireless communication[J]. IEEE Transactions on Wireless Communications, 2019, 18(8):4157–4170.
43. ABEYWICKRAMA S, ZHANG R, YUEN C. Intelligent reflecting surface: Practical phase shift model and beamforming optimization[C]. 2020 IEEE International Conference on Communications (ICC) , 2020.
44. XIE H, XU J, LIU Y F. Max-Min fairness in IRS-aided multi-cell MISO systems via joint transmit and reflective beamforming[C]. 2020 IEEE International Conference on Communications (ICC), 2020.
45. GAO Y, YONG C, XIONG Z, et al. Reconfigurable intelligent surface for MISO systems with proportional rate constraints[C]. 2020 IEEE International Conference on Communications (ICC) , 2020.
46. ZHOU G, PAN C, REN H, et al. Intelligent reflecting surface aided multigroup multicast MISO communication systems[J]. IEEE Transactions on Signal Processing, 2019, 68: 3236–3251.
47. LI H, LIU R, LI M, et al. IRS-enhanced wideband MU-MISO-OFDM communication systems[C]. 2020 IEEE Wireless Communications and Networking Conference (WCNC), 2019.
48. H. Han, J. Zhao, D. Niyato, M. D. Renzo and Q. -V. Pham, "Intelligent Reflecting Surface Aided Network: Power Control for Physical-Layer Broadcasting," ICC 2020 - 2020 IEEE International Conference on Communications (ICC), Dublin, Ireland, 2020, pp. 1–7, https://doi.org/10.1109/ICC40277.2020.9148827.
49. WU Q, ZHANG R. Intelligent reflecting surface enhanced wireless network via joint active and passive beamforming[J]. IEEE Transactions on Wireless Communications, 2019, 18(11): 5394–5409.
50. ZHOU G, PAN C, REN H, et al. A framework of robust transmission design for IRS-aided MISO communications with imperfect cascaded channels[J]. IEEE Transactions on Signal Processing, 2020, 68: 5092–5106.

51. M. Shao, Q. Li and W. -K. Ma, "Minimum Symbol-Error Probability Symbol-Level Precoding With Intelligent Reflecting Surface," in IEEE Wireless Communications Letters, vol. 9, no. 10, pp. 1601–1605, Oct. 2020, https://doi.org/10.1109/LWC.2020.2997653.
52. XU Y, GAO Z, WANG Z, et al. RIS-Enhanced WPCNs: Joint Radio Resource Allocation and Passive Beamforming Optimization[J]. IEEE Transactions on Vehicular Technology, 2021, 70(8): 7980–7991.
53. ZHAO D, LU H, WANG Y, et al. Joint Passive Beamforming and User Association Optimization for IRS-assisted mmWave Systems[C]. 2020 IEEE Global Communications Conference, 2020:1–6.
54. MA X, GUO S, ZHANG H, et al. Joint Beamforming and Reflecting Design in Reconfigurable Intelligent Surface-Aided Multi-User Communication Systems[J] IEEE Transactions on Wireless Communications, 2021, 20(5): 3269–3283.
55. Z. Li, M. Hua, Q. Wang and Q. Song, "Weighted Sum-Rate Maximization for Multi-IRS Aided Cooperative Transmission," in IEEE Wireless Communications Letters, vol. 9, no. 10, pp. 1620–1624, Oct. 2020, https://doi.org/10.1109/LWC.2020.2999356.
56. YAN W, HE Z Q, KUAI X. IRS-aided large-scale MIMO systems with passive constant envelope precoding[J]. arXiv, 2019.
57. X. Yu, D. Xu and R. Schober, "Optimal Beamforming for MISO Communications via Intelligent Reflecting Surfaces," 2020 IEEE 21st International Workshop on Signal Processing Advances in Wireless Communications (SPAWC), Atlanta, GA, USA, 2020, pp. 1–5, https://doi.org/10.1109/SPAWC48557.2020.9154337.
58. ZHOU G, PAN C, REN H, et al. Robust Beamforming Design for Intelligent Reflecting Surface Aided MISO Communication Systems[J]. IEEE Wireless Communications Letters, 2020, 9(10): 1658–1662.
59. QIAN X, RDNZO M. D, LIU J, et al. Beamforming Through Reconfigurable Intelligent Surfaces in Single-User MIMO Systems: SNR Distribution and Scaling Laws in the Presence of Channel Fading and Phase Noise[J]. IEEE Wireless Communications Letters, 2021, 10(1): 77–81
60. DI B, ZHANG H, LI L, et al. Practical hybrid beamforming with limited-resolution phase shifters for reconfigurable intelligent surface based multi-user communications[J]. IEEE Access, 2020.
61. B. Di, H. Zhang, L. Song, Y. Li, Z. Han and H. V. Poor, "Hybrid Beamforming for Reconfigurable Intelligent Surface based Multi-User Communications: Achievable Rates With Limited Discrete Phase Shifts," in IEEE Journal on Selected Areas in Communications, vol. 38, no. 8, pp. 1809–1822, Aug. 2020, https://doi.org/10.1109/JSAC.2020.3000813.
62. X. Mu, Y. Liu, L. Guo, J. Lin and N. Al-Dhahir, "Capacity and Optimal Resource Allocation for IRS-Assisted Multi-User Communication Systems," in IEEE Transactions on Communications, vol. 69, no. 6, pp. 3771–3786, June 2021, https://doi.org/10.1109/TCOMM.2021.3062651.
63. GAO J, ZHONG C, CHEN X, et al. Unsupervised Learning for Passive Beamforming[J]. IEEE Communications Letters, 2020, 24(5):1052–1056.
64. SONG H, ZHANG M, GAO J, et al. Unsupervised Learning-Based Joint Active and Passive Beamforming Design for Reconfigurable Intelligent Surfaces Aided Wireless Networks[J]. IEEE Communications Letters, 2021, 25(3): 892–896.
65. HU C, DAI L, HAN S, et al. Two-Timescale Channel Estimation for Reconfigurable Intelligent Surface Aided Wireless Communications[J].IEEE Transactions on Communications, 2021, 69(11): 7736–7747.
66. SHEN D, DAI L. Dimension Reduced Channel Feedback for Reconfigurable Intelligent Surface Aided Wireless Communications[J]. IEEE Transactions on Communications, 2021, 69(11): 7748–7760.
67. WEI X, SHEN D, DAI L. Channel Estimation for RIS Assisted Wireless Communications—Part II: An Improved Solution Based on Double-Structured Sparsity[J]. IEEE Communications Letters, 2021, 25(5): 1403–1407.

68. Jian M, Zhao Y. A modified off-grid SBL channel estimation and transmission strategy for RIS-assisted wireless communication systems[C]//2020 International Wireless Communications and Mobile Computing (IWCMC). IEEE, 2020: 1848–1853.
69. SHI X, WANG J, SONG J. Triple-Structured Sparsity-Based Channel Feedback for RIS-Assisted MU-MIMO System[J]. IEEE Communications Letters, 2022, 26(5): 1141–1145.
70. HE J, WYMEERSCH H, JUNTTI M. Channel Estimation for RIS-Aided mmWave MIMO Systems via Atomic Norm Minimization[J]. IEEE Transactions on Wireless Communications, 2021, 20(9): 5786–5797.
71. SINGH C, SINGH K, LIU K H. Fast Beam Training for RIS-Assisted Uplink Communication[J]. arXiv preprint arXiv:2107.14138, 2021.
72. HE J, WYMEERSCH H, SANGUANPUAK T, et al. Adaptive Beamforming Design for mmWave RIS-Aided Joint Localization and Communication[C]. 2020 IEEE Wireless Communications and Networking Conference Workshops (WCNCW), 2020: 1–6.
73. CHEN W, WEN C K, LI X, et al. Adaptive Bit Partitioning for Reconfigurable Intelligent Surface Assisted FDD Systems With Limited Feedback[J]. IEEE Transactions on Wireless Communications, 2022, 21(4): 2488–2505.
74. TAHA A, ALRABEIAH M, ALKHATEEB A. Enabling Large Intelligent Surfaces With Compressive Sensing and Deep Learning[J]. IEEE Access, 2021, 9: 44304–44321.
75. JIN Y, ZHANG J, ZHANG X, et al. Channel Estimation for Semi-Passive Reconfigurable Intelligent Surfaces With Enhanced Deep Residual Networks[J]. IEEE Transactions on Vehicular Technology, 2021, 70(10): 11083–11088.
76. SCHROEDER R, HE J, JUNTTI M. Passive RIS vs. Hybrid RIS: A Comparative Study on Channel Estimation[C]. 2021 IEEE 93rd Vehicular Technology Conference (VTC2021-Spring), 2021: 1–7.
77. ARDAH K, GHEREKHLOO S, ALMEIDA A L F. D. TRICE: A Channel Estimation Framework for RIS-Aided Millimeter-Wave MIMO Systems[J]. IEEE Signal Processing Letters, 2021, 28: 513–517.
78. LIU H, YUAN X, ZHANG Y J A. Matrix-Calibration-Based Cascaded Channel Estimation for Reconfigurable Intelligent Surface Assisted Multiuser MIMO[J]. IEEE Journal on Selected Areas in Communications, 2020, 38(11): 2621–2636.
79. LIU H, ZHANG J, WU Q, et al. ADMM Based Channel Estimation for RISs Aided Millimeter Wave Communications[J]. IEEE Communications Letters, 2021, 25(9): 2894–2898.
80. MA S, SHEN W, AN J, et al. Wideband Channel Estimation for IRS-Aided Systems in the Face of Beam Squint[J]. IEEE Transactions on Wireless Communications, 2021, 20(10): 6240–6253.
81. LIU Y, ZHANG S, GAO F, et al. Cascaded Channel Estimation for RIS Assisted mmWave MIMO Transmissions[J]. IEEE Wireless Communications Letters, 2021, 10(9): 2065–2069.
82. ZEGRAR S E, AFEEF L, ARSLAN H, Reconfigurable intelligent surface (RIS): Eigenvalue Decomposition-Based Separate Channel Estimation[C]. 2021 IEEE 32nd Annual International Symposium on Personal, Indoor and Mobile Radio Communications (PIMRC), 2021: 1–6.
83. ZHOU Z, GE N, WANG Z, et al. Joint Transmit Precoding and Reconfigurable Intelligent Surface Phase Adjustment: A Decomposition-Aided Channel Estimation Approach[J]. IEEE Transactions on Communications, 2021, 69(2): 1228–1243.
84. ARAÚJO G T D, ALMEIDA A L F D, BOYER R. Channel Estimation for Intelligent Reflecting Surface Assisted MIMO Systems: A Tensor Modeling Approach[J]. IEEE Journal of Selected Topics in Signal Processing, 2021, 15(3): 789–802.
85. MAO Z, PENG M, LIU X. Channel estimation for reconfigurable intelligent surface assisted wireless communication systems in mobility scenarios[J]. China Communications, 2021, 18(3): 29–38.
86. XU M, ZHANG S, MA J, et al. Deep Learning-Based Time-Varying Channel Estimation for RIS Assisted Communication[J]. IEEE Communications Letters, 2022, 26(1): 94–98.
87. SUN S, YAN H. Channel Estimation for Reconfigurable Intelligent Surface-Assisted Wireless Communications Considering Doppler Effect[J]. IEEE Wireless Communications Letters, 2021, 10(4): 790–794.

88. Y. Zhao, "Cascaded Channel Decoupling Based Solution for RIS Regulation Matrix," in Intelligent and Converged Networks, vol. 5, no. 1, pp. 19–27, March 2024, https://doi.org/10.23919/ICN.2024.0002.
89. HE J, WYMEERSCH H, JUNTTI M. Leveraging Location Information for RIS-Aided mmWave MIMO Communications[J]. IEEE Wireless Communications Letters, 2021, 10(7): 1380–1384.
90. KEYKHOSRAVI K, KESKIN M F, GRANADOS G S, et al. SISO RIS-Enabled Joint 3D Downlink Localization and Synchronization[C]. 2021 IEEE International Conference on Communications, 2021: 1–6.
91. HU X, ZHONG C, ZHANG Y, et al. Location Information Aided Multiple Intelligent Reflecting Surface Systems[J]. IEEE Transactions on Communications, 2020. 68(12): 7948–7962.
92. KUNDU N K, MCKAY M R. Channel Estimation for Reconfigurable Intelligent Surface Aided MISO Communications: From LMMSE to Deep Learning Solutions[J]. IEEE Open Journal of the Communications Society, 2021, 2:471–487.
93. SHTAIWI E, ZHANG H, ABDELHADI A, et al. RIS-Assisted mmWave Channel Estimation Using Convolutional Neural Networks[J]. 2021 IEEE Wireless Communications and Networking Conference Workshops (WCNCW), 2021:1–6.
94. ZHANG S, LI M, JIAN M, et al. AIRIS: Artificial intelligence enhanced signal processing in reconfigurable intelligent surface communications[J]. China Communications, 2021, 18(7):158–171.
95. LIN S, ZHENG B, ALEXANDROPOULOS G C, et al. Progressive Channel Estimation and Passive Beamforming for RIS-Assisted OFDM Systems[C]. 2020 IEEE Global Communications Conference, 2020: 01–06.
96. LIU Y, LIU E, WANG R, et al. Channel Estimation and Power Scaling of Reconfigurable Intelligent Surface with Non-Ideal Hardware[C]. 2021 IEEE Wireless Communications and Networking Conference (WCNC), 2021:1–6.
97. PAPAZAFEIROPOULOS A, PAN C, KOURTESSIS P, et al. Intelligent Reflecting Surface-Assisted MU-MISO Systems With Imperfect Hardware: Channel Estimation and Beamforming Design[J]. IEEE Transactions on Wireless Communications, 2022, 21(3): 2077–2092.
98. ZHOU G, PAN C, REN H, et al. A Framework of Robust Transmission Design for IRS-Aided MISO Communications With Imperfect Cascaded Channels[J]. IEEE Transactions on Signal Processing, 2020, 1(68):5092–5106.
99. JUNG M, SAAD W, JANG Y et al. Performance Analysis of Large Intelligent Surfaces (LISs): Asymptotic Data Rate and Channel Hardening Effects[J]. IEEE Transactions on Wireless Communications, 2020, 19(3): 2052–2065.
100. ZHAO Y J, JIAN M N, Applications and challenges of Reconfigurable Intelligent Surface for 6G networks (In Chinese), Radio Communications Technology, 2021, 47(06): 679–691.
101. Zhao, Y. Reconfigurable intelligent surfaces for 6G: applications, challenges, and solutions. Front Inform Technol Electron Eng 24, 1669–1688 (2023). https://doi.org/10.1631/FITEE.2200666
102. ZHAO Yajun. Reconfigurable Intelligent Surface Constructing 6G Near-Field Networks[J]. Mobile Communications, 2024,48(4): 2–11.
103. Jiayi Zhang, Jiying Xiang, Bo Ai, Mengnan Jian, Yajun Zhao. "6G Multiple Antennas and Reconfigurable Intelligent Surfaces," Publishing House of Electronics Industry, 2023.
104. H. Xu, Y. Zhao, L. Mo, C. Huang and B. Sun, "Inter-cell antenna calibration for coherent joint transmission in TDD system," 2012 IEEE Globecom Workshops, Anaheim, CA, USA, 2012, pp. 297–301, https://doi.org/10.1109/GLOCOMW.2012.6477586.
105. ZHAO Y J, LV X. Network Coexistence Analysis of RIS-Assisted Wireless Communications[J]. IEEE Access, 2022, 10: 63442–63454.
106. ZHANG J Y, LIU H, WU Q Q, et al. RIS-aided next-generation high-speed train communications-challenges, solutions, and future directions [J]. IEEE Wireless Communications, 2021, 28(6): 145–151.

107. Zhao Y J, Zhang J Y, Ai B. Applications of Reconfigurable Intelligent Surface in Smart High Speed Train Communications(In Chinese). ZTE Communications, 2021, 27(04): 36–43.
108. Zhao, Yajun. "RIS assisted wireless networks: Collaborative regulation, deployment mode and field testing." *IET Communications* (2024).
109. Ma H B, Zhang P, Yang F et al. Reflections on Development Reconfigurable Intelligent Surface. ZTE Communications, 2022, 28(03): 70–77.
110. RENZO M D, DANUFANE F H, XI X, et al. Analytical modeling of the path-loss for reconfigurable intelligent surfaces - Anomalous mirror or scatterer?[J]. IEEE International Workshop on Signal Processing Advances in Wireless Communications (SPAWC), 2020: 1–5.

Chapter 9
AI-Enabled Air Interface

In recent years, artificial intelligence (AI) technology has developed rapidly, due to the motivation of the rapid development of deep learning algorithms, the explosive growth of data volume and data sources, as well as the continuous improvement of hardware computing power. Breakthroughs has been made in many fields such as natural language processing, speech recognition, and image recognition, etc. The development and applications of AI extend far beyond the computer industry. It has also been widely used in other fields such as finance, medical, transportation, education, industrial manufacturing, resulting in revolutionary changes and driving the ongoing evolution and development of these industries.

In the research of wireless communication, many earlier academic studies have applied neural networks, such as introducing neural networks in the algorithm design of mobility management and wireless resource allocation. However, the application of AI in 4G or earlier commercial wireless communication systems is not common. With the maturity of 5G and corresponding evolution versions as well as the rapid development of potential 6G technologies, it can be foreseen that physical layer related technologies will undoubtedly become more and more complex, and the corresponding algorithms running on the complex hardware systems are also required to adapt to these changes. Meanwhile, the potentials of AI have been continuously explored in recent years, and many AI technologies such as machine learning, deep learning, and reinforcement learning has been proposed. The advantages of utilizing AI technologies in dealing with complex, difficult modeling and solving problems have been widely recognized. Thus, both the academic and industrial communities have gradually reached a consensus on using AI to solve complex physical layer design problems in the future wireless communications, and AI technologies are also frequently used in various air interface technology development processes.

This chapter will focus on explaining typical AI-enabled air interface technologies and introducing the latest research progress.

G. Liu et al., *6G Migrates from Communication to XaaS*,
https://doi.org/10.1007/978-981-96-7579-1_9

9.1 AI-Enabled Air Interface Technology and Standardization Research

Traditional air interface physical layer design is typically optimized by decomposing the system to several modules according to different functions and optimizing different modules separately to ensure each module to be optimal under specific conditions. Such method cannot guarantee the optimality under complex conditions, let alone the global optimality of the whole system. However, AI provides us with a new design methodology for air interface physical layer technologies.

9.1.1 AI Enabled Physical Layer Technology

According to recent years' research, it is shown that the integration of AI and physical layer air interface technology can be reflected at different levels. First of all, applying AI to optimize or even directly replace of a certain physical layer function module can be considered as the first level of basic AI-enabled air interface technology. For example, some scenarios such as interference detection, uplink/downlink channel heterogeneity, channel prediction, etc., are difficult to handle since non-linearity of the problem hinders the accurate modeling. However, based on AI technologies, the problem design can be solved by replacing a function module with a deep neural network. The second level of AI-enabled air interface technology applies AI to jointly optimize multiple physical layer modules, such as coding, modulation and waveform. In traditional air interface designs, these modules are generally optimized separately in traditional air interfaces designs due to the high complexity of joint optimization process at the receiver. However, with the aid of AI technologies, it is possible to jointly optimize several physical layer modules, thereby eliminating the independent optimization of each module through data-driven machine learning. The top level of AI-enabled air interface is to achieve end-to-end system designs by utilizing the AI technologies. By adopting this end-to-end system design method, the transmitters and receivers can even be obtained through real gradient and approximate gradient iteration under the scenarios with the unknown channel models. Moreover, since the physical layer functions in the end-to-end design can be implemented using the same hardware architecture and accelerator, the system efficiency can be improved and the future expansion and deployment are also simplified.

For each AI-enabled air interface technology study, researchers often need to first find specific problems which are needed to be optimized and also are suitable for AI to solve, determine whether the AI methods can achieve higher performance than traditional algorithms, then further identify under which scenarios and conditions the advantages can be achieved. After confirming the performance advantages brought by AI, it is also necessary to determine the cost of obtaining these advantages, e.g., the availability of data, the complexity of AI implementation (i.e., training and inference complexity), the difficulty of AI implementation on hardware, etc.

Further, balancing performance and cost, the corresponding AI-enabled air interface technology is designed and implemented. In the following parts of this chapter, we will detail the research progress and results in the three typical fields, i.e., AI-enabled channel estimation, AI-enabled channel state information feedback, and AI-enabled millimeter wave beam management based on the trade-offs of performance and cost.

9.1.2 AI Enabled Air Interface Technology Standardization Evolution

It is worth noting that the vast majority of AI-enabled air interface technology research focuses on the implementation optimization, which means that the UE or base station side is implemented through the AI-based algorithm design to achieve performance optimization, energy saving effects, etc. Further, to better utilize AI-enabled air interface technology, the wireless communication standardization research work has also gradually gained attention from the industry. In the 5G standardization process of the International Standardization Organization 3GPP, the 3GPP SA2 working group introduced a new network function—Network Data Analytics Function (NWDAF) in Rel-15 to analyze and process network data for related intelligent applications, and the function was continuously enhanced in subsequent versions. On the access network side, the 3GPP RAN3 working group studied the improved data collection related content in Rel-16 and Rel-17. By analyzing the relevant use cases and functional architecture of intelligent wireless access networks, they investigated the impact of introducing AI technologies on the current NG-RAN network nodes and interfaces [1, 2]. On the other hand, the 3GPP's work of improving the physical layer air interface using AI technologies began with the research topic of Rel-18 in 2022 [3].

Specifically, by comparing with the traditional algorithms, the AI or machine learning (ML) based NR air interface research in 3GPP Rel-18 aims to develop the AI or ML based physical layer air interface algorithms to achieve performance improvement, complexity or cost reduction. After discussions, three typical AI-enabled air interface technologies will be studied as use cases firstly.

(1) Enhanced Channel State Information (CSI) feedback, such as cost reduction, accuracy enhancement, and CSI prediction.
(2) Beam Management [4], for example, beam prediction is carried out in the time and spatial domain to reduce cost and delay and enhance the accuracy of beam selection.
(3) Enhanced positioning accuracy in specific scenarios, including the scenarios with a large number of Non-Line Of Sight (NLOS) links.

Referring to the above three typical use cases, 3GPP aims to come up with a widely applicable framework for future AI/ML research. The framework is expected to support the AI/ML model analysis at different stages, such as the generation stage of AI/ML models (including model training, validation, and testing), the network structure, implementation details (input/output, pre/post-processing procedures, etc.) and the complexity analysis in the inference stage of AI/ML models. In addition, the level of AI cooperation between UE and gNB is also an important part of this research framework, where AI cooperation level is determined by analyzing the AI cooperation required by a certain AI-enabled air interface technology. Besides, the studies of the lifecycle management of AI/ML models (including model generation and training, model deployment, model prediction, model supervision and model update), large datasets, the unification of corresponding terms and symbols will create solid foundation for emerging AI-enabled air interface technologies in future B5G and 6G.

9.2 AI-Enabled Channel Estimation

Channel estimation is one of the earliest research areas that has been widely investigated by the academic community among AI-based communication physical layer technology topics. In communication systems, the Demodulation Reference Signal (DM-RS) is used for channel estimation at the receiver in either uplink or downlink transmission stage before the demodulation of data channel, control channel or broadcast channel information. In the NR communication system, the transmitter will transmit pilot symbols using the pre-agreed wireless time–frequency resources for the sake of receiver channel estimation. Generally, the transmitter will modulate these DM-RS pilot signals on a small number of Resource Elements (RE) in the wireless time–frequency resource in order to calculate the wireless channel for the entire time–frequency resource, i.e., the impulse response of this wireless channel. The whole process is called the channel estimation problem, which is also one of the most basic problems in wireless communication systems.

9.2.1 Traditional Channel Estimation

The classic channel estimation algorithm in the NR wireless communication system is the non-blind channel estimation performed through the known DM-RS pilots at the transmitter and receiver. In practice, the typical DM-RS configuration on the wireless Resource Block (RB) is shown in Fig. 9.1a, where the grey resource units carry the DM-RS pilot signal (the position of the NR system pilot point in the wireless resource block is determined by 3GPP standard which includes multiple configuration modes, controlling the pilot density in the time domain and frequency domain [5]). Before the demodulation of the received data, the receiver needs to estimate the channels of the entire wireless resource block based on the received DM-RS at a few positions. The channel estimation over the entire wireless resource based on the received DM-RS is shown in Fig. 9.1b.

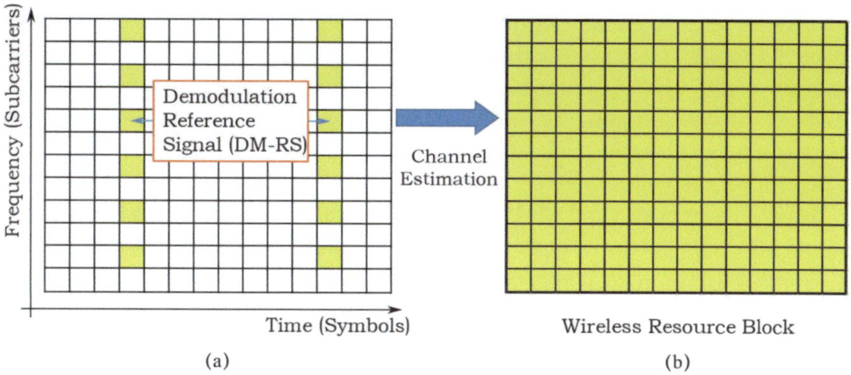

Fig. 9.1 Pilot-based channel estimation in NR system

For the given transmitted signal $\boldsymbol{X}$, the received signal at the receiver side $\boldsymbol{y}$, is generally denoted as

$$\boldsymbol{y} = \boldsymbol{X}\boldsymbol{h} + \boldsymbol{z} \tag{9.1}$$

where $\boldsymbol{h}$ is the wireless channel to be estimated, z is the additive Gaussian white noise at the receiver, whose elements follow the complex Gaussian distribution $CN\left(0, \sigma_z^2\right)$.

For traditional wireless communication algorithms, the channel estimation is generally performed in two steps.

In the first step, the channel at the location of the DM-RS is estimated based on the received pilot signal. The estimation can be implemented using either least squares (LS) method or minimum mean square error (MMSE) method [6]. For the LS method, the estimation criterion is to optimize the estimated channel $\hat{h}$ so that the error term $\|\boldsymbol{y} - \boldsymbol{X}\hat{\boldsymbol{h}}^2\|$ is minimized. Based on the criterion, the estimated channel for the LS method is derived as

$$\hat{\boldsymbol{h}}_{\mathrm{LS}} = \left(\boldsymbol{X}\boldsymbol{X}^H\right)^{-1}\boldsymbol{y}\boldsymbol{X}^H \tag{9.2}$$

The biggest advantage of the LS method is that the channel estimation can be directly implemented based on the received signal and the known transmitted signal. The calculation involved in the estimation is simple and no additional information is required. The other MMSE estimation method is more accurate, whose criterion is to minimize the term $\|E\boldsymbol{y} - \boldsymbol{X}\hat{\boldsymbol{h}}^2\|$ instead. Based on this criterion, the estimated channel is obtained as

$$\hat{\boldsymbol{h}}_{\mathrm{MMSE}} = \boldsymbol{R}_h\left(\boldsymbol{X}^H\boldsymbol{X}\boldsymbol{H}\boldsymbol{X}\boldsymbol{R}_h\boldsymbol{R}_h + \sigma_z^2\boldsymbol{I}\right)^{-1}\boldsymbol{X}^H\boldsymbol{y} \tag{9.3}$$

where $\boldsymbol{R}_h = \mathrm{E}\boldsymbol{h}\boldsymbol{h}^H$ represents the second-order statistics of the real channel. In fact, this process is to minimize the impact of random noise on channel estimation, so it can also be seen as a denoising process. It can be proven that the error of MMSE channel estimation is always less than that of LS channel estimation. Although the accuracy of the LS channel estimation is lower, this method is generally used in practical receivers because the MMSE method requires accurate channel statistical information as well as the power of noise signal, and the matrix inversion calculation in the MMSE method is more complex.

Besides, there is also an LMMSE channel estimation method which lies between LS and MMSE methods. It can be regarded as a simplified version of MMSE method by eliminating the high complexity inversion operation [7]. The LMMSE algorithm replaces the matrix $\boldsymbol{X}^H\boldsymbol{X}$ with the statistical characteristics of the transmitted symbols $\mathrm{E}(\boldsymbol{X}^H\boldsymbol{X})$ during the derivative calculation stage. Thus, the estimated channel is simplified to

$$\hat{\boldsymbol{h}}_{\mathrm{LMMSE}} = \boldsymbol{R_h}\left(\boldsymbol{R_h} + \frac{\beta}{SNR}\boldsymbol{I}\right)^{-1}\hat{\boldsymbol{h}}_{\mathrm{LS}} \tag{9.4}$$

where β is a constant related only to the modulation method of the transmitted symbol (e.g., $\beta_{QPSK} = 1$, $\beta_{16QAM} = 1.8889$, $\beta_{64QAM} = 2.6857$). Despite this, it is still very difficult to know the exact channel statistics and signal-to-noise ratio (SNR) in practice.

The second step is to infer the channel over the entire wireless resource block based on the channel information obtained at the DM-RS points. Different interpolation algorithms are used to recover the channel information depending on the specific situation. Thus, this step can also be regarded as an interpolation process. There are many interpolation algorithms and they often estimate the received SNR first. If the time–frequency offset of the received signal is considered, a series of operations such as time offset compensation, frequency offset compensation, smoothing, and reverse compensation are needed in the interpolation process to assist the channel estimation at the receivers. Generally, different receiving algorithms has different processing procedures, and the detailed procedures will not be elaborated here. In summary, the channel estimation process in the cellular wireless communication system can be divided into two main parts: denoising and interpolation, which are shown in Fig. 9.2.

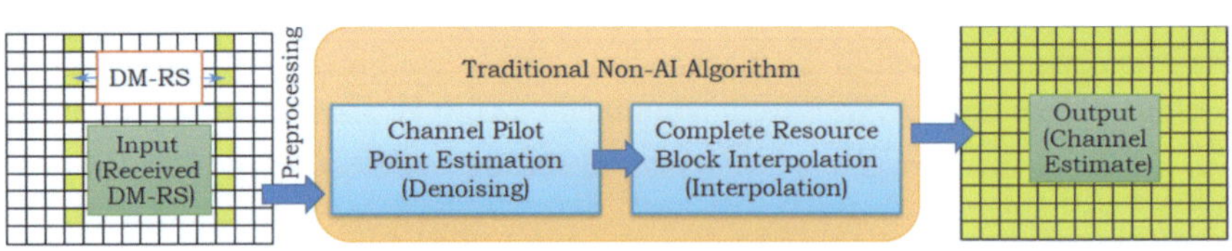

Fig. 9.2 Channel estimation process in wireless communication system

9.2.2 Algorithm Framework

After the emerging of deep neural networks, a large number of deep learning models and algorithms have been applied to image processing, including image denoising, generation, and recovery, etc. The convolutional neural network (CNN) based deep learning models, which is identified by local features focusing and weight sharing, are often used to process the image-like serialized data.

The super-resolution problem has been widely investigated in the image field for a long time, which aims to restore low-resolution images to the images with higher resolution using algorithms [8]. This kind of problem shares a lot common points with wireless channel estimation. Both of these two problems need to recover or generate detailed information based on limited prior knowledge. Generally, the signal processing technique cannot increase the amount of information. Thus, the machine learning algorithms become one of the most suitable processing methods. In addition, supervised machine learning algorithms are easy to design as end-to-end structures to achieve specific tasks, which can further reduce the cumulative errors introduced by the traditional cascaded algorithms.

To use supervised learning algorithms, the input information, output information and corresponding model of the algorithm should be determined, then the model will go through the training and the prediction stages. The input information is usually known data and its features, while the output data generally depends on the task to be performed and it should contain real output data as the label for model training. For the channel estimation task, the machine learning models can be devised as an end-to-end deep learning model, and the channel (channel impulse response) over the entire wireless resource is estimated based on the pilot signal at a few points.

According to the above description, we will first introduce the input information, output information and corresponding model for channel estimation task. Although the task is implemented using machine learning algorithms, the known information is same with that in the traditional designs. Therefore, the input information of the machine learning algorithm is also the received signal $\boldsymbol{y}$ at the DM-RS position and the pilot signal $\boldsymbol{X}$ transmitted by the transmitter. For a certain received symbol and transmitted symbol at a pilot point, the input signal (input feature) of channel estimation can be the channel estimation value based on the LS criterion at each pilot point $\hat{\boldsymbol{H}}(f_0, t_0) = \boldsymbol{y}(f_0, t_0)/\boldsymbol{x}(f_0, t_0)$, i.e., the ratio of the received and transmitted signal vectors at the DM-RS point (f_0, t_0), where f_0 and t_0 represent the time and frequency indicators to be estimated which are generally subcarrier sequence number and OFDM symbol number, respectively. Using the LS estimate as input information, the differences brought by different pilot can be ignored. In fact, the input data includes the information of the actual channel model at this point and the noise information.

For the channel estimation algorithm, since its goal is to recover the channel over the entire wireless resource, the model should output the channel estimation result over the entire time–frequency wireless resource $\hat{\boldsymbol{H}}(f, t)$. Considering the timeliness required by tasks such as demodulation, the scale of estimation in the time domain will not be too long, which is usually a slot. Whereas the estimation in the frequency domain mainly depends on the bandwidth of the channel.

During the training phase of learning, in order to accurately restore the complete channel, the model output should be the entire, real channel impulse response. Therefore, the labeled data during training should be the impulse response of the real channel over the entire resource block given the pilot signal point estimate input. This can be seen as a classic supervised learning regression task in machine learning, in which the output data is fitted according to the input data to minimize the difference between the fitted data and the real data. A typical set of input and output data visualization is shown in Fig. 9.3, where the wireless resource block is composed of 96 subcarriers and 14 OFDM symbols, and the pilot symbol is configured on every two subcarriers of the 4th and 12th symbols.

For regression tasks, it is important to measure the difference between the output and the ideal output during algorithm design. Typically, minimal Normalized Mean Square Error (NMSE) criterion is an important standard for measuring the differences between two different channel impulse responses. It can also be used to determine the accuracy of channel estimation in channel estimation problems.

$$NMSE = \mathrm{E}\left\{\frac{\|\hat{\boldsymbol{H}} - \boldsymbol{H}\|_2^2}{\|\boldsymbol{H}\|_2^2}\right\} = \mathrm{E}\left\{\frac{\mathrm{MSE}(\hat{\boldsymbol{H}}, \boldsymbol{H})}{\|\boldsymbol{H}\|_2^2}\right\} \tag{9.5}$$

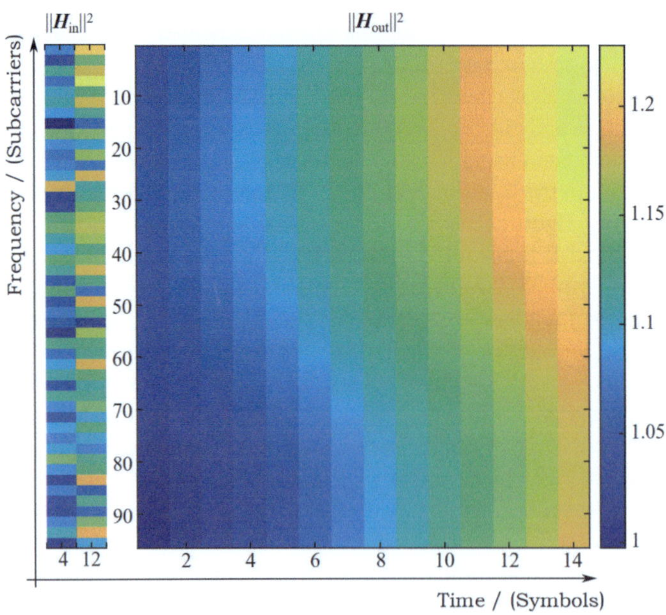

Fig. 9.3 Input $\boldsymbol{H}_{\text{in}}$ and output $\boldsymbol{H}_{\text{out}}$ visualization

Taking Fig. 9.3 as an example, NMSE can be written as (measuring NMSE in dB):

$$NMSE = 10lg\left(\frac{\sum_{t=0}^{N_{symb}^{slot}-1}\sum_{f=0}^{N_{RB}\cdot N_{subc}^{RB}-1}\|\boldsymbol{H}_{out}(f,t)-\boldsymbol{H}_{ideal}(f,t)\|^2}{\sum_{t=0}^{N_{symb}^{slot}-1}\sum_{f=0}^{N_{RB}\cdot N_{subc}^{RB}-1}\|\boldsymbol{H}_{ideal}(f,t)\|^2}\right) \tag{9.6}$$

where N_{symb}^{slot} represents 14 symbols in the time domain; $N_{RB} \cdot N_{subc}^{RB}$ represents 96 subcarriers in the frequency domain (8 RBs each of which contains 12 subcarriers); $\boldsymbol{H}_{out}$ is the output channel estimate; $\boldsymbol{H}_{ideal}$ is the actual channel.

Once the inputs and outputs of the learning algorithm are determined, an appropriate learning model can be chosen to complete this task. Although many learning models perform well on regression tasks in supervised learning problems, deep neural network models undoubtedly have the best performance.

9.2.3 Model Selection

Since the channel estimation problem has many similarities with image problems such as super-resolution problem, deep learning network models that can be used to handle image problems are introduced to solve channel estimation problem. For example, in [9], the authors cascaded the image super-resolution network SRCNN and the image recovery network DnCNN to implement pilot-based wireless channel estimation, which is shown in Fig. 9.4. The functions of the two different CNN are interpolation and denoising, respectively.

Although the features of image data and channel data in time–frequency domain have many similarities, there are still differences. Natural images usually have high local correlation, while for the channel responses, metadata between rows or columns are often highly correlated and the correlations between rows and columns are independent. In fact, these correlations depend on the time correlation (Doppler spread characteristics) and the frequency correlation (delay spread characteristics) of channels. Thus, it is different from the image data. Based on this, a carefully designed

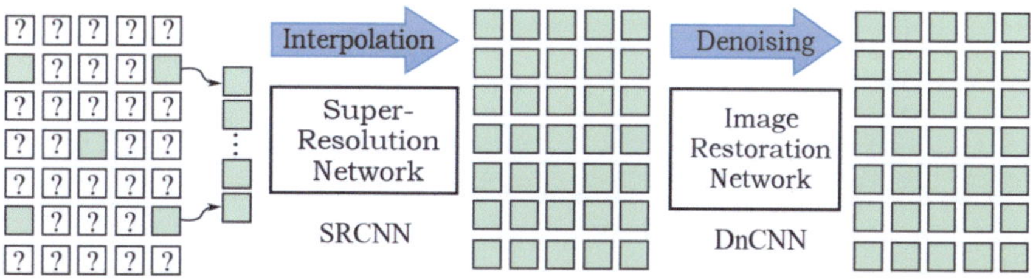

Fig. 9.4 Channel estimation using classic CNN structure

neural network made up of a large number of fully connected layers is more suitable for the feature extraction in time–frequency domain channel.

According to the Mixer network proposed by Google in 2021 [10], a fully connected layers based deep learning algorithm can achieve better NMSE performance for channel estimation task. The Mixer network is composed of stacked fully connected layers (also known as multilayer perceptrons, MLP) and operators for data scaling and dimension transforming. It can achieve comparable performance to CNN when handling image tasks. However, benefitting from the small computational scale of the fully connected layer, the network can be directly implemented through matrix multiplication, and thus under the same performance, it has higher training and execution efficiency.

The structure of the Mixer network makes it suitable for three-dimensional (height, width, channel) serialized data input such as images. It will perform fully connected layer operations on row and column dimensions respectively. The Mixer layer operation on the image is essentially the same as CNN, which can also be regarded as a local connection method. The CNN has local and spatial invariant connection feature, while the Mixer layer operation is to perform local connections on the data of a certain dimension separately. As is shown in Fig. 9.5, the network first performs a fully connected operation (inner product operation) for the data of a certain dimension, and then performs scale and dimension transformation (i.e., row and column transpose).

Therefore, the Mixer network can be understood as data features extracting method by performing correlation operations on single-dimensional data, rather than extracting the correlation between different dimensions directly. This connection method is more suitable for channel data with relatively low direct correlation between rows and columns.

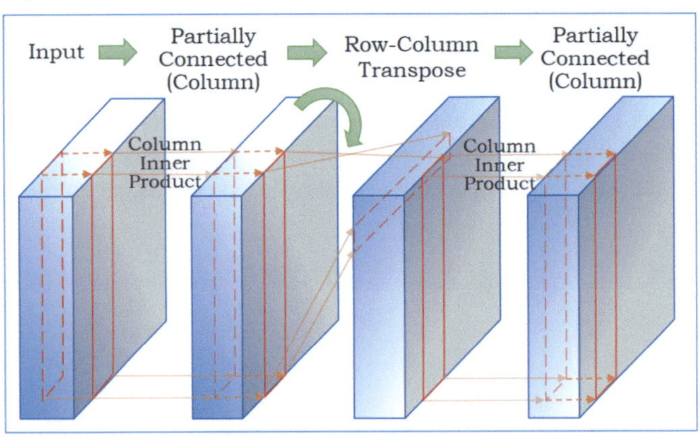

Fig. 9.5 Neuron connection method in Mixer layer operation

9.2.4 Algorithm Design

Based on the Mixer network structure and the aforementioned analysis, in the following part, we will describe a typical deep learning channel estimation algorithm design.

The input data for the deep learning model of the channel estimation task is the baseband channel estimation values based on the LS criterion which are related to the wireless time–frequency resource block composed of 96 subcarriers and 14 OFDM symbols. The input data is organized in a $48 \times 2 \times 2$ input data matrix form and is related to a total of 48×2 pilot DM-RS symbols. The last dimension of the input matrix is the real and imaginary parts of the channel estimation values, i.e., a total of 192 real numbers. The model's output is a $96 \times 14 \times 2$ data matrix, representing the entire channel estimation results. During training, the data labels are the real complete channel matrix, with the same dimensions of $96 \times 14 \times 2$. The loss function for training is the aforementioned NMSE function, measuring the difference between the estimate and the real channel. The network is composed of a stack of multiple Mixer layers, which is shown in Fig. 9.6.

By using the Mixer network structure, this network gradually expands multiple frequency point input data to the entire time–frequency resource blocks. Further, based on the previous discussion, some simplifications and improvements can be made to the overall deep learning model. In some cases, the SNR of the receiver can be obtained during data inputting, so the SNR can also be input into the network as an additional data feature. This method improves the performance, but the SNR is not a necessary input feature. Furthermore, the SNR, as a prior knowledge, can also

Fig. 9.6 Network structure

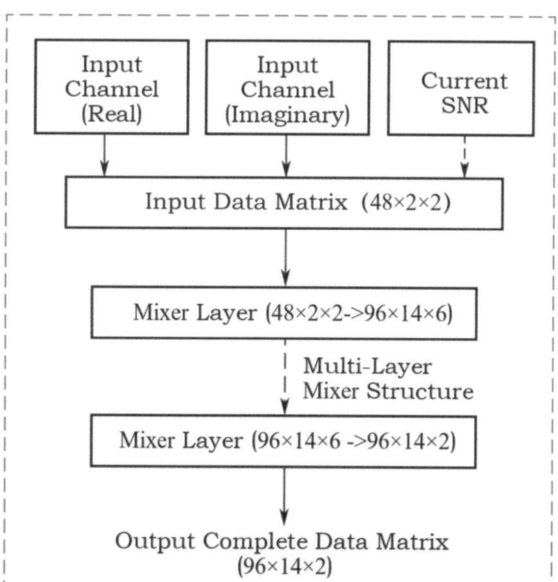

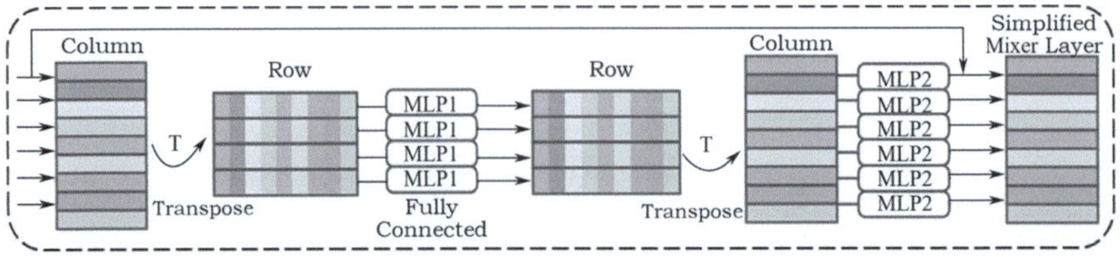

Fig. 9.7 Diagram of the simplified Mixer layer operation

be added to the loss function for further algorithm optimization. At the same time, the Mixer layer operation can also be simplified to reduce the number of computations and further reduce the computational complexity. The simplified Mixer layer operation diagram is shown in Fig. 9.7.

In addition, more optimizations can be made in the network structure design and training process according to the characteristics of the problem, which are not detailed here.

9.2.5 *Performance Analysis and Discussion*

The simulation evaluation dataset is a public dataset obtained through link-level simulation. The simulation configuration parameters are listed in Table 9.1. This dataset has been used for the channel estimation track of the second international wireless AI competition [11].

It is shown that compared to other algorithms, the trained Mixer deep neural network model achieves significantly superior channel estimation performance on data not involved in the training. Figure 9.8 shows the performance comparison of different channel estimation algorithms under the same pilot overhead conditions. For the benchmarks, the CNN-based neural network can refer to the ChannelNet proposed in [9]; the ideal LMMSE based algorithm uses the ideal statistics of the channel and noise for estimation, which can refer to Sect. 9.2.1 for the detailed LMMSE channel estimation description.

Table 9.1 Simulation configuration parameters

Parameters	Values	Parameters	Values
Channel model	Mixed CDL model (TS38.901 CDL-A/B/C/D/E)	UE velocity	0–60 km/h
Antenna configuration	32 Tx, 1 RX	Time selay spread	0–300 ns
Carrier frequency	3.5 GHz	Bandwidth	96 subcarriers
SNR	0–20 dB (1 dB interval)	Dataset size	210 K (10 K per SNR level)

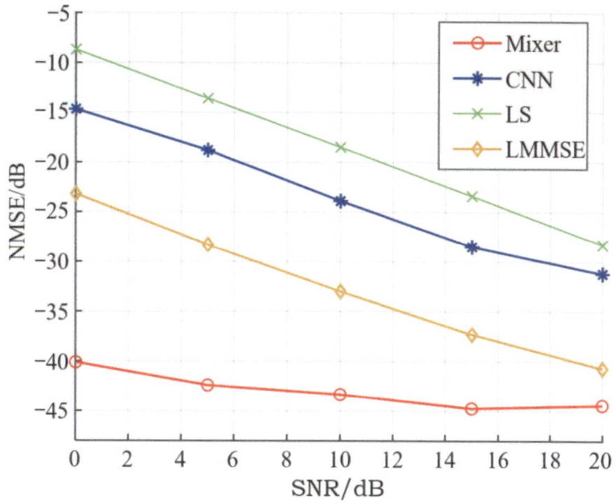

Fig. 9.8 Performance comparison of different channel estimation algorithms under the same pilot overhead conditions

From the experiment results, it can be seen that, overall, the increase of the SNR will influence the performance of channel estimation. This is because the negative influence of noise is removed during the estimation of the pilot points, or it can be explained that the negative impact of the noise estimation error on the overall channel estimation result has reduced. However, the overall NMSE of channel estimation mainly depends on the quality of the interpolation algorithm of the majority of non-pilot points. As can be seen from Fig. 9.8, under low SNR (0 dB) conditions, the Mixer deep neural network algorithm significantly outperforms other traditional algorithms and CNN-based deep learning algorithms, which demonstrates its significant advantage in denoising performance. While under high SNR (20 dB) conditions, the denoising performance of the ideal LMMSE algorithm is closer to the Mixer deep neural network algorithm due to the reduced negative impact of noise, but there is still a performance gap.

As the SNR continues to increase, the slopes of the LS and ideal LMMSE curves almost remain the same, while the slope of the Mixer curve gradually flattens. The performance of Mixer algorithm at 15–20 dB is almost the same. This indicates that when the SNR reaches around 15 dB, the channel estimation performance of the Mixer deep neural network is almost not limited by the noise any more. Further, when the SNR is 15 dB or higher, the algorithm has exerted all its denoising performance, and increasing the SNR cannot further improve the channel estimation performance. In fact, the slopes of the performance curves of traditional linear algorithms are all very close to − 1, which means that the denoising performance is completely proportional to the SNR of the signal during channel estimation, while the performance loss caused by interpolation is constant and less affected by the SNR. However, neural network-based algorithms consider both interpolation and denoising issues, which is equivalent to jointly optimizing the two processes and achieving a global optimal solution, rather than two independent optimal solutions for separate processes. Thus, it can achieve a significant performance improvement.

Although the high computational complexity of deep learning algorithms has been widely criticized, the conclusion is drawn from the image processing deep networks which are generally composed multi-layer CNN stacking. Thus, the corresponding computational complexity is significantly higher than that of the traditional algorithms. However, due to the limited size of input data and obvious data features, the computational complexity of a general channel estimation network is generally 2–3 orders of magnitude lower than that of common image deep learning networks [12]. Thus, the Mixer network structure also has a clear advantage in computational complexity compared to the CNN and other network structures. Therefore, whether on existing network equipment or mobile terminal devices, the real-time requirement for neural network-based channel estimation can be guaranteed.

The robustness of deep learning algorithms largely depends on the coverage of training samples. The network algorithm performance discussed in this section is based on a certain DM-RS pilot configuration, and the practical network pilot configurations, such as location and density, have many possibilities. Therefore, it may be necessary to design multiple neural networks to cope with different pilot configurations, or to train a unified network with more input information to adapt to channel estimation under different pilot configurations.

The channel estimation problem described in this section is one of the most basic problems that the cellular wireless communication receiver must deal with. Through end-to-end design, the deep neural networks-based AI channel estimation algorithm combines the denoising and interpolation steps of traditional channel estimation method and achieves significantly better channel estimation performance than traditional LS-based and other algorithms. Among them, the Mixer network structure extracts features in the time and frequency domains of the channel separately and obtains significantly better estimation performance than other network models. Deep learning channel estimation algorithms usually use the LS estimate on the pilot points as the network input, and output the estimation results of the entire time–frequency domain channel. During the training stage, the data labels are given by the obtained entire time–frequency domain real channel response. This section uses the public channel estimation dataset in the second international wireless AI competition, and compares various different algorithms. The simulation results show that the Mixer-based deep learning algorithm significantly outperforms other algorithms in terms of NMSE performance metric for channel estimation task. Therefore, the deep learning-based channel estimation algorithms have great application potential in future receiver channel estimation designs. However, the timeliness and robustness of the deep learning-based algorithm still need to be investigated in practical usage.

9.3 AI-Enabled Channel State Information Feedback

In MIMO systems, obtaining accurate CSI is crucial. The CSI currently defined by 3GPP includes Rank Indicator (RI), Channel Quality Indicator (CQI), and Precoding Matrix Indicator (PMI). As the number of antennas increases, the overhead of CSI feedback also increases, especially the PMI which represents channel matrix information. How to use AI technology to efficiently feedback CSI is an important research topic.

9.3.1 Terminal Channel State Information Feedback

1. Traditional CSI Feedback

Traditional CSI feedback is based on codebooks that is a set of pre-defined precoding matrix candidates. Based on the channel estimation, the UE evaluates the transmission performance, selects the precoding matrix that can bring the highest performance, and feeds back its codebook index to the BS. The core of CSI feedback mechanism is information extraction and codebook design. 3GPP has currently standardized Type 1 and Type 2 codebooks, which are designed under the assumption that the angles of arrival and departure of signals obey uniform distribution [13]. However, in the practical environment, the angles of arrival and departure may not be uniformly distributed and the statistical distributions are different for different BSs. This characteristic makes it possible for the future optimizations. For instance, for some angle ranges with high probability, the codebook should be designed more densely, and vice versa. In addition, Type 1/2 codebooks have different performance and usage ranges. Type 1 codebook is simpler but less accurate, which is designed for single-user transmission. Type 2 codebook is more accurate, which can be used for multi-user transmission, but its overhead is too large.

2. AI-based CSI Compression Feedback

The AI-based CSI feedback architecture is shown in Fig. 9.9. The AI-based CSI feedback is much different from traditional feedback mechanisms. After channel estimation, the UE adopts an AI model to extract and compress the features of the input channel matrix $\boldsymbol{h}$, then map the features to a floating-point feature vector z. Afterwards, the UE will feedback the quantized bit stream $\hat{z}$ to the BS. Similarly, the BS first dequantizes $\hat{z}$ into a floating-point vector z', then uses an AI model to restore the channel matrix $\boldsymbol{h}'$ from z' as accurate as possible. In general, the AI model on the UE side is referred to as Encoder, and the AI model at the BS side is known as Decoder. This architecture corresponds to the autoencoder model framework in the deep learning field. The optimization goal of training is to make $\boldsymbol{h}$ and $\boldsymbol{h}'$ as close as possible.

The introduction of AI technology can optimize for a specific base station or channel environment. In complex real-world scenarios, the BS can obtain more complete channel information and design more suitable precoding matrix for transmission. In addition, it can train different AI models according to different scenarios and requirements to achieve CSI feedback with arbitrary number of bits and arbitrary precision requirements.

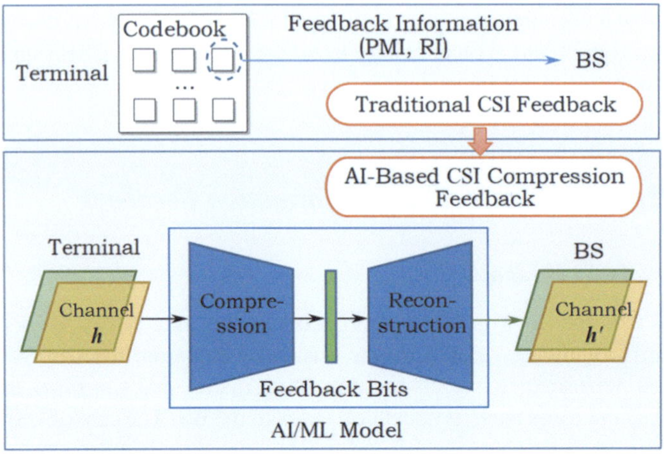

Fig. 9.9 AI-based CSI feedback architecture

9.3.2 Evaluation Metrics

For CSI feedback, the overall design expectation includes: smaller feedback overhead, more accurate the information that the BS can get, lower computational complexity. Therefore, AI-based CSI feedback has the following evaluation metrics.

1. Compression ratio

This can be directly assessed by the number of bits of CSI feedback by the air interface.

2. Feedback performance

This part has two levels of evaluation.

(1) CSI reconstruction accuracy: Evaluate the performance of the AI model itself and measure the difference between $\boldsymbol{h}$ and $\boldsymbol{h}'$ on the testing dataset. The optional metrics include MSE, NMSE. Besides, if the transmit precoding feedback compression is used, cosine similarity can be adopted for measurement.
(2) System gain: Evaluate the performance gain brought by the AI-based CSI feedback scheme in communication systems. The current practice is to deploy the AI model on the simulation platform and compare block error rate (BLER), throughput, and other performance metric with traditional CSI feedback methods through link-level or system-level simulation.

3. Complexity

The complexity of the AI model is generally evaluated by the number of floating-point operations (FLOPs) and the size of the model. Since the CSI feedback have stringent real-time requirements, the complexity of the AI model needs to be limited within a feasible range. on the other hand, the computing power of the chip is generally represented by the number of FLOPS. Thus, the processing delay can be estimated by setting a reference computing power. Especially, the computing power on the UE side will directly limit the network scale of the Encoder.

9.3.3 Channel Information for Feedback

In wireless communication systems, CSI feedback data includes various forms and categories. The AI-based CSI feedback can also have various scenarios, including feedback of the complete time–frequency domain channel, feedback of MIMO transmit precoding, etc.

Currently, there are two types of data used for AI-based CSI feedback method [14, 15].

(1) Full channel information feedback data

Full channel information is usually fed back in the form of power delay spectrum on different antenna pairs, which includes complete power delay information corresponding to several subcarriers/clusters and transmit/receive antenna sets. It retains the complete channel data information.

(2) Channel characteristic information feedback data

Channel characteristic information is derived from the singular value decomposition (SVD) of the full channel information, which can be used for precoding at the transmitter. Compared with full channel information, the SVD-based characteristic information aims to provide reference for transmit precoding at the BS in MIMO channels. The characteristic information is compressed, fed back, received, and reconstructed during the feedback process. The dimension of feedback transmit precoding is determined by the precoder of each transmit antenna after channel decomposition for different sub-carriers.

For example, after receiving the reference signal, the terminal obtains the full channel information through channel estimation, then obtains the channel characteristic information using SVD, and feeds back the information to the BS in a codebook-based or AI compression-based manner. Therefore, the channel characteristic information-based compression and feedback are more suitable for the current 3GPP standards and can be directly compared with the Type 1 and Type 2 feedback mechanisms in 3GPP standard.

The CSI channel data used for AI model training can come from simulation or field sampling. Currently, the channel data used for training in the industry-related research is mainly generated from channel model simulation in 3GPP 38.901 [16].

(3) Link-level channel

The typical channels include CDL-A, CDL-B, CDL-C, CDL-D, CDL-E, etc. Each channel has its own channel characteristics in terms of LOS/NLOS, transmission/arrival angle, multipath, delay spread, and other influencing factors.

Relying on the link-level channels, the model design scheme for the CSI feedback problem can be initially verified, and the comparison of performance gain can be analyzed. Since the channel scenario corresponding to the link-level channel is relatively single, there are some shortcomings in the model generalization. Thus, the model can be trained through a mixed process method by combining multiple channels to construct the training dataset.

(4) System-level channel

The typical channels include various types of channels in Urban Macro (UMa), Urban Micro (UMi), Indoor Hotspot (InH), Rural Macro (RMa), and other common communication scenarios. Compared with link-level channels, the system-level channels consider more influential factors and scenario characteristics. They can accurately describe the diversity feature of practical wireless channels, and is able to generate datasets for further model performance and generalization capability verifications.

9.3.4 Algorithm Design

In summary, AI-based CSI feedback can be described as a procedure that a terminal obtains the compressed CSI information through an AI model, quantized it into a bit stream, and fed back to the BS through the air interface. Then, the BS also uses an AI model to recover the CSI, and obtains the reconstructed CSI for data transmission at the transmitter. During this process, the channel information will undergo preprocessing, compression, quantization, and finally decompression. The compression and decompression parts usually implemented by a deep neural network-based autoencoder. The following context provides a detailed explanation of each part of the algorithm and discusses possible directions for the further optimization.

1. Data preprocessing

The channel data to be compressed should be preprocessed according to data characteristics firstly, and different CSI may have different processing methods.

(1) Sparsity-based data trimming

For the full channel information feedback, if the channel data is the power delay spectrum in the antenna-delay domain, as shown in Fig. 9.10, the overall energy of the channel information is distributed on a few delay paths, which means that the main channel information is also concentrated on these paths. Therefore, during the channel information compression, the delay paths with extremely low energy is ignored and only the paths with high energy is selected. At the same time, since the distribution of high-energy paths is changing, it is also necessary to effectively compress and feedback the positions of these paths during the compression so that the receiver can correctly recover the feedback information.

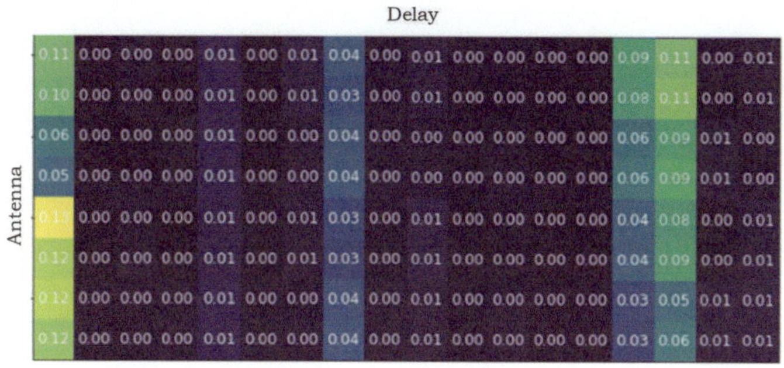

Fig. 9.10 Power delay spectrum in the antenna-delay domain

(2) Input data transformation

Apart from the sparsity in the antenna-delay domain, the channel may also show sparsity under some channel conditions when transforming the channel from antenna domain into angle domain using the Discrete Fourier Transform (DFT). In this way, it can also be revealed that the energy of channel is concentrated in certain beam directions. In addition, transforming the time delay domain into the frequency domain will also reveal some implicit characteristics of the channel, and corresponding processing can be done according to the characteristics of the practical training dataset.

Furthermore, the baseband MIMO channel is usually a complex matrix, which can be processed into two feature maps consisting of real and imaginary parts. In wireless communications, the complex form of amplitude-phase is more common, and it has specific physical meaning. Therefore, the real-imaginary part feature map can either be transformed into an amplitude-phase feature map or used as additional input of encoders, as shown in Fig. 9.11.

(3) Data augmentation

In the computer vision field, data augmentation can be achieved by processing sample images in terms of space and color, such as flipping, brightness adjusting, etc. By increasing training samples, the model generalization ability is improved, and overfitting problem can be prevented. For channel information, there are no image-color-like attributes. However, if the feedback information is regarded as an image, spatial processing methods can also be used on the training data during training to improve algorithm performance.

For the CSI compression feedback problem, another data augmentation method is to construct a large-scale training dataset from a known small-scale channel dataset based on an AI-based data generation scheme. Studies [17, 18] have shown that, under the limited training sample size conditions, using the large-scale samples constructed by AI-based data generation method can significantly improve the performance of the CSI compression feedback model, as shown in Fig. 9.12. For the dataset constructed based on real environment collection, its sample size is often limited by the difficulty

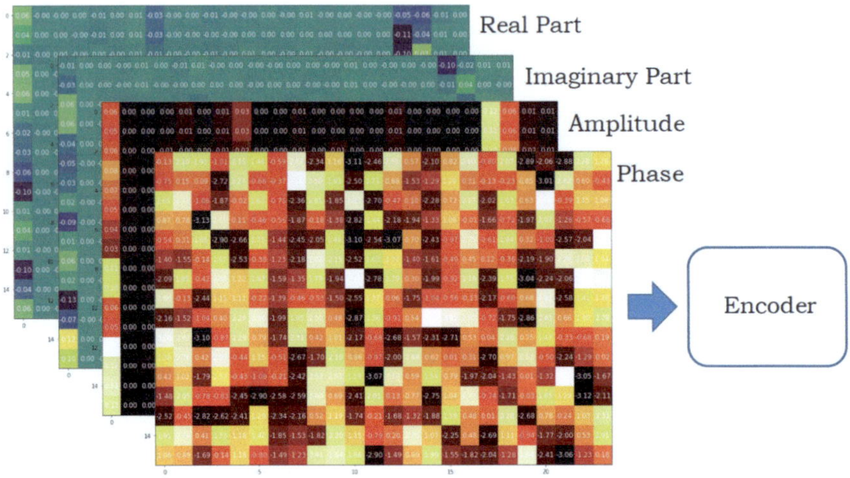

Fig. 9.11 Increase the dimension of input channel data

of collection. The AI-based data augmentation scheme is particularly suitable for these scenarios.

2. Network model

The characteristics of the data to a large extent determine the selection of AI network models. Channel data has spatial features like images, and temporal features like natural language. In the computer vision and natural language processing fields, deep learning has achieved great success, and there emerge many mature models to draw on.

(1) CNN

Many studies have used CNN as the basic network model for CSI compression feedback [19, 20]. The channel information is fed into the autoencoder in an image-like manner, and the channel characteristics can be extracted using convolution kernels.

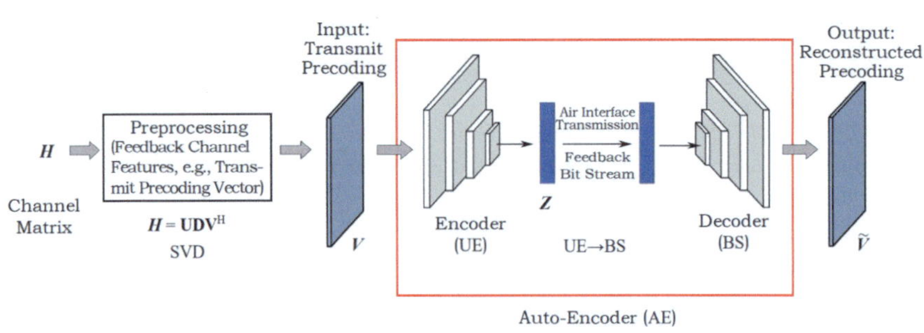

Fig. 9.12 AI autoencoder-based transmit precoding compression feedback and reconstruction

Correspondingly, in the decoder, upsampling is performed and convolution kernel is also used for feature recovery. Appropriate convolution kernel size and image size allows CNN to capture more spatial correlations. For example, for information between different antenna pairs on a path, a matrix with (N_t, N_r) form is better for a CNN to extract the correlation between data than a vector with $(N_t \times N_r, 1)$ format.

As the number of CNN layers increases, the network can perform more complex feature extraction. At the same time, the network also faces with the increased training difficulty and overfitting problems. Therefore, the residual structure-based ResNet [21] becomes a common component in network design. This structure enables residual learning by providing short-circuit connections between layers.

(2) Transformer model

Different from the CNN-based network models, the Transformer based network models [22] treat channel data of each path or each antenna pair as the sequential information input similar to natural language processing field, which is shown in Fig. 9.13.

There are three advantages to use the Transformer model for CSI compression feedback: First, the network structure of the Transformer model is suitable

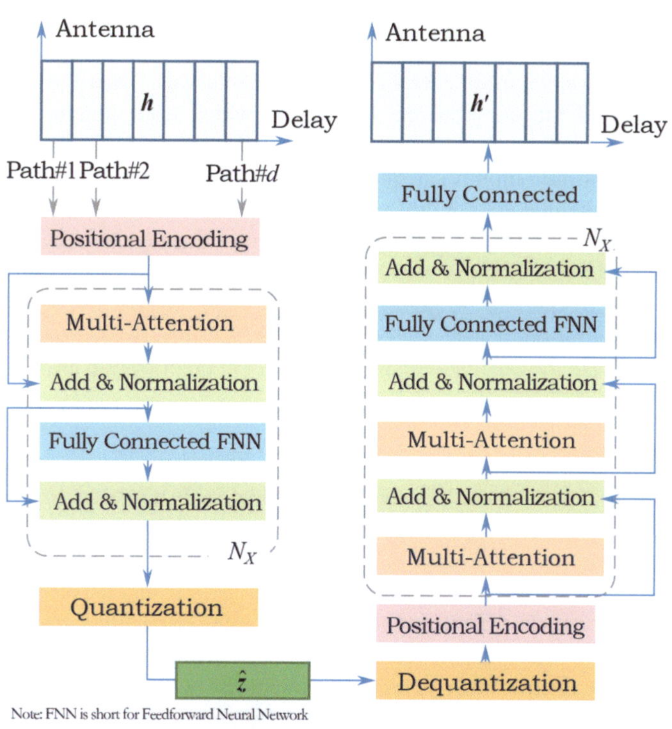

Fig. 9.13 CSI feedback compression model based on Transformer

for the autoencoder structure, because it originates from solving translation problems between different languages, which is similar to the overall architecture of CSI compression feedback, and is very suitable to be deployed on UE and BS. Second, regarding channel data as sequential input has certain physical significance in the wireless communication. Based on the relevant knowledge, the network can be optimized to better extracting the channel characteristics. Third, the Transformer model contains a self-attention module. For the CSI compression feedback problem, this module can better extract the correlation information between different paths or different antenna pairs.

3. Quantization

Since UEs need to feedback the channel information to the BS in a form of bitstream, the last layer of the encoder is generally a quantization layer to implement the data conversion which transform the floating-point feature vectors to bitstreams. For the decoder, a dequantization layer is required at the beginning of the network. Quantization will inevitably introduce errors. Different quantization strategies have a significant impact on the performance of CSI compression feedback. Common quantization methods are summarized as follows.

(1) Uniform quantization and non-uniform quantization

The simplest method is uniform quantization, which divides the distribution interval of floating-point values into several equal-length intervals. Then the quantization layer uses several bits to label the interval to which each floating-point value belongs. The dequantization layer then selects the floating-point value at the center of this interval as the output of dequantization. Since the floating-point values of the compressed channel are not uniform distributed, non-uniform quantization may achieve better performance by splitting floating-point value region to non-equal intervals. However, the quantization intervals and corresponding dequantization method is difficult to give directly. A possible solution is to analyze the distribution of compressed floating-point data after fixing other parameters of the network, and iteratively optimize the quantization interval until the quantization error converges.

(2) Vector quantization

Vector quantization is widely adopted in CSI compression feedback techniques. There are many types of vector quantization methods. It has been proved that the vector quantization in the VQ-VAE neural network model for quantizing the feature vectors after compression is more mature. The quantization codebook composed of central vectors can be mapped by an embedding layer and shared between the encoder and decoder. The bitstream after quantization is actually the indexes in the quantization codebook. The size and structure of the codebook also have influence on the number of feedback bits. The vector quantization of a training codebook is shown in Fig. 9.14.

Apart from the above two basic methods, there are some optimization techniques for quantization. For example, the channel information for a path can be quantized into different numbers of bits according to the importance of the path. In general,

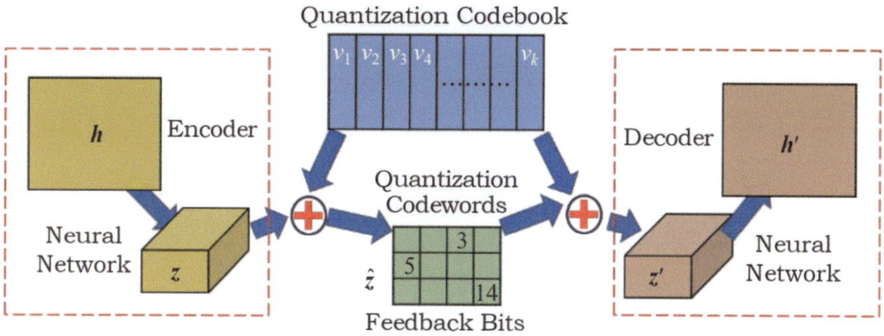

Fig. 9.14 Vector quantization of the training codebook

the high the energy of the path corresponds to the high important of the path, and the path will have more quantization bits, which means the quantization granularity of important path is more refined. In addition, after dequantization, the quantization error can be corrected utilizing the overall characteristics of the data [23], then the information matrix can be further restored.

9.3.5 Performance Results

The throughput comparison for different CSI feedback algorithms is shown in Fig. 9.15. This figure compares spectral efficiency of different CSI feedback algorithms when the system adopts corresponding transmission precoding. It can be seen that when the reconstruction accuracy of CSI feedback, i.e., NMSE > 0.85, the performance of AI-based CSI full channel information feedback method is significantly better than the codebook-based Type 1 and Type 2 precoding methods provided in the current standard.

The IMT-2020 5G and AI Fusion Research Task Group have organized two wireless AI competitions. The competitions are conducted to develop AI-based CSI compression feedback schemes based on full channel information data [15] and channel characteristic information data [16], respectively. The competition results provide more authoritative performance of AI-based CSI compression feedback algorithms.

In the first competition, the CSI compression feedback question provided the full channel information data and required feedback bit overhead can be minimized under the condition of the CSI reconstruction accuracy NMSE < 0.1. The results showed that using deep learning algorithms, the antenna-time delay domain CSI channel sample of 768 floating-point real numbers could be compressed to 286–384 bits while still meeting the reconstruction accuracy conditions.

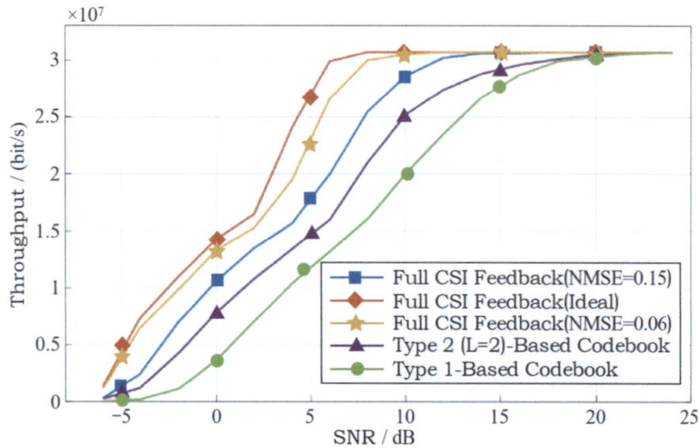

Fig. 9.15 Throughput comparison for different CSI feedback algorithms

In the second competition, the CSI feedback question provided the channel characteristic information data and required to maximize the accuracy of CSI reconstruction under the condition of the fixed feedback bit number. The results showed that on the testing dataset with 48 bits fixed feedback overhead, the squared cosine similarity after reconstruction could reach 0.84–0.87. On the testing dataset with 128 bits fixed feedback overhead, the squared cosine similarity after reconstruction could reach 0.92–0.94. Compared with the best eType-II codebook in the 3GPP standard, with the same number of feedback bits, the AI-based scheme can bring a 10–20% reconstruction accuracy improvement and for the same reconstruction accuracy, the AI-based scheme can save 50–60% of feedback overhead.

9.4 AI-Enabled Millimeter Wave Beam Management

In the new generation of wireless communication systems, in order to cope with the growing business demands, the millimeter wave band, known as its abundant bandwidth resources, is recognized as an indispensable key for the future wireless communication technologies. However, the higher frequency of the millimeter wave suffers from greater path loss, leading to rapid signal attenuation. To overcome the significant attenuation of the millimeter wave band channel, the multi-antenna array technology has been adopted to provide high-gain beamforming and compensate for the high path loss. However, due to high gain, high directivity, and narrow bandwidth of beams, each beam only has a relatively small coverage area in three-dimensional space. Thus, multiple beams need to be designed at both the transmitter and receiver sides to cover the entire three-dimensional space as much as possible. Therefore, the communication system can ensure that there is always a suitable beam in each direction to produce a larger signal gain, and certain methods can be used to select the appropriate beam for transmission or reception. How to select and manage these beams is the major research topic of this section.

On the one hand, due to changes in the environment, such as the movement, rotation, and obstruction of the user equipment (UE), the signal arrival direction on the UE side will change at any time. Therefore, the UE needs to continuously perform receiving beam scanning, measure the Reference Signal Received Power (RSRP), in order to find the optimal receiving beam and complete the update of the UE active beam.

On the other hand, due to the movement of the UE, the signal transmission direction on the BS side also needs to be adjusted in time. The BS determines the optimal transmission beam according to the measured results of multiple reference signals reported by the UE (sent by the BS to the UE, representing multiple transmission directions), and completes the update of the BS active beam.

This section will introduce AI technologies to assist the beam selection and beam management (BM) problems on both the terminal and the BS sides. Specifically, the AI-enabled millimeter wave beam management technology for both the terminal and the BS sides is introduced, and the performance of the terminal side AI beam management algorithm is verified based on the hardware platform.

9.4.1 Terminal Side AI Beam Management

1. Background Introduction

In the fifth-generation and the future sixth-generation mobile communication system (5G and 6G), both the BS and the terminal will adopt MIMO technology, using antenna panels with multiple antennas to form beams in specified directions. By controlling the phase of each antenna unit, different beam directions can be formed [25].

The existing standard provides a beamforming weight formula model, which is used to calculate the weight factor of each antenna unit (n, m) when forming a certain beam i, see formula (9.7) [26, 27].

$$w = \frac{1}{\sqrt{N_H N_V}} \exp\left(i \cdot 2\pi \cdot \left(n \cdot \frac{d_V}{\lambda} \cdot \sin\left(\theta_{i,\text{etilt}}\right) - m \cdot \frac{d_H}{\lambda} \cdot \cos\left(\theta_{i,\text{etilt}}\right) \cdot \sin\left(\varphi_{i,\text{escan}}\right)\right)\right) \tag{9.7}$$

where $n = 0, 1, \cdots, N_V - 1$ and $m = 0, 1, \cdots, N_H - 1$ are the antenna unit indexes in the vertical and horizontal directions, respectively. N_{V} and N_{H} are the number of antenna units in the vertical and horizontal directions, respectively. d_{V} and d_{H} are the distances between adjacent antenna units in the vertical and horizontal directions, respectively. $\theta_{i,\text{etilt}}$ and $\phi_{i,\text{escan}}$ are the vertical and horizontal angles of the i-th beam, respectively, and λ is the wavelength.

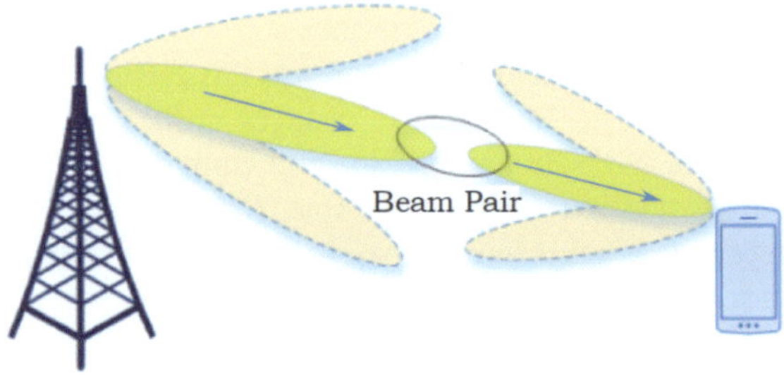

Fig. 9.16 Beam management: optimal beam pair matching at the transmitter and receiver

The goal of beam management is to find an appropriate beam pair by measuring the relevant reference signals, so that the transmission and the receive beam directions are matched, which is shown in Fig. 9.16.

On the one hand, the BS periodically sends reference signal bursts. Each reference signal burst contains multiple reference signals, and each reference signal is sent through different transmit beams to complete the scanning of the transmit beam (Transmit Beam Sweeping). On the one hand, the UE generates different receive beams at different times to measure multiple reference signals sent by the BS and obtains the quality of each reference signal. Then the UE reports the reference signal with the strongest RSRP to the BS. At the same time, the UE also records the receive beam number corresponding to the strongest RSRP. Thus, when the BS uses the same transmit beam for transmission in the future, the UE can use the corresponding receive beam for signal reception. The naive receive beam scanning method described above periodically triggers receive beam scanning. In other words, the UE traverses each receive beam in turn, measures the power of the reference signal for the transmit beam, then uses the receive beam with the maximum receiving power for subsequent data transmission.

The transmission of the reference signal is periodic. For example, a typical transmission cycle is 20 ms, i.e., the BS generates a reference signal burst every 20 ms. However, the UE can only use one receive beam to measure a certain reference signal within a reference signal burst, therefore, the UE needs to use different receive beams in turn within multiple reference signal cycles to complete the RSRP measurement. It can be seen that the UE needs to take a long time to complete a round of RSRP measurement, which leads to a large beam selection delay. Moreover, since the beam selection delay is large, the optimal beam may change after the beam scanning is completed, i.e., the optimal beam during the scanning process is no longer the optimal beam at the current moment. This phenomenon leads to a low accurate beam selection problem and downgrades the beam transmission performance.

In response to the shortcomings and problems of the naive receive beam scanning method, this section proposes an AI-enabled beam selection and beam management scheme that can timely and effectively select the appropriate receive beamforming plan, thereby improving the beam transmission performance.

2. Introduction of AI

During the receive beam selection at the terminal, the reference signal received by the UE can be used to identify the optimal beam, but the prediction information of the optimal beam cannot be directly obtained from the measured RSRP. Our goal is to timely select the optimal beam from all candidates as the terminal receive beam for data signal reception. When the UE is moving or rotating, it is difficult to quickly and accurately identify the optimal receive beam based on manual rules. In this case, AI will be a powerful tool in UE beam management that can effectively capture the required characteristics. A Beam Management (BM) scheme based on the Q-learning Network can achieve intelligent UE beam management. Next, we will further elaborate this scheme (hereinafter referred to as the BMQN scheme).

3. BMQN Scheme

Figure 9.17 shows the AI-enabled UE beam management scheme. In this scheme, the UE only uses the current active beam and a wide beam to measure the reference signal in a certain order. Then the UE uses a deep neural network to predict the optimal receive beam in the next moment in advance. Therefore, it can achieve the receive beam switching in time.

Before beam prediction, the machine learning model needs to be trained using a carefully designed Reinforcement Learning (RL) framework, where the training data includes both simulation data of 5G NR system and measurement data of the millimeter wave communication hardware platform.

4. RSRP Data Format

In the BMQN scheme, the UE alternately measures the RSRP value using the current active beam and the corresponding wide beam. Then it generates the input data of the machine learning model in the form of a sliding window. The input data format is shown in Fig. 9.18.

The sliding window contains two RSRP values at each step, which are the RSRP value measured using the currently activated beam and the RSRP value measured using the wide beam. A total of M (such as $M = 4$) measures of the RSRP pair forms a $2 \times M$ dimensional data table, where the two rows are the RSRP values measured using the currently activated beam and the wide beam, respectively.

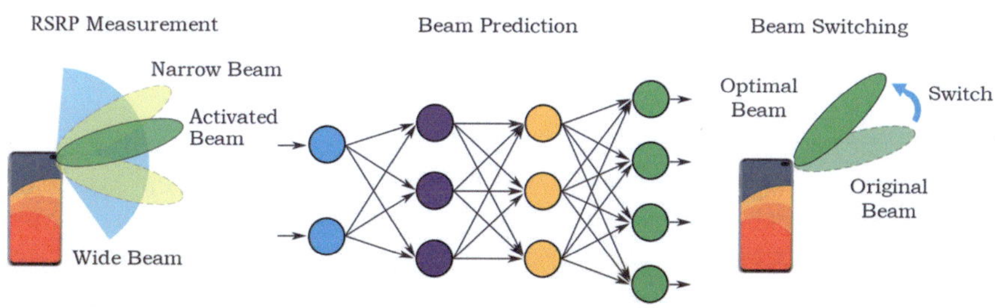

Fig. 9.17 AI-enabled UE beam management scheme

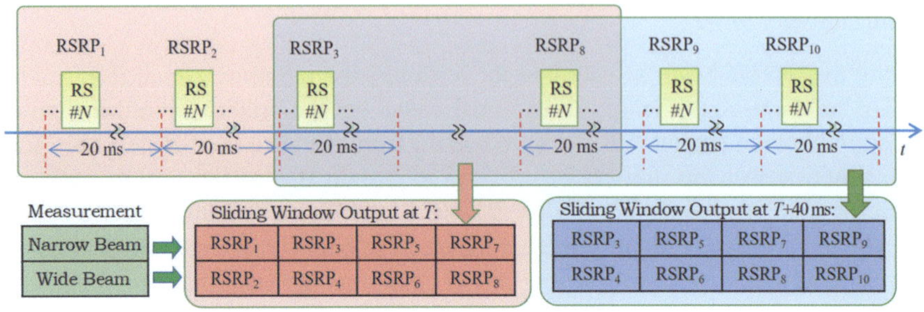

Fig. 9.18 Input data format of the machine learning model

Before fed into the ML model, the $2 \times M$ dimensional data table is reorganized into an $M \times 2$ dimensional data table, and further expanded to an $M \times (N + 1)$ dimensional data table, where N is the number of all narrow beams of the UE (such as $N = 16$). There are $(N + 1)$ data in each row, including N narrow beams RSRP values and a wide beam RSRP value. However, among N narrow beam RSRP values, only the RSRP value corresponding to the currently activated beam is actually measured, and the other $(N - 1)$ RSRP values are set to default values. The organized RSRP data table will carry the implicit information of the receive beam that the RSRP value corresponds to. This information is helpful during the neural network model training and predicting.

In addition, in order to facilitate the model training, the input RSRP data table needs to be further preprocessed. Linearly scale the original RSRP value and add a certain bias to make data roughly uniformly distributed around zero. The processing formula can be summarized as $E = \frac{1}{g}(\text{RSRP} + b)$, where RSRP is the RSRP value before preprocessing, E is the RSRP value after preprocessing, g is the scaling factor, and b is the bias.

5. Q-learning algorithm design

In the BMQN scheme, the Q-learning algorithm is the core of the AI algorithm. The Q-learning algorithm is a value-based algorithm in reinforcement learning. The Q denotes the value $Q(s, a)$, that is a reward by executing an action a ($a \in A$) in a specific state s ($s \in S$) at a certain moment. The environment will feedback a rewards r according to the agent's actions, and obtain the state at the next moment s'. The main idea of the algorithm is to build a Q-table containing states and actions to store Q-values, and select the action that can obtain the maximum reward based on Q-values.

For the beam management on the UE side, the UE acts as an agent. The state s is the RSRP data table preprocessed from the RSRP value, and the optimal beam selection corresponds to the action a. The change of the action will influence the subsequent RSRP values and get corresponding reward r.

In the basic Q-learning algorithm, Q-value is stored in a Q-table, whose dimension is consistent with the possible number of states. In the proposed scheme, the state corresponds to the RSRP data table measured by the UE, and its possible number could be infinite. So, it is impossible to store Q-value in the traditional Q-table way. Thus, the Q-function is come up to replace Q-table. In order to better extract useful information from the historical RSRP values, the deep neural network model is used to implement this Q-function.

In order to select the optimal beam (action, Action), the goal of the Q-learning algorithm is to maximize the long-term reward Q-value, which is expressed as the following formula

$$Q^*(s, a) = \max_{\pi} E(r_t + \gamma r_{t+1} + \cdots | s_t = s, a_t = a, \pi) \quad (9.8)$$

where r_t, γr_{t+1} are the instantaneous rewards at time t and $t + 1$, respectively. γ is the weight factor and π is the mapping strategy from state s to action a, i.e., $a = \pi(s)$ [28].

Based on Q-value, the UE will choose an action a_t that maximizes the $Q(s_t, a_t; \theta)$ value at time t, where s_t represents the environment state at time t, and θ is the parameter of the Q-function (including weights and biases). The action a is selected from the set A, which includes all UE receiving beam candidates. Therefore, the goal of the UE beam management is to select the optimal action at time t, i.e., a_t^*:

$$a_t^* = \arg\max_{a_t \in A} Q(s_t, a_t; \theta) \quad (9.9)$$

In summary, the neural network-based Q-function obtains the Q-value for each action (candidate beam) based on the input data of the current state (time-continuous RSRP values), then compares the Q-values of each action to select the optimal receiving beam, which is shown in Fig. 9.19.

In this Q-learning algorithm, the Q-function network calculates the Q-value. Generally, a good neural network structure design not only needs to have high prediction accuracy, but also needs to have good interpretability. Since the historical RSRP information in the input data is a time-dependent series, the Recurrent Neural Network (RNN) is good weapon to process time-dependent sequences and extract key feature information from it. Thus, the RNN model is adopted for the data analysis and the optimization as well as implementing the Q-function.

Specifically, an improved RNN model is utilized, namely the Long Short Term Memory (LSTM) [29], to implement the Q-function. It extracts features related to beam management from historical RSRP information, such as UE movement,

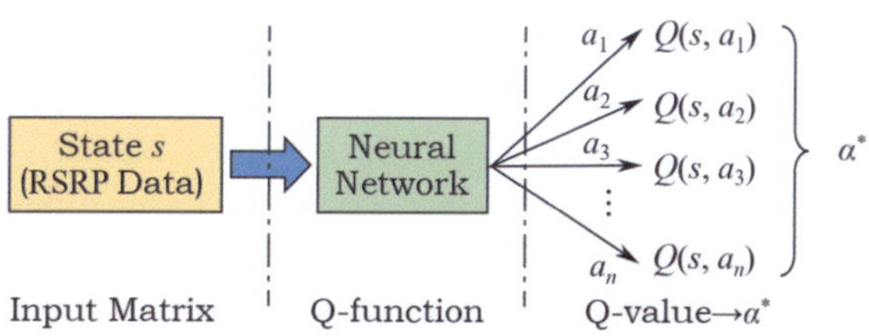

Fig. 9.19 Optimal action selection of the Q-learning algorithm

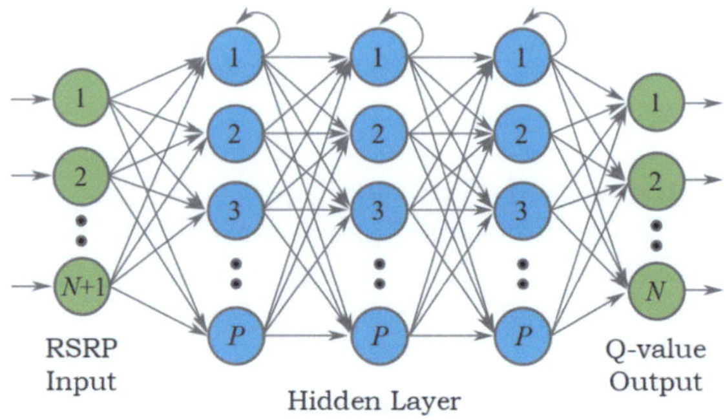

Fig. 9.20 Q-function: LSTM model-based deep neural network

changes of the wireless environment, actual beam distribution patterns, etc. To illustrate the network, a schematic diagram of a three-layer LSTM model is shown in Fig. 9.20, in which the network takes $(N + 1)$ RSRP data as input, has three LSTM hidden layers, and outputs N Q-values corresponding to different actions. Based on the output of the network the optimal beam can be decided.

In this scheme, the specific model parameters are set as follows: time step length $M = 4$, which means the network takes 4 activated beam measurement values and 4 wide beam measurement values as input. The number of received beams $N = 16$, the number of hidden layers $L = 3$, the number of neurons in each layer $P = 156$, and the output of the output layer is 16 Q-values, which corresponds to 16 received beams.

6. *Training Framework*

Before using the BMQN scheme for beam prediction, neural network model first needs to be trained using a large amount of training data. The training data comes from either 5G NR system-level simulation or measurement of the beam management PoC hardware prototype. The following part will mainly describe the BMQN training framework design based on system-level simulation data, and the PoC hardware prototype verification of the millimeter wave beam management scheme will be discussed in Sect. 9.4.3.

First, the collection of original data is introduced. The system-level simulation adopts the channel based on the 3D channel model of 3GPP TR 38.901. The BS is equipped with a 16×16 antenna array where a sector can form 64 beams. The UE installs two antenna panels, each of which uses a 2×2 antenna configuration, forming 8 beams. Thus, the two panels of the UE can form 16 beams in total.

There are 3 movement states for the UE, namely stationary, moving in random directions, and rotating in a fixed place.

During data collection, data is recorded every 20 ms. At each recording point, there exist an optimal beam pair, i.e., the optimal transmit beam and receive beam (the ideal optimal beam pair can be obtained in the simulation). For the optimal transmit beam, the RSRP measured values of all 16 receive narrow beams and the two wide beams of the UE are recorded.

By recording a large amount of RSRP measurements for different UE movement states, the original training data for the neural network model is obtained and the dataset is formed.

For conventional reinforcement learning training, since only the situation for the chosen action can be obtained in the current state, it is impossible to know the situations for all actions. Thus, record data cannot contain every situation. However, our original simulation data includes all *N* receiving beam and two wide beam RSRP measurements, so for any state *s*, all possible actions and the corresponding next state *s'* can be obtained. This means a complete training dataset can be obtained based on the original simulation data. However, during the training and prediction in the proposed BMQN scheme, only one RSRP value can be obtained at a given time. This means that only 1 of the 16 receiving beams is the real measured RSRP value, and the other 15 values are default values. Moreover, the measurement of narrow beam RSRP and wide beam RSRP must be arranged in different measurement periods.

For the Q-learning algorithm, the input data for network training is a four-tuple, denoted as {*s, a, r, s'*}. Randomly select a current state *s* and an action *a*, the network can obtain the next moment state *s'*, and calculate the corresponding reward *r*.

During the training process of reinforcement learning, when non-linear functions (such as neural networks) is used to represent Q-value, the network may be instable or divergent. This instability mainly comes from the correlation between observed sequences. In fact, even a slight update on the Q-value can cause significant changes in the policy as well as the correlation between the Q-value and the target value. Therefore, the double network is proposed, where two separate mapping functions (i.e., the main network and the target network) are trained using separate experiences in a symmetrical way. This architecture makes the training easier and more stable.

During the training process, batch training data are generated dynamically. The model is trained using the backpropagation (BP) algorithm so as to minimize the difference between the two networks, and the parameters of the two models, i.e., θ and θ', are updated until the two networks converge (see Fig. 9.21). The loss function optimized during the training can be given by:

$$L(\theta) = E_{(s,a,r,s')\sim U(D)}\left(r_t + \gamma \max_{a'} Q\left(s', a'; \theta'\right) - Q(s, a; \theta)\right)^2 \tag{9.10}$$

The final optimization objective is:

$$Q^*(s, a; \theta) \leftarrow E_{s'}\left[r + \gamma \cdot \max_{a'} Q^*\left(s', a'; \theta\right)|s, a\right] \tag{9.11}$$

Differentiating the loss function with respect to θ yields the following gradient:

$$\nabla_\theta L(\theta) = E_{(s,a,r,s')}\left[\left(r_t + \gamma \max_{a'} Q\left(s', a'; \theta'\right) - Q(s, a; \theta)\right)\nabla_\theta Q(s, a; \theta)\right] \tag{9.12}$$

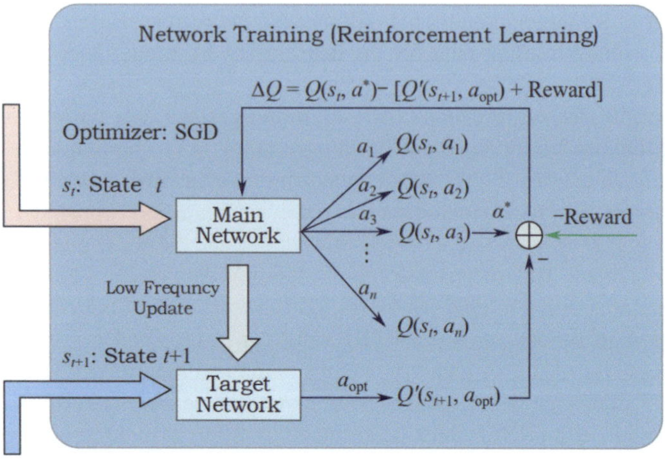

Fig. 9.21 Training process of the dual network

By using the Stochastic Gradient Descent (SGD) optimizer [30], the above gradient is backpropagated through the LSTM model, then the model can eventually reach a stable state.

The key to network training lies in the reward design. A good reward design mechanism helps to improve the network prediction accuracy. This is also the key to reinforcement learning training. The most direct reward design is based on the action selection results. Therefore, the reward calculation method is first designed as follows:

$$r_t = \mathrm{RSRP}_{t+1} - \mathrm{RSRP}_{\max} + C \tag{9.13}$$

where r_t is the reward value at time t. RSRP_{t+1} denotes the RSRP value measured at $t+1$ moment by the UE, and $\mathrm{RSRP}_{\max}$ is the maximum RSRP value in all received beam measurements. C is a given constant, which is used to balance the ratio of positive and negative rewards.

However, considering the practical situation, the UE is not able to measure the RSRP value of all received beams at the same time, so the above reward calculation method's $\mathrm{RSRP}_{\max}$ is unobtainable. Therefore, another reward calculation method is adopted, which considers the difference between the RSRP value measured by the current narrow beam and the RSRP value measured by the wide beam. The reward is calculated as follows:

$$r_t = \mathrm{RSRP}_s - \mathrm{RSRP}_w + C \tag{9.14}$$

where RSRP_s is the RSRP value measured by the current narrow beam and RSRP_w The RSRP value obtained for wide beam measurement. Replacing the aforementioned ideal reward calculation method with (9.14), the online network training process can be realized (i.e., training the model using the data from the prediction tasks, without the need for additional information collection). According to test results, this reward calculation method can achieve almost the same network performance.

9.4.2 BS Side AI Beam Management

1. Background introduction

As is mentioned earlier, both the BS and UE can adopt MIMO technology and use multiple antenna arrays to form high-gain, highly directional narrow beams. However, the deployment of antenna arrays requires more space, and would lead to increased implementation complexity. So, it is easier to implement MIMO technology on the BS side than on the terminal side. When the terminal moves, the BS needs to update the transmission beam in time, adjust the direction of the transmission signal, and complete beamforming.

There are two major solutions for BS side beam selection traditionally.

One solution is a naive full transmission beam scanning scheme, where the BS configures corresponding reference signal (e.g., SSB) resources for each narrow beam and triggers the terminal to measure and report all configured reference signal resources. Then based on reported RSRP value of the terminal, the BS chooses the narrow beam with the highest RSRP as the optimal beam for data transmission at the next moment.

The other solution is a hierarchical structure-based transmission beam scanning scheme, which combines SSB resources-based wide beams and CSI-RS resources-based narrow beams in a hierarchical structure. The procedure of hierarchical structure-based transmission beam scanning can be explained as follows: The BS first triggers the terminal to measure the wide beam RSRP on the SSB resources. Then based on the optimal wide beam, the BS configures CSI-RS resources for all narrow beams covered by the optimal wide beam and triggers the terminal to measure the narrow beam RSRP. Finally, the BS selects the optimal narrow beam based on the feedback narrow beam RSRP values for data transmission at the next moment.

However, both of the above traditional schemes suffer from problems. The BS needs to occupy a large amount of SSB resources or CSI-RS resources for RSRP measurement, which wastes a lot of resources. Moreover, the terminal needs to measure RSRP for a large amount of SSB resources or CSI-RS resources in a short time, which makes the terminal suffer from a great measurement burden and huge battery power consumption.

Specifically, for the hierarchical structure-based transmission beam scanning scheme, the BS can only choose the narrow beam covered by the optimal wide beam. If the terminal moves at a high speed, it could move out of the coverage of the optimal wide beam during the measurement and report time of the terminal. Then the BS cannot choose the actual optimal beam at the current time, which may lead to a low transmission quality, or even the link interruption.

In response to the problems of the traditional schemes, an AI-enabled beam selection and beam management scheme under the BS hierarchical beam structure is proposed, which can enable the BS to select the appropriate transmission beam for beamforming timely and effectively, and thus improve beam transmission performance.

2. Scheme process

Figure 9.22 shows the AI-enabled beam selection procedure under a hierarchical BS structure. The left-hand side shows that the BS and UE transmits data through the currently activated narrow beam before determining the new beam. When beam selection is required, the BS first triggers the UE to perform RSRP measurement based on the SSB wide beam and selects the two wide beams with the highest RSRP values among all wide beams. Then the BS selects a small number of narrow beams from the narrow beams covered by the two wide beams for future measurement and triggers the UE to measure RSRP of the selected narrow beams. Afterwards, the AI prediction model is used to predict the optimal beam at the target time where multiple AI prediction models can be used to predict the beam. Finally, when the target time is reached, the BS will switch to the target narrow beam, that is, switch from the currently used narrow beam to the predicted target narrow beam, as shown in the right-hand side of Fig. 9.22.

3. AI model and input data

The AI-based beam prediction framework on the BS side also adopts the Q-learning algorithm in reinforcement learning, where the neural network model implementing the Q-function is a 3-layer bidirectional long short-term memory network model (Bi-LSTM), which is similar to the AI-enabled beam management scheme on the terminal side.

The input data format of the AI model on the BS side is illustrated in Fig. 9.23. It is shown that at time step *i*, the input data of LSTM includes the $M = 4$ wide beam RSRP (assuming that the BS is configured with 4 wide beams) and N_2 narrow beam RSRP (assuming that the BS needs to select N_2 narrow beams covered by

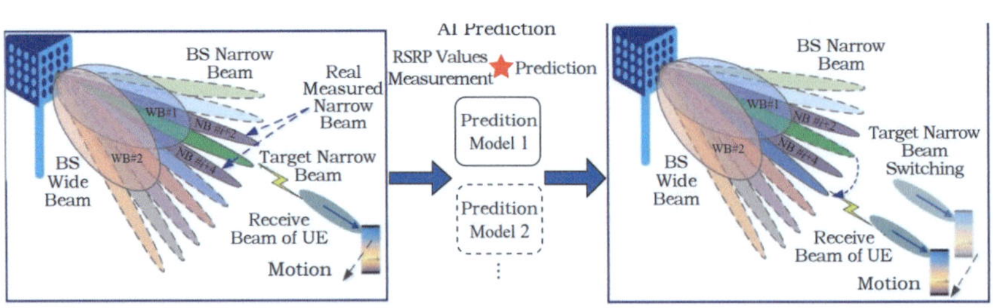

Fig. 9.22 AI-enabled BS beam selection process

Fig. 9.23 The input data format of the AI model on the BS side

$\mathrm{WRSRP}_{1,i-1}$	$\mathrm{WRSRP}_{1,i}$
$\mathrm{WRSRP}_{2,i-1}$	$\mathrm{WRSRP}_{2,i}$
$\mathrm{WRSRP}_{3,i-1}$	$\mathrm{WRSRP}_{3,i}$
$\mathrm{WRSRP}_{4,i-1}$	$\mathrm{WRSRP}_{4,i}$
$\mathrm{NRSRP}_{1,i-1}$	$\mathrm{NRSRP}_{1,i}$
$\mathrm{NRSRP}_{2,i-1}$	$\mathrm{NRSRP}_{2,i}$
$\mathrm{NRSRP}_{3,i-1}$	$\mathrm{NRSRP}_{3,i}$
...	...
$\mathrm{NRSRP}_{N_2,i-1}$	$\mathrm{NRSRP}_{N_2,i}$

the two highest RSRP wide beams for further measurement) at the current time step i, which can be denoted as: $\mathrm{WRSRP}_{1,i}$, $\mathrm{WRSRP}_{2,i}$, $\mathrm{WRSRP}_{3,i}$, $\mathrm{WRSRP}_{4,i}$, $\mathrm{NRSRP}_{1,i}$, $\mathrm{NRSRP}_{2,i}$, $\mathrm{NRSRP}_{3,i}$,..., $\mathrm{NRSRP}_{N_2,i}$. The input data also includes the 4 wide beam RSRP and N_2 narrow beam RSRP at the previous time step $i-1$ (historical information, i.e., the last received RSRP information), which can be denoted as: $\mathrm{WRSRP}_{1,i-1}$, $\mathrm{WRSRP}_{2,i-1}$, $\mathrm{WRSRP}_{3,i-1}$, $\mathrm{WRSRP}_{4,i-1}$, $\mathrm{NRSRP}_{1,i-1}$, $\mathrm{NRSRP}_{2,i-1}$, $\mathrm{NRSRP}_{3,i-1}$, ..., $\mathrm{NRSRP}_{\mathrm{N2},\mathrm{i}-1}$. These RSRP data are placed into a two-dimensional input matrix, with the size of $2 \times (M + N_2)$.

4. Performance results

A system-level simulation is conducted to verify the performance of this beam prediction algorithm. To implement performance simulation, a system-level simulation platform and an AI model algorithm platform are developed. The system-level simulation platform is used to collect a large amount of RSRP measurement data and generate training data and test data. The AI model algorithm platform is used for the AI algorithm design, the AI model training, and the beam selection prediction.

The traditional hierarchical structure-based transmit beam scanning algorithm is set as the baseline, and compared with the AI-enabled beam selection algorithm. The performance comparison of the two algorithms is shown in Table 9.2.

From the above simulation result, it can be seen that compared with the baseline, the AI-enabled beam selection algorithm occupies fewer CSI-RS resources, and the number of measurements will be reduced by half when the UE performs a narrow beam RSRP measurement. In addition, the AI-enabled algorithm is slightly more

Table 9.2 Performance comparison of two algorithms

Algorithm	Top1 (%)	Top1–2 (%)	Top1–3 (%)	Top1–4 (%)	Top1–5 (%)	RSRP difference from ideal	Number of CSI-RS resources
Baseline	54.30	74.16	83.26	87.06	89.11	3.623	16
AI-enabled Algorithm	56.68	76.88	85.46	88.64	90.32	3.216	8

accurate than the baseline during beam selection. The difference between the theoretical RSRP upper bound and the simulated RSRP is also about 0.4 dB smaller than the baseline, which indicates that the AI-enabled beam selection algorithm is more likely to select the actual optimal transmit beam.

9.4.3 Verification with Hardware Platform for Terminal Beam Management Scheme

In order to better verify the advantages of the BMQN scheme, the algorithm is deployed to a hardware platform, where the millimeter wave beam management is required. Figure 9.24 shows the structure of the millimeter wave beam management hardware platform and also the downlink data transmission process from the BS to the UE.

The mmPSA in Fig. 9.24 is a millimeter wave transceiver antenna array, operating at the frequency of 28 GHz millimeter wave band. The configuration of antenna units is set as 1×8, using a 6-bit phase shifter to generate 16 beams in the horizontal direction. Each beam covers a range from 8° to 16°. Thus, the antenna array can cover half of the spatial domain. On the other hand, the wide beam can be formed by one of the antenna units, with the 3 dB beam width of about 120°.

Before up-conversion to millimeter wave, the signal needs to go through a USRP-RIO for baseband signal processing. The USRP is short for universal software radio peripheral which is generally used for baseband processing. The USRP in the platform implements NR frame structure and RSRP measurement and other baseband

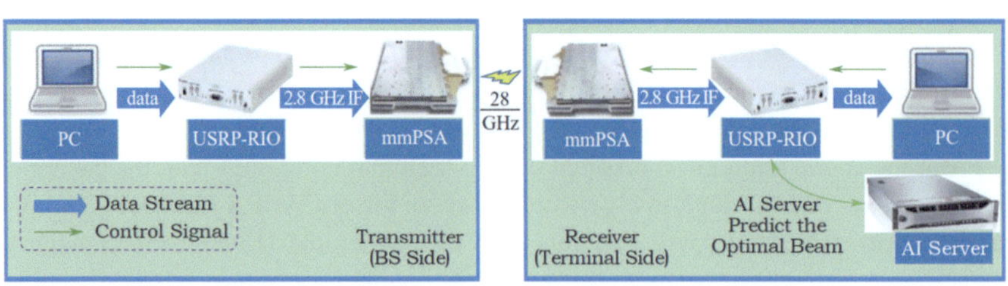

Fig. 9.24 Structure of the millimeter wave beam management hardware platform

functions and is connected to a laptop, which acts as the controller. At the receiver side, there is an AI server connected to the USRP, which is used to predict the optimal receive beam in real time using BMQN scheme and inform the receiver about the result.

To verify the algorithm performance using the hardware platform, we should first collect practical data for model training. During the data collection, we place the antenna panel (phase shifter) on a remote-controlled robot chassis so that the data collection under different UE motion states can be simulated.

The hardware layout is shown in Fig. 9.25. When the robot chassis moves and rotates in a predetermined manner, the AI server automatically records the RSRP values of all 16 narrow beams and one wide beam every 40 ms. By setting different moving and rotating speeds for the robot chassis, more measurement data can be collected.

After collecting the complete dataset, the model is trained offline base on the aforementioned Q-learning algorithm. The model trained with practical data will be deployed on the AI server. The UE manages the receive beam according to the real-time prediction results of this server. In order to better illustrate the performance of the BMQN scheme, several rule-based benchmark algorithms have been proposed and four algorithms are listed below.

(1) Naive Sweeping (NaS): Continuously scan all receive beams and select the beam with the highest RSRP value.
(2) Binary Search (BS): A hierarchical measurement that continuously compares and groups beams, then further executes grouping for the better beam groups.

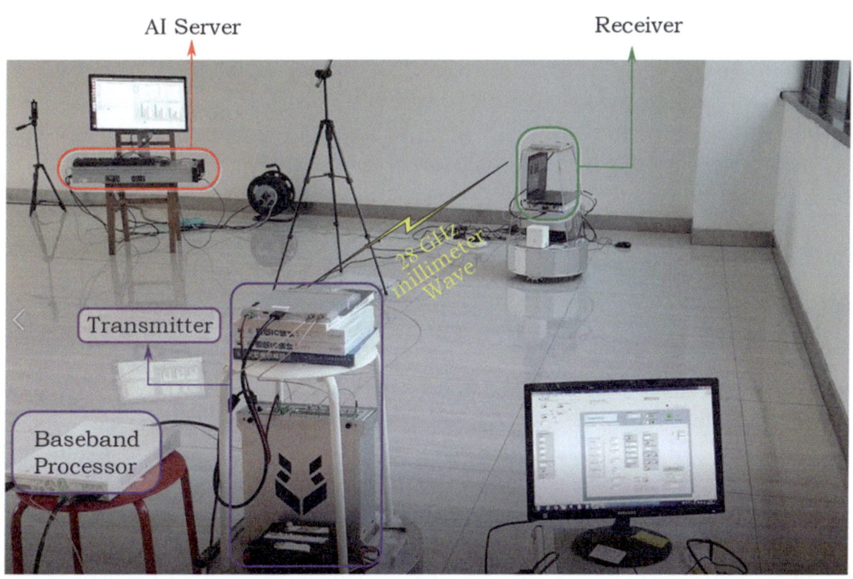

Fig. 9.25 Hardware layout

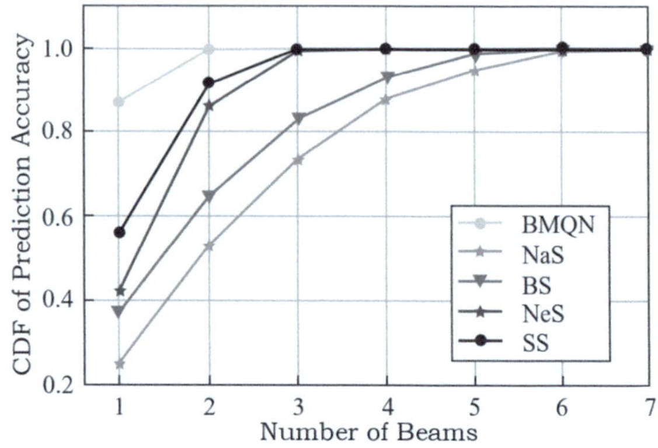

Fig. 9.26 Comparison of beam prediction accuracy

(3) Sampling Sweeping (SS): Select representative beams to measure and choose the beam with the highest RSRP value.
(4) Neighbor Search (NeS): Only measure the current active beam and the adjacent beams, and select the beam with the highest RSRP value.

The direct performance criterion of the algorithm is beam prediction accuracy. The performance comparison of the algorithm is shown in Fig. 9.26. The top-N rankings of beam prediction accuracy for the BMQN scheme and other four traditional algorithms are demonstrated in the form of a cumulative distribution function (CDF).

It can be seen that the top-1 beam selection accuracy of the BMQN scheme is above 0.85, and the top-2 beam selection accuracy is close to 0.99, which is significantly superior to the other four traditional rule-based algorithms. Generally, the complete prediction process on the hardware platform is only 3 to 5 ms, which can fully satisfy the real-time requirement of beam selection. Considering the computational power of practical terminal devices and the GPU computing scenarios, the computational delay can be significantly reduced. In addition, the scenario of beam tracking recovery after link interruption has also been examined. It is verified that the time of recovering to the optimal beam is much less than that of traditional rule-based algorithms.

9.5 Summary

In recent years, the development of AI-enabled air interface technology has attracted increasing attention from the industry and academia. Meanwhile, it should be noted that there are still many challenges when applying AI technology to mature air interface physical layer. The challenges include the collection and acquisition of big data, the generalization performance of AI algorithms, the complexity control of AI algorithms on different stages, and the standardization of AI-enabled air interfaces, etc. In the early stage of 6G research, it is beneficial to consider the AI-enabled air interface technologies that can achieve better performance, and combine it with the standardization achievement of the 5G and B5G. This chapter introduced the evolution and standardization progress of AI-enabled air interface technology. By illustrating three important AI-enabled air interface technology research (i.e., channel estimation, CSI feedback, and millimeter wave beam management), this chapter present current research results and the future potentials of AI-enabled air interface technology.

References

1. 3GPP. Study on RAN-centric Data Collection and Utilisation for NR: RP-181456[S]. 3GPP work item description, 2018.
2. 3GPP. Study on further enhancement for data collection: RP-201304[S]. 3GPP work item description, 2020.
3. 3GPP. Study on Artificial Intelligence (AI)/Machine Learning (ML) for NR Air Interface: TR 38.843[S]. 3GPP work item description, 2022.
4. ZHANG J, HUANG Y, WANG J, et al. Intelligent beam training for millimeter-wave communications via deep reinforcement learning[C] in Proceedings GLOBECOM'19, Waikoloa, HI, USA, 2019: 1–7.
5. 3GPP. NR; Physical channels and modulation: TS 38.211[S]. 3GPP. v17.0.0, 2022.
6. COLERI S, ERGEN M, PURI A, et. al. Channel estimation techniques based on pilot arrangement in OFDM systems[J]. IEEE Transactions Broadcast, 2002, 48: 223–229.
7. SAVAUX V, LOUËT Y. LMMSE channel estimation in OFDM context: a review[J]. IET Signal Process, 2017, 11: 123–134.
8. DONG C, CHEN C L, HE K, et. al. Image Super-Resolution Using Deep Convolutional Networks[J]. IEEE Transactions on Pattern Analysis and Machine Intelligence, 2016, 38: 295–307.
9. SOLTANI M, MIRZAEI A, POURAHMADI V, et. al. Deep learning-based channel estimation[J]. IEEE Communications Letters, 2019.
10. TOLSTIKHIN I, HOULSBY N, KOLESNIKOV A, et. al. Mlp-mixer: An all-mlp architecture for vision[J]. Advances in Neural Information Processing Systems, 2021, 34.
11. IMT-2020(5G), CAICT, VIVO. AI-based Channel Estimation, Second Wireless AI Competition (2nd WAIC), [OL]. [2022–12–08]. DataFountain website.
12. VIVO, 3GPP RAN Technical Standards Working Group. Study on AI/ML based Air Interface Enhancement in Rel-18: RP-210170[S]. 2021.
13. IMT-2020(5G) Promotion Group. Research Report on AI-based MIMO Technology[R].2021.
14. DataFountain. 2nd wireless communication AI competition (2nd WAIC) [OL]. [2022–12–08]. DataFountain website.
15. IMT-2020(5G) Promotion Group. 5G and AI Fusion Basic Theory and Dataset Research Report[R].2021.
16. DataFountain. 1st wireless communication AI competition (1st WAIC) [OL]. [2022–12–08]. DataFountain website.
17. IMT-2020(5G) Promotion Group, OPPO. Discussion on R18 study on AIML-based 5G enhancements[R]. 2021.
18. XIAO H, TIAN W, LIU W, et al. ChannelGAN: Deep Learning-Based Channel Modelling and Generating[J]. IEEE Wireless Communications Letters, 2022, 11: 650–654.
19. XIAO H, WANG Z, TIAN W, et al. AI enlightens wireless communication: Analyses, solutions and opportunities on CSI feedback[J]. China Communications, 2021,18: 104–116.
20. LIU W, TIAN W, XIAO H, et al. EVCsiNet: Eigenvector-Based CSI Feedback Under 3GPP Link-Level Channels[J]. IEEE Wireless Communications Letters, 2021, 10: 2688–2692.

21. HE K, ZHANG X, REN S, et al. Deep Residual Learning for Image Recognition[J]. 2016 IEEE Conference on Computer Vision and Pattern Recognition (CVPR), 2016: 770–778.
22. VASWANI A, SHAZEER N, PARMAR N, et al. Attention is All you Need[J]. ArXiv, 2017.
23. CHEN T, GUO J, JIN S , et al. A Novel Quantisation Method for Deep Learning-Based Massive MIMO CSI Feedback[J]. 2019 IEEE Global Conference on Signal and Information Processing (GlobalSIP), 2019: 1–5.
24. LU Z, WANG J, SONG J. Multi-resolution CSI Feedback with Deep Learning in Massive MIMO System[J]. 2020 IEEE International Conference on Communications (ICC), 2020: 1-6.
25. ZHANG Y, REN P, DU Q ,et al. Antenna tilt assignment for three-dimensional beamforming in multiuser systems[J]. 2015 IEEE Global Communications Conference,2015: 1-6.
26. 3GPP. Study on channel model for frequencies from 0.5 to 100 GHz: TR38.901 V14.2.0[S]. 2017.
27. 3GPP. Study on new radio access technology: Radio Frequency (RF) and co-existence aspects: TS38.803 V14.1.0[S]. 2017.
28. MNIH V, KAVUKCUOGLU K, SILVER D, et al. Human-level control through deep reinforcement learning[J] Nature, 2015, 518(7540): 529–533.
29. HOCHREITER S, SCHMIDHUBER J. Long short-term memory[J]. Neural computation, 1997, 9:1735–1780.
30. BOTTOU L, BOUSQUET O. The tradeoffs of large scale learning[J]. Advances in Neural Information Processing Systems, 2008: 161–168.

Chapter 10
Advanced Duplex

This chapter explains the necessity of researching Advanced Duplex technology from the perspective of driving factors, and provides an overview of key issues, main challenges, and applicable scenarios. For future high-power full-duplex cellular base stations, this chapter discusses the design criteria for joint self-interference cancellation and self-interference cancellation technologies in various domains, outlines network cross-interference and candidate mechanisms, and provides some reference viewpoints on interference cancellation schemes from an engineering practice perspective.

10.1 Driving Factors and Trends

Currently, 3GPP has released the 5G NR (New Radio) Rel-17 standard technical specifications. Compared with 4G LTE, one of the features of 5G NR is that it supports higher carrier frequencies and larger bandwidths. Most of the newly allocated 5G spectrum worldwide has higher frequency than 4G with unpaired spectrum for Time-Division Duplex (TDD), such as the 3.3–3.8 GHz, 28 GHz, and 39 GHz bands. The initial deployment of 5G NR systems will face coverage limitations due to unavoidable larger path losses. Compared with Frequency Division Duplex (FDD), TDD's coverage capability significantly decreases due to limited frequency band allocation and slot resources, which restricts transmission repetition resources. The coverage comparison of 5G is shown in Fig. 10.1, which brings challenges to deployment in terms of co-station and construction costs. To address the shortcoming of TDD coverage, the industry has conducted some research [1], one of which is the Supplementary Uplink (SUL) technology adopted by 3GPP. It provides scheduling flexibility and can schedule terminals with poor coverage on 5G high-frequency band carriers to low-frequency uplink carriers, to improve coverage to

G. Liu et al., *6G Migrates from Communication to XaaS*,
https://doi.org/10.1007/978-981-96-7579-1_10

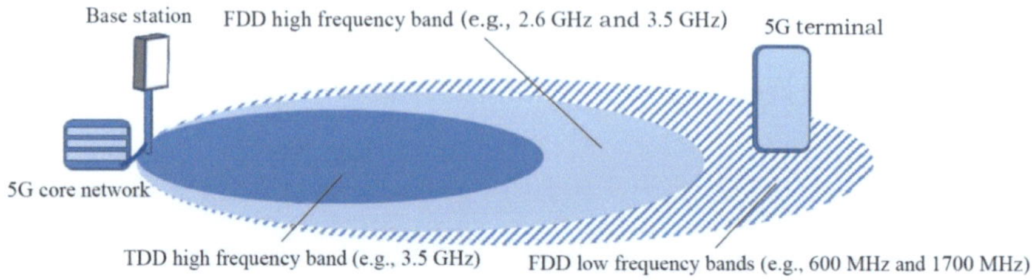

Fig. 10.1 Comparison of 5G coverage

some extent at the cost of additional auxiliary spectrum resources and more complicated transceivers. However, it does not fundamentally solve the problem of TDD coverage, but only avoids new spectrum for 5G when coverage issues arise. Another solution is to increase the deployment density of 5G base stations, which requires addressing more sites. From the perspective of mobile operators, this is not attractive.

Furthermore, as emerging services such as ultra-high-definition video, AR/VR/XR, holography, and digital twins become increasingly popular, global mobile service volume will continue to explode. It is estimated that by 2030 (the 6G era), there will be more than 80 times growth compared to 2020 (the 5G era) (approximately 5000 EB/m) [2]. However, the spectrum is always limited. The available spectrum with good propagation characteristics in the low-frequency band is nearly exhausted. 5G has taken the lead in developing frequencies above 6 GHz, including millimetre wave bands, indicating that the scarcity of spectrum is one of the key issues to be solved in mobile communications. To meet the growing demand for service volume and data rates, on the one hand, exploratory research is needed on the feasibility of using higher frequency bands (such as terahertz, visible light, etc.) with abundant bandwidth resources for 6G mobile communications. On the other hand, efforts need to be made to improve the efficiency of existing spectrum use.

In addition, although TDD has certain advantages over FDD in effectively supporting asymmetric services and channel reciprocity, the time-division configuration of uplink and downlink slots in the TDD frame structure and the corresponding Hybrid Automatic Repeat Request (HARQ) feedback timing have led to a hard limit on the air interface delay of the TDD system. The rise of ultra-low latency services has prompted mobile communications to formulate schemes with air interface delays less than 10 ms from 5G NR onwards. Future 6G will need to further reduce delays.

Based on the demand for enhanced coverage, increased spectral efficiency, and reduced latency, the evolution of duplex modes has gradually been discussed since the early stages of 5G. Many studies are dedicated to realising a new type of duplex technology that combines the advantages of both TDD and FDD, namely Full Duplex technology. Currently, 3GPP Rel-18 has initiated a research project on new duplex technology, exploring the feasibility of evolving duplex technology from 5G-Advanced. However, the scope of this research project is currently limited to Sub-band Non-overlapped Full Duplex, Sub-band Overlapped Full Duplex, and Full-band

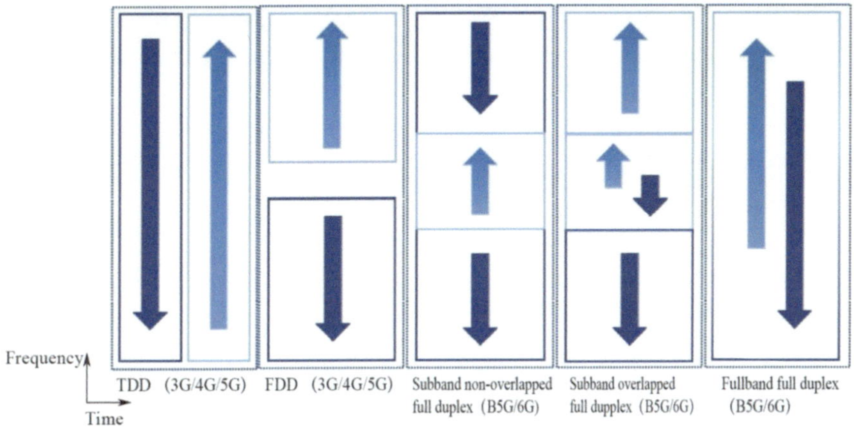

Fig. 10.2 Differences between the three new duplex modes and traditional TDD/FDD

Full Duplex (also referred to as In-band Full Duplex or co-time co-frequency Full Duplex in some literature).

The concept of Sub-band Non-overlapped Full Duplex allows simultaneous uplink transmission and downlink reception, or uplink reception and downlink transmission on non-overlapping frequency sub-bands within a carrier bandwidth. The concept of Sub-band Overlapped Full Duplex allows simultaneous uplink transmission and downlink reception, or uplink reception and downlink transmission on the same frequency sub-bands within a carrier bandwidth. The concept of Full-band Full Duplex allows simultaneous uplink transmission and downlink reception, or uplink reception and downlink transmission on all bandwidth resources of the same carrier. It can be seen that the former two can be considered as a special case of the latter, but they face different technical challenges and are suitable for different application scenarios. The differences between the three new duplex modes and traditional TDD/FDD are shown in Fig. 10.2.

Theoretically, Full-band Full Duplex has the potential to double spectral efficiency, have the same latency as FDD, and enhance uplink coverage. Sub-band Full Duplex has similar potential for latency and coverage enhancement as Full-band Full Duplex, but it does not maximize spectral usage efficiency compared to Full-band Full Duplex. The coverage enhancement capability here means Full Duplex able to provide more available continuous resources, allowing users at the cell edge to perform retransmission based signal combination on these continuous resources to improve coverage. In addition, Sub-band Non-overlapped Full Duplex is least technically challenging to implement, Sub-band Overlapped Full Duplex is more, while Full-band Full Duplex faces the most challenges. Therefore, Sub-band Non-overlapped Full Duplex was the first to be studied by 3GPP for 5G NR duplex evolution research. The general view of 3GPP companies is to prioritise non-overlapping sub-band full duplex on the base station side, and study its feasibility for 5G-Advanced. The future success of Sub-band Non-overlapped Full Duplex will become the foundation for the research of the other two full duplex techniques.

10.2 Main Technical Issues, Challenges and Application Scenarios

Full Duplex technology was proposed as early as the 1990s [3]. In the past decade, the academic and industrial sectors have continuously explored duplex evolution, making great progress or breakthroughs on some key issues [4–10]. However, to this day, it is difficult to achieve a mobile communication standard. For a long time, the common understanding in the industry was that simultaneous transmission at the same frequency in actual mobile communication was not feasible due to a series of well-known serious interference problems caused by the high transmission power of base stations. To apply Full Duplex in mobile cellular systems, it is necessary to develop effective and cost-effective solutions to solve the following two interference problems.

(1) Self-Interference (SI)

This includes self-interference caused by near-field coupling of transmitting and receiving antennas, and self-interference received by the receiving antenna from electromagnetic waves reflected back by nearby reflectors. The schematic diagram of Full Duplex self-interference is shown as follows: As in Fig. 10.3, it demonstrates a multipath channel that includes a direct path due to antenna coupling and multiple strong reflection paths. Considering the radiation power level of the base station in mobile cellular network, this self-interference can be very strong. If the self-interference cannot be effectively removed, the desired useful signal will not be successfully received. The higher the transmission power, the greater the antenna coupling self-interference; the lower the frequency band, the greater the transmission self-interference, and the more complex the self-interference channel.

① Non-overlapping subband full duplex uses different subbands in the operating frequency band for uplink and downlink. Theoretically, it is orthogonal for uplink and downlink transmission, but due to non-ideal factors of the device in practice, mainly including the power amplifier (Power Amplifier, PA) non-linearity and/or frequency deviation, the transmission on different subbands are not orthogonal, and there will be self-interference caused by adjacent channel leakage. The cases of self-interference under different duplex types are shown in Fig. 10.4.

② The self-interference of co-frequency full duplex (including overlapping subband full duplex and full band full duplex) is more severe than that of non-overlapping subband full duplex. The self-interference in the overlapping band is much stronger than the adjacent channel leakage self-interference, which includes linear interference and in-band non-linear interference caused by non-ideal factors of the device.

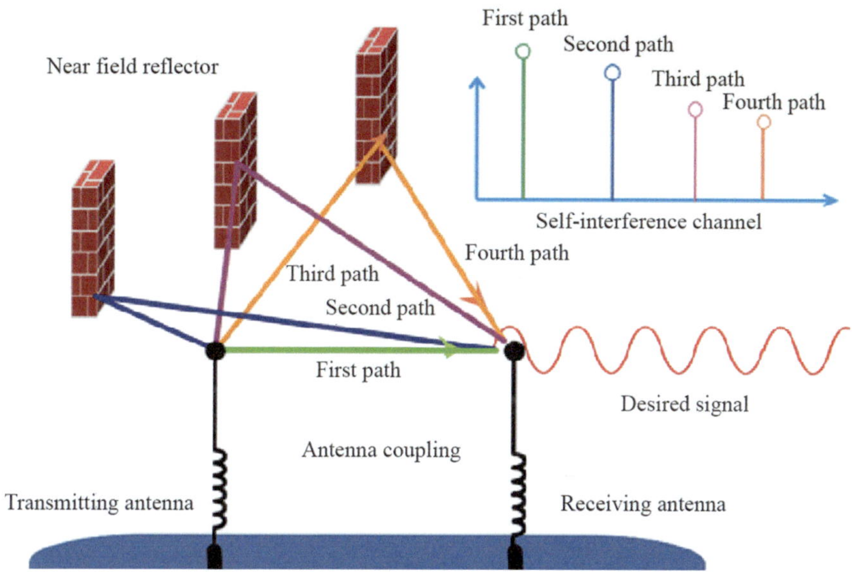

Fig. 10.3 Full duplex self-interference schematic

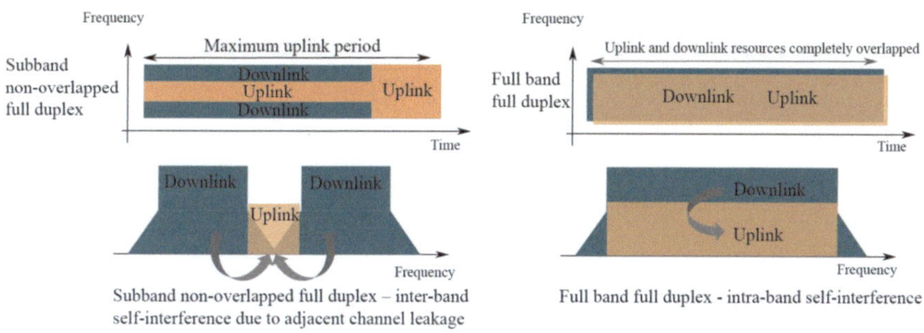

Fig. 10.4 Self-interference cases of different duplex types

(2) Network cross-interference

Similar to the TDD system, if full duplex is applied in the mobile cellular network, it may cause cross-interference from downlink transmission of one base station to the uplink reception of another adjacent base station, or from uplink transmission of one terminal to downlink reception of another adjacent terminal (Cross Link Interference, CLI) [6, 7]. Such CLI will bring serious network performance degradation to the neighboring operators deployed at adjacent frequencies, superimposing the inherent co-frequency inter-cell interference. The full duplex network interference is shown in Fig. 10.5. The interference situation is more complicated than the current TDD

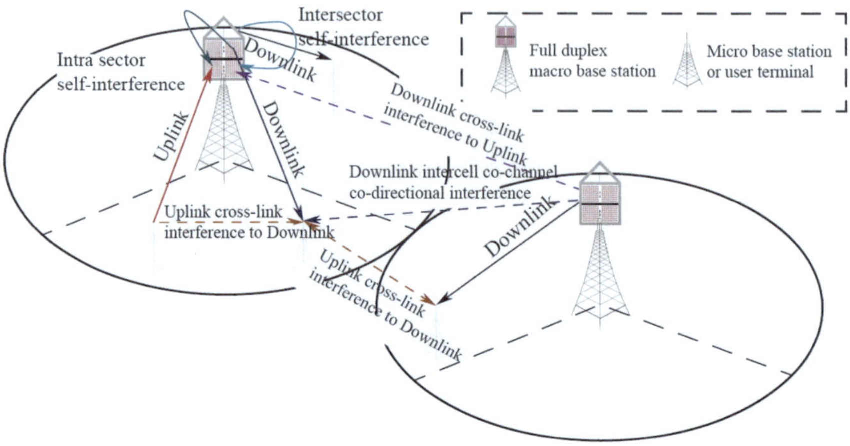

Fig. 10.5 Full-duplex networking interference

system. Besides the traditional inter-cell interference (Inter-Cell Interference, ICI), it also introduces the self-interference and cross-interference between sectors.

① For Non-overlapping subband full duplex or overlapping subband full duplex, the network cross-interference between neighboring operators can be alleviated to a certain extent by assigning proper frequency resources for uplink transmission. For example, if both operators place the uplink frequency band in the middle of the carrier bandwidth, then the adjacent channels are all for downlink. Such adjacent channel interference is similar to current negligible channel interference between operators, which doesn't need to be handled. This is also the main motivation to still consider subband full duplex deployment even in full band full duplex. In the current spectrum allocation for mobile cellular network, it is still relatively common that in the same area the spectrum allocated to one operator is adjacent to that of another operator, such as the frequency 3.42–3.5 GHz, 3.5–3.6 GHz and 3.6–3.7 GHz in the 3GPP band n78. These three adjacent sections of the spectrum are for national 5G network coverage in the same country, and can be allocated to different operators. Assuming that one operator deploys full band full duplex in the second spectrum section, then the cross-interference to the networks of other two operators due to the adjacent channel leakage is extremely difficult to handle, which limits the use of full band full duplex, and makes subband full duplex more attractive. In current TDD networking, three operators can negotiate uplink and downlink frequency allocation. Similar concept can be reused for operators negotiating the frequency position of the uplink subband in subband full duplex, thereby avoiding serious cross-interference between operators.

② Sub-band non-overlapping full-duplex has slightly less cross-interference than sub-band overlapping full-duplex networking, similar to the interference situation with dynamic TDD that has been supported by 3GPP. However, Compared with full-band full-duplex, such an alleviation on interference situation is obtained by sacrificing the theoretical twice frequency usage efficiency with full-duplex.

From the above explanations on self-interference and cross-interference, it can be understood that full-band full-duplex is difficult to be used in scenarios with uncontrollable cross-interference, such as densely deployed access networks in urban areas, and is more suitable for backhaul links, isolated deployments and e.t.c., while sub-band full-duplex has advantages in the scenarios with adjacent frequency deployments for different operators. Table 10.1 briefly compares several full duplex types.

Table 10.1 Comparison on full duplex types

Type contrast	Duplex				
	FDD	TDD	Subband nonoverlapping full duplex	Subband overlapping full duplex	Full band full duplex
Frequency utilization expectation	Low	Low	Medium	Medium	Highest
Latency	Low	High	Low	Low	Low
Uplink Coverage Expectation	Highest	Low	High	High	High
SI	Low, negligible	None	Medium	High	Highest
CLI	None	Controllable	Higher than TDD	High	Highest and most complex
Interference negotiation between adjacent frequency operators	Not required	Required, uplink and downlink allocation negotiation	Required, uplink and downlink allocation negotiation	Required, uplink and downlink allocation negotiation	Hard
Applicable scenarios	Flexible, all scenarios	Flexible, all scenarios	Secondary flexibility, basically all scenarios	Generally flexible, slightly smaller than full band full duplex limit	Restricted, CLI controllable scenarios: such as return link (especially 3GPP IAB scenario), island, indoor low-power independent deployment, millimeter wave cell

10.3 Key Technology for Self-interference Cancellation

For the local self-interference caused by full-duplex operation, a series of studies have shown that self-interference suppression/cancellation (Self-Interference Cancellation, SIC) is practically feasible [5–15]. SIC design can be jointly performed in multiple domains, generally divided into antenna domain SIC, RF domain SIC, and digital domain SIC. It is particularly worth noting that for the high radiation power of base stations in cellular network, it is unrealistic to achieve the desired SIC performance solely by operation on a certain domain. If the antenna domain and/or RF domain SIC does not have enough cancellation capability, the saturation in receiver is likely to happen, and causes signal distortion that makes SIC in digital domain impossible to implement.

Literature [14] also points out that the ability of digital domain SIC is only capable to process signals within the dynamic range of the receiver. The dynamic range of the receiver is constrained by the RF module of the receiver, mainly the Lower Noise Amplifier (LNA) and the Analog to Digital Converter (ADC). Unlike the Moore's law of digital devices, the industrial development trend of these analogue devices over the past 20 years has shown that the improvement of commercial devices is slow and limited [16]. Therefore, it is difficult to greatly enhance the dynamic range of the receiver by replacing hardware, and it is also difficult to rely on an efficient and high-quality SIC in digital domain to reduce self-interference below the receiver's noise floor.

Joint SCI among multiple domains is inevitable for the full-duplex implementation of high-power base stations. As shown in Fig. 10.6, the SIC receiver needs to ensure that receiving signal after SIC in antenna domain and RF domain not exceed the dynamic range before inputting to the LNA and ADC, so that digital domain SIC can further effectively remove residual self-interference. This requires the antenna domain or RF domain to have a certain self-interference suppress capability. The residual interference after digital domain SIC is expected to be lower than the receiver's noise floor, to avoid the receiver becoming insensitive due to residual self-interference. Thereby it ensures that the demodulation performance of the desired signal is comparable to the current TDD, especially for high-order modulation transmission. Otherwise, the expected full-duplex gain will be greatly reduced.

The requirement on SIC can be simply estimated by the transmission power level and the receiver's noise floor, based on the principle as residual self-interference being reduced lower than the noise floor. It can avoid over insensitivity of receivers compared to the receivers for TDD/FDD that doesn't suffer from self-interference. Figure 10.6 shows an example of SIC requirement for a high-power base station with a bandwidth of 100 MHz. The Equivalent Isotropically Radiated Power (EIRP) of 5G base station can be 54 dBm, assuming the array gain of multi-antenna is 22 dBi, the total output power (TRP) of the power amplifier is 32 dBm, the noise factor is assumed to be 7 dB, then the noise floor at 100 MHz is -87 dBm. If the residual self-interference is suppressed lower than noise floor, then SIC needs to achieve 141 dB based on EIRP

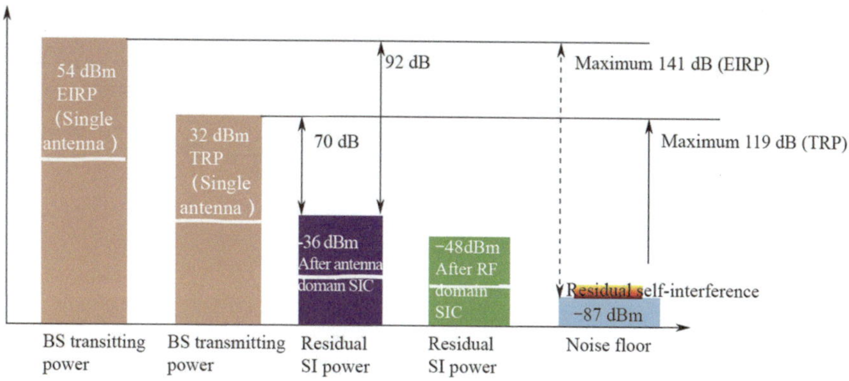

Fig. 10.6 SIC requirements and examples

calculation, and 119 dB based on TRP calculation. Note that both requirements on SIC are equivalent, based on different observation point. Assuming that the spatial domain isolation and antenna coupling isolation can achieve 70 dB (calculated as 92 dB according to EIRP), then the SIC capability at full-duplex receiver needs to reach 49 dB. If the front-end isolation is less than 70 dB, measures must be taken to ensure that the receiver is not saturated, e.g., using high-quality RF components in receiver, and the overall SIC requirement and the reasonable allocation for each stage of SIC in different domain need to be considered in the beginning.

10.3.1 Full-Duplex Transceiver Framework

In order to achieve a SIC below noise floor, the transceiver design for high-power full-duplex base station faces great challenges. From the perspective of engineering implementation, a transceiver design framework with joint SIC across antenna domain, RF domain, and digital domain is more practical [15], as shown in Fig. 10.7. The framework design guidelines are as follows.

(1) Under the two limiting factors of acceptable antenna size and radiation pattern, maximize the SIC capability in the antenna domain, and reduce the direct coupling self-interference between the transmitting and receiving antennas to avoid receiver saturation, thereby reducing the burden of RF domain SIC. Increasing the SIC capability in the antenna domain can also effectively eliminate the transmitter noise in the first stage.

(2) The purpose of RF domain SIC is to match the strongest few paths for cancellation only in the case that near-field reflection of self-interference exists, to avoid strong reflection causing receiver saturation at a lower cost. In some cases, RF domain SIC may not be needed, by minimizing the requirement on RF domain SIC.

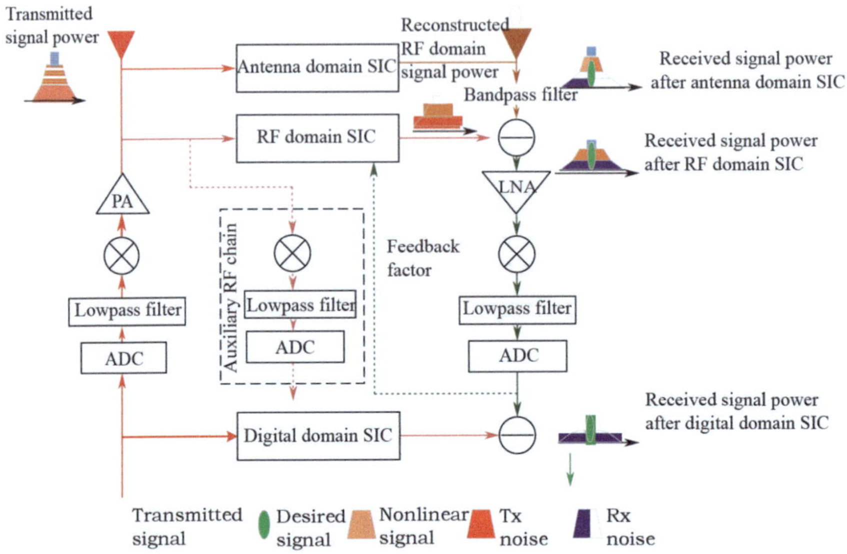

Fig. 10.7 Multi-domain joint SIC transceiver framework

(3) On the conditions that the receiver is not saturated with the first two stages of SIC, and the transmitter noise is suppressed to negligible level, residual self-interference can be further minimized in digital domain by using efficient digital signal processing. Among them, to achieve SIC requirement below the receiver noise floor, effective cancelation on non-linear residual self-interference and accurate self-interference channel estimation are crucial. Non-linear self-interference processing can be achieved through non-linear modelling or auxiliary RF feedback links.

Based on the above framework design, the subsequent content first briefly describes the self-interference signal model, and then elaborates on the SIC schemes in these three domains.

10.3.2 Self-interference Signal Model

As shown in Fig. 10.7, in the full-duplex transceiver, a known transmitted self-interference signal $s(t)$ is added with emission noise and the cascading non-linear effects of various devices after passing through active radio frequency components, due to the non-ideal factors of the components, especially power amplifier. Based on the Polynomial Hammerstein model [18], the transmission signal outputted by the power amplifier $s_{PA}(t)$ can be represented as

$$s_{\text{PA}}(t) = \sum_{p}^{P} f_{\text{PA},t} \otimes s(t)|s(t)|^{p-1} + n_{\text{TX}}(t) \tag{10.1}$$

where, P is the maximum non-linear order of odd numbers; p represents the p-th odd order component; $f_{\text{PA},t}$ represents the nonlinear response; $\otimes$ is the convolution operator; $n_{\text{TX}}(t)$ is the emission noise.

The above nonlinear signal passes through the complex self-interference channel shown in Fig. 10.3, and is represented by $h_{\text{SI}}(t)$ which represents the equivalent channel response of the self-interference signal from transmitter to receiver (i.e., the coupling response from PA to LNA, including nonlinear response, near-field antenna coupling, and air interface reflection). The self-interference signal at the receiver can be represented as

$$s_{RX,SI}(t) = h_{SI}(t) \otimes s_{PA}(t)) = \sum_{p}^{P} \underbrace{h_{p,t} \otimes f_{PA,t}}_{h_{SI,p,t}} \otimes s(t)|s(t)|^{p-1} + n'_{TX}(t) \tag{10.2}$$

where, $h_{\text{SI},p,t}$ is the unknown equivalent channel response of the p-th component. The full-duplex received signal is

$$r(t) = s_{\text{RX,SI}}(t) + y_{\text{wanted}}(t) + n_{\text{RX}}(t) \tag{10.3}$$

where, $y_{\text{wanted}}(t)$ is the desired signal to be received; $n_{\text{RX}}(t)$ is the noise at the receiver.

It can be seen that, to remove $s_{\text{RX,SI}}(t)$ from $r(t)$, is actually to use the estimated equivalent self-interference channel response $\hat{h}_{\text{SI},p,t}$ for reconstructing the received self-interference signal $\hat{s}_{\text{RX,SI}}(t)$, that is,

$$\hat{s}_{\text{RX,SI}}(t) = \sum_{p}^{P} \hat{h}_{\text{SI},p,t} \otimes s(t)|s(t)|^{p-1} \tag{10.4}$$

The more accurate the channel estimation is, the closer $\hat{s}_{\text{RX,SI}}(t)$ approaches to $s_{\text{RX,SI}}(t)$, , and the smaller the residual self-interference will be. Then a larger the Single to Interference Noise Ratio (SINR) of the received signal after SIC $r_{\text{SIC}}(t)$ can be obtained, which approaches to that of non-full duplex.

$$r_{\text{SIC}}(t) = \underbrace{\left(s_{\text{RX,SI}}(t) - \hat{s}_{\text{RX,SI}}(t)\right)}_{\text{Residual SI}} + y_{\text{wanted}}(t) + n_{\text{RX}}(t) \tag{10.5}$$

10.3.3 Antenna Domain SIC

Antenna domain SIC, also known as spatial suppression, is the first stage in the overall SIC framework. It aims to minimize the self-interference power received by the receiver antenna, to avoid receiver saturation to the greatest extent. There are generally two types of methods: antenna physical isolation technology [4, 15, 20] or antenna interference cancellation [21–23]. The latter has drawbacks as high cost, implementation difficulty, and hard to guarantee cancellation performance in practice. The basic idea is to use auxiliary antennas or auxiliary phase networks to specifically construct a received signal with a phase difference of 180°, thereby cancelling the interference. The principle of this method is to use antenna distance and/or phase shifting, that is, the electromagnetic wave coupling path, to achieve 180° difference for electromagnetic wave cancellation. However, in reality, it is impossible for large bandwidth systems to achieve a precise 180° difference for all frequency signals, because the path construction can only determine a path difference based on the wavelength of a certain frequency point. The 180° difference is only effective for that precise frequency point component, besides this method can potentially create null areas of the received signal. Considering the controversy on the practical feasibility of the latter method, in this section we mainly discuss the first method of antenna physical isolation technology and introduces a more practical method of antenna domain SIC. From the perspective of the signal model, antenna physical isolation technology does SIC by attenuating the signal $s_{\mathrm{RX,SI}}(t)$ in Eq. (10.2), that is, the signal after antenna isolation is equivalent to $\beta s_{\mathrm{RX,SI}}(t)$, and the self-interference signal power is reduced according to the isolation degree β^2. As can be seen from Eq. (10.2), the emission noise is also suppressed, avoiding the impact on subsequent digital domain processing.

The near-field self-interference coupled between the transmitting and receiving antennas can be largely attributed to the surface wave coupling caused by surface currents and the spatial coupling of electromagnetic waves radiated in the near field. Polarization, circulators, shielding cavities, defective grounding structures, electromagnetic bandgap structures, and absorbing materials can all achieve a certain level of transmit-receive isolation, but considering the power level of base station, none of them can independently achieve the desired SIC performance to avoid receiver saturation (such as greater than 60 dB of self-interference transmit-receive isolation) and some of them may have large impact on the antenna radiation pattern. In addition, in actual systems, both polarized ports will be used for transmission, hence it is not practical to use polarized ports for full-duplex self-interference isolation. Besides, the polarization isolation in the near field inside the antenna shield cannot be expected to be as good as that in the far field, which can be observed in some full-duplex antenna domain SIC tests [15]. For high-power base stations, it is not realistic to use circulators for transmitter-and-receiver shared antennas at present. In this section we only discuss self-interference suppression under separated transmitting and receiving antennas.

Mobile base stations usually use multiple antennas, and the design of antenna domain SIC is preferred to be applied to all the scenarios with different number of antennas from an engineering perspective. The physical separation of transmitting-receiving antennas with additional choke structures is more practical, with acceptable implementation difficulty and good interference suppression effect.

Early research on choke structure [28–30] proved from the theory of electromagnetic wave transmission that the impedance of the corrugated surface at a height of $\lambda/4$ (where λ is the wave length) is infinite, which means that surface wave transmission is perfectly suppressed (or understood as surface current does not conduct). Thus there is no coupling interference between the transmitting and receiving antennas. Note that this proof is only for theoretical single-frequency signals. In practice, electromagnetic wave signals are transmitted with a certain bandwidth, so the corrugated surface at a height of $\lambda/4$ designed for the centre frequency of electromagnetic waves cannot perfectly suppress surface waves. Nevertheless, this principle on choke provides good theoretical support to the design of antenna domain SIC.

Based on the above corrugated surface theory, we strive for design on self-interference suppression antennas for large bandwidth signals. Given the limitations of antenna size and radiation pattern, it is necessary to specifically design the parameters of the corrugated surface to maximize the effect of interference suppression [4, 15]. The self-interference suppression antenna based on the high-impedance corrugated surface structure is shown in Fig. 10.8, which is a 2T2R dual-polarization antenna based on the above principle. Multiple choke walls can construct cascaded choke slots, and such a corrugated surface has a better choking effect than a single slot, because the theoretical height of choke wall in corrugated surface is different for different frequencies within wide system bandwidth, and multi-wall structure can better suppress electromagnetic surface waves at different frequency segments. The height h, thickness t and spacing d of each choke wall must be selected under certain boundary conditions, according to factors such as bandwidth, frequency band, size, etc., to achieve best self-interference cancellation performance.

From the perspective of surface impedance, it can be proven that when the height h is in the range $\lambda/4 < h < \lambda/2$, where λ is the wavelength of the centre frequency of the electromagnetic wave, the corrugated surface has a good suppression effect on the surface wave. In addition, according to literature [28], in order to present only non-evanescent waves (Nonevanescent Mode, i.e., transverse electric field surface waves) in the corrugated surface choke slot for easy suppression of surface wave coupling, the thickness and spacing of the choke plate need to satisfy $t \ll d + t < \lambda/2$.

The design concept of full-duplex self-interference suppression antenna scheme based on corrugated surface can be extended to large-scale antenna arrays. It only needs to adopt an appropriate choke scheme between the transmitting and receiving antenna arrays for effectively suppressing near-field coupled self-interference. Of course, the specific design varies according to the actual situation, but the concept is the same. This method is low-cost and easy to combine with other antenna coupling suppression technologies, such as electromagnetic bandgap structures, which is suitable for the implementation of integrated full-duplex antennas.

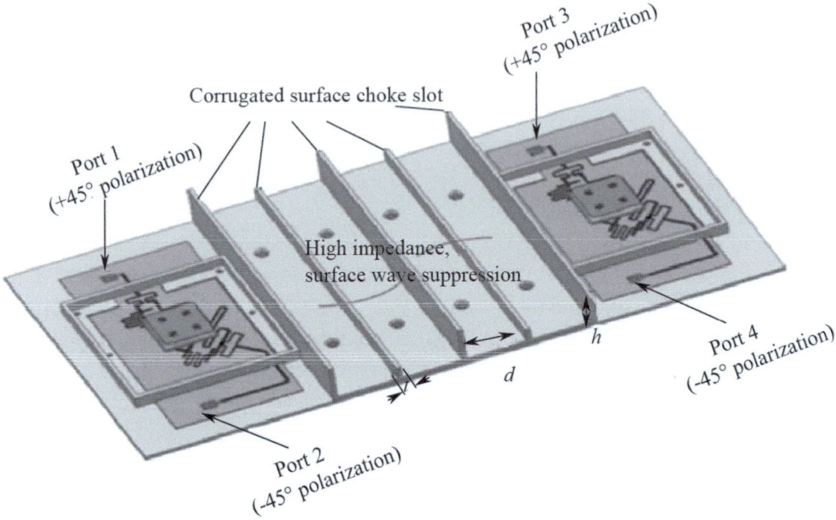

Fig. 10.8 Self-interference suppression antenna based on high-impedance corrugated surface structure

10.3.4 RF Domain SIC

In general, RF domain SIC can be regarded as an analogue filtering process, where the filtering coefficients include delay, amplitude, and phase to match the self-interference channel. To avoid saturation of the subsequent receiver, the analogue self-interference signal can be reconstructed according to dominant strong paths.

For better compatible to the multipath self-interference channel, RF domain SIC can use a multi-tap adjustable filtering network for self-interference reconstruction. Each tap consists of a delay device and an amplitude-phase adjuster, as shown in Fig. 10.9. All the delay, amplitude, and phase support a certain adjustable range, thereby approximating the N strong paths in the self-interference channel for elimination.

The reconstruction of self-interference is usually a searching process to match the self-interference channel, based on the known analogue signal obtained from the transmitter through the coupler, by adjusting the delay, amplitude, and phase parameters of the filtering network to approximate the self-interference channel as much as possible, thereby obtaining the reconstructed signal of formula (10.4), and then performing the cancellation operation with the received self-interference. Specifically, the filtering parameter optimization problem can be given by (10.6)

$$\min_{(\omega_1,\cdots,\omega_N,\nu_1,\cdots,\nu_N)} \left(r(t) - \sum_{i=1}^{N} \mathrm{Re}\left\{\omega_i e^{j2\pi f_0 t} s(t-\nu_i)\right\}\right)^2 \tag{10.6}$$

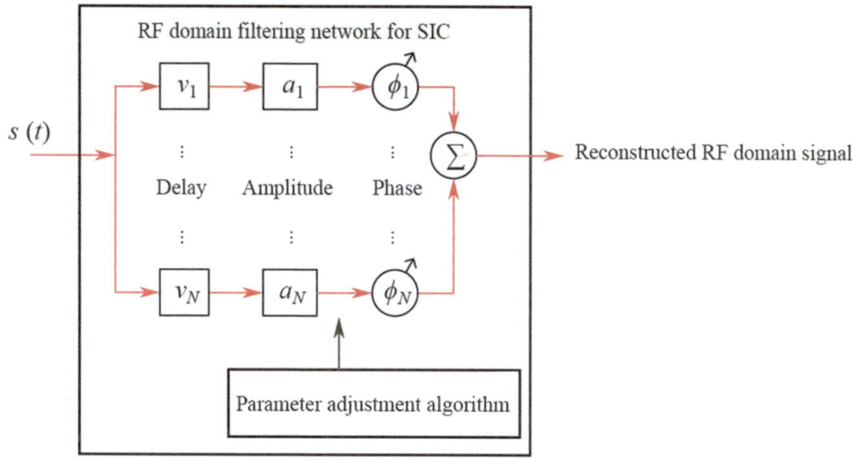

Fig. 10.9 RF domain multi-tap adjustable filtering network for self-interference reconstruction

where, $\omega_1, \ldots, \omega_N$, $\nu_1, \ldots, \nu_N$ are respectively the complex coefficients (amplitude and phase) and tap delays of each tap; $r(t)$ is the received signal; $s(t)$ is the known transmitted signal. Although the gradient descent algorithm [31] is suggested for solving Eq. (10.6), non-ideal factors of the RF circuit may cause equation in (10.6) to be non-convex, which makes solutions based on this algorithm cannot achieve global optimality.

The following improved search algorithm is more effective and worth considering in engineering practice, as described in the following.

Firstly, the initial search point is obtained based on the residual self-interference power P_{res} calculation.

$$\frac{P_{\text{res}}}{B} = R + P^H W + W^H P + W^H Q W \tag{10.7}$$

where, B is the bandwidth; $W = [\omega_1, \cdots, \omega_N]^T$ is the complex vector of self-interference cancellation gains.

$$R = \sum_{k=1}^{K} \sum_{m=1}^{K} a_k^* a_m \text{sinc}[B(\tau_m - \tau_k)] \exp\{-j2\pi F_c(\tau_m - \tau_k)\}$$

$$Q = \begin{bmatrix} \frac{\text{sinc}[B(\nu_1 - \nu_1)]}{e^{j2\pi F_c(\nu_1 - \nu_1)}} & \cdots & \frac{\text{sinc}[B(\nu_1 - \nu_N)]}{e^{j2\pi F_c(\nu_1 - \nu_N)}} \\ \vdots & & \vdots \\ \frac{\text{sinc}[B(\nu_N - \nu_1)]}{e^{j2\pi F_c(\nu_N - \nu_1)}} & \cdots & \frac{\text{sinc}[B(\nu_N - \nu_N)]}{e^{j2\pi F_c(\nu_N - \nu_N)}} \end{bmatrix}$$

$$P = \begin{bmatrix} \sum_{k=1}^{K} a_k \mathrm{sinc}[B(\tau_k - \nu_1)]e^{-j2\pi F_c(\tau_k - \nu_1)} \\ \vdots \\ \sum_{k=1}^{K} a_k \mathrm{sinc}[B(\tau_k - \nu_N)]e^{-j2\pi F_c(\tau_k - \nu_N)} \end{bmatrix}$$

where, F_c represents the carrier frequency; K is the number of channel paths in the actual self-interference channel; k is the channel path index; N is the number of taps in the filtering network, corresponding to the number of channel paths to be cancelled in the RF domain. As can be inferred from the aforementioned "delay, amplitude, and phase support a certain adjustable range, thereby approximating the N strong paths in the self-interference channel to be eliminated", N is the number of dominant paths in the K path of the actual self-interference channel. The meaning of the above formula is the autocorrelation function of the self-interference channel. Therefore, k should traverse the number of channel paths K in the actual self-interference channel, which is not relevant to N. a_k and τ_k are complex coefficient and delay of the *k-th* channel path. ω_k and ν_k are complex coefficient and delay of the *k-th* tap in the RF SIC filtering network. The optimal $\boldsymbol{W}$ value for minimizing residual self-interference can be obtained from Eq. (10.8).

$$W_{\mathrm{opt}} = -Q^{-1}P \tag{10.8}$$

Here Q is a matrix determined based on delays in the RF domain SIC circuit, and P can be obtained through the following steps.

(1) Randomly generate a new self-interference gain vector.
(2) Based on the vector in step (1), perform RF domain SIC.
(3) Calculate the residual self-interference.
(4) Obtain a new SIC observation vector.

Repeat the above steps $2N + 1$ times to obtain Eq. (10.9).

$$\begin{gathered} XY = G \\ X = \begin{bmatrix} \omega_{1,\mathrm{re}}^{(1)} & \omega_{1,\mathrm{im}}^{(1)} & \cdots & \omega_{N,\mathrm{re}}^{(1)} & \omega_{N,\mathrm{im}}^{(1)} & 1 \\ \vdots & \vdots & & \vdots & \vdots & \vdots \\ \omega_{1,\mathrm{re}}^{(2N+1)} & \omega_{1,\mathrm{im}}^{(2N+1)} & \cdots & \omega_{N,\mathrm{re}}^{(2N+1)} & \omega_{N,\mathrm{im}}^{(2N+1)} & 1 \end{bmatrix} \\ Y = \begin{bmatrix} p_{1,\mathrm{re}} & p_{1,\mathrm{im}} & \cdots & p_{N,\mathrm{re}} & p_{N,\mathrm{im}} & \frac{R}{2} \end{bmatrix} \end{gathered} \tag{10.9}$$

where, X is a square matrix containing $2N + 1$ gain vectors, $\omega_{k,\mathrm{re}}^{(i)}$, $\omega_{k,\mathrm{im}}^{(i)}$ is the real and imaginary part of the k th tap's i th coefficient. $p_{i,\mathrm{re}}$ and $p_{i,\mathrm{im}}$ are the real and imaginary parts of the $\boldsymbol{P}$'s i-th element, $\boldsymbol{G}$ is the observation matrix with self-interference removed. Accordingly, $\boldsymbol{Y}$ obtained through $Y = X^{-1}G$ can be used to calculate W_{opt} in Eq. (10.8) with the required $\boldsymbol{P}$.

Once the optimal W_{opt} is obtained, it can be used as the initial search point for the second search stage, so that the traditional gradient descent algorithm can be used in the second search stage to quickly obtain the locally optimal coefficients for RF domain SIC.

In addition, as shown in Fig. 10.7, digital self-interference channel estimation assisted feedback can also be used to optimize the coefficients of the RF domain SIC circuit, generally by initially cancelling the RF domain self-interference to a certain extent so that receiver saturation doesn't happen when receiving pilot signal, and then the digital domain channel estimation can be used to further assist in adjusting the coefficients of the multi-tap filtering network required for RF domain self-interference reconstruction. Assuming that the pilot signal reception is not saturated, and the digital domain channel estimation is valid, then the optimization problem in Eq. (10.6) is simplified to a self-interference channel matching optimization problem in Eq. (10.10).

$$\min_{(\omega_1,\cdots,\omega_N,\nu_1,\cdots,\nu_N)} \sum_i^N \left| h_{\text{SI},\tau_i} - \underbrace{\omega_i \delta(t-\nu_i)}_{\hat{h}_{\text{SI},\tau_i}} \right|^2 \tag{10.10}$$

This can avoid the use of search algorithms, and ensure the searching complexity of the RF domain multi-tap filter network same as that with traditional digital domain channel estimation. h_{SI,τ_i} is the equivalent channel response of the i th path of the self-interference channel. This channel estimation is performed in the digital domain, and the optimization formula can assist the RF domain multi-tap filter network ω_i and ν_i parameter settings, that is, the digital-assisted RF domain reconstruction of the $\hat{h}_{\text{SI},\tau_i}$ channel and the self-interference channel h_{SI,τ_i} Optimization is carried out to minimise self-interference. The difficulty in implementing this method is how to ensure the pilot reception unsaturated. To use low-power pilot design is one option, but it will reduce the estimation accuracy and increase system overhead, thereby reducing the gain of full duplex. Another option is to remove RF self-interference on the initial direct coupling path by pre-setting the RF tap coefficient to avoid saturation when receiving the high-power pilot. However, in complex environments with strong reflections, it is impossible to guarantee the performance. Therefore, digital self-interference channel estimation assisted RF interference cancellation has significant limitations in practical use, and combining it with search algorithms may be a better choice.

RF domain self-interference cancellation can be understood as adding a multi-tap filtering network equivalent to increasing the spatial path of electromagnetic waves, except that the electromagnetic waves are transmitted through the multi-tap filtering network rather than through the air. These increased degrees of freedom are used for self-interference cancellation, similar to the principle of antenna interference cancellation mentioned we discussed earlier, with the implementation cost becoming the addition of RF circuits composed of couplers, delay devices, and amplitude and phase modulators, rather than physical antennas. The more spatial paths of the self-interference channel, the more degrees of freedom required, and therefore the more taps in the filtering network. Therefore, it is quite difficult for actual large-bandwidth multi-antenna systems to achieve a satisfied performance with RF domain SIC. For example, a 5G base station with 64 transmitting and 64 receiving antennas, considering only one path, to optimize the performance of RF domain SIC, requires 4096 taps to reconstruct 4096 direct coupling paths of transmitting and receiving antennas. Considering the number of paths in a 100 MHz bandwidth system in a real environment, the number of taps needs to be multiplied, making the hardware complexity of RF domain SIC unrealistic. This is why one of the design principles proposed in Sect. 10.3.1 is to minimize the expectation of RF domain SIC. Starting from the complexity that engineering implementation can bear, we can use dimension reduction design to implement a limited number of taps for RF domain SIC to ensure that the receiver is not saturated, and the rest can be handed over to efficient digital domain processing.

10.3.5 Digital Domain SIC

After the first two stages of self-interference cancellation, as long as the receiver is not saturated, the digital domain can carry out efficient digital signal processing, reconstruct the residual self-interference in digital domain, effectively delete the residual self-interference, and effectively handle complex multi-antenna, multi-path self-interference channels. The goal of digital domain self-interference cancellation is to further delete the residual self-interference to below the receiver's base noise, making the SINR of the useful signal comparable to the FDD/TDD SNR, which is a key indicator for full duplex to approach the maximum theoretical gain.

For sub-band non-overlapping full duplex, as mentioned earlier, despite theoretically the transmission and reception are orthogonal, both out-of-band leakage due to non-linearity and non-orthogonality caused by frequency offset can result in significant non-linear self-interference. Furthermore, full-band full duplex and sub-band overlapping full duplex will also have in-band non-linear self-interference caused by non-linearity. Proactively addressing the transmitter non-linearity can reduce non-linear self-interference from the beginning, thereby reducing the burden of digital domain and RF domain SIC at the receiver. In base stations equipped with Digital Pre-distortion (DPD), the nonlinear self-interference generated by the transmitter can be initially mitigated using DPD. The non-linearity of the transmitting channel is equivalently modelled as shown in Fig. 10.10. The non-linear signal output by the power amplifier $x_{n\mathrm{PA}}$ is obtained after the coupler and the auxiliary RF feedback link's ADC to get the reference sample $\tilde{x}_{n\mathrm{PA}}$. Based on the reference sample, the Indirect Learning Architecture with Least-Square (ILA-LS) method [19] is used to train the inverse parameters relative to the non-linear response, allowing for the pre-distortion of the transmitted signal $\boldsymbol{x}$ to yield the signal $\tilde{x}$. $\tilde{x}$, after undergoing the response of active device at transmitter $f_{\mathrm{PA},t}$, can maximizes the linearity of the transmitted signal

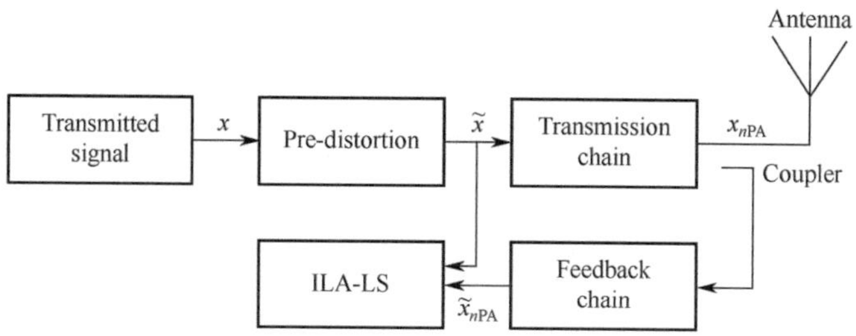

Fig. 10.10 Non-linear self-interference improvement system based on transmitter DPD

$\boldsymbol{x}$, because pre-distortion serves as an inverse operation counteracting the non-linear response of the transmitter $f_{\text{PA},t}$.

Although utilizing DPD can alleviate non-linear self-interference to a certain degree, practical constraints such as fractional delay precision alignment limit its effectiveness. It remains challenging to actively eliminate non-linear self-interference at the transmitter end down to a negligible level. Furthermore, some implementations either lack support for DPD functionality or face difficulties implementing it. Regardless of the full duplex mode adopted, high-power mobile cellular networks require digital domain SIC capabilities, not only to deal with residual self-interference, but also to handle non-linear self-interference (e.g., when the previous SIC stage fails to suppress non-linearity lower than the noise floor).

The core aspect of digital domain SIC is the estimation of the self-interference channel, based on which the self-interference is reconstructed and finally deleted. Either pilot signal or data signal can be used for self-interference channel estimation, as long as uncontaminated transmission is ensured. To enhance the digital domain SIC, power boosting can be applied to the pilot. As for self-interference reconstruction, non-linearity issue should be taken into account. As discussed earlier, the auxiliary RF link shown in Fig. 10.7 can be used to introduce a non-linear transmitted signal replica for reconstruction, simplifying nonlinear processing to linear processing with minimal complexity without additional hardware costs for base stations equipped with DPD functionality, as the DPD auxiliary RF feedback link can be directly reused. However, for devices without this feedback link, the baseband transmitted signal without non-linearity must be used for self-interference reconstruction, necessitating a modelling based approach similar to DPD. At the receiver end, non-linearity needs to be re-modelled, which is more difficult than DPD since reconstruction also needs to estimate the air interface channel of the non-linear component.

The concept behind digital domain SIC is easy to understand. Referring to the signal model in Sect. 10.3.2, once the self-interference channel is estimated using conventional channel estimation algorithms, the self-interference reconstruction and cancellation operation can be performed according to Eq. (10.4). Emphasis should be placed on the signal quality of the pilot and the accuracy of the self-interference channel estimation. It is essential to consider the effects of oscillator phase noise, frequency offset, and quantization error on digital domain interference cancellation in practical applications. Necessary link optimization needs to be conducted to ensure the accuracy of the reconstructed self-interference signal. Full-duplex system designs leveraging an auxiliary RF feedback link exhibit optimal performance in practice, as the aforementioned adverse error impacts can be entirely eliminated through link optimization during engineering implementation.

10.4 Network Cross-link Interference Management Mechanism

For dynamic TDD systems, the 3GPP standard protocol has supported a network cross-link interference (CLI) management mechanism. Since CLI in full-duplex network has some similarities with that in dynamic TDD, the same mechanism can be reused or further enhanced for full duplex. As early as 3GPP LTE-Advanced Release 12, enhanced Interference Mitigation and Traffic Adaptation (eIMTA) was standardized to enable different cells and operators to utilize different uplink and downlink ratios in certain scenarios. In the 3GPP 5G NR standardization process, enhanced schemes were initially investigated in Release 15 to support duplex flexibility and LTE-NR coexistence. Further evaluation and research were conducted in Release 16 for CLI [32, 33] and then necessary measurement, reporting and information exchange mechanisms were standardized subsequently.

In full-duplex networking, apart from self-interference, there exists CLI as described in Sect. 10.2 of this book. Similar to dynamic TDD networking, this interference includes CLI from downlink transmission of one base station to uplink reception of its neighbouring base station (DL-to-UL CLI) and CLI from uplink transmission of one terminal to downlink reception of its neighboring terminal (UL-to-DL CLI). Note that DL-to-UL CLI is crutial to networking.

DL-to-UL CLI arises when neighboring base stations apply different transmission directions (DL or UL) on the same time–frequency resource. Cellular base stations typically have a high Equivalent Isotropically Radiated Power (EIRP), making it likely for one base station' downlink transmission to overpower the uplink signal of a nearby base station, which results in receiver blocking or desensitization, hindering normal uplink reception, especially when base stations are deployed at height. DL-to-UL CLI is further classified into co-channel CLI and Adjacent Channel CLI. The former pertains to CLI on the same frequency of the same operator, while the latter occurs between different operators using adjacent frequency carriers.

Typically, it is impossible for operators to manage DL-to-UL Adjacent Channel CLI through coordination between base stations. Instead, using adjacent channel leakage suppression or isolation is more effective. Sub-band full duplex offers flexibility to isolate adjacent channel CLI. As shown in Fig. 10.11, adjacent channel interference between operators can be avoided by allocating the uplink sub-band away

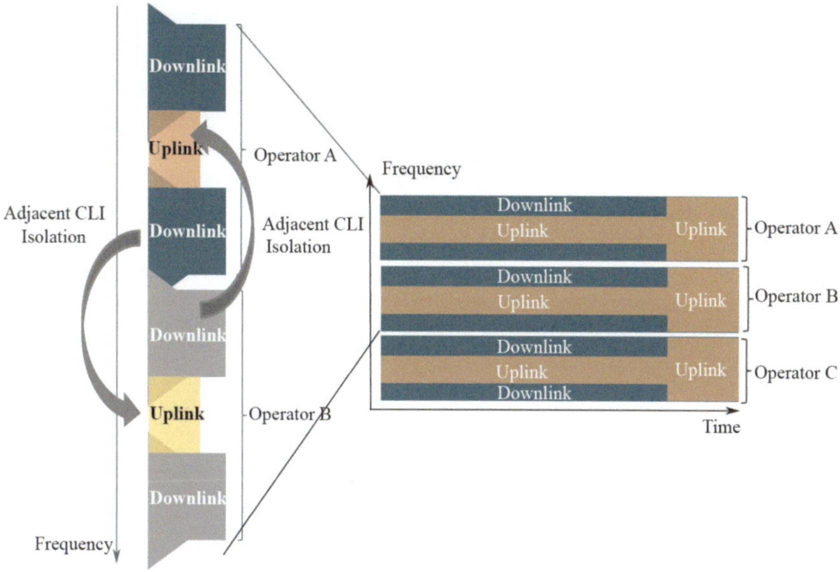

Fig. 10.11 Mechanism for adjacent channel interference avoidance between operators

from the downlink band of the neighboring operator, i.e., isolate adjacent channel leakage.

DL-to-UL Co-channel CLI, which occurs within the same operator's network, allows for base station cooperation. Effective methods to control this type of CLI include precise downlink beamforming, cooperative scheduling, cooperative beam management, and power control. In some specific deployment scenarios, adjusting the antenna height and tilt, utilizing beam alignment and beam nulling techniques [34] can effectively reduce the DL-to-UL CLI caused by full duplex, so that CLI will not limit the achievable throughput gains of full duplex. For example, in Integrated Access and Backhaul (IAB) networks employing full duplex, proper antenna deployment planning can be expected because the backhaul link is stable. Besides, large-scale antenna array can be used to enhance point-to-point transmission, providing narrow beams to precisely align beam directions and reduce interference between devices. Furthermore, it also provides flexibility to suppress sidelobes and nullify in beam directions that produces severe interference, which make CLI manageable.

UL-to-DL CLI occurs when neighboring terminals employ distinct transmission directions (e.g., DL, UL) on the same time–frequency resources, either within the same cell or between neighboring cells. Because maximum transmitting power of terminal generally lower than 26 dBm, much smaller than base stations, the power level of UL-to-DL CLI is much lower than DL-to-UL CLI. It usually does not totally obstruct downlink reception, but can considerably affect downlink performance. The current mechanisms studied by 3GPP [33] on UL-to-DL CLI management can be effective to full duplex networks, including cooperative scheduling, cooperative beamforming, terminal power control, and link adaptive adjustment. These mechanisms rely on CLI measurement, reporting mechanisms, and essential information exchange to offer cooperative references for network decision-making, enabling efficient management of this type of CLI.

Reference [3] presents a comprehensive summary and evaluation results of candidate solutions for CLI management, from 3GPP. Those solutions serve as valuable references for designing CLI handling strategies in future full duplex networks.

10.5 Concept Verification

Reference [24] summarizes initial verification results on full-band full-duplex SIC. However, indicates that achieving sufficient SIC capability for high-power full-duplex systems suitable for mobile communication systems remains challenging, with only 100–110 dB [25, 26] being achieved. This falls short of the required SIC capability, considering that base stations typically operate at powers exceeding 1W (30 dBm) and full-band full-duplex necessitates at least 120 dB of SIC capability to prevent receiver desensitization.

Industry leaders have recently showcased successful full-duplex prototype verification systems, demonstrating the feasibility of full-band full-duplex for mobile cellular base stations. Reference [27] utilizes joint antenna domain, RF domain and digital domain SIC, achieving an overall SIC capability of 119.4 dB in the low frequency band by significantly improving performance of RF domain SIC. This study verifies the potential implementation of full-duplex at base stations operating at powers beyond 1W in realistic reflective environments. Another study, reference [15], adopts similar SIC techniques, including the corrugated surface choke technique, multi-tap filter network RF domain SIC technique, and auxiliary RF feedback link digital domain SIC technique. This approach enhances the overall SIC capability to 122.5 dB (Fig. 10.12 shows the field test results of SIC with 2.4 GHz prototype), allowing error-free simultaneous bidirectional transmission at the same frequency for a base station power amplifier output power of 32 dBm.

Reference [4] presents the world's first sub-band non-overlapping full-duplex concept verification test bench (operating at 2.4 GHz with a 100 MHz bandwidth), which implements a three-stage SIC with transmitter DPD non-linear self-interference suppression, antenna domain isolation and receiver non-linear SIC. This three-stage SIC design demonstrates that multi-stage SIC transceivers can suppress sub-band leakage non-linear self-interference below the receiver's noise floor when the power amplifier outputs 30 dBm. Test results indicate that sub-band non-overlapping full-duplex SIC degrades signal quality by just 1 dB compared to TDD's self-interference-free signal quality. This enables sub-band full-duplex to potentially double uplink coverage by combining retransmissions with 5 times of the uplink slot resources with traditional TDD under a given modulation order. Figure 10.13 illustrates the sub-band non-overlapping full-duplex test bench and SIC testing process.

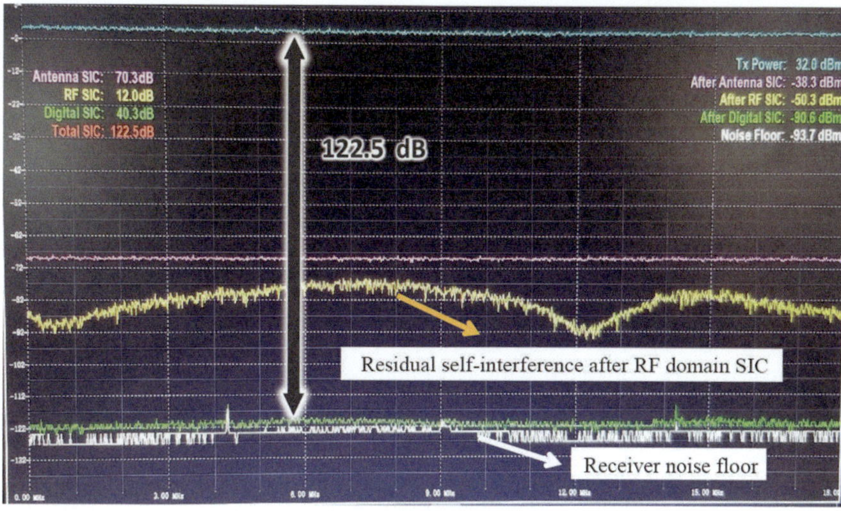

Fig. 10.12 Field test results of SIC with 2.4 GHz prototype

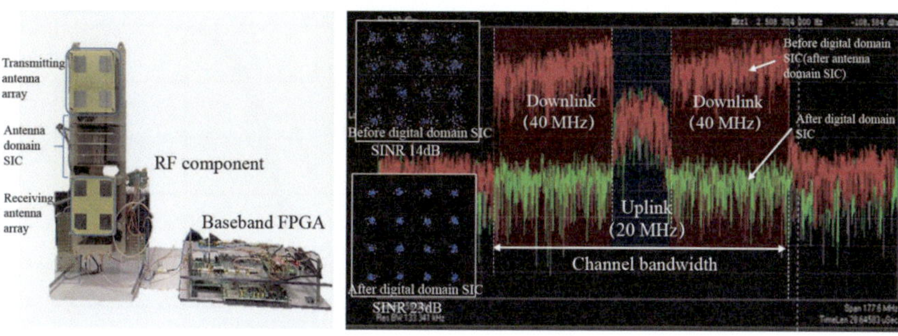

Fig. 10.13 Sub-band non-overlapping full-duplex test bench and SIC test

To further validate the feasibility of full-band full-duplex for integrated access and backhaul (IAB) networks, reference [17] measured self-interference levels in a developed millimeter wave full-duplex base station, which includes self-interference between the same sector and the same station sector. Then system-level evaluation analysis is conducted based on the measurement results. The analysis adopted evaluation method of 3GPP under IAB topology structure and considered all types of network interference. It indicated that network interference is manageable with aid of precise control on narrowband beams, and full-band full-duplex can achieve objective gains under multi-cell networking. It also compared the SINR of half-duplex and full-band full-duplex, and analyzed the SINR loss rate caused by full-duplex self-interference and CLI, and showed that network interference is also manageable in the IAB scenario. The SINR distribution and loss rate post-SIC are presented in Fig. 10.14. By calculating the SINR loss due to self-interference and CLI relative to half-duplex without interference, the loss rate was determined, which is defined as the Cumulative Distribution Function (CDF) difference between TDD and full-duplex at the SINR threshold required for demodulating 64 QAM and 256 QAM with the highest channel coding rate. This loss rate represents the portion of transmission failure due to full-duplex interference when utilizing the highest modulation coding scheme. Comparison results suggest that effective SIC prevents significant SINR losses from full-duplex interference. Moreover, comparing the contributions of self-interference to SINR loss with and without SIC confirms the controllability of CLI under millimeter wave narrow beams, limiting its impact on SINR.

Reference [17] further compared the cell throughput under full-duplex and half-duplex IAB networking. Following effective SIC, full-duplex can increase the throughput by 70% compared with half-duplex, which once again shows that the networking interference is controllable.

Reference [17] did not present the link performance test results of the developed millimeter wave full-duplex concept test bench. However, during a keynote speech at the 2022 IEEE Consumer Communications & Network Conference (CCNC), a demonstration of the millimeter wave full-duplex test bench in an outdoor setting at a

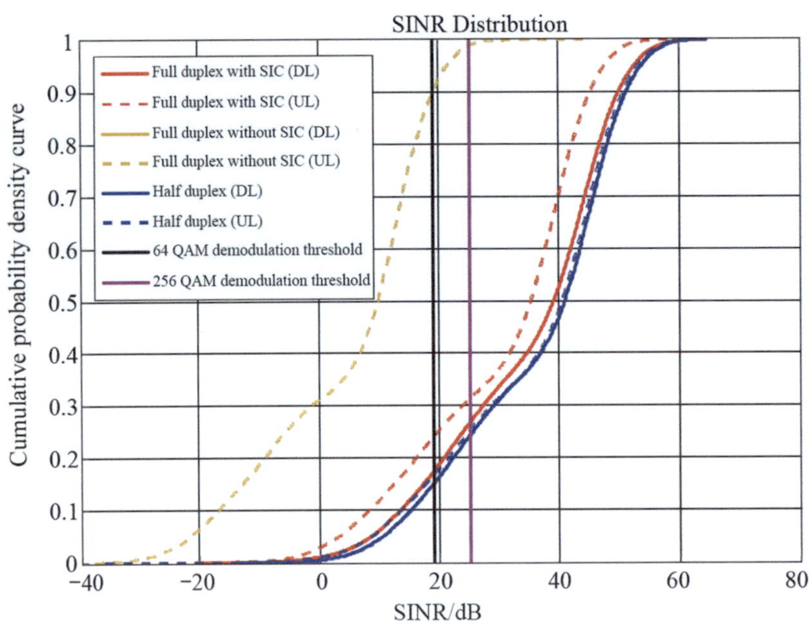

	Downlink	Uplink
Average SINR loss	1.6 dB	5.2 dB
Loss rate (up to 64 QAM)	2.3%	7.3%
Loss rate (up to 256 QAM)	1.9%	5.0%

Fig. 10.14 SINR distribution and loss rate after SIC

distance of 100 m was showcased. Under 64 QAM modulation, full-duplex achieved twice the throughput. This serves as evidence for the feasibility of obtaining high-performance SIC with high power and large antenna arrays, bolstering confidence in the adoption of full-duplex technology in the future.

The industry's concept verification efforts extend beyond these examples, with the focus shifting towards practical large-scale antenna array high-power systems. We anticipate further advancements, facilitating easier engineering implementation of SIC schemes, CLI management schemes, and real-world verification results. Figure 10.15 compares the average throughput of full-duplex and half-duplex cells.

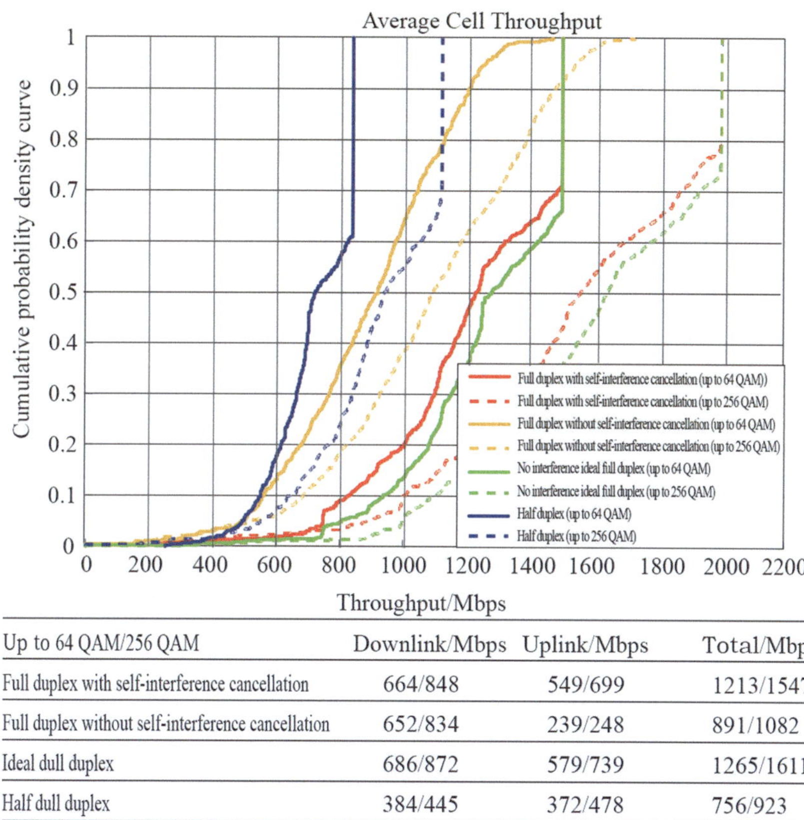

Up to 64 QAM/256 QAM	Downlink/Mbps	Uplink/Mbps	Total/Mbps
Full duplex with self-interference cancellation	664/848	549/699	1213/1547
Full duplex without self-interference cancellation	652/834	239/248	891/1082
Ideal dull duplex	686/872	579/739	1265/1611
Half dull duplex	384/445	372/478	756/923

Fig. 10.15 Comparison on average throughput of full-duplex and half-duplex cells

10.6 Summary

Duplex technology has been actively researched since the early stages of 5G development. Over the past decade, it has matured and evolved to provide networking flexibility while avoiding interference. Sub-band non-overlapping full-duplex led the way in 3GPP project research, and full-band full-duplex, as a promising enabling technology for 6G, requires accelerated practical exploration to fulfill its potential on the 6G stage.

This chapter explores the driving forces behind new duplex research, highlighting key issues, main challenges, and applicable scenarios. For future cellular high-power full-duplex base stations, joint SIC design criteria are discussed, along with outlining SIC technologies for each domain. Networking cross-link interference and candidate mechanisms are addressed, providing insights from an engineering practice perspective for interference cancelation schemes.

Finally, this chapter shows representative concept verification results, reflecting the current research progress and technological maturity. The new duplex technology is anticipated to become a reality, yet it requires collaborative efforts across industries and continuous advancement in key technologies.

References

1. 3GPP. Study on NR Coverage Enhancements (Release 17), V17.0.0, document TR 38.830[S]. 2020.
2. ITU. IMT Traffic Estimates for 2020 to 2030: R M.2370[S]. 2015.
3. CHEN S, BEACH M A, MCGEEHAN J P. Division-free duplex for wireless applications[J]. IEEE Electronics Letters, 1998, 34(2): 147–148
4. JI H, KIM Y, MUHAMMAD K, et al, Extending 5G TDD Coverage with XDD: Cross Division Duplex[J]. IEEE Access, 2021, 9: 51380–51392.
5. HUA Y, LIANG P, MA Y, et al. A method for broadband full-duplex mimo radio[J]. IEEE Signal Processing Letters, 2012, 19:793–796.
6. KIM D, LEE H, HONG D. A Survey of In-Band Full-Duplex Transmission: From the Perspective of PHY and MAC Layer[J]. IEEE Communications Surveys & Tutorials, 2015, 17(4): 2017–2046.
7. AHMED E, ELTAWIL A M. All-digital self-interference cancellation technique for full-duplex systems[J]. IEEE Transactions on Wireless Communications, 2015, 14(7): 3519–3532.
8. XIA X, XU K, WANG Y, et al. A 5G-enabling technology: Benefits, feasibility, and limitations of in-band full-duplex mMIMO[J]. IEEE Vehicular Technology Magazine, 2018, 13(3): 81–90.
9. CIRIK A C, BISWAS S, TAGHIZADEH O, et al. Robust transceiver design in full-duplex MIMO cognitive radios[J]. IEEE Transactions on Vehicular Technology, 2018, 67(2): 1313–1330.
10. SINGH U, BISWAS S, SINGH K, et al. Kanaujia and C. -P. Li, Beamforming Design for In-Band Full-Duplex Multi-Cell Multi-User MIMO LSA Cellular Networks[J] IEEE Access, 2020, 8: 222355–222370.
11. ALVES H, RIIHONEN T, SURAWEERA H A. Full-Duplex Communications for Future Wireless Networks[M]. Springer, 2020.
12. SADJINA S, MOTZ C, PAIREDER T, et al. A Survey of Self-Interference in LTE-Advanced and 5G New Radio Wireless Transceivers[J]. IEEE Transactions on Microwave Theory and Techniques, 2020, 68(3): 1118–1131.

13. MOTZ C, PAIREDER T, PRETL H, et al. A Survey on Self Interference Cancellation in Mobile LTE-A/5G FDD Transceivers[J]. IEEE Transactions on Circuits and Systems II: Express Briefs, 2021.
14. SABHARWAL A, SCHNITER P, GUO D, et al. In-band full-duplex wireless: challenges and opportunities[J]. IEEE Journal Selected Areas in Communications, 2014.
15. YU B, QIAN C, PENG L, et al. Full Duplex Communication with Practical Self-Interference Cancellation Implementation[J]. IEEE ICC, 2022: 1100–1105.
16. CORCORAN J, POULTON K. Analog-to-digital converters – 20 years of progress in "Agilent" oscilloscope[J]. Agilent Measurement Journal, 2007(1): 34–40.
17. KIM S, LEE K, YU B, et al. Measurement and System-level Evaluation of Inter-Sector Self-Interference on Full Duplex in mmWave Band[J]. IEEE ICC 2022, 2022: 21991146.
18. KIAYANI A, ANTTILA L, VALKAMA M. Modelling and dynamic cancellation of TX-RX leakage in FDD transceivers[J]. 2013 IEEE 56th International Midwest Symposium on Circuits and Systems (MWSCAS), 2013.
19. HUSSEIN M A, BOHARA V A, VENARD O. On the system level convergence of ILA and DLA for digital predistortion[J]. Proceedings International Symposium on Wireless Communication Systems (ISWCS), 2012: 870–874.
20. HANEDA K, KAHRA E, WYNE S, et al. Measurement of loop-back interference channels for outdoor-to-indoor full-duplex radio relays[J]. Proceedings of the Fourth European Conference on Antennas and Propagation (EuCAP), 2010.
21. KNOX M E. Single antenna full duplex communications using a common carrier[J]. IEEE Wireless Microwave Technology Conference, 2012.
22. NIKOLAOU S, BAIRAVASUBRAMANIAN R, LUGO C, et al. Pattern and frequency reconfigurable annular slot antenna using PIN diodes[J]. IEEE Transactions on Antennas and Propagation, 2006.
23. Choi J, JAIN M, SRINIVASAN K, et al. Achieving single channel, full duplex wireless communication [C]//ACM Mobicom, 2010.
24. KOLODZIEJ K E. In-band Full-Duplex Wireless Systems Overview[J]. IEEE ICC 2021, 2021: 1–6.
25. SARRET M G, BERARDINELLI G, MAHMOOD N H, et al. Analysing the potential of full duplex in 5G ultra-dense small cell networks[J]. EURASIP Journal on Wireless Communications and Networking, 2016, 284: 1–16.
26. BHARADIA D, MCMILIN E, KATTI S. Full duplex radios[J]. ACM SIGCOMM, 2013: 375–386.
27. CHEN F, LEE H H, MORAWSKI R, et al. RF/Analog Self-Interference Canceller for 2x2 MIMO Full-Duplex Transceiver[J]. IEEE ICC 2017, 2017:1–6.
28. OLVER A D, XIANG J. Wide angle corrugated horns analysed using spherical modal--matching[J]. Proceedings of the IEEE, 1988, 135(1).
29. KINDAL P. Artificially soft and hard surfaces in electromagnetics[J]. IEEE Transactions on Antenna and Propagation, 1990, 38(10).
30. HANNIEN I, NIKOSKINEN K. Implementation of method of moments for numerical analysis of corrugated surfaces with impedance boundary condition[J]. IEEE Transactions on Antenna and Propagation, 2008, 56(1).
31. WANG J, ZHAO Z, TANG Y. A RF adaptive least mean square algorithm for self-interference cancellation in co-frequency co-time full duplex systems[J]. IEEE ICC 2014, Sydney, 2014: 5622–5627.
32. 3GPP. Cross Link Interference (CLI) handling and Remote Interference Management (RIM) for NR: Tech. Rep. 38.828, v. 16.1.0[S]. 2019.
33. 3GPP. Study on New Radio Access Technology Physical Layer Aspects: Tech. Rep. 38.802, v. 14.2.0 [S]. 2018.
34. CHOI Y, SHIRANI M H. Simultaneous Transmission and Reception: Algorithm, Design and System Level Performance[J]. IEEE Trans. Wireless Commun.2013, 12(12): 5992–6010.

Chapter 11
Orthogonal Time–Frequency Space Modulation

Orthogonal frequency division multiplexing (OFDM) has been widely employed in 4G LTE and 5G NR. The OFDM communication system possesses outstanding advantages such as excellent resistance to multipath fading, simple channel estimation and equalization, easy implementation of receiver hardware, and tight coupling with MIMO technology. However, OFDM also exhibits some inherent drawbacks, such as high adjacent channel leakage ratio (ACLR), weak inter carrier interference (ICI) performance, and high peak to average power ratio (PAPR). These drawbacks can be mitigated through various engineering methods, allowing OFDM to continue its use in both 4G and 5G protocols. Nonetheless, this also presents some challenges and costs, such as low spectrum utilization, stringent synchronization requirements, and excessive feedback overhead. Faced with the increasingly demanding technical requirements of 6G, traditional OFDM waveform appear inadequate. Therefore, it is imperative to explore new waveform for 6G, in the hope of finding a highly flexible, spectrum-efficient, and unified waveform for both uplink and downlink, in order to better meet the demands of future ultra-high-speed, large-scale connectivity, and high-speed mobile scenarios.

11.1 Background

With the rapid development of technologies such as high-speed railways, smart highways, and drones, high-speed mobile scenarios are poised to become one of the critical environments in 6G. The Doppler frequency shift effect in high-mobility communications disrupts the orthogonality between subcarriers in OFDM systems, causing distortion and blurring in the time-domain signals at the receiver side, significantly impacting symbol detection performance. The current approach to address this issue involves trading reliability for overhead, sacrificing a portion of resource allocation for reference signals to conduct frequency offset estimation and compensation based

G. Liu et al., *6G Migrates from Communication to XaaS*,
https://doi.org/10.1007/978-981-96-7579-1_11

on measurement-feedback mechanisms. The paradox lies in the fact that in high-speed scenarios, where the channel's coherence time is shorter, frequent utilization of reference signals and triggering measurement-feedback mechanisms introduce a significant amount of resources and feedback overhead, ultimately reducing the efficiency of the communication system. Consequently, the performance of OFDM in high-speed mobile scenarios is greatly constrained.

From the perspective of the wireless channel, the transmission signals in high-mobility scenarios exhibit dual dispersion in both the time and frequency domains. First, due to multipath propagation, the non-coherent superposition of multipath signals at the receiver side causes a rapid change in signal amplitude, known as the fading effect. Secondly, as user mobility increases or carrier frequency rises, the channel's impulse response undergoes frequency shifts due to the Doppler effect, resulting in a blurred version of the transmission signal spectrum, known as frequency dispersion. In traditional time–frequency domain channel analysis, the channel's delay and Doppler characteristics are modeled as distortions to the time-domain waveform. In engineering practice, only rough estimates of the channel can be obtained through limited reference signals, making it impossible to acquire precise channel information at the granularity of sampling points. Since the OFDM technology modulates symbols in the time–frequency domain, the demodulation of symbols depends on the accurate estimation of the rapidly time-varying channel in the time–frequency domain, thus inherently limiting performance.

It is worth noting that the rapidly time-varying channel in the delay Doppler domain exhibits stable characteristics. The delay Doppler domain channel model directly characterizes the signal propagation characteristics in high-speed mobile environments, explicitly reflecting the delay, Doppler frequency shift, amplitude, and phase variations of each path in the physical scattering environment. Due to the limited number of scatterers in the physical channel and the slow relative positional changes between the transmitter–receiver nodes and scatterers, the delay Doppler domain channel demonstrates sparsity and stability.

The coherent region of the time–frequency domain channel and the stationary region of the delay-Doppler domain channel are depicted in Fig. 11.1. Typically, we use coherence to denote the variation characteristics of the time–frequency domain channel. Within a coherent region, the complex gain of the channel remains approximately constant. In contrast, for the delay-Doppler domain channel, we can define stationarity to represent the variation characteristics. Within a stationary region, the delay and Doppler of the channel remain approximately constant. The stationarity of the delay-Doppler domain channel depends on slowly changing parameters, namely the physical channel topology and relative velocity, while the coherence of the time–frequency domain channel is affected by rapidly changing parameters, such as signal phase variations, resulting in the stationary region lasting much longer than the coherent region.

Orthogonal time frequency space (OTFS) modulation technology leverages the aforementioned characteristics of the delay-Doppler domain channel by modulating symbols in the delay-Doppler domain. This transforms a received signal that exhibits fading and frequency shift relative to the original signal in the time–frequency domain

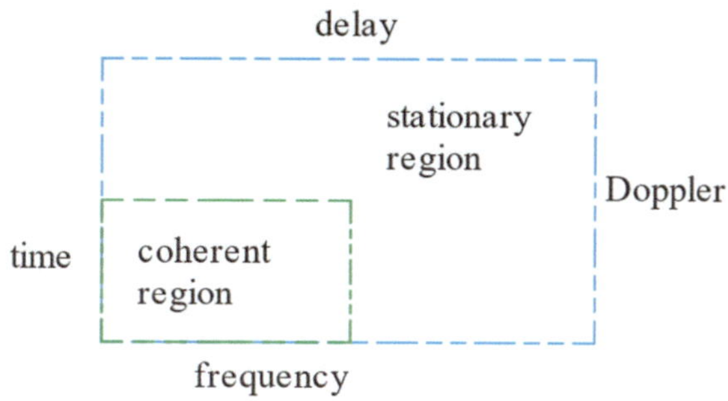

Fig. 11.1 Coherent area of the time–frequency domain channel and stable area of the delay-Doppler domain channel

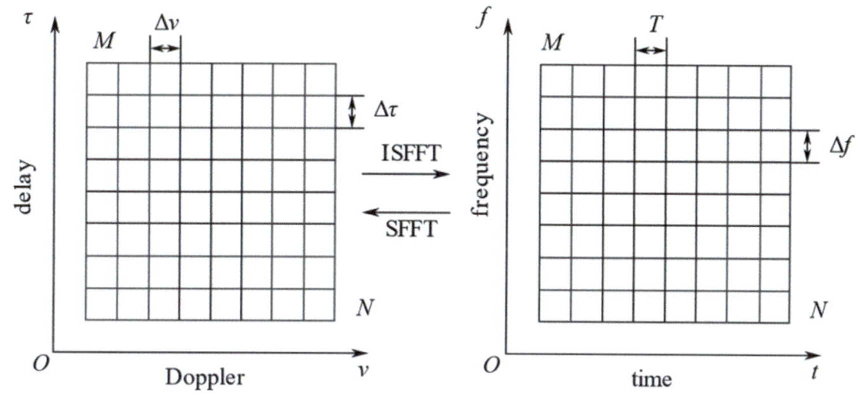

Fig. 11.2 Transformation between delay Doppler domain and time–frequency domain

into a superposition of impulse responses of various paths in the delay-Doppler domain. By utilizing the distinct delay and Doppler characteristics of different path components, complete separation of multipath signals is achieved, enabling precise estimation of the delay-Doppler domain channel response for each path. OTFS technology harnesses the stability of the delay-Doppler domain channel to effectively reduce the pilot and feedback overhead for channel estimation. Additionally, when the modulation symbols mapped in the delay-Doppler domain are transformed and transmitted in the time–frequency domain through a two-dimensional transformation, it increases both the temporal and spectral diversity of the signal, thereby mitigating the impact of channel fading on demodulation performance.

11.1.1 Application Scenarios

1. High Mobility Scenarios

In the context of 6G communication systems, which encompass a richer set of services and demands, the corresponding networks exhibit a high degree of heterogeneity. Among these, high-speed mobility scenarios, driven by the rapid development of high-speed transportation, have gradually become an indispensable aspect of mobile communication systems. Whether it's communication between high-speed trains and trackside base stations or between vehicles on highways and roadside units, the relative motion between transceivers generates significant Doppler effects, posing a considerable challenge for reliable communication. Especially in harsh channel conditions such as vehicle-to-everything (V2X), driven by safety considerations, there is a need for ultra-reliable and low-latency communication. Therefore, the exploration of new waveform and coding techniques is essential. Literature [1] evaluated current mainstream waveform schemes combined with advanced channel coding and found that under various fading channel conditions, OTFS outperforms other waveforms in terms of both spectral efficiency and bit error rate, exhibiting excellent resistance to multipath interference and Doppler effects, making it a strong candidate waveform for high-mobility scenarios in 6G. In literature [2], the authors utilized OTFS channel modeling in a high-speed train scenario and designed a method using reconfigurable intelligent surfaces (RIS) to improve the channel quality for distant users. The reliable channel estimation and Doppler tolerance of OTFS significantly reduced the overhead required for RIS state adjustments, and the waveform itself demonstrated outstanding performance in high-speed communication.

2. Millimeter Wave Bands

The millimeter wave bands have been designated as optional frequency bands by various communication protocols, and the demand for large bandwidth in 6G will further drive the spectrum requirements for millimeter wave bands. However, the high carrier frequency of millimeter wave bands leads to significant Doppler frequency shifts, and waveform design faces challenges posed by phase noise issues stemming from hardware limitations. Literature [3] compared the performance of OTFS and OFDM technologies under millimeter wave channel models, and the results show that OTFS exhibits significant gains over OFDM in both channel estimation accuracy and bit error rate. Literature [4] studied the impact of oscillator phase noise on OTFS modulation performance in millimeter wave communications. In the adopted equivalent channel model, the phase noise of the oscillator and the Doppler frequency shift of the channel caused significant frequency dispersion in the transmitted signal. The study employed an improved message passing (MP) algorithm for simulation verification and concluded that OTFS exhibits stronger tolerance to oscillator phase noise compared to OFDM, achieving two to three times better bit error rate performance in the evaluated 28 GHz system.

3. Satellite Communications

Satellite communication systems, including low earth orbit (LEO) satellites, are potential enablers for achieving seamless global coverage in 6G. Satellite communication channels exhibit three main characteristics: (1) significant path loss, mainly dominated by line-of-sight paths; (2) high relative velocities between space and ground, resulting in high Doppler frequency shifts in the channel; (3) long propagation distances, leading to significant propagation delays. Due to the high Doppler spread brought about by high mobility, the channel between space and ground experiences dramatic variations in the time–frequency domain, and the large propagation delay is not conducive to using traditional measurement-feedback mechanisms for Doppler compensation. Therefore, the inherent Doppler-resistant properties of OTFS modulation technology are advantageous for its application in satellite communications. In a wide-area coverage large-scale IoT network based on satellite communications, literature [5] utilized the aforementioned characteristics to design a non-scheduled uplink non-orthogonal multiple access (NOMA) transmission mechanism based on OTFS modulation, achieving low-complexity device activity detection and channel estimation.

4. Underwater Acoustic Communications

Underwater acoustic communication systems utilize sound waves for communication underwater. Due to the low speed of sound waves, the multipath propagation delay can reach several hundred milliseconds. Additionally, the continuous vibration of the propagation medium results in severe Doppler spread and frequency shift effects on the signal. This leads to the underwater acoustic channel exhibiting faster time-varying characteristics compared to traditional wireless channels. Literature [6] used OTFS modulation technology to transform the rapidly time-varying underwater acoustic channel into a relatively steady delay-Doppler domain for analysis, facilitating more precise channel estimation and compensation. This approach better mitigates subcarrier interference, leading to improved bit error rate, spectral efficiency, and PAPR performance.

5. Integrated Sensing and Communications (ISAC) Scenarios

Future ISAC devices will integrate the dual functions of communication and sensing, enabling hardware sharing, spectrum sharing, and protocol interoperability between communication and sensing signals. The unified waveform design between communication and sensing signals is critical for enhancing performance. The key to ISAC's unified waveform design lies in minimizing the interference between communication and sensing signals, meeting the requirements of both communication and sensing functions, and improving spectral efficiency while ensuring system performance. The traditional multi-carrier integrated waveforms are mainly based on the OFDM waveform, which boasts advantages such as a simple signal generation process, all-digital processing, and extremely high time–frequency processing gains [7]. However, the cyclic prefix (CP) of OFDM symbol is useless overhead in sensing detection. Furthermore, OFDM waveforms exhibit poor adaptability to high Doppler scenarios and can only detect Doppler frequency shifts of up to 10% of subcarrier spacing, limiting their performance in tracking fast-moving objects.

As a sensing waveform, OTFS overcomes the aforementioned drawbacks of OFDM waveforms. Firstly, consider the same resource block with M subcarriers and N symbols for both OTFS and OFDM waveform. When transformed into the time–frequency domain, an OTFS signal only needs to add a CP before N OFDM symbols to avoid interference. However, a traditional OFDM signal requires a CP before each symbol, resulting in N CPs for the same aforementioned resource block. Therefore, the overhead of OTFS is significantly reduced, which means that it can allocate more energy to the sensing signal itself, thereby enhancing target detection and tracking capabilities. Secondly, due to the natural tolerance of OTFS signals to high Doppler shifts, OTFS radar exhibits excellent detection capabilities for high Doppler frequency shifts in high-speed scenarios, outperforming OFDM radar.

11.1.2 Industry Research Status

Ronny Hadani, a mathematics professor at the University of Texas, first proposed a new modulation technique designed in the delay Doppler domain in 2017, known as Orthogonal Time Frequency Space modulation [8]. His article detailed the signal processing process of OTFS modulation and demodulation, pointing out that OTFS, as a new type of waveform, has unique advantages in channel estimation, diversity gain, and precoding.

During the study item stage of 5G NR, OTFS was once discussed as a candidate waveform for 5G NR at the 3GPP TSG RA WG1 during meetings #84 to #87. Led by Cohere Technologies, it was requested that OTFS modulation technology and the corresponding reference signal design to be included in the evaluation of the 5G system air interface.

In literature [9], the author pointed out that OTFS modulation technology has significant spectral efficiency advantages in high-order MIMO and high Doppler scenarios, reference signal efficiency, and channel estimation and prediction. It proposed that all reference signals and data are mapped in the delay Doppler domain, experiencing the same delay Doppler domain channel response within the designed transmission slot, and obtaining diversity gain in time and frequency through two-dimensional transformation.

Subsequently, in literature [10], the author evaluated the performance of OTFS and OFDM in high mobility scenarios (such as high-speed rail) under the 3GPP extended vehicular A (EVA) channel environment. The results showed that OTFS has better performance than OFDM. Due to the inherent anti-Doppler characteristics of OTFS technology, it can achieve high throughput in high mobility communication systems without the need to increase the subcarrier spacing and the CP overhead. In literature [11], the authors evaluated the performance of OTFS in SISO and MIMO scenarios under TDL-C channels, and compared it with OFDM. The results consistently demonstrated the superiority of OTFS.

In literature [12], the authors investigated the performance of OTFS under different subcarrier spacing in high-speed mobile scenarios. The simulation results indicated the superiority of OTFS over OFDM across all subcarrier spacing conditions. By utilizing the proposed delay Doppler domain pilot superimposition scheme in the article, coupled with a non-linear iterative Turbo receiver, a channel estimation iterative algorithm that achieves carrier interference elimination without increasing subcarrier spacing was implemented. It is evident that the proposed pilot superimposition scheme has a certain gap with the performance of ideal channel estimation. However, due to the inability to obtain perfect channel state information in practice, the proposed channel estimation method demonstrates a certain level of effectiveness, and reduces the pilot overhead to a negligible extent.

11.2 Overview of OTFS Technology

The OTFS technology maps the information in a data packet of size $M \times N$, , such as quadrature amplitude modulation (QAM) symbols, onto an $M \times N$ grid in the 2D delay Doppler domain plane, with each grid point modulating a QAM symbol from the data packet. Subsequently, by designing a set of orthogonal 2D basis functions, the dataset on the $M \times N$ delay Doppler domain is transformed into the $N \times M$ time–frequency domain. This transformation is mathematically referred to as the inverse symplectic finite fourier transform (ISFFT). Correspondingly, the transformation from the time–frequency domain to the delay Doppler domain is called the symplectic finite fourier transform (SFFT). Analysis in delay Doppler and time–frequency domain can be converted through the aforementioned ISFFT and SFFT. The transformation relationships described above are illustrated in Figure Orthogonal frequency division multiplexing (OFDM) has been widely employed in 4G LTE and 5G NR. The OFDM communication system possesses outstanding advantages such as excellent resistance to multipath fading, simple channel estimation and equalization, easy implementation of receiver hardware, and tight coupling with MIMO technology. However, OFDM also exhibits some inherent drawbacks, such as high adjacent channel leakage ratio (ACLR), weak inter carrier interference (ICI) performance, and high peak to average power ratio (PAPR). These drawbacks can be mitigated through various engineering methods, allowing OFDM to continue its use in both 4G and 5G protocols. Nonetheless, this also presents some challenges and costs, such as low spectrum utilization, stringent synchronization requirements, and excessive feedback overhead. Faced with the increasingly demanding technical requirements of 6G, traditional OFDM waveform appear inadequate. Therefore, it is imperative to explore new waveform for 6G, in the hope of finding a highly flexible, spectrum-efficient, and unified waveform for both uplink and downlink, in order to better meet the demands of future ultra-high-speed, large-scale connectivity, and high-speed mobile scenarios (Fig. 11.2).

The delay Doppler domain channel model directly reflects the channel's delay Doppler response characteristics caused by the geometric properties of the relative positions of the reflecting bodies between the transmitter and receiver in the wireless link, including the delay, Doppler, and channel gain on each separable path, as shown in Eq. (11.1).

$$h(\tau, \nu) = \sum_{i=1}^{P} h_i \delta(\tau - \tau_i)\delta(\nu - \nu_i) \tag{11.1}$$

The delay Doppler domain channel exhibits the following characteristics:

(1) Invariance of channel coupling status: The signal's delay and Doppler, reflecting the direct impact of reflecting objects in the physical channel, are solely determined by the relative velocity and position of the reflecting objects. Therefore, on the timescale of wireless frames, the signal's delay and Doppler response can be considered invariant.
(2) Separability of channel coupling status: In the frequency response of the delay Doppler domain channel, all diversity paths are represented as separate impulse responses, completely separable, while QAM symbols traverse all diversity paths.
(3) Orthogonality of channel coupling status: The impulse response of the delay Doppler domain channel is confined to a single delay Doppler domain resource element. If a waveform satisfying the bi-orthogonality assumption can be found, there will be no inter-delay interference (IDeI) and inter-Doppler interference (IDoI) at the receiving end.

Due to these characteristics, delay Doppler domain analysis eliminates the difficulties of tracking time-varying fading characteristics in traditional time–frequency domain analysis. Instead, it extracts all diversity properties of the time–frequency domain channel by analyzing the time-invariant delay Doppler domain channel. Furthermore, through the transformation relationship between the delay Doppler domain and the time–frequency domain, the time–frequency domain channel can be computed. In practical systems, the number of delay paths and Doppler frequency shifts is much smaller than the number of time and frequency domain responses of the channel, making the representation of the channel in the delay Doppler domain more concise. Therefore, utilizing OTFS technology for analysis in the delay Doppler domain can enable a tighter and more flexible encapsulation of reference signals, particularly beneficial for supporting large-scale MIMO systems with large antenna arrays.

11.2.1 Principle and Implementation

1. Transformation Relationship Between Channels

The OTFS modulation transforms symbols in the delay Doppler domain to the time–frequency domain for transmission. After experiencing the channel, the receiving

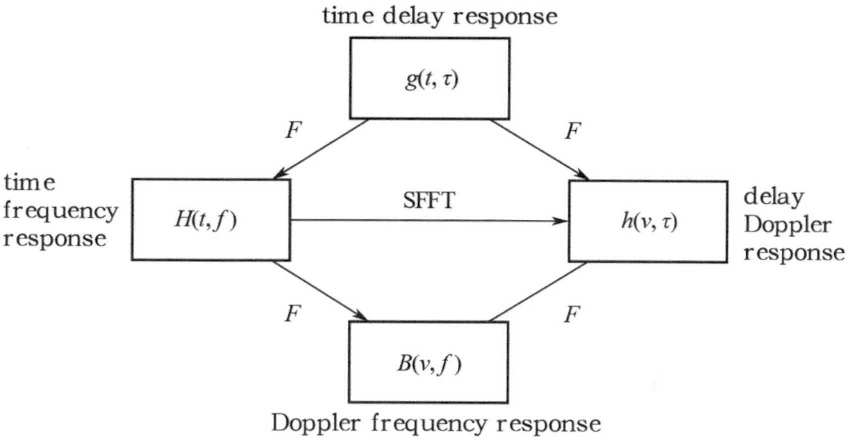

Fig. 11.3 Transformation relationship between channels

end still needs to return to the delay Doppler domain for symbol decision. Therefore, analysis methods for wireless channel responses in the delay Doppler domain can be introduced. When the signal passes through a linear time-varying channel, the transformation relationship between its channels is as shown in Fig. 11.3.

In Fig. 11.3, the SFFT transformation formula is

$$h(\tau, \nu) = \iint H(t, f) \mathrm{e}^{-\mathrm{j}2\pi(\nu t - f\tau)} \mathrm{d}\tau \mathrm{d}\nu \tag{11.2}$$

Correspondingly, the ISFFT transformation formula is

$$H(t, f) = \iint h(\tau, \nu) \mathrm{e}^{\mathrm{j}2\pi(\nu t - f\tau)} \mathrm{d}\tau \mathrm{d}\nu \tag{11.3}$$

When the signal passes through a linear time-varying channel, let the time-domain received signal be $r(t)$, its corresponding frequency domain received signal is $R(f)$, and $r(t) = \mathcal{F}^{-1}\{R(f)\}$. $r(t)$ can be represented in the following form

$$r(t) = s(t) * h(t) = \int g(t, \tau) s(t - \tau) \mathrm{d}\tau \tag{11.4}$$

As shown by the relationship in Fig. 11.3, $g(t, \tau)$ can be expressed as

$$g(t, \tau) = \int h(\nu, \tau) \mathrm{e}^{\mathrm{j}2\pi\nu t} \mathrm{d}\nu \tag{11.5}$$

Substituting Eq. (11.5) into Eq. (11.4), $r(t)$ can be further expressed as

$$r(t) = \iint h(\nu, \tau) s(t - \tau) \mathrm{e}^{\mathrm{j}2\pi\nu t} \mathrm{d}\tau \mathrm{d}\nu \tag{11.6}$$

Based on the relationship shown in Fig. 11.3, the classical Fourier transform theory, and Eq. (11.6), we know

$$\begin{aligned} r(t) &= \iint h(\nu,\tau)\Big(\int S(f)\mathrm{e}^{\mathrm{j}2\pi f(t-\tau)}\mathrm{d}f\Big)\mathrm{e}^{\mathrm{j}2\pi\nu t}\mathrm{d}\tau\mathrm{d}\nu \\ &= \int\Big(\iint h(\nu,\tau)\mathrm{e}^{\mathrm{j}2\pi(\nu t-f\tau)}\mathrm{d}\tau\mathrm{d}\nu\Big)S(f)\mathrm{e}^{\mathrm{j}2\pi ft}\mathrm{d}f \\ &= \int H(t,f)S(f)\mathrm{e}^{\mathrm{j}2\pi ft}\mathrm{d}f \\ &= \mathcal{F}^{-1}\{R(f)\} \end{aligned} \tag{11.7}$$

Equation (11.7) indicates that in the context of an OTFS system, the analysis of the delay Doppler domain can be achieved by leveraging the existing communication framework established in the time–frequency domain, and by introducing additional signal processing at the transmitter and receiver. Importantly, this supplementary signal processing consists solely of Fourier transforms and can be entirely implemented using existing hardware, without the need for additional modules. This high level of compatibility with the current hardware infrastructure significantly facilitates the application of OTFS systems. In practical systems, OTFS technology can be realized using the front-end and back-end processing modules of a filtering OFDM system, thereby demonstrating excellent compatibility with multi-carrier systems under the existing NR technology architecture.

2. Combination of OTFS and Multi-carrier Systems

When OTFS is combined with existing OFDM multi-carrier systems, it can be considered as a precoding process in the digital domain. Firstly, placing QAM symbols on the resource grid in the delay Doppler domain, they are transformed into a transformed symbol set in the time–frequency domain through a two-dimensional ISFFT, and then, after the Heisenberg transform, they are transmitted as time-domain sampling points. Unlike traditional OFDM, there is no need to add a cyclic prefix before each time-domain sampling point of an OFDM symbol, only a single cyclic prefix before the entire time-domain sampling points is required. Figure 11.4 shows the transmission side signal processing flow of an OTSF system with $N = 128, M = 2048$.

The receiver of the OTFS system is the inverse process of the transmitter. After the receiver receives the time-domain sampling points, they undergo a Wigner transform to obtain the received symbol set in the time–frequency domain. Subsequently, they are transformed into the received symbol set in the delay Doppler domain through a two-dimensional SFFT, and then processed, including channel estimation, equalization, demodulation, and decoding, among others. The processing flow of the transmitter and receiver of the OTFS system described above can be referenced in Fig. 11.5.

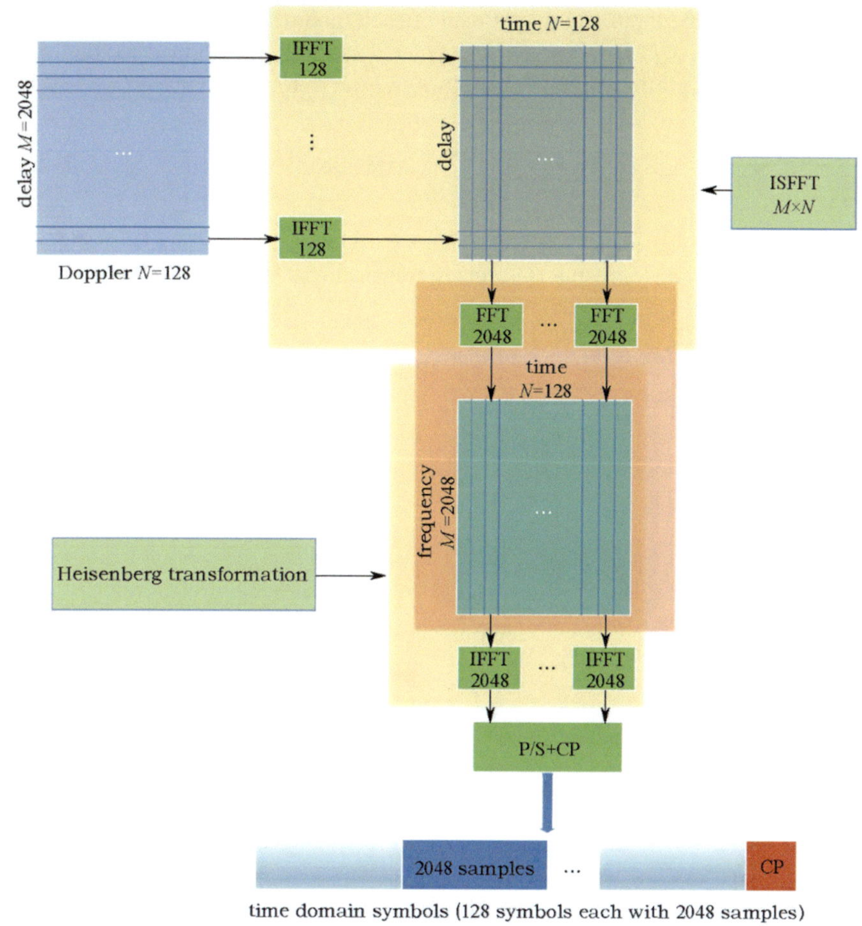

Fig. 11.4 OTFS system transmission side signal processing flow

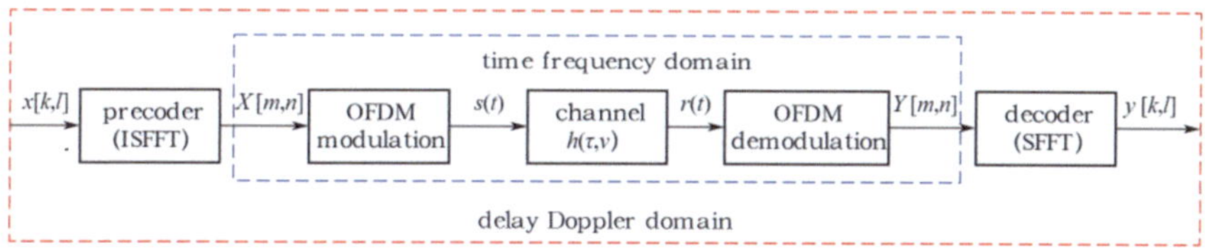

Fig. 11.5 OTFS system transmitting and receiving end processing flow

In Fig. 11.5, the input and output data of each module are described as follows. First, the symbol $x[k, l]$ in the delay doppler domain is transformed to the time–frequency domain through ISFFT, as shown in Eq. (11.8).

$$X[n, m] = \frac{1}{\sqrt{NM}} \sum_{k=0}^{N-1} \sum_{l=0}^{M-1} x[k, l] e^{j2\pi\left(\frac{nk}{N} - \frac{ml}{M}\right)} \tag{11.8}$$

Then, the time–frequency domain signal $X[n, m]$ is transformed into a continuous time-domain signal $s(t)$ by the Heisenberg transform, as shown in Eq. (11.9).

$$s(t) = \sum_{n=0}^{N-1} \sum_{m=0}^{M-1} X[n, m] g_{\text{tx}}(t - nT) e^{j2\pi m\Delta f(t-nT)} \tag{11.9}$$

where, $g_{\text{tx}}(t)$ represents the transmission pulse. The continuous time-domain signal $s(t)$ passes through a time-varying channel to reach the receiver. The time-domain received signal $r(t)$ is

$$r(t) = \iint h(\nu, \tau) s(t - \tau) \mathrm{e}^{\mathrm{j}2\pi\nu(t-\tau)} \mathrm{d}\tau \mathrm{d}\nu \tag{11.10}$$

At the receiver, the cross-ambiguity function (CAF) is calculated using a matched filter, as shown in Eq. (11.11).

$$Y(t,f) = A_{\text{gx},r}(t,f) = \int g_{\text{rx}}^{*}(t' - t) r(t') \mathrm{e}^{-\mathrm{j}2\pi f(t'-t)} \mathrm{d}t' \tag{11.11}$$

By sampling $Y(t,f)$, the output of the matched filter is obtained as

$$Y[m, n] = Y(t,f)|_{t=nT, f=m\Delta f} \tag{11.12}$$

The combination of Eqs. (11.11) and (11.12) is called the Wigner transform. $Y[n, m]$ is the transformed from the time frequency domain to the delay doppler domain by SFFT, given by (11.13),

$$y[k, l] = \frac{1}{\sqrt{NM}} \sum_{k=0}^{N-1} \sum_{l=0}^{M-1} Y[m, n] e^{-j2\pi\left(\frac{nk}{N} - \frac{ml}{M}\right)} \tag{11.13}$$

After the above process, the final input–output relationship of the signal in the delay doppler domain at the receiving end is

$$y[k, l] = \sum_{i=1}^{P} h_i x[[k - k_{\nu_i}]_N, [l - l_{\tau_i}]_M] = h[k, l] \circledast x[k, l], \tag{11.14}$$

where $\circledast$ is the symbol for circular convolution.

11.2.2 Waveform Design

The OTFS system utilizes symbol multiplexing in the delay Doppler domain, resulting in a two-dimensional convolution in the input–output relationship under the assumption of bi-orthogonal waveforms. However, bi-orthogonal waveforms are impractical in reality and OTFS systems typically employ rectangular pulses as shaping waveforms. OTFS using rectangular pulses provides maximum diversity gain, but suffers from severe out of band (OOB) leakage, making it unsuitable for multi-user scenarios. In order to achieve the best balance in terms of OOB, PAPR, and demodulation performance, other shaping waveforms, such as elongated ellipsoidal waveforms, circular pulse shaping, and methods like windowing functions and filtering operations to improve the characteristics of the shaped signal have been proposed.

In reference [13], the vectorized input–output relationship in the delay Doppler domain under arbitrary waveforms is derived and can be expressed as follows:

$$y = (F_N \otimes G_{\mathrm{rx}})H\left(F_N^H \otimes G_{\mathrm{tx}}\right)x + (F_N \otimes G_{\mathrm{rx}})w = H_{\mathrm{eff}}x + \tilde{w} \tag{11.15}$$

In the formula, $\otimes$ represents the Kronecker product. H_{eff} is the equivalent channel matrix under any waveform, which is represented as

$$H_{\mathrm{eff}} = (F_N \otimes G_{\mathrm{rx}})H\left(F_N^H \otimes G_{\mathrm{tx}}\right) \tag{11.16}$$

In the formula, H is the time domain channel matrix, i.e.

$$H = \sum_{i=1}^{P} h_i \Pi^{l_i} \Delta^{k_i},\ \tau_i = \frac{l_i}{M\Delta f},\ \nu_i = \frac{k_i}{NT} \tag{11.17}$$

where, $\Delta f$is the subcarrier spacing; Π^{l_i} and Δ^{k_i} corresponding to the delay and Doppler parameter matrix of the i-th path, whose details can be found in literature [18]. Specifically, when using a rectangular waveform, the equivalent channel matrix is

$$H_{\mathrm{eff}}^{\mathrm{rect}} = (F_N \otimes I_M)H\left(F_N^H \otimes I_M\right) \tag{11.18}$$

Based on the equivalent channel matrix of rectangular waveform as shown in (11.18), the equivalent channel matrix under any waveform can be expressed as

$$\begin{aligned} \boldsymbol{H}_{\mathrm{eff}} &= (\boldsymbol{I}_N \otimes \boldsymbol{G}_{\mathrm{rx}})(\boldsymbol{F}_N \otimes \boldsymbol{I}_M)\boldsymbol{H}(\boldsymbol{F}_N^{\mathrm{H}} \otimes \boldsymbol{I}_M)(\boldsymbol{I}_N \otimes \boldsymbol{G}_{\mathrm{tx}}) \\ &= (\boldsymbol{I}_N \otimes \boldsymbol{G}_{\mathrm{rx}})\boldsymbol{H}_{\mathrm{eff}}^{\mathrm{rect}}(\boldsymbol{I}_N \otimes \boldsymbol{G}_{\mathrm{tx}}). \end{aligned} \tag{11.19}$$

Since the sparsity of the channel is only determined by $H_{\mathrm{eff}}^{\mathrm{rect}}$, the sparsity of the equivalent channel matrix holds whether it is a rectangular pulse waveform or not.

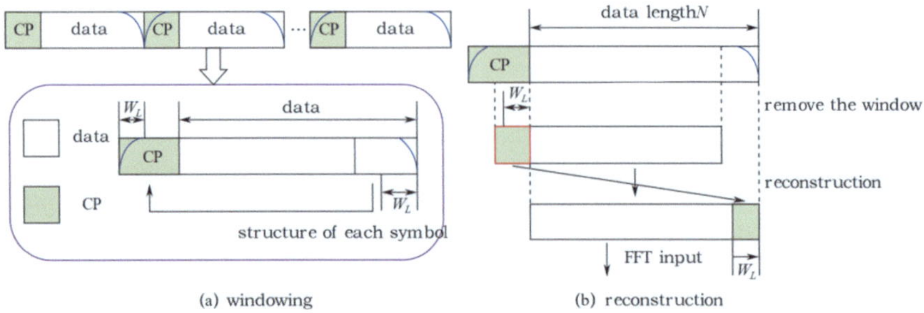

Fig. 11.6 Windowing and reconstruction of the signal

Therefore, the detection based on MP algorithm with lower complexity can be used for detection.

1. OTFS waveform design to reduce OOB

In literature [14], the authors have introduced a technique known as windowing and restructuring OTFS (WR-OTFS) to mitigate the emission of OOB power. The proposed WR-OTFS leverages temporal windowing at the transmitter to effectively diminish OOB power emission. At the receiver, redundant samples from the CP are utilized for signal reconstruction, ensuring symbol detection performance. The windowing and reconstruction of the signal are shown in Fig. 11.6.

The evaluation shows that the longer the window length used, the more obvious the OOB improvement. When BER is 10^{-3}, the proposed WR-OTFS can reduce the OOB power by 66 dB, while gaining 10 dB over OFDM. However, the cost is that a CP needs to be added before each OFDM symbol, and the CP needs to be more than twice the window length, which reduces the spectral efficiency.

In literature [15], the authors proposed a circular pulse shaping framework for OTFS to reduce OOB. The expression for circular pulse shaping is

$$g_{(n,m)}(t) = g(t - nT)_{T_f} e^{j2\pi m\Delta f(t-nT)} \tag{11.20}$$

The expression for the time-domain transmitted signal after circular pulse shaping is

$$s(t) = \sum_{n=0}^{N-1} \sum_{m=0}^{M-1} X[n, m] g\big((t - nT) \text{mod} T_f\big) e^{j2\pi m\Delta f(t-nT)} \tag{11.21}$$

The author has designed a low-complexity transmitter for this system, using frequency domain cyclic Dirichlet pulses for pulse shaping. Simulation shows that compared to traditional OTFS systems, the use of frequency domain cyclic Dirichlet pulses can significantly reduce OOB leakage (by about 50 dB). The proposed pulse-shaped OTFS has a lower peak-to-average power ratio without BER leakage.

2. OTFS Waveform Design to Reduce PAPR

Multi-carrier systems such as OFDM and OTFS both have the concern of high PAPR. Understanding the PAPR characteristics under different modulation systems and devising waveform configurations to reduce PAPR is a pivotal concern.

In reference [16], the authors analyzed the peak-to-average power ratio of OTFS waveforms and derived the upper limit of PAPR and its complementary cumulative distribution function when OTFS signals are shaped using rectangular pulses. The conclusion drawn was that the maximum PAPR in OTFS systems increases linearly with the OTFS parameter N (i.e., the Doppler number). The paper simulated the complementary cumulative distribution function (CCDF) of PAPR in OTFS and compared it with OFDM and generalized frequency division multiplexing (GFDM), indicating that OTFS can exhibit superior PAPR performance compared to OFDM and GFDM.

In practical systems, the insertion of high-power pilot pulses for channel estimation within OTFS data frames can deteriorate the PAPR performance. In such scenarios, PAPR reduction can be achieved through time-domain scrambling. For instance, when a single-point pulse with PSR = 17 dB is inserted in an OTFS with N = 16 and M = 256. The comparison of the PAPR of the time-domain sequence before and after scrambling is shown in Fig. 11.7.

The system diagram of the OTFS after time-domain scrambling is shown in Fig. 11.8. In addition, this problem can be avoided by changing the design of the reference signal, for example, using the pilot sequence in the DD domain to replace the single-point pulse pilot for channel estimation. The specific technology and analysis of advantages and disadvantages will be presented later.

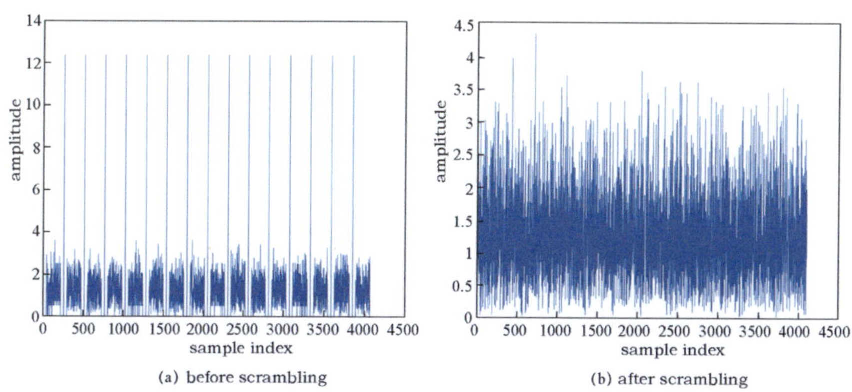

Fig. 11.7 Comparison of PAPR of time-domain sequence before and after scrambling

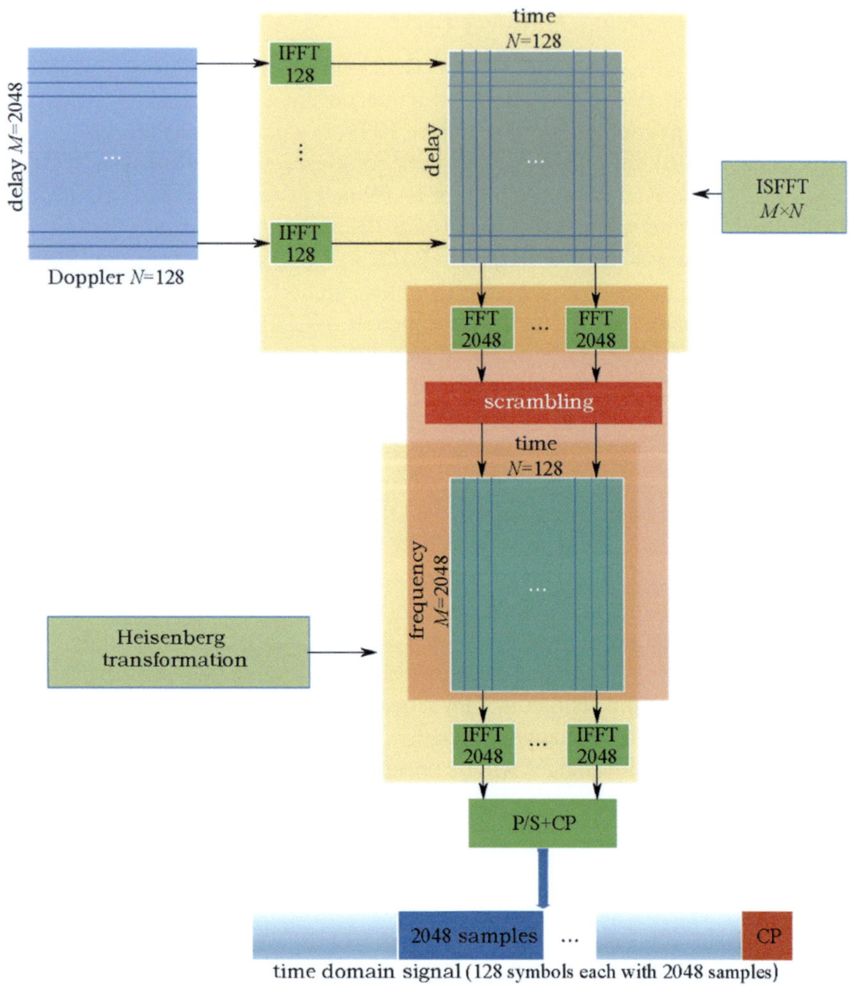

Fig. 11.8 System diagram of OTFS after time-domain scrambling

11.2.3 Pilot Design

Due to the limited number of reflecting objects in the actual physical channel, the number of delay and Doppler states determined by the reflecting objects in the delay-Doppler domain is much smaller than the number of time and frequency states caused by signal phase variations in the time–frequency domain. In other words, the channel matrix in the delay-Doppler domain exhibits sparsity, which is advantageous for the targeted design of pilot mapping and corresponding channel estimation schemes.

At the transmitter, using OTFS modulation transforms the rapidly fading channel in the time–frequency domain between the transmitter and receiver into a slowly varying and smooth channel in the delay-Doppler domain. Therefore, it can be approximated that each information symbol transmitted in a time slot experiences the same static channel response.

At the receiver, employing unique pilot pulse designs allows the OTFS receiver to resolve distinguishable paths and their channel responses in the physical channel through delay-Doppler imaging, and coherent combining of energy from different reflected paths using a receive equalizer. This effectively provides a non-fading static channel response.

Specifically, channel estimation in OTFS systems can be accomplished as follows. The transmitter maps pilot pulses onto the delay-Doppler domain, and the receiver utilizes delay-Doppler imaging of the pilots to estimate the delay-Doppler domain channel response. Subsequently, the time–frequency domain channel response expression can be obtained based on the relationship shown in Fig. 11.3, facilitating the application of existing time–frequency domain techniques for signal analysis and processing. Reference [17] indicates that, compared to traditional reference signal sequences mapped in the time–frequency domain, pilot pulses in the delay-Doppler domain offer the following advantages: they can optimally support the simultaneous estimation of channels with different delay and/or Doppler spreads, and pilot mapping provides greater flexibility, as pilots can be placed at any position in the continuous delay-Doppler plane, rather than the limited choices in the time–frequency plane, making this flexibility more conducive to the optimization of the system design.

1. Pilot Mapping of OTFS

The pilot mapping of the delay-Doppler channel can be carried out as shown in Fig. 11.9.

In Fig. 11.9, the transmitted signal is a single-point pilot located at $\left(l_p, k_p\right)$ (black grid in the figure), surrounded by an area of $(2l_\tau + 1)(4k_\nu + 1) - 1$ protection symbols

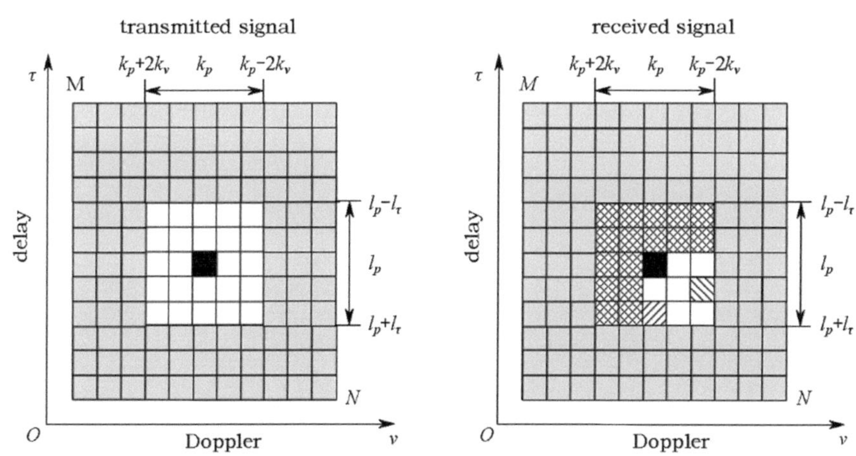

Fig. 11.9 Pilot mapping of the delay-Doppler channel

(white grid in the figure), and $MN - (2l_\tau + 1)(4k_\nu + 1)$ data symbols (grey grid in the figure). At the receiver, two additional symbols corresponding to the pilot symbol (grid with slash in Fig. 11.9) appear in the protection area, indicating that there are secondary paths with different delays and Dopplers other than the main path. By measuring the amplitude, delay, and Doppler parameters of all secondary paths, the delay doppler domain expression of the channel, i.e., $h(\nu, \tau)$ is obtained. Specifically, to prevent the contamination of the pilot symbols by the data on the received signal grid points, the area of the protection symbols should satisfy the following conditions:

$$l_\tau \geq \tau_{\max} M \Delta f, k_\nu \geq \nu_{\max} N \Delta T \tag{11.22}$$

In the formula, $\tau_{\max}$ and $\nu_{\max}$ are the maximum delay and maximum Doppler frequency shift of all paths in the channel, respectively.

The detection of pilot frequency positions on the receiver side is shown in Fig. 11.10. The receiver sequentially demodulates and transforms the received time–frequency domain sample points into QAM symbols in the delay-Doppler domain using OFDM demodulation and SFFT transformation. The position of the pilot pulse is then determined using threshold-based signal power detection. It is worth noting that, because the transmission of the pilot signal usually involves power boosting, the power of the pilot pulse at the receiver side is much greater than the data power, making it easy to determine the pilot frequency position using power detection.

The example shown in Fig. 11.9 corresponds to a single-port scenario, that is, only one set of reference signals needs to be transmitted. In multi-antenna systems, we often utilize multiple antenna ports to simultaneously transmit multiple streams of data, thereby fully exploiting the spatial degrees of freedom of the antennas to achieve spatial diversity gain or to enhance system throughput. When multiple antenna ports are present, multiple pilot signals need to be mapped in the delay-Doppler plane, which leads to the mapping diagram shown in Fig. 11.11.

In Fig. 11.11 15 antenna ports correspond to 15 pilot signals. Each pilot signal uses the method shown in Fig. 11.9, that is, a central impulse signal plus protection symbols. The number of delay-Doppler domain resource elements (REs) occupied by a single pilot symbol is $(2l_\tau + 1)(4k_\nu + 1)$. If there are P antenna ports, the total resource overhead of the pilot tones is $[P_1(l_\tau + 1) + l_\tau][P_2(2k_\nu + 1) + 2k_\nu]$ assuming

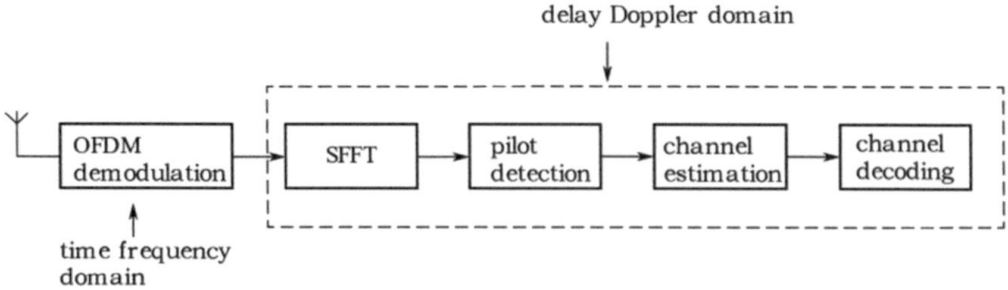

Fig. 11.10 Pilot position detection on the receiver side

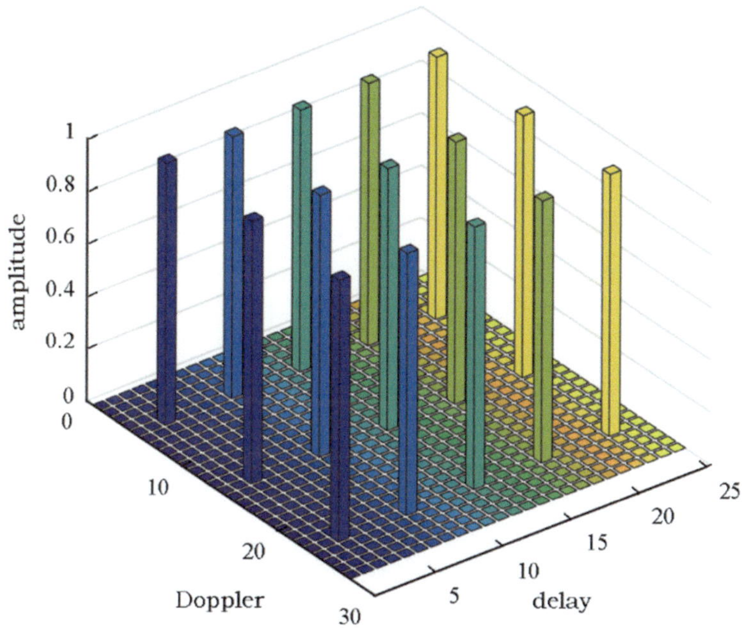

Fig. 11.11 Schematic diagram of the mapping of multi-port reference signals in the delay-Doppler domain

that the guard bands of adjacent antenna ports can be reused, the placement of pilot tones is in the delay dimension as P_1, and in the Doppler dimension as P_2, and satisfying $P = P_1P_2$.

It can be seen that, while single-port transmission offers the advantage of minimal resource utilization and simple detection algorithms, for communication systems with multiple antenna ports, the single-point pilot plus guard band scheme cannot achieve resource reuse, leading to a linear increase in overhead. Therefore, as depicted in Fig. 11.12, a pilot mapping scheme that effectively reduces pilot overhead has been proposed for multi-antenna systems, which enables resource reuse in the delay-Doppler domain.

In Fig. 11.12, rather than a pilot realized by a single -point pulse, a pilot sequence constructed by a pseudo-random noise (PN) sequence generated in a specific way, and mapped on the two-dimensional resource grid on the delay Doppler plane according to specific rules. The black grids in the middle of the left figure in Fig. 11.12 illustrate the pilot sequence. In the following text, we call the black grids (i.e., the resource position occupied by the pilot sequence) the pilot resource block. The white part next to the pilot resource block is the pilot guard area, composed of blank resource elements that do not transmit any signals/data. Similar to the single-point pilot in Fig. 11.9, we also have a guard area around the pilot resource block to avoid interference with the data. The overhead of the guard area is basically the same as that in the single-point pilot mapping in Fig. 11.9. The difference is that in the resource area

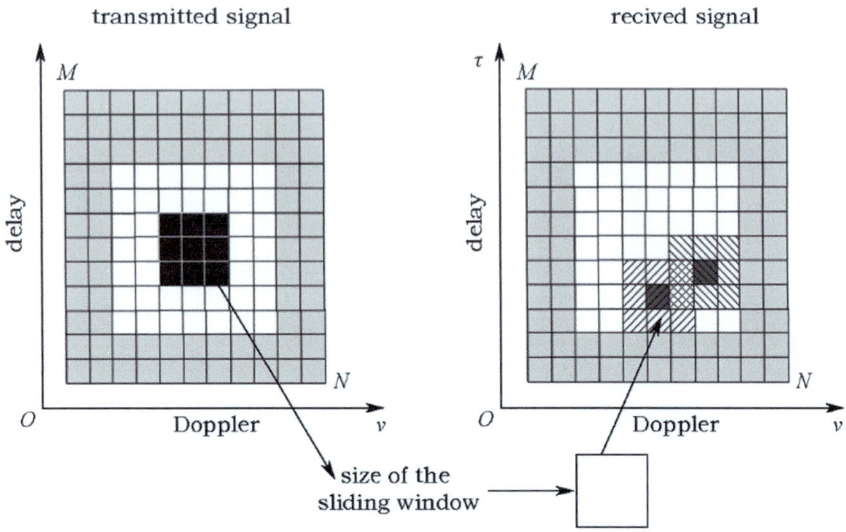

Fig. 11.12 Pilot resource reuse in the delay Doppler domain

where the pilot sequence is mapped, the pilot sequences of different ports can select low-correlation pilots, superimpose and map on the same block of resources. Then, the pilot sequence at the receiver can be detected through a specific algorithm, so as to distinguish the pilots corresponding to different antenna ports. Since full reuse of resources is carried out at the transmitter, the pilot overhead under the multi-antenna port system can be greatly reduced.

Similar to the scenario in Fig. 11.9, at the receiver, since different paths are with different delays and Doppler frequency shifts, the received pilot signal block shift to grids with slash in the right subfigure of Fig. 11.12. At this time, the size of the known transmitted pilot (the black grids in the left subfigure of Fig. 11.12) is used as the sliding window size at the receiver end, and sliding window detection is performed in the delay Doppler domain in the right subfigure. The result derived from the sliding window detection is given by $M(R, S)[\delta, \omega]$. When $N_p \to +\infty$, the detection results have the following properties (the probability of the following formula tends to 100%):

$$\begin{aligned} M(R, S)[\delta, \omega] &= 1 + \varepsilon'_{N_p}, \text{ if } (\delta, \omega) = (\delta_0, \omega_0) \\ &= \varepsilon_{N_p}, \text{ if } (\delta, \omega) \neq (\delta_0, \omega_0) \end{aligned} \tag{11.23}$$

where, $\left|\varepsilon_{N_p}\right| \leq \frac{1}{\sqrt{N_p}}$, $\left|\varepsilon_{N_p}\right| \leq \frac{C+1}{\sqrt{N_p}}$, $C > 0$ is a certain constant, and decreases as the SINR of the signal increases; (δ, ω) and (δ_0, ω_0) are the current location of the sliding window and the position that the pilot signal block is shifted to in the received signal, respectively. It can be seen that only when $(\delta, \omega) = (\delta_0, \omega_0)$, then can we obtain a value near 1. Otherwise, the result of the sliding window detection is a very small value. Therefore, when the sliding window coincides with the shifted pilot signal block, the detection result will be an energy peak. Using this method, as long as it is ensured to have sufficient length, the receiver can obtain the correct pilot position based on the value of $M(R, S)$, that is, obtain the delay and Doppler information of the channel. At the same time, the amplitude of the channel is given by the value of $1 + \varepsilon_{N_p}{}'$ obtained from the detection calculation.

The pilot sequence scheme in Fig. 11.12 and the pilot pulse scheme in Fig. 11.11 each have their advantages and disadvantages, as shown in Table 11.1.

In certain scenarios, the constraints on the overhead of pilot guard intervals may be insufficient to fully cover the potential delay and Doppler shifts in the channel. In such

Table 11.1 Comparison of various pilot schemes

	Delay doppler domain		Time frequency domain
	Pilot pulse	Pilot sequence	DMRS
Pros	The receiver employs energy detection, a straightforward method	Multi−port/user multiplexing allows for the reuse of delay−Doppler domain resources, reducing overhead	The implementation is easy and straightforward
	It can utilize blind detection of pilot position to carry information	It can utilize blind detection of pilot position and sequences to carry information	Information can be carried using blind detection of pilot sequences
	With a sufficiently large total number of grid points in the two−dimensional delay−Doppler plane, it offers high detection accuracy and accurately reconstructs channel information	Sequence detection offers high precision, and its accuracy can be flexibly adjusted using sequence length	The precision of pilot detection and resource overhead can be flexibly balanced through signaling configuration
	The detection performance can be flexibly adjusted through power enhancement	This approach saves on protection symbol overhead, enabling the acquisition of approximate results even when the overhead is insufficient	Resource mapping of the pilots allows for adjustments at the symbol/subcarrier level, offering strong flexibility
cons	The protection interval overhead of multi−port pilot pulses is relatively high	Sequence correlation/matching detection has relatively high complexity	However, detection precision depends on time–frequency domain interpolation, resulting in relatively lower accuracy
	The PAPR of the time−domain signal is also high	Accuracy is constrained by the sequence length	Furthermore, in high Doppler scenarios, the system's ability to resist ICI and phase noise is limited

instances, the pilot sequence scheme continues to demonstrate acceptable performance, while the pilot pulse scheme experiences significant performance degradation. The performance comparison of the two pilot design schemes under different pilot overhead conditions is depicted in Fig. 11.13. The curves marked with diamonds and circles in the figure represent the performance curves of the pilot sequence scheme based on different detection algorithms, while the curve marked with squares represents the performance curve of the pilot pulse scheme. It is evident that in the specific scenario illustrated in the figure (where the delay and Doppler shifts of the channel are significant), even when the pilot overhead reaches 60%, the performance of the pilot pulse scheme is still far inferior to that of the pilot sequence scheme.

2. OTFS Channel Estimation Algorithm

Accurate channel estimation in communication systems is a hot research topic. Due to the slow-varying nature of the channel in the delay-Doppler domain, it is advantageous for OTFS systems to utilize advanced channel estimation algorithms in the delay-Doppler domain to achieve performance superior to traditional time–frequency domain channel estimation algorithms.

In references [18] and [19], an embedded pilot-assisted channel estimation scheme for OTFS is proposed. In each OTFS frame, pilots, guards, and data symbols are arranged in the delay-Doppler plane to appropriately avoid interference between pilot and data symbols at the receiver. Such symbol arrangement has been developed for OTFS on multipath channels with both integer and fractional Doppler frequency

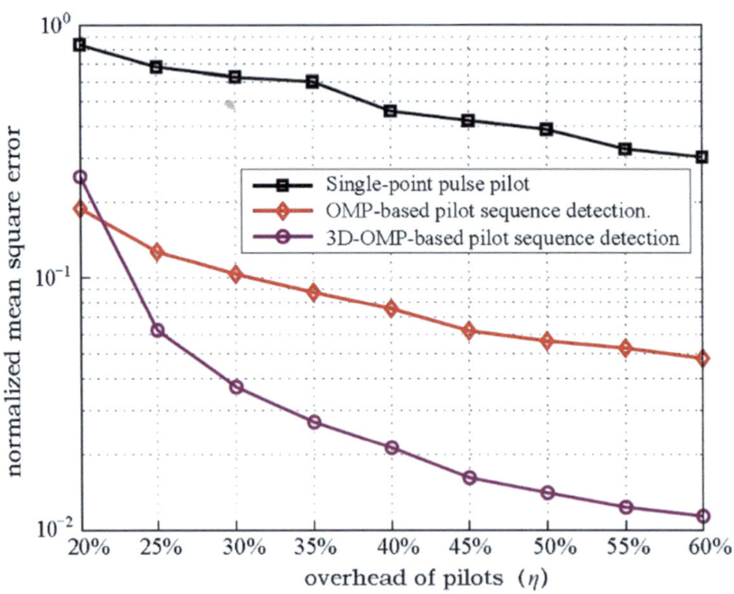

Fig. 11.13 Performance comparison of two pilot design schemes under different pilot overhead conditions

shifts. At the receiver, channel estimation is performed using a threshold-based method, and the estimated channel information is utilized for data detection through the MP algorithm. Due to the specific embedded symbol arrangement, channel estimation and data detection can be executed with minimal overhead within the same OTFS frame.

The pilot design and symbol detection under integer Doppler frequency shifts are illustrated in Fig. 11.14. In Fig. 11.15, a simulation compares the channel estimation performance for integer and fractional Doppler frequency shifts. It can be observed that, at high pilot signal-to-noise ratios, there is only minimal performance loss, demonstrating the reliability of the proposed channel estimation method.

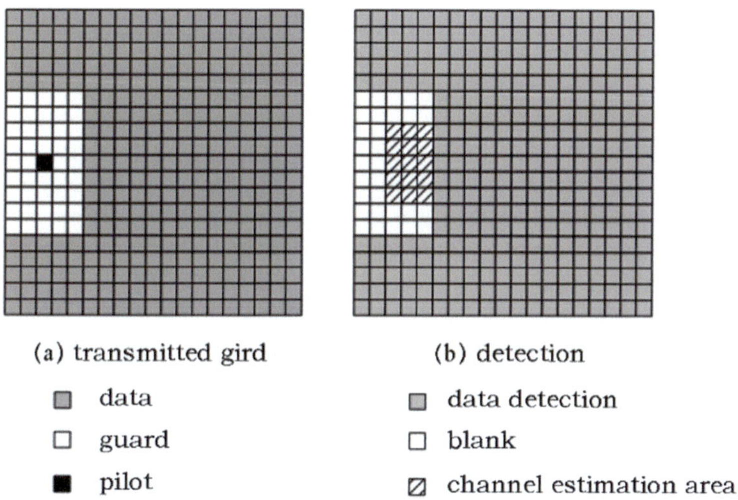

Fig. 11.14 Pilot design and symbol detection under integer Doppler frequency shifts

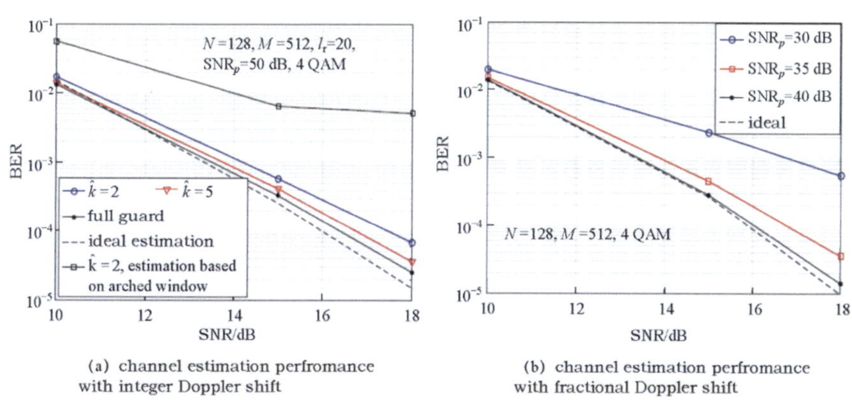

Fig. 11.15 Channel estimation performance under integer and fractional Doppler shifts

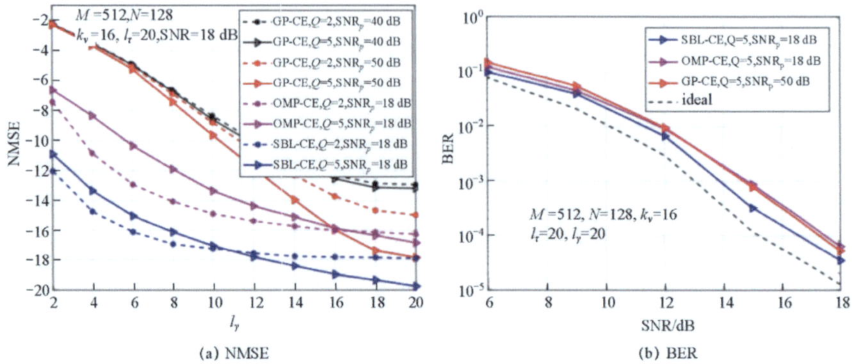

Fig. 11.16 NMSE and BER performance comparison of channel estimation

In reference [20], a problem is addressed concerning the need for higher pilot overhead and pilot power to estimate channel state information due to the effects of Doppler frequency shifts. Based on the sparsity of the channel in the delay-Doppler domain, a channel estimation algorithm is proposed using pilot patterns and sparse Bayesian learning. There are no guard pilots in the pilot pattern.

Under the new pilot pattern, the received pilot signal in the delay-Doppler domain is interfered by data, requiring a more complex approach for path separation and gain estimation. This method first transforms the channel estimation problem into a sparse signal recovery problem. Then, a sparse Bayesian learning framework is introduced, and a sparse signal prior model is constructed as a hierarchical Laplace prior model. Finally, the expectation–maximization algorithm is used to update the parameters in the previous model. Numerical simulations validate the advantages of this algorithm in terms of pilot overhead, pilot power consumption, and resistance to noise interference. The performance comparison of channel estimation NMSE and BER is shown in Fig. 11.16.

3. OTFS Channel Estimation Performance Evaluation

We have compared the performance of delay-Doppler domain and time–frequency domain channel estimation under TDL-A and CDL-A channels through simulation. The delay-Doppler domain channel estimation adopted the algorithm in literature [19], while the time–frequency domain channel estimation used the comb-shaped pilot and MMSE algorithm, the pilot pattern of which is shown in Fig. 11.17.

The relevant simulation parameters for time–frequency domain channel estimation and delay-Doppler domain channel estimation under TDL-A channel are as follows: $N = 128, M = 1024$, carrier frequency $f_c = 4$ GHz, speeds v are 3 km/h, 120 km/h, 240 km/h, and 500 km/h respectively; the pilot signal-to-noise ratio for delay-Doppler domain channel estimation $\text{SNR}_p = 45$ dB, the pilot interval of time–frequency domain channel estimation is 8 subcarriers. The comparison of the

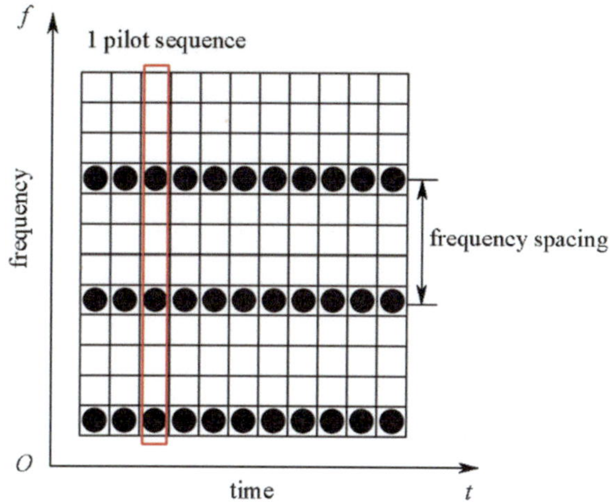

Fig. 11.17 Time–frequency domain pilot pattern

performance of time–frequency domain and delay-Doppler domain channel estimation is shown in Fig. 11.18. The channel estimation performance under the change of root mean square delay spread with $v = 500$ km/h is shown in Fig. 11.19.

In the CDL-A channel, time–frequency domain and delay-Doppler domain channel estimation are performed. The pilot interval of the time–frequency domain

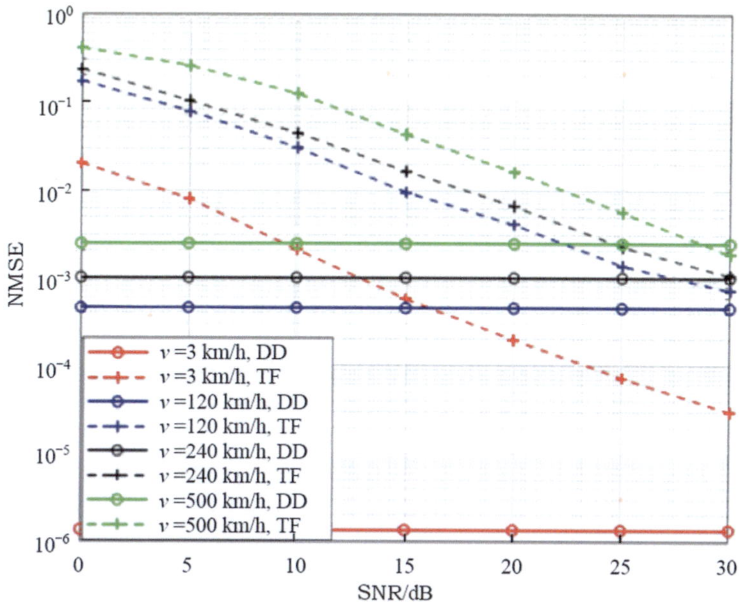

Fig. 11.18 Comparison of time–frequency domain and delay-Doppler domain channel estimation performance

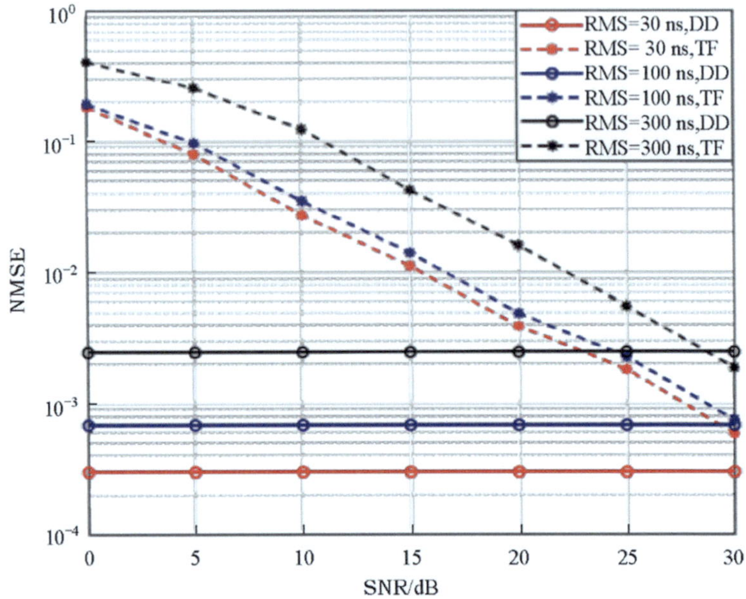

Fig. 11.19 $v = 500$ km/h change in root mean square delay spread under channel estimation performance

channel estimation is 2 subcarriers, and the pilot signal-to-noise ratio of the delay-Doppler domain is $\text{SNR}_p = 45$ dB, the performance comparison is shown in Fig. 11.20.

It can be seen that the channel estimation performance in the delay Doppler domain is significantly better than that in the time–frequency domain under the CDL-A channel model at four speeds. In the following simulation, the signal-to-noise ratio of the data symbols is fixed at $\text{SNR}_\text{d} = 20$ dB. After aligning the pilot power in the time–frequency domain and the delay-Doppler domain, the equivalent pilot power data in the delay-Doppler domain is shown in Table 11.2. Based on the configuration in Table 11.2, the comparison of channel estimation between the time–frequency domain and the delayed Doppler domain is illustrated in Fig. 11.21.

It can be seen that when the pilot power is the same, the channel estimation performance in the delay Doppler domain is better, or in other words, the cost of channel estimation in the delay Doppler domain is lower at the same channel estimation performance. In summary, we can conclude that the channel estimation performance in the delay Doppler domain is better and the cost is lower under the TDL-A and CDL-A channel models.

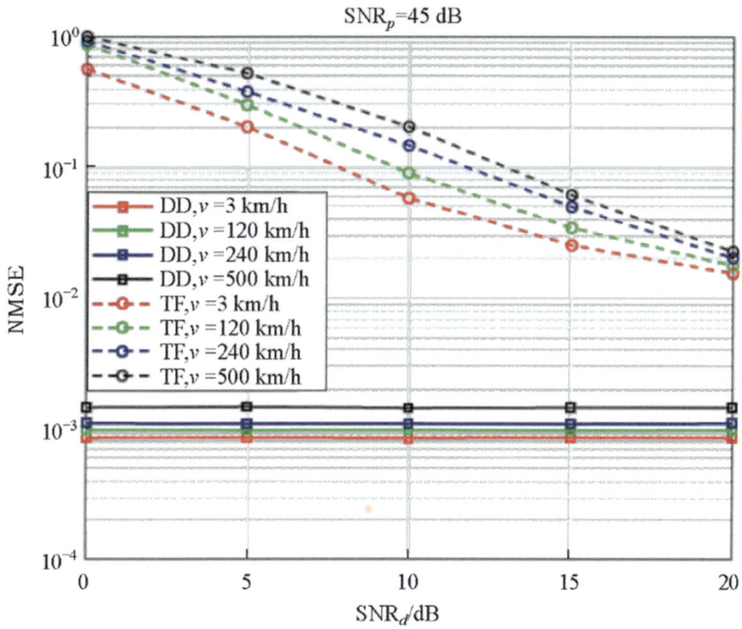

Fig. 11.20 Comparison of channel estimation performance at different speeds under CDL-A channel

Table 11.2 Equivalent pilot power of delay Doppler domain corresponding to time–frequency domain pilot interval

TF domain pilot interval	2	4	8	16	12
DD domain pilot power/Db	51.1751	48.1648	45.1545	42.1442	39.1339

11.2.4 Receiver Algorithm

The OTFS system presents significant challenges for symbol detection due to the two-dimensional convolution input–output relationship in the delay-Doppler domain. To reduce the complexity of OTFS detection, low-complexity linear and nonlinear OTFS receiver strategies have been successively proposed. The design of linear receivers in OTFS systems mainly focuses on the structure of equalization matrices, including the double-cyclic structure of matrices in the delay-Doppler domain, as well as the sparse band structure in the time domain, which are characterized by simple implementation. The nonlinear receiver of the OTFS system features an iterative structure, which can provide excellent detection performance. The nonlinear receiver in the OTFS system employs an iterative structure, which can offer excellent detection performance, including detection based on MP and related improved algorithms; iterative Rake receiver after zero padding (ZP); detection algorithm based on parallel interference

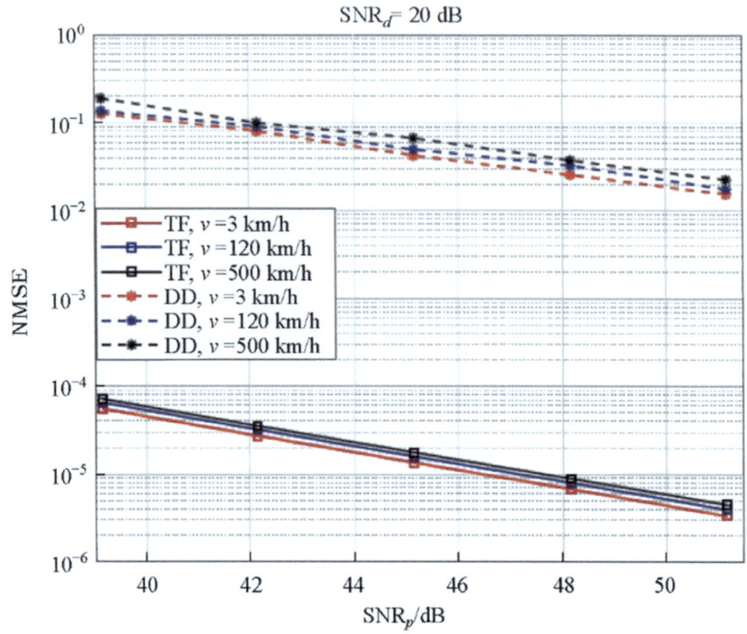

Fig. 11.21 Comparison of channel estimation performance under different pilot powers

cancellation; maximum a posteriori (MAP) algorithm based on Markov chain-Monte Carlo approximation, etc.

1. Low complexity linear equalization

Literature [21] proposes a low-complexity OTFS linear equalizer, which uses the structure of the effective channel matrix in the delay Doppler domain. Th proposed low complexity linear equalizer can provide accurate minimum mean square error (MMSE) and zero-forcing (ZF) solutions. For an $N \times M$ OTFS system, the time complexity of the proposed method is $\mathrm{O}(MN\lg(MN))$, where N and M are respectively the number of grid points in the Doppler and delay dimensions. In comparison, the MMSE and ZF solutions using the traditional matrix inversion method require $\mathrm{O}(M^3N^3)$ time complexity. For channel matrices that do not satisfy the block cyclic structure, the proposed method can provide a low-complexity initial solution for local search techniques to achieve enhanced bit error rate performance. The related simulation results are shown in Fig. 11.22, it can be seen that the proposed low-complexity linear equalizer can obtain accurate ZF and MMSE solutions. Furthermore, when the block cyclic matrix structure is not satisfied, using the proposed method as the initial solution for likelihood ascent search (LAS) can achieve better performance than the MP detection algorithm.

Literature [22] studied a low-complexity linear minimum mean square error receiver, in which the solution of the linear MMSE receiver in the OTFS system was

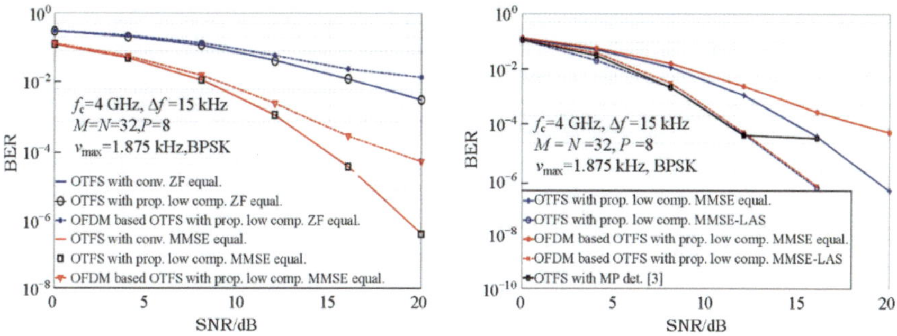

Fig. 11.22 Bit error rate performance of the proposed low-complexity MMSE and ZF equalizer

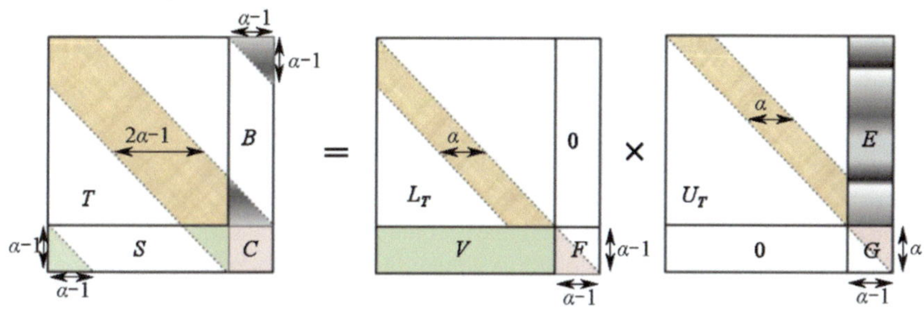

Fig. 11.23 Structure and LU decomposition of matrix $\boldsymbol{\Psi}$

first derived as $\hat{\boldsymbol{d}} = (\boldsymbol{HA})^{\dagger}\left[(\boldsymbol{HA})(\boldsymbol{HA})^{\dagger} + \frac{\sigma_n^2}{\sigma_d^2}\boldsymbol{I}\right]^{-1}\boldsymbol{r}$. Then, the receiver utilizes the sparsity and quasi-band structure of the matrix involved in the demodulation process to perform the corresponding LU decomposition to obtain $\boldsymbol{\Psi} = \left[\boldsymbol{HH}^{\dagger} + \frac{\sigma_n^2}{\sigma_d^2}\boldsymbol{I}\right]$, as shown in Fig. 11.23.

Finally, the proposed detection method in [22] is with the complexity of logarithmic linear order, and simulations have found that the error rate performance will not damages.

2. Low Complexity Non-linear Equalization

Nonlinear equalization provides better detection performance than linear equalization, but it comes with relatively higher complexity. In the context of OTFS data detection, numerous linear equalization algorithms with lower computational complexity have been proposed to address the high complexity brought by two-dimensional convolution in the delay-Doppler domain.

(1) MP and its improved algorithms

In references [23] and [24], an explicit input–output relationship describing OTFS modulation and demodulation is first derived. Ideal pulse shaping waveforms that satisfy the bi-orthogonality condition and rectangular waveforms that do not satisfy

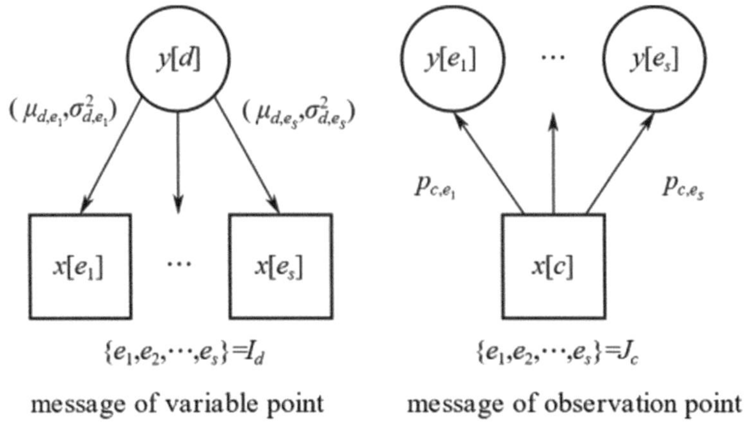

Fig. 11.24 The message passing process in the MP algorithm between observation (factor) nodes and variable nodes

the bi-orthogonality condition are analyzed. The results indicate that in the former case, only IDoI exists, while in the latter case, additional ICI and inter symbol interference (ISI) occur. Next, the interference is characterized, and a low-complexity and efficient MP algorithm is developed to achieve joint interference cancellation and symbol detection, as shown in Fig. 11.24. Although ICI and ISI are mitigated through proper phase adjustments, IDI can be alleviated by adjusting the MP algorithm to consider only the largest interference term. The MP algorithm effectively compensates for various channel Doppler spreads. Simulation results demonstrate that using rectangular waveforms with OTFS can achieve performance comparable to that of ideal but unrealizable pulse-shaped OTFS. Finally, simulation results prove that the proposed uncoded OTFS outperforms OFDM under various channel conditions.

From the simulation results in this literature, it can be observed that the performance of the Message Passing algorithm using rectangular waveforms is almost identical to that using ideal waveforms. Therefore, the use of the MP algorithm can effectively reduce the impact of ICI, making it practical to use rectangular waveforms.

In reference [25], a low-complexity MP algorithm based on Doppler compensation precoding and an improved stopping criterion for data detection in OTFS is proposed. Additionally, taking advantage of principal component analysis (PCA), an improved approximate MP algorithm based on covariance processing is introduced, which has almost the same complexity as the traditional AMP algorithm but exhibits better bit error rate (BER) performance.

(2) Iterative Rake Receiver

In the OTFS system, references [26] and [27] propose a linear complexity iterative Rake receiver. The basic idea is to use maximal ratio combining (MRC) to extract and coherently combine the received multipath components of the transmitted symbols in the delay Doppler grid, to improve the SNR of the combined signal. By placing

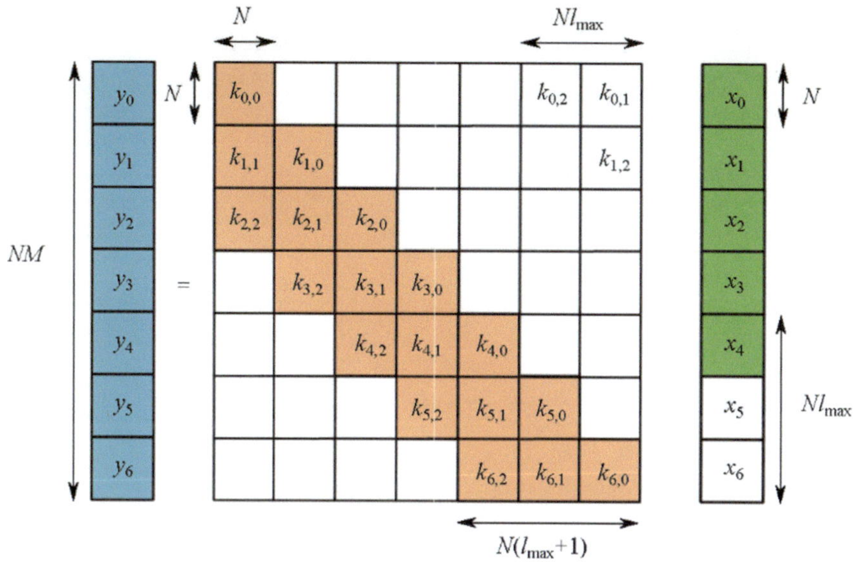

Fig. 11.25 The delay Doppler domain channel matrix $\boldsymbol{H}$ after adding null symbols

guard symbols in the delay Doppler grid and utilizing the cyclic nature of the channel matrix block, the OTFS input–output relationship is reconstructed in vector form, and the delay Doppler domain channel matrix $\boldsymbol{H}$ after adding null symbols only contains shadow blocks, as shown in Fig. 11.25.

In the OTFS system, references [26] and [27] propose a linear complexity iterative Rake receiver. The fundamental concept involves utilizing MRC to extract and coherently combine the received multipath components of the transmitted symbols in the delay-Doppler grid, in order to enhance the signal-to-noise ratio (SNR) of the combined signal. By placing pilot symbols in the delay-Doppler grid and leveraging the cyclical properties of the channel matrix blocks, the input–output relationship of OTFS is reconstructed in vector form. As shown in Fig. 11.25, the delay Doppler domain channel matrix $\boldsymbol{H}$ only contains shadow blocks after adding null symbols.

Using the new vector input–output relationship, references [26] and [27] introduce a low-complexity MRC-based decision feedback equalizer (DFE). The performance and complexity of the proposed equalizer are comparable to existing MP algorithms. Building upon this, the literature also presents a Gauss–Seidel-based over-relaxed Rake receiver to enhance the performance and convergence speed of iterative detection. Furthermore, based on the proposed maximal ratio combining strategy in the delay-Doppler domain, an equivalent low-complexity delay-time domain symbol detection algorithm is further proposed to reduce detection complexity. In order to enhance the reliability of the proposed algorithms, the literature also considers Turbo channel coding and introduces the iterative Rake Turbo receiver, as illustrated in Fig. 11.26.

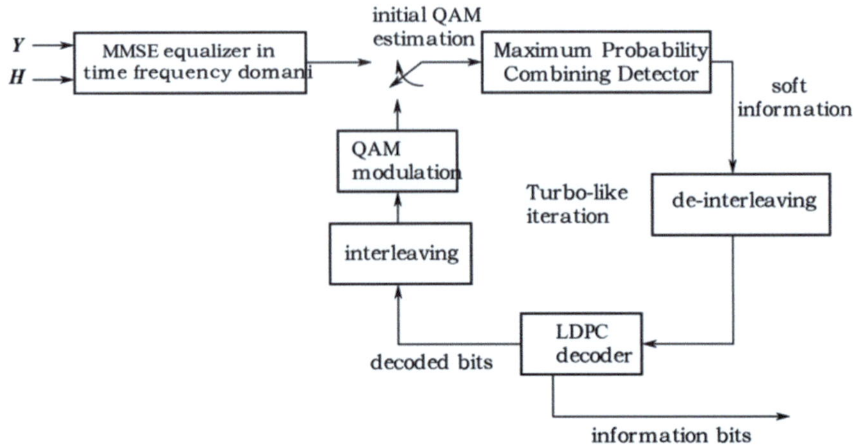

Fig. 11.26 Iterative Rake Turbo Receiver

Figure 11.27 shows the bit error rate comparison of uncoded MRC, MPA and MMSE-OFDM. It can be seen that the performance of the proposed algorithm for OTFS with good initial values is optimal compared to other MPA memory MMSE-OFDM.

Figure 11.28 shows the frame error rate (FER) comparison of MRC and bit interleaved coded (BIC) MMSE-OFDM (OFDM-BICM) under Turbo channel coding. It can be seen that the successive over relaxed (SOR) Turbo Rake receiver achieves optimal performance after three iterations.

(3) Linear MMSE based parallel interference cancellation (LMMSE-PIC)

The literature [28] discusses the design of a low-complexity iterative equalizer for encoding OTFS systems. In order to combat time frequency interference, the

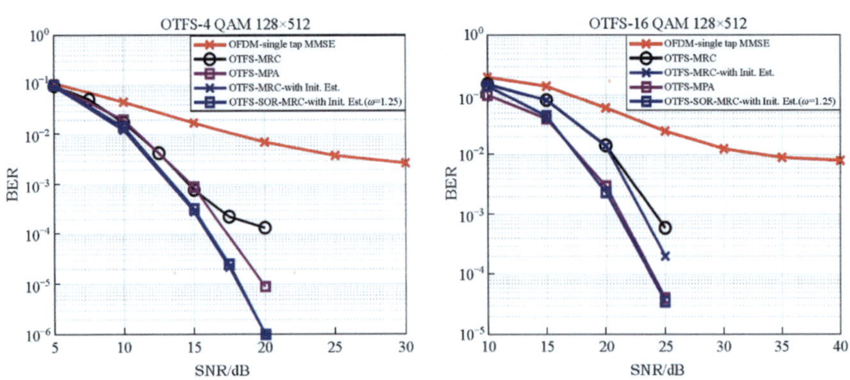

Fig. 11.27 Bit error rate comparison of uncoded MRC, MPA and MMSE-OFDM

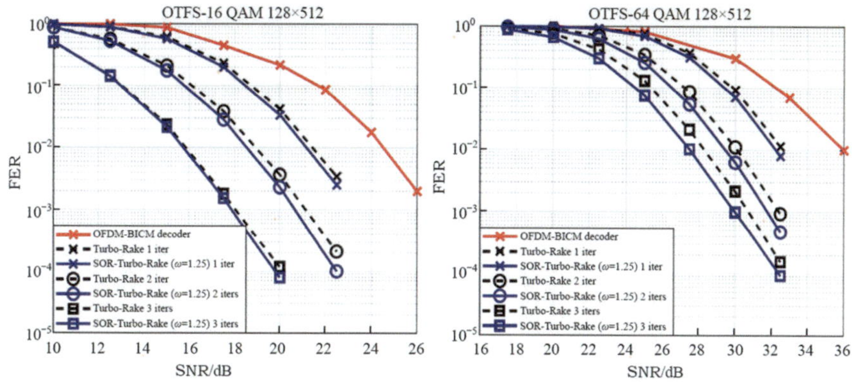

Fig. 11.28 Frame error rate comparison of MRC and OFDM-BICM under Turbo channel coding

LMMSE-PIC is adopted as the equalizer for OTFS. A first-order Neumann series is employed to approximate the matrix inversion involved in the LMMSE-PIC of OTFS, which reduces the complexity of the equalizer to near-linear, thereby decreasing the total number of transmitted symbols. The LMMSE-PIC equalizer framework proposed in literature [28] is illustrated in Fig. 11.29.

The transmission of external information and the simulation results of the coded bit error rate show that the proposed LMMSE-PIC equalizer is superior to the recently proposed OTFS PIC method. Furthermore, it also achieves considerable gains compared to OFDM, as shown in Fig. 11.30.

(4) MCMC

The literature [29] investigates OTFS from the perspectives of signal detection and channel estimation, and proposes a detection scheme based on Markov Chain Monte

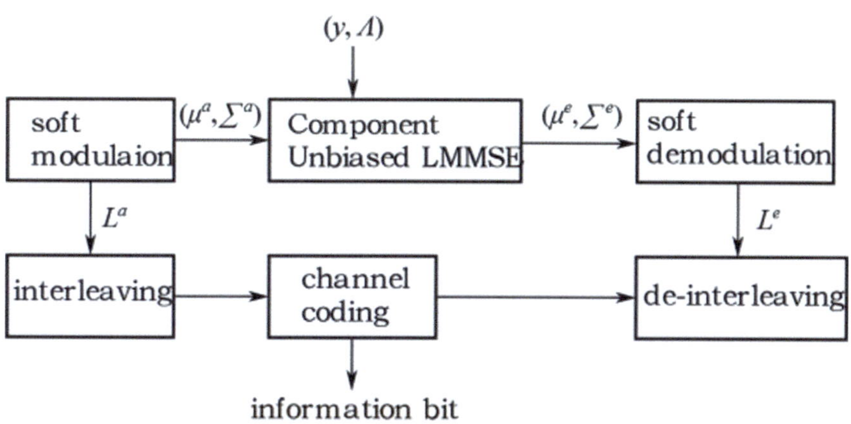

Fig. 11.29 LMMSE-PIC Equalizer Framework

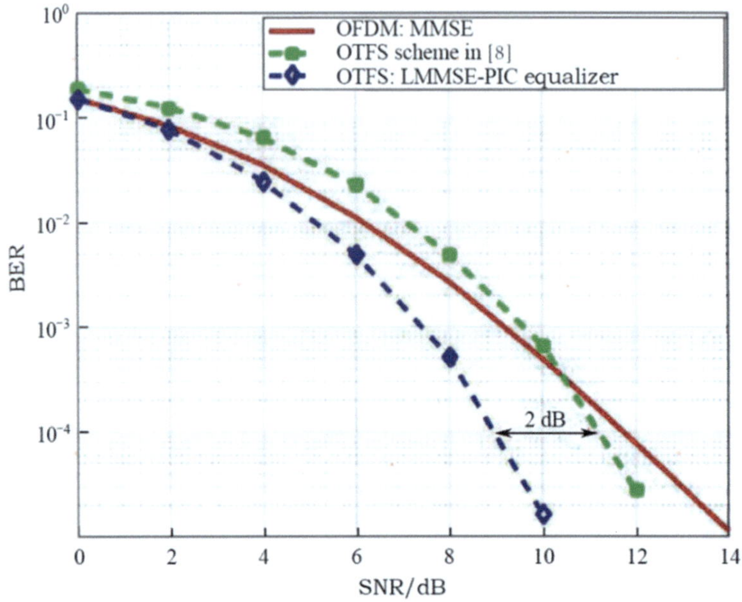

Fig. 11.30 Bit error rate comparison of LMMSE-PIC and OFDM

Carlo (MCMC) sampling. Assuming equiprobable transmission symbols based on the input–output relationship in the delay-Doppler domain, the maximum likelihood (ML) detection of the received signal in the delay-Doppler domain yields the solution of $\hat{\boldsymbol{x}}_{\mathrm{ML}} = \underset{\boldsymbol{x}\in\mathbb{A}^{NM}}{\operatorname{argmin}} \|\boldsymbol{y} - \boldsymbol{H}\boldsymbol{x}\|^2$. The detection complexity is extremely high, growing exponentially with the total number of grid points NM in the delay-Doppler two-dimensional plane. Therefore, in literature [29], MCMC is employed to conduct random Gibbs sampling to approximate the optimal solution of the ML algorithm. The reliability of the proposed symbol detection algorithm was verified through simulations, as illustrated in Fig. 11.31.

11.3 MIMO-OTFS System Design

The design of an OTFS-based communication system needs to consider the combination with MIMO systems. The combination of OTFS and MIMO can multiply the wireless signal transmission rate in high-speed mobile scenarios. However, interference from space, delay, and Doppler will pose a severe challenge to MIMO-OTFS symbol detection. To achieve stable data transmission in the MIMO-OTFS system, an efficient precoding mechanism is needed. Depending on the requirement of channel information, MIMO precoding can be divided into open-loop MIMO precoding and closed-loop MIMO precoding, which are introduced separately below.

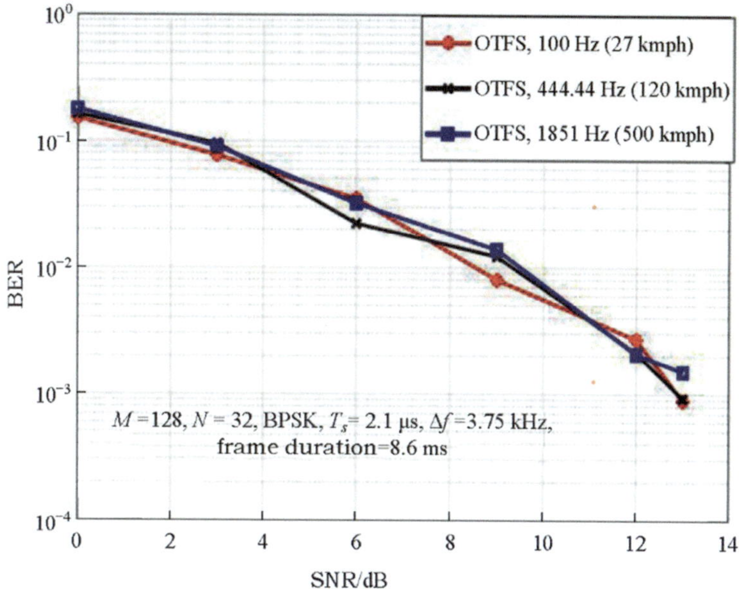

Fig. 11.31 Performance of random Gibbs sampling detection under three types of Doppler frequency shifts

11.3.1 Open-Loop MIMO Precoding

Open-loop MIMO precoding is a precoding scheme that does not depend on channel feedback information. Literature [30] proposes an open-loop diversity coding scheme under the assumption that the delay-Doppler domain channel remains unchanged between two consecutive frames.

Consider a MIMO-OTFS system with layer number of N_v, the transmit antenna ports number of N_t, and the number of receive antenna ports N_r. The transmission is in units of time slots, and the number of symbols transmitted in each time slot is N, and the number of subcarriers is M. Consider a 2×1 MIMO-OTFS system with $N_v = N_t = 2$, $N_r = 1$, the open-loop diversity scheme is shown in Table 11.3, where P is a permutation matrix, determined by the following equation:

$$P = P_N \otimes P_M \tag{11.24}$$

Table 11.3 Open-loop diversity scheme

	1st frame	2nd frame
Antenna 1	$\boldsymbol{x}_1$	$-(\boldsymbol{Px}_2)^*$
Antenna 2	$\boldsymbol{x}_2$	$(\boldsymbol{Px}_1)^*$

In the Eq. (11.24), $\otimes$ represents the Kronecker product; $P_M \in \mathbb{R}^{M\times M}$ and $P_N \in \mathbb{R}^{N\times N}$ represent the shift matrices of order M and N respectively. A shift matrix of order L is represented as

$$P_L = \begin{bmatrix} 1\,0 \ldots 0 \\ 0\,0 \ldots 1 \\ 0\,0 \cdots 0 \\ \vdots\ \vdots \quad \vdots \\ 0\,1 \cdots 0 \end{bmatrix}_{L\times L} \tag{11.25}$$

The purpose of introducing the permutation matrix P is to ensure the orthogonality of codewords to obtain the coding gain and diversity gain of the space–time code. Therefore, in the above open-loop diversity coding scheme, when transmitting the first frame, the vectorized data in delay Doppler domain for antenna 1 isx_1, the vectorized data in delay Doppler domain for antenna 2 is x_2; when the second frame is transmitted, the vectorized data in delay Doppler domain for antenna 1 is$-(Px_2)^*$, the vectorized data in delay Doppler domain for antenna 2 is$(Px_1)^*$. The received signals on the two frames at the receiver are given by

$$y_1 = H_1x_1 + H_2x_2 + v_1 \tag{11.26}$$

and

$$y_2 = -H_1(Px_2)^* + H_2(Px_1)^* + v_2 \tag{11.27}$$

respectively, where v_1 and v_2 represent the additive Gaussian white noise of the first and second frames. At the receiver, multiplying y_2 by the permutation matrix P and then taking the complex conjugate, it derives

$$\begin{aligned} \begin{bmatrix} \boldsymbol{y}_1 \\ (\boldsymbol{P}\boldsymbol{y}_2)^* \end{bmatrix} &= \begin{bmatrix} \boldsymbol{H}_1 & \boldsymbol{H}_2 \\ (\boldsymbol{P}\boldsymbol{H}_2)^*\boldsymbol{P} & -(\boldsymbol{P}\boldsymbol{H}_1)^*\boldsymbol{P} \end{bmatrix} \begin{bmatrix} \boldsymbol{x}_1 \\ \boldsymbol{x}_2 \end{bmatrix} + \begin{bmatrix} \boldsymbol{v}_1 \\ (\boldsymbol{P}\boldsymbol{v}_2)^* \end{bmatrix} \\ &= \begin{bmatrix} \boldsymbol{H}_1 & \boldsymbol{H}_2 \\ \boldsymbol{H}_2^{\mathrm{H}} & -\boldsymbol{H}_1^{\mathrm{H}} \end{bmatrix} \begin{bmatrix} \boldsymbol{x}_1 \\ \boldsymbol{x}_2 \end{bmatrix} + \begin{bmatrix} \boldsymbol{v}_1 \\ (\boldsymbol{P}\boldsymbol{v}_2)^* \end{bmatrix} \end{aligned} \tag{11.28}$$

Let $\overline{y} = \begin{bmatrix} y_1 \\ (Py_2)^* \end{bmatrix}$, $\overline{H} = \begin{bmatrix} H_1 & H_2 \\ H_2^H & -H_1^H \end{bmatrix}$ and $\overline{v} = \begin{bmatrix} v_1 \\ (Pv_2)^* \end{bmatrix}$, it can be expressed as

$$\overline{y} = \overline{H}\begin{bmatrix} x_1 \\ x_2 \end{bmatrix} + \overline{v} \tag{11.29}$$

Therefore, you can directly use MMSE or MP equalization to eliminate the impact of $\overline{H}$ and detect the transmitted symbols x_1 and x_2. It should be noted that the dimension of $\overline{H}$ is $2MN \times 2MN$, the complexity of equalization will increase as well.

The open-loop diversity coding scheme in literature [30] requires the channel to remain unchanged for two frames. However, when the relative speed between the transmitter and receiver is large or the dimensions (M and N) of a frame are large, there may be some deviation in the channel between the two frames. In response to this, literature [31] proposes an open-loop diversity coding scheme implemented within a single frame, as shown in Fig. 11.32. Firstly, the delay Doppler domain frames on each antenna are divided into two half-frames in the Doppler direction, and the data of two layers are respectively mapped onto these two half-frames. At this point, two issues need to be addressed: the first issue is to ensure that the two half-frames of the same antenna do not interfere with each other, and the second issue is to ensure that the data of the two half-frames of the same antenna experience the same equivalent channel. The first issue can be addressed by setting a guard interval between the two half-frames. As for the second issue, it needs to be discussed separately: for an ideal wave system, the equivalent channels experienced by the two half-frames are the same, so no additional processing is required; however, for a system based on rectangular waves, the assumption of orthogonality does not hold, leading to subcarrier interference in the time–frequency domain and symbol interference, which manifests as phase offset in the delay Doppler domain. Further research has revealed that only the data at the tail end along the delay direction in the two half-frames will experience different channels. Due to the cyclic nature of the delay Doppler domain equivalent channel, the data at corresponding positions in the two half-frames will still experience the same channel. Therefore, in a system based on rectangular waves, a suitable guard interval can be set at the tail end in the delay direction to ensure that the data of the two half-frames experience the same channel.

The coding scheme of the transmitter is shown in Fig. 11.32, the permutation matrix $P = \tilde{P}_{\frac{N}{2}} \otimes \tilde{P}_M$, where:

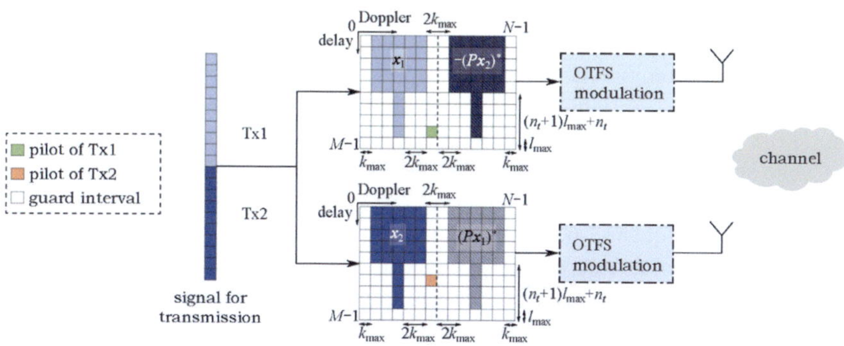

Fig. 11.32 Open-loop diversity coding scheme implemented within a frame

$$\tilde{\boldsymbol{P}}_{\frac{N}{2}} = \begin{bmatrix} \boldsymbol{I}_{k_{max}} & \boldsymbol{0} & \boldsymbol{0} \\ \boldsymbol{0} & \boldsymbol{P}_{\frac{N}{2}-2k_{max}} & \boldsymbol{0} \\ \boldsymbol{0} & \boldsymbol{0} & \boldsymbol{I}_{k_{max}} \end{bmatrix}_{\frac{N}{2}\times\frac{N}{2}} \tag{11.30}$$

$$\tilde{P}_M = \begin{bmatrix} P_{M-l_{\max}} & 0 \\ 0 & I_{l_{\max}} \end{bmatrix}_{M\times M} \tag{11.31}$$

The aforementioned modifications serve to accomplish the adaptation from a full frame to half frames, while also avoiding the confusion of pilot positions caused by permutation matrices.

The receiver multiplies the received signal of the second half frame by the permutation matrix P on the left, and then takes the complex conjugate to perform MMSE or MP equalisation. At this time, the equivalent channel dimension used for equalization is $MN \times MN$, and the complexity of equalization is lower than the scheme in literature [30].

In addition to performing open-loop diversity coding, diversity gain can also be obtained by cyclic shifting in the delay doppler domain. Specifically, for the same data, different antennas perform different cyclic shifts along the delay direction or along the Doppler direction to achieve diversity of multiple antennas.

Cyclic shifting along the delay direction means that, on the l-th antenna ($l = 1, 2, \cdots, L$), the last d_l elements of $\boldsymbol{X}$ along the delay direction can be placed at the beginning of the sequence, with the remaining $M - d_l$ elements successively shifted by d_l positions; alternatively, the first d_l elements of $\boldsymbol{X}$ along the delay direction can be placed at the end of the sequence, with the remaining $M - d_l$ elements successively shifted forward by d_l positions. The transmitted signal after cyclic shifting is denoted as $\boldsymbol{X}_l$, as shown in Fig. 11.33.

The cyclic shift is performed in the Doppler direction, that is, on the l th antenna ($l = 1, 2, \cdots, L$), X can be placed at the beginning of the last p_l bits in the Doppler direction, and the remaining $N - p_l$ bits are moved back p_l bits in sequence; X can also be placed at the end of the first p_l bits in the Doppler direction, and the remaining $N - p_l$ bits are moved forward p_l bits in sequence.

p l p_{l} position, the remaining $N - p_l$ positions move forward p_l positions in sequence, the emission lattice after cyclic shift is represented as X_l, as shown in Fig. 11.34.

Cyclic shifting along the Doppler direction means that, on the l-th antenna ($l = 1, 2, \cdots, L$), the last p_l elements of $\boldsymbol{X}$ along the Doppler direction can be placed at the beginning of the sequence, with the remaining $N - p_l$ elements successively shifted by p_l positions; alternatively, the first p_l elements of $\boldsymbol{X}$ along the Doppler direction can be placed at the end of the sequence, with the remaining $N - p_l$ elements successively shifted forward by p_l positions. The transmitted signal after cyclic shifting is denoted as $\boldsymbol{X}_l$, as shown in Fig. 11.33.

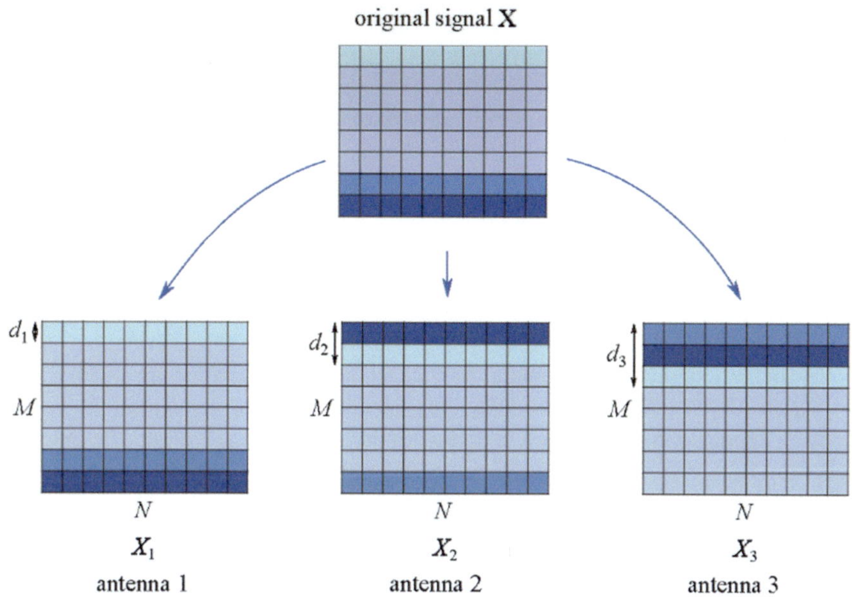

Fig. 11.33 Cyclic shift in the delay direction

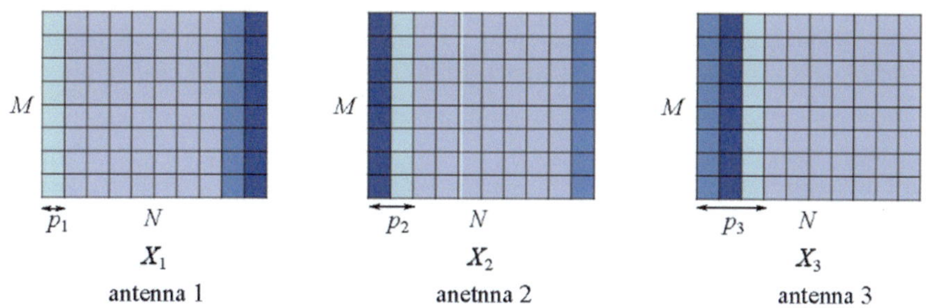

Fig. 11.34 Cyclic shift along the Doppler direction

11.3.2 Closed-Loop MIMO Precoding

The use of channel state information provided by the user for MIMO precoding at the transmitter is referred to as closed-loop MIMO precoding. Closed-loop MIMO precoding can be further divided into linear MIMO precoding and nonlinear MIMO precoding.

1. Linear MIMO Precoding

Linear MIMO precoding is a widely used precoding technique. In the context of multi-user large-scale MIMO-OTFS systems, a maximal ratio combining precoder was proposed in reference [32]. The transmitter performs precoding on symbols by multiplying them by the Hermitian values of the equivalent channel. Despite the lower computational complexity of precoding in reference [33], interference between users is not completely eliminated, thus reducing the detection performance of OTFS.

In high-speed scenarios, the limited number of scatterers between the base station and users results in minimal beam angle expansion, causing the multipath signals passing through scatterers to occupy only a small portion of the angle-domain channel. Consequently, the angle-domain channel of the MIMO-OTFS system exhibits sparsity. Figure 11.35 depicts the sparsity of a user's delay-Doppler-angle domain, showing confinement within a small range in the dimensions of delay, Doppler, and angle [33].

The delay-Doppler-angle domain channel is obtained by performing a DFT transformation on the multi-antenna delay-Doppler channel (also known as the delay-Doppler spatial domain channel, a three-dimensional channel) along the antenna direction. Reference [33] first performs a DFT transformation on the multi-antenna delay-Doppler channel (also known as the delay-Doppler spatial domain channel, a three-dimensional channel) to obtain the delay-Doppler-angle domain channel, and analyzes that the channel exhibits sparsity in the directions of delay, Doppler, and angle. Therefore, beamforming can be carried out based on the sparsity of different users in the angle domain to differentiate between the signals of different users.

Firstly, through scheduling, users transmitting signals simultaneously are ensured not to overlap in the angle domain. Secondly, based on the uplink pilot estimation of the channel information for each user (including information in the dimensions of delay, Doppler, and angle), the signals of different users are detected at different

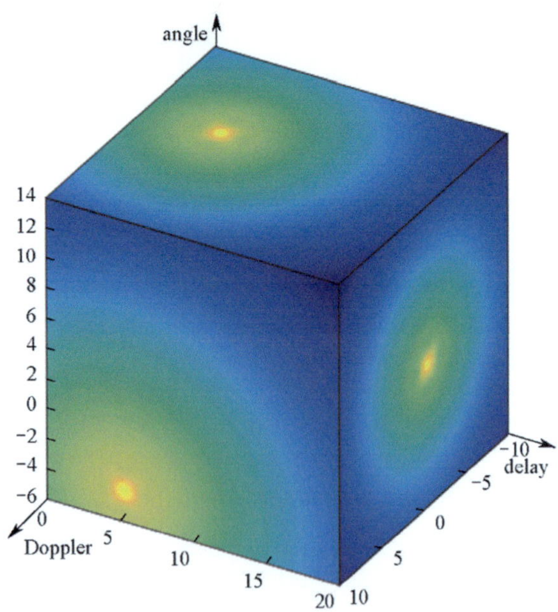

Fig. 11.35 Delay doppler angle domain sparsity

angles. Lastly, based on the estimated angle direction for each user, their respective beamforming vectors are formed to avoid interference between multiple users in the downlink. Since beamforming can only achieve alignment within a certain angle range, there are instances where multiple users' paths exist within this range. In such cases, resource allocation in the delay-Doppler domain or the design of NOMA schemes can be implemented for interference elimination based on the different characteristics of the angle sets of different users (or non-completely overlapping characteristics).

This approach combines user scheduling and the sparsity of the multi-user angle domain, and utilizes the spatial DFT transformation and beamforming of multiple antennas to address the issues of uplink and downlink multi-user interference. However, under this approach, it is still not possible to achieve multi-stream transmission for a single user.

Literature [34] proposes to perform linear precoding based on the delay Doppler domain equivalent MIMO matrix, such as delay Doppler domain ZF or MMSE precoding. However, the dimension of the delay Doppler domain equivalent MIMO matrix $\boldsymbol{H}$ is $N_rMN \times N_tMN$ (where N_{r} represents the number of antennas at the receiving end, N_{t} represents the number of antennas at the transmitting end, M represents the dimension of the delay domain, N represents the dimension of the Doppler domain), performing inversion operations for ZF or MMSE precoding on such a large dimension matrix is not feasible in real systems. In response to this, reference [34] proposed a low-complexity scheme for the special characteristics of the delay-Doppler domain equivalent MIMO matrix in a 2×2 MIMO system. This scheme can reduce the complexity of the matrix inversion operations in the ZF and MMSE algorithms with large-dimensional matrices. However, this low-complexity scheme cannot be extended to other antenna configurations, thus limiting its applicability.

Literature [35] proposes a low-complexity ZF MIMO-OTFS scheme under the assumption of ideal wave. In the case of ideal wave, the equivalent MIMO matrix $\boldsymbol{H}$ It has block cyclic characteristics. At this point, firstly, the equivalent MIMO matrix $\boldsymbol{H}$ is subjected to eigenvalue decomposition, and the inversion of the matrix $\boldsymbol{H}$ is replaced by the inversion of the eigenvalue matrix. Then, the eigenvalue matrix is divided into blocks (iteratively, each time a $MN \times MN$ block is divided out), and the inverse matrix of the eigenvalue matrix is constructed with the inverse matrix of the sub-block.

Reference [36] proposes the use of Neumann Series Expansion to approximate the inversion of large-dimensional matrices, and then constructs the inverse of the target channel matrix using the inverse of regular matrices to reduce complexity. However, such methods have certain requirements for the characteristics of the channel matrix and are not universally applicable to arbitrary channel matrices.

2. Non-linear MIMO Precoding

Compared to linear MIMO precoding, nonlinear MIMO precoding incurs higher complexity but offers better interference cancellation. Due to the transmission characteristics of OTFS signals, the received signal in the delay-Doppler domain is a cyclic convolution of the transmitted signal and the channel response, rather than a direct product of the two. As a result, the relationship between the transmitting and receiving signals exhibits certain nonlinear characteristics. In this regard, Hadani, a pioneer in the OTFS field at Cohere, believes that nonlinear MIMO precoding is better suited for OTFS technology.

Tomlinson-Harashima precoding (THP) is a well-known nonlinear precoding scheme that can also be used in closed-loop MIMO-OTFS systems. In practice, implementing THP is akin to performing pre-equalization at the transmitter to eliminate the effects of the channel. For MIMO-OTFS systems, the receiver needs to apply THP in the delay-Doppler domain to obtain time–frequency two-dimensional diversity gain. However, the channel in the delay-Doppler domain needs to be convolved with the transmitted symbols in the delay-Doppler domain to obtain the received symbols, and as such, the channel in the delay-Doppler domain cannot be decomposed into individual delays or Dopplers for the THP process. In MIMO-OTFS systems, it is common to decompose the equivalent channel matrix in the delay-Doppler domain to directly eliminate inter-symbol interference across different layers.

It is possible to consider using ZF/MMSE-QR-THP for MIMO-OTFS systems, as shown in Fig. 11.36. For QR-THP under these two criteria, the matrix design and the transmission and reception processes for THP are consistent; the only difference lies in the matrices used for QR decomposition.

Consider a MIMO-OTFS system with layer number of N_v, the transmit antenna ports number of N_t, and the number of receive antenna ports N_r. The transmission is in units of time slots, and the number of symbols transmitted in each time slot is N, and the number of subcarriers is M. Considering the scenario of ideal channel estimation, where the transmitter is aware of the equivalent channel matrix $H_{DD} \in \mathbb{C}^{N_r MN \times N_t MN}$ in the delay-Doppler domain, and further designs the feedback matrix and precoding matrix based on H_{DD}. The specific process of ZF/MMSE-QR-THP in the delay-Doppler domain is illustrated in Fig. 11.37, where, $s \in \mathbb{C}^{N_v MN \times 1}$ represents the data of all layers in the delay-Doppler domain and $s = [{s_1}^T, {s_2}^T, \cdots, {s_{N_v}}^T]^T$; $p \in \mathbb{C}^{N_t MN \times 1}$ represents the data of all transmitting antenna ports in the delay Doppler domain

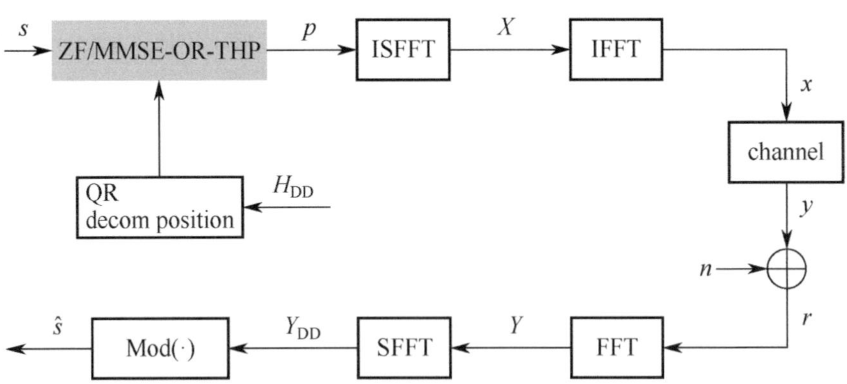

Fig. 11.36 Using ZF/MMSE-QR-THP in MIMO-OTFS systems

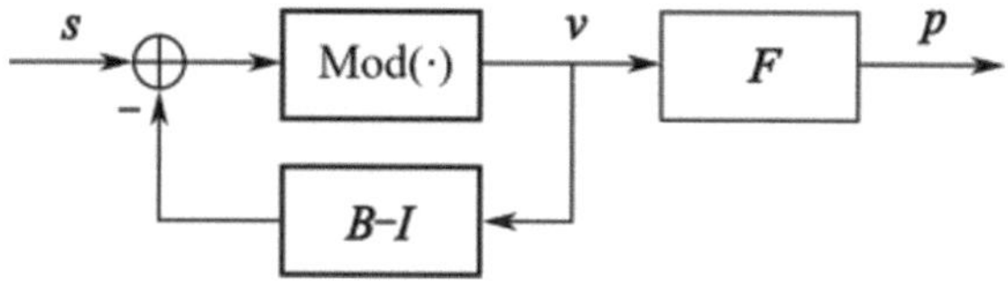

Fig. 11.37 The specific process of ZF/ MMSE-QR-THP

and $p = [p_1{}^T, p_2{}^T, \cdots, p_{N_t}{}^T]^T$. The feedback matrix is $B \in \mathbb{C}^{N_v MN \times N_v MN}$ and the pre-coding matrix is $F \in \mathbb{C}^{N_t MN \times N_v MN}$.

At first, QR decomposition is conducted to the equivalent channel matrix H_{DD}, given by

$$H_{\text{DD}}{}^H = QR \tag{11.32}$$

Then, the feedback matrix B and the pre-coding matrix F can be designed as

$$F = QG \tag{11.33}$$

and

$$B = R^H G \tag{11.34}$$

Respectively. Here, G is a diagonal matrix given by

$$G = \text{diag}\{r_{11}, r_{22}, \cdots, r_{ll}\}^{-1} \tag{11.35}$$

where, $r_{11}, r_{22}, \cdots, r_{ll}$ are all the diagonal elements of the matrix R^H.

In the delay Doppler domain, if the impact of noise is not considered, it is equivalent to pre-equalizing in the delay Doppler domain at the transmitting end, eliminating the impact of the delay Doppler domain channel.

$$H_{\text{DD}} F B^{-1} = I \tag{11.36}$$

Since the THP design based on the ZF criterion may amplify noise, the THP matrix can be designed according to the Minimize Mean Square Error (MMSE) criterion. The $\boldsymbol{Q}$ matrix and $\boldsymbol{R}$ matrix are obtained according to the MMSE criterion.

$$H_{\text{DD}}{}^{-1}\left(H_{\text{DD}} H_{\text{DD}}{}^H + \frac{\sigma_n{}^2}{\sigma_v{}^2} I\right) = QR \tag{11.37}$$

The design of other matrices is the same as Eqs. (11.33)–(11.36).

In the MIMO-OTFS system, the use of THP follows a similar basic principle to that in MIMO-OFDM. However, due to the particularity of the delay-Doppler domain channel matrix, pre-equalization can only be performed jointly across all subcarriers. As a result, during the computation process, the matrix dimensions are larger, leading to higher computational complexity. Nevertheless, the system benefits from time–frequency domain two-dimensional diversity gain, resulting in significantly improved performance at high speeds compared to MIMO-OFDM systems.

3. Equivalence of time-frequency domain and delay Doppler domain MIMO precoding

In the MIMO-OFDM system, codeword-based precoding effectively maps the time–frequency domain data streams onto different antenna ports in the spatial domain. Therefore, in the MIMO-OTFS system, it is also possible to consider precoding in the time–frequency domain, with the actual operation being identical to the codeword-based precoding process in the MIMO-OFDM system. However, precoding can be performed in the delay-Doppler domain at the receiver. As the delay-Doppler domain channel changes relatively slowly, the channel estimation in the delay-Doppler domain will be more accurate. Hence, it is feasible to equivalently carry out the time–frequency domain precoding process in the delay-Doppler domain. The following presents the proof of the equivalence relationship between time–frequency domain and delay-Doppler domain MIMO precoding.

For ease of derivation, consider a downlink MIMO-OTFS system with a number of transmitter antenna ports $N_t = 2$, a number of receiver antenna ports $N_r = 2$, and a number of data streams $N_v = 2$. The number of subcarriers for each transmitted data stream is M, and the number of symbols is N. $D_1 \in \mathbb{C}^{M \times N}$, $D_2 \in \mathbb{C}^{M \times N}$ are the transmitted symbols in the delay-Doppler domain. The delay-Doppler domain symbols are converted to the time–frequency domain through ISFFT:

$$T_1 = F_M D_1 {F_N}^H \tag{11.38}$$

$$T_2 = F_M D_2 {F_N}^H \tag{11.39}$$

According to the properties of matrix vectorization:

$$\text{vec}(ABC) = \left(C^T \otimes A\right)\text{vec}(B) \tag{11.40}$$

where, A, B, C are matrices of any dimension; $\otimes$ represents the Kronecker product. According to Eq. (11.40), T_1, T_2 can be represented in vector form as

$$\text{vec}(T_1) = \left({F_N}^* \otimes F_M\right)\text{vec}(D_1) \tag{11.41}$$

$$\text{vec}(T_2) = \left({F_N}^* \otimes F_M\right)\text{vec}(D_2) \tag{11.42}$$

Let

$$T = [\text{vec}(T_1), \text{vec}(T_2)]^T \tag{11.43}$$

$$D = [\mathrm{vec}(D_1), \mathrm{vec}(D_2)]^T \tag{11.44}$$

$$\mathrm{vec}\left(X^T\right) = [t_1^1, t_1^2, t_2^1, t_2^2, \cdots, t_{MN}^1, t_{MN}^2] \tag{11.45}$$

where $T \in \mathbb{C}^{2\times MN}$, $X \in \mathbb{C}^{MN\times 2}$, $D \in \mathbb{C}^{2\times MN}$ represent the time–frequency domain symbols before and after pre-coding, and the delay Doppler domain symbols before pre-coding, respectively; t_i^j represents the symbol on j-th antenna and i-th subcarrier after pre-coding.

Considering that the precoding matrix on each subcarrier is different, the time–frequency domain precoding matrix can be represented as

$$W_{\mathrm{TF}} = \begin{pmatrix} W_1 & 0 & \cdots & 0 \\ 0 & W_2 & \cdots & 0 \\ \vdots & \vdots & & \vdots \\ 0 & 0 & \cdots & W_{MN} \end{pmatrix} \tag{11.46}$$

where $W_i \in \mathbb{C}^{2\times 2}, i = 1, 2, \cdots, MN$, representing the precoding matrix on each subcarrier. From this, the precoded time–frequency domain symbols can be obtained

$$\mathrm{vec}\left(X^T\right) = W_{\mathrm{TF}}\mathrm{vec}(T) \tag{11.47}$$

$$= W_{\mathrm{TF}}\mathrm{vec}\{[\left(F_N{}^* \otimes F_M\right)D^T]^T\} \tag{11.48}$$

$$= W_{\mathrm{TF}}\mathrm{vec}[D\left(F_N{}^H \otimes F_M{}^T\right)] \tag{11.49}$$

$$= W_{\mathrm{TF}}[\left(F_N{}^* \otimes F_M\right) \otimes I_2]\mathrm{vec}(D) \tag{11.50}$$

Suppose the time–frequency domain pre-coding is equivalent to the delay Doppler domain, the pre-coding process in the delay Doppler domain is represented as

$$\mathrm{vec}\left(\tilde{D}\right) = W_{\mathrm{DD}}\mathrm{vec}(D) \tag{11.51}$$

where $\tilde{D} \in \mathbb{C}^{2\times MN}$ represents the delay Doppler domain symbols after equivalent delay Doppler domain pre-coding, $W_{\mathrm{DD}} \in \mathbb{C}^{2MN\times 2MN}$ represents the equivalent precoding of the time–frequency domain in the delay-Doppler domain. The delay Doppler domain symbols after equivalent delay Doppler domain pre-coding are transformed into the time–frequency domain by ISFFT, which can be represented as

$$\tilde{T}_1 = F_M\tilde{D}_1F_N{}^H \tag{11.52}$$

$$\tilde{T}_2 = F_M \tilde{D}_2 F_N{}^H \tag{11.53}$$

Similarly,

$$\mathrm{vec}(\tilde{T}_1) = (F_N{}^* \otimes F_M)\mathrm{vec}(\tilde{D}_1) \tag{11.54}$$

$$\mathrm{vec}(\tilde{T}_2) = (F_N{}^* \otimes F_M)\mathrm{vec}(\tilde{D}_2) \tag{11.55}$$

$$\tilde{T} = [\mathrm{vec}(\tilde{T}_1), \mathrm{vec}(\tilde{T}_2)]^T \tag{11.56}$$

$$\tilde{D} = [\mathrm{vec}(\tilde{D}_1), \mathrm{vec}(\tilde{D}_2)]^T \tag{11.57}$$

Therefore,

$$\mathrm{vec}(X^T) = \tilde{T} = [(F_N{}^* \otimes F_M) \otimes I_2]\mathrm{vec}(\tilde{D}) \tag{11.58}$$

Then, according to formula (11.28) and formula (11.36), we have

$$[(F_N{}^* \otimes F_M) \otimes I_2]\mathrm{vec}(\tilde{D}) = W_{\mathrm{TF}}[(F_N{}^* \otimes F_M) \otimes I_2]\mathrm{vec}(D) \tag{11.59}$$

$$\mathrm{vec}(\tilde{D}) = [(F_N{}^* \otimes F_M) \otimes I_2]^{-1} W_{\mathrm{TF}}[(F_N{}^* \otimes F_M) \otimes I_2]\mathrm{vec}(D) \tag{11.60}$$

Then, according to Eq. (11.29), we have

$$W_{\mathrm{DD}} = [(F_N{}^* \otimes F_M * \otimes F_M) \otimes I_{N_t}] \tag{11.61}$$

If extended to arbitrary number of transmit and receive antenna ports and data layers, then Eq. (11.61) becomes

$$W_{\mathrm{DD}} = [(F_N{}^* \otimes F_M) \otimes I_{N_t}]^{-1} W_{\mathrm{TF}}[(F_N{}^* \otimes F_M) \otimes I_{N_v}] \tag{11.62}$$

It can be observed that equivalent transformation can be performed between time–frequency domain precoding and delay-Doppler domain precoding, achieving the same effect. As the channel estimation accuracy in the delay-Doppler domain is higher than that in the time–frequency domain, MIMO precoding can be directly carried out in the delay-Doppler domain based on the channel estimation in the delay-Doppler domain, without the need to transform the channel estimation results to the time–frequency domain for time–frequency domain MIMO precoding.

11.4 Advantages and Challenges of OTFS

11.4.1 Advantages

OTFS, as a highly competitive new waveform candidate technology, demonstrates its advantages primarily in the following aspects:

(1) OTFS resolves the physical channel's reflections through delay Doppler imaging and coherently combines the energy from different reflected paths using a receiver equalizer. This effectively provides a fading-free static channel response. Leveraging this static channel characteristic, OTFS does not require closed-loop channel adaptation like OFDM to cope with rapidly changing channels, thereby enhancing system robustness and reducing system design complexity.

(2) Due to the significantly smaller number of delay-Doppler states in the delay Doppler domain compared to the time–frequency states in the time–frequency domain, the channel in OTFS can be expressed in a highly compact form. This reduces the channel estimation overhead and enhances precision.

(3) The advantages of OTFS are also evident in highly Doppler-sensitive channels. By appropriately analyzing the delay Doppler imaging with signal processing parameters, the Doppler characteristics of the channel are fully displayed, which is advantageous for signal analysis and processing in Doppler-sensitive scenarios (such as high-speed mobility and millimeter-wave).

11.4.2 Challenges

The current research on OTFS technology generally assumes that the modulation waveform of OTFS satisfies the double orthogonality assumption. When processing digital signals, OTFS can be seen as a precoder for the OFDM system, where the transmitted constellation symbols are mapped onto time–frequency domain resources after undergoing ISFFT, and then converted into a time-domain signal for transmission using traditional OFDM modulation. Based on this architecture, the OTFS system exhibits a concise two-dimensional convolution form in the delay Doppler domain input–output relationship, leading to accurate channel estimation and outstanding demodulation performance. However, as the ideal waveform satisfying the double orthogonality assumption cannot be achieved in practical systems, rectangular pulses, raised cosine pulses, and similar waveforms are commonly used as shaping pulses. In the presence of non-ideal waveforms, the concise input–output relationship in the delay Doppler domain is disrupted, introducing Doppler and delay interferences between symbols, compromising the orthogonality among received symbols in the delay Doppler domain, resulting in biased channel estimation. Currently, there are two main approaches for mitigating fractional delay and Doppler: one is to increase the delay and Doppler resolution for more granular channel estimation, and the other is to apply filtering or windowing to the time–frequency domain signal at the transmitter to counteract the fractional delay and Doppler effects. However, both approaches have significant drawbacks. The former significantly increases processing delay and complexity, which is not conducive to system implementation, while the latter cannot completely eliminate the effects of fractional delay and Doppler, and may introduce signal distortion, affecting receiver performance and potentially leading to an error floor in certain scenarios. Therefore, optimizing the transmitted waveform at the system implementation level to realize the superior performance of OTFS in theory is a pressing challenge in future communication systems. If a waveform suitable for OTFS can be found that is both close to the ideal waveform satisfying the double orthogonality assumption and easy to implement in engineering, OTFS is more likely to shine in 6G communication systems.

Another advantage of OTFS is the innovative introduction of delay Doppler domain signal analysis in communication systems, using unique pilot pulse design to enable more refined channel estimation without the interpolation errors of time–frequency domain channel estimation. However, the received signal in the delay Doppler domain is the result of the two-dimensional convolution of the transmitted signal and the channel response, inevitably introducing interference from neighboring resource grid points, leading to suboptimal performance of low-complexity linear equalization. Practical OTFS typically employs message passing algorithms for iterative demodulation, which has high complexity and its performance is greatly affected by channel sparsity. Current low-complexity receiver algorithms often use redundant resources (such as adding CP/ZP) to improve demodulation performance, sacrificing spectral efficiency and losing some advantages over OFDM systems. Finding a delay Doppler domain receiver algorithm that achieves a better balance between complexity and performance, and exploring the potential advantages of cross-domain joint equalization and detection algorithms, are important technical challenges to further drive the deployment of OTFS.

The combination of OTFS and multi-antenna systems remains an area that has not been fully explored. Linear precoding transmission based on MIMO-OTFS has been proven to achieve the same capacity boundary as linear precoding transmission of MIMO-OTFS. Nonlinear precoding based on MIMO-OTFS is expected to achieve a higher capacity boundary than the same precoding method based on MIMO-OTFS. The reason for this is that nonlinear precoding is essentially a pre-equalization of the channel at the transmitter, and its effectiveness heavily depends on the accuracy of the channel CSI. However, in practical systems, we can only use the measured channel CSI results from the previous time slot to determine the design of the precoding matrix for the current time slot. Therefore, for rapidly changing time–frequency domain channels in high-mobility scenarios, the obtained precoding matrix may not match the current time slot's channel well. In contrast, the delay Doppler domain channel exhibits temporal stability, providing inherent advantages in channel estimation and prediction, which can better unleash the performance of nonlinear precoding. However, current nonlinear precoding techniques are overly complex, and designing a MIMO-OTFS nonlinear precoding technique that balances performance and implementation complexity is an urgent problem to solve.

Additionally, in an ISAC system based on MIMO-OTFS, the geometric position information of the physical channel and the motion status of communication nodes can be measured by the received signals. Because the delay Doppler domain channel model directly reflects the geometric relationships of nodes in the physical channel, it is conducive to the implementation of perception-assisted communication. For example, predicting changes in the geometric topology of the physical channel through perception-based ranging and speed measurements, and then exploiting the relationship between the geometric topology of the physical channel and the wireless channel to assist in predicting changes in the wireless channel response and designing corresponding pre-equalization algorithms to reduce the cost of channel measurement and feedback. The algorithm development and process design related to the above issues are also one of the challenges in expanding the application of OTFS waveforms.

References

1. ANWAR W, KRAUSE A, KUMAR A, et al. Performance analysis of various waveforms and coding schemes in V2X communication scenarios[C]//2020 IEEE Wireless Communications and Networking Conference (WCNC). IEEE, 2020: 1–8.
2. LIN J, WANG G, XU R, et al. Channel and phase shift estimation for TM-aided OTFS railway communications[C]//2021 IEEE/CIC International Conference on Communications in China (ICCC Workshops). IEEE, 2021: 444–448.
3. HADANI R, RAKIB S, MOLISCH A F, et al. Orthogonal time frequency space (OTFS) modulation for millimeter-wave communications systems[C]//2017 IEEE MTT-S International Microwave Symposium (IMS). IEEE, 2017: 681–683.
4. SURABHI G D, RAMACHANDRAN M K, CHOCKALINGAM A. OTFS modulation with phase noise in mmWave communications[C]//2019 IEEE 89th Vehicular Technology Conference (VTC2019-Spring). IEEE, 2019: 1–5.
5. ZHOU X, GAO Z. Joint active user detection and channel estimation for grant-free NOMA-OTFS in LEO constellation Internet-of-Things[C]//2021 IEEE/CIC International Conference on Communications in China (ICCC). IEEE, 2021: 735–740.
6. FENG X, ESMAIEL H, WANG J, et al. Underwater acoustic communications based on OTFS[C]//2020 15th IEEE International Conference on Signal Processing (ICSP). IEEE, 2020, 1: 439–444.
7. RAVITEJA P, PHAN K T, HONG Y, et al. Orthogonal time frequency space (OTFS) modulation based radar system[C]//2019 IEEE Radar Conference (RadarConf). IEEE, 2019: 1–6.
8. HADANI R, RAKIB S, TSATSANIS M, et al. Orthogonal time frequency space modulation[C]//2017 IEEE Wireless Communications and Networking Conference (WCNC). IEEE, 2017: 1–6.
9. 3GPP TSG RA WG1. Overview of OTFS waveform for next generation RAT. Meeting #84-bis R1 162929 [R]. 2016.
10. 3GPP TSG RA WG1. Performance Results for OTFS Modulation. Meeting #85 R1 165053 [R]. 2016.
11. 3GPP TSG RA WG1. Performance evaluation of OTFS waveform in single user scenarios. Meeting #86 R1 167593 [R]. 2016.

12. 3GPP TSG RA WG1. OTFS-Performance-in-High-Doppler-with-Varying-Subcarrier Spacing. Meeting #86bis R1 1609825 [R]. 2016.
13. RAVITEJA P, HONG Y, VITERBO E, et al. Practical pulse-shaping waveforms for reduced-cyclic-prefix OTFS[J]. IEEE Transactions on Vehicular Technology, 2018, 68(1): 957–961.
14. HOSSAIN M N, SUGIURA Y, SHIMAMURA T, et al. Waveform design of low complexity WR-OTFS system for the OOB power reduction[C]//2020 IEEE Wireless Communications and Networking Conference Workshops (WCNCW). IEEE, 2020: 1–5.
15. TIWARI S, DAS S S. Circularly pulse-shaped orthogonal time frequency space modulation[J]. Electronics Letters, 2020, 56(3): 157-160.
16. SURABHI G D, AUGUSTINE R M, CHOCKALINGAM A. Peak-to-average power ratio of OTFS modulation[J]. IEEE Communications Letters, 2019, 23(6): 999–1002.
17. 3GPP TSG RA WG1. On UL RS for CSI Estimation. Meeting #87 R1 1612632 [R]. 2016.
18. RAVITEJA P, PHAN K T, HONG Y. Embedded pilot-aided channel estimation for OTFS in delay–Doppler channels[J]. IEEE Transactions on Vehicular Technology, 2019, 68(5): 4906-4917.
19. RAVITEJA P, PHAN K T, HONG Y, et al. Embedded delay-Doppler channel estimation for orthogonal time frequency space modulation[C]//2018 IEEE 88th Vehicular Technology Conference (VTC-Fall). IEEE, 2018: 1–5.
20. ZHAO L, GAO W J, GUO W. Sparse Bayesian learning of delay-Doppler channel for OTFS system[J]. IEEE Communications Letters, 2020, 24(12): 2766–2769.
21. SURABHI G D, CHOCKALINGAM A. Low-complexity linear equalization for OTFS modulation[J]. IEEE Communications Letters, 2019, 24(2): 330–334.
22. TIWARI S, DAS S S, RANGAMGARI V. Low complexity LMMSE receiver for OTFS[J]. IEEE Communications Letters, 2019, 23(12): 2205-2209.
23. RAVITEJA P, PHAN K T, JIN Q, et al. Low-complexity iterative detection for orthogonal time frequency space modulation [C] //2018 IEEE Wireless Communications and Networking Conference (WCNC). IEEE, 2018: 1–6.
24. RAVITEJA P, PHAN K T, HONG Y, et al. Low-complexity iterative detection for orthogonal time frequency space modulation [C]//IEEE Transactions on Wireless Communications. IEEE, 2018, 17(10): 6501–6515.
25. LI L, LIANG Y, FAN P, et al. Low complexity detection algorithms for OTFS under rapidly time-varying channel[C]//2019 IEEE 89th Vehicular Technology Conference (VTC2019-Spring). IEEE, 2019: 1–5.
26. THAJ T, VITERBO E. Low complexity iterative rake detector for orthogonal time frequency space modulation[C]//2020 IEEE Wireless Communications and Networking Conference (WCNC). IEEE, 2020: 1–6.
27. ZEMEN T, HOFER M, LOESCHENBRAND D, et al. Iterative Detection For Orthogonal Precoding In Doubly Selective Channels[C]// IEEE International Symposium on Personal, Indoor and Mobile Radio Communications (PIMRC). IEEE, 2017.
28. LONG F, NIU K, DONG C, et al. Low complexity iterative LMMSE-PIC equalizer for OTFS[C]// 2019 IEEE International Conference on Communications (ICC). IEEE, 2019: 1–6.
29. MURALI K R, CHOCKALINGAM A. On OTFS modulation for high-Doppler fading channels[C]//2018 Information Theory and Applications Workshop (ITA). IEEE, 2018: 1–10.
30. AUGUSTINE R M, SURABHI G D, CHOCKALINGAM A. Space-Time Coded OTFS Modulation in High-Doppler Channels[C]//IEEE Vehicular Technology Conference. IEEE, 2019.
31. WANG D, SUN B L, WANG F G, et al. Transmit diversity scheme design for rectangular pulse shaping based OTFS[J]. China Communications, 2022, 19(3): 116–128.
32. PANDEY B C, MOHAMMED S K, RAVITEJA P, et al. Low complexity precoding and detection in multi-user massive MIMO OTFS downlink [J]. IEEE Transactions on Vehicular Technology, 2021, 70(5): 4389–4405.
33. LI M Y, ZHANG S, GAO F F, et al. A new path division multiple access for the massive MIMO-OTFS networks [J]. IEEE Journal on Selected Areas in Communications, 2021, 39(4): 903–918.

34. SURABHI G D, CHOCKALINGAM A. Low-complexity linear equalization for 2×2 MIMO-OTFS signals [C]//IEEE 21st International Workshop on Signal Processing Advances in Wireless Communications. IEEE, 2020: 1–5.
35. SINGH P, MISHRA H B, BUDHIRAJA R. Low-Complexity Linear MIMO-OTFS Receivers[C]// 2021 IEEE International Conference on Communications Workshops (ICC Workshops). IEEE, 2021.
36. FENG J, NGO H Q, FLANAGAN M F, et al. Performance Analysis of OTFS-based Uplink Massive MIMO with ZF Receivers[C]// 2021 IEEE International Conference on Communications Workshops (ICC Workshops). IEEE, 2021.
37. RAVITEJA P, PHAN K T, HONG Y, et al. Interference cancellation and iterative detection for orthogonal time frequency space modulation [J]. IEEE Transactions on Wireless Communications, 2018, 17(10): 6501–6515.

Chapter 12
Integrated Sensing and Communication

With the development of information technology and the expansion of communication network functions, new services such as ubiquitous intelligence, human–machine interaction, immersive XR, digital twins, etc., are constantly emerging. In addition to communication capabilities, future 6G systems will have the ability to sense the physical world, and provide richer services and applications based on the information of target objects, events or environments. This chapter introduces how the communication and sensing can be integrated to extend the network capabilities and empower the new service, application and scenarios.

12.1 Background of Integrated Sensing and Communication (ISAC)

From an implementation perspective, sensing can be categorized into radio frequency (RF) sensing and non-RF sensing. Radar, as a typical RF sensing method, has been widely used. Its main operational mode is to detect targets and measure their distances by transmitting radio waves and receiving echoes. With the development of radar technology, radar can not only measure the distance of the target, but also their velocity, azimuth, elevation angle, and extract more information about the targets from the above information, including their size and shape. Radar technology was initially used in the military field to detect targets such as aircraft, missiles, vehicles, and ships. With the development of radar technology, it is increasingly used in civil applications, with typical applications such as weather radar measuring the echoes of meteorological targets such as clouds and rain to determine the position and intensity of the weather for weather forecasting. With the vigorous development of the electronic information industry, the Internet of Things, communication technology, and so on, radar technology is beginning to be applied in people's daily lives, greatly improving the convenience and safety of work and life. For example, automotive radar

G. Liu et al., *6G Migrates from Communication to XaaS*,
https://doi.org/10.1007/978-981-96-7579-1_12

provides early warning information for vehicle driving by measuring the distance and relative velocity between vehicles, between vehicles and surrounding environmental objects, and between vehicles and pedestrians, greatly improving the safety of road traffic. In terms of terminal applications, Google Pixel 4 smart phone uses a miniature motion-sensing Soli radar chip to recognize gestures, making it convenient for users to operate in the air [1].

Non-RF sensing involves using various sensors to collect environmental information and obtain sensing results. Typical methods include capturing images or video information based on cameras, or obtaining specific sensing information through other sensors such as temperature sensors, barometers, accelerometers, gyroscopes, pedometers, etc. With the development of image recognition technology, camera-based sensing has been widely applied in fields such as transportation and public safety. However, camera-based sensing technology has certain limitations, including dependence on lighting conditions and Line Of Sight (LOS) conditions, limited coverage distance, severe impact from adverse weather in outdoor scenes, and privacy and security issues. Additionally, sensor equipment often requires specialized deployment and maintenance, or needs to be carried, and there are power supply issues, lacking advantages in convenience and cost. In contrast, RF sensing utilizes existing wireless infrastructure, without the need for dedicated sensing devices, offering high convenience and low cost, and is not affected by weather and lighting conditions. It effectively overcomes the shortcomings of non-RF sensing in these aspects and can complement each other, providing higher quality sensing services. Furthermore, using widely distributed wireless communication facilities for sensing greatly enhances the coverage and ubiquity of sensing.

In fact, wireless signals are influenced by the surrounding environment during propagation, causing changes in signal amplitude, phase and other characteristics, i.e., the environment information has a modulating effect on the signals. By analyzing the wireless signals at the receiver, not only can the information carried by the transmitter be obtained, but also the information reflecting the characteristics of the propagation environment can be extracted. In other words, electromagnetic waves have inherent sensing capabilities. Currently, the academic community has conducted a large amount of exploration and verification on the sensing capabilities of Wi-Fi systems, involving various applications such as vital sign monitoring, intrusion detection, trajectory tracking, gesture recognition, etc. IEEE 802.11 also established the IEEE 802.11bf working group in September 2020, targeting indoor positioning, automation control, security, health and other fields, to explore the potential of wireless local area network sensing and promote the implementation of related wireless sensing applications. There has been extensive research on utilizing cellular communication systems including LTE and 5G NR (New Radio) for sensing functions. Apart from various indoor applications similar to Wi-Fi sensing, cellular communication systems can also be used for outdoor applications such as weather or air quality monitoring. As a typical 3GPP sensing service, standard research on positioning was initiated during the LTE phase. NR wireless technology offers unique advantages and can provide additional value in enhancing positioning capabilities, and standardization work for NR positioning has been initiated in NR Release-16. By measuring the delay, angle, and power information of the channel between the user equipment (UE) and base station, the position information of the UE is calculated. In Release-17, positioning technology has been further enhanced to meet the demands for higher accuracy, lower latency and higher efficiency.

In the past, communication and sensing were strictly distinguished due to differences in research objects and focus, and communication systems and sensing systems were studied separately in most scenarios. In fact, communication and RF sensing have many similarities in working principles, system architecture and frequency bands. Both communication systems and sensing systems are based on electromagnetic wave theory, using the transmission and reception of electromagnetic waves to acquire and transmit information. Both communication systems and sensing systems have antennas, transmitters, receivers, signal processors and other hardware structures, with a large overlap in hardware resources. with the development of technology, there is an increasing overlap in working frequency bands. And there are similarities in key technologies such as signal modulation and detection, waveform design. For example, civil radar and communication systems have been evolving towards miniaturization and higher frequency bands for a long time, and the spectrum expansion of 6G towards high frequency bands will promote the integration of communication technology with imaging, sensing and other applications. Due to its characteristics of short wavelengths and large bandwidth, not only will data rate be improved, but also angular resolution and ranging accuracy in imaging and radar applications [2]. In addition to spectrum resources, existing radar and communication systems have a high degree of similarity in hardware architecture, channel characteristics and signal processing methods. Therefore, integrated design for communication and sensing functions has become a trend. The integration of communication and sensing systems can bring many advantages, such as cost savings, size reduction, power consumption reduction, spectrum efficiency improvement, interference reduction, etc., thereby improving the overall performance of the system. From the perspective of interaction and impact of communication and sensing systems, the following different degrees of integration levels can be distinguished [3].

(1) Coexistence: Communication systems and sensing systems share space–time-frequency resources, but do not share information or eliminate mutual interference. Without considering physical integration design and optimization, when perfect isolation cannot be achieved, interference between the two systems leads to a decrease in their individual performance.
(2) Cooperation: The communication system and sensing system achieve the goal of reducing interference with each other by sharing partial information. Furthermore, through mutual information interaction, more enriched functionality can be achieved or the communication and sensing performance of the system can be enhanced.
(3) Joint design and optimization: Communication and sensing functions are considered together during system design. From the perspectives of hardware architecture, waveform design, protocol design, received signal processing, etc.,

the tight integration of communication and sensing functions is achieved to improve spectrum efficiency and optimize overall system performance.

Considering the research results of the academic and industrial sectors, the technical route of integrated sensing and communication research can be divided into three main categories: communication-centric technical route, sensing-centric technical route, and a route that balances communication and sensing [4]. In the early stages of research, technical research is primarily conducted from one side of the communication system or sensing system. The communication-centric technical route requires the design of integrated sensing and communication from the perspective of the communication system. The system performance assessment indicators are mainly communication performance indicators such as spectrum efficiency, channel capacity, signal-to-noise ratio, and bit error rate performance. To support sensing functions on the premise of maximizing communication performance, sensing can be introduced based on the existing mobile cellular network architecture, which can adopt either uplink sensing or downlink sensing. For uplink sensing, the main method is for the base station to use the uplink signal of the user equipment (UE) to realize sensing functions, and the current communication system architecture can be implemented without any changes. For downlink sensing, it is usually for the UE to receive the downlink signal from the base station for sensing, or for the base station to receive the echo of its own downlink communication signal for sensing. To ensure communication efficiency, continuous waveforms are preferred over pulse waveforms, requiring the transceiver to work simultaneously. High transmission efficiency communication waveforms such as Orthogonal Frequency Division Multiplexing (OFDM) waveforms are prioritized for waveform selection. Sensing signals mainly reuse existing reference signals, synchronization signals, or data of the communication system. To ensure sensing performance, certain improvements such as increasing bandwidth, duration, sampling frequency, etc. can be made without significantly impacting communication performance. The sensing-centric technical route involves integrated design from the perspective of the sensing system with minimal changes to achieve communication functions. The performance assessment indicators of the system primarily focus on sensing indicators, emphasizing sensing performance such as ranging, velocity measurement, angle measurement, positioning accuracy, detection, or recognition probability. The research focuses on achieving communication functions with minimal impact on sensing performance. Common waveforms in sensing systems, such as Frequency Modulated Continuous Wave (FMCW) waveforms, are mainly used for waveform selection. Communication information embedding can be achieved through methods such as index modulation, which is suitable for low communication rate, low hardware cost, and high sensing accuracy requirements in IoT applications. An important technical route for communication and sensing integrated design is the route that balances communication and sensing, which requires a focus on deep-level joint design and optimization. The two major functions of communication and sensing need to be pre-arranged for future deep integration on the technical evolution route, reflecting the foresight of integrated design. Currently, there has been preliminary research in the academic community on the fusion of communication and sensing, including research on integrated waveforms and signal processing, research on joint performance boundary, performance evaluation indicators design, integrated hardware platform design, etc. However, in general, it is still in a preliminary stage, and continuous theoretical and hardware research is needed.

12.2 Typical Application Scenarios and Requirements of ISAC

Sensing is a fundamental service that 6G can provide. More precisely, 6G systems can offer sensing as a basic service to entities outside the 6G system. In fact, the service object of sensing can also be the 6G system itself. Categorized by the scope of service objects, sensing can be divided into two types. One is external services, where the service object is entities outside the 6G system, i.e., network for sensing. The other is internal services, where the service object is the 6G system itself, i.e., sensing for network. External services of sensing include a variety of new vertical application scenarios, represented by smart cities, intelligent transportation, industrial automation, environmental monitoring, smart living, and intelligent manufacturing. Internal services of sensing include using sensing information to enhance the performance of communication systems.

12.2.1 External Services of Sensing

According to the coverage of sensing, typical sensing functions and application scenarios can be classified, as shown in Table 12.1.

As can be seen, there are many types of use cases related to wireless sensing, and the deep integration of communication and sensing is expected to support more new applications, creating more value for social development and industrial upgrading. The following is an analysis of typical scenarios.

(1) Intelligent transportation: With the popularization of systems such as autonomous driving and Internet of Vehicles in future, it is necessary to measure the positions, velocities, and distances between vehicles, detect road environments, and identify emergencies. Vehicles act as communication nodes and sensing nodes, working in coordination with integrated base stations deployed on roadsides to sense the environment while performing communication tasks, to achieve real-time obstacle avoidance, route selection, hazard detection, and compliance with traffic regulations. For intelligent transportation, in addition to basic sensing functions such as vehicle ranging, velocity measurement, positioning and tracking, simultaneous localization and mapping (SLAM) based on wireless signals is also required. SLAM technology has been widely applied in unmanned aerial vehicles, unmanned

Table 12.1 Typical sensing functionalities and application scenarios

Sensing functionalities	Application scenarios
Monitoring of weather conditions and air quality	Meteorology, agriculture, daily life services
Traffic and pedestrian flow detection, quantity statistics, intrusion detection	Intelligent transportation, commercial services, security monitoring
Target positioning and tracking, range/velocity/angle measurement	Various applications of traditional radar
Environmental reconstruction	Intelligent driving and navigation, smart cities, network planning and optimization
Motion/posture/facial recognition	Smart interaction, gaming, smart homes
Monitoring of vital signs such as heartbeat/respiration	Health care, medical care
Imaging, material detection, ingredient analysis	Security inspection, industrial applications, biomedicine

submarines, planetary rovers, domestic robots, virtual reality (VR), augmented reality (AR), and life assistance, including behavior monitoring, fall detection and real-time assistance [5]. Traditional SLAM based on stereo cameras and integrated sensors cannot automatically scan environments, and has poor flexibility and high costs. The use of wireless signal-based SLAM can address these issues, providing strong support for vehicle navigation and route planning. Additionally, responding to and making decisions regarding emergencies on road requires the use of artificial intelligence (AI) technology to achieve real-time recognition and judgment of the surrounding traffic environment.

(2) Industrial automation: With the development of digitalization, networking, and intelligence in factories, more and more devices with sensing and execution capabilities are being introduced into industrial systems. Information acquisition, transmission, processing and control functions interact with each other, and communication and sensing technologies are highly integrated to form autonomous industrial systems to maximize production process efficiency. Similar to intelligent transportation, basic sensing functions such as ranging, velocity measurement and positioning are also required in industrial manufacturing scenarios. These functions are used for sensing machine or equipment components and operators, intelligent planning of assembly line production operations, product tracking and counting, and combined with action and event recognition to achieve automated control, anomaly detection and emergency handling.

(3) Environmental monitoring: Monitoring weather changes such as rain, snow, fog, humidity, temperature and air quality is of great significance for weather forecasting, climate modeling, agricultural and water conservancy engineering, as well as for personal daily travel and health protection. Taking rainfall monitoring as an example, traditional methods involve rain gauges, weather radar and satellites. Monitoring rainfall using communication signals has advantages such as a large number of

sensing nodes, wide spatial distribution, and high temporal resolution, which effectively improves the cost efficiency. Rainfall monitoring can utilize Received Signal Strength (RSS), where the Probability Density Function (PDF) of RSS of no rain, light rain, moderate rain and heavy rain has different mean and variance characteristics [6]. Utilizing instantaneous values, mean and variance of RSS for rainfall level classification requires normalization of the RSS reference value of no rain to eliminate the influence of other attenuation factors. AI methods can be employed for rainfall classification.

(4) Smart living: Using integrated sensing and communication technology will provide more personalized and convenient services for people's daily lives. Typical application cases include breathing and heartbeat monitoring, gesture and posture recognition, intrusion detection, indoor positioning and tracking. Breathing and heartbeat rate act as an important vital sign and can effectively reflect the current health status. Using wireless signals for human body breathing monitoring can be applied to medical bed monitoring and daily home sleep monitoring. Compared with traditional wearable sensor devices, it has the advantages of non-contact and low cost. Intrusion detection, indoor positioning and tracking of individuals are of great significance for home safety, public area safety, and the care of children or elderly living alone. The method based on visual surveillance has privacy issues and is affected by light and other environmental conditions. Using wireless communication signals to monitor the presence and movement of people in the area not only reduces installation time and cost, but also has the advantages of all-weather and wide coverage. Non-contact gesture and action recognition is one of the commonly used methods in human–computer interaction. Smart homes, wearable devices, smart cars, and VR/AR can support gesture recognition control functions. It can also be combined with the indoor positioning and tracking mentioned earlier, which is an important aspect of the application of integrated sensing and communication technology in smart living scenarios.

The introduction of sensing into mobile communication networks greatly enriches the functions of communication systems, expands their applications in different scenarios, and also puts forward new requirements for the key performance indicators of existing systems. In the past, communication systems typically aimed to optimize throughput and transmission reliability, focusing on performance indicators such as spectrum efficiency, channel capacity, Signal-to-Noise Ratio (SNR), Signal to Interference plus Noise Ratio (SINR), Bit Error Rate (BER), Block Error Rate (BLER). However, integrated sensing and communication systems need to consider sensing performance and even integrated performance. Moreover, in different application scenarios, the evaluation criteria and performance requirements for integrated sensing and communication may vary. According to the typical scenarios mentioned earlier, the requirements for sensing performance indicators need to be added on top of the existing communication performance indicators. The definition of the performance indicators related to the sensing service is shown in Table 12.2.

To support the aforementioned rich sensing services, the receiving node of the sensing signal first needs to receive the sensing signal, then obtain basic measurements based on the received signal or channel response. Basic measurements include

Table 12.2 Definition of performance indicators related to the sensing service

Parameter	Definition
Sensing accuracy	Sensing accuracy reflects the degree of deviation between the true result and the sensing result at a certain confidence level, and can be characterized by sensing errors, such as root mean square error. Examples include range accuracy, velocity accuracy, angle accuracy, rainfall accuracy, respiratory rate accuracy, etc.
Sensing resolution	Sensing resolution refers to the ability to distinguish multiple targets from different dimensions, such as range resolution (including horizontal and vertical directions), velocity resolution, angle resolution, humidity resolution, imaging resolution, etc.
Sensing range	Sensing range refers to the range of sensing under the premise of meeting certain performance requirements, such as the range of sensing distance, the range of sensing velocity, the range of sensing angle, etc.
Sensing update rate	The reciprocal of the time interval between two consecutive sensing results
Detection probability	The ability to judge the presence or absence of a sensing target or the occurrence or non-occurrence of a sensing event, usually the probability of judging as existing under the assumption that the target exists
False alarm probability	The ability to judge the presence or absence of a sensing target or the occurrence or non-occurrence of a sensing event, usually the probability of judging as existing under the assumption that the target does not exist
Sensing latency	Sensing latency is used to quantitatively describe the real-time requirements of sensing services, such as the maximum latency from the reception of a sensing service request to the response to that request by the sensing function node

delay, Doppler, angle, strength and their multi-dimensional combination representation. The receiving node of the sensing signal or the computing node determines the characteristics of the sensing target based on the basic measurements, including range, velocity, direction, position, acceleration, whether the target exists, trajectory, action, expression, vital signs, quantity, imaging results, weather, air quality, shape, material, composition, etc. Table 12.3 provides the sensing information at different levels.

Table 12.3 Sensing information at different levels

Sensing information	Contents of sensing information
Received signal or raw channel information	Complex results of the received signal or channel response
Sensing measurement quantity	Delay, Doppler, angle, strength, and their multi-dimensional combined representation
Sensing results/sensing intermediate results	Presence or absence of target, distance, velocity, orientation, acceleration, position, trajectory, movement, expression, respiration rate, heart rate, imaging results, weather, air quality, material and composition, etc

12.2.2 *Internal Services of Sensing*

Internal services of sensing, or sensing-assisted communication, involve the acquisition of channel information through environmental sensing, which can be utilized to enhance the performance of communication systems regarding UE positioning, obstruction avoidance, non-line-of-sight detection, and channel estimation [7]. In the case of millimeter-wave communication systems with a large codebook, the beam training process is time-consuming. By employing sensing technology to obtain position information of UE, the overhead of beam training can be reduced, facilitating rapid beamforming and tracking in mobile scenarios. Furthermore, this approach allows for the positioning and tracking of communication targets, as well as the prediction of potential obstacles in the channel environment. Consequently, beam switching can be performed to prevent communication link failures caused by obstruction. Additionally, sensing different targets in the surrounding environment helps distinguish line-of-sight and non-line-of-sight channels, addressing the issue of high misidentification rates during analysis based on channel characteristics. Furthermore, the channel parameters obtained through sensing can also enhance the accuracy of communication channel estimation.

12.3 Key Technologies for Air Interface of ISAC

12.3.1 *Waveform and Signal Design*

Waveform design is a key focus of research on integrated sensing and communication technology. The design approach involves reusing existing communication waveforms or sensing waveforms, utilizing time division, frequency division, and spatial division to achieve diversity transmission of communication and sensing waveforms. This method has low resource efficiency. Alternatively, by modifying existing waveforms or designing new waveforms, communication and sensing functions can be integrated into one waveform to achieve integrated design. The key to waveform design is to minimize the interference between communication signals and sensing signals, meet the requirements of communication and sensing, improve spectrum efficiency while ensuring system performance, and consider whether to prioritize communication functions or sensing functions, or both, to find a balance in performance. Below is an analysis of several common waveforms that can be used for integrated sensing and communication.

(1) Single carrier waveforms: These are typically combined with spread spectrum technology for integrated sensing and communication, such as single carrier integrated sensing and communication waveform based on Direct Sequence Spread Spectrum (DSSS), Time Hopping Spread Spectrum (THSS), or Chirp Spread Spectrum (CSS) [8]. Spread spectrum technology has been widely used in communication systems to achieve higher security and robustness, providing stronger resistance

to interference and interception, and different spread spectrum sequences can be chosen to enable multi-user access. To ensure sensing performance, it is necessary to use spread spectrum sequences and spread factors with good auto-correlation and cross-correlation properties to adapt to different sensing accuracies and dynamic range requirements. In addition, the training sequences in single-carrier communication systems, such as the short training field (STF) and channel estimation field (CEF) composed of complementary Gray sequences in the frame structure of IEEE 802.11ad using single carrier frequency domain equalization (SC-FDE) waveforms, can be used for synchronization, frequency offset estimation, and channel estimation in communication systems, and can also be used for target detection, ranging, and velocity measurement in radar systems [9].

(2) Multi-carrier waveforms: OFDM waveforms are widely used in 4G and 5G mobile communication systems due to their high spectral efficiency, flexible bandwidth resource allocation and ability to meet high throughput demands. Additionally, due to its inherent advantages in sensing, OFDM waveforms have also been applied in radar systems. Using OFDM waveforms for parameter estimation has no distance-Doppler coupling effect, and the detection algorithm based on Fourier transform at the receiver is simple and efficient. Moreover, the transmission mechanisms of radar and communication systems based on OFDM waveforms are highly similar, which is conducive to the integrated design of communication and sensing. However, OFDM waveforms also have challenges such as high peak-to-average power ratio (PAPR) and sensitivity to Doppler and phase noise, and optimization is required. For example, constant envelope orthogonal frequency division multiplexing (CE-OFDM) is used to overcome the issue of high peak-to-average power ratio in traditional OFDM, which leads to non-linear distortion in high-power amplifiers [10]. In addition, the use of filter bank-based multi-carrier (FBMC) is considered to replace OFDM as the fusion waveform [11]. FBMC can reduce the overlap between subcarriers through filtering, and it has the characteristics of concentrated spectrum and excellent side-lobe suppression. Furthermore, this waveform does not require the insertion of a cyclic prefix (CP) in the time domain, thus avoiding the redundancy issues and the impact on cognitive radio accuracy caused by CP. As a potential new waveform for 6G, orthogonal time frequency space (OTFS) waveform can obtain the full diversity gain of sensing signal processing by spreading the transmitted symbols to the whole time–frequency domain, thus performing well in range, Doppler estimation and tracking rate. Existing research and simulations show that its performance in radar estimation characteristics is comparable to that of OFDM. Due to the absence of CP redundancy, OTFS achieves greater spectral efficiency at the cost of higher complexity [12] than OFDM.

(3) Radar waveforms: The concept of integrated communication and radar based on pulse groups carrying communication information has emerged as early as the 1960s. It can use the different positions of pulses within the group relative to a reference pulse to achieve modulation of the communication data [13], and can also use different pulse repetition frequencies to carry the communication information [14]. Pulse radar emits simple pulse signals within the time range given by the duty cycle in each pulse period, receives echoes in the remaining time of the pulse period, and

processes the echoes for sensing. This waveform does not require the system to have full-duplex capability, it is relatively simple to implement, but there is a near-distance blind zone. Moreover, due to the influence of the duty cycle, a larger peak transmission power is required during pulse emission, and the discontinuity of the signal in time lead to inefficient resource utilization. Simple pulse waveforms are usually used for long-distance sensing applications, for example, military applications for monitoring the intrusion of flying objects. FMCW radar emits frequency linearly changing chirp signals while simultaneously receiving echoes for sensing. The FMCW waveform has advantages such as a large time-bandwidth product, constant envelope, good auto-correlation properties, and a simple transceiver architecture and signal processing flow. It can also carry communication information by loading communication modulation data onto the FMCW waveform to achieve integration of communication and sensing. However, FMCW radar requires full-duplex capabilities [8], higher requirements for signal isolation, and is suitable for short-range sensing applications with low transmission signal power, such as automotive radar in autonomous driving applications. Ultra-Wide Band (UWB) waveform refers to a waveform with an absolute bandwidth greater than 500 MHz or a relative bandwidth greater than 0.2 relative to the center frequency. It's usually a nanosecond or sub-nanosecond ultra-narrow pulse waveform, which obtains the range information of the target by measuring the flight time of the ultra-narrow pulse. The system structure and implementation of this waveform are simple, with low cost and power consumption. Additionally, due to its low signal power, it is typically suitable for short-range detection applications such as vital signs monitoring and gesture recognition [15]. Furthermore, the design of integrated sensing and communication based on radar waveforms can consider the combination with spatial modulation or generalized spatial modulation. For example, using different transmitting antennas to carry communication information, employing spatial or generalized spatial modulation can avoid altering the characteristics of the sensing signal and minimize the impact on sensing performance. The waveform design can also be combined with the concept of Multiple Input Multiple Output (MIMO) radar, where signals associated with different transmitting antennas are mutually orthogonal, which is advantageous for demodulating the carried communication information.

For MIMO radar, the orthogonality between the signals from different antennas can be achieved through Time Division Multiplexing (TDM), Frequency Division Multiplexing (FDM), Doppler Division Multiplexing (DDM), and Code Division Multiplexing (CDM) [16]. The orthogonal signals of MIMO radar can be divided into fast-time orthogonal signals and slow-time orthogonal signals. The fast-time orthogonal signals separate the signals from each transmitting antenna in the fast-time domain or fast-time frequency domain, such as fast-time FDM signals and CDM signals. The slow-time orthogonal signals separate the signals from each transmitting antenna in the slow-time domain or Doppler domain, such as slow-time TDM signals and DDM signals. TDM signals achieve orthogonality in the slow time domain by transmitting signals at different slow times. DDM signals can be seen as a special case of slow time CDM signals or slow time FDM signals, which achieve orthogonality of the transmitted waveform in the Doppler domain by applying different linear phase modulation to the signals on different transmitting array elements and shifting the signals to different frequencies in the Doppler domain [17]. For each transmitting antenna of MIMO radar, FMCW is usually used as the basic waveform.

12.3.2 Sensing Signal Detection Theory

Sensing signal detection is an important component of sensing signal processing, the purpose of which is to judge whether the sensing target exists in the presence of noise, interference and clutter. This section begins with the fundamentals of hypothesis testing and Neyman-Pearson criteria, and focuses on the widely used Constant False Alarm Rate (CFAR) detection in practical engineering, as well as several specific implementation methods of CFAR detection.

1. Hypothesis Testing

Sensing signal detection is a binary hypothesis testing problem. At any given time, one of the following two hypotheses holds [18].

Hypothesis 1 (denoted as H_0): Noise, interference and clutter exist.

Hypothesis 2 (denoted as H_1): Noise, interference, clutter exist, and the target signal exist.

For sensing signal processing, signal detection is usually carried out in units of resolution cells, for example, dividing the range measurement range into several range resolution cells, dividing the velocity measurement range into several velocity resolution cells, and dividing the angle measurement range into several angle resolution cells. In the given resolution cell, if it is judged to be H_0, it is assumed that there is no target in this resolution cell, and if it is judged to be H_1, it is assumed that there is a target in this resolution cell.

Detection probability and false alarm probability are important metrics for evaluating the performance of target detection as follows.

(1) Detection probability: The probability of correctly determining the presence of a target within the detection resolution cell, typically denoted as $P(H_1|H_1)$.

(2) False alarm probability: The probability of incorrectly determining the presence of a target within the detection resolution cell when there is no target present, typically denoted as $P(H_1|H_0)$.

Let $P(x|H_0)$ and $P(x|H_1)$ be the probability density functions of the echo signal power of the cell under test in the absence and presence of a target, respectively. Then, the detection probability and false alarm probability are expressed as

$$\begin{aligned} P_d &= P(H_1|H_1) = P(x > x_0|H_1) \\ &= \int_{x_0}^{\infty} P(x|H_1)\,\mathrm{d}x P_f = P(H_1|H_0) \end{aligned}$$

$$= P(x > x_0|H_0) = \int_{x_0}^{\infty} P(x|H_0)\,dx \quad (12.1)$$

where x_0 represents the detection threshold.

2. Neyman-Pearson criterion

Theoretically, the goal of signal detection is to maximize the detection probability P_d and minimize the false alarm probabilityP_f. However, in practice, it is not possible to achieve both simultaneously. In engineering, the Neyman-Pearson criterion is commonly used to optimize the detection performance to a certain extent [19]. The Neyman-Pearson criterion states that by imposing a certain constraint on the false alarm probability P_f, the detection probability P_d can be maximized. This criterion on one hand limits the entry of false targets into the system, thus affecting the efficiency of the system, and on the other hand, ensures that real targets are detected with the highest probability.

The mathematical description of the Neyman-Pearson criterion is to minimize the following expression under the constraint that P_f is a constant

$$\begin{aligned} F &= 1 - P_{\mathrm{d}} + \eta(P_{\mathrm{f}} - \alpha) \\ &= \eta(1-\alpha) + \int_{x<x_0} [p(x|H_1) - \eta p(x|H_0)]\mathrm{d}x \end{aligned} \quad (12.2)$$

where η is the Lagrange multiplier. Obviously, minimizing F requires the integral term in the equation to be 0, thus obtaining the threshold detection that satisfies the Neyman-Pearson criterion as

$$\frac{p(x|H_1)}{p(x|H_0)} \underset{H_1}{\overset{H_0}{\gtrless}} \eta \quad (12.3)$$

Equivalently, the expression for the power of the echo signal corresponding to the cell under test is given by

$$x \underset{H_1}{\overset{H_0}{\gtrless}} x_0 \quad (12.4)$$

where x_0 is determined by $P(x|H_0)$, $P(x|H_1)$ and η.

3. CFAR detection

Neyman-Pearson criterion provides an ideal CFAR detection method under known $P(x|H_0)$ and $P(x|H_1)$. In practical engineering applications, where $P(x|H_0)$ and $P(x|H_1)$ are unknown and complex, the engineering implementation of CFAR

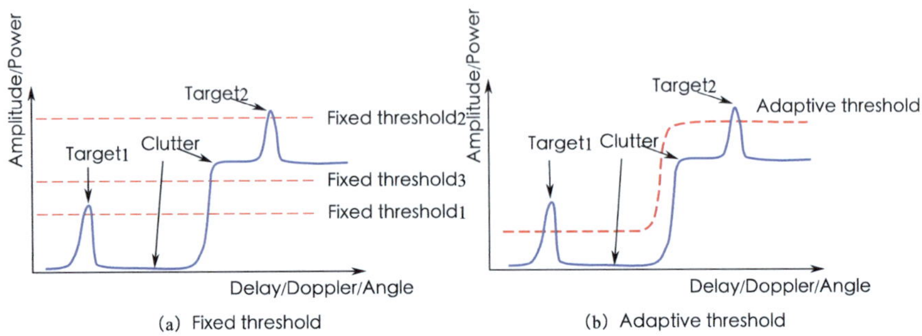

Fig. 12.1 Comparison of fixed threshold and adaptive threshold

detection adapts the detection threshold based on the Neyman-Pearson criterion to maintain a certain level of false alarm probability for the detection results [20].

The comparison between the fixed threshold and the adaptive threshold is shown in Fig. 12.1. In an environment including noise, interference and clutter, the use of fixed threshold detection can lead to a significant variation in false alarm probability or detection probability over a wide range. For example, in Fig. 12.1a, fixed threshold 1 exhibits a high false alarm probability, fixed threshold 2 has a low detection probability, and fixed threshold 3 simultaneously has a high false alarm probability and a low detection probability. However, with reference to the changing background environment, applying CFAR detection based on the Neyman-Pearson criterion can maintain relatively stable false alarm probability and detection probability when using adaptive threshold detection, as shown in Fig. 12.1b.

The fundamental concept of CFAR detection is to estimate the background noise level at the location of the cell under test by estimating the signal levels from neighboring cells in the range, velocity or angle dimension. This estimated background noise level is then used to set the detection threshold for the cell under test. The block diagram of CFAR detection is illustrated in Fig. 12.2, where D represents the cell under test, and $(x_1, x_2, \ldots, x_n)$ and $(x_{n+1}, x_{n+2}, \ldots, x_{2n})$ represent the leading and lagging reference cells, used for estimating the background noise level at the cell under test. To prevent the influence of the target at the location of the cell under test on the estimation of the background noise level, a guard cell is placed between the cell under test and the reference cells. The core of CFAR detection is to estimate the background noise level Z at the cell under test using the signal sequence within the reference cells. Finally, the product of the background noise level Z and the threshold factor T, denoted as S, serves as the detection threshold for the cell under test.

Depending on the estimation method of the background environment level Z, the commonly used CFAR detection methods are mainly divided into four categories: Mean-based CFAR, ordered statistic CFAR, adaptive CFAR, and clutter map CFAR. Each will be introduced separately below.

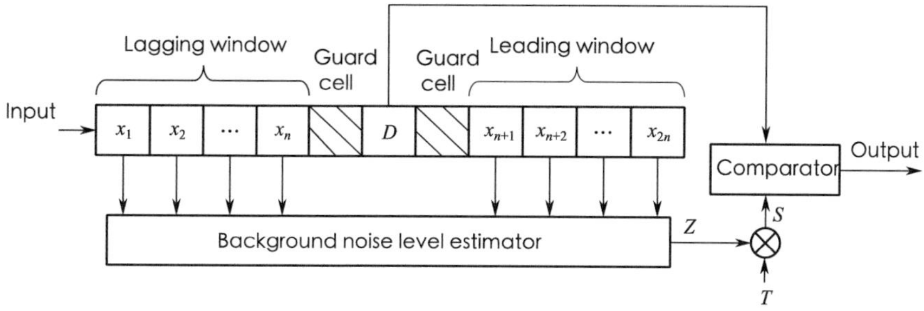

Fig. 12.2 Block diagram of CFAR detection

(1) Mean-based CFAR

In the background environment level estimation process of the mean-based CFAR method, the signal sequences of the leading reference half-window and the trailing reference half-window are first averaged to obtain

$$\overline{x} = \frac{\sum_{i=1}^{n} x_i}{n} \tag{12.5}$$

$$\overline{y} = \frac{\sum_{i=n+1}^{2n} x_i}{n} \tag{12.6}$$

The three typical methods of mean-based CFAR are as follows.

(1) Cell Averaging Constant False Alarm Rate (CA-CFAR): The average of the mean values $\overline{x}$ and $\overline{y}$ from the leading and lagging reference windows is calculated, and the result is used as the background noise level for the cell under test, denoted as $Z = (\overline{x} + \overline{y})/2$.
(2) The Greatest Of-CFAR (GO-CFAR): The larger value between the averages of the leading and lagging reference windows, denoted as $\overline{x}$ and$\overline{y}$, is selected as the background noise level for the cell under test, denoted as $Z = \max\{\overline{x}, \overline{y}\}$.
(3) The Smallest Of-CFAR (SO-CFAR): The smaller value between the averages of the leading and lagging reference windows, denoted as $\overline{x}$ and $\overline{y}$ is selected as the background noise level for the cell under test, denoted as $Z = \min\{\overline{x}, \overline{y}\}$.

(2) Ordered Statistics CFAR

In the context of Ordered Statistic CFAR (OS-CFAR) for estimating background environment levels, the process involves sorting all the sample values in the leading and lagging reference windows in descending order to obtain $X = [x(1), x(2), \cdots, x(2n)]$. Then, the background environment level estimation is based on the sorted results. The following are two typical methods.

(1) Ordered Statistics Constant False Alarm Rate (OS-CFAR): The kth sample value is selected as the background level for the cell under test, i.e., $Z = X(k)$.

(2) Censored Mean Level Detector Constant False Alarm Rate (CMLD-CFAR) [21]: Delete the largest r sample value, with the remaining $2n$ - r sample values' mean as the background environment level for the cell under test, i.e., $Z = \sum_{i=r+1}^{2n} x(i)/(2n-r)$.

(3) Adaptive CFAR

The methods of obtaining the background environment level of the mean-based CFAR and the ordered statistic CFAR mentioned above are fixed, and their performance is usually poor when the background environment changes dramatically. The adaptive CFAR method estimates the background environment level by choosing to use the mean-type CFAR method or the ordered statistic-type CFAR method based on the analysis of real-time sample values. A typical method is the Variable Index-CFAR (VI-CFAR) [22].

In the VI-CFAR method, computations on the leading and lagging reference windows yield the second-order statistic variable VI and the mean ratio MR.

$$\begin{aligned} \mathrm{VI} &= 1 + \frac{\hat{\sigma}^2}{\hat{\mu}^2} = 1 + \frac{1}{n-1}\sum_{i=1}^{n}\frac{(x_i-\overline{x})^2}{(\overline{x})^2} = n\frac{\sum_{i=1}^{n}(x_i)^2}{(\sum_{i=1}^{n}x_i)^2} \\ \mathrm{MR} &= \frac{\overline{x}_A}{\overline{x}_B} = \frac{\sum_{i\in A} x_A}{\sum_{i\in B} x_B} \end{aligned} \tag{12.7}$$

Then make the following judgement based on VI and MR

$$\begin{cases} VI \le K_{VI} \Rightarrow Non-uniformbackgroundenvironment \\ VI > K_{VI} \Rightarrow Uniformbackgroundenvironment \end{cases} \tag{12.8}$$

$$\begin{cases} K_{MR}^{-1} \le MR \le K_{MR} \Rightarrow Samemeanvaluesofthereferencewindows \\ MR \le K_{MR}^{-1}\ Or\ MR > K_{MR} \Rightarrow Differentmeanvaluesofthereferencewindows \end{cases} \tag{12.9}$$

where K_{VI} and K_{MR} are the predetermined thresholds.

The CFAR method is selected according to the above judgement as shown in Table 12.4.

(4) Clutter Map CFAR

The clutter map is obtained by updating and iteratively processing the echoes from the detection unit over several scans to estimate the clutter's strength and storing the results. The clutter map stores the estimated clutter power at each clutter cell, including one or more resolution cells, and then detection thresholds are set based on the clutter power of the clutter cell corresponding to the cell under test [20]. The

Table 12.4 CFAR method selection

Whether the background environment is uniform		Whether the mean values of the leading and lagging reference windows are the same	CFAR method
The leading reference window	The lagging reference window		
Yes	Yes	Yes	CA-CFAR
Yes	Yes	No	GO-CFAR
Yes	No	-	CA-CFAR
No	Yes	-	CA-CFAR
No	No	-	SO-CFAR

clutter establishment methods can be divided into point clutter map technology and area clutter map technology.

(1) Point clutter map technology: The clutter map is established using the echo signals of the clutter cell itself in each scan. The point clutter map only iterates the clutter power estimation in the time domain, that is, the echo signal scanned at times $1, 2, 3, \cdots, t-1$ is used to estimate the clutter power at time t through cyclic iteration, and then the threshold judgement is made based on the estimated clutter power. If the judgement result at time t is no target, then the t moment's echo signal power is incorporated into the iterative cycle of clutter power estimation. If the judgement result at the t moment is that there is a target, then the clutter power data of that clutter cell is not updated.

(2) Surface clutter map technology: The clutter map is established using the echo signals from several adjacent cells in each scan of the clutter cell, and the surface clutter map combines time domain and spatial domain to iteratively estimate clutter power, as shown in Fig. 12.3. At the $1, 2, 3, \cdots, t$ moment, the clutter power at the cell under test D is estimated using $MN - mn$ clutter cells (excluding guard cells) adjacent to D. In the process of establishing the clutter map, the neighboring samples can be selected by methods such as maximum selection, minimum selection, averaging and ordered statistics.

At the t moment, the sample value input into the clutter map iterative process is q_t, then the power clutter stored in the iterated clutter cell is

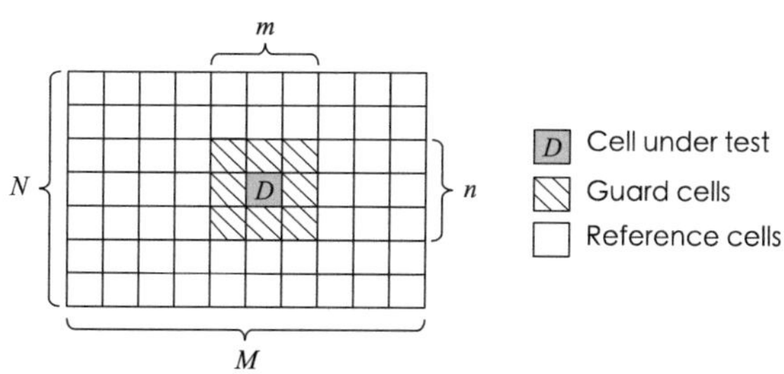

Fig. 12.3 Schematic diagram of surface clutter map

$$\begin{aligned}\hat{p}_t(k) &= (1-w)\hat{p}_{t-1}(k) + wq_t(k) \\ &= w\sum_{i=0}^{t}(1-w)^i q_{t-i}(k)\end{aligned} \tag{12.10}$$

where w is the forgetting factor, taking values ranging from 0 to 1, the larger w is, the faster the clutter map converges, and the value of w should not be excessively large. In point clutter map technology, q_t is the echo signal power of the current clutter cell at the t moment. Whereas in surface clutter map technology, q_t is the value determined in the reference cell by methods such as maximum selection, minimum selection, averaging and ordered statistics depicted in Fig. 12.3.

12.3.3 Sensing Parameter Estimation Algorithm

Sensing tasks include detecting targets or estimating dynamic parameters such as range, angle and velocity. It also includes the recognition of sensing targets, such as detection, recognition and classification of target features or behaviors. Parameter estimation related to sensing is a non-linear problem, making it challenging to apply most classical linear estimators. Common parameter estimation methods include: Periodogram methods, such as two-dimensional or three-dimensional Fast Fourier Transform (FFT); spatial beamformers, such as Bartlett beamformers and Minimum Variance Distortionless Response (MVDR) beamformers; subspace-based spectral analysis techniques, such as Multiple Signal Classification (MUSIC); Estimating Signal Parameters via Rotational Invariance Techniques (ESPRIT); Maximum Likelihood (ML) methods; Compressed Sensing (CS) techniques; Tensor Decomposition techniques, etc. [23]. These techniques have different performances and complexities and have been widely used in the fields of communication, radar and image processing.

The three-dimensional FFT parameter estimation technique is illustrated with the common FMCW multi-antenna radar as an example. The signal model for a single-target FMCW radar is given by [24]

$$r(l,n,p) = \alpha\exp\left\{j2\pi\left[\left(\frac{2\mu R}{c}+f_d\right)\frac{n}{f_s}+\frac{ld\sin\theta}{\lambda}+f_d pT_0+\frac{2R}{\lambda}\right]\right\}+\omega(l,n,p) \tag{12.11}$$

where r is the radar echo, l, n and p are the indices of the antenna, chirp signal fast time and slow time respectively, α is the amplitude of the echo signal, μ is the slope of the chirp signal, R is the distance between the target and the radar, c is the speed of light, f_d is the Doppler frequency, f_s is the sampling frequency of the echo signal, d is the spacing between the receiving antenna array elements, θ is the arrival angle of the target echo, λ is the signal wavelength, T_0 is the total time of a single chirp duration plus the echo delay, ω represents noise. If two of the parameters l, n, p are fixed, then

$$r(l_0, n, p_0) = \alpha \exp\left\{j2\pi\left[\frac{l_0 d \sin\theta}{\lambda} + f_d p_0 T_0 + \frac{2R}{\lambda}\right]\right\}\exp\left[2\pi\left(\frac{2\mu R}{c}\right)\frac{n}{f_s}\right] + \omega \tag{12.12}$$

$$r(l_0, n_0, p) = \alpha \exp\left\{j2\pi\left[\frac{l_0 d \sin\theta}{\lambda} + \left(\frac{2\mu R}{c}\right)\frac{n_0}{f_s} + \frac{2R}{\lambda}\right]\right\}\exp(2\pi f_d T_0 p) + \omega \tag{12.13}$$

$$r(l, n_0, p_0) = \alpha \exp\left\{j2\pi\left[\left(\frac{2\mu R}{c} + f_d\right)\frac{n_0}{f_s} + f_d p_0 T_0 + \frac{2R}{\lambda}\right]\right\}\exp\left(2\pi\frac{l d \sin\theta}{\lambda}\right) + \omega \tag{12.14}$$

The organized equation is as follows.

$$r(l_0, n, p_0) = s_{lp}\exp\left[2\pi\left(\frac{2\mu R}{cf_s}\right)n\right] + \omega \tag{12.15}$$

$$r(l_0, n_0, p) = s_{\text{ln}}\exp\left[2\pi(f_d T_0)p\right] + \omega \tag{12.16}$$

$$r(l, n_0, p_0) = s_{np}\exp\left[2\pi\left(\frac{d\sin\theta}{\lambda}\right)l\right] + \omega \tag{12.17}$$

According to the signal model of FMCW radar, the parameters to be estimated, including range, Doppler and angle, are contained in the angular frequency components of the corresponding dimensions, and can thus be directly obtained using 2D or 3D FFT.

In array signal processing, a fundamental issue is the estimation of the direction of arrival (DOA) of spatial signals, which is also a crucial task in many fields such as radar and sonar. The earliest attempt to estimate the angle through the array was the "Beamformer" method, with representative algorithms being the Bartlett Beamformer and the MVDR (also known as Capon) Beamformer [25]. For the Bartlett Beamformer, the basic idea is to "steer" the array in a certain direction, i.e., linearly combining received signals, and measure the array gain in that direction. The direction with the maximum gain is the estimated direction of the incoming wave. The total power of the combined array received signal $\boldsymbol{x}(t)$ is given by

$$P(\boldsymbol{w}) = \frac{1}{N}\sum_{t=1}^{N}|y(t)|^2 = \frac{1}{N}\sum_{t=1}^{N}\boldsymbol{w}^H\boldsymbol{x}(t)\boldsymbol{x}^H(t)\boldsymbol{w} = \boldsymbol{w}^H\hat{\boldsymbol{R}}\boldsymbol{w} \tag{12.18}$$

where $\boldsymbol{w}$ is the filter coefficient for the Bartlett beamformer, $\hat{\boldsymbol{R}} = \frac{1}{N}\sum_{t=1}^{N}\boldsymbol{x}(t)\boldsymbol{x}^H(t)$ is the correlation matrix of the measured signal. The optimization objective of the

Bartlett beamformer is

$$\max_{w} E\{w^{\mathrm{H}} x(t) x^{\mathrm{H}}(t) w\} = \max_{w} w^{\mathrm{H}} E\{x(t) x^{\mathrm{H}}(t)\} w$$
$$= \max_{w}\{E|s(t)|^2 \left|w^{\mathrm{H}} a(\theta)\right|^2 + \sigma^2 |w|^2\}, \text{ s.t. } |w| = 1 \tag{12.18}$$

where $\boldsymbol{a}(\theta)$ is the array steering vector in the direction θ. The solution to the above equation gives the spatial spectrum and the optimal beamformer coefficients as

$$P_{BBF}(\theta) = \frac{\boldsymbol{a}^H(\theta)\hat{\boldsymbol{R}}\boldsymbol{a}(\theta)}{\boldsymbol{a}^H(\theta)\boldsymbol{a}(\theta)}, \boldsymbol{w}_{BF} = \frac{\boldsymbol{a}(\theta)}{\sqrt{\boldsymbol{a}^H(\theta)\boldsymbol{a}(\theta)}} \tag{12.19}$$

The peak position of the angular power spectrum corresponds to the direction of arrival of the signal. Since the Bartlett beamformer seeks the direction with the maximum combined power, the actual received signal includes interference and noise components, resulting in a wide main lobe and low resolution. The MVDR beamformer considers the impact of interference and noise, its core idea is to maintain constant power in the observation direction while minimizing the power of noise and interference from directions other than the observed one. Therefore, it can be regarded as a sharp spatial band-pass filter, with its optimization objective being

$$\min_{w} P(\boldsymbol{w}), s.t.\ \boldsymbol{w}^H \boldsymbol{a}(\theta) = 1 \tag{12.21}$$

wherein, the spatial spectrum and the optimal beamformer coefficients are obtained through the Lagrange multiplier method.

$$P_{CAP}(\theta) = \frac{1}{\boldsymbol{a}^H(\theta)\boldsymbol{R}^{-1}\boldsymbol{a}(\theta)}, \boldsymbol{w}_{CAP} = \frac{\boldsymbol{R}^{-1}\boldsymbol{a}(\theta)}{\boldsymbol{a}^H(\theta)\boldsymbol{R}^{-1}\boldsymbol{a}(\theta)} \tag{12.22}$$

Based on the classic spectral estimation methods and spatial beamforming techniques, such as the periodogram-based methods, there are limitations in terms of resolution. In contrast, subspace-based spectral estimation methods, such as MUSIC [26], ESPRIT [27], Space-Alternating Generalized Expectation–maximization (SAGE) [28], compressed sensing [29], and tensor decomposition [30], often achieve higher parameter estimation accuracy and resolution. It should be noted that these representative parameter estimation methods also have some limitations. For example, MUSIC and ESPRIT require a certain number of measurement samples to ensure that the signal subspace and noise subspace can be accurately separated; SAGE can greatly reduce the computational complexity of the expectation maximization process through subspace alternation, but the time efficiency is low in the case of multi-parameter estimation; compressed sensing does not require signal equidistant measurement, but it needs to meet the signal sparsity condition, and the performance of the parameter estimation problem with continuous values will significantly decrease; tensor decomposition often needs to be combined with the aforementioned

ESPRIT, CS and other methods, so it also has some limitations corresponding to these methods. Due to limited space, a detailed discussion of the above-mentioned parameter estimation methods is not provided here.

The performance of the parameter estimation algorithm is closely related to the overall performance of the integrated sensing and communication system, and directly affects the quality of sensing services. Due to the numerous 6G sensing scenarios and complex wireless channel environments, signal processing in integrated sensing and communication scenarios will face huge challenges. Low complexity parameter estimation, real-time high-precision parameter estimation, parameter estimation based on discontinuous data signals, and cooperative parameter estimation of multiple sensing nodes are potential research directions for integrated sensing and communication parameter estimation technology.

12.3.4 MIMO ISAC Technologies

A typical use case of sensing is to measure the azimuth and position of the sensing target, which is naturally closely related to radar. In the late 1950s, the Lincoln Laboratory at the Massachusetts Institute of Technology conducted a series of research on phased-array radar, including theoretical analysis, application research, hardware design, device manufacturing and system testing [31]. Early radars with multiple radar array elements were mainly phased array radars, which achieved radar beamforming and rapid beam steering by adjusting the phase of the transmission signals of each antenna element in the radar array. Around 2003, with further advancements in signal processing technology, MIMO radar emerged. The MIMO concept was generally believed to have been initially used in the field of wireless communication (around 1990) and was later adopted in radar. The distinctive feature of MIMO radar is that the signals transmitted by each antenna in the array are orthogonal to each other, which can be achieved through time division multiplexing, frequency division multiplexing, Doppler frequency division multiplexing, code division multiplexing, or combinations of the above. Studies have shown that MIMO radar, utilizing waveform diversity and virtual array characteristics, can achieve higher detection or estimation resolution, a higher maximum number of identifiable targets and better clutter suppression ability [32]. MIMO radar can be divided into co-located MIMO radar [33] and distributed MIMO radar [32] according to the antenna deployment position. In distributed MIMO radar, the antenna spacing is typically much larger than half the wavelength of the transmitted signal, and the signal model of this type of radar often needs to consider the near-field assumption, in which signals arriving at the receiving array are spherical waves. Compared with co-located MIMO radar, distributed MIMO radar can also better utilize the radar cross-section (RCS) diversity of the target to improve detection performance [32]. Co-located MIMO radar can estimate the target angle on the array side with consideration of the far-field assumption, while distributed MIMO radar generally directly estimates the target's azimuth coordinates with high estimation accuracy. From an implementation perspective, precise synchronization between antennas in distributed MIMO radar is a challenge. Subsequently, new types of radar systems such as phased MIMO radar [34] and beamspace MIMO radar [35] have emerged, providing more flexible radar signal and antenna configuration schemes, enabling the radar to simultaneously take advantage of both phased-array radar and MIMO radar, and possess excellent detection and parameter estimation performance. Future MIMO ISAC systems are likely to have both MIMO communication and MIMO radar functions. The improvement in system sensing accuracy utilizes the concept of virtual arrays in MIMO radar, which is briefly introduced below. Consider that the total number of antennas in the MIMO radar transmitting array is M, and the position coordinates of each transmitting antenna are $\boldsymbol{x}_{T,m}, m = 0, 1, \cdots, M-1$. And the total number of receiving array antennas is N, the coordinates of each receiving antenna are $\boldsymbol{x}_{R,n}, n = 0, 1, \cdots, N-1$. Assuming that the signals transmitted by each transmitting antenna are orthogonal, then

$$\int \phi_m(t)\phi_k^*(t)dt = \delta_{mk} \tag{12.23}$$

Each receiving antenna uses M matched filters to separate the transmitted signals, so the receiver obtains a total of NM received signals. Considering a far-field point target, the response of the target obtained by the n-th receiving antenna's m-th matched filter can be represented as

$$y_{n,m}^{(t)} = \alpha^{(t)}\exp\left[j\frac{2\pi}{\lambda}\boldsymbol{u}_t^T\left(\boldsymbol{x}_{T,m} + \boldsymbol{x}_{R,n}\right)\right] \tag{12.24}$$

where $\boldsymbol{u}_t$ is a unit vector pointing from the radar transmitter to the point target, $\alpha^{(t)}$ is the reflection coefficient of the point target. It can be seen that the phase of the reflected signal is jointly determined by the transmitting antenna and the receiving antenna. Equivalently, the target response of Eq. (12.14) is exactly the same as the target response obtained by an array with NM antennas, which is referred to as a virtual array (VA). And the equivalent array antenna position coordinates are

$$\{\boldsymbol{x}_{T,m} + \boldsymbol{x}_{R,n} | m = 0, 1, \cdots, M-1; n = 0, 1, \cdots, N-1\} \tag{12.25}$$

Figure 12.4 shows a MIMO radar antenna configuration with $M = 3$ and $N = 4$, and the corresponding virtual array diagram. When deploying MIMO radar, by reasonably setting the positions of the transmitting antenna elements and receiving antenna elements, an array containing NM non-overlapping virtual antennas can be constructed with only $N + M$ physical antennas. Since the virtual array can often form a larger array aperture, it can achieve a higher angular resolution.

With mature hardware implementation and signal processing methods, phased array radar sensing technology is currently available. Phased array radar utilizes the entire array for beamforming, creating high-gain, high-directivity narrow beams, which is conducive to improving the sensing SNR. However, the beam width of the phased array radar determines the angular resolution. When the sensing area is large,

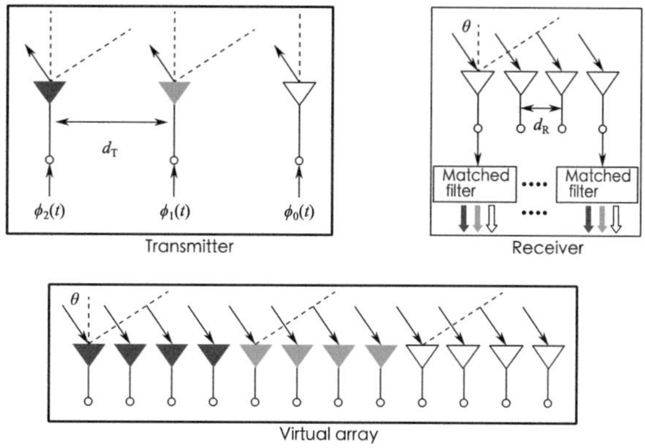

Fig. 12.4 MIMO radar antenna configuration and corresponding virtual array diagram

beam scanning is required, and multiple targets within a distance smaller than the beamwidth cannot be distinguished, limiting the maximum number of detectable targets. MIMO radar transmits signals from different antennas independently (quasi-orthogonal or orthogonal), and by deploying antennas properly, it can form a large aperture virtual array with the same number of antennas, significantly improving angular resolution. Additionally, MIMO radar exhibits strong clutter suppression capabilities. However, due to the orthogonal waveforms of each transmitting antenna, the beamwidth is wider and the beamforming gain is limited.

It is worth mentioning that with the development of millimeter-wave (mmWave) and massive MIMO technologies, the hybrid architecture of massive MIMO is becoming a trend [36]. This hybrid architecture includes fully-connected structures, partially-connected structures, and a combination of both. One or more antenna array elements are connected to an analogue phase shifter to achieve analogue phase control, and a group of antennas connected to multiple analogue phase shifters form a subarray. A subarray is connected to an analogue-to-digital converter (ADC) / digital-to-analogue converter (DAC) to achieve digital amplitude/phase control. This hybrid architecture significantly reduces the number of RF chains required, reducing implementation costs. Currently, the academic community has conducted extensive research on hybrid beamforming for this structure [37], but research on integrated sensing and communication under this architecture is still in the exploratory stage [38].

In future joint sensing and communication scenarios, sensing based on radar technology, such as the positioning and trajectory tracking of pedestrians, motor vehicles, drones, etc., often requires sensing of one or more targets or events in a certain area. Prior to this, detection may be required for a large area with wide angle coverage to identify the approximate location of the targets. Unlike traditional radar scenarios, in joint sensing and communication scenarios, the business coverage distance is generally tens to hundreds of meters, and the surrounding environment and objects easily generate significant clutter, seriously affecting sensing performance. In joint sensing and communication scenarios, signal multipath propagation can increase capacity for communication, but the situation is more complex for sensing, some will become clutter, and some may also help improve sensing performance [39]. It can be foreseen that hybrid architecture large-scale MIMO will still be a key technology for future high-capacity and high-reliability communication, and this structure is expected to achieve variable antenna/antenna subarray topology and flexible configuration of transmitted signals, which provides a solid foundation for integrated sensing and communication applications. Using a considerable number of antenna array elements, it is possible to finely control the sensing beam and improve sensing performance.

12.4 Key Technology of ISAC Network Architecture

12.4.1 Different Sensing Modes

According to the different transmitting and receiving nodes of sensing signals, there are six sensing modes, as shown in Fig. 12.5. It is worth noting that each sensing mode in Fig. 12.5 is illustrated by a sensing signal transmitting node and a sensing signal receiving node. In actual systems, different sensing modes can be selected according to different sensing requirements. Each sensing mode can have one or more transmitting and receiving nodes, and the actual sensing system can include multiple different sensing modes. The sensing objects in Fig. 12.5 are exemplified by people and vehicles, and the sensing objects of the actual system will be more diverse.

The definition of the six basic sensing modes are as follows.

(1) Base station simultaneously transmits and receives sensing signals: In this mode, Base station 1 transmits the sensing signal and obtains the sensing result by receiving the echo of the sensing signal.

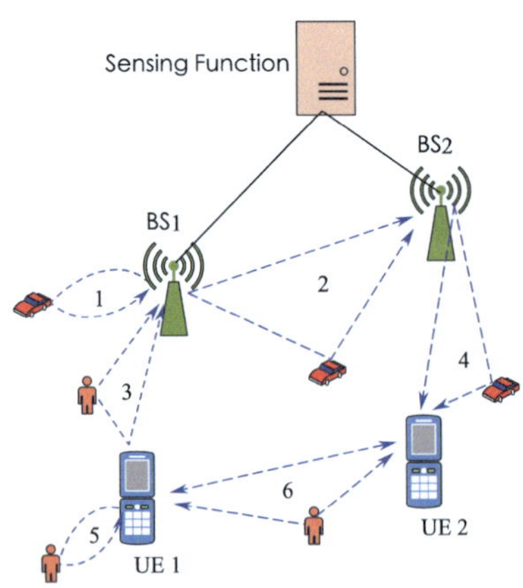

Fig. 12.5 Six basic sensing modes

(2) Base station 1 transmits and base station 2 receives sensing signals: In this mode, Base station 2 receives the sensing signal transmitted by Base station 1 to obtain the sensing result.
(3) UE transmits and base station receives sensing signals: In this mode, Base station 1 receives the sensing signal transmitted by UE 1 to obtain the sensing result.
(4) Base station transmits and UE receives sensing signals: In this mode, UE 2 receives the sensing signal transmitted by Base station 2 to obtain the sensing result.
(5) UE simultaneously transmits and receives sensing signals: In this mode, UE 1 transmits the sensing signal and obtains the sensing results by receiving the echo of the sensing signal.
(6) UE 1 transmits and UE 2 receives sensing signals: In this mode, UE 2 receives the sensing signal transmitted by UE 1 and obtains the sensing result.

12.4.2 Wireless Sensing Function and Network Architecture

1. End-to-end functionality explanation

The integrated sensing and communication network architecture is a system architecture that supports wireless communication and wireless sensing functions and services from an end-to-end architecture perspective. Wireless sensing services refer to mobile networks receiving sensing service requests for application scenarios, selecting appropriate sensing function nodes, triggering sensing business, receiving the sensing results from the selected sensing function nodes, and responding to sensing service requests. From the perspective of wireless sensing functions, the potential functions are as follows:

(1) The aforementioned selected sensing function node receives the sensing service request, determines the required sensing measurement quantity according to the sensing service request. Correspondingly, it receives the sensing measurement result, i.e., the value of the sensing measurement, generates the sensing result, and responds to the sensing service request.
(2) Control of quality of service (QoS), that is, controlling sensing-related nodes to meet the sensing QoS requirements.
(3) Determine the sensing signal transmitting or receiving node, including network equipment (such as base stations) and user equipment (such as mobile phones) in the mobile communication system.
(4) Determine the sensing links or sensing modes, where the sensing modes include the previously mentioned six types of sensing modes.
(5) Determine the sensing signal. Potential sensing signals include reference signals and data signals, where the reference signal can be a communication reference signal or a sensing-specific reference signal.
(6) Determine the time–frequency resources used for sensing. Potential sensing resources include time–frequency resources not used in communication (such

as guard bands), shared time–frequency resources already used in communication (such as reference signals or data signals), sensing-specific time–frequency resources. Furthermore, it is necessary to determine the configuration of the sensing signal, potential configurations include the time–frequency and spatial resource information of the sensing signal. If the node determining the sensing time–frequency resources is not the sensing signal transmitting node, then the configuration of sensing signals is transmitted to the sensing signal transmitting node by the node determining the sensing time–frequency resources.

(7) Determine the configuration of the sensing measurement quantity. Potential configurations include the indication of the sensing signal to be measured, the quantity or time window of the sensing signal to be measured, the indication of the measurement result report, etc. If the node determining the configuration of the sensing measurement quantity is not the receiving and measurement node of the sensing signals, then the configuration of sensing measurement quantities is transmitted to the sensing signal receiving node.

(8) Determine and configure the transmission channel for reporting sensing measurement results, including establishing, modifying or releasing the transmission channel, etc.

(9) The sensing signal transmitting node transmits the sensing signal, the sensing signal receiving node receives the sensing signal and performs measurement and reports the sensing measurement result.

In addition to the aforementioned wireless sensing services and sensing capabilities, the registration and interaction of sensing capabilities, sensing security and privacy, sensing billing and sensing policies are also important components. The registration and interaction of sensing capabilities can provide sensing capability information of sensing function nodes, sensing signal transmitting nodes, or sensing signal receiving nodes, thereby supporting the selection of sensing function nodes, sensing signal transmitting nodes, sensing signal receiving nodes, sensing links or sensing modes, etc. Sensing security and privacy must ensure that the privacy information of the environment or sensing target obtained by wireless sensing is used within the authorization scope of its owner, and also ensure the security and authentication of the end-to-end process of wireless sensing services, avoiding attacks or tampering, etc. Sensing billing is an important part of integrated sensing and communication business, with potential content including billing objects, billing standards and billing schemes, where the billing objects need to include charges for the demanders of sensing services, and also consider payments for UE that serve as sensing information transmitting or receiving nodes. Sensing policies can provide relevant policies for the selection of sensing function nodes, sensing signal transmitting/receiving nodes, and the transmission channels of sensing measurement results.

Considering the ISAC network architecture from the perspective of standardization, further discussions are needed on the definition of network functions, the interfaces and flow design between network functions. The definition of network functions refers to dividing or combining the aforementioned functions to form the network functions of the core network, wireless access network or UE. Whether the sensing network functions of the core network or wireless access network need to be defined separately for control (C)/user (U), which protocol layers make up the sensing control or data protocol stack, and which network functions of the core network, wireless access network, or UE can produce sensing results are issues that still need further industry discussion. According to the analysis of typical application scenarios, the demanders of sensing services could be external network functions such as application functions, or internal network functions such as UE, base stations, etc. Therefore, in the two situations of external sensing services and internal sensing services, further industry discussion is needed on which network functions, interfaces and processes between network functions can be uniformly designed or reused.

2. Sensing QoS

High QoS guaranteed service is one of the important features of mobile communication networks, therefore the wireless sensing QoS mechanism is one of the important contents of the integrated sensing and communication architecture. Existing communications define communication QoS based on the quality requirements of user plane data transmission for different services, using characteristic parameters such as Resource Type, Priority Level, Packet Delay Budget, Packet Error Rate (PER), etc. From the perspective of wireless sensing service quality requirements, wireless sensing QoS can be defined by characteristic parameters such as sensing accuracy, sensing resolution, sensing range, sensing update rate, sensing latency, etc.

In scenarios such as high-speed highway traffic monitoring of vehicles' velocities and traffic flow in smart transportation, the aforementioned sensing QoS parameters mainly involve sensing accuracy, sensing resolution, sensing range, and sensing update rate. Among these, sensing accuracy or resolution mainly refers to the accuracy or resolution of range, velocity, and angle, and sensing range encompasses distance range and velocity range. In scenarios such as respiratory monitoring in smart living, the aforementioned sensing QoS parameters mainly involve sensing accuracy, sensing resolution, sensing range, sensing latency and sensing update rate. Among them, sensing accuracy is the accuracy of respiratory rate, sensing resolution is the resolution of respiratory rate, sensing range refers to the distance range from the target to the sensing signal receiving node, and sensing latency is the maximum latency required for each output of respiratory monitoring results. Therefore, the above-mentioned sensing QoS is more suitable for representing the quality requirements of the sensing service, and the number of sensing QoS parameters and the specific meanings of some characteristic parameters vary for different application scenarios. Furthermore, it is necessary to consider enhancing the compatibility of sensing QoS parameters for different sensing services, as well as the ease of parsing, interacting and guaranteeing the required sensing QoS for sensing-related network functions, such as sensing function nodes, sensing signal transmitting nodes, sensing signal receiving nodes in mobile communication networks, the sensing QoS parameters can also be defined from the dimensions of quality requirements for sensing signals, quality requirements for sensing measurements, or quality requirements for the transmission of sensing measurement results. Defining sensing QoS feature parameters from dimensions such as the quality of sensing signals, the quality of sensing measurements, and the quality of sensing measurement results contributes to decoupling from specific sensing services and improving the efficiency of network function coordination. For example, based on the relationship between sensing range resolution and signal bandwidth, as well as the relationship between maximum unambiguous velocity and the repetition period of sensing signals, sensing QoS can be defined through parameters such as the minimum bandwidth of sensing signals, the maximum repetition period of sensing signals, and the minimum duration of sensing signals. Considering the quality requirements of sensing signals, sensing QoS can be defined through parameters such as the minimum bandwidth, the maximum repetition period and the minimum time-domain duration of the sensing signal. Sensing QoS can also be defined according to the relationship between sensing update rate and the reporting of sensing measurement results, as well as the relationship between sensing accuracy and the quality of sensing measurement results. Considering the quality requirements of sensing measurements, the sensing QoS can be defined by parameters such as the minimum number of sensing signals or the shortest time to be measured, the minimum time of the sensing measurements, such as the time interval between receiving the sensing signals and the reporting of the results, and the maximum time for reporting of the sensing results. Additionally, QoS parameters can also be defined based on the quality requirements and transmission mode of sensing measurement results, such as reusing communication user-plane transmission channels or using new sensing transmission channels. If sensing measurement results are transmitted through the existing communication user-plane channel, sensing QoS can consider integrating the existing communication QoS parameters with sensing data transmission requirements for comprehensive definition. Therefore, in order to better ensure the quality of wireless sensing services, the definition and interaction mechanism of sensing QoS parameters need to consider the requirements and efficiency of both sensing services and sensing network functions.

3. Examples of Wireless Sensing Function and Architecture

To facilitate a better understanding of the integrated sensing and communication architecture and flows, and considering the ongoing discussions about 6G network architecture and protocols, the following example is based on the 5G protocol. Currently, there are various options for the interaction of potential sensing configuration information and sensing measurement results. The following flow is a brief illustration of providing wireless sensing services externally, using the sensing mode in which the base station transmits sensing signals and the UE receives sensing signals as an example.

(1) The Application Function (AF) sends a sensing service request to the Network Exposure Function (NEF). This request may include the sensing target or area, sensing business information, such as velocity measurement, respiratory monitoring, etc., and sensing service QoS requirements, including sensing accuracy, sensing resolution and other QoS parameters.

(2) Based on the sensing service request and information from the Sensing Function (SF), the NEF selects the appropriate sensing function node, and sends a sensing service request to the SF.
(3) The SF receives the sensing service request and determines the required sensing measurements based on the information of the sensing service request. If the aforementioned definition of sensing QoS parameters that enhance sensing business compatibility and network function resolution is adopted, and the sensing service request only includes the QoS parameters of the sensing service, then the SF can convert the sensing service QoS parameters to generate QoS feature parameters for sensing signal, sensing measurement, or sensing measurement result transmission, which are used for sensing QoS control among various sensing-related functions in mobile communication networks.
(4) Assuming the SF determines the sensing signal transmitting or receiving nodes, as well as the sensing link or sensing mode. The SF selects a group or multiple suitable combinations of sensing signal transmitting or receiving nodes, as well as corresponding sensing links or sensing modes, based on the sensing service request information and the sensing capabilities of each node. When the received sensing service request is aimed at a larger geographical area, the SF may need to divide the request into multiple smaller geographical areas based on the sensing range of the sensing transmitting or receiving nodes, hence the received sensing service request corresponds to multiple sets of sensing signal transmitting or receiving nodes.
(5) Assuming the base station determines the sensing signal, the time–frequency resources used for sensing, and the configuration of sensing measurements. The base station determines the aforementioned sensing configuration information based on the received sensing QoS information, such as sensing signal, sensing measurement, and sensing measurement result transmission related QoS parameters. If the base station is not the transmitting node or receiving node for the sensing signal, it sends the sensing signal configuration to the transmitting node and the sensing measurement configuration to the receiving node.
(6) As previously described, this example assumes that the UE is the receiving node for the sensing signal and uses existing user plane channels to transmit sensing measurement results. Therefore, when the SF selects the UE as the sensing receiving node, it also triggers the protocol data unit (PDU) session establishment or modification process, so that the UE, as the sensing signal receiving node, can transmit the sensing measurement results through this PDU session.
(7) The sensing signal transmitting node, such as the base station, transmits the sensing signal, the UE as the sensing signal receiving node receives and measures the sensing signal, and reports the sensing measurement results through the aforementioned PDU session.
(8) The SF receives the sensing measurement results, i.e., the values of the sensing measurements, generates the sensing results and sends them to the NEF.
(9) The NEF responds to the AF's sensing service request.

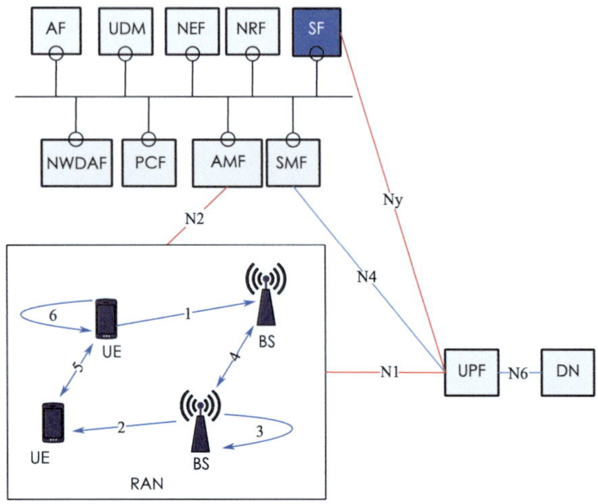

Fig. 12.6 Network architecture of integrated sensing and communication

The network architecture of the integrated sensing and communication is shown in Fig. 12.6.

12.5 Future Research Directions and Challenges

In order to implement the application of integrated sensing and communication technology, in addition to the key technologies in the previous chapters related to air interface and network architecture, there are still some challenges that need to be addressed, such as integrated waveform design and performance evaluation metrics, sensing channel modeling, and the impact of hardware non-idealities. Furthermore, the integration of sensing with other cutting-edge technologies as a future research direction for integrated sensing and communication technology also requires further exploration and research.

12.5.1 Integrated Waveform and Performance Metrics

To achieve a deep integration of communication and sensing systems and to efficiently utilize limited spectrum resources, it is necessary to balance waveform parameters based on performance metrics for both communication and sensing functions, in order to achieve highly integrated waveform design. Previous chapters have proposed some research on integrated waveforms, mainly based on the modification of commonly used communication or sensing waveforms to simultaneously possess communication and sensing functionalities. However, there is currently a lack of standardized theoretical guidance for waveform design. Some research has proposed using metrics such as mean square error and estimation-information rate as unified performance evaluation metrics for communication-sensing systems, or introducing the concept of capacity-distortion functions based on rate-distortion theory into integrated sensing and communication systems, and using maximization of mutual information weighted by communication and sensing as criteria for waveform design [40–43]. Typically, communication and sensing systems have different emphases on performance metrics, as well as different signal processing methods and objectives. A unified standard for measuring the performance of integrated sensing and communication is of significant importance in guiding the design of integrated waveforms. Additionally, considering the constraints and mutual interference between communication and sensing, in cases where optimal communication-sensing performance cannot be simultaneously achieved, compromises need to be made in waveform design based on different application scenarios.

12.5.2 Backscatter Technology-Assisted Sensing

ISAC has the potential to give rise to a range of new 6G applications. For instance, sensing based on low-power communication devices is expected to become a significant application scenario for 6G. Wireless sensing based on Radio Frequency Identification (RFID) and backscatter communication technology, as compared to device-free wireless sensing, not only achieves basic sensing functions but also obtains additional target information, thereby potentially enhancing the performance of sensing. Due to the low cost, low power consumption and suitability for large-scale deployment of RFID and backscatter, the use of RFID/backscatter-based sensing is expected to gain widespread applications in 6G networks. For example, in vehicular networks, roadside units (RSUs) with sensing capabilities can communicate with vehicles on the road using integrated sensing and communication signals, enabling vehicle localization, velocity measurement and trajectory tracking. When vehicles are equipped with RFID or backscatter tags, these tags can provide additional information such as vehicle identification and current status. By receiving the reflected signals from the tags, the RSUs can achieve high-precision vehicle localization [44], velocity measurement, and trajectory tracking, as well as accurate identification and differentiation of different vehicles, thereby expanding the sensing capabilities.

Currently, there have been numerous studies on wireless sensing based on RFID. For example, by adding tags to intravenous drip bags in hospitals, real-time monitoring of drip rates for multiple patients can be achieved [45]. Deploying multiple tags around sensing targets can enable target orientation recognition [46], material identification and 2D cross-sectional imaging [47]. Compared to widely used RFID technology, backscatter communication offers faster communication data rates and longer communication distances [48], making it a promising implementation method for integrated sensing and communication. The corresponding key technologies, such as tag-based integrated sensing and communication waveform design, frame structure design, and sensing scheme and algorithm design require future research.

Tag devices can be installed on sensing targets or deployed near sensing targets, and the receiver can detect the tag signals to complete the sensing process. RFID-based sensing offers the following advantages:

(1) By installing different tags on different sensing targets, different sensing targets can be distinguished according to the different characteristics of the reflected signals from different tags.
(2) To a certain extent, tag devices exhibit more stable signal reflection characteristics compared to sensing targets, resulting in higher signal reflection strength and better coverage and interference suppression capabilities, thereby enhancing sensing performance.
(3) Tags are cost-effective, have low power consumption, and are compact. Deploying multiple tag devices around sensing targets and performing joint sensing processing on signals from multiple tags in the environment can expand the sensing range and improve sensing performance.

12.5.3 Reconfigurable Intelligent Surface-Assisted Sensing

Wireless sensing applications have different requirements for wireless channels compared to traditional wireless data services. In traditional wireless data services, wireless networks utilize the multipath characteristics of wireless channels to support multi-user multi-stream data transmission, maximizing frequency utilization. On the other hand, wireless sensing applications, such as wireless localization, require the accurate estimation of the position information of sensing targets using the line-of-sight paths between sensing nodes and sensing targets. The multipath characteristics of wireless channels have a negative impact on the detection performance of wireless sensing applications. The presence of non-line-of-sight paths mixed with line-of-sight paths in the wireless channel results in a decrease in the detection performance of line-of-sight paths. Additionally, in complex wireless environments, there are often cases where line-of-sight paths are obstructed, presenting new challenges for traditional channel detection algorithms based on line-of-sight paths. Therefore, the deployment of integrated sensing and communication networks needs to consider the needs of both communication and sensing, balancing the spectrum efficiency of multi-user communication and the performance of wireless sensing.

Future wireless sensing applications need to explore sensing schemes in non-line-of-sight signal environments. The premise of utilizing non-line-of-sight sensing in multipath channels is to identify useful paths in the multipath channel or to reconstruct the multipath propagation environment, allowing sensing nodes to extract path information of sensing targets. 6G networks can utilize Reconfigurable Intelligent Surfaces (RIS) to enhance the performance of wireless sensing networks. RIS deployed in multipath propagation environments ensures the existence of line-of-sight paths between RIS and sensing nodes, as well as between RIS and sensing targets. By adjusting the reflection beams of RIS, the signal quality of the transmission path from sensing nodes to RIS and then to sensing targets can be maximized, further improving the performance of wireless sensing. Given the known positions of sensing nodes and RIS, the wireless network can obtain information about the sensing targets based on the channel information between sensing nodes, RIS, and sensing targets, as well as the forwarding beams of RIS. RIS-assisted sensing networks can help sensing nodes overcome the effects of non-line-of-sight signal environments. Meanwhile, in line-of-sight propagation environments, the transmission path from sensing nodes to RIS and then to sensing targets can serve as auxiliary information for line-of-sight sensing, further enhancing the performance of sensing applications.

RIS-assisted sensing networks also present some challenges in deployment. RIS beamforming for sensing applications is a key technology to enable RIS-assisted sensing. As a low-cost passive device, RIS can be deployed on a large scale in the environment. Determining the forwarding beams of RIS in the absence of information about sensing targets is a highly challenging problem. For active sensing targets, the sensing targets can determine the optimal RIS forwarding beams through beam training based on RIS [49]. For passive sensing targets, the sensing network can consider using machine learning to infer the information of sensing targets based on the correlation between different RIS forwarding beams and the variations in wireless channels. The large-scale deployment of RIS brings about sensing scenarios based on near-field assumption. Traditional sensing algorithms typically approximate signal processing based on far-field assumptions. Therefore, traditional channel modeling and signal processing algorithms need to be modified to match RIS-assisted sensing based on near-field assumption.

12.5.4 Channel Modeling for ISAC

The theoretical research and establishment of 6G channel models support early-stage technology development and standardization work for 6G, providing reliable theoretical and empirical evidence for the assessment and standardization promotion of candidate technologies. Additionally, it can be applied to the design and optimization of communication systems, playing a fundamental role in reducing the testing and network planning costs of actual system deployments.

Research on integrated sensing and communication involves aspects such as integrated waveform design, sensing algorithm design, and performance evaluation of communication and sensing, all of which rely on the support of channel models that accurately reflect the wireless propagation environment experienced by actual systems. For channel modeling in integrated sensing and communication scenarios, research is needed on how to construct joint representation models for communication and sensing. Traditional communication channels and radar channels have some noticeable differences. The wireless communication channel environment is complex and rich in multipath characteristics, which is advantageous for the spatial multiplexing and spatial diversity of MIMO systems. In contrast, the radar channel is often considered sparse channels, with few or no multipath, and reflections caused by other reflectors in the environment apart from the target are referred to as clutter, considered harmful. Additionally, the transmission distance of communication system is generally shorter than the detection distance of radar, and the transmission power of communication is much lower than that of radar. Furthermore, communication channels consider both line-of-sight and non-line-of-sight (NLOS) propagation, while traditional radar channels primarily consider line-of-sight signal propagation between the radar and the target. For example, in the case of sensing in a mobile network, the considerations for channel models differ depending on the different sensing methods. The reflection and scattering characteristics of the sensing target need to be considered in the channel model for base station self-sending and self-receiving sensing signals or integrated sensing and communication signals to achieve sensing. For echo signals, the round-trip path loss and additional reflection loss need to be considered, and the reflection loss is related to the RCS of the sensed object. For multi-target sensing and communication, the channels of all the communicating targets as well as the echo channels of all the sensing targets need to be accurately and jointly modelled, i.e., the reflective and scattering characteristics of all the sensing targets need to be reflected in the channel model.

Currently, the mainstream channel models for 5G mainly include Stochastic Model, Deterministic Model, and Quasi-deterministic Mode [51]. Stochastic models are mostly based on actual channel measurements and have been reasonably simplified to balance complexity and credibility. Additionally, their channel characteristics are controlled by a set of key channel parameters, with different parameter sets corresponding to different scenarios, providing flexible configuration. The standardized channel models in 3GPP TR 38.901 [52] belong to stochastic models, which are convenient for later expansion. Deterministic models include full-wave analysis models, ray tracing (RT) models, and hybrid models of both. Full-wave analysis models are based on the finite difference time domain (FDTD) method, directly simulating the distribution of electric or magnetic fields, with high accuracy but high computational complexity. Ray tracing models are based on electromagnetic theory, geometric optics theory, and geometric diffraction theory, describing all paths from the transmitter to the receiver using rays, offering precision and ease of programming. Hybrid models combining ray tracing and FDTD use ray tracing models to predict electromagnetic wave propagation in most indoor and outdoor areas, while FDTD is used to predict the electromagnetic wave propagation characteristics in some complex indoor structures. The advantage of deterministic models lies in their strong physical foundation, enabling accurate prediction of electromagnetic wave propagation based on environmental details and deterministic analysis of point-to-point electromagnetic wave propagation, making them applicable to a wide range of complex environments. In [53], the authors verified the velocity and range measurement performance of a vehicle radar algorithm based on OFDM signals using an RT channel model. Semi-deterministic models combine the characteristics of stochastic and deterministic models, with some parameters generated using stochastic modeling methods and others borrowing from deterministic modeling methods. Currently representative semi-deterministic models include Millimetre-Wave Evolution for Backhaul and Access (MiWEBA) [54], IEEE 802.11ay [55], Quasi-Deterministic Radio Channel Generator (QuaDRiGa) [56] models, etc. In

[57], the authors verified the feasibility of integrated sensing in 5G networks using compressive sensing based on the QuaDRiGa model and 3GPP model. Currently, whether there are some new electromagnetic wave propagation mechanisms in the integrated sensing and communication scenario, which modeling method is used in the model, which type of channel model it belongs to, and whether there exists a more accurate and efficient channel modelling method, etc. are all issues that need to be studied and solved.

12.5.5 Impact of Non-Idealities on Sensing Performance

In integrated sensing and communication scenarios, obtaining accurate measurement information is particularly important, and non-idealities of devices and hardware circuits can significantly affect measurement accuracy. For bistatic sensing between base stations and terminals, the sensing information is usually obtained from the processing of the channel information. Therefore, obtaining high-quality sensing channel information is particularly important, and channel measurement errors caused by some non-idealities can significantly affect the accuracy of sensing.

Literature [58] summarizes the impact of non-idealities at the receiver on the acquisition of channel information, as follows:

(1) Power Amplifier Uncertainty (PAU). Due to the resolution limitation of hardware such as Low Noise Amplifiers (LNA) and Programmable Gain Amplifiers (PGA), there is an unknown offset for the amplitude of the measured channel information.
(2) IQ path imbalance. The orthogonality of the baseband signals is destroyed, resulting in distortion of the channel information.
(3) Frequency deviation, including Carrier Frequency Offset (CFO) and Carrier Phase Offset (CPO). The residual frequency deviation leads to a time-dependent phase shift of the channel information between sampling points, which affects the Doppler measurement accuracy.
(4) Timing deviation, including Sampling Time Offset (STO) and Sampling Frequency Offset (SFO). Timing deviation affects the accuracy of delay estimation and thus the ranging accuracy.

In [59], the authors summarized methods such as shared reference clocks, cross-correlation of multiple antennas in a single station, and joint elimination of timing errors across multiple stations, and explained that the impact of clock deviation on sensing can be addressed by improving the Global Positioning System (GPS) clock, relaxing the sensing requirements of a single node, and performing multi-node measurements for the target. Literature [60] summarized the impact of transmitter processing on the acquisition of channel information, which mainly includes windowing, precoding, beamforming and other receiver-agnostic processing leading to the inability of the receiver to acquire the actual channel information.

References

1. DIETER BOHN. Google Pixel 4 and 4 XL hands-on: this time, it's not about the camera[EB/OL]. The Verge, 2019.
2. JUNTTI M, KANTOLA R, KYSTI P, et al. Key Drivers and Research Challenges for 6G Ubiquitous Wireless Intelligence[R]. 2019.
3. PAUL B, CHIRIYATH A R, BLISS D W. Survey of RF Communications and Sensing Convergence Research[J]. IEEE Access, 2016, 5: 252–270.
4. IMT-2030(6G) Promotion Group. Integrated Sensing and Communication Technology Research Report[R]. 2021.
5. WITRISAL K, MEISSNER P, LEITINGER E, et al. High-Accuracy Localization for Assisted Living: 5G systems will turn multipath channels from foe to friend[J]. IEEE Signal Processing Magazine, 2016, 33(2):59–70.
6. BERITELLI F, CAPIZZI G, LO S G, et al. Rainfall Estimation Based on the Intensity of the Received Signal in a LTE/4G Mobile Terminal by using a Probabilistic Neural Network[J]. IEEE Access, 2018: 1–1.
7. ALI A, GONZALEZ-PRELCIC N, HEATH R W, et al. Leveraging Sensing at the Infrastructure for mmWave Communication[J]. IEEE Communications Magazine, 2020, 58(7): 84–89.
8. HAN L, WU K. Joint wireless communication and radar sensing systems-state of the art and future prospects[J]. IET Microwaves, Antennas & Propagation, 2013, 7(11): 876–885.
9. PREETI, KUMARI, JUNIL, et al. IEEE 802.11ad-Based Radar: An Approach to Joint Vehicular Communication-Radar System[J]. IEEE Transactions on Vehicular Technology, 2017.
10. HUANG Y X, HU S, MA S Y, et al. Constant envelope OFDM Rad-Com fusion system[J]. EURASIP Journal on Wireless Communications and Networking, 2018, 2018 (1): 1–15.
11. KOSLOWSKI S, BRAUN M, JONDRAL F K. Using filter bank multicarrier signals for radar imaging[C]//2014 IEEE/ION Position, Location and Navigation Symposium-PLANS 2014. IEEE, 2014: 152–157.
12. GAUDIO L, KOBAYASHI M, BISSINGER B, et al. Performance analysis of joint radar and communication using OFDM and OTFS[C]//2019 IEEE International Conference on Communications Workshops (ICC Workshops). IEEE, 2019: 1–6.
13. MEALEY R M. A Method for Calculating Error Probabilities in a Radar Communication System [J]. IEEE Transactions on Space Electronics and Telemetry, 1963: 37–42.
14. ILLIAN H. Fiden, Donald Walter Czubiak. Radar-Compatible Data Link System: 7298313 B1[P]. 2007–11–20.
15. KHAN F, SEONG K L, SUNG H C. Hand-based gesture recognition for vehicular applications using IR-UWB radar[J]. Sensors, 2017, 17(4): 833.
16. SUN H, BRIGUI F, LESTURGIE M. Analysis and comparison of MIMO radar waveforms[C]//2014 International Radar Conference. IEEE, 2014: 1–6.
17. RABIDEAU D J. Doppler-offset waveforms for MIMO radar[C]//2011 IEEE RadarCon (RADAR). IEEE, 2011: 965–970.
18. Xing Mengdao, Wang Tong. Fundamentals of Radar Signal Processing[M]. Beijing: Electronics Industry Publishing House, 2008.
19. Thomas Schönhofer, Arthur Jordan. Signal Detection and Estimation - Theory and Application[M]. Guan Xin, Yang Aiping, Bai Yu, et al. trans. Beijing: Electronics Industry Publishing House, 2012.
20. He You, Guan Jian, Meng Xiangwei, et al. Radar Target Detection and Constant False Alarm Rate Processing[M]. Beijing: Tsinghua University Press, 2011.
21. RITCEY J A. Performance analysis of the censored mean-level detector[J]. IEEE Transactions on Aerospace and Electronic Systems, 1986 (4): 443–454.
22. SMITH M E, VARSHNEY P K. VI-CFAR: A novel CFAR algorithm based on data variability[C]//Proceedings of the 1997 IEEE National Radar Conference. IEEE, 1997: 263–268.

23. ZHANG J A, RAHMAN M L, WU K, et al. Enabling joint communication and radar sensing in mobile networks—A survey[J]. IEEE Communications Surveys & Tutorials, 2021, 24(1): 306–345.
24. PATOLE S M, TORLAK M, WANG D, et al. Automotive radars: A review of signal processing techniques[J]. IEEE Signal Processing Magazine, 2017, 34(2): 22–35.
25. VEEN B V, BUCKLEY K M. Beamforming: a versatile approach to spatial filtering[J]. IEEE Assp Magazine, 2002, 5(2):4–24.
26. SCHMIDT R. Multiple emitter location and signal parameter estimation[J]. IEEE Transactions on Antennas and Propagation, 1986, 34(3): 276–280.
27. ROY R, KAILATH T. ESPRIT-estimation of signal parameters via rotational invariance techniques[J]. IEEE Transactions on acoustics, speech, and signal processing, 1989, 37(7): 984–995.
28. FESSLER J A, HERO A O. Space-alternating generalized expectation-maximization algorithm[J]. IEEE Transactions on Signal Processing, 1994, 42(10): 2664–2677.
29. HADI M A, ALSHEBEILI S, JAMIL K, et al. Compressive sensing applied to radar systems: an overview[J]. Signal, Image and Video Processing, 2015, 9: 25–39.
30. CHEN H, AHMAD F, VOROBYOV S, et al. Tensor decompositions in wireless communications and MIMO radar[J]. IEEE Journal of Selected Topics in Signal Processing, 2021, 15(3): 438–453.
31. FENN A J, TEMME D H, DELANEY W P, et al. The development of phased-array radar technology[J]. Lincoln laboratory journal, 2000, 12(2): 321–340.
32. HAIMOVICH A M, BLUM R S, CIMINI L J. MIMO radar with widely separated antennas[J]. IEEE Signal Processing Magazine, 2007, 25(1): 116–129.
33. LI J, STOICA P. MIMO radar with colocated antennas[J]. IEEE Signal Processing Magazine, 2007, 24(5): 106–114.
34. HASSANIEN A, VOROBYOV S A. Phased-MIMO radar: A tradeoff between phased-array and MIMO radars[J]. IEEE Transactions on Signal Processing, 2010, 58(6): 3137–3151.
35. KHABBAZIBASMENJ A, HASSANIEN A, VOROBYOV S A, et al. Efficient transmit beamspace design for search-free based DOA estimation in MIMO radar[J]. IEEE Transactions on Signal Processing, 2014, 62(6): 1490–1500.
36. ALKHATEEB A, MO J, GONZALEZ-PRELCIC N, et al. MIMO precoding and combining solutions for millimeter-wave systems[J]. IEEE Communications Magazine, 2014, 52(12): 122–131.
37. MOLISCH A F, RATNAM V V, HAN S, et al. Hybrid beamforming for massive MIMO: A survey[J]. IEEE Communications Magazine, 2017, 55(9): 134–141.
38. LIU F, MASOUROS C, PETROPULU A P, et al. Joint radar and communication design: Applications, state-of-the-art, and the road ahead[J]. IEEE Transactions on Communications, 2020, 68(6): 3834–3862.
39. XU Z, FAN C, HUANG X. MIMO radar waveform design for multipath exploitation[J]. IEEE Transactions on Signal Processing, 2021, 69: 5359–5371.
40. KUMARI P, NGUYEN D H N, HEATH R W. Performance trade-off in an adaptive IEEE 802.11 ad waveform design for a joint automotive radar and communication system[C]//2017 IEEE International Conference on Acoustics, Speech and Signal Processing (ICASSP). IEEE, 2017: 4281–4285.
41. CHIRIYATH A R, PAUL B, JACYNA G M, et al. Inner bounds on performance of radar and communications co-existence[J]. IEEE Transactions on Signal Processing, 2015, 64(2): 464–474.
42. KOBAYASHI M, CAIRE G, KRAMER G. Joint state sensing and communication: Optimal tradeoff for a memoryless case[C]//2018 IEEE International Symposium on Information Theory (ISIT). IEEE, 2018: 111–115.
43. YUAN X, FENG Z, NI W, et al. Waveform optimization for MIMO joint communication and radio sensing systems with imperfect channel feedbacks[C]//2020 IEEE International Conference on Communications Workshops (ICC Workshops). IEEE, 2020: 1–6.

44. Han Kaifeng, Liu Tiezhi. Backscatter communication assisted vehicular positioning technology with ultra-high accuracy[J]. Telecommunications Science, 36(7): 107–117.
45. LIN Y, XIE L, WANG C, et al. Dropmonitor: Millimeter-level sensing for RFID-based infusion drip rate monitoring[J]. Proceedings of the ACM on Interactive, Mobile, Wearable and Ubiquitous Technologies, 2021, 5(2): 1–22.
46. WEI T, ZHANG X. Gyro in the air: tracking 3d orientation of batteryless internet-of-things[C]// Proceedings of the 22nd Annual International Conference on Mobile Computing and Networking. 2016: 55–68.
47. WANG J, XIONG J, CHEN X, et al. TagScan: Simultaneous target imaging and material identification with commodity RFID devices[C]//Proceedings of the 23rd Annual International Conference on Mobile Computing and Networking. 2017: 288–300.
48. DASKALAKIS S, GEORGIADIS A, KIMIONIS J, et al. A printed millimeter-wave modulator and antenna array for low-complexity Gigabit-datarate backscatter communications[J]. Nature Electronics, 2021(4): 439–446.
49. SHAO X, YOU C, MA W, et al. Target sensing with intelligent reflecting surface: Architecture and performance[J]. IEEE Journal on Selected Areas in Communications, 2022, 40(7): 2070–2084.
50. LI L, SHUANG Y, MA Q, et al. Intelligent metasurface imager and recognizer[J]. Light: science & applications, 2019, 8(1): 97.
51. WANG C X, BIAN J, SUN J, et al. A survey of 5G channel measurements and models[J]. IEEE Communications Surveys & Tutorials, 2018, 20(4): 3142–3168.
52. 3GPP. TR 38.901, V16.1.0–2019 Study on channel model for frequencies from 0.5 to 100 GHz(Release 16) [S]. 3rd Generation Partnership Project (3GPP), 2019.
53. SIT Y L, STURM C, REICHARDT L, et al. Verification of an ofdm-based range and doppler estimation algorithm with ray-tracing[C]//2011 IEEE-APS Topical Conference on Antennas and Propagation in Wireless Communications. IEEE, 2011: 808–811.
54. MALTSEV A, PUDEYEV A, BOLOTIN I, et al. Channel Modeling and Characterization, V1.0, document FP7-ICT-608637/D5.1[R].2014.
55. MALTSEV A, PUDEYEV A, GAGIEV Y, et al. Channel Models for IEEE 802.11ay, Document IEEE 802.11–15/1150r9 [R]. 2017.
56. JAECKEL S, RASCHKOWSKI L, BÖRNER K, et al. QuaDRiGa: A 3-D multi-cell channel model with time evolution for enabling virtual field trials[J]. IEEE Transactions on Antennas and Propagation, 2014, 62(6): 3242–3256.
57. RAHMAN M L, ZHANG J A, HUANG X, et al. Joint communication and radar sensing in 5G mobile network by compressive sensing[J]. IET Communications, 2020, 14(22): 3977–3988.
58. ZHUO Y, ZHU H, XUE H, et al. Perceiving accurate CSI phases with commodity WiFi devices[C]//IEEE INFOCOM 2017-IEEE Conference on Computer Communications. IEEE, 2017: 1–9.
59. ZHANG J A, WU K, HUANG X, et al. Integration of radar sensing into communications with asynchronous transceivers[J]. IEEE Communications Magazine, 2022, 60(11): 106–112.
60. TADAYON N, RAHMAN M T, HAN S, et al. Decimeter ranging with channel state information[J]. IEEE Transactions on Wireless Communications, 2019, 18(7): 3453–3468.

Chapter 13
Backscatter Communication

With the development of social informatization and digitalization, the Internet of Things (IoT) has become an important direction for the future development of the mobile communication industry. As a result, the industry has established numerous IoT technology standards, such as enhanced Machine-Type Communication (eMTC), Narrow Band Internet of Things (NB-IoT), Reduced Capability (RedCap) 5G new air interface, and Radio Frequency Identification (RFID), which can to some extent meet the low-cost and low-power consumption requirements of the IoT. In view of the "digital twin, ubiquitous intelligence" 6G vision and the "carbon peaking and carbon neutrality" goal, ultra-low power consumption, and even zero power consumption IoT solutions have become an important part of 6G research. This chapter will introduce the application scenarios, basic principles, and key technologies of backscatter communication technology for zero power communication.

13.1 Introduction

According to the IoT market forecast report released by IoT Analytics in September 2021 [1], even with the impact of the global COVID-19 pandemic and global chip shortage, there will be more than 27.1 billion IoT device connections globally by 2025. The massive number of future IoT devices poses higher requirements from the perspective of cost and power consumption. Compared with current NB-IoT/eMTC devices, future IoT devices will require lower deployment costs, with a single module costing less than 0.1 cents, further reduced power consumption, such as at the level of hundreds of microwatts or even zero power consumption, and enhanced functionality. In addition to supporting traditional communication functions, they will also support sensing and positioning functions, thus enabling ubiquitous interconnection and deep integration between the physical world and the digital world.

G. Liu et al., *6G Migrates from Communication to XaaS*,
https://doi.org/10.1007/978-981-96-7579-1_13

Towards 2030, 6G will support the connectivity requirements of trillions of wide-area IoT devices, where the IoT devices have different data rates, bandwidths, and hardware capabilities. Future IoT applications are expected to evolve in two distinct directions. The first one is to transform narrowband IoT into broadband IoT for accommodating the scenarios such as vehicle video surveillance, live education, and park logistics monitoring that demand wide-area coverage, numerous connections, and high-bandwidth transmission. The other one involves the shift from active IoT to passive IoT to support extreme industrial environments that necessitate battery-free operation, such as the scenarios with ultra-low/high temperatures, high humidity, high pressure, high radiation, or specific scenarios such as the miniaturization of implantable medical devices in the health and medical field, or the demand for cost-competitive IoT devices in logistics warehousing. For passive IoT, which aims for low cost, low power consumption, wide coverage, and numerous connections, a cellular evolution based on the current RFID system will be a crucial focus for future 6G development, which can achieve extensive connections and broad coverage, meanwhile ensuring low cost and low power consumption.

Backscatter communication (BSC), also known as backscattering communication, is deemed as one of the most representative technologies for achieving passive IoT. In recent years, new types of backscatter communication technologies have emerged in the industry, including bistatic backscatter, environmental backscatter, full-duplex backscatter, and cross-technology backscatter. These technologies provide medium/low-rate communication while ensuring low-cost flexible deployment, ultra-low power consumption, or even zero power consumption. As a result, they have become key technologies of interest in 6G network design and "green" IoT, and an important means to achieve "ubiquitous intelligence". Additionally, backscatter communication provides additional multipath for the wireless channel environment, which promotes the emergence of "symbiotic communication" technology. Specifically, based on special signal modulation methods, backscatter communication devices can not only use the radio frequency (RF) signals in the environment to transmit their own information but also can provide gains for environmental signals, thus achieving mutualism [2]. Because of the technical advantages and broad market prospects, backscatter communication has attracted widespread attention in academia and industry.

Currently, academia mainly focuses on the application scenarios of backscatter communication and the research of key technologies at the physical layer, including modulation technology, multiple access technology, and interference cancellation technology. Globally renowned 6G projects and major companies are conducting research on backscatter communication for zero-power communication. In 2021, Georgia Tech and Bell Labs developed a printed millimeter-wave modulator and antenna array for backscatter communication [3], achieving Gbps data rate transmission. Additionally, in December 2021, OPPO led the completion of the "Zero Power Communication" white paper [4] at the Future Mobile Communication Forum, and in the same year, released the "Zero Power Communication" white paper [5]. In terms of standardization, the IEEE Standards Group established the IEEE 802.15 RFID Research Group in 2008, conducting the research on the physical layer technology, data link layer technology, networking technology, interference suppression technology, positioning, and typical applications of active/semi-passive/passive RFID tags or sensors. In February 2021, China submitted a proposal to the ITU-R WP5D meeting, suggesting that backscatter communication should be included in the ITU-R's future technology trends report [6]. Moreover, in February 2022, the 3GPP TSG-SA1#97e meeting approved the project of Ambient power-enabled IoT [7], which focuses on the scenarios and requirements of passive IoT. In September 2022, the Ambient IoT SID project was approved at the 3GPP TSG-RAN#97 meeting [8], following the scenario research of SA1 and focusing on the discussion of the radio access technology enabling passive IoT.

13.2 Typical Application Scenarios

Due to the advantages of backscatter communication, such as low-cost deployment, ultra-low power consumption, support for large connections, and wide coverage, as well as its applicability in traditional IoT scenarios and logistics warehousing and other asset inventory scenarios, it also has a wide range of potential applications in other fields in the future. Generally speaking, the typical application scenarios of backscatter communication can be divided into two categories (see Fig. 13.1): wide-area coverage scenarios, including logistics warehousing, smart agriculture, industrial IoT, railway operation and maintenance, Unmanned Aerial Vehicle (UAV) inspection, etc.; and local area coverage scenarios, including smart home, human health monitoring, smart wearables, and implantable medical devices.

1. Industrial IoT

The Industrial Internet of Things (IIoT) refers to the integration of various sensors with sensing and monitoring capabilities, mobile communication technology, big

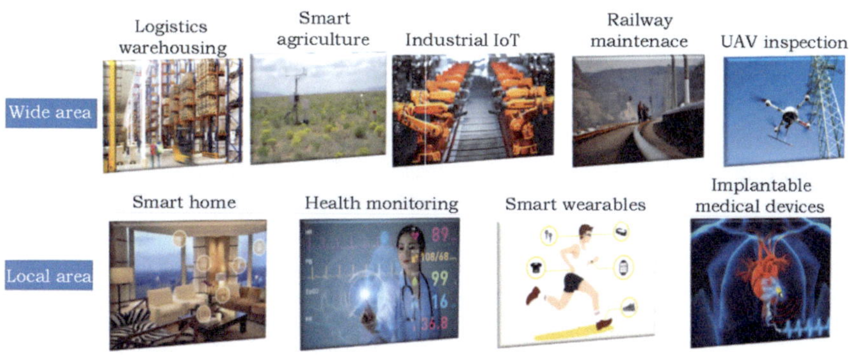

Fig. 13.1 Typical application scenarios of backscatter communication

Fig. 13.2 Example of the application of backscatter communication in the IIoT

data analysis, and other intelligent analysis technologies into all aspects of industrial production. This integration aims to improve production quality and efficiency, as well as reducing production costs and resource consumption. By collecting and analyzing various types of data through IoT sensing devices, manufacturers can improve control, enhance supply chain efficiency, assist technicians in determining the degree of equipment damage and equipment maintenance, and monitor infrastructure and asset tracking in industrial operations. The IIoT is applied in industries such as manufacturing, oil industry, chemical industry, steel industry, etc. In these industries, IoT sensing devices often need to work in extremely harsh environments. System failures and equipment out of control may lead to high risks and even endanger property and life. Traditional IoT devices powered by lithium batteries cannot function in extreme environments, such as high temperature, ultra-low temperature, high humidity, high speed, and high radiation. Additionally, the maintenance cost of battery-powered IoT devices is high. For instance, replacing batteries for a large number of IoT devices in extreme environments consumes a lot of manpower and cost. Backscatter communication devices powered by environmental energy harvesting can achieve large connections and wide coverage communication without the need for battery power. This solution addresses the applicability of IoT devices in extreme environments and reduces subsequent maintenance costs, thereby expanding the application scenarios of the IIoT and reducing deployment and maintenance costs. An example of the application of backscatter communication in the IIoT is shown in Fig. 13.2.

2. Logistics and Warehousing

In recent years, the rapid growth of e-commerce has led to an explosive increase in the demand for logistics and warehousing. To manage logistics transportation, warehousing, packaging, loading and unloading, circulation processing, distribution, and information links effectively, enterprises need to reduce industry transportation costs and improve transportation efficiency. Traditional logistics and warehousing use printed QR (Quick Response) codes to track goods, requiring manual scanning and data input, which is inefficient. Additionally, warehouse locations are sometimes not clearly divided, leading to chaotic stacking and missing goods. Integrating IoT technology into traditional logistics and warehousing to create an intelligent warehousing management system can enhance the efficiency of goods handling, reduce manual labor intensity and costs, and enable real-time monitoring of goods in and out, thereby improving delivery accuracy.

Fig. 13.3 Example of the application of backscatter communication in smart logistics and smart warehousing

Given the immense scale of the logistics and warehousing market, the cost of deploying IoT devices needs to be highly competitive. The cost of existing NB-IoT single modules typically ranges from 4 to 8 dollars, and RedCap modules are even more expensive, ranging from 13 to 27 dollars, which does not meet the cost requirements for logistics and warehousing scenarios. While traditional RFID tags are inexpensive, their signals may be easily obstructed, leading to reduced recognition accuracy and speed. Additionally, the reader's scanning range is limited to about 3 m, requiring a high deployment cost to meet large-scale batch asset inventory needs. In contrast, backscatter communication devices powered by environmental energy are extremely low-cost, with module costs less than 0.2 cents due to their lack of complex RF devices. Their wide coverage and large connection technical advantages enable rapid batch reading and large-scale reading and writing, significantly improving the efficiency of logistics and warehousing management while reducing operating costs. Furthermore, the positioning technology based on backscatter communication allows for rapid positioning of goods, greatly enhancing the efficiency of goods sorting and asset tracking. The backscatter communication tags in smart logistics and smart warehousing are depicted in Fig. 13.3.

3. Smart Wearables

The smart wearable scene is consumer-centric, wirelessly connecting various devices worn by consumers through IoT technology and has been applied in multiple fields, such as health monitoring, activity recognition, assisted living, mobile sensing, smart clothing, and indoor positioning. Currently, the mainstream product forms include watches (including watches and wristbands), shoes (including shoes, socks, or other products worn on the legs in the future), and glasses (including glasses, helmets, headbands). Additionally, there are various other product forms such as smart clothing, backpacks, walking sticks, and accessories.

Battery-powered smart wearable devices often have a short battery life. When more functions are turned on, power consumption further increases. Users often need to charge the device frequently during the daily use, which seriously affects the user experience. Backscatter communication IoT terminals have excellent characteristics, such as extremely low cost, extremely small size, extremely low power consumption (battery-free), flexibility, foldability, and washability. These characteristics make them particularly suitable for smart wearable scenes and are easy to be accepted by consumer-related industries, such as kindergartens and clothing factories. On one

(a) Health monitoring (b) Tracking and positioning (c) Portable payment

Fig. 13.4 Application of backscatter communication in smart wearables

hand, backscatter communication devices obtain energy through energy harvesting and do not need batteries, fundamentally solving the problem of short battery life. On the other hand, backscatter communication devices are low-cost, small in size, and soft in material, washable and foldable, effectively improving the comfort of wearing and user experience. The applications of backscatter communication in the wearable field are shown in Fig. 13.4.

The typical applications of the smart wearable scene include the following categories:

(1) *Health monitoring*: Backscatter communication devices integrated with sensors are embedded in wristbands, shoes, socks, and other wearable products for health monitoring. They provide timely feedback on a person's physical condition and monitor and collect data such as sleep status, weight information, heart rate, blood pressure, etc.
(2) *Location and tracking*: Backscatter communication devices can be combined with location services to provide care for the elderly, children, or hospital patients, offering location and tracking in case of loss. More comfortable materials can optimize the wearing experience, while the feature of passive ultra-low power consumption can greatly extend the usage time.
(3) *Portable payment*: Tied to personal information, backscatter communication devices can be used for convenient payments for bus rides, subway rides, shopping, etc.

4. Implantable Medical Devices

As medical devices continue to miniaturize, a large number of miniature devices and equipment can now be implanted in the human body for indicator detection or function maintenance. Examples include pacemakers, insulin pumps, various chips, catheter stents, interventional devices, artificial organs, assistive devices, and miniature robots. These devices contain numerous electronic components to enable remote data exchange and control. The backend system can obtain the operation status of the device and relevant health data from the human body in a timely or regular manner.

While these implantable medical devices significantly improve the quality of human life, they also pose challenges. The primary concern is the power supply of the device. Most functional implantable devices currently require space for batteries,

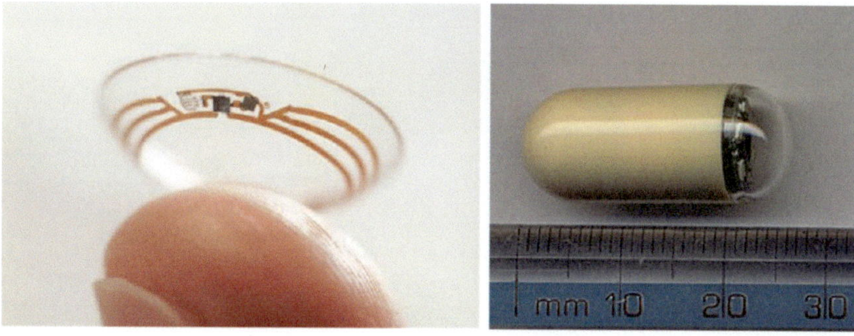

(a) Implantable lens (b) Implantable capsule endoscope

Fig. 13.5 Application of backscatter communication technology in the field of medical health

which increases the device's volume and can impact the ease of implantation. Additionally, the human body often exhibits immune rejection to larger implanted devices, leading to issues such as muscle redness and necrosis. Furthermore, as the battery power is limited, even if it can last for ten years, it will eventually need replacement through a second implantation surgery, increasing the discomfort of the patients. Backscatter communication technology, based on environmental RF power supply, can reduce the battery consumption of these medical devices during communication, thereby extending the device's working life. When in operation, the device only requires external environmental RF power supply and transmission control commands to return the stored data to the controller, enabling the monitoring of human health data. Additionally, because backscatter communication devices have no RF devices, the hardware circuit is simple, enabling further miniaturization and reducing the risk of immune rejection. The application of backscatter communication technology in the field of medical health is illustrated in Fig. 13.5.

5. Smart Home

Smart homes utilize residences as a platform, connecting various devices in the home through the Internet of Things to create an efficient and comfortable living system. They employ automatic control of home appliances, lighting, temperature, anti-theft and alarm systems, and other functions to make the home environment safer, more convenient, and comfortable. Sensors and small devices in smart homes can communicate using backscatter technology [5].

Backscatter communication can achieve battery-free operation, eliminating the need for charging, greatly extending the life of the devices in smart homes, and reducing maintenance costs. Additionally, due to the ultra-low cost, extremely small size, washability, and flexible/foldable form factors, it can be deployed very flexibly in smart homes, such as being embedded in walls, ceilings, and furniture, or attached to keys, passports, clothes, and shoes. Based on these advantages, backscatter communication can expand the application scenarios of smart homes and has great appeal

Fig. 13.6 Application of backscatter communication in smart homes

in the field of smart homes. The application of backscatter communication in smart homes is illustrated in Fig. 13.6.

Typical application scenarios of smart homes include the following categories:

(1) *Item finding*: Extremely small, washable, flexible, and foldable backscatter communication devices can be attached to items that are easy to lose, such as keys, passports, bank cards, wallets, etc. When these items need to be found, they can be quickly located, and the lost items can be found.
(2) *Environmental monitoring and alarms*: Backscatter communication devices integrated with sensors are used to monitor the temperature, humidity, etc. of the house and can also be used for alarms in emergency situations such as gas leakage. The battery-free feature of backscatter communication devices can greatly extend the usage time of the device and achieve maintenance-free operation.
(3) *Intelligent control*: Backscatter communication devices integrated with sensors can achieve intelligent control of household appliances, such as controlling the on and off of washing machines, air conditioners, televisions, curtains, etc., and can also provide more precise control by embedding/attaching labels on doors and furniture for home robots to navigate [6].

13.3 Fundamentals of Backscatter Communication

13.3.1 Backscatter Communication System Architecture

In general, a backscatter communication system consists of a backscatter communication transmitter and a backscatter communication receiver. The backscatter communication transmitter is commonly known as a backscatter communication device, which is a device that modulates the information to be transmitted on the RF signals transmitted by other network nodes. Therefore, backscatter communication literally uses the RF signals transmitted by other network nodes for signal modulation, regardless of whether its power supply comes from an energy harvesting circuit or battery power.

Figure 13.7a depicts the backscatter communication transmitter, and the following introduces its basic constituent modules and main functions:

(1) Antenna: This is used to receive RF signals and control commands, and it also transmits modulated backscatter signals.

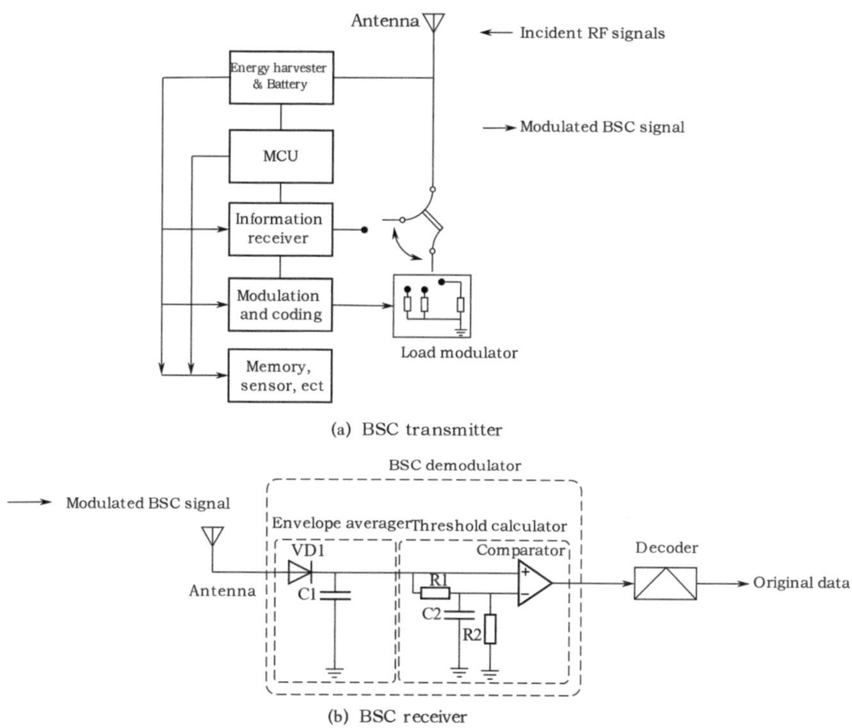

Fig. 13.7 Backscatter communication system architecture

(2) Energy harvester or battery module: The energy harvesting module is responsible for RF energy harvesting for the backscatter communication device, or other energy harvesting methods, including but not limited to solar energy, kinetic energy, mechanical energy, thermal energy, etc. Additionally, it may also include a battery power supply module, with which the backscatter communication device can be deemed as a semi-passive device. The energy harvesting module or power supply module powers all other modules in the device.
(3) Microcontroller (MCU): The MCU controls baseband signal processing, energy storage or data scheduling status, switching of circuits, system synchronization, etc.
(4) Signal receiving module: This is used for demodulating control commands or data sent by the backscatter communication receiver or other network nodes.
(5) Encoding and modulation module: Under the control of the controller, this module performs channel encoding and signal modulation, and selects different load impedances through switching of circuits to achieve modulation.
(6) Storage or sensor module: This is used for storing the identification (ID) information, location information, or sensor data of the device.

In addition to the above typical modules, future backscatter communication transmitters may even integrate reflection amplifier modules, low noise amplifier modules, etc., to enhance the receiver sensitivity and transmission power of the transmitter.

Figure 13.7b depicts the backscatter communication receiver, which in a traditional RFID system is the reader. The following introduces its basic modules and main functions:

(1) Antenna: This is used for transmitting RF signals or receiving modulated backscatter signals.
(2) Demodulation module: This includes envelope detection and threshold comparison for detecting the backscatter signal transmitted by the transmitter. It involves Amplitude Shift Keying (ASK) detection, Phase Shift Keying (PSK) detection, Frequency Shift Keying (FSK) detection, or Quadrature Amplitude Modulation (QAM) detection, etc. After detection, the threshold comparison module, composed of the mean circuit and comparator, outputs the demodulated signals.
(3) Decoding module: This module decodes the demodulated information to recover the original data.

13.3.2 Modulation Technique for Backscatter Communication Backward Link

The purpose of backscatter communication is to transmit signaling and data between backscatter communication devices and other network nodes. The link from the network node to the backscatter communication device is called the downlink or forward link. The network node can provide the target RF carrier, RF energy, or downlink data, or command, for the backscatter communication device through the forward link. The link from the backscatter communication device to the network node is called the uplink or backward link. In backward link, backscatter communication device uses the target RF carrier to modulate its own information for backscatter transmission. This section mainly introduces the modulation technique for backscatter communication backward link.

The modulation method is one of the key features that distinguish backscatter communication from traditional communication. The purpose of modulation is to transform the transmitted analog signal or digital signal into signals suitable for transmission. Digital baseband signals often have rich low-frequency energy, so it is necessary to modulate the carrier with the digital baseband signal, rather than directly transmit the digital baseband signal, to make the transmission signal match the characteristics of the wireless channel. In backscatter communication, the modulation method not only needs to match the forward channel or backward channel but also needs to consider the modulation characteristics based on load impedance switching, demodulation capability, energy storage capacity, clock synchronization capability, etc., in the backscatter communication device. Therefore, designing a modulation method suitable for the forward link and backward link of the backscatter communication system requires considering interference suppression characteristics, energy efficiency and spectral efficiency, demodulation gain, etc., as well as the ability to provide clock synchronization, the ability to provide energy, etc., which is different from other communication systems.

In backward link or uplink from the backscatter communication device to the backscatter communication receiver, considering the cost and energy consumption limitations of backscatter communication device, backscatter signals are usually modulated in low-order ASK, PSK, or FSK modulation. The principle of modulation is mainly achieved by switching impedance components, changing the amplitude, phase, or frequency of the incident RF signals. Figure 13.8 shows the principles of ASK, PSK, and FSK modulation [9]. For ASK modulation, at least two load impedances are required. Next, the working principle of 2ASK modulation, i.e., On–Off Keying (OOK) modulation, is explained as an example.

Assuming the antenna impedance is Z_A, the load impedance is Z_L, and the phase of the antenna impedance is equal to the phase of the load impedance, both are θ_A, then

$$Z_A = |Z_A|e^{j\theta_A} \tag{13.1}$$

$$Z_L = |Z_L|e^{j\theta_L} = |Z_L|e^{j\theta_A} \tag{13.2}$$

Then, assuming the reflection coefficient is Γ, the expressions for the magnitude of the reflection coefficient $|\Gamma|$ and phase θ can be given by

$$|\Gamma| = \left|\frac{Z_L - Z_A}{Z_L + Z_A}\right| = \frac{|Z_L|^2 + |Z_A|^2 - 2|Z_A||Z_L|\cos(\theta_A - \theta_L)}{|Z_L|^2 + |Z_A|^2 + 2|Z_A||Z_L|\cos(\theta_A - \theta_L)}$$

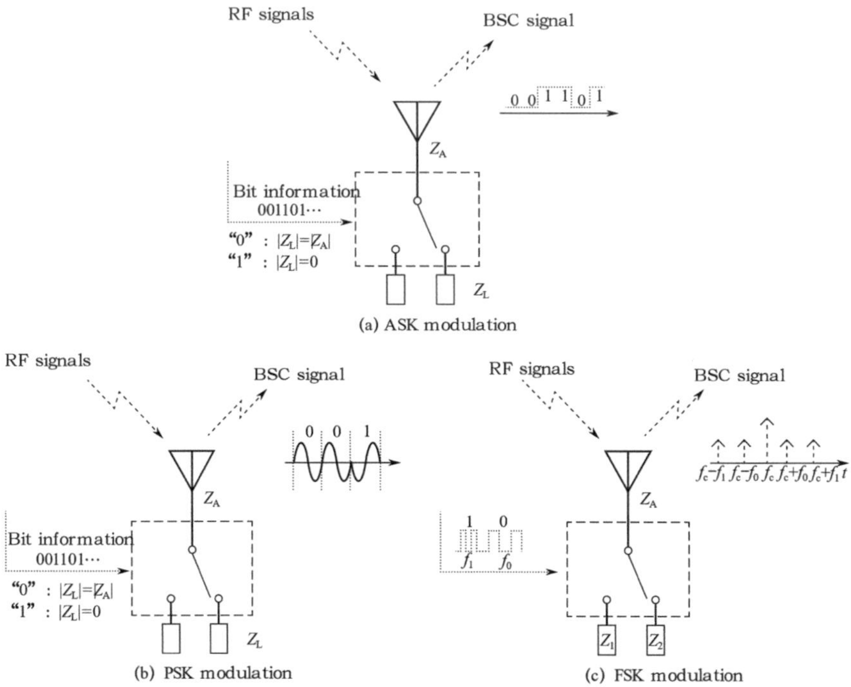

Fig. 13.8 Principles of ASK, PSK and FSK modulation in backscatter communication

$$= \frac{|Z_L|^2 + |Z_A|^2 - 2|Z_A||Z_L|}{|Z_L|^2 + |Z_A|^2 + 2|Z_A||Z_L|} \tag{13.3}$$

$$\theta = \arctan\left(\frac{2|Z_A||Z_L|\sin(\theta_A - \theta_L)}{|Z_A|^2 - |Z_L|^2}\right) = \arctan[\alpha \sin(\theta_A - \theta_L)] = 0 \tag{13.4}$$

where, α represents a constant.

According to Eq. (13.4), it can be seen that in order to make the phase of the reflection coefficient be 0, it is assumed that the phase of the load impedance is equal to the phase of the antenna impedance. If the backscatter communication device transmits a bit of “1”, the antenna should be connected with a load impedance to be 0, making the reflection coefficient be 1, and the signal is modulated by the backscatter communication device. If the backscatter communication device transmits a bit of “0”, the antenna should be connected with a load impedance equal to the antenna impedance, making the reflection coefficient be 0, and the backscatter communication device absorbs the incident carrier frequency signal.

The principle of PSK modulation is similar to that of ASK modulation, and PSK modulation can be achieved by selecting different load impedances. Taking Binary Phase Shift Keying (BPSK) modulation as an example, in order to ensure that the amplitude of the reflection coefficient does not affect the modulation performance of

BPSK, it is necessary to ensure that the phase of the load impedance is 90° different from the phase of the antenna impedance, so that the amplitude of the reflection coefficient is always 1, maximizing the power of the backscatter signal.

For FSK modulation, its modulation principle is different from the ASK modulation and PSK modulation. Taking 2FSK modulation as an example, if a backscatter communication device wants to implement 2FSK modulation, it needs to generate different frequency square wave signals of bit "1" and bit "0" through the MCU and control the on–off frequency of the RF switch associated with the load impedance. 2FSK can be implemented by a single load impedance or by two load impedances. If the impedance values of the two load impedances are at the open circuit point and short circuit point of the Smith chart, that is, the impedance values of the two load impedances are infinite and 0, the absolute value of the reflection coefficient can always be 1, ensuring the power of the backscatter signal is maximized [9]. Assuming the impedance value of Z_1 is infinite, the impedance value of Z_2 is 0, and Δf represents the frequency difference, then the reflection coefficient can be expressed as:

$$\Gamma(t) = \begin{cases} 1,\ Z_1 = \infty t \in \frac{n}{\Delta f}, \frac{2n+1}{2\Delta f}\Big) n \text{ is an integer} \\ -1,\ Z_2 = 0 t \in \frac{2n+1}{2\Delta f} \frac{n+1}{\Delta f}\Big) n \text{ is an integer} \end{cases} \tag{13.5}$$

Assume the incident signal is $S_{in} = \mathrm{Re}\left(e^{j2\pi f_c t}\right)$, after 2FSK modulation, the backscattered signal S_{bs} can be represented as

$$S_{bs} = \Gamma(t) \cdot S_{in} \tag{13.6}$$

Since the square wave function is a periodic function, it can be represented by a Fourier series. Taking the first harmonic component of the square wave function, Eq. (13.6) can be transformed into

$$\tilde{S}_{bs} = \frac{4}{\pi} \sin(2\pi f_c t) \cdot \sin(2\pi \Delta f t) = \frac{2}{\pi}\left\{\cos[2\pi (f_c - \Delta f)t\right] - \sin[2\pi (f_c + \Delta f)t]\right\} \tag{13.7}$$

According to Eq. (13.7), it is evident that the backscatter signal transforms into a double-sideband signal with a frequency offset of Δf.. Furthermore, if the RF signal source is a Long Range Radio (LoRa) signal, Chirp Spread Spectrum (CSS) modulation can be achieved through the linear change of the switching frequency, which can be controlled by the MCU or implemented by a voltage-controlled oscillator controlled by a Digital to Analog Converter (DAC). From an information-carrying perspective, CSS is an extension of FSK modulation.

ASK provides continuous transmission power, and the demodulation of the backscatter signal is relatively easy to implement. However, when using ASK, the backscatter signal is easily affected by interference and noise. Therefore, some literature proposes differential modulation technology to overcome the impact of interference and noise, but differential modulation has the disadvantage of a high Bit Error Rate (BER) [10]. FSK also has a strong adaptability to interference and noise, but the spectrum utilization of FSK is only half that of ASK.

13.3.3 Modulation Technique for Backscatter Communication Forward Link

In the forward link or downlink from other network nodes to backscatter communication device, consider employing joint coding and modulation to achieve low complexity synchronization and demodulation of backscatter communication device. The joint coding and modulation can be designed according to the following principles [11]:

(1) The signal after joint coding and modulation should be at the high level for as long time as possible and should not interrupt the energy supply to the backscatter communication device. This necessitates the baseband coding to have flip-flop characteristic between every two adjacent data bits. Such a coding scheme ensures continuous energy supply to the backscatter communication device even when "0" appears continuously, and also facilitates the extraction of clock information from the received code. Therefore, a coding scheme with flip-flop characteristic needs to be selected.
(2) To ensure the reliability of the forward link, the coding should provide error detection or verification protection, enabling the backscatter communication device to determine whether a bit error or signal conflicts have occurred.
(3) Since the backscatter communication device uses a low-cost and unstable crystal for synchronization, it is desirable to extract the clock from the code stream transmitted from the network node. Therefore, the signal after joint coding and modulation should facilitate the backscatter communication device in extracting clock information.
(4) During joint coding and modulation, a reasonable Signal-to-Noise Ratio (SNR) should be provided, and after modulation, a balance between spectral efficiency and energy efficiency should be achieved.

Considering the above joint coding and modulation principles, the coding method for the forward link can adopt Pulse Interval Encoding (PIE) or Manchester encoding, and the modulation method can adopt ASK modulation. Next, taking ISO/IEC 18000-6C as an example, the corresponding joint coding and modulation method is introduced, as shown in Table 13.1.

The modulation methods include ASK and its derived modulation schemes, such as Double Sideband Amplitude Shift Keying (DSB-ASK), Single Sideband Amplitude Shift Keying (SSB-ASK), and Phase Reversal Amplitude Shift Keying (PR-ASK). ASK modulation has relatively low spectral efficiency, and for a given data

Table 13.1 ISO/IEC 1800-6C Standard's three forward link coding and modulation methods

Technical features	Type A	Type B	Type C
Frequency	860–960 MHz	860–960 MHz	860–960 MHz
Modulation	ASK	ASK	DSB-ASK SSB-ASK PR-ASK
Coding	PIE	Manchester	PIE

rate, it occupies a relatively large bandwidth. Compared with SSB-ASK and PR-ASK, the bandwidth efficiency of DSB-ASK is the lowest, at only 0.2 bps/Hz. Under the same BER and 100% modulation depth, DSB-ASK and SSB-ASK require the minimum Carrier-to-Noise Ratio (CNR), but they also provide the smallest amount of RF energy to the backscatter communication device. PR-ASK can maximize the reduction of the carrier-to-noise ratio within a narrow band range, and consequently, the RF energy that can be provided to the backscatter communication device increases accordingly, making it more suitable for narrowband and long-distance transmission. Figure 13.9 shows the modulation principles of the above three modulation methods [12]. As can be seen from the figure, DSB-ASK modulation is the simplest of the three modulation methods, only needing to control the carrier on and off through the baseband signal, or only one of the In-Phase Quadrature (IQ) baseband signals is needed, and a DAC can drive the modulator. In the SSB-ASK modulation model, the IQ signals are distinguished by the Hilbert transformer, resulting in a 90° phase difference of the IQ signals, and the output is a locally oscillated suppressed single sideband RF signal. However, the Hilbert transformer increases the power consumption of the digital processor. Compared with SSB-ASK, the two-path IQ signal of PR-ASK does not differ, does not require a Hilbert transformer, and is relatively easy to implement. In summary, PR-ASK is more suitable for long-distance forward link transmission.

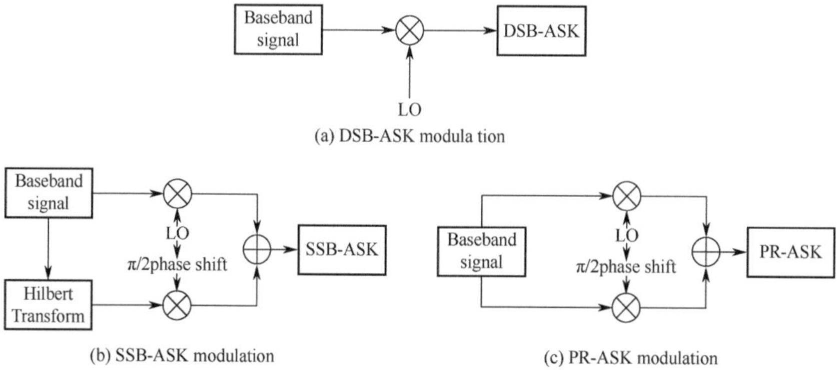

Fig. 13.9 DSB-ASK, SSB-ASK and PR-ASK modulation principles

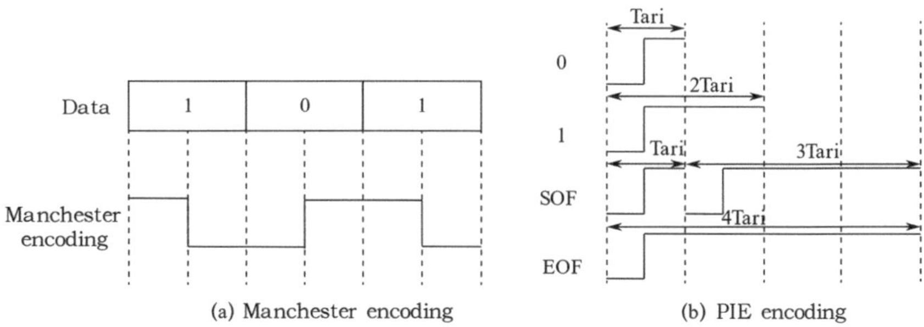

Fig. 13.10 Principle of Manchester encoding and PIE encoding

The main coding methods include Manchester encoding and PIE, etc., and the principle is shown in Fig. 13.10. The spectral efficiency of Manchester encoding is only 50%, using the phase of voltage jumps to distinguish between "0" and "1", where a negative jump from high to low represents "1", and a positive jump from low to high represents "0". The characteristics of Manchester encoding are as follows: since each jump occurs in the middle of a code element, the backscatter communication device can conveniently use Manchester encoding as a bit synchronous clock. The code elements of Manchester encoding are divided into two sub-code elements, the second sub-code element can be regarded as a check bit, thus providing a check function. Pulse width encoding is a coding method where "0" and "1" have different time intervals. Based on a continuous pulse with a fixed interval of Tari, the pulse repetition period varies according to "0" and "1". By defining different time widths between the falling edges of the pulses, 4 symbols are represented: 0, 1, frame start symbol (Start of Frame, SOF) and frame end symbol (End of Frame, EOF), their corresponding encoding durations are Tari and 2 Tari, followed by Tari with 3 Tari, 4 Tari. Here, Tari represents the time interval between pulses, indicating the width between two adjacent pulses of the symbol "0". The characteristics of pulse width encoding are as follows:

(1) The bit time of pulse width encoding can be flexibly changed;
(2) Because the encoding signal has a large number of pulse signals, errors in binary code data can be easily detected;
(3) When there is no information to transmit, there is a large amount of available idle time in the slot for power supply and calculation. Since pulse width encoding has more slots for power supply than other encodings, it can provide a higher energy conversion level for backscatter communication devices, thus supporting longer distance transmission coverage.

13.3.4 Energy Harvesting

Backscatter communication devices have the advantage of being able to operate without the need for traditional batteries or power sources. These devices require only a few hundred microwatts or even a few microwatts, which can be provided by energy harvester. Electromagnetic wave energy, solar energy, light energy, vibration energy, thermal energy, and other environmental sources can all be utilized as energy sources for backscatter communication devices. This section introduces the environmental energy harvesting technology of backscatter communication devices, with an example of harvesting electromagnetic wave energy from environmental RF signals.

Based on the principle of electromagnetic wave transmission, backscatter communication devices collect electromagnetic wave energy from environmental RF signals through an energy harvesting circuit, as shown in Fig. 13.11. The energy harvesting module mainly consists of a rectifying antenna and a boost regulation module. The rectifying antenna includes a receiving antenna, a matching network, and a rectifying circuit. Firstly, the receiving antenna collects electromagnetic wave energy from the environment and converts it into an Alternating Current (AC) signal. The matching network ensures impedance matching between the antenna and the rectifying circuit, maximizing the absorption of energy by the rectifying circuit, thereby improving the energy harvesting efficiency. The rectifying circuit then converts the high-frequency AC signal from the antenna into a Direct Current (DC) signal. Since the rectifier only outputs a weak forward voltage, which is insufficient to reach the startup voltage of the controller, a boost module is needed to increase the input voltage to reach the required startup voltage of the backscatter communication device. The regulation module mainly addresses the issue of unstable input signals affecting the working state of the backscatter communication device circuit. The receiving antenna, matching network, and rectifying circuit within the rectifying antenna are crucial components of the energy harvesting circuit, determining the overall energy conversion efficiency of the backscatter communication device.

To reduce the constraints on the number of receiving antennas and improve the energy harvesting efficiency of backscatter communication devices, while also

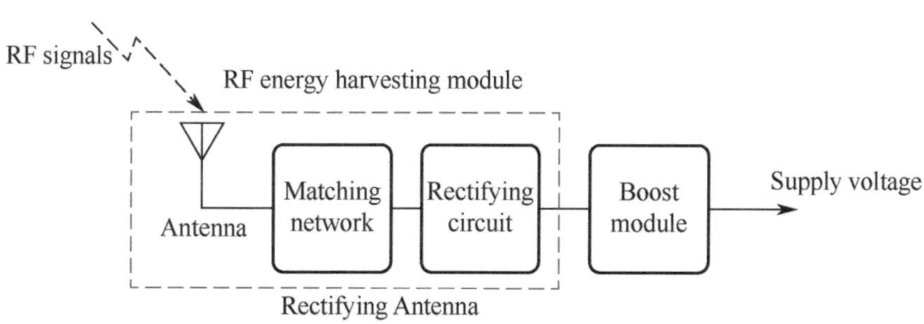

Fig. 13.11 Principle of RF energy harvesting of backscatter communication devices

addressing concerns related to energy consumption, hardware complexity and cost, there has been a rapid development in the field of miniaturized, highly integrated, wideband and multi-frequency antennas. The wider the bandwidth and the more frequency bands that the receiving antenna supports, the higher energy conversion efficiency of the energy harvester can be achieved. However, there is often an inherent trade-off between bandwidth and antenna gain, and the working frequency band of the antenna needs to be consistent with the matching network and rectification circuit, necessitating a comprehensive consideration of the expansion of the receiving antenna's bandwidth. The electrical parameters that measure the bandwidth of the antenna mainly include return loss, which is related to the input impedance of the antenna and the characteristic impedance of the transmission line. Additionally, since the polarization direction of the RF signal in the environment is uncertain, i.e., the RF signal may be a linearly polarized signal that is horizontally or vertically polarized, the polarization mode of the antenna also affects the efficiency of energy harvesting. Compared with linearly polarized antennas, circularly polarized antennas have better receiving sensitivity for energy harvesting and can receive any linearly polarized incoming wave signal. Therefore, the development of miniaturized, highly integrated, wideband, and multi-frequency antennas has the potential to significantly enhance the energy harvesting capabilities of backscatter communication devices across multiple communication scenarios.

During energy harvesting, it is essential to ensure that the impedance of the rectification circuit matches the impedance of the receiving antenna to maximize the absorption of energy. Additionally, using a matching network in power distribution can help reduce the amplitude and phase errors of the current [13]. The matching network typically consists of inductors and capacitors, and can have one or more components. Its forms include basic Γ shape, extended T shape, and Π shape, as illustrated in Fig. 13.12, where Z_A represents the impedance of the antenna, Z_L represents the load impedance, L represents the inductor, and C represents the capacitor. By taking into account the characteristics of the rectification circuit and the hardware capabilities of the backscatter communication device, selecting a suitable matching network can enhance the efficiency of energy harvesting and ensure stable performance. When designing a matching network, the inductive and capacitive characteristics of the matching network can be analyzed in conjunction with the Smith Chart, and the response frequency of the matching network can be adjusted by varying the capacitance or inductance value.

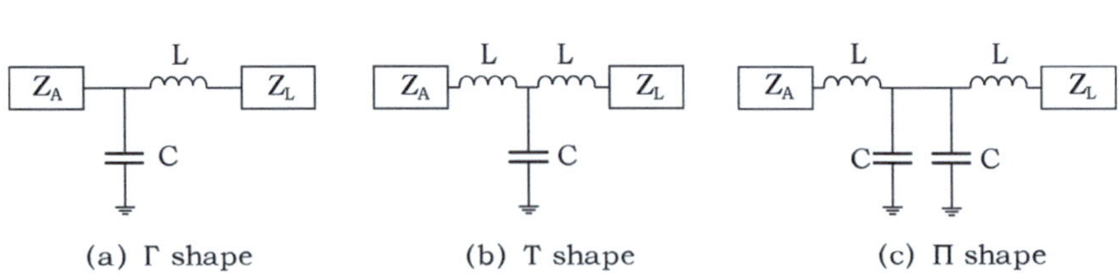

(a) Γ shape (b) T shape (c) Π shape

Fig. 13.12 Several forms of matching networks

The rectification circuit in backscatter communication devices contains non-linear devices, leading to the generation of a large number of harmonic signals when the environmental RF signal passes through the rectifier. This results in energy loss and affects the energy conversion efficiency. The rectification efficiency of the circuit is influenced by factors such as circuit topology, frequency and amplitude of the input RF signal, and output load [14]. There exists a fundamental trade-off between the rectification efficiency and the input power of RF signals. When the input power of RF signals is below − 10 dBm, the rectification efficiency decreases with decreasing power, and when the input power is less than − 30 dBm, the rectification efficiency is generally less than 10% [15]. When designing a rectification circuit, key design indicators to consider include energy conversion efficiency, stability time, bandwidth, and sensitivity. Energy conversion efficiency determines the ability of the rectification circuit to convert the power of RF signal into a DC level power, while stability time refers to the duration of the output voltage of the rectification circuit maintaining a relatively stable level. To improve rectification efficiency, researchers have explored the use of advanced diodes such as spin diodes, heterojunction reverse tunnel diodes, and Schottky diodes [15]. However, even with these high-efficiency diodes, there are significant limitations to improving rectification efficiency, particularly when backscatter communication devices operate at low input power levels. In recent years, researchers have designed rectifier circuits using Complementary Metal Oxide Semiconductor (CMOS) technology, achieving a rectification efficiency of 40% when the RF signal power is − 20 dBm [16]. This represents a significant advancement in improving rectification efficiency under low power conditions.

13.4 Backscatter Communication Air Interface Technology

Traditional backscatter communication has limitations when it comes to transmission distance. Cellular backscatter communication, however, extends the reach of backscatter communication from short-range local-area communication to wide-area cellular wireless network applications. This expansion provides customers with standardized and convenient cellular network functions, thereby stimulating new applications and market directions. To achieve the integration of backscatter communication with cellular networks, improvements are needed in its backward link data rate, coverage, connection, and synchronization, as well as other air interface technologies. The corresponding physical layer key technologies include channel modeling and capacity analysis, channel coding and decoding, data rate enhancement, coverage improvement, synchronization technology, multi-access technology, channel detection and estimation, interference cancellation, and Multi-Input Multi-Output (MIMO) technology, etc.

13.4.1 Backward Link Data Rate Improvement Technology

Due to the hardware limitations of energy storage capacity and modulation circuits in backscatter communication devices, they typically use low-order modulation. Additionally, due to the limited stability of the oscillator, the synchronization performance of these devices is not good, resulting in lower communication data rate. Therefore, one of the key challenges for practical implementation of backscatter communication is to further improve the backward link data rate and achieve data rate of kbps or even Mbps.

High-order modulation is an effective way to improve spectral efficiency. In recent years, high-order modulation, including QAM technology, has been successfully applied to backscatter communication [17]. For instance, the paper [18] demonstrates the use of a combination of transistors and power dividers in backscatter communication devices to achieve 16 QAM modulation. However, the power consumption of QAM modulation using Positive Intrinsic Negative (PIN) diodes is high. This issue can be effectively addressed by replacing the current-controlled PIN diode with a voltage-controlled transistor. To minimize the higher normalized power consumption caused by high-order QAM modulation, researchers have proposed an unequal error protection based coded modulation scheme that combines QAM with unequal error protection [19].

The principle of QAM modulation is illustrated in Fig. 13.13. The RF signal is split into two paths by a power equal division 3 dB Wilkinson power divider, corresponding to the I path and Q path in Fig. 13.13. The two branches have a phase difference of 45°, resulting in the phase of the reflected wave of each branch being 90° different from the phase of the reflected wave of the other branch [9]. The reflected wave b in Fig. 13.13 is the sum of the reflected waves of the I path and Q path. Combined with the properties of the Wilkinson power divider, the reflected wave b can be represented as

$$b = \frac{b_1}{\sqrt{2}} + \frac{b_2}{\sqrt{2}} = \frac{a_1 \Gamma_1 e^{j\frac{\pi}{2}}}{\sqrt{2}} + \frac{a_2 \Gamma_2}{\sqrt{2}} \tag{13.8}$$

where, Γ_1 and Γ_2 represent the reflection coefficients of the I path and Q path respectively, and the corresponding load impedances are Z_1 and Z_2, which can be specifically represented as

$$\Gamma_i = \frac{Z_i - Z_0}{Z_i + Z_0},\ i = 1, 2 \tag{13.9}$$

Therefore, the reflection coefficient of the backscatter communication device can be represented as

$$\Gamma = \frac{b}{a} = \frac{\Gamma_1 \frac{1}{\sqrt{2}}}{\sqrt{2}} e^{j\frac{\pi}{2}} + \frac{\Gamma_2 \frac{1}{\sqrt{2}}}{\sqrt{2}} = \frac{\Gamma_2}{2} + j\frac{\Gamma_1}{2} \tag{13.10}$$

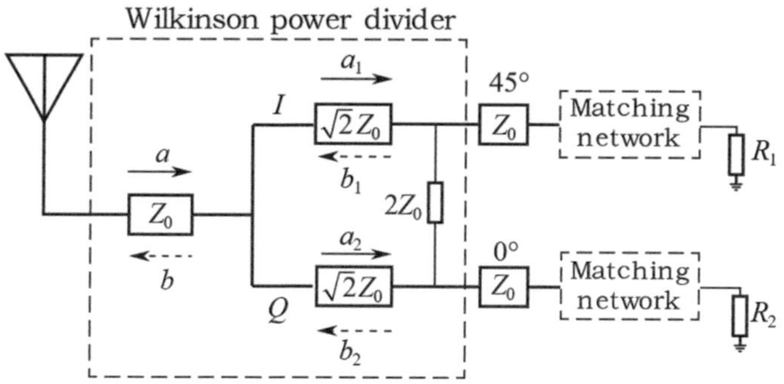

Fig. 13.13 Principle of QAM modulation

As seen in Eq. (13.10), to implement 4 QAM modulation in the backscatter communication device, it is necessary to connect two different load impedances after the power divider in order to generate four different reflection coefficients.

In addition to advanced modulation techniques, millimeter-wave communication, full duplex, MIMO, and other technologies have recently been applied in backscatter communication to increase data rate. A printed millimeter-wave antenna array for backscatter communication has been developed [3], as shown in Fig. 13.14. This array achieved a data rate of 2 Gbps at a cost of only 0.17 pJ per bit. The RF front end of the system comprises a microstrip patch antenna array and a single high-frequency single crystal transistor that supports advanced modulation. This setup reduces the volume and cost issues associated with configuring multiple stacked transistors in traditional millimeter-wave devices. Using a single crystal transistor for analog modulation compresses the front-end complexity into a single component, significantly reducing the required number of transistors and eliminating the need to add additional mixers in the millimeter-wave transceiver. Full duplex is also an effective method for increasing data rate for backscatter communication [20]. Researchers have designed a full duplex backscatter communication system [21] and have effectively addressed the in-band full duplex self-interference problem through spread spectrum and other methods.

13.4.2 Communication Coverage Enhancement Technology

Limited by the transmission power of network nodes, round-trip signal attenuation, energy conversion efficiency and energy storage capacity, the sensitivity of backscatter communication equipment, antenna gain and signal interference, both forward and backward link coverage of backscatter communication need to be further improved. Specifically, for the forward link from the network node to the backscatter communication device, the driving power consumption needs to be a few microwatts to tens of microwatts, and the allowable minimum signal reception strength or receiver sensitivity of the backscatter communication device receiving

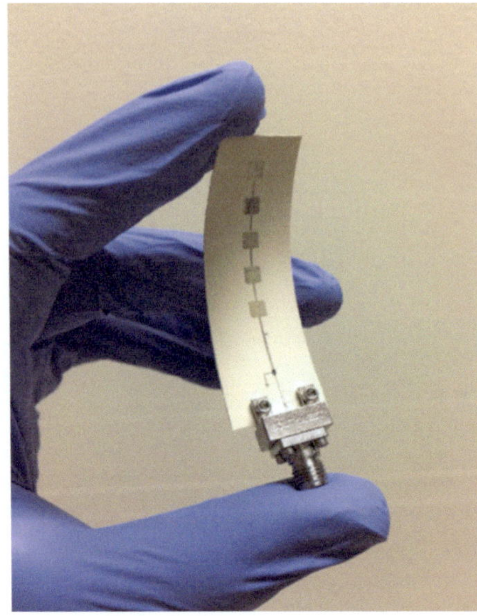

Fig. 13.14 Printed millimeter wave antenna array for backscatter communication [3]

the RF carrier signal is about −20 dBm, while the sensitivity of traditional terminal equipment receivers is about −100 dBm. If the backscatter communication device has energy storage capabilities, the receiver sensitivity can be relaxed to − 30 dBm. Additionally, limited by the energy conversion efficiency of the RF energy harvesting circuit, when the power of the input RF signal is below −23 dBm, it is difficult for the energy harvesting circuit to effectively collect the signal and rectify it into usable DC power. The lower the power of the input signal, the lower the energy conversion efficiency. In the backward link from the backscatter communication device to the network node, since part of the signal energy needs to be used for circuit power supply, the signal strength of the backscatter is 3–5 dB lower than the incident RF carrier signal strength. Additionally, the antenna gain of backscatter communication devices is generally not too large, at 0–2 dBi.

Table 13.2 presents the link budget for different frequency bands, transmission power, network node and device antenna gain, backscatter transmission loss, and the use of a Low Noise Amplifier (LNA) [5]. From this table, the following conclusions can be drawn as following.

(1) Limited by the receiver sensitivity of the backscatter communication device and the transmission loss of the backscatter communication, both the forward link and the backward link encounter weak signal coverage problems in certain scenarios, with the forward link becoming the bottleneck of network coverage.
(2) Enhancing the receiver sensitivity of the backscatter communication device can effectively improve the coverage of the forward link, but it will reduce the communication distance of the backward link.

Table 13.2 Backscatter communication link budget [5]

System parameters	Case 1	Case 2	Case 3	Case 4	Case 5	Case 6	Case 7
Frequency/GHz	2.4	0.7	0.7	0.7	0.7	0.7	0.7
Network node							
Transmission power/dBm	36	36	36	36	36	36	36
Antenna gain/dBi	8	8	8	8	8	8	8
Sensitivity/dBm	−100	−100	−100	−100	−100	−100	−100
Maximum backward link communication distance/m	176.89	606.48	191.78	1917.84	606.48	60.65	606.48
Zero−power tag							
Antenna gain/dBi	2	2	2	12	12	2	2
Sensitivity/dBm	−20	−20	−30	−20	−30	−40	−40
Maximum forward link communication distance/m	19.85	68.05	215.19	215.19	680.48	680.48	680.48
Backscatter transmission loss/dB	5	5	5	5	5	5	5
LNA/dB	0	0	0	0	0	0	20

Note The maximum backward link communication distance in the table represents the scenario in which the backscatter communication device is operating at the receiver sensitivity threshold. At this point, the received signal power is just enough to reach the start-up voltage of the energy harvesting circuit

(3) The integration of an LNA can effectively enhance the receiver sensitivity of backscatter communication devices and extend the communication distance of the backward link.
(4) Constrained by the maximum transmission power regulations of the unauthorized frequency band, the transmission power allowed in the authorized frequency band is at least 10 dB higher than that in the unauthorized frequency band. Hence, the authorized frequency band is more conducive to building a cellular network for backscatter communication.

The bistatic separation architecture is considered as one of the most effective ways to enhance backscatter communication coverage. In this architecture, the RF signal source or carrier generator is physically separated from the backscatter communication receiver, effectively avoiding the problem of round-trip signal attenuation and self-interference in monostatic backscatter communication. By separately placing the RF signal source and the backscatter communication receiver, or even deploying a dedicated RF signal source for RF power supply, the transmission coverage of backscatter communication can be effectively improved (FIg. 13.15).

Additionally, the use of beamforming technology to concentrate the energy of the RF carrier signal, in combination with high-efficiency energy harvesting circuits, can also effectively address the problem of limited coverage. By maximizing the energy collection of backscatter communication devices and implementing hybrid beamforming of the RF signal source along with passive beamforming in the backscatter

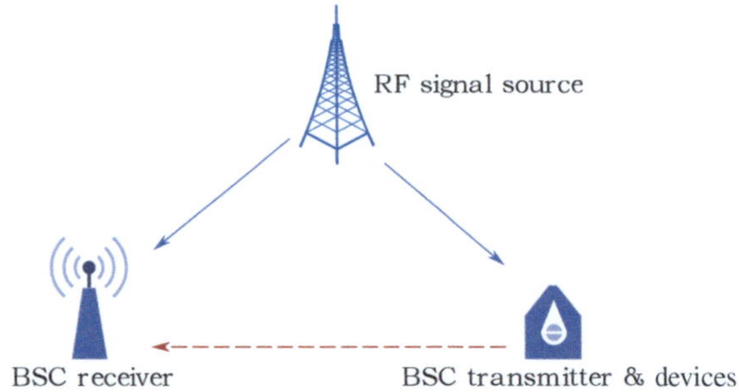

Fig. 13.15 Bistatic backscatter communication

communication device, a joint beamforming scheme can be formed to enhance the forward link coverage. However, it is more challenging to enhance the coverage of the backward link through MIMO precoding due to the hardware complexity and power consumption of backscatter communication devices. Nevertheless, the load impedance switching mode of the backscatter communication device can change the amplitude and phase of the signal, providing the possibility for the backscatter communication device to implement precoding technology by configuring multiple antennas. Each antenna of the backscatter communication device needs to select the corresponding load impedance connection according to the target direction of beamforming. It is important to note that while having more impedances can improve the accuracy of beamforming, excessive load impedances may increase the hardware complexity and power consumption of the backscatter communication device.

In addition, designing a new hardware architecture for backscatter communication devices is also an effective technique for increasing communication coverage. In recent years, researchers have explored the integration of tunnel diodes in backscatter communication devices to amplify the signal power of backscatter signals. Figure 13.16 depicts the tunnel diode and circuit schematic integrated in backscatter communication [25]. This hardware architecture design allows for a signal gain of 40 dB with only an additional 29 μW power consumption. Furthermore, researchers have been able to further improve the transmission coverage of backscatter communication by designing high-gain antennas for backscatter communication devices [26] and integrating LNAs into the system [27].

The impedance matching in backscatter communication is crucial for achieving the desired modulation depth of the backscatter signal. However, due to transmission line losses, insertion losses of the switch and other factors, it is often challenging to achieve the theoretical level of impedance matching, which can affect the modulation depth of the backscatter signal. This modulation depth, in turn, impacts the strength of the backscatter signal, ultimately affecting the data rate and detection performance of the signal. To address this problem, the precision of hardware design needs to be improved, and the impedance matching network should be optimized. Research has indicated that the use of negative impedance can amplify the incident signal. However, to achieve negative impedance, additional bias sources are required, such as tunnel diodes or CMOS.

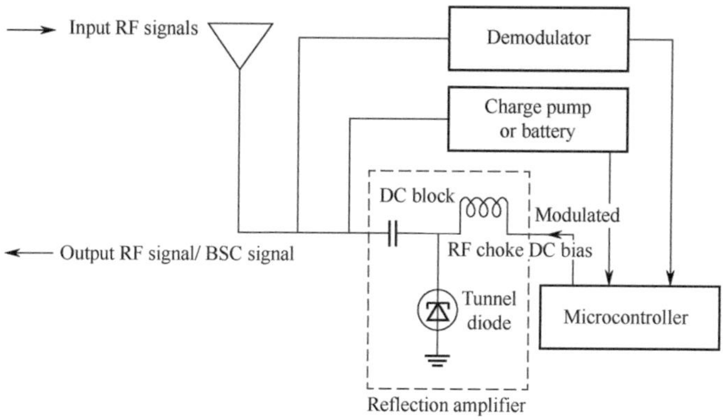

Fig. 13.16 Tunnel diode and circuit schematic integrated in backscatter communication

When representing negative impedance as $Z_L = -R_L + jX_L, R_L > 0$, where R_L represents the resistance of the negative impedance and X_L represents the reactance of the negative impedance, and the antenna impedance as $Z_A = -R_A + jX_A, R_A > 0$, where R_A represents the resistance of the antenna and X_A represents the reactance of the antenna, the definition of the reflection coefficient can be applied to address these challenges.

$$|\Gamma_i|^2 = \frac{(R_L + R_A)^2 + (X_L + X_A)^2}{(R_L - R_A)^2 + (X_L + X_A)^2} > 1 \tag{13.11}$$

Based on Eq. (13.11), it is evident that the reflection coefficient is always greater than 1, amplifying the incident signal.

13.4.3 Reliable Communication Transmission Technology

Due to the round-trip signal attenuation in backscatter communication and the impact of strong direct link interference or self-interference, there are significant technical challenges in achieving reliable transmission. Currently, the RFID system only supports four data rates, thereby limiting the system's ability to adapt to the link. Additionally, the Bi-Phase Space Coding (FM0), deformed bi-phase code (also known as Miller code), Manchester coding, etc., used only have error detection capabilities but no error correction capabilities, resulting in poor performance.

In dynamic wireless channel environments, using channel coding can effectively improve communication reliability and service quality. However, backscatter communication devices have high requirements for power consumption, cost and implementation complexity. Therefore, it is necessary to innovate existing channel coding schemes, design the best code structure and channel coding scheme for backscatter business characteristics, channel conditions, and usage scenarios. Table 13.3 presents several typical application scenarios and performance requirements [7], from which it can be seen that the packet size in general backscatter communication is usually a few dozen bytes, or even less than 100 bits. Most existing channel coding schemes are designed for long codes, and there are fewer designs for short codes (less than 200 bits). Although the channel coding scheme used by the current 5G data plane is not difficult to implement, there is still a 1–2 dB gap between the actual decoder and the finite length limit performance. Therefore, the coding and decoding technology for 6G backscatter communication needs to further narrow the performance gap, consider hardware capability constraints, simplify the complexity of coding and decoding, and improve the energy efficiency of coding and decoding. To reduce the 1 dB performance difference with the theoretical error correction limit of finite length coding, Polar Codes, short algebraic codes such as Reed-Muller Codes (RM Codes), Bose–Chaudhuri–Hocquenghem Codes (BCH Codes), and Reed-Solomon Codes (RS Codes) can be considered [29]. Among them, Polar codes are decoded by SCL decoders. In short code scenarios, by optimizing the code structure through list decoders, modifying the polarization kernel function to improve the polarization rate, and cascading the outer code structure, the best balance between performance and complexity can be achieved. Additionally, optimized algebraic codes can also be adapted to backscatter communication, with design principles including researching how to efficiently and systematically construct short channel codes to maintain good minimum code distance characteristics, researching code rate and code length adaptive methods for high-performance short codes to adapt to different packet sizes, and designing new decoding algorithms to achieve low-complexity decoding.

Table 13.3 Several typical application scenarios and performance requirements

Scenarios	Maximum allowable end-to-end delay	User experience data rate	Message size	Device density	Distance
Medical equipment inventory management	Hundreds of ms	2 kbps	176 bits	1000/km2	Indoor 50 m outdoor 200 m
Automobile manufacturing	>100 ms	<1 kbps	96 bits	1.5×106/km2	Indoor 30 m
Automated warehousing	1 s	100/128 bps	96/128 bits	—	Indoor 30 m
Smart agriculture	>1 s	<1 kbps	<1000 bits	1/m2	30–100 m
Smart home sensor	20 s	—	8~96 bits	–	Indoor 10–30 m

In digital communication systems, joint coding and modulation typically involves the optimization design between channel encoding and modulation to maximize the system's reliability and efficiency. FM0 code, Miller code, and Manchester code are typical codes with joint coding and modulation design, which aimed at achieving low cost and low power consumption of the device. In backscatter communication, the concept of joint design of coding and modulation can be extended by simultaneously designing channel codes with higher coding efficiency and signal waveforms to ensure low cost and low power consumption of the hardware. Recently, researchers have proposed a balanced grouping code based on FM0 [30], which optimizes the codeword structure and achieves 50% higher coding efficiency than FM0 while ensuring low implementation complexity. For bistatic backscatter communication, researchers have proposed a method called short group length cyclic channel coding [31], which can be achieved through simple shift registers, enabling long-distance transmission. Additionally, some scholars have proposed a low-power μ Code encoding technology [32] to enhance the transmission distance and parallel transmission of backscatter communication. Unlike traditional encoding that uses pseudorandom sequences to represent information bits, μ Code uses periodic signals to characterize information bits, allowing the receiver to demodulate and decode signals without the need for phase and carrier synchronization.

Space Time Block Code (STBC) is an effective and reliable transmission technique. By introducing signal redundancy in the spatial and temporal domains and rationally constructing block coding transmission matrices, diversity gain can be achieved without increasing bandwidth [33]. Specifically, STBC is divided into Orthogonal Space Time Block Code (O-STBC) [34] and Quasi-Orthogonal Space Time Block Code (QO-STBC) [35], among which Alamouti Space Time Block Code is a typical O-STBC. Traditional STBC is designed for active RF communication systems and do not consider the load modulation characteristics and implementation complexity of backscatter communication devices. Since backscatter communication device controls the amplitude or phase of the signal by changing the load impedance, considering the non-ideal factors of the modulation circuit, the amplitude or phase of the output signal may suffer errors. However, as long as these signal errors are within a distinguishable range, they have no impact on signal demodulation. Therefore, the less the load impedance that needs to be changed on each antenna, the higher fault tolerance, and the smaller the probability of error detection. To solve the above problems, some scholars have designed a new STBC structure by combining the characteristics of impedance matching modulation [36], namely the Hsinchin Space Time Block Code, which ensures diversity gain while reducing the number of impedance matching types on each antenna, thereby reducing the implementation complexity of the backscatter modulation module. Tables 13.4 and 13.5 compare the number of load impedance types required by these two types of STBC with two antennas. As can be seen from the two tables, based on the Alamouti Space Time Block Code, antenna 1 and antenna 2 require four types of load impedance $|\Gamma|e^{j\theta}$, $|\Gamma|e^{j(\theta+\pi)}$, $|\Gamma|e^{-j\theta}$ and $|\Gamma|e^{-j(\theta+\pi)}$. Based on the Hsinchin STBC, antenna 1 only needs two types of load impedances $|\Gamma|e^{j\theta}$ and $|\Gamma|e^{j(\theta+\pi)}$; antenna 2 also needs two types of load impedances $|\Gamma|e^{-j\theta}$ and $|\Gamma|e^{-j(\theta+\pi)}$ as well.

Table 13.4 Alamouti STBC two-antenna transmit diversity encoding table (BPSK modulation)

00	Antenna 1	Antenna 2	01	Antenna 1	Antenna 2
T	$\lvert\Gamma\rvert e^{-j\theta}$	$\lvert\Gamma\rvert e^{-j\theta}$	t	$\lvert\Gamma\rvert e^{-j\theta}$	$\lvert\Gamma\rvert e^{j(\theta+\pi)}$
$t+T$	$\lvert\Gamma\rvert e^{j(\theta+\pi)}$	$\lvert\Gamma\rvert e^{-j\theta}$	$t+T$	$\lvert\Gamma\rvert e^{-j\theta}$	$\lvert\Gamma\rvert e^{j(\theta+\pi)}$
10	Antenna 1	Antenna 2	11	Antenna 1	Antenna 2
T	$\lvert\Gamma\rvert e^{j(\theta+\pi)}$	$\lvert\Gamma\rvert e^{-j\theta}$	t	$\lvert\Gamma\rvert e^{j(\theta+\pi)}$	$\lvert\Gamma\rvert e^{j(\theta+\pi)}$
$t+T$	$\lvert\Gamma\rvert e^{j(\theta+\pi)}$	$\lvert\Gamma\rvert e^{j(\theta+\pi)}$	$t+T$	$\lvert\Gamma\rvert e^{-j\theta}$	$\lvert\Gamma\rvert e^{j(\theta+\pi)}$

Table 13.5 Hsinchin STBC two-antenna transmit diversity encoding table (BPSK modulation) [36]

00	Antenna 1	Antenna 2	01	Antenna 1	Antenna 2
T	$\lvert\Gamma\rvert e^{-j\theta}$	$\lvert\Gamma\rvert e^{-j\theta}$	t	$\lvert\Gamma\rvert e^{-j\theta}$	$\lvert\Gamma\rvert e^{j(\theta+\pi)}$
$t+T$	$\lvert\Gamma\rvert e^{-j\theta}$	$\lvert\Gamma\rvert e^{j(\theta+\pi)}$	$t+T$	$\lvert\Gamma\rvert e^{j(\theta+\pi)}$	$\lvert\Gamma\rvert e^{j(\theta+\pi)}$
10	Antenna 1	Antenna 2	11	Antenna 1	Antenna 2
t	$\lvert\Gamma\rvert e^{j(\theta+\pi)}$	$\lvert\Gamma\rvert e^{-j\theta}$	t	$\lvert\Gamma\rvert e^{j(\theta+\pi)}$	$\lvert\Gamma\rvert e^{j(\theta+\pi)}$
$t+T$	$\lvert\Gamma\rvert e^{-j\theta}$	$\lvert\Gamma\rvert e^{-j\theta}$	$t+T$	$\lvert\Gamma\rvert e^{j(\theta+\pi)}$	$\lvert\Gamma\rvert e^{-j\theta}$

Traditional STBC uses coherent detection for decoding, requiring the receiver to have knowledge of the Channel State Information (CSI) from the transmit antenna to the receive antenna. However, in backscatter communication, the power consumption and communication resources limit the feasibility of channel estimation through transmitting pilot signals, and the cost of estimating CSI without pilot signals at the receiver is even higher. Differential STBC provides a scheme where neither the transmitter nor the receiver needs to know the CSI information. The encoding and decoding processes are simple, and diversity gain can be achieved [37, 38]. Moreover, by combining the structure of differential STBC and Hsinchin STBCs, it is possible to reduce the number of load impedance types on each antenna while ensuring diversity gain without pilot signals. Additionally, traditional differential STBC is only suitable for constant modulus modulation, such as Multiple Phase Shift Keying (MPSK). For non-constant modulus modulation, such as Amplitude Phase Shift Keying (APSK) and QAM, the energy of each constellation symbol varies, making the traditional differential STBC encoding and decoding methods unusable. However, researchers have extended differential STBC to non-constant modulus modulation [39], considering the constellation shaping gain brought by non-constant modulus modulation compared to constant modulus modulation, and the difference in amplitude modulation and phase modulation capabilities based on impedance matching modulation circuits, thereby further improving the performance of the system.

13.4.4 Multi-access Technology

The future 6G network will need to support a massive number of passive IoT devices, represented by backscatter communication devices, to access the network. This will require a connection density which is 10 to 100 times higher than that of 5G. Traditional Orthogonal Multiple Access (OMA) methods, based on collision avoidance and scheduling, are challenging to apply to the large-scale access of backscatter communication devices. Therefore, new multiple access technologies must be designed to meet the connection needs of a massive number of backscatter communication devices.

OMA avoids interference through orthogonal resource allocation, effectively reducing the complexity of decoding at the communication receiver. However, it also limits the number of simultaneous connections allowed by the communication system. Considering the large-scale connection demand in the mMTC business scenario targeted by the backscatter communication system, Non-Orthogonal Multiple Access (NOMA) with non-orthogonal resource allocation has greater application potential. NOMA allows multiple connections to reuse the same resources in the time, frequency and code domains, and superimpose transmission signals. At the receiver, Successive Interference Cancellation (SIC) is used to decode the superimposed signals. Information theory has also proven that NOMA can achieve higher spectral efficiency than OMA [40].

Power Domain Non-Orthogonal Multiple Access (PD-NOMA) is one NOMA technology and has the potential to be used for backscatter communication systems. Its basic idea is to use the power differences of different backscatter communication devices at the receiving end to achieve superimposed transmission [41]. Figure 13.17 shows its working principle using a simple scenario with a single carrier and two backscatter communication devices as an example. Consider that both backscatter communication devices 1 and device 2 use constant envelope phase modulation, meaning the impedance used by the modulated signal has the same reflection coefficient but different phases, represented by ξ_1 and ξ_2. Additionally, the distances of the two backscatter communication devices from the RF signal source and the receiver are also different, represented by h_1 and h_2, to represent the channel gains of backscatter communication devices 1 and 2 to the receiver, while P_1^{in} and P_2^{in} represent the incident RF signal power, respectively. If $s_1(t)$ and $s_2(t)$ are used to represent the modulated signals of backscatter communication device 1 and device 2, then the receiver will receive the superimposed signal

$$y(t) = \xi_1 h_1 \sqrt{P_1^{in}} s_1(t) + \xi_2 h_2 \sqrt{P_2^{in}} s_2(t) + w(t) \tag{13.12}$$

where, $w(t)$ is Additive White Gaussian Noise (AWGN).

In order to recover the original signals transmitted by the two devices from the superimposed signal, the receiver needs to decode based on SIC, the receiver first sorts the Signal to Interference Noise Ratio (SINR) of the two signals; then decodes the signal with the higher SINR, at which point the signal with the lower SINR will

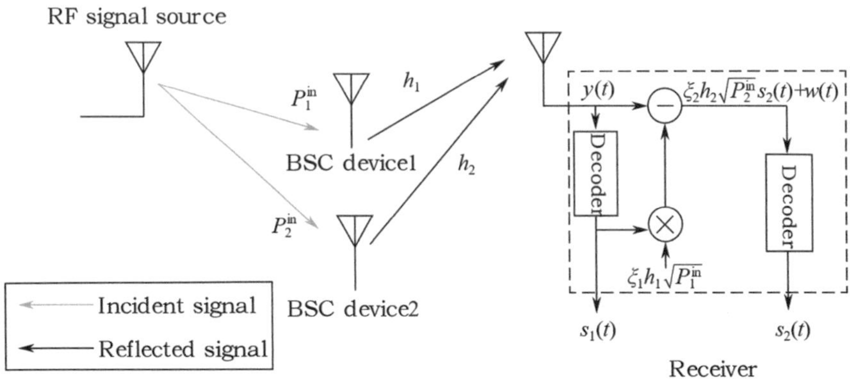

Fig. 13.17 PD-NOMA working principle

be regarded as noise; then eliminates the interference caused by this signal from the received signal; and finally decodes the signal with the lower SINR. In order to ensure that the SINR of the two signals can reach the minimum threshold Γ, the reflection coefficient can be adjusted by solving the following set of inequalities:

$$\begin{cases} \frac{(\xi_1 h_1)^2 P_1^{in}}{(\xi_2 h_2)^2 P_2^{in} + \sigma^2} > \Gamma \\ \frac{(\xi_2 h_2)^2 P_2^{in}}{\sigma^2} > \Gamma \end{cases} \tag{13.13}$$

Here it is assumed that $h_1\sqrt{P_1^{in}} > h_2\sqrt{P_2^{in}}$, and vice versa.

When extending PD-NOMA from two backscatter communication devices to any number of backscatter communication devices, some backscatter communication devices may not be able to achieve a sufficiently high SINR within the range of reflection coefficient adjustment. In such cases, these backscatter communication devices that cannot use PD-NOMA can access via Time Division Multiple Access (TDMA) [41]. Additionally, PD-NOMA can also be combined with Space Division Multiple Access (SDMA) to achieve higher spectral efficiency [42]. Specifically, all devices are first divided into multiple device groups according to their positions, and then the receiver equipped with multiple antennas uses beamforming to align each device group for directional reception. In other words, devices between groups use SDMA to reduce mutual interference, while devices within a group use PD-NOMA (Fig. 13.18).

In addition to the ultra-large number of connections, another important feature of backscatter communication services is short packet length, low transmission frequency, and sporadic spontaneous transmission, which is difficult to match with the current user access method based on network authorization. On the one hand, users need to interact with the network through a series of control signaling before transmitting data, obtain access authorization, and related resource configuration, and the signaling overhead caused may even be much higher than the actual data payload

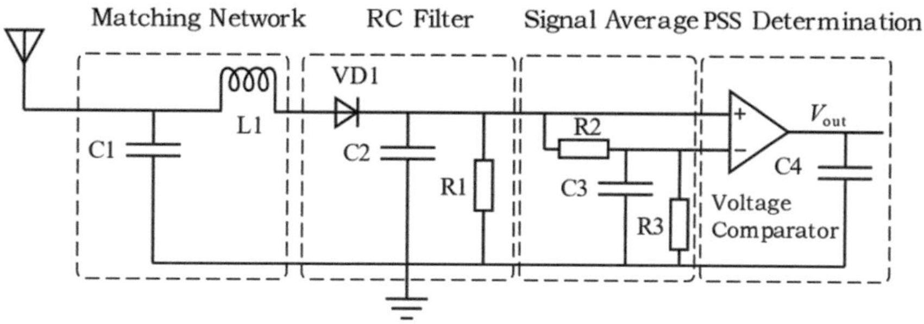

Fig. 13.18 Low-power synchronization circuit used for BSC devices

to be transmitted. On the other hand, although Semi-Persistent Scheduling (SPS) and Configured Grant (CG) can optimize and reduce signaling overhead for periodic services, they are difficult to fully match the service characteristics of backscatter communication. More importantly, the cost and power consumption of backscatter communication devices are extremely low, and it is difficult to implement the complex protocols and signaling processes required for user access based on network authorization. Therefore, the development of random access or grant free NOMA technology that can transmit data at any time without network authorization has important significance and value for backscatter communication systems.

However, grant free means that users can reuse the same resources to transmit information. Therefore, NOMA technology, which allows non-orthogonal use of resources, is the evolutionary basis of grant free random access technology. Compared with NOMA technology, grant free random access also needs to solve problems such as active user blind detection and user identification. One solution is to design a codebook containing many low-correlation sequences for users to choose as extension sequences, and the receiver determines the user and decodes the data based on the blind detection results of the sequence, such as the multi-user shared multiple access based on grant free random access [43, 44]. Additionally, some studies have proposed to use the sparse characteristics of active users to design algorithms based on compressed sensing, directly obtaining the number of active users, channel estimation and transmitted signals [45]. Currently, as an emerging access technology, grant free random access technology is still in its infancy, and technologies suitable for backscatter communication systems are worth further exploration and development.

13.4.5 Synchronization Technology

Synchronization is a fundamental requirement for a communication system to achieve multiple access, data transmission and scheduling. However, the built-in crystal oscillator in the backscatter communication devices is often limited by cost and have inaccurate sampling time and time drift, which significantly impacts the performance and network capacity of backscatter communication systems. Mainstream crystal oscillators are typically divided into two types: passive crystal oscillators and active crystal oscillators. Passive crystal oscillators work through the built-in oscillator in the MCU chip, with a frequency generally ranging from 1 Hz to 1 MHz. The signal quality and accuracy of the crystal oscillator are not good, and it needs to be precisely matched with peripheral circuits such as inductance, capacitance and resistance to maintain stability. On the other hand, active crystal oscillators provide nominal frequencies from tens of megahertz to hundreds of megahertz through the crystal oscillator and can offer high-precision frequency reference, but this comes at the cost of increased power consumption, cost and volume. Currently, most backscatter communication devices use passive crystal oscillators due to cost and power consumption considerations. The challenge lies in solving the synchronization problem caused by poor crystal oscillator stability without increasing the cost and power consumption of backscatter communication devices. In practical terms, this is a problem that urgently needs to be addressed for backscatter communication to become feasible.

In some systems, such as Long Term Evolution (LTE) and New Radio (NR) systems, gNB and UE synchronization is achieved by periodically transmitting synchronization signals with fixed transmission periods and signal formats. In backscatter communication systems that use LTE signals as RF signals, researchers have utilized the fixed transmission period and signal format characteristics of the LTE signal's Primary Synchronization Signal (PSS). Special low-power circuits have been designed to detect LTE signals and synchronize with them. The low-power synchronization circuit is used to maximize the incident signal power through an impedance matching network, employ a Resistor-Capacitance (RC) filter for envelope detection, average the RF filtered signal using a signal averaging circuit, and utilize a voltage comparator to determine the arrival of the PSS signal. By incorporating an additional low-power synchronization circuit in the device, the timing error in backscatter communication can be reduced to less than 40 μs with a 90% probability, meeting the requirement of a hundred kbps data rate. This approach addresses the synchronization challenges in backscatter communication systems.

Additionally, to achieve high-precision synchronization performance while ensuring low complexity, one can consider the design of a preamble sequence for forward and backward link synchronization, instead of solely relying on adding an extra dedicated synchronization circuit. Furthermore, for the cost-, volume-, and power consumption-constrained backscatter communication devices, reducing the device's sensitivity to synchronization accuracy can be achieved through an asynchronous transmission mechanism. This involves utilizing the spatio-temporal Poisson model of the device in the network, developing a rational decentralized and asynchronous transmission method, and optimizing the transmission slot length and energy storage time to enable asynchronous transmission of multiple devices [47].

13.4.6 Interference Cancellation Technology

In monostatic backscatter communication systems and bistatic backscatter communication systems, the received signal comprises a superposition of useful backscatter signal and leaked self-interference signal or direct link interference signal. The strength of the self-interference signal and direct link interference signal may be much greater than that of the backscatter signals. As a result, the cancellation of strong direct link interference signal and self-interference signal is a technical prerequisite for achieving improved data rate, coverage, reliable transmission, and large-scale multi-access in backscatter communication system. The schematic illustrating strong signal interference in backscatter communication is depicted in Fig. 13.19.

In monostatic backscatter communication, the reader emits RF carrier signal to supply energy and target carriers for the backscatter communication device, while simultaneously receiving the backscatter signals transmitted by the backscatter communication device. This half-duplex working mechanism results in carrier leakage at the front end of the reader, which is referred to as the self-interference signal. The self-interference signal strength is significantly greater than the backscatter signal. Specifically, three reasons contribute to the self-interference signal caused by carrier leakage:

(1) Limited isolation between transmission and reception, leading to carrier leakage from the transmitting front end to the receiving front end.
(2) Mismatch of the backscatter device antenna, causing the carrier signal to reflect to the receiving front end.
(3) Reflection of the carrier signal by the environment, re-entering the receiving antenna.

To mitigate the self-interference signal caused by carrier leakage, the reader transmission and reception channels can be isolated in monostatic backscatter communication receiver. This can be achieved by adopting a dual antenna structure with transmission and reception antenna isolation, using a multi-port circulator, or employing a directional coupler, among other methods [48].

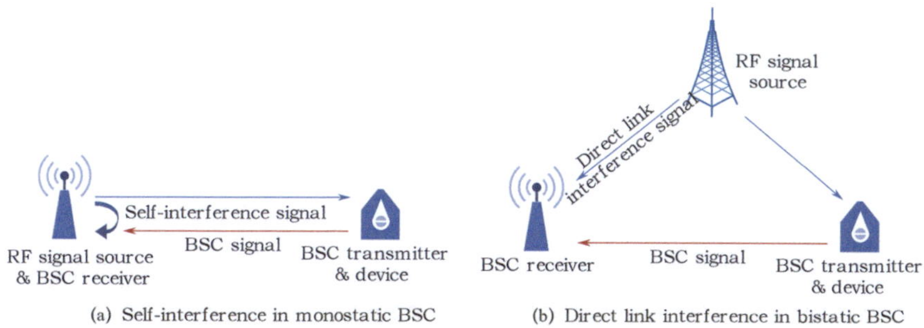

Fig. 13.19 Schematic of strong signal interference in backscatter communication

The use of transmission and reception antenna isolation is the most effective among the isolation methods, reducing carrier leakage through 25–30 dB isolation between the receiver and transmitter antennas. As shown in Fig. 13.20a, assuming the power of the reader RF signal is 36 dBm, and the strength of the useful backscatter signal received by the receiving antenna is −55 dBm, with 28 dB isolation between the transmitting and receiving antennas and an input reflection coefficient of −15 dB, the signal received by the receiver after the band-pass filter includes an 8 dBm self-interference signal generated by carrier leakage and a −55 dBm useful backscatter signal. The advantage of the dual antenna is its good isolation, but the disadvantage is that the system requires two antennas, which increases the cost and complexity of implementation.

A circulator is a multi-port device that utilizes the different magnetic conductivities of ferrite in a constant electric field to select the conduction port and isolate the transmission and reception channels. As depicted in Fig. 13.20b, assuming the power of the reader's RF signal is 36 dBm, and the strength of the useful scattering signal received by the receiving antenna is −55 dBm. At this point, the 36 dBm output signal of the transmitter reaches the antenna radiation power of 35 dBm after 1 dB attenuation by the circulator. Assuming the antenna input reflection coefficient is −15 dB, the strength of the carrier leakage signal reflected by the antenna is 20 dBm. This signal reaches the receiver with a strength of 19 dBm after another 1 dB attenuation by the circulator. Additionally, the transmitter's signal can also reach the receiver through the isolation port of the circulator, with an isolation leakage signal strength of 16 dBm. Since the strength of the isolation leakage signal is less than the strength of the carrier leakage caused by antenna mismatch, the strength of the self-interference signal of the receiver's carrier leakage is 19 dBm. Simultaneously, the −55 dBm reverse scattering signal received by the receiving antenna reaches the receiver with a signal strength of −56 dBm after 1 dB attenuation by the circulator. The advantage of using a circulator as an isolator is that only a single antenna is needed for the transmitting and receiving antennas, but the disadvantage is that the size is larger, and the cost is higher.

A directional coupler is another commonly used isolation component, which achieves direct, coupled, or isolated connections between different ports through power coupling or distribution components with directionality. As depicted in Fig. 13.20c, assuming the power of the reader's RF signal is 36 dBm, and the strength of the useful reverse scattering signal received by the receiving antenna is − 55 dBm. The input reflection coefficient of the antenna is −15 dB. Assuming that the loss from the output to the input of this directional coupler is 0 dB, the isolation from the output to the coupling end is 30 dB, and the coupling from the input to the coupling end is 10 dB, then the carrier signal strength of the antenna at this time is 30 dBm, the strength of the carrier leakage signal reflected by the antenna is 21 dBm, and the strength of the carrier leakage signal coupled to the receiver is 11 dBm. Additionally, the strength of the signal directly coupled from the transmitter's carrier to the receiver is 6 dBm. Since the strength of the isolation leakage signal is less than the coupling leakage of the antenna reflection, the strength of the self-interference signal of the receiver's carrier leakage is 11 dBm, and the strength of the useful backscatter

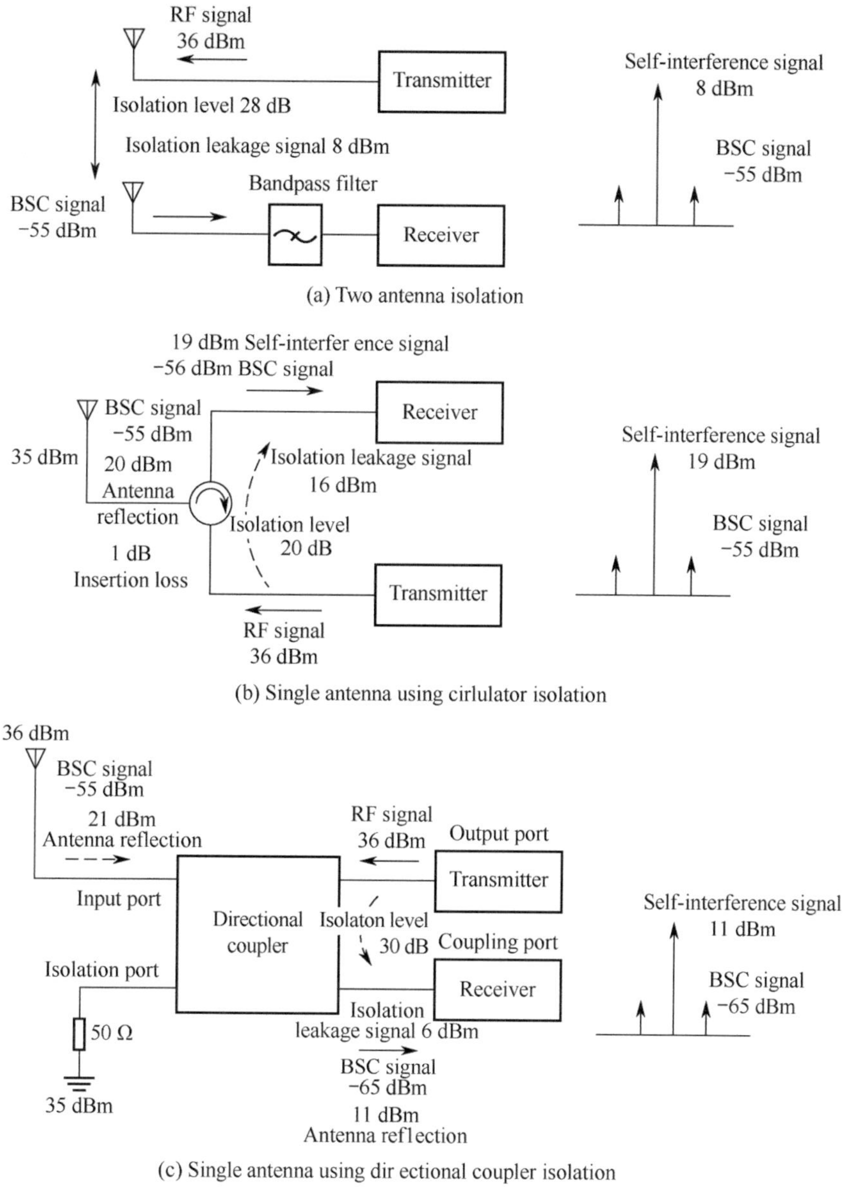

Fig. 13.20 Isolation methods in reader

signal is − 65 dBm. The isolation effect of the coupler is similar to that of the circulator, but it is smaller in size and lower in cost. For the already leaked carrier, carrier cancellation technology or self-interference cancellation technology can be further used to eliminate carrier leakage, thereby improving the sensitivity of the receiver.

According to the circuit structure, mainstream carrier leakage cancellation methods can be divided into three major categories: the receiving dual-path cancellation method, negative feedback loop method, and dead zone amplifier cancellation method [48]. The receiving dual-path cancellation method for carrier leakage cancellation operates by setting up two RF paths at the receiving front end, one is a linear RF path, and the other is a non-linear amplitude-limited RF path. The signals on the two paths are subtracted to retain the useful backscatter signal and cancel the leaked self-interference signal. The negative feedback loop method for carrier leakage cancellation is the most successful method in the market. The typical negative feedback loop circuit generally consists of four circuit modules: carrier cancellation reference source, phase and amplitude adjustment, detection circuit, and control unit, as shown in Fig. 13.21. The carrier cancellation reference source is typically obtained from the transmitter output through methods such as directional couplers. The detection circuit is used to detect the residual carrier leakage signal, and the control unit adjusts the amplitude and phase of the carrier cancellation reference signal according to the output of the detection circuit. This adjustment makes it the same as the carrier leakage signal amplitude and opposite in phase, ultimately eliminating the carrier leakage signal through vector addition. Depending on the different implementation methods of phase and amplitude circuits, the negative feedback loop method can be further subdivided into the I/Q orthogonal cancellation negative feedback loop method and the amplification and phase shift cancellation negative feedback loop method. The dead zone amplifier cancellation method utilizes the characteristics of the dead zone amplifier to attenuate the carrier leakage signal in the dead zone and amplify the useful signal in the envelope, thereby achieving the purpose of suppressing the carrier leakage signal and amplifying the useful backscatter signal.

In bistatic backscatter communication, unlike the self-interference signal in monostatic backscatter communication, the interference signal is the direct link interference signal from the RF signal source to the backscatter communication receiver.

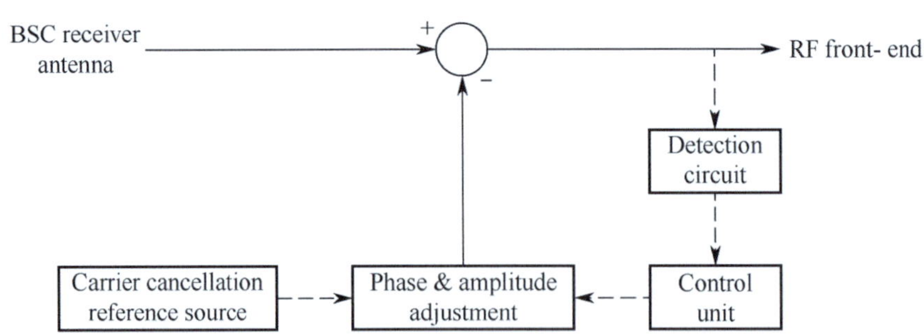

Fig. 13.21 Negative feedback loop method for carrier leakage cancellation

This direct link interference signal may be a modulated signal, and the backscatter communication receiver generally does not know the modulation characteristics of the direct link signal, making the challenge of eliminating direct link interference signal greater. To effectively eliminate the strong direct link interference signal from the RF signal source, researchers have based their efforts on the time domain structure and frequency domain structure characteristics of the RF carrier signal and the joint design of the baseband signal in backscatter communication. Through this approach, the receiver can effectively eliminate the strong direct link interference signal. For instance, when the RF carrier signal is the Orthogonal Frequency Division Multiplexing (OFDM) signal waveform widely used in LTE and NR systems, researchers have found that they can effectively eliminate strong direct link interference signal when the channel delay does not exceed the Cyclic Prefix (CP) length based on the repeated structure characteristics of the CP in the OFDM signal. This is achieved by jointly designing the differential class baseband modulation signal in the backscatter communication device [49]. In addition to utilizing the repetitive structure in the time domain of OFDM, interference signal can also be eliminated by using the guard band in the frequency domain of OFDM. This is done by modulating the signal through moving the equivalent frequency of the baseband signal to different guard bands [50]. The same design concept can also be extended to unmodulated single sine wave RF signals and differential chaos modulated signals [51, 52].

13.5 Backscatter Communication Protocol Stack and Network Architecture Design

During the communication process, backscatter communication performance is constrained by the power of the incident RF signal, memory size and RF hardware capabilities. Therefore, it is necessary to design a lightweight and minimalist protocol stack to reduce power consumption, signaling overhead and computational complexity during protocol stack processing. Furthermore, despite the limited software and hardware capabilities of backscatter communication devices, the needs for reliable access and secure data transmission still remain critical. Therefore, it is necessary to study low-cost secure access schemes and secure transmission mechanisms that align with the capabilities of backscatter communication devices. Additionally, as backscatter communication supports monostatic backscatter communication architecture, bistatic backscatter communication architecture and ambient backscatter communication architecture, flexible network deployment is needed to meet different requirements.

13.5.1 Lightweight Protocol Stack Design

When backscatter devices access the cellular network, the network can provide secure data reading and transmission functions for these devices. It can select different data transmission channels based on the specific needs of the backscatter devices, effectively utilizing network resources. For instance, for small packet data services, a data transmission method optimized for the control plane can be used to conserve user plane resources. The core network of the cellular network, which backscatter devices access, can store the context of these devices, manage their mobility and achieve real-time location tracking.

Due to the low cost of backscatter devices, it is essential to design a protocol stack with lower complexity for them. A new air interface protocol stack, called IoT Access Network Protocol, can be developed for communication between backscatter devices and normal UE. Rather than having a separate protocol stack between backscatter devices and the core network, normal UE can act as proxies to facilitate the interaction between backscatter devices and the core network.

Figure 13.22 depicts a protocol stack illustrating a terminal device acting as a proxy for a backscatter device to access the core network. Once the UE reads the identity of the backscatter device through the IoT Access Network Protocol, it registers the backscatter device in the core network using the Non-Access Stratum (NAS) protocol stack.

In Fig. 13.23, the protocol stack illustrates a terminal device acting as a proxy for a backscatter device to transmit data. After reading the data of the backscatter device through the IoT Access Network Protocol, the UE transmits it through the session channel of the UE. For small packet IoT data, the UE can establish a control plane optimized session channel to carry the data of the backscatter device to the core network through NAS messages.

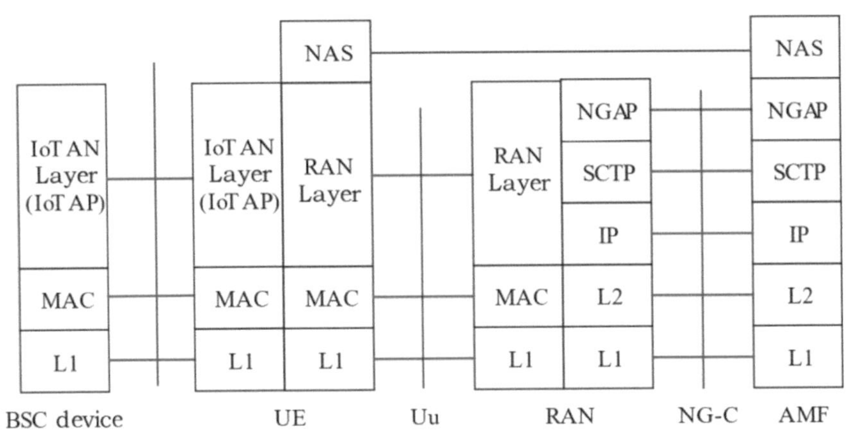

Fig. 13.22 Protocol stack for UE acting as a proxy for backscatter device to access the core network

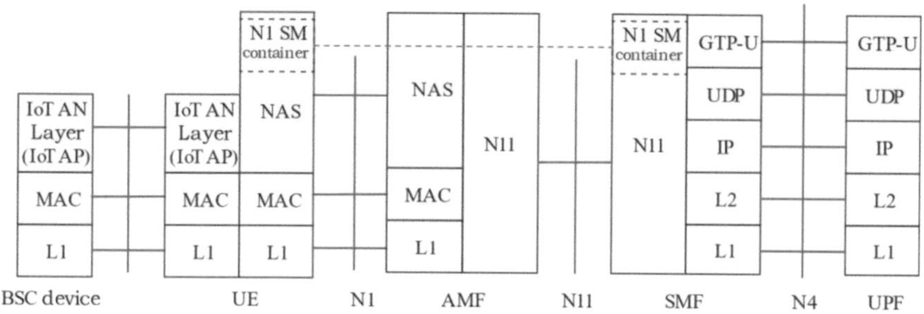

Fig. 13.23 Protocol stack for UE acting as a proxy for backscatter device to transmit data

13.5.2 Minimalist Network Architecture

Cellular-based backscatter communication helps to extend backscatter communication from traditional short-range applications to wide-area cellular wireless network applications, providing customers with convenient and standardized cellular network functions and stimulating new applications and market directions. Depending on the differences in the uplink/backward link and downlink/forward link of the backscatter communication device, coexistence or integration of backscatter communication and cellular communication can be implemented by the three modes shown in Fig. 13.24 [53]. Mode 1 is the base station direct connection mode, where the backscatter device directly and bidirectionally communications with a base station. This deployment architecture requires high reception sensitivity for both the base station and the backscatter device, but the network deployment is simple. The communication between the base station and backscatter device includes IoT data or signaling. Mode 2 is the UE/relay assisted mode, where the backscatter device transmits data/signaling to a base station, and receives data/signaling from the assisting node; or the backscatter device receives data/signaling from a base station and transmits data/signaling to the assisting node. In this mode, the assisting node can be a relay, an integrated access and backhaul (IAB) node, or a repeater. Mode 3 is the UE direct connection mode, where the backscatter device communication bidirectionally with a UE. The communication between UE and backscatter device includes IoT data and/or signaling. This type of deployment scenario is mainly for scenarios without base station coverage or scenarios that require local storage and processing for data privacy and security. Table 13.6 summarizes the self-interference cancellation capability requirements and downlink reception sensitivity requirements of different sub-mode devices.

In practical deployment, the power supply device, control device, target carrier or excitation signal providing device, and data transmission and reception device of the backscatter device can be the same device or different devices. The three modes shown in Fig. 13.24 can be further divided into 6 sub-modes to achieve flexible network deployment. It is important to note that Table 13.6 assumes that the device providing energy for the backscatter device and the device providing the carrier signal

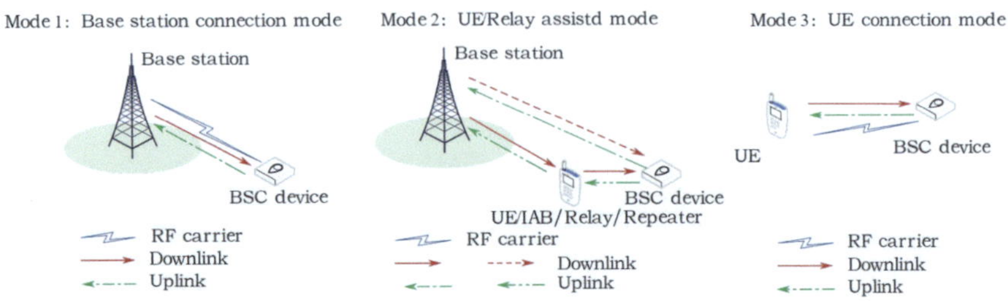

Fig. 13.24 Three connectivity topologies for cellular-based backscatter communication

Table 13.6 Self-interference cancellation capability requirements and receiver sensitivity requirements of three connectivity topologies

Mode	Downlink	Uplink	Energy source	Self-interference cancellation capability	Receive sensitivity requirements
1–1	BS→BSC Device	BSC Device→BS	BS	Low	High
2–1	BS→BSC Device	BSC Device→UE→BS	BS	Low	High
2–2	BS→BSC Device	BSC Device→UE→BS	UE	High	High
2–3	BS→UE→BSC Device	BSC Device→BS	BS	Low	Low
2–4	BS→UE→BSC Device	BSC Device→BS	UE	Low	Low
2–5	BS→UE→BSC Device	BSC Device→UE→BS	UE	High	Low
2–6	BS→UE→BSC Device	BSC Device→UE→BS	BS	Low	Low
3–1	UE→BSC Device	BSC Device→UE	UE	High	Low

source are the same device. This device can be a base station, terminal, or relay, or other independent network elements or nodes. Additionally, the backscatter device can also obtain energy from multiple different types of devices to improve energy collection efficiency. Similarly, the device receiving the backscatter signal can be a base station, terminal, or relay, or other independent network elements or nodes. The backscatter signal can be jointly received by multiple devices.

The cellular backscatter communication has the following three possible network architectures, in which the IoT devices are the backscatter devices.

Architecture 1: Transparent architecture for mobile network operators is shown in Fig. 13.25. In this architecture, the UE proxies the data transmission of the IoT device, and the IoT device interacts with the UE using unlicensed spectrum. The

Fig. 13.25 Architecture 1: Transparent architecture for mobile network operators

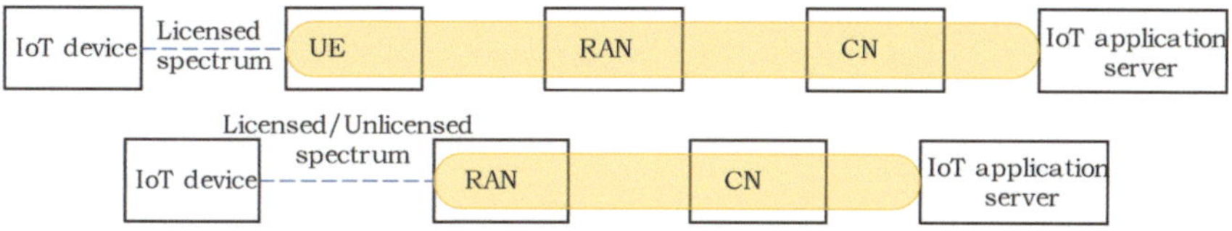

Fig. 13.26 Architecture 2: Mobile operator provides the receiver, IoT devices do not access the core network

mobile network does not need to authorize the UE as a receiver and charges the UE according to the existing billing method. Terminal users, IoT device manufacturers, and IoT service providers form a business ecosystem together.

Architecture 2: In this architecture, mobile operators provide receivers, and IoT devices do not access the core network, as shown in Fig. 13.26. The mobile operator provides UEs or access network devices as receivers and proxies the data transmission of the IoT devices. Here, the IoT device interacts with the UE using licensed spectrum, and the IoT device interacts with the access network device using licensed or unlicensed spectrum. The mobile network needs to authorize the UE or configure the access network to use licensed spectrum and support receiver functions. The mobile operator charges the IoT service provider, and the billing method can be based on the number of reads and writes to the IoT device or a monthly subscription.

Architecture 3: The mobile operator provides the receiver, and proxy IoT devices access the mobile core network, as shown in Fig. 13.27. This architecture inherits the functions of Architecture 2, such as the use of licensed spectrum by the mobile network, and the authorization of receiver functions, etc. On this basis, the mobile network can also provide location tracking for IoT devices, remote access to IoT devices by other UEs, differential services for Quality of Service (QoS) according to the business needs of IoT devices, identification and management of IoT devices, etc.

Although flexible network deployment can adapt to different backscatter communication needs, complex network architectures also bring new challenges to the operation, power consumption and cost of backscatter communication devices. Therefore, it is necessary to design simplified network architectures suitable for backscatter communication devices. Specifically, the design can be carried out from the following aspects:

(1) *Mobility management*: For static and semi-static backscatter communication devices, such as those used for warehouse inventory, sensors, monitors and other types of backscatter devices, the network does not need to obtain or track their location in real-time. For mobile backscatter devices, such as those

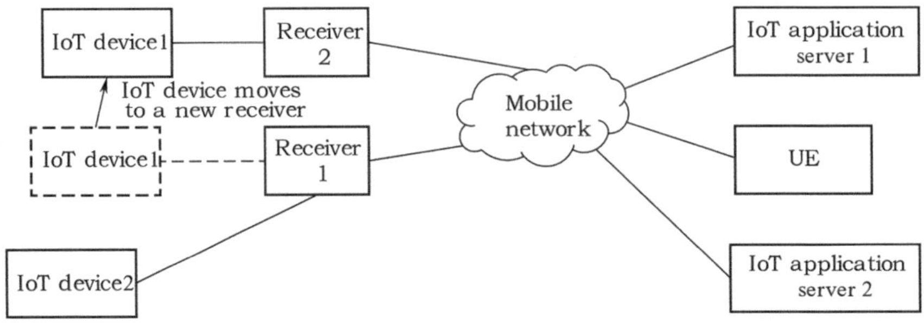

Fig. 13.27 Architecture 3: Mobile operator provides the receiver, proxy IoT devices access the mobile core network

used for real-time monitoring IoT devices (location tracking), the network needs to know their location in real-time. The network can determine the appropriate mobility management mechanism according to the use scenario of the backscatter device. The status management of backscatter devices can be simplified, such as merging the registration status and connection status.

(2) *Security*: Bidirectional authentication between backscatter devices and mobile networks, ensuring that readers securely obtain the identity information and data of backscatter devices, and avoid the risk of transmitting data to illegal readers.
(3) *Session management*: The receiver obtains the session channel establishment parameters controlled by the network and establishes the session channel on behalf of the backscatter communication device according to these parameters. Session channel parameters include session channel granularity, such as establishing a session channel at the granularity of backscatter devices or groups; the corresponding data network name and slice of the session; session channel type, such as establishing a control plane optimized session channel.
(4) *Opening of IoT device management capabilities*: The mobile network provides the corresponding Application Layer Interface (API) to the IoT device application server, supporting the IoT device application server's management, control, and information acquisition capabilities for IoT devices.
(5) *Location services for IoT devices*: Design location services for IoT devices based on the use scenario of IoT devices, requests from the application server, capabilities of IoT devices, etc.
(6) *Billing*: Design different billing strategies according to the use scenario of IoT devices, such as billing based on IoT device groups, billing based on the number of IoT devices served by the mobile network, the number of communications of IoT devices, etc.

13.6 Summary

In future, low-cost and low-power backscatter communication devices will be ubiquitous, enabling smart cities and homes, smart healthcare, environmental monitoring and other applications. They will become the simplest bridge between the physical world and the digital world in the "digital twin world" vision. This chapter provides a detailed introduction to the technologies in backscatter communication, including RF energy harvesting, backscatter device hardware architecture, physical layer air interface technology, minimalist protocol stack, network connectivity topologies and security mechanisms. While ensuring low cost, low power consumption, and even zero power consumption, backscatter communication is expected to achieve the ability to transmit for hundreds of meters at a data rate of hundreds of kbps. This will truly enable the vision of 6G wide-area IoT and the transmission of more bits with less wattage. Future sensors will have more accurate, reliable and powerful sensing capabilities. By integrating energy harvesting technology with sensors, it is possible to achieve passive sensors. With their low cost and massive deployment characteristics, ubiquitous sensing can be achieved, thereby meeting the integrated sensing requirements in 6G while enhancing the functionality of backscatter communication devices.

References

1. IOT ANALYTICS. Cellular IoT & LPWA Market Tracker (Q3 2021)[R]. 2021.
2. LONR R, LIANG Y, GUO H, et al. Symbiotic radio: A new communication paradigm for passive Internet of Things[J]. IEEE Internet of Things Journal, 2019, 7(2): 1350–1363.
3. KIMIONIS J, GEORGIADIS A, DASKALAKIS S, et al. A printed millimetre-wave modulator and antenna array for backscatter communications at gigabit data rates[J]. Nature Electronics, 2021, 4(6): 439–446.
4. Future Mobile Communication Forum. Zero Power Consumption Communication[R]. 2022.
5. OPPO. Zero Power Consumption Communication[R]. 2022.
6. RADIO SECTOR OF ITU. Future Technology trends of terrestrial IMT systems towards 2030 and beyond[R]. 2022.
7. 3GPP. TR22.840: Study on Ambient power-enabled Internet of Things[R]. 2022.
8. 3GPP. RP-222664: New SID Study on Ambient IoT[R]. 2022.
9. ABBASI Q, ABBAS H, ALOMAINY A, et al. Backscattering and RF Sensing for Future Wireless Communication[M], John Wiley & Sons, 2021.
10. QIAN J, GAO F, WANG G. Signal detection of ambient backscatter system with differential modulation[C]//2016 IEEE International Conference on Acoustics, Speech and Signal Processing (ICASSP). Piscataway: IEEE, 2016: 3831–3835.
11. Huang Yulan. Internet of Things: Detailed Explanation of RFID Core Technology[M]. Beijing: People's Posts and Telecommunications Publishing House, 2016.
12. Gao Ruyun, Lu Manru, Zhang Qimin, et al. Communication Electronic Circuits[M]. 4th Edition. Xi'an: Xi'an University of Electronic Science and Technology, 2016.
13. Zeng Xingwen. Principles and Analysis of High Frequency Circuits[M]. Xi'an: Xi'an University of Electronic Science and Technology Press, 2006.
14. VULLERS R, SCHAIJK R V, VISSER H J, et al. Energy harvesting for autonomous wireless sensor networks[J]. IEEE Solid-State Circuits Magazine, 2010, 2(2): 29–38.

15. GU X, GUO L, HEMOUR S, et al. Optimum temperatures for enhanced power conversion efficiency (PCE) of zero-bias diode-based rectifiers[J]. IEEE Transactions on Microwave Theory and Techniques, 2020, 68(9): 4040–4053.
16. CUI Q, ZHANG C, GUO Y, et al. A 1.8 V Output RF Energy Harvester at Input Available Power of− 20 dBm[C]//2018 IEEE 3rd International Conference on Integrated Circuits and Microsystems (ICICM). Piscataway: IEEE, 2018: 33–37.
17. THOMAS S, REYNOLDS M S. QAM backscatter for passive UHF RFID tags[C]//2010 IEEE International Conference on RFID (IEEE RFID 2010). Piscataway: IEEE, 2010: 210–214.
18. CORREIA R, BOAVENTURA A, Carvalho N B. Quadrature amplitude backscatter modulator for passive wireless sensors in IoT applications[J]. IEEE Transactions on Microwave Theory and Techniques, 2017, 65(4): 1103–1110.
19. BOYER C, ROY S. Coded QAM backscatter modulation for RFID[J]. IEEE Transactions on Communications, 2012, 60(7): 1925–1934.
20. SABHARWAL A, SCHNITER P, GUO D, et al. In-band full-duplex wireless: Challenges and opportunities[J]. IEEE Journal on Selected Areas in Communications, 2014, 32(9): 1637–1652.
21. LIU W, HUANG K, ZHOU X, et al. Full-duplex backscatter interference networks based on time-hopping spread spectrum[J]. IEEE Transactions on Wireless Communications, 2017, 16(7): 4361–4377.
22. VAN H N, HOANG D T, LU X, et al. Ambient backscatter communications: A contemporary survey[J]. IEEE Communications Surveys & Tutorials, 2018, 20(4): 2889-2922.
23. HE C, WANG Z, LEUNG C. Unitary Query for the M × L × N MIMO Backscatter RFID Channel[J]. IEEE Transactions on Wireless Communications, 2015, 14(5): 2613–2625.
24. YANG G, WEI T, LIANG Y. Joint hybrid and passive beamforming for millimeter wave symbiotic radio systems[J]. IEEE Wireless Communications Letters, 2021, 10(10): 2294–2298.
25. AMATO F, PETERSON C W, AKBAR M B, et al. Long range and low powered RFID tags with tunnel diode[C]//2015 IEEE International Conference on RFID Technology and Applications (RFID-TA). Piscataway: IEEE, 2015: 182–187.
26. CORREIA R, CARVALHO N B, KAWASAKI S. Continuously power delivering for passive backscatter wireless sensor networks[J]. IEEE Transactions on Microwave Theory and Techniques, 2016, 64(11): 3723–3731.
27. SONG C, DING Y, EID A, et al. Advances in wirelessly powered backscatter communications: From antenna/RF circuitry design to printed flexible electronics[J]. Proceedings of the IEEE, 2021, 110(1): 171–192.
28. AMATO F, PETERSON C W, DEGNAN B P, et al. Tunneling RFID tags for long-range and low-power microwave applications[J]. IEEE Journal of Radio Frequency Identification, 2018, 2(2): 93–103.
29. TONG W, ZHU P Y. The new journey of 6G wireless communication: Crossing human and machine connectivity, towards intelligent connectivity of everything[M]. Beijing: China Machine Press, 2021.
30. DURGIN G D, DEGNAN B P. Improved channel coding for next-generation RFID[J]. IEEE Journal of Radio Frequency Identification, 2017, 1(1): 68–74.
31. FASARAKIS-HILLIARD N, ALEVIZOS P N, BLETSAS A. Coherent detection and channel coding for bistatic scatter radio sensor networking[J]. IEEE Transactions on Communications, 2015, 63(5): 1798–1810.
32. PARKS A N, LIU A, GOLLAKOTA S, et al. Turbocharging ambient backscatter communication[J]. ACM SIGCOMM Computer Communication Review, 2014, 44(4): 619–630.
33. ALAMOUTI S M. A simple transmit diversity technique for wireless communications[J]. IEEE Journal on Selected Areas in Communications, 1998, 16(8): 1451–1458.
34. TAROKH V, JAFARKHANI H, CALDERBANK A R. Space-time block codes from orthogonal designs[J]. IEEE Transactions on Information Theory, 1999, 45(5): 1456–1467.
35. JAFARKHANI H. A quasi-orthogonal space-time block code[J]. IEEE Transactions on Communications, 2001, 49(1): 1–4.
36. LIU H, LIN W, LIN M, et al. Passive UHF RFID tag with backscatter diversity[J]. IEEE Antennas and Wireless Propagation Letters, 2011, 10: 415–418.

37. TAROKH V, JAFARKHANI H. A differential detection scheme for transmit diversity[J]. IEEE Journal on Selected Areas in Communications, 2000, 18(7): 1169–1174.
38. LIU W, SHEN S, TSANG D H K, et al. Enhancing ambient backscatter communication utilising coherent and non-coherent space-time codes[J]. IEEE Transactions on Wireless Communications, 2021, 20(10): 6884–6897.
39. HWANG C S, NAM S H, CHUNG J, et al. Differential space time block codes using nonconstant modulus constellations[J]. IEEE Transactions on Signal Processing, 2003, 51(11): 2955–2964.
40. TSE D, VISWANATH P. Fundamentals of wireless communication[M]. Cambridge: Cambridge University Press, 2005.
41. GUO J, ZHOU X, DURRANI S, et al. Design of non-orthogonal multiple access enhanced backscatter communication[J]. IEEE Transactions on Wireless Communications, 2018, 17(10): 6837–6852.
42. LI L, HUANG X, FANG Y. Hierarchical Multiple Access for Spectrum-Energy Opportunistic Ambient Backscatter Wireless Networks[J]. IEEE Transactions on Mobile Computing, 2022.
43. YUAN Z, YAN C, YUAN Y, et al. Blind multiple user detection for grant-free MUSA without reference signal[C]//2017 IEEE 86th Vehicular Technology Conference (VTC-Fall). Piscataway: IEEE, 2017: 1–5.
44. YUAN Z, HU Y, LI W, et al. Blind multi-user detection for autonomous grant-free high-overloading multiple-access without reference signal[C]//2018 IEEE 87th Vehicular Technology Conference (VTC Spring). Piscataway: IEEE, 2018: 1–7.
45. JIANG S, YUAN X, WANG X, et al. Joint user identification, channel estimation, and signal detection for grant-free NOMA[J]. IEEE Transactions on Wireless Communications, 2020, 19(10): 6960–6976.
46. CHI Z, LIU X, WANG W, et al. Leveraging ambient LTE traffic for ubiquitous passive communication[C]//Proceedings of the Annual conference of the ACM Special Interest Group on Data Communication on the applications, technologies, architectures, and protocols for computer communication. New York: ACM Press, 2020: 172–185.
47. YANG Q, WANG H, ZHENG T, et al. Wireless powered asynchronous backscatter networks with sporadic short packets: Performance analysis and optimisation[J]. IEEE Internet of Things Journal, 2018, 5(2): 984–997.
48. GAN QUAN. Internet of Things UHF RFID technology, products and applications[M]. Beijing: Tsinghua University Press, 2021.
49. YANG G, LIANG Y, ZHANG R, et al. Modulation in the air: Backscatter communication over ambient OFDM carrier[J]. IEEE Transactions on Communications, 2017, 66(3): 1219–1233.
50. ELMOSSALLAMY M A, PAN M, JÄNTTI R, et al. Noncoherent backscatter communications over ambient OFDM signals[J]. IEEE Transactions on Communications, 2019, 67(5): 3597–3611.
51. TAO Q, LI Y, ZHONG C, et al. A novel interference cancellation scheme for bistatic backscatter communication systems[J]. IEEE Communications Letters, 2021, 25(6): 2014–2018.
52. LUO R, YANG H, SHI H, et al. Ambient Backscatter Communication Design over DCSK Signals[J]. International Journal of Bifurcation and Chaos, 2021, 31(13): 1–12.
53. Future Mobile Communication Forum. Terminal friendly 6G technology white paper[R]. 2022.

Chapter 14
6G Integrated Satellite-Terrestrial Network

Facing the needs of social development beyond 2030, mobile communication networks need to provide more comprehensive three-dimensional coverage, and satellite networks will play a more important role. Therefore, 6G will be an integrated network deeply integrating satellite networks and terrestrial networks, providing more ubiquitous, intelligent, secure, and trustworthy public mobile information basic service capabilities [1–3]. Compared with terrestrial mobile communication systems, the satellite communication system has the advantages of wide coverage, large communication capacity, small terrain impact, high flexibility, and adaptability to various services. Based on an overview of the development state of integrated Satellite-Terrestrial networks (ISTNs), this chapter looks forward to the development path of the ISTN, and analyzes several key technologies adopted by the ISTN for 6G.

14.1 Current Development of ISTN

Although 5G technology has been commercialized around the world, providing users with high bit rate, low latency, high capacity, multiple new services and vertical application communication services [4–6]. It is limited by geographical environment and service models, which makes it unable to ensure network coverage in remote oceanic and land areas. To break through geographical restrictions, integrating satellite networks and terrestrial networks into a global seamless ISTN has become a hot topic of research in the academic and industrial world. The ISTN is based on the terrestrial network and extended by the satellite network, covering space, sky, land, sea and other natural spaces, providing information assurance for activities of all types of users such as space-based, air-based, land-based, sea-based users [7, 8]. The construction of the ISTN is a long-term gradual advancement and continuous

G. Liu et al., *6G Migrates from Communication to XaaS*,
https://doi.org/10.1007/978-981-96-7579-1_14

improvement process, and scientific and reasonable system architecture and network model design are the basis and starting point of the research [9, 10].

International organizations such as 3GPP and ITU have set up corresponding working groups to carry out standardization research on the ISTN, and the China Communications Standards Association (CCSA) has also set up the Aerospace Communication Technology Committee (TC12) in 2019 to carry out research on satellite-terrestrial integration. Among them, the Non-Terrestrial Networks (NTN) project, initiated by 3GPP, is committed to integrating satellite communication with 5G to solve the key issues of New Radio (NR) technology supporting NTN [11]. With the continuous evolution of 5G standardization, key technology breakthroughs for 6G have also been launched, and the ITU 6G standard plan has initially taken shape. These works have laid a technical foundation for the research of the ISTN for 6G.

14.1.1 The Rise of Satellite Communication

Terrestrial networks can only set up base stations in limited areas, and it is difficult to set up base stations in areas with harsh natural conditions and high economic costs (such as deserts, oceans). Therefore, due to economic costs and technical factors, the existing terrestrial mobile communication network only covers about 20% of the earth's land area, less than 6% of the earth's surface area, and covers about 70% of the total population. Satellite networks, with their wide coverage area, high-speed and large-capacity data transmission, and independence from geographical factors, can well make up for the shortcomings of terrestrial networks.

After experiencing a development downturn in the early twenty-first century, satellite communication systems have made some progress with the rapid development of the mobile internet. Nowadays, a new generation of satellite communication systems, mainly characterized by broadband interconnection, has gradually developed and is demonstrating a trend of acceleration. The new generation of satellite communication systems has the following characteristics.

(1) It has evolved from narrowband voice to broadband voice data transmission, from pipeline services to mobile internet and mobile IoT, and from industry applications to general services.
(2) Represented by low-orbit satellites, giant constellations consisting of thousands to tens of thousands of satellites have entered the planning and construction phase.
(3) Through satellite link interconnection, a globally covered interconnected space network is formed.
(4) Mass-produced and factory-based low-cost satellites, terminals, and rocket launch technologies have significantly reduced the cost of satellite network deployment.

Table 14.1 Existing international typical LEO satellite plans

	Oneweb (USA)	Starlink (USA)	LeoSat (France)	Telesat (Canada)	Kuiper (USA)
Number of satellites	588	4425–42,000	108	117	3236
Service link frequency band	Ku, Ka	Ku, Ka, V	Ka	Ka	Ka, V
Inter-satellite link	No	Yes	Yes	Yes	Yes
Beam type	Fixed beams	Flexible scanning beams	Spot beams	Flexible scanning beams	/
Construction state	In commercial	In commercial	/	Experimental satellites in orbits	/

Major technology powers have begun to launch their own satellite launch plans. There are currently 14 constellation plans announced abroad, including 9 in the United States, and 1 each in Russia, Canada, India, South Korea, and the Netherlands. As shown in Table 14.1, the existing international typical LEO satellite plans include the United States' OneWeb, Starlink and Kuiper, France's LeoSat, and Canada's Telesat. China is also keeping pace, and since 2018 has begun to build the Hongyan Constellation and Hongyun Project and launched some backbone network satellites. After the Hongyan Constellation is expected to be deployed, it will provide users with narrowband communication, IoT, and broadband internet services. The Hongyun Project is expected to provide broadband mobile communication services with seamless global coverage after deployment, and to build a comprehensive information platform integrating "communication, navigation, and remote sensing" for various users [12]. In 2021, the China Satellite Communication Network Group (referred to as "Satellite Network") was officially established to coordinate the development of satellite internet at the national level.

14.1.2 Satellite Communication System Standards

The degree of standardization of the existing satellite communication system and technical protocols is low, and the protocol standards lag behind the development and construction of the communication system, resulting in a small application range for the satellite communication system. Among them, the standards for the narrowband satellite communication system are drawn from those of the terrestrial communication system, as shown in Table 14.2; the broadband satellite communication system standards modify the original broadband multimedia standard protocol, as shown in

Table 14.3, and are mainly used for satellite internet services. At present, there are several shortcomings in the broadband satellite communication system.

(1) The existing protocol cannot support the coexistence of various different services, and the satellite communication system protocol is prone to evolving into a "chimney" protocol, with small industry scale and high equipment prices.
(2) Although the existing protocols have standardized the underlying protocols for basic air interface transmission, they rarely involve system upper-layer protocol standards, system construction, testing, etc., and lack the necessary capabilities to support network switching, flexible transmission architecture, inter-satellite interfaces, network security, etc.
(3) The air interface design of the existing protocol lacks flexibility in resource scheduling, and the efficiency and reliability of resource scheduling are low, and the control of service quality is weak. For instance, QoS cannot be guaranteed, and complex service transmission cannot be supported.

Table 14.2 Narrowband satellite communication system standards

Satellite communication system	First launching year	Frequency band	Protocol standard
MAST	1995	L	AMPS
ACeS	2000	L	Similar with GSM
Thuraya	2000	L	GMR-3G
Iridium	1996	L, Ka	Similar with GSM
GlobalStar	1998	L, S	Similar with IS-95
TerreStar	2009	S	WCDMA(3G)
TT-1	2016	S	Similar with 3G

Table 14.3 Broadband satellite communication system standards

	DVB-RCS	IPoS	DOCSIS-S
Background	From ETSI	From TIA， IP	From USA
Protocol standard	PHY, MAC	PHY, MAC, RLC	PHY, MAC
Forward link transmission	DVB-S/DVB-S2, > 100Mbps	DVB-S/DVB-S2, > 100Mbps	DVB-S/DVB-S2, 72Mbps,52Mbps
Backward link transmission	MF-TDMA, < 6Mbps	MF-TDMA, < 2.5Mbps	MF-TDMA, < 10Mbps
Standard maturity	high	medium	low
Application scale	Wide application	normal	normal

Note IPoS The full English name is Internet Protocol over Satellite, that is, Satellite Internet Protocol

14.1.3 New Trends in the Development of Satellite-Terrestrial Integration

Low Earth Orbit (LEO) satellites, with their low cost, low latency, high speed, and large capacity, play a crucial role in building satellite internet. LEO satellite networks, as a supplement to terrestrial 5G/6G networks, have significant advantages and are the mainstream trend for the satellite communication industry in the 5G/6G era. The outlook for ISTNs, with LEO satellites as the main satellites, can be summarized as follows.

(1) LEO satellites operate in low earth orbits of 500–1500 km, are lightweight, small in size, and low in manufacturing cost. The cost of traditional high-orbit satellite is about 1.04 billion RMB each, while the cost of LEO satellite is only one percent of that. At the same time, constellation networking of LEO satellites can achieve global seamless coverage. LEO satellites can be put into commercial use with lower launch costs and higher use value.
(2) Satellites supplement terrestrial networks with their vast coverage area. i.e., base stations cover densely populated areas, leveraging capacity advantages to satisfy multiple-user connections. Satellite coverage is used in remote areas where base stations cannot cover, leveraging the coverage advantage of satellites and saving the cost of base station construction.
(3) In the 5G/6G era, the industry expects leveraging the synergy between high-capacity, high-bandwidth satellites and terrestrial networks to support enhanced Mobile Broadband (eMBB) and massive Machine Type Communications (mMTC) related applications in the three major application scenarios.
(4) Expanding wireless coverage paths in a satellite-terrestrial integrated manner, deploying high-gain antennas on satellites, and deploying base station processing devices compatible with mobile communication systems on the ground can enable direct communication between terrestrial terminals and LEO satellites within the medium and low frequency bands [13].

14.1.4 Progress and Trends of ISTN Standards

Satellites will become an important component of the 5G system and play a crucial role, which has been widely recognized by the industry. ITU has launched the NGAT SAT project, proposing in ITU-R M.2083 that "the next-generation mobile communication network should meet the user's need for access to services anytime, anywhere". ITU-R M.2460 analyzes the key factors of integrating satellite systems into the next-generation access technology and proposes typical application scenarios for satellite networks: relay broadband transmission services, data backhaul and distribution services, broadband mobile communication services, and hybrid multimedia services.

In June 2017, Europe established the SaT5G alliance to explore feasible solutions for integrating satellites into 5G networks, with the main tasks including:

(1) Implementing NFV and SDN technologies in satellite/5G networks;
(2) Researching satellite/5G multi-link and heterogeneous transmission technologies;
(3) Integrating the control plane and data plane of satellite networks and 5G networks;
(4) Management and operation technologies for integrated satellite/5G networks.

3GPP started researching satellite communications from Rel-14, aiming to achieve broader coverage by complementing the advantages of satellite networks and terrestrial networks to meet user access service requirements. In TR 38.913 "Scenarios and Requirements for Next Generation Access Technologies" starting in January 2016, 3GPP included satellite access technology as one of the basic access technologies for 5G networks. The TR22.822 "Research on Satellite Access to 5G" that started in 2018 presented some research results on the use of satellite access in 5G. In March 2017, 3GPP launched a new plan to study the role of satellites in 5G and has completed two Study Items (SIs). After two years of research, 3GPP has approved NTN as a new key feature of 5G, and the Work Item (WI) has been running since January 2020 [14]. At the same time, 3GPP has defined the three main service categories of the NTN-5G system as service continuity, service ubiquity, and service scalability [15].

(1) Service Continuity: Service continuity aims to improve the reliability of 5G services by providing continuous access to services granted by the 5G system for 5G users (such as cars, trains, airborne platforms, sea vessels). While terrestrial networks can provide reliable coverage in urban areas, they cannot meet coverage needs in certain special geographical areas such as mountains, deserts, oceans, etc. At this time, satellite access is used to ensure service continuity worldwide.
(2) Service Ubiquity: This type of service aims to provide 5G services to users in areas without service or insufficient service through satellite access networks. For example, some rural and remote areas may not be able to provide terrestrial coverage, or terrestrial network services may be temporarily interrupted or even completely destroyed by natural disasters.
(3) Service Scalability: This type of service will effectively utilize the data multicast and broadcast capabilities of satellite networks, such as the distribution of ultra-high-definition television content to support the scalability of 5G networks.

Rel-15 NTN SI studied NTN deployment scenarios and channel models, analyzed the impact factors of 5G air interface, and released TR 38.811. Rel-16 NTN SI proposed technical solutions for supporting NTN networks based on the 5G air interface and carried out link and system-level performance evaluations. Rel-17 NTN carried out three standardization projects, including NR over NTN, IoT over NTN, 5G ARCH_SAT, and completed the first satellite bent pipe transparent forwarding technology standard based on 5G in Rel-17. At the current Rel-18 phase, 3GPP is continuing to study the characteristics of satellite and air interface transmission link enhancement on the User Plane Function (UPF) for NTN. 5G standards are still evolving, and standardization work for 5G Pro has begun. With the release of higher planning (Release) version standards, 5G continues to improve existing performance characteristics while also introducing some new features. At the same time, 6G air interface technology needs to meet the capability requirements of 6G, and 6G needs to enhance its capabilities in air interface access and three-dimensional coverage. Currently, the ISTN for 6G is focusing on the integration and unification of satellite and terrestrial network air interface design schemes, and the key technology breakthroughs for 6G have been carried out, and the ITU 6G technology standards have initially taken shape [16].

14.2 Development Path of ISTNs

The ISTN has multi-layered, dynamically time-varying characteristics. The dynamism, complexity, and scalability of its topology, the high-speed mobility of space nodes, and the limited storage and processing capabilities make existing space information transmission technologies unable to meet the communication needs under the new space network architecture. The following will analyze the ISTN from the technical development paths of service integration, standard integration, and system integration.

14.2.1 Past: Service Integration

Traditional terrestrial network technology is mature and rich in resources, but it is greatly affected by the geographical environment. The satellite relay nodes in satellite networks are distributed in space, providing higher flexibility and larger coverage. Therefore, people began to attempt to complement the advantages of satellite and terrestrial networks early on, that is, preliminary integration: the ISTN achieves interconnection and intercommunications through intermediate gateways, enhancing mutual service, but the satellite network and the terrestrial network remain independent.

The increase in traffic volume and diversified demands has brought problems to traditional mobile networks. Firstly, the surge in mobile traffic volume and the development of new generation wireless access technology have shifted the bottleneck of mobile networks from the radio interface to backhaul and core networks [17–19]. Secondly, networks are independent of each other, and the technical standards are incompatible. Thirdly, the interconnection of satellite and terrestrial network services is achieved through a gateway, and the types of satellite network services are single. Finally, resources cannot be coordinated uniformly, resulting in low network efficiency.

14.2.2 Present (5G): Standard Integration

Due to the many differences between satellite communication and terrestrial wireless communication in terms of deployment environment, coverage range, and channel transmission characteristics, achieving deep integration between the two systems faces some challenges. The second stage is standard integration, where the satellite communication system adopts a communication standard similar to or the same as the terrestrial system, and the spectrum resources are shared with the terrestrial system [20].

With the integration and transformation of satellite-terrestrial networks, the current architecture of the ISTN is the satellite-terrestrial complementary network and the satellite-terrestrial hybrid network. Under the architecture of the satellite-terrestrial complementary network, the terrestrial system and the satellite system share a network management center, but their access networks, core networks, and frequency bands used remain independent. Under the architecture of the satellite-terrestrial hybrid network, the air interface part is as unified as possible. In terms of transmission and technical integration design, the two networks adopt same or similar standards and key technologies, making full use of the rich industrial chain foundation of the terrestrial network to improve R&D efficiency. In complex 5G scenarios, all participants (terrestrial and satellite operators, 5G vertical industries, and infrastructure providers) participate in an ecosystem composed of various resources (terrestrial radio systems, satellites, cloud/edge computing and transmission), and the unification of the air interface standard becomes even more important [21, 22], which is also the key to standard integration.

At present, from the perspective of the service composition of satellite networks, interconnection with terrestrial networks still occupies the main share, and the research on the architecture of satellites and 5G is actually still only achieved through gateways for the interconnection of satellite and terrestrial network services. The types of satellite network services are single, resources cannot be coordinated and managed uniformly, frequency sharing and coordinated management have not been achieved, resulting in low network efficiency.

14.2.3 Future (6G): System Integration

According to the current planning of China's spatial information network and the development trend in related fields abroad, the future of the ISTN for 6G is shown in Fig. 14.1. It moves from satellite networks to spatial information networks, from space-earth integration to space-air-earth-sea integrated spatial information networks, using various spatial platforms (geosynchronous satellites, medium–low orbit satellites, stratospheric balloons, and aircraft, drones, etc.) as carriers, making satellites and terrestrial networks function as a whole. The access points, frequencies, access networks, and core networks of the entire system are completely unified in planning and design, providing users with consistent services without perception, adopting coordinated resource scheduling, consistent service quality, and seamless roaming between satellite and earth. Compared with the past and present 5G ISTN, the future system integration process for 6G will be divided into seven levels.

(1) Standard integration: Unify air interface standard by using the same design scheme, the same transmission and switching technology on the air interface layer structure.
(2) Network integration: A unified network architecture across the entire network and a unified TCP/IP protocol allows all kinds of IP-based services to inter-communicate, such as data networks, telephone networks, video networks, all of which can be integrated together.
(3) Management integration: Unify resource scheduling and management.
(4) Spectrum integration: Frequency sharing and coordinated management.

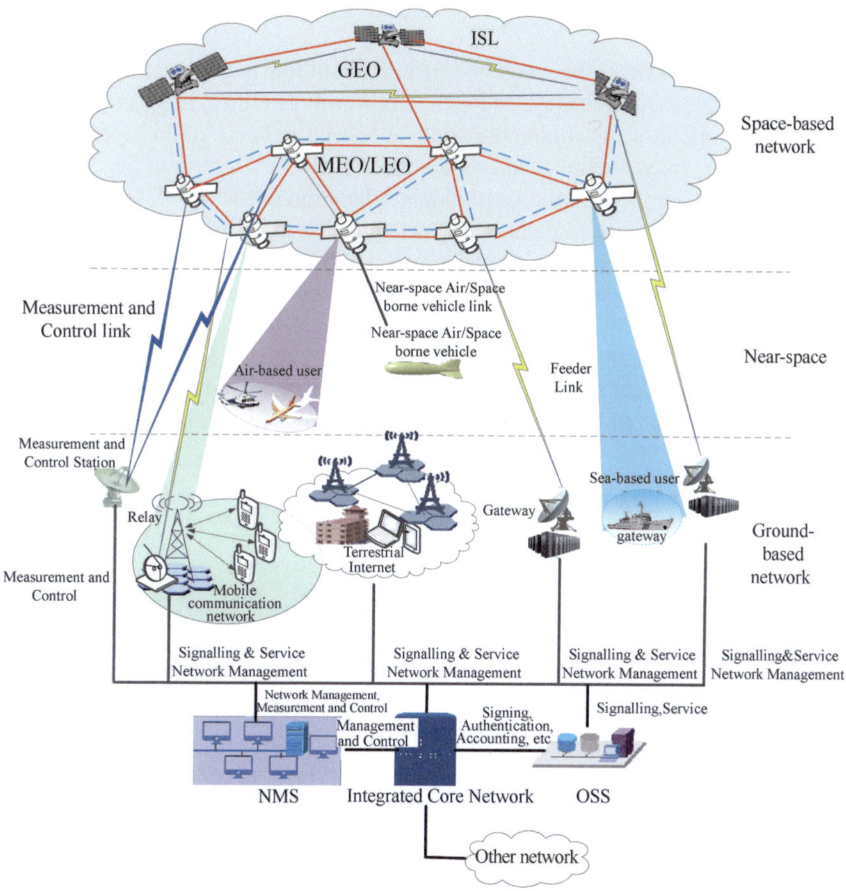

Fig. 14.1 Future of the ISTN for 6G

(5) Service integration: Unify service support and scheduling.
(6) Platform integration: The network platform adopts an integrated design.
(7) Terminal integration: Unify terminal identification and access methods and user terminals, gateway stations or satellite payloads can largely adopt the achievements of terrestrial network technology.

The integration of satellites and terrestrial networks will expand network coverage, improve the utilization of network frequency resources, and achieve shared use of satellite-terrestrial frequencies; at the same time, it will enhance service support capabilities and transmission efficiency through satellite-terrestrial collaborative transmission, building a green, efficient, and energy-saving network communication environment.

14.3 Integration of Satellite-Terrestrial Networks

The future ISTN for 6G is the integrated fusion of multiple heterogeneous networks, including space-based multi-layer subnets (high-orbit satellites, medium–low orbit satellites and monitoring equipment) and terrestrial cellular multi-layer subnets (macro cellular, micro cellular and pico cellular) [23], as shown in Fig. 14.2 for the large spatial–temporal scale cross-domain heterogeneous 6G ISTN. However, multi-layer complex cross-domain networking will lead to difficulties in network architecture design, large-scale spatial propagation environment will lead to low transmission efficiency, and the high-speed movement of satellites will lead to high dynamic changes in network topology, which in turn will make it difficult to guarantee service quality. These are all huge challenges faced by the 6G ISTN. To solve these problems, key technological breakthroughs need to be achieved from the aspects of network architecture, air interface transmission, networking methods, and frequency management of satellite-terrestrial integration [24, 25].

14.3.1 Network Architecture of Satellite-Terrestrial Integration

In terms of network architecture, the unified design of satellite and terrestrial cellular communication architecture is research. A flexible, reconfigurable network architecture and an efficient multi-domain, multi-dimensional network management architecture is designed to achieve flexible partitioning of network functions between satellite-terrestrial network nodes and improve resource management efficiency in ISTNs [26–29].

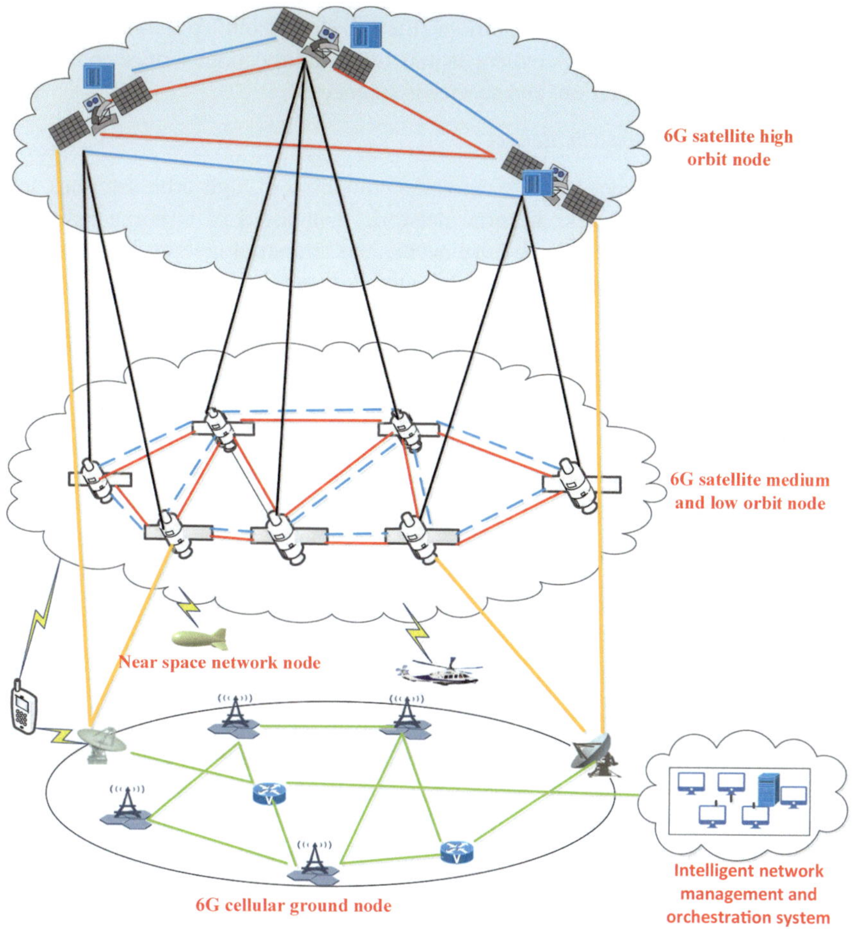

Fig. 14.2 Large spatial–temporal scale cross-domain heterogeneous 6G ISTN

1. Challenges Faced by the ISTN Architecture

Traditional space-based satellite mobile networks and terrestrial cellular networks have developed independently, with isolated networking and poor interoperability between systems, making it difficult to meet the needs for information integration and comprehensive utilization, and unable to achieve the development goals of 6G ISTNs. It is necessary to deeply integrate and unify the standard between space-based networks and mobile cellular networks to form a unified and efficient intelligent network, achieve multi-network integration, improve network transmission efficiency and performance, reduce network construction and maintenance costs, and enhance the flexibility of network deployment and the quality of service transmission.

Compared to the original satellite networks or terrestrial networks, due to fundamental changes in network structure (from planar to three-dimensional large spatiotemporal span) and node capabilities (transmission, mobility, routing, coverage capabilities, etc.), ISTNs face challenges in network architecture, service continuity, network capability adaptation, and service assurance.

(1) Network Architecture Challenge

The ISTN for 6G includes satellite networks composed of high-orbit satellites and medium–low orbit satellites, airborne networks composed of airborne balloons, airships and drones to meet temporary needs, and terrestrial cellular networks. Its network nodes have characteristics such as three-dimensional multi-level distribution, high-speed movement, and uneven functions and networking characteristics of each node. Therefore, the ISTN faces challenges such as large differences in network composition structure, difficulty in integrating network resources uniformly, inability to form efficient service guarantees; significant differences in communication protocols, high protocol conversion overhead during information transmission; dispersed network management and control functions, low network management efficiency, high complexity; uneven distribution of users and services, dynamic changes in network topology and architecture. How to efficiently organize heterogeneous networks, build a service-oriented network architecture that includes unified air interface protocols and networking protocols, modular functions, intelligence, user-centricity, capable of meeting different deployment scenarios and diverse service requirements is a challenge faced by the ISTN.

(2) Service Continuity Challenge

The ISTN faces the high dynamics of network topology caused by the high-speed movement of space-based and airborne network nodes, large differences in transmission delay between different network layers, large changes in network communication link performance, and poor robustness of satellite-terrestrial, inter-satellite, and airborne links, among other multi-scale heterogeneous network problems. At the same time, due to the rapid movement of low-orbit satellites, the phenomenon of frequent changes in service satellites will frequently occur for a fixed area. When there are many users in the area, there will be a problem of inter-satellite switching for group users. The ISTN faces complex, efficient, reliable, and secure routing and forwarding issues for multiple types, large capacity, and unevenly distributed services, and faces huge challenges in the network routing establishment and maintenance. To ensure the quality of 6G service continuity services and form reliable connections of seamless coverage, how to break through the closed nature of mobility management between heterogeneous networks and ensure the continuity of efficient service cooperation between terrestrial and satellite will be a challenge faced by the ISTN.

(3) Network Capability Adaptation and Service Assurance Challenge

The ISTN will provide all-scenario services, and the network should be able to provide matching network service capabilities according to different service requirements, such as in some service scenarios where the terminal needs to simultaneously access multiple types of satellite access networks and/or terrestrial access networks.

At the same time, the ISTN faces a series of factors that are not conducive to service quality assurance, such as large link delay jitter, frequent switching between users and feeder links, dynamic changes in inter-satellite/satellite-terrestrial network topology, etc. Therefore, how to perceive service requirements and provide corresponding network services, and how to achieve service quality assurance for ISTNs, are challenges faced by ISTNs in terms of service support.

2. Requirements for ISTN Architecture

The 6G network architecture will continue to evolve towards cloudification, making the network simpler and more efficient. Network functions can be quickly and frequently built, released, and deployed based on cloud-native technologies. In the process of network architecture evolution, it should comply with the basic requirements of service-oriented network architecture such as control and forwarding separation, and on-demand statelessness. The service-oriented network architecture will not only include the core network control plane, but also the core network user plane and base stations. For different application scenarios, network functions will be further decoupled, simplified, and customized to adapt to a variety of service requirements in the future.

Due to the characteristics of satellite networks, such as rapid dynamic topology changes, limited transmission power, limited information processing capabilities, and transmission delays much greater than terrestrial networks, the integration of satellite and terrestrial networks is complex. Simple service intercommunication and standard integration cannot meet the intelligent, minimalist, and on-demand customization requirements of 6G networks. Therefore, the ISTN for 6G will be a system-integrated network, deeply integrating satellite networks and terrestrial networks in network architecture, network functions, and air interface transmission, as well as wireless resource management and scheduling, through unified user management and security mechanisms to achieve unified network management across the entire network [29].

The ISTN should have characteristics of simplicity, agility, resilience, and intensification.

(1) Simplicity: The hierarchy, types, quantities, and interfaces of the ISTN should be as few as possible to reduce the complexity and cost of operation and maintenance. Utilize technologies such as distributed artificial intelligence, SDN, NFV, etc., to establish a distributed network with on-demand adjustment, elastic scalability, and self-organization and self-evolution capabilities.
(2) Agility: The ISTN should have programmable capabilities, network resources should have elastic scalability, enabling the network to be flexible and controllable, integrated and evolvable, as well as elastic and customizable, facilitating rapid deployment and assurance of networks and services, making the network more intelligent and flexible, with higher adaptability and flexible elasticity..
(3) Resilience: The ISTN, through its endogenous elastic scalable architecture and technologies such as software-defined networks and virtualization, builds a flexible and efficient network capability resource pool that can be used as needed, achieving on-demand customization, dynamic deployment, and

elastic scalability of network capabilities, meeting the reliability and robustness requirements of the ISTN.

(4) Intensification: The ISTN should have the ability to deploy network functions or services as needed under a unified network architecture, and the ability to dynamically orchestrate and schedule resources as needed. Network resources should be able to be planned and deployed uniformly, changing the high-cost, low-efficiency situation under scattered and segmented conditions.

3. Elastic and reconfigurable network architecture

The ISTN for 6G will be centered on service/users, serving multiple communication scenarios. In order to provide efficient network services and better user experience in different scenarios, the ISTN architecture should have flexible adaptability and be able to construct service networks at multiple scales according to different scenarios and service requirements. The ISTN can enhance network vitality, reduce operating costs, and promote network openness and service innovation through network function reconstruction. It adopts a modular function design pattern and constructs networks that meet the needs of different application scenarios through the combination of "function components". The architecture of the ISTN is shown in Fig. 14.3.

The ISTN for 6G is a large spatial–temporal scale multi-layer heterogeneous network composed of high-orbit satellites, medium and low-orbit satellites, and terrestrial communication systems.

(1) Terrestrial communication systems

The terrestrial communication system, composed of a terrestrial network and an intelligent network management and control system (INMCS), is the core of the ISTN. The ISTN manages the terrestrial communication system and the satellite communication system in a unified manner by constructing a global view of the network. In order to unify and coordinate the network resources of satellites and terrestrial access points, and to construct an ISTN as needed, the intelligent network control platform in the terrestrial communication system is used to manage and control the network topology between the satellite network and the terrestrial network.

(2) Low and medium orbit satellites

Low and medium orbit satellites not only support ordinary access network functions, but can also deploy customized core network functions according to service requirements, such as mobility management, session management, and user plane network functions. By deploying network functions, each satellite can form an independent service network/service terminal, or it can serve as an access network node connecting other satellites/terrestrial core networks. When a single satellite forms its own system, the links between satellites will only carry service data, which can avoid using the feeder link of the ground gateway station in the control plane, reduce the

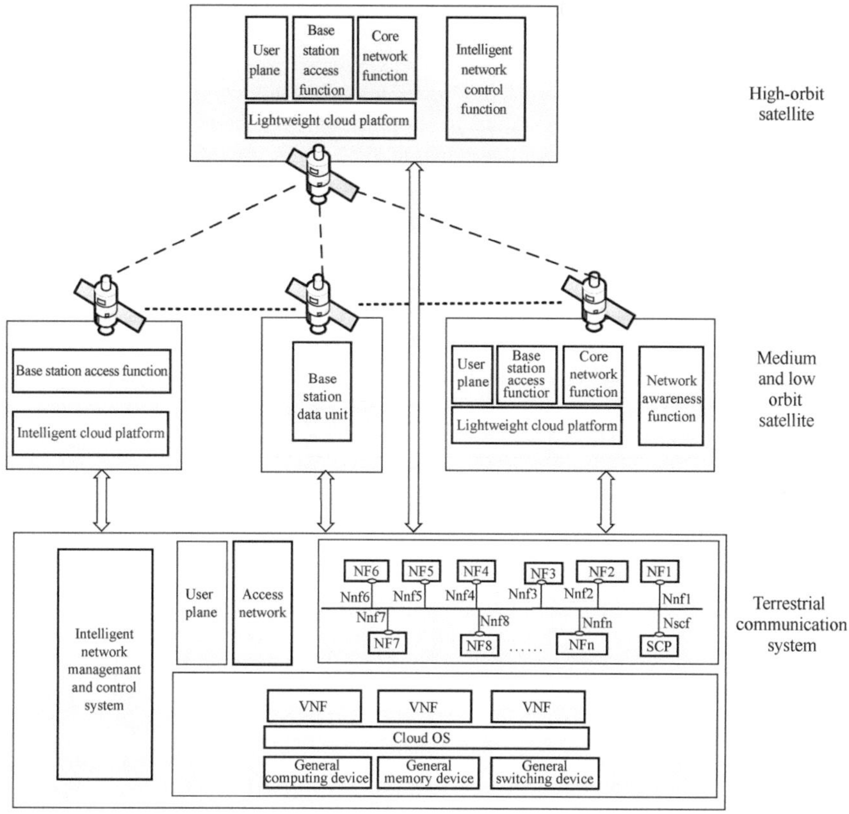

Fig. 14.3 Architecture of the ISTN

latency of the control plane, and also reduce the impact on the inter-satellite interface protocol, allowing the inter-satellite link to adopt a unified interface, providing flexibility for inter-satellite routing.

(3) High orbit satellites

High orbit satellites, as the backbone nodes of the space-based network, can manage the inter-satellite network topology of the low and medium orbit satellites within their management area and form a single autonomous domain in the satellite network. Therefore, high orbit satellites not only support the core network functions of low and medium orbit satellites, but also support intra-domain management and control functions. The intra-domain management and control platform on high orbit satellites is mainly responsible for receiving abnormal status notifications of satellite nodes or links within their autonomous domain, thereby dynamically adjusting the network structure based on service characteristics, network topology, network load, etc. In case of emergencies, it can also restructure the network based on the instructions of the terrestrial control center and the availability, security, and load of the intra-domain satellite network nodes.

The architecture of the ISTN can be divided into management plane, control plane, and data forwarding plane according to its functions, as shown in Fig. 14.4. The data forwarding plane includes high-orbit satellite nodes, medium–low orbit satellite nodes, and the data plane of the terrestrial communication system, mainly used for data forwarding. The control plane includes two parts: network routing control and network function control. Network routing control is mainly achieved through a programmable satellite network distributed architecture to implement dynamic routing planning and load balancing of the ISTN. The network routing control function is mainly responsible for real-time collection of node status, network topology and status, as well as real-time collection of network link traffic; based on the global network and traffic view, it performs centralized calculation and real-time deployment of end-to-end paths for key services; it achieves rapid recovery of controllers in case of failure, online deployment and upgrading of key functions; it realizes the synchronicity of status data between the terrestrial network management and control platform and the satellite network management and control platform, ensuring the rapid recovery of related configuration data and strategy data after cold start outside the equipment. Network function control is mainly achieved through implementing network functions of mobile communication on the virtualization platform, including access control, mobility management, session management and other core network functions and access network functions. The INMCS belongs to the management plane of the ISTN, mainly responsible for dynamic planning of network routing and dynamic reconstruction of network functions according to service requirements and network resource conditions. For example, the INMCS can deploy different network element functions on satellite nodes according to the network formation, service requirements, and network resource conditions, achieving customization and on-demand reconstruction of network functions. Through the INMCS, flexible partitioning of network element functions is achieved between satellite-based and terrestrial-based, between different satellite subnets, and between different satellite nodes in the same layer network. For example, the terrestrial-based network has stronger processing capabilities and can implement complete network functions; the space-based network, according to network deployment and service requirements, implements lightweight, tailorable network element functions. For satellite nodes with strong processing capabilities, such as high-orbit or some low-orbit satellites, they can become network control nodes by deploying control network element functions; for some satellite nodes with weak processing capabilities, only data forwarding plane functions are deployed for data processing and transmission.

4. Architecture of Intelligent Network Management and Control Systems

To meet the demand for an elastic and reconfigurable architecture in ISTNs, the intelligent network management and control system of ISTNs will be based on artificial intelligence technology. It will intelligently analyze network capabilities that match service requirements based on user demand data, network operation data, network topology data, ephemeris data, etc. The INMCS can manage and schedule

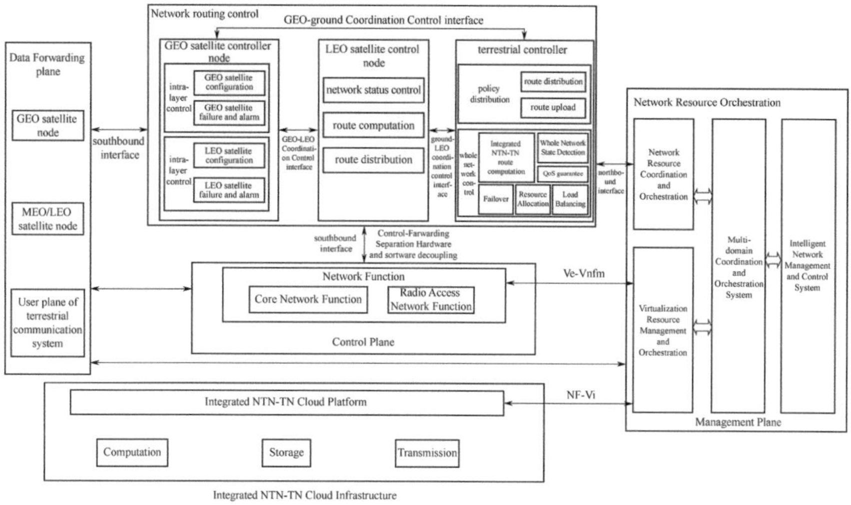

Fig. 14.4 Architecture of the ISTN

network resources on-demand and automatically according to service requirements. Combined with the enhanced network control, analysis, and collection capabilities in the ISTNs, a real-time autonomous closed-loop system of dynamic cross-domain resources is formed. This system achieves intelligent analysis and orchestration of the network, avoids wasting network resources, quickly adjusts network service capabilities, and ensures service needs of key users. Furthermore, the intelligent network and control system should also have end-to-end network policy management capabilities, capable of making intelligent decisions and policy issuance, in combination with user service needs, network status, and resource status.

The architecture of the INMCS is shown in Fig. 14.5. This system includes an intelligent network control engine and a network resource coordination and orchestration system. The INMCS can be deployed either as a terrestrial-based platform for managing and controlling networks intelligently or as an intra-domain network management and control platform onboard satellites. Through the network resource coordination and orchestration system, the INMCS coordinates and schedules resources in ISTNs, thereby completing the management of end-to-end networks and services management.

The INMCS leverages a multi-dimensional intelligent perception data model and a multi-party negotiation/sharing mechanism to support the deep integration of computation, data, algorithms, and networks. Utilizing distributed artificial intelligence algorithms, it achieves analysis of both satellite-based and terrestrial network situations, establishing a global network view that integrates both satellite-based and terrestrial networks. Through intelligent management and control, it achieves zero-touch networking. It dynamically selects appropriate network levels based on network status and service requirements. This process is accompanied by automatic configuration of network resources and intelligent generation of networking

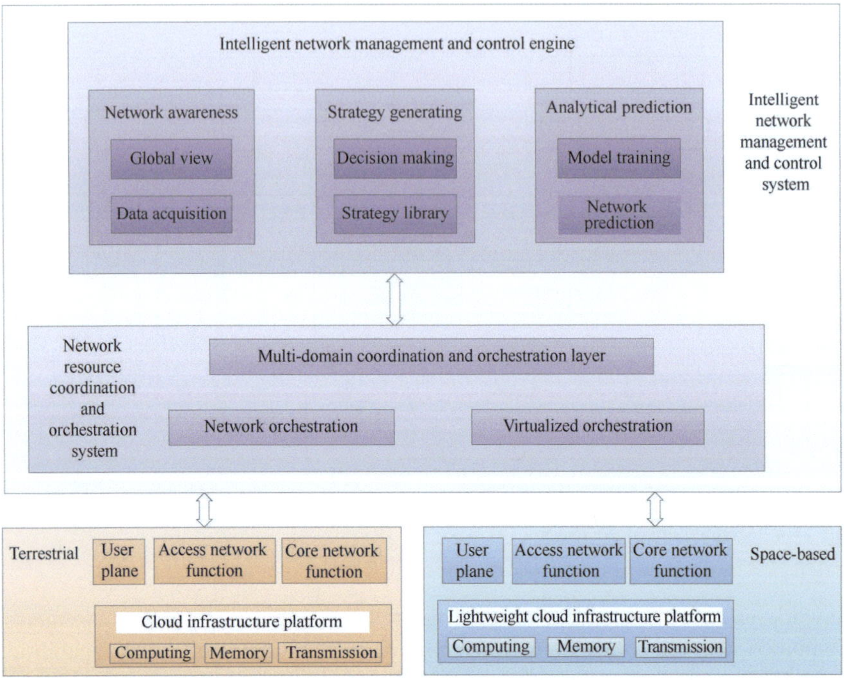

Fig. 14.5 Architecture of the INMCS

strategies. For instance, for priority users, it can configure high QoS level network connections to meet their service requirements. In instances where low-orbit satellite connections are congested, the system autonomously allocates network resources, prioritizing to ensure this user's network connection. Some delay-insensitive services or low-priority users can choose to transmit via high-orbit satellites. Furthermore, this system can span physical and virtual networks, enabling true end-to-end service management and facilitating the smooth evolution of existing siloed networks and network management systems towards 6G intelligent management and digital twin systems. Through this system, the automation management of network architecture can be achieved, resulting in significant operating cost savings and reduced time to service initiation. Furthermore, this system supports a flexible information model that leverages abstraction of products, services, and resources to shield the complexity of the network, enabling rapid data source integration, service virtualization, service assurance and analysis, service process management and other functions.

To achieve intelligent management and orchestration of network resources for diverse scenarios and services, it is necessary to solve the problem of heterogeneous network resource management and scheduling within ISTNs. The network resource collaboration and orchestration system achieves unified orchestration of both satellite-based and terrestrial network resources through cross-domain collaborative orchestration., and can configure different network element functions in low-orbit and high-orbit satellites according to network conditions and service requirements. Concurrently, it ensures synergy in computing resources, security policies, application management, service management, etc., of the satellite-based and terrestrial-based networks. This is achieved in accordance with the network resource deployment strategy and service strategy of the INMCS. Consequently, the system accomplishes on-demand end-to-end network resource configuration and orchestration, and achieving efficient utilization of heterogeneous network resources.

5. Virtualization Technology of ISTNs

The 6G network architecture will continue its evolution towards cloud-based development. This approach, rooted in cloud-native software design, enables the rapid and frequent construction, release, and deployment of network functions. The service-oriented network architecture will include not only the core network control plane, but also the user plane and base stations. In response to varied application scenarios, network functions will be further decoupled, simplified and customized to accommodate a variety of future service requirements. The architecture of the cloud-based platform for ISTNs is shown in Fig. 14.6.

The diverse applications of satellite-terrestrial integration will transmit massive amounts of data, necessitating robust processing capabilities in computing, storage, and transmission for support. Importantly, these capabilities must be achieved with the lowest possible payload and power consumption. Furthermore, beyond the fundamental requirement of adapting to harsh space environments, space-based networks face challenges such as the difficulty of maintenance and the impracticality of replacing software and hardware. Hence, it is crucial that even if some devices experience abnormalities, failures or damage, the network should still guarantee the continuation of information processing service. Therefore, lightweight, highly reliable and high fault isolation virtualization technology in space-based networks is a key technology for the development of ISTNs towards cloudification.

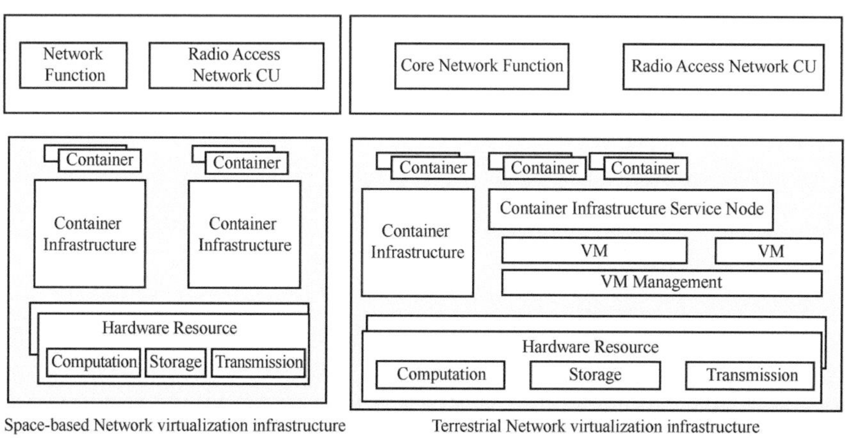

Fig. 14.6 Architecture of the cloud-based platform for ISTNs

Containers, a type of operating system-level virtualization technology, isolate different processes through operating system isolation techniques. Compared with NFV, containers are lighter and more manageable. Additionally, as a technology for application packaging, containers define standardized application release formats, significantly facilitating the development, deployment, and portability of applications. Containers can promote the micro service transformation of network elements through technologies such as layered images and centralized image repositories, thereby accelerating software development and deployment.

Therefore, the virtualization technology of ISTNs is based on containers constructing and operating elastically scaled network functions through cloud-native technology. The use of containerized packaging facilitates the reuse of virtualized components, simplifying the maintenance of the virtualization platform. Network element functions run within containers, serving as independently deployed units to achieve high-level resource isolation. Moreover, a centralized orchestration scheduling system dynamically manages and schedules virtualized resources.

Cloud-native virtual network functions (VNFs) employ a highly scalable architecture, supporting distributed state and asynchronous message processing. They utilize efficient, lightweight containers to provide on-demand capacity and fault switching as necessary. The collaboration with a highly automated intelligent network orchestration system enhances the agility of overall service.

In the space-based network, the use of micro services to construct and deploy cloud-native virtual network functions within a container environment enables rapid initiation of network functionalities. Additionally, this approach ensures a high degree of fault isolation when part of the network functionality fails. Furthermore, it allows for the addition of micro services as needed to handle increased capacity or backup requirements, facilitating flexible deployment and enhancing network reliability.

In the context of ISTNs, it is imperative that network elements achieve micro service decoupling. This begins with vertical decoupling, which involves the separation of the control plane from the data plane to ensure statelessness. Subsequently, horizontal decoupling is performed by breaking down monolithic structures into multiple distributed, loosely coupled micro services. These micro services can be independently developed and upgraded, fostering rapid service iteration.

14.3.2 Air Interface Transmission for Satellite-Terrestrial Integration

In terms of air interface transmission [30, 31], it is necessary to study the unified air interface design scheme for both satellite and terrestrial cellular communication. The design supports various types of service transmissions, ensuring that terminals can access the most suitable node of ISTNs.

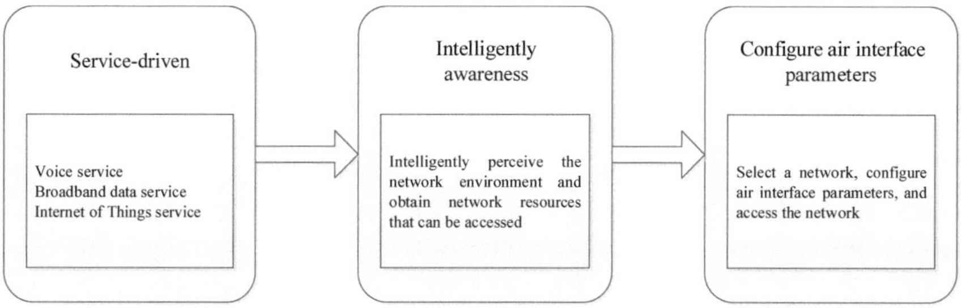

Fig. 14.7 Primary design concept for unified air interface in ISTNs

1. Design Concept for Unified Air Interface in ISTNs

The design for the unified air interface in ISTNs is based on key links such as service-driven, network awareness, and the configuration of variable parameter sets. The primary design concept is illustrated in Fig. 14.7 [32].

The design for the unified air interface in ISTNs is driven by service requirements and employs intelligent network environment sensing to determine the currently available network resources for accessing. Subsequently, based on the constraints of both services and network resources, configurable parameters are selected from a parameter set. These constraints encompass the Quality of Service (QoS) requirements such as bandwidth and latency needed for the current service, as well as the access resources provided by the currently available network. For terminals, there is no need to distinguish between satellite networks and terrestrial networks. They only need to assess whether the currently accessible network resources can meet their service demands, and then select the appropriate air interface parameters for configuration, access the network.

The core of the unified air interface design lies in the configuration technology of variable parameter sets, which can adaptively match the air interface standard for users based on their usage scenarios and types of services, and select the appropriate network access. For both satellite and terrestrial air interfaces, the same design scheme can be adopted with different parameters configured. For example, the satellite air interface may be configured with large bandwidth, large subcarrier interval, more HARQ (Hybrid Automatic Repeat Request) processes, etc. From the user's perspective, the distinguish between satellite and terrestrial networks becomes unnecessary. Instead, they access the network by dynamically configuring different air interface parameters. This achieves seamless switching and roaming through unified network resource configuration and management, realizing truly seamless ISTNs.

2. Variable Parameter Set

In the unified air interface standard for ISTNs, the design of the variable parameter set is central and crucial. The set comprises the primary technical parameters of the air interface, which is applicable to both satellite and terrestrial cellular communications. For different usage scenarios of ITSC, the values of the parameters within this set vary. Air interface parameters can be flexibly configured to adapt to the requirements of diverse application scenarios, facilitating opportunistic and on-demand access between satellite and terrestrial networks.

The primary parameters of the unified air interface's variable parameter set may include: transmission bandwidth, modulation method, coding method, transmission waveform, subcarrier interval, pilot format, HARQ configuration, orthogonal access method, random access method, control channel format, etc.

Table 14.4 provides the primary parameters list for the air interface's variable parameter set. The unified air interface's variable parameter set for ISTNs includes, but is not limited to, the parameters in this table.

3. Selection of waveform and modulation method

The rapid movement characteristics of low-orbit satellites cause serious Doppler effects, impacting the selection of subcarrier spacing. Due to the limited satellite power, it is generally necessary to maximize satellite power utilization. Choosing the appropriate waveform can reduce the impact of the signal PAPR (Peak to Average Power Ratio) on satellite power amplifiers. In terrestrial communications, high-order modulation mainly utilizes QAM (Quadrature Amplitude Modulation). To mitigate the impact of PAPR, satellite communications commonly employ PSK (Phase Shift Keying) modulation. Therefore, it is also necessary to choose an appropriate high-order modulation method based on specific requirements. In addressing the large delay and significant Doppler characteristics of satellite scenarios, it is necessary to choose appropriate sub-band bandwidth, subcarrier spacing and cyclic prefix, etc. For example, in satellite scenarios, it is recommended to use 120 kHz, 240 kHz subcarrier spacing to combat strong Doppler effects. To reduce the impact of waveform PAPR on satellite power amplifiers, the modulation waveform for the satellite air interface can use the single carrier form such as DFT-S-OFDM, or employ techniques like peak clipping to pre-process CP-OFDM signals to reduce the impact of PAPR effects. The following presents the PAPR statistical characteristics of DFT-S-OFDM and CP-OFDM signals, as shown in Fig. 14.8.

Additionally, regarding the issues of modulation methods, the current 5G system employs $\pi/2$-BPSK, QPSK, 16 QAM, 64 QAM and 256 QAM modulation methods. For satellite communications, DVB-S2X includes modulation methods such as $\pi/2$-BPSK, QPSK, 8 PSK, 16 APSK, 32 APSK, etc. It is recommended to continue using $\pi/2$-BPSK, QPSK for low-order modulation, while adopting 8 PSK and high-order APSK modulation for high-order modulation to reduce the impact of PAPR on satellite power amplifiers.

4. Time–frequency synchronization technology

Due to the remote distance, high velocity (especially for low earth orbit satellite, LEO), and power limitation of the satellite, the impact of low signal-to-noise ratio (SNR), large Doppler frequency shift, and significant time delay on time–frequency synchronization must be considered in the design of the unified air interface for ISTNs. For LEO satellites at an orbital height of 600 km, with a carrier frequency of 20 GHz, the motion introduces a maximum Doppler frequency shift of $\pm$ 480 kHz

Table 14.4 Primary parameters list for the air interface's variable parameter set

Main parameters	Range of values	Note
Bandwidth	180 kHz-1 GHz	Adapt to IoT, voice and wideband services
Waveform	DFT-S-OFDM or CP-OFDM	Satellite communication prefers single carrier type waveform, while terrestrial communication can support OFDM type directly
Modulation type	π/2-BPSK, BPSK, QPSK, 8PSK, 16QAM, 16APSK, 32APSK,64APSK, 64QAM, 128APSK, 256APSK, 256QAM, etc	Adapt to most cases for terrestrial communication and satellite communication
Coding type	Convolution, Turbo, Polar, LDPC, etc	Extensional to any other coding types
Sub-carrier space (SCS)	15 kHz, 30 kHz, 60 kHz, 120 kHz, 240 kHz	Provide variable SCSs
Pilot format	Continuous pilot, comb-like pilot	Extensional to any other types of pilot
HARQ	$\leq$ 16, 64 or128 process Or disable HARQ	Support scenarios with different delay ranges
Multiple access	OFDMA, PDMA, etc	Support orthogonal or non-orthogonal multiple access types
Time–frequency synchronization	Pre-compensation based on UE location and ephemeris	Provide high precision compensation, to support signal detection in low SNR case
Beam coordination transmission	Support SU-MIMO or MU-MIMO with multiple beams from a single satellite, or from different satellites, or from satellite and terrestrial access points	Enhance system capacity with achieved diversity gain
Control channel format	Specific control channel or dynamic control channel	Can be selected flexibly according to scenarios
Random access	On-demand random access, simplified access	Select access type according to service requirement and network state
Handover	Intelligent handover	Support variable handover types such as location based, UE based or AI assisted

and a maximum Doppler change rate of $\pm$ 5.44 kHz/s, which significantly affects frequency synchronization. Additionally, the one-way satellite-to-earth delay varies from several milliseconds to tens of milliseconds, impacting time synchronization between satellite and earth. The long transmission distance between satellite and

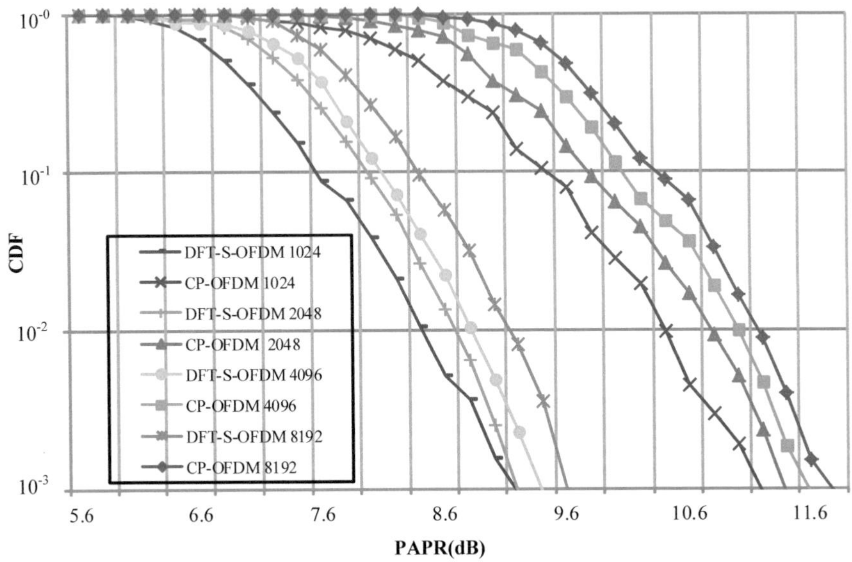

Fig. 14.8 PAPR statistical characteristics of DFT-S-OFDM and CP-OFDM signals

earth results in significant signal attenuation and the low SNR further increase the difficulty of signal detection.

To address the characteristics of significant Doppler frequency shift effect and large delay introduced by LEO satellites, a pre-compensation mechanism for uplink and downlink time–frequency synchronization based on terminal GNSS assistance can be employed. User terminals utilize GNSS to acquire their own position information and simultaneously obtain ephemeris information broadcasted by satellites. Based on the ephemeris information, terminals can estimate the position and velocity of the satellite, and then calculate the delay and frequency shift from the user to the satellite, and perform pre-compensation accordingly.

The schematic diagram of time–frequency pre-compensation is shown in Fig. 14.9. For the frequency offset, the satellite can compensate for the common frequency offset during the transmission of downlink signals and the reception of uplink signals, ensuring that the user terminal at point C experiences a frequency offset of 0. As for the time offset, the satellite can broadcast a common time offset to calculate the total transmission delay, while the Timing Advance (TA) value used by the user terminal when transmitting uplink signals is determined solely by the timing deviation relative to a reference point. During compensation, the common delay can also be accounted for, eliminating the need for further network processing. This significantly reduces the range of frequency offsets that each user terminal needs to handle, reducing the computational load for searching reference signal in downlink synchronization and simplifying the time–frequency offset handling in uplink synchronization.

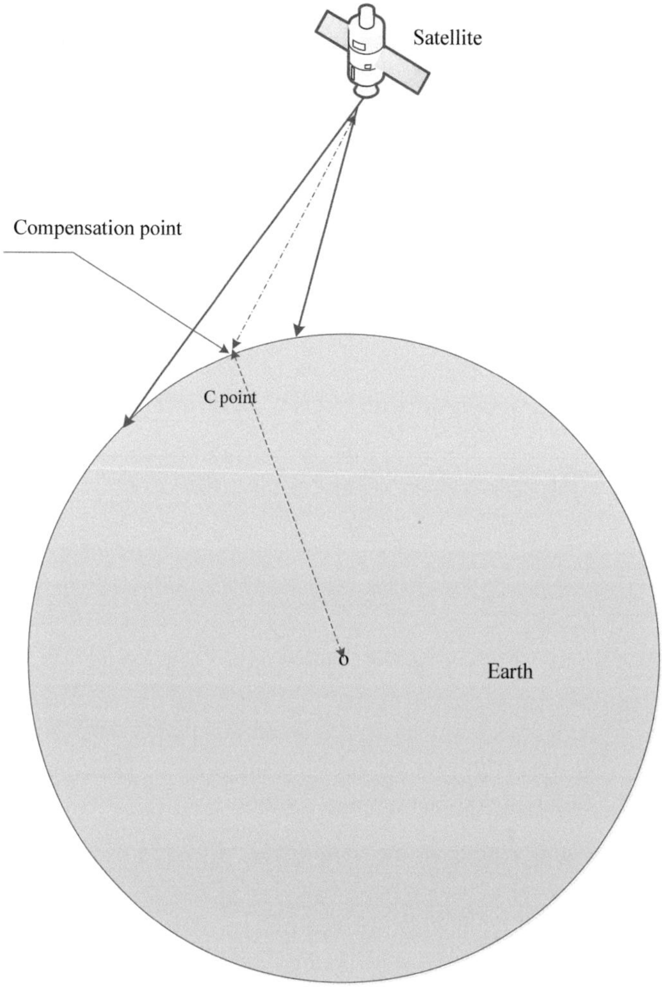

Fig. 14.9 Schematic diagram of time–frequency pre-compensation

In response to characteristics of low signal strength of satellite uplink and downlink signals, techniques such as repetitive transmission and multi-symbol combining detection can be employed to improve the success rate of signal detection. Taking the downlink PBCH as an example, this section evaluates the impact of multiple symbol combinations on timing synchronization under different SNRs, with the results illustrated in Fig. 14.10.

The frequency synchronization simulation results are shown in Fig. 14.11. By employing 8 times PBCH combined detection, at SNR −10 dB, the probability that the downlink timing synchronization range is less than ± 100 ns exceeds 95%. The simulation results indicate that multiple combinations can significantly enhance

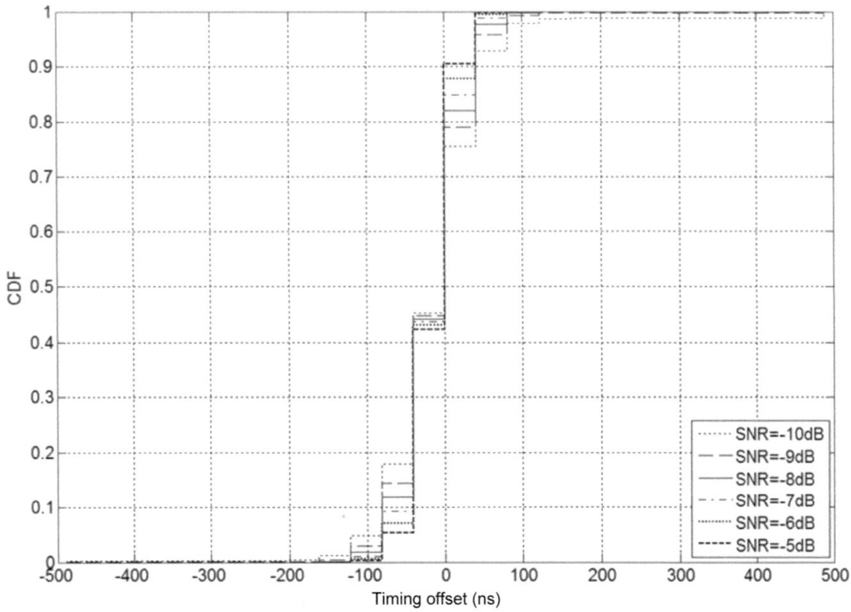

Fig. 14.10 Simulation results of Timing synchronization

the frequency synchronization accuracy, with residual frequency offset of only a few kilohertz under the 120 kHz subcarrier configuration. The simulation results demonstrate the feasibility of the proposed solution.

5. Access and mobility management technology

Due to the extended transmission delay inherent in satellite communications, the traditional four-step access method suffers from a significant latency, necessitating a simplified design of the access scheme. Furthermore, owing to the rapid movement of LEO satellites, terminals will frequently undergo inter-satellite handovers. Concurrently, conventional handover decision-making processes based on signal RRM measurements struggle to meet requirements in satellite cells where the SNR distribution is relatively uniform.

To address the access issue, it is necessary to optimize the design of the access procedure by reducing the steps involved and enhancing the success rate, thus facilitating the "ultra-simplified" access. It is recommended to adopt a two-step access method, as illustrated in Fig. 14.12.

To address the handover issue, a joint handover decision-making approach based on terminal position, satellite ephemeris, and RRM measurement results has been designed and introduced. The primary procedure involves: the terminal calculates its current coverage position based on GNSS positioning and ephemeris information. If entering an overlapping coverage area, the terminal reports its current location and RRM measurement results (optional). After receiving the information reported

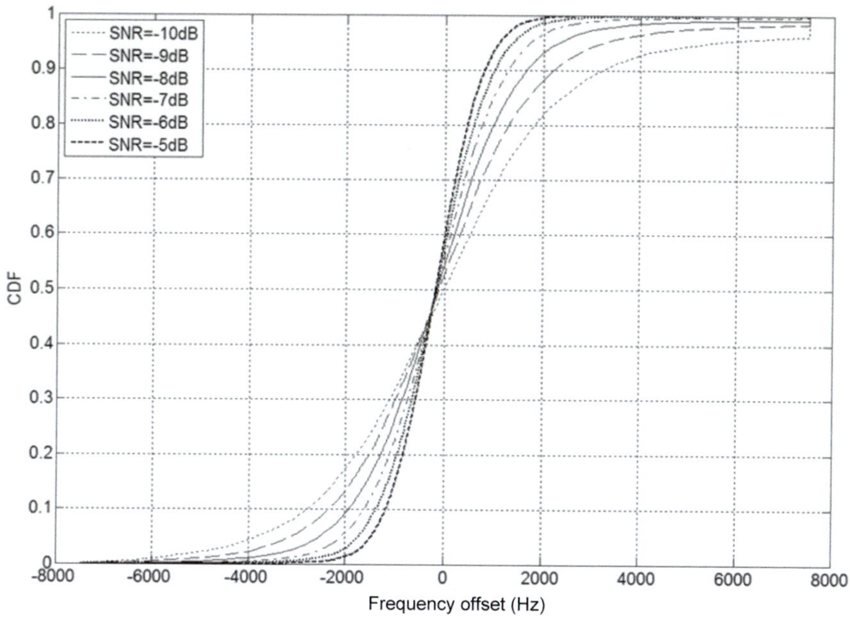

Fig. 14.11 Simulation results of Frequency synchronization

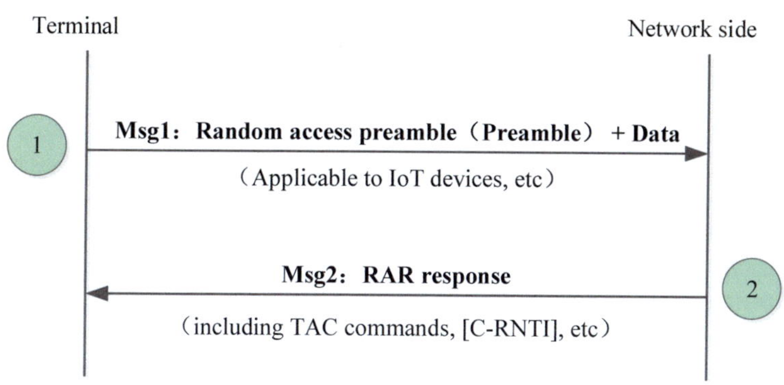

Fig. 14.12 Two-step access method procedure

by the terminal, the network side processes it comprehensively and determines the necessity of a handover. If a handover is required, the network side initiates a handover request such as RRC reconfiguration process, the terminal then executes the handover according to the established procedures. Figure 14.13 illustrates the inter-satellite handover process based on a joint decision using terminal location and ephemeris, taking the Xn interface as an example [33].

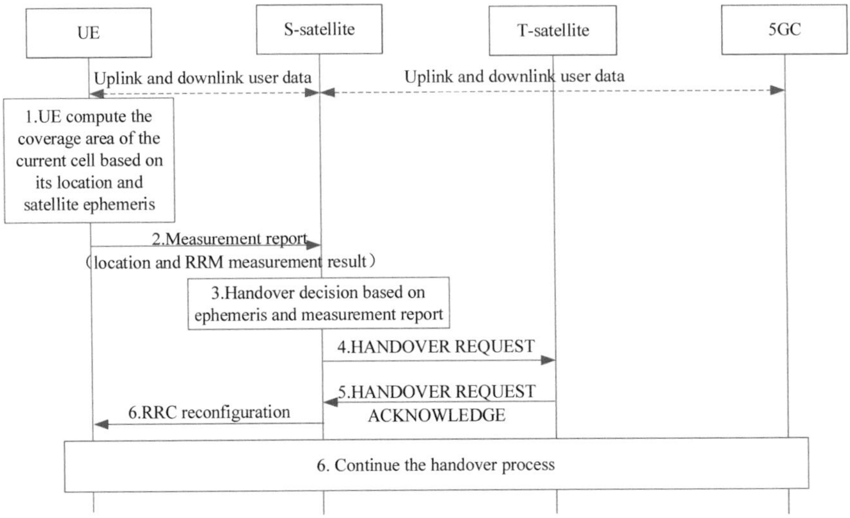

Fig. 14.13 Inter-satellite handover process based on joint decision-making using terminal location and ephemeris (Xn interface)

Additionally, a terminal-initiated handover scheme can be designed, where the handover decision is initiated from the terminal side, to achieve fast "ultra-intelligent" handover, etc.

6. Multi-beam cooperative transmission technology

Traditional satellite communications generally utilize a single satellite and a single beam to serve one user, which to a certain extent limits the user's data transmission rate and the utilization efficiency of satellite resources [34]. To further utilize the spatial transmission characteristics of satellites, a terminal can be simultaneously connected to multiple satellites in a formation. These satellites collaborate to achieve joint data transmission, thereby obtaining transmission diversity gain or multiplexing gain.

Multi-satellite and multi-beam cooperative transmission technology, a candidate technology for improving satellite transmission rates, involves the concurrent transmission of data to the same user via multiple beams from multiple satellites or multiple polarized beams from a single satellite, within the same spectrum resources. This technology poses high demands on the system, and its primary research contents and technical challenges are outlined as follows.

(1) Research on the theory and model of multi-satellite and multi-beam cooperative transmission based on line-of-sight (LOS) multiple-input multiple-output (MIMO).
(2) Research on transmission schemes, synchronization technology and codebook design for multi-satellite and multi-beam cooperative transmission.

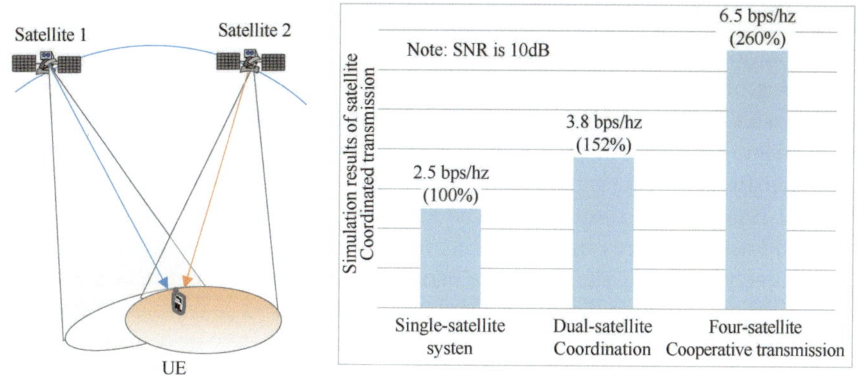

Fig. 14.14 Schematic and simulation results of multi-satellite and multi-beam cooperative transmission

(3) Support distributed and centralized cooperative signal processing for on-board processing and transparent forwarding mechanisms.

Figures 14.14 presents the schematic and simulation results of multi-satellite and multi-beam cooperative transmission. The simulation results indicate that through the collaborative transmission of multi-beams from four satellites, a more than 150% enhancement in transmission rate for a single user can be achieved.

14.3.3 Networking Method for ISTNs

In terms of networking methods, the primary research areas include inter-cell frequency planning, adaptive routing and seamless handover between multi-layer networks, as well as task migration of multi-level edge computing in ISTNs.

1. Efficient unified management of heterogeneous cross-domain resources between satellite and terrestrial networks

Utilizing the temporal and spatial characteristics of diverse services in the satellite-terrestrial heterogeneous network environment, the research is conducted on unified representation techniques for cross-domain multi-dimensional resources. Under the framework of multi-level edge intelligent control driven by overall service behavior, the networking and intelligent collaboration mechanisms of multiple edge control groups are explored. Utilizing the characteristics of heterogeneous cross-domain satellite-terrestrial networks and the system information capacity theory, a flexible deployment and dynamic mapping method for virtualized resource service function chains in ISTNs is proposed.

(1) Efficient unified representation and resource correlation graph construction of heterogeneous cross-domain resources in ISTNs

In satellite-terrestrial heterogeneous networks, the lack of system-level representation for various types of resources poses challenges to the unified management and scheduling of resources across the entire network. Consequently, there is a need to develop a unified method for the generalized representation of system resources. By distinguishing the types of network resources and their distribution in heterogeneous cross-domain networks, unified identification and representation can be applied to different transmission resources, including access resources, spectrum resources, link resources and processing forwarding resources. This involves qualitatively and quantitatively recording their attribute parameters and status parameters. Based on the correlation among multi-dimensional resources, a resource correlation graph is created and maintained. Internal attribute correlation parameters are defined to reflect the degree of mutual constraints among resources. This allows for the rational allocation of resource combinations to provide guaranteed end-to-end services, laying the foundation for collaborative resource management.

(2) Intelligent collaboration mechanism of multi-level edge computing in ISTNs

In response to the integration and interconnection requirements of terrestrial and satellite networks, an integrated network is constructed as a cross-domain multi-level edge computing intelligent collaboration network. The study investigates the synergistic effect among various elements in multi-resource collaboration and proposes a multi-level edge computing intelligent collaboration strategy to achieve the collaborative optimization of edge network communication, computation, storage, and user perception capabilities. Furthermore, user behavior-driven collaborative methods for resource-constrained multi-level edge computing networks are explored, providing guiding methodologies for network resource configuration and transmission mechanism strategies and enabling collaborative sensing between satellite and terrestrial networks, which lays a theoretical foundation for efficient resource configuration methods.

(3) Mechanism for joint coordination and deployment of heterogeneous cross domain service function chain resources in ISTNs

Here, it is necessary to study the service function chain mapping mechanism for ISTNs, including the design of corresponding service function chain definitions, virtual network function deployment, and service function chain mapping. Considering the service requirements, network environment and other characteristics in ISTNs, a service function chain mapping method is designed for this specific network type. This involves dynamic deployment of the number of function chains, their functionalities, and the selection of entity nodes within these chains. The study also focuses on optimizing signaling interaction methods and dynamic configuration control techniques for function chains, thereby enhancing the dynamic configuration and optimization capabilities of service function chains.

2. Elastic, efficient dynamic routing and mobility management

In ISTN, mobility management faces technical challenges such as large spatial–temporal scales, highly time-varying topologies, extremely frequent handovers, and cross-domain multi-path. Research efforts are directed towards satellite-terrestrial collaborative location management, elastic and efficient dynamic routing, seamless handover for large-scale users, aiming to break through key technologies of elastic, efficient dynamic routing and mobility management.

(1) Low complexity dynamic adaptive routing for ISTNs

The ISTN is characterized by its heterogeneous multi-dimensional nature, with a highly dynamic network topology. The relationships between nodes are complex and subject to frequent changes, resulting in slow routing convergence and poor adaptability. To address these challenges, research on SDN-based low-cost dynamic adaptive switching routing has been conducted, proposing dynamic adaptive routing and rapid convergence methods to achieve low-cost, high-reliability routing. Additionally, studies on real-time network topology analysis methods for multi-dimensional information integration have been carried out, introducing an active collaborative routing approach that aggregates capabilities. This is tailored to different network transmission capabilities and Quality of Service (QoS) assurances, achieving on-demand routing. Furthermore, based on differentiated reliable routing of network intelligence, a reliability situation assessment model for multi-hop transmission links between satellites and the earth is conducted. The model aims to achieve multi-level routing that fulfills spatiotemporal constraints, adapts to mission requirements, and optimizes resources.

(2) Heterogeneous low-latency efficient mobility management in ISTNs

In satellite networks, mobility management confronts challenges such as high handover frequencies and diverse types of handovers (horizontal/vertical, etc.), which makes it difficult to implement the selection of access points, measurement and decision-making for handovers, and path redirection, leading to intricate signaling processes. Research on low-latency conditional handovers based on intelligent mobility prediction has been conducted. An integrated coverage model based on dynamic graph has been proposed, introducing multi-granularity location registration and paging for satellite-terrestrial collaboration. Additionally, conduct a stochastic optimization model for access point selection under user fairness and resource constraints, supporting a seamless handover decision mechanism without user perception. Furthermore, research into the group characteristics and deterministic attributes (such as latency sensitivity) of user terminals and service applications leads to the proposal of a low-latency group handover method through satellite-terrestrial collaboration, aiming to achieve efficient on-demand group handovers.

3. On-demand deterministic service under high dynamic topology environment

The disparity between satellite network environments and terrestrial network environments is significant. The characteristics of satellite networks such as the high error rate, long latency and high dynamic nature mean that traditional terrestrial network service mechanisms cannot meet the service requirements of satellite networks, such as low latency, high bandwidth, high stability, etc. Research into new on-demand deterministic service strategies has been conducted, proposing high deterministic traffic control methods.

(1) High deterministic on-demand service layered architecture

To achieve deterministic services in high dynamic environments, it is necessary to achieve comprehensive perception of various services, real-time monitoring of network conditions, and precise management of traffic. Therefore, designing high deterministic on-demand service layered architecture is crucial. The layered architecture of the high dynamic topology environment is shown in Figs. 14.15. The SDN-based system is divided into the application layer, control layer, and data layer. The collaborative service mechanism is designed to facilitate centralized, real-time, and precise management of service flows in high dynamic topology environments, and provide support for deterministic services.

(2) High deterministic traffic control method based on queue scheduling

The constraints on satellite network resources render best-effort routing methods inadequate in meeting service QoS requirements. Moreover, the complex and sudden physical environment of satellite networks, coupled with their high dynamic topology, means that optimal resource allocation for current service flows may not yield satisfactory service outcomes subsequently. To address this issue, research on high deterministic traffic control methods based on queue scheduling has been conducted. By perceiving application-layer service demands and collecting real-time network conditions, this approach aims to optimize multi-dimensional QoS indicators through route optimization and queue configuration. It implements fine-grained, high deterministic queue settings to achieve on-demand deterministic services.

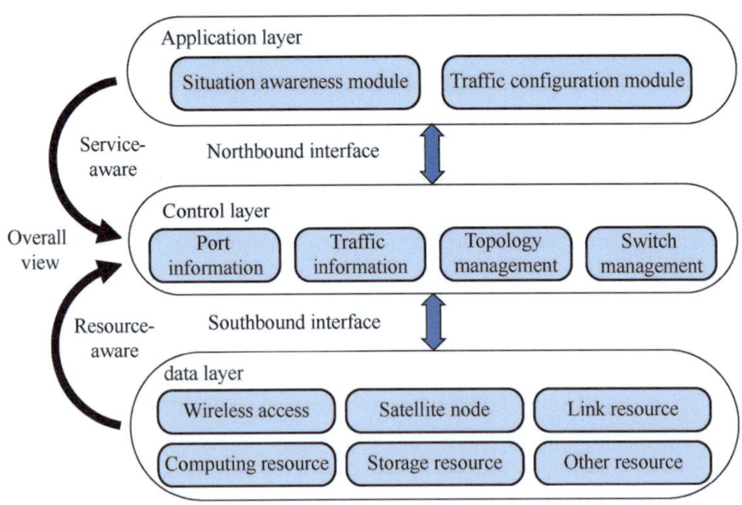

Fig. 14.15 Layered architecture for high dynamic topology environments

(3) Protection mechanism for SDN control flow

The high dynamic environment of satellite networks causes regular interruptions and frequent unexpected incidents in communication links, which can easily lead to control link failures in SDN networks. Consequently, switches may be unable to handle unknown traffic, leading to packet loss, and controllers may fail to obtain real-time status of the data plane, degrading overall performance. To handle these challenges, research on SDN control flow protection mechanisms [35–37] aims to achieve recovery and rerouting of control flows in case of topology failures, ensuring elastic connections between the data plane and control plane, and supporting deterministic guarantees for QoS.

14.3.4 Frequency Management for ISTNs

As user service demands increase, spectrum resources are becoming increasingly scarce. 6G aims to cover all scenarios in space, air, and ground, and its frequency requirements are anticipated to significantly exceed those of 5G. If traditional frequency rigid segmentation is used in ISTNs, it will lead to a decrease in transmission efficiency. However, flexible spectrum sharing can allow 6G to perform refined spectrum management, achieving spectrum resource sharing in multiple dimensions such as frequency domain, time domain, and spatial domain.

For the mid to high frequency bands that are favored by both satellite and terrestrial communications, flexible spectrum sharing can shift the competitive relationship between the two into a collaborative one, significantly enhancing the spectrum utilization efficiency of the 6G ISTN system. Given the spatial distribution differences in wireless transmission links between satellite-based and terrestrial-based systems, spatial reuse and interference avoidance become crucial means for flexible spectrum sharing in ISTNs. With the aid of AI, terminals can better distinguish between satellite and terrestrial communication signals based on signal direction, achieving spatial reuse and interference avoidance. To improve the utilization efficiency of frequency resources, it is necessary to study the signal transmission characteristics of multi-layer spatial networks, explore soft frequency reuse methods for ISTNs by leveraging the differences in beams and coverage. Furthermore, through interference prediction and resource coordination, research on technologies and methods for dynamic frequency sharing and reuse should be conducted to reduce cell edge interference and improve cell edge transmission efficiency. By introducing machine learning-based frequency situation prediction methods, dynamic frequency sharing strategies for satellite-terrestrial heterogeneous systems are proposed [38–40]. The principles of spectrum coordination and sharing between satellite and terrestrial networks in the mid to low-frequency bands are shown in Fig. 14.16, and the principles of interference avoidance and sharing between satellite and terrestrial networks in high frequency bands are illustrated in Fig. 14.17.

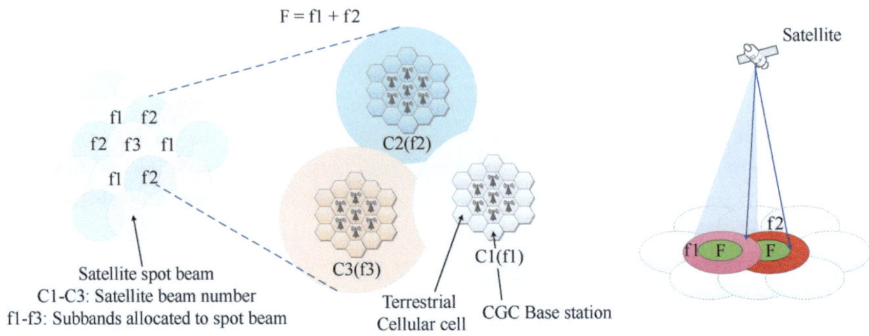

Fig. 14.16 Principles of spectrum coordination and sharing between satellite and terrestrial networks in the mid to low-frequency bands

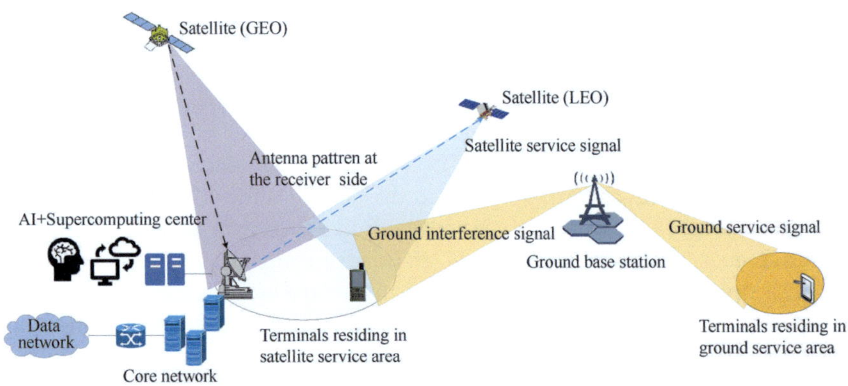

Fig. 14.17 Principles of interference avoidance and sharing between satellite and terrestrial networks in high frequency bands

Taking millimeter waves, the potential frequency band of 6G, as an example, emerging giant constellations are vying to use Ku, Ka frequency bands to enhance the capacity of single satellite, while terrestrial communications have also embarked on millimeter waves-based standardization and industrial advancement since 5G. The flexible spectrum sharing between satellite and terrestrial communications enables both domains to utilize millimeter wave bands, transforming their relationship from frequency competition to frequency collaboration. Due to the significant difference in propagation directions between satellite communication links and terrestrial communication links, terminals can distinguish between satellite communication signals and terrestrial communication signals based on the signal direction, thereby avoiding interference between satellites and terrestrial access points.

14.4 Summary

Satellite network and terrestrial network technologies are rapidly developing in their respective fields, and the deep integration of satellites and terrestrial networks has become the direction for the future development of 6G network technologies, which has been widely recognized by the industry. Through complementary advantages and close integration between satellite networks and terrestrial networks, the network coverage will be expanded, the overall efficiency of the network will be improved, and a global seamless three-dimensional coverage network will be realized. The integration of satellite and terrestrial networks will follow the industry consensus on the three-stage technical development path: service integration, standard integration, and system integration. This involves integrating in aspects such as frequency utilization, network architecture, resource management, air interface and service support, ultimately forming a unified network that provides imperceptible services through satellite-terrestrial integration.

References

1. JIANG W, HAN B, HABIBI M A, et al. The road towards 6G: a comprehensive survey[J]. IEEE Open Journal of the Communications Society, 2021, 2: 334–366.
2. LIU G Y, HUANG Y H, LI N, et al. Vision, requirements and network architecture of 6G mobile network beyond 2030[J]. IEEE China Communications, 2020, 17(09): 92–104.
3. ZHU J K, ZHAO M, ZHANG S H, et al. Exploring the road to 6G: ABC—foundation for intelligent mobile networks[J]. IEEE China Communications, 2020, 17(06): 51–67.
4. CHEN S Z, SUN S H, KANG S L. System integration of terrestrial mobile communication and satellite communication —the trends, challenges and key technologies in B5G and 6G[J]. IEEE China Communications, 2020, 17(12): 156–171.
5. CHEN S Z, SUN S H, WANG Y M, et al. A comprehensive survey of TDD-based mobile communication systems from TD-SCDMA 3G to TD-LTE(A) 4G and 5G directions[J]. IEEE China Communications, 2015, 12(02): 40–60.
6. QIAO X Q, REN P, NAN G S, et al. Mobile web augmented reality in 5G and beyond: challenges, opportunities, and future directions[J]. IEEE China Communications, 2019, 9(16): 141–154.
7. LIU J, DU X Q, CUI J H, et al. Task-Oriented intelligent networking architecture for the Space–Air–Ground–Aqua integrated network[J]. IEEE Internet of Things Journal, 2020, 7(06): 5345–5358.
8. XU K, QU Y, YANG K. A tutorial on the internet of things: from a heterogeneous network integration perspective[J]. IEEE Network, 2016, 30(02): 102–108.
9. KODHELI O, LAGUNAS E, MATURO N, et al. Satellite communications in the new space era: a survey and future challenges[J]. IEEE Communications Surveys & Tutorials, 2021, 23(01): 70-109.
10. GABER A, EIBAHAAY M A, MOHAMED A M, et al. 5G and satellite network convergence: survey for opportunities, challenges and enabler technologies[C]// Novel Intelligent and Leading Emerging Sciences Conference (NILES). Giza: IEEE, 2020: 366-373.
11. 3GPP. 3GPP TR 38.811: Study on new radio (NR) to support non-terrestrial networks V15.0.0 (Release 15) [R]. 2018.06.
12. Xu Xiaotao, Pang Jiangcheng, Li Chao. Recent developments and insights in constellation satellite mobile communication systems[J]. Defence Technology, 2021, 42(01): 100–105.

13. XU X T, PANG J C, LI C. Recent developments in satellite constellation networks and lessons learned[J]. Defence Technology, 2021, 42(01): 100-105.
14. GUIDOTTI A, VANELLI-CORALLI A, CAUS M, et al. Satellite-enabled LTE systems in LEO constellations[C]// International Conference on Communications Workshops (ICC Workshops). Paris: IEEE, 2017: 876–881.
15. 3GPP. 3GPP TR 38.821: Solutions for NR to support non-terrestrial networks (NTN) V1.0.0 (Release 16) [R]. December 2019.
16. NIEPHAUS C, KRETSCHMER M, GHINEA G. QoS provisioning in converged satellite and terrestrial networks: a survey of the state-of-the-art[J]. IEEE Communications Surveys & Tutorials, 2016, 18(04): 2415–2441.
17. CHEN S Z, LIANG Y C, SUN S H, et al. Vision, requirements, and technology trend of 6G: how to tackle the challenges of system coverage, capacity, user data-rate and movement speed[J]. IEEE Wireless Communications, 2020, 27(02): 218–228.
18. MA L, WEN X M, WANG L H, et al. An SDN/NFV based framework for management and deployment of service based 5G core network[J]. IEEE China Communications, 2018, 15(10): 86–98.
19. JIN X N, ZHANG P Y, YAO H P. A communication framework between backbone satellites and ground stations[C]//International Symposium on Communications and Information Technologies (ISCIT). Qingdao: IEEE, 2016, 479–482.
20. KODHELI O, GUIDOTTI A, VANELLI-CORALLI A. Integration of satellites in 5G through LEO constellations[C]// Global Communications Conference (GLOBECOM). Singapore: IEEE, 2017, 1–6.
21. BI Y G, HAN G J, XU S, et al. Software defined Space-Terrestrial integrated networks: architecture, challenges, and solutions[J]. IEEE Network, 2019, 33(01): 22–28.
22. KHALILI H, KHODASHENAS P S, SIDDIQUI S. On the orchestration of integrated satellite components in 5G networks and beyond[C]// International Conference on Transparent Optical Networks (ICTON). Bari: IEEE, 2020: 1–4.
23. ZHAO B K, FEI C J, MAO X L, et al. Networking in space terrestrial integrated networks[C]/ / International Conference on Optical Communications and Networks (ICOCN). Huangshan: IEEE, 2019: 1–3.
24. Chen Shanzhi. Analysis and development suggestions on low orbit satellite communication[J]. Telecommunications Science, 2020(6).
25. Sun Shaohui, Dai Cuiqin, Xu Hui, et al. Research on integrated networking of star-earth for 6G[N]. Journal of Chongqing University of Posts and Telecommunications, 2021(12).
26. Xu Hui, Miao Deshan, Kang Shaoli, et al. Satellite network architecture and key transmission technologies for space-earth integration[J]. Integrated Information Network of Heaven and Earth, 2020, 1(02): 2–10.
27. Wang Hucheng, Xu Hui, Sun Shaohui. Research on 5G network technology integrating satellite communication[J]. Radio Communication Technology, 2021.
28. ABDULGHAFFAR A, MAHMOUD A, ABU-AMARA M, et al. Modelling and evaluation of software defined networking based 5G core network architecture[J]. IEEE Access, 2021, 9(01): 10179–10198.
29. Xu Hui, Sun Shaohui. Research on the network architecture of heaven and earth integration for 6G[J]. Integrated Information Network of Heaven and Earth, 2021, 2(4):2–9.
30. Xu Changzhi, Jin Yi, Li Li, et al. Wireless transmission technology for satellite-terrestrial integration towards 6G[J]. Journal of Electronics & Information Technology, 2021, 43(01): 28–36.
31. XU C Z, JIN Y, LI L, et al. Wireless transmission technology of satellite-terrestrial integration for 6G mobile communication[J]. Journal of Electronics & Information Technology, 2021, 43(01): 28–36.
32. Hou Liming, Han Bo, Miao Deshan, et al. Air interface transmission technology based on 5G and its evolution for satellite-terrestrial integration[J]. Information Communication Technology and Policy, 2021(9).

33. SUN S HUI, HOU L M, MIAO D S. Beam Switching Solutions for Beam-Hopping Based LEO System[J]. IEEE VTC Workshop, 2021.
34. GOTO D, YAMASHITA F, SUGIYAMA T, et al. Broadband Multi-Satellite/Multi-Beam system with single frequency reuse[C]// IEEE Vehicular Technology Conference (VTC Spring). Glasgow: IEEE, 2015, 1–5.
35. PENG F, SHOU G C, HU Y H, et al. Lightweight edge computing network architecture and network performance evaluation[C]// Asia Communications and Photonics Conference (ACP). Hangzhou: IEEE, 2018: 1–3.
36. DONG P, GAO M, TANG F L, et al. Multi-Layer and heterogeneous resource management in SDN-Based Space-Terrestrial integrated networks[C]//International Conference on High Performance Computing and Communications; International Conference on Smart City; International Conference on Data Science and Systems (HPCC/SmartCity/DSS). Yanuca Island: IEEE, 2020: 377–384.
37. MENESES F, CORUJO D, NETO A, et al. SDN-based End-to-End flow control in mobile slice environments[C]// Network Function Virtualization and Software Defined Networks (NFV-SDN). Verona: IEEE, 2018: 1–5.
38. PARK J M, OH D S, PARK D C. Coexistence of mobile-satellite service system with mobile service system in shared frequency bands[J]. IEEE Transactions on Consumer Electronics, 2009, 55(3): 1051–1055.
39. MATINMIKKO-BLUE M, YRJÖLÄ S, AHOKANGAS P. Spectrum management in the 6G era: the role of regulation and spectrum sharing[C]// Wireless Summit (6G SUMMIT). Levi: IEEE, 2020: 1–5.
40. YAN S, CAO X Y, LIU Z L, et al. Interference management in 6G space and terrestrial integrated networks: challenges and approaches[J]. IEEE Intelligent and Converged Networks, 2020, 1(03): 271–280.

Chapter 15
6G Network Architecture

From the voice services of the 2G era, to the data services of the 3G and 4G eras, to the digitalisation of industries in the 5G era, the architecture of mobile communication networks has undergone revolutionary changes. Looking towards 6G, where should the architecture of mobile communication networks go? Mobile communication networks can be seen as a platform providing universal services, and we need to design the network architecture natively around the universal services provided. This chapter provides a detailed introduction to the overall architecture design of the 6G network.

15.1 The Evolution of Mobile Communication Networks Accompanied by Architectural Iterations

15.1.1 The History of Mobile Communication Network Evolution

The development of mobile communication networks has evolved from 2 to 5G.

(1) In the 2G era, mobile communication networks mainly provided voice and short message services. 2.5G added support for data services based on Circuit Switch (CS) architecture, supporting mobile data services at a small cost. At this time, the terminal was in a "semi-intelligent" form, which can be said to have a certain data service capability.
(2) 3G ushered in the real era of mobile internet, where users could surf the internet anytime, anywhere using mobile terminals. However, due to the limitations of mobile terminals, the initial network usage experience was not good. The emergence of smart terminals represented by the iPhone completely solved the bottleneck of the terminal, and the applications and content that followed

G. Liu et al., *6G Migrates from Communication to XaaS*,
https://doi.org/10.1007/978-981-96-7579-1_15

also experienced explosive development. Gradually, the wireless bandwidth provided by the 3G network could no longer meet the usage requirements of mobile applications, and the network gradually became the bottleneck, under which 4G was born.

(3) 4G was designed entirely for data services, and the mobile broadband (MBB) connection services it provided solved the bottleneck of the network, enabling the mobile internet to flourish and become an indispensable part of people's lives. Mobile communication networks have also increasingly become indispensable infrastructure like water, electricity, and coal.

(4) 5G was designed based on the historic opportunity of industry digitalisation. Based on the deployment method of the 4G network, deploying a single network can no longer meet the differentiated network needs of different industries. Networks from 2 to 4G were all built using dedicated hardware devices. If each vertical industry deploys a full set of dedicated mobile communication networks, the cost would be enormous. In addition, the unified deployment of the control plane and user plane is not conducive to the low-latency reliable transmission of data packets. Benefiting from the development of NFV technology, core network elements can be flexibly deployed on general servers through IT software, that is, service-oriented architecture. Based on the control plane implementation method of service-oriented network functions, the mobile network can be logically sliced into any network (network slicing technology) to meet the communication needs of different industry users. The complete separation of the user plane and control plane also brings unprecedented low-latency high-reliability transmission characteristics to 5G. 5G not only brings about a revolution in communication technology but also promotes the revolution of the digital industry.

Looking back at the development of networks from 2 to 5G, the air interface technology and network architecture of mobile networks are constantly evolving, and the native design of the architecture for services is the key to achieving universal services. The 2G network architecture was natively designed for voice services and began to introduce superimposed data services; data services made many optimisation designs in 3G, but from an architectural perspective, it still cannot be considered a complete native data service design; 4G is truly designed around data services [1], at this time, voice services have become a subsystem on data services, or it can be said to be an application. And 5G, with its flexible customisation requirements for industry digitalisation, introduces slicing and service-based architecture (Service-Based Architecture, SBA) at the architectural level [2, 3], it can also be said that a certain degree of native design has been carried out.

Although the native design of the architecture is not the only way to implement new features (in fact, more features are implemented by patching and overlaying the basic infrastructure), it is clear that the native design of the architecture can better, more efficiently, and more thoroughly implement new features.

15.1.2 Architecture is the Foundation of the System

We know that architecture has a significant impact on both the functional and quality attributes of a system and is the foundation of the entire system. Good architecture can make logical functions and services run more efficiently, bringing characteristics such as high cohesion, low coupling, easy expansion, agile evolution, and developer-friendly. However, doing native design at the architectural level usually comes at a high cost and requires sufficient driving force. In the 6G era, from 2C to 2B, from air interface to network, from KPI to business model, there are huge driving forces for change, which are important opportunities for architectural innovation.

In addition, extending the existing architecture often introduces additional complexity, either explicitly or implicitly, while adding features. When accumulated to a certain critical point, it can easily cause systemic problems, affecting functional and quality attributes, as well as the implementation of new features and the robustness of the overall system, among other key attributes.

A most typical example is the tragedy of the Boeing 737-MAX: Under the original design of the 737's short-legged architecture, considering new requirements for energy saving and long-range, the diameter of the aircraft engine is getting larger and larger, but the original 737 architecture's wing-to-ground distance cannot accommodate the new engine. Boeing's final approach was to adjust the engine position and shape through various patching measures, rather than designing a completely new body architecture like Airbus to natively support the new engine. As a result, whether it's the oddly shaped flattened engine on the Boeing 737-NG, or the engine's constant forward movement throughout the history of the Boeing 737, after further forward and upward movement, the aircraft will automatically raise its head in high angle of attack flight conditions. Boeing had to introduce a new mechanism to revise such a head raise, and the overall system complexity was too high. The pilots did not have time to undergo comprehensive training to deal with such abnormal conditions, leading to several disasters of aircraft stalling and crashing. If the Boeing 737-MAX could have undergone an architectural change before its debut, I believe there would have been a completely different outcome. The design evolution of Boeing aircraft is shown in Fig. 15.1.

15.1.3 6G the "Intelligent Inclusiveness" of the Network Requires Native Architectural Design

Looking at the evolution of mobile communication network architecture, as shown in Fig. 15.2, 2G to 5G each provide different types of inclusive basic services, all of which are inseparable from native architectural design.

(1) The inclusive basic service of 2G is voice service, and the matching 2G end-to-end network architecture is natively designed for voice.

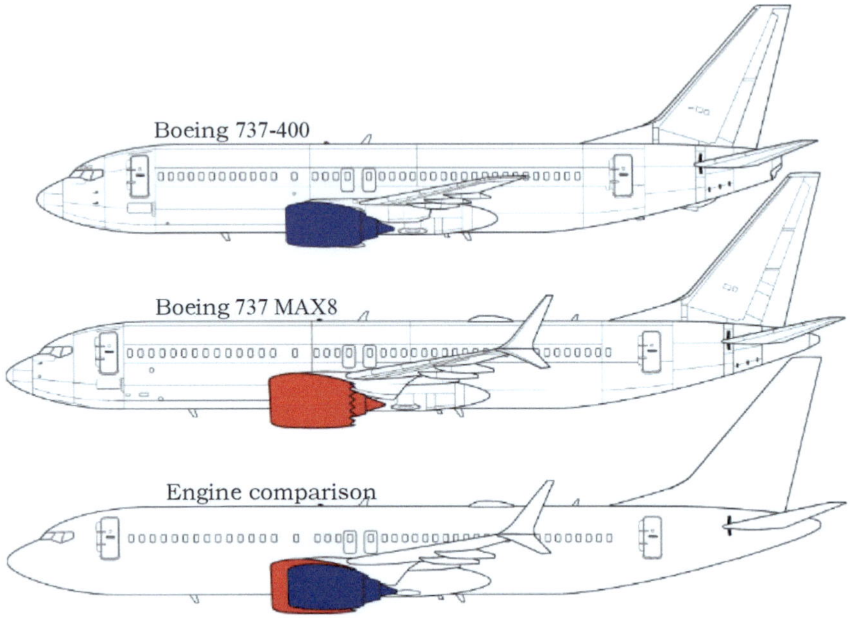

Fig. 15.1 Design evolution of Boeing aircraft [4]

(2) By the time of 3G and 4G, the inclusive basic service is data service, but the 3G architecture cannot be fully considered as a native data architecture, and more often data services are overlaid on the traditional network base; the 4G architecture is completely based on data services for native design, and traditional services such as voice must be provided based on basic data services (Voice Over LTE, VoLTE).

(3) The basic service provided by 5G is the Internet of Everything [2], from MEC to slicing (uRLLC, mMTC and eMBB), to non-public networks (None Public Network, NPN), etc., 5G has designed many architectural capabilities to better serve the industry, and is still on the path of evolution.

The current industry consensus on 6G is that the core vision of 6G is to achieve "universal intelligence". In the field of intelligence, relying on the NWDAF of the 5G architecture to add functions or to provide cloud AI (Cloud AI) in a standalone manner, will face many challenges, such as difficulty in data acquisition, difficulty in ensuring data quality, insufficient generalisation of AI models, and lack of effective verification and guarantee methods for application effects, etc. These factors have led to the performance and efficiency of artificial intelligence being lower than expected.

In addition, how to provide more convenient service capabilities for 2B, how to release the new value of the air interface and the network, how to provide a more secure and trustworthy network service environment, etc., these key issues that affect the network usage experience, all need to be natively designed at the architectural level.

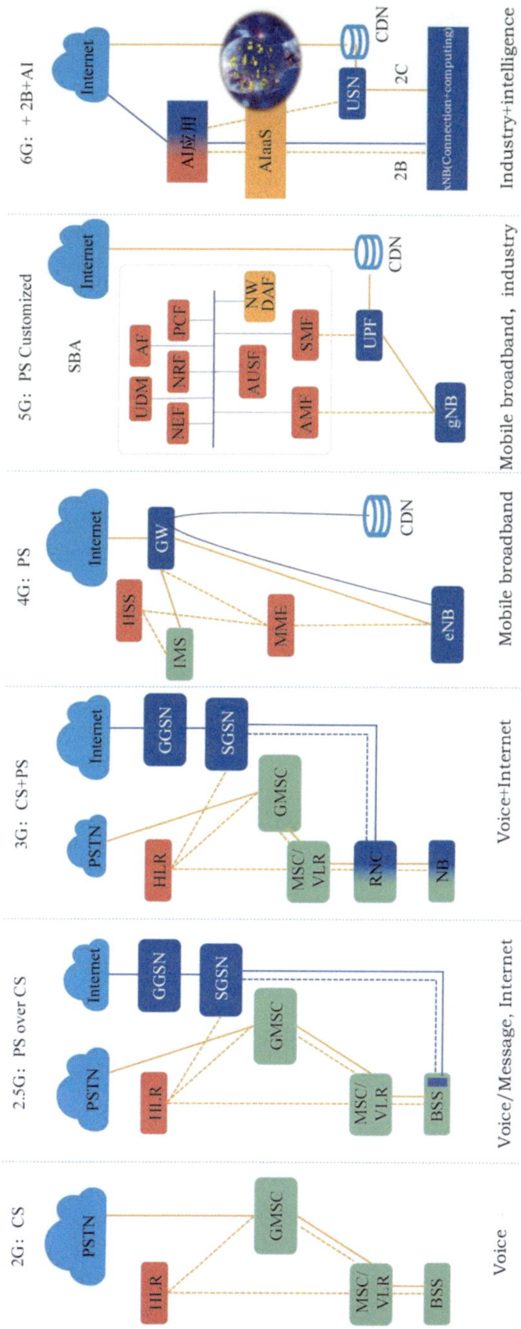

Fig. 15.2 Evolution of wireless communication network architecture

15.2 The Driving Force of 6G Network Architecture Transformation

15.2.1 The Existing 2C Business Model Has Reached a Bottleneck

5G networks already widely support high-speed, large-bandwidth applications, driving the further rapid development of the mobile Internet, achieving high-speed growth in mobile traffic every year. In the 6G era, more innovative applications will emerge, such as extended reality (Extended Reality, XR), tactile Internet, holographic communication, etc., may all become mainstream applications, which also stimulates the demand for network capacity. However, from a business driving force perspective, it is difficult for 6G networks to bring substantial revenue growth to operators by simply providing more traffic. At the same time, the future new applications' demands for the network are not just about low-latency high-bandwidth connections, especially in terms of computing power, data processing capabilities, etc., there are new demands. Therefore, it is necessary for 6G networks to further explore new service capabilities, such as the universal environment of AI, data and enhanced industry capabilities, etc.

These new service capabilities will go beyond the traditional business scope of operators and involve all walks of life. The 6G network architecture needs to provide various new capabilities in the form of XaaS, which constitutes the business driving force for the transformation of the 6G network architecture.

15.2.2 2B the Full Arrival of Business Scenarios

Starting from 5G, mobile networks have begun to consider providing technical support for 2B scenarios, providing them with customisable high-reliability communication services [2]:

(1) Network edge deployment (MEC);
(2) Network slicing;
(3) Non-public network (Non-Public Network, NPN).

Regardless of the above technologies, although they can provide certain near-deployment and isolation evolution capabilities, they are essentially still controlled by network service providers. For 2B users, the control and management range of the network is quite limited and cannot achieve real-time network control.

With the full arrival of 2B scenarios, 6G networks need more 2B-friendly capabilities, such as:

(1) Flexible and fully customisable network space;
(2) Diversified user capabilities beyond connection (connection + data + computing);
(3) Network operability rights of users/tenants/integrators;

(4) Integration and cost advantages.

Through the construction of these capabilities, each participant can own their own network space at a lower cost, and influence the network space directly and efficiently to serve themselves, thereby growing a more prosperous ecosystem for thousands of industries.

15.2.3 New Business Scenario—Metaverse

With the continuous evolution of applications, mobile networks will welcome more and more new business scenarios, including the hot metaverse scenario in recent years. As an emerging concept that affects the entire industry chain, the metaverse is increasingly affecting the technical development and business decisions of all relevant parties. A mature metaverse ecosystem includes not only the upper layer of experience, discovery, and creator economy, but also the lower layer of spatial computing, decentralisation, human–machine interaction, and the corresponding infrastructure layer. Due to its special ability attributes, mobile networks can build capabilities at the lower level and become the base of the metaverse. The specific reasons are introduced below.

1. Mobile networks are associated with digital humans

Mobile networks can perceive people and businesses, and provide differentiated services based on the identity of users and the type of sessions, which is a prototype of digital humans. In the real world, mobile numbers and SIM cards are increasingly tied to a large number of digital applications and social services, and are constantly evolving, becoming a cornerstone of trust. This shows that using the identity of the mobile network as the basis for the identity of the digital human is widely existing and can be a natural choice for the path of the metaverse.

2. Mobile networks are associated with natural humans

Natural humans must be associated with digital humans through terminals, and the terminals and corresponding wireless access are the core concerns of mobile networks. As the metaverse application is getting closer, the "perception" of digital humans to natural humans will become more complex and real, from location to imaging, from single sense to holographic perception, a large number of natural human agents—sensors will be spread and controlled by mobile networks.

3. Ubiquitous access of mobile networks

The core advantage of mobile networks is ubiquitous access. No matter where the user is, such as the ground, ocean, sky or even the universe, 6G networks can provide network access services for them, keeping the connection between the user and the application server at all times. At the same time, the wireless attribute also frees users from the constraints of cables.

In summary, new business scenarios put higher demands on mobile networks, in turn, 6G networks will also use their own favourable conditions to provide soil for the growth of new businesses.

15.2.4 Release of the Value of New Air Interface Capabilities

Achieving "ultimate connection" is one of the visions of 6G [5, 6], more and more access technologies will be applied to the 6G access network, including terahertz access and visible light communication. Traffic growth usually requires more wireless spectrum, and cellular network infrastructure tends to use low-frequency spectrum to achieve ubiquitous coverage. After several generations of wireless network evolution, more and more spectrum resources have been used for network upgrades. 6G will not only be deployed in the millimetre wave band, but will also be applied to terahertz and even visible light for the first time. This means that in order to achieve the ultimate connection, any available frequency band may be used.

In order to reasonably apply various access technologies, innovative wireless signal coverage solutions are needed to meet the requirements of 6G networking and capacity. Low Earth Orbit satellites can enhance ground-to-space communication coverage, and high-altitude platform stations can also be deployed as temporary solutions, both of which are beneficial supplements to the wireless signal coverage infrastructure. In this way, 6G can provide almost unlimited capacity, enabling 6G wireless connections to reach unprecedented speeds.

In addition, besides serving as a communication network, the 6G network will also become a sensor network. Based on the design of integrated sensing and communication, wireless networks can provide new sensing services and enhance wireless communication capabilities. Traditional sensing is an independent function, including a complete set of dedicated devices, such as ordinary radar, lidar, computer tomography, magnetic resonance imaging, etc. With the development of mobile terminal positioning technology, mobile phone positioning in mobile communication systems has also become a basic type of sensing capability. Mobile terminals use air interface signals and internal measurement technology to achieve precise positioning. However, using millimetre waves and terahertz bands with larger bandwidth and smaller wavelengths, sensing functions can be integrated into 6G networks that are strongly related to these two bands.

Whether it's the ultimate connection of 6G or the new sensing capabilities, they need to be provided through the native design at the architectural level to provide new dimensions and operating modes, thereby truly releasing the capabilities and value of the new 6G air interface.

15.2.5 Breakthrough in Network Pipeline Capabilities

Since entering the mobile Internet era, the mobile network itself has always focused on doing the pipeline well. Operators support a variety of colourful mobile Internet applications by building differentiated connection pipelines. Such division of labour has contributed to the vigorous development of the mobile Internet, truly changing all aspects of people's lives..

But in the era of 6G intelligence for all, is the evolution of mobile networks still limited to connection pipelines? From the consensus of current industry research, the answer is no. For example, AI is essentially supported by three basic elements: data, computing power, and algorithms, all of which are indispensable. These basic elements may be ubiquitous in networks, terminals, and applications, and cannot be separated. Communication, computing, data, control, etc. must be closely integrated.

From another perspective, in future networks, the deep integration of information technology, communication technology, and data technology [Information Technology (IT), Communication Technology (CT), Data Technology (DT), ICDT] is a common consensus in the industry. If we continue to maintain the boundary division of end-pipe-cloud, it will be difficult to flexibly and uniformly schedule data, computing power, and algorithms in the entire ICDT system, and will face huge challenges in technology and ecology.

The evolution of wireless network positioning is shown in Fig. 15.3. From the history of wireless network evolution, the wireless pipeline must move towards endogenous intelligence. During the evolution from 2 to 4G, with the end-pipe-cloud each deriving their own independent ecosystems, relatively independent evolution was achieved. This is an inevitable result of the transition from voice to mobile Internet. Behind it is the fact that the fixed Internet has led the development for more than ten years, the mode of mobile Internet has been integrated by the fixed Internet, and the connection is very thin (funnel, IP thin waist). By 5G, due to the trend of industry digitisation, the mobile network has expanded its pipeline capabilities, but the end-pipe-cloud model has not been rethought, and 5G has also encountered problems in industry applications. What the industry really needs is a complete solution of connection + computing + intelligence, which may only be truly solved by 6G. 6G is no longer a pure "pipeline", but will break through the original boundaries and become an ubiquitous, distributed, endogenous intelligent innovation network.

15.2.6 Comprehensive Improvement of Network KPIs

Different services will have different network requirements. The design goals of 6G will not only be reflected in the key performance indicators (KPIs) of the new air interface, but also in new indicator dimensions in the network architecture. Traditional transmission services pursue high bandwidth, low latency, high reliability, and

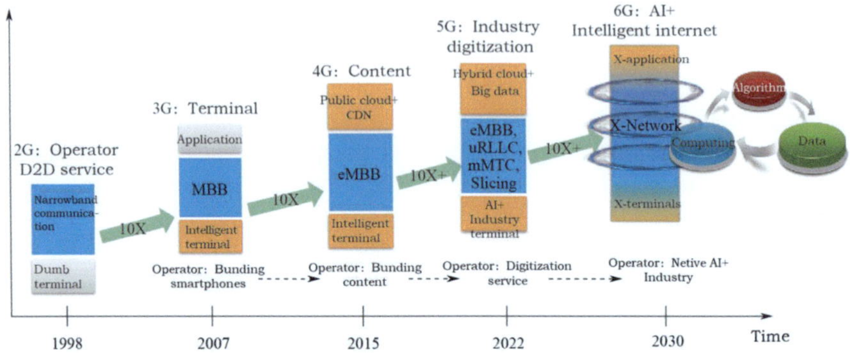

Fig. 15.3 Evolution of wireless network positioning

massive access performance indicators, while future 6G networks will pursue excellent network performance in more dimensions, including KPI design in computing power, data processing ability, perception ability, network deployment flexibility, and security.

For example, computing services need to meet the demands for computing time and resources; data services need to meet the demands for data capacity and data processing speed; perception services need to meet the demands for perception accuracy and range; trustworthiness needs to meet the demands for trust level classification; intelligence needs to meet the demands for automation service level classification and algorithm capability. In addition, the ease of network deployment and the degree of customisation are also KPIs for whether the future network can flexibly meet user needs. The reasonable customisation of these KPIs is a necessary condition for the 6G network to meet different applications in the future. The KPIs of the 6G network architecture are shown in Fig. 15.4.

The future 6G network requires the joint design of operators, equipment vendors, application providers, and users. In order to build an inclusive and efficient 6G network [5], the industry consensus on new KPIs will play a crucial role in the future standardisation of 6G.

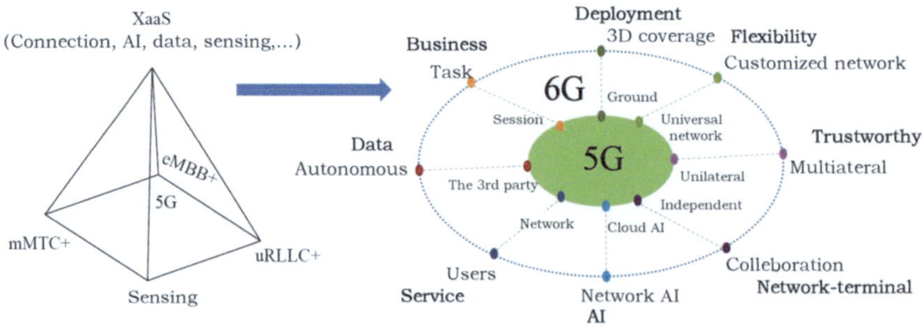

Fig. 15.4 KPIs of 6G network architecture

From historical experience, the network can achieve small gains in KPIs through patch-like technology enhancements, such as the control plane and user plane separation scheme in the later stage of 4G, and the distributed AI improvement technology in the later stage of 5G. However, breakthrough improvements in KPIs must be considered at the beginning of network architecture definition, and revolutionary changes should be made to the network architecture at the beginning of each generation, such as the 5G SBA architecture greatly increasing the flexibility of network deployment through software-based network function design.

15.3 6G Network Architecture Technical Requirements

Based on the business and technical trends facing the 6G network architecture, the requirements for the network architecture can be extracted from the following aspects.

15.3.1 Endogenous AI, Universal Intelligence

In recent years, technological innovations such as the Internet, big data, cloud computing, artificial intelligence, and blockchain, as well as the continuous emergence of new business forms, new scenarios, and new models, have accelerated the development of the digital economy, promoted the generation of massive data, and made various industries more urgent in their demands for communication and computing. Communication and computing have become the cornerstone of the digital transformation of the whole society, directly determining the height of social intelligence development. As a conduit for connecting users and transmitting data, mobile networks can perceive computing, supporting the efficient use of diverse distributed computing resources, such as edge computing deployed within mobile networks to reduce end-to-end latency and enhance service experience.

As a current hot topic in academia, the most discussed aspect of AI at this stage is how to use AI to optimise air interface links and network performance, improve network operation and maintenance automation, and user experience, etc. These may only be one of the applications of endogenous AI [6] in the network.

The future 6G network should achieve universal intelligence, and should not be limited to using AI to solve the network's own problems. It should also provide better architectural support for third-party AI applications such as industry digitisation. Therefore, how to define endogenous AI at the future architecture level to natively support various types of AI applications is something we need to consider.

The essence of architecture-level endogenous AI is to provide a complete AI environment in the 6G network, including AI infrastructure, AI workflow logic, data and model services, etc., to support all types of AI scenarios in 6G with a unified framework, which we call Network AI [7, 8], the specific supported application scenarios are introduced below.

(1) AI for Network (AI4NET): Enhancing the performance, efficiency and user service experience of the mobile network through AI. The main research of AI4NET includes using AI to optimise traditional algorithms (such as air interface channel coding, modulation), optimise mobile network functions (such as mobility optimisation, session management optimisation), optimise mobile network operation and maintenance management (such as resource management optimisation, planning management optimisation), etc.
(2) Network for AI (NET4AI): Providing various support capabilities for AI through the mobile network, making AI training/inference more efficient, more real-time, or enhancing data security and privacy protection, etc. NET4AI extends the traditional network range from connection services to computing power, data, algorithms, etc.
(3) AI as a Service (AIaaS): Building AI application service capabilities in the mobile network infrastructure, AI applications include AI for network use or new AI businesses, AI applications can be deployed by operators or third parties.

Network AI coexists and complements existing cloud AI, but at the same time, Network AI can also evolve and develop independently of the cloud.

15.3.2 Network is Flexible and User-Friendly

In previous mobile networks, users (including end users, tenants, etc.) often only acted as consumers of network services provided by operators, and could not directly influence the design and operation of the network. For users, the network was neither flexible nor friendly. Although 5G introduced the concept of slicing, allowing slice tenants to customise and manage network slices through the management plane, there are still many problems with network slicing, such as the inability to support tenants to deploy independent data processing and storage, inefficient implementation of security isolation, inability to flexibly customise networks according to the way they want, inability to design and control security policies within the enterprise organisation, and more importantly, lack of flexibility in business models, such as difficulty in leveraging professional industry integrators to provide better services for enterprises.

Therefore, from the user's perspective, in the 6G network, the mobile network needs to provide users with flexible network customisation methods to meet users' needs for network manageability and controllability, which can specifically include the following points.

(1) Flexibly customised user space. The 6G network can provide users with network space services, in addition to connection services based on slice logic, user space can also provide users with data processing and storage capabilities. User spaces can be independent of each other and adopt independent security mechanisms based on user needs. In addition, user space can also provide network services such as AI, data governance or perception based on user needs.

(2) User-autonomous asset governance. Users have autonomous management capabilities for user space, and the network assets in the user space (such as data stored or generated by computing in the user space, connection or computing capabilities services in the user space, etc.) are managed by the user, who can choose to open them to the operator or other users. For example, the network can give users the ability to allow other users or devices to access the network space owned by the user, and the operator needs to use the capability services or data of the user space through the user's authorisation.
(3) Efficient user control methods. In the current network, when a user (tenant) needs to adjust the network, they often need to negotiate with the operator, who operates the network through the management plane, which is very inefficient and complex. Users of the 6G network can operate the assets of the user space in a more direct and diverse way (such as control plane and user plane instructions, etc.), thereby enabling professional industry integrators to provide services for enterprises.

15.3.3 Safe and Reliable, Trusted by Many

6G clearly faces the threat of quantum computers, and the traditional cryptography security based on computational complexity has great hidden dangers. In addition, the widespread application of new technologies such as AI and blockchain (Block Chain, BC) is a double-edged sword. These new technologies are used to enable some features of the 6G network, but they may also bring more security risks to the network. It can be foreseen that the security attacks faced by 6G will be more diverse and intelligent. Therefore, for the security of the 6G network, it is necessary to consider how to form a consensus in the dimensions of architecture and standards, define architecture-level endogenous trustworthy security mechanisms (not just authentication and encryption), and avoid attributing safety and trust to the engineering trust of the manufacturer's product implementation.

Trustworthy functions independently carry safety and trust capabilities. The difference between trustworthy functions and the security capabilities of 5G and before mainly lies in the following four features.

(1) Security evolves towards trust. With the development of security technology, security capabilities will not be limited to the categories of confidentiality, integrity and availability (Confidentiality, Integrity and Availability, CIA), but will further evolve towards trust capabilities. Security introduces physical layer security technology, blockchain, artificial intelligence and other technologies, which will further enhance the trust capabilities of 6G. With the help of physical layer security technology, the security of the air interface can be enhanced. With the help of blockchain technology, a new type of consensus-based decentralized trust model can be built, which complements traditional centralized trust and endorsement trust. Based on artificial intelligence technology, intelligent learning and decision-making of security policies can be achieved.

(2) Trust capabilities evolve independently. From the perspective of 3GPP standardization, the security capabilities of mobile networks have always been strongly coupled with other network capabilities, with security messages embedded in control plane messages, such as terminal-side security capability reporting embedded in terminal capability reporting sets, private network authentication embedded in private network access processes, etc. The strong coupling of security and communication capabilities ensures the economical use of network resources, the establishment of communication and security simultaneously, and a method is needed to resolve the contradiction that security technology needs to serve mobile networks deeply and evolve independently of mobile networks. We can consider adopting the independent evolution of security capabilities and network capabilities to solve the above contradiction.

(3) Trust is equal. The centralized architecture of the secure mobile network establishes a network message flow mode of the dominant party and the passive party between the operator and the user. Among them, the security-related messages of the control plane are initiated by the network side, such as access network authentication, security context carry transfer, etc.; the security capabilities are completely defined and decided by the network side, such as encryption algorithm priority, authentication timing and session key update cycle, etc. Enhancing the user's voice in the customization of trust capabilities, enhancing the access network's advisory power and execution ability in the execution of trust capabilities, reducing the core network's decision-making power in establishing the trust relationship between the two ends of communication, making it possible for 6G users to actively demand, choose and decide on the trust capabilities they need; making it possible for the access network to provide special trust services for the user end, reducing the transmission and request decision cycle of the security context.

(4) Trust as a Service. As part of the 6G network architecture, the trust function, along with the management plane, computing function, and data function, provides the ability for Trust as a Service. Blockchain is part of the trust-inducing enabling technology. The 6G blockchain differs from traditional blockchain, it is dynamically built based on 6G end-to-end network infrastructure, serving 6G business needs, executing 6G unified protocol stack and communication protocols, flexibly deployed according to 6G network characteristics, and established using cryptographic algorithms. It supports various chain structures, block structures, and consensus algorithms, and is one of the capabilities of 6G Trust as a Service. The blockchain enabling module executes transaction consensus, link communication, and smart contracts, and has the ability to open features, allowing customisers to build business blockchains oriented to specific business logic by calling the enabling module, such as Decentralized Public Key Infrastructure (DPKI) chain, Identity Management (IDM) chain, network behaviour recording chain, etc.

15.3.4 Data Collaboration, Regulatory Compliance

Data is generated, flows, and is consumed in mobile networks, playing a significant role in the operation of the economy and society, and is considered the "new oil" of a data-driven society. The 6G network will become the infrastructure of various key infrastructures. As the network scale, new technologies, applications, etc. develop, the data in the network is increasing and becoming more important. Efficient use of data requires a complete service architecture, spanning the entire lifecycle from data generation, collection, storage, transmission, processing, analysis, exchange, and sharing. Data services are based on a framework of data collection, preprocessing, distribution, publication, and analysis, meeting the requirements of data laws and regulations, balancing data sharing and security, and providing data as a service product.

From the perspectives of data value discovery, regulatory requirements, technology development trends, and business monetisation models, the demand for data services in the 6G era will be more urgent, and different from the traditional monolithic data service architecture, a unified data service architecture is needed to meet these needs. The specific requirements are introduced below.

(1) Massive data collection and mining. Network regulation and optimisation, fault warning, and user experience improvement require operators to continuously monitor the network, thereby generating a large amount of uninterrupted network status and behaviour data, and conducting in-depth analysis to achieve fault warning, network optimisation, and operational value-added. The network is a complex system that requires unified coordination from the architectural level and mining the hidden correlations from massive real-time data.

(2) Meet the legal and regulatory requirements for data supervision. With the widespread and deep application of ICT, the continuous disclosure of data security and privacy leakage incidents, people are increasingly aware of the importance of privacy and data ownership. Major countries and organisations have introduced relevant laws and regulations to regulate the use of data, such as China's Personal Information Protection Law (PIPL), the European Union's General Data Protection Regulation (GDPR), etc., clarifying users' control over personal data. Data subjects should be able to independently decide whether to monetise, share, or provide personal data for AI model training. Therefore, in the 6G era, whether a trustworthy data service can be provided will become an important prerequisite for the monetisation of data value. And the processing of data, from data collection to the final result feedback, requires full-process compliance detection, which requires end-to-end data services, to achieve closed-loop control, and meet the end-to-end supervision requirements for data processing.

(3) Data-centric distributed heterogeneous processing. Existing networks are mostly designed according to specific types of applications, designing data service architectures, providing application-centric services, and application expansion capabilities are limited. The network is increasingly generating and

consuming massive data, and in the 6G era, Extended Reality (XR), Integrated Sensing And Communications (ISAC), Digital Twin Network (DTN) are emerging. New terminals, new businesses, and new functions such as Delay Tolerant Networks (DTN) and AI generate a large amount of data in the 6G environment and may process third-party industry data. A data-centric approach requires an architecture designed based on the distributed characteristics of data and the demand for data services. At the same time, the generation and consumption of data span different ends, wireless access networks (Radio Access Network, RAN), transport networks (Transport Network, TN), core networks (Core Network, CN), etc., presenting multi-source and heterogeneous characteristics. It is necessary to unify cross-domain data services, manage data from a global perspective, and fully and efficiently mine the value of massive data in the 6G network.

(4) The demand for data sharing/exchange and monetisation. Data, as the fifth major factor of production, is a new type of productivity and bears the mission of prospering emerging markets. However, the differences in data types, quality, attributes, etc., make a huge demand gap between the generation and consumption of data, making it difficult to directly share and exchange, and unable to support data monetisation. Therefore, it is necessary to lead the global data service, eliminate the difference between data production and consumption through global data collaboration.

15.3.5 Open Cooperation to Expand the Ecosystem

In the technical requirements of the 6G network, whether it is AI, data, or security and trust, there are issues of ecosystem synergy.

1. Network Co-Management Ecosystem

In the process of network planning, construction, and use, it often involves multi-party co-construction of network infrastructure, multi-party maintenance of network resources, and multi-party co-management of network services. The constructors, maintainers, and operators of the network may not be the same unit, and each part may be jointly responsible by units from multiple fields. This requires enabling a richer ecosystem at the operation and maintenance level.

2. Data Sharing and Trading Ecosystem

The real value of data depends on its liquidity and confidentiality. Open sharing of data is an important mechanism to realise data flow and reflect its value; the owner or provider of data must provide data in the form of a service to the data consumer to realise value monetisation. In a sense, the network itself needs to support a data market with data security and privacy protection.

3. Industry Private Network Ecosystem

In 2B scenarios, the function of the network participates in the production environment of the industry, and some network information needs to interact with industry knowledge (such as AI algorithms and models, application strategies, etc.). These industry knowledge often can only come from specialised industry organisations, which requires an open ecosystem to connect the network with the industry.

4. Integrator Ecosystem

Beyond issuing cards and numbers, operators will wholesale the network to integrators for users/tenants/integrators, and integrators will carry out secondary processing (such as data adaptation, network capability adjustment, policy application, etc.) according to their respective industry fields. Then a large number of integrators will bring the network to thousands of industries. This method draws on the successful experience of digital technology landing in traditional industries, introduces new roles through changes in business models and control models, and truly activates the 2B market.

5. Developer Ecosystem

The 6G network has inherent connection, data, and computing capabilities, but how to associate complex network capabilities with end users (industry terminals/applications, Application, App) is not enough to rely on equipment vendors and operators themselves, often also requires a developer ecosystem. In the 6G network architecture, developers should be allowed to orchestrate and assemble network capabilities, define and deploy network capabilities as needed, flexibly load applications and data, to get closer to the industry and prosper network applications.

15.4 6G Network Functional Architecture

15.4.1 Overall Functional Architecture

In order to meet the aforementioned technical requirements and better meet the demands of 6G network development in a native architecture manner, a comprehensive functional view needs to be defined to express the logical functions of the network.

The overall architecture of the 6G network is shown in Fig. 15.5. The complete overall architecture mainly includes task orchestration scheduling and control, connection function (control plane), connection function (user plane), computing function, data function, and trust function, which will be introduced one by one in the following content.

From the perspective of meeting technical requirements, "Endogenous AI, intelligent inclusiveness" is mainly reflected in the computing function and data function; "Network flexibility, user-friendly" is mainly reflected in how users (tenants) can

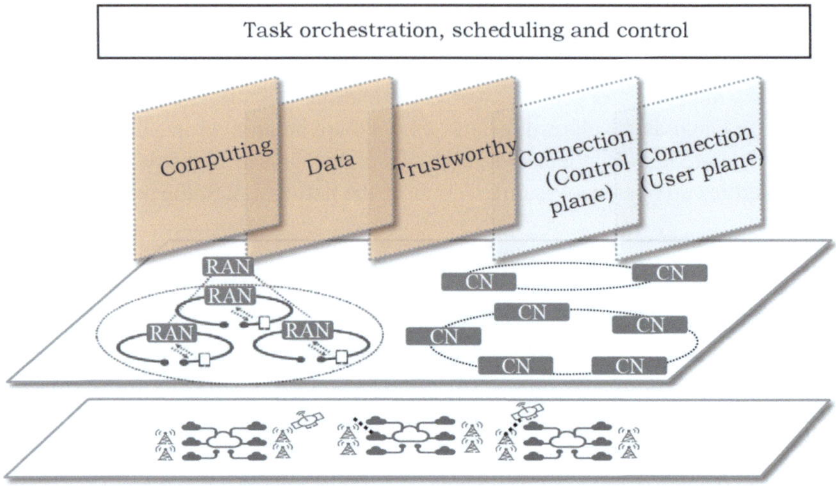

Fig. 15.5 Overall architecture of the 6G network

conveniently obtain and access connection functions, as well as the services provided by the computing function and data function; "Secure and reliable, multi-party trust" is mainly reflected in the trust function; "Data collaboration, regulatory compliance" is mainly reflected in the data function; "Open cooperation, enlarging the ecosystem" will affect all functions.

The aforementioned architecture involves a large number of different nature function sets, which can either run as enhanced network capabilities on top of existing network functions, or exist as independent network functions; specifically, we also explore considering them as an independent "plane", encapsulating service functions across multiple entities through the plane, and having a layered (protocol) structure internally, such as connection plane, computing plane, data plane, trust plane, etc.

To become a "plane" in the network architecture, the following conditions need to be met.

(1) Provide a type of basic functional service, such as session control, data transmission services, etc.
(2) Have an independent protocol stack, such as connection protocols (such as NAS and RRC), computing protocols, etc.
(3) The resources and functions involved in the services provided are multiple entities, such as the base station and UPF of the user plane, etc.

15.4.2 *Task Orchestration Scheduling and Control*

The 6G network goes beyond pipeline connection services, providing XaaS, and needs to introduce a business orchestration and management system. Based on the previous discussion, the vision of 6G to achieve intelligent inclusiveness, more XaaS services can essentially be considered as a kind of intelligent service, that is, providing endogenous artificial intelligence capabilities and services in the 6G network infrastructure. Therefore, a new task orchestration scheduling and control architecture needs to be introduced to complete the functions of business opening, execution scheduling, and resource unified management. The service system of XaaS is shown in Fig. 15.6.

The task orchestration scheduling and control architecture mainly has three layers, as shown in Fig. 15.7.

(1) Interface layer: Mainly facing developers and network operation deployers. Developers can submit intelligent business requirements by inputting program code, intelligent models or logical images, and can perform flexible and efficient computing on the cloud edge of the 6G network. Network operation deployers, by describing inference services, training services, or general graph computing services, can provide intelligent services flexibly and efficiently on the cloud edge of the 6G network.
(2) Scheduling layer: Mainly provides the allocation management functions of images, models, data, and tasks. Management includes storage, forwarding, allocation, orchestration, scheduling, performance monitoring. It receives the computing task requirements obtained from the interface layer, coordinates computing storage resources, completes the mapping of computing tasks to computing power, and objects to storage. At the same time, in order to improve network resource efficiency or user experience, the management layer will endogenously generate some intelligent computing tasks. The scheduling algorithm can be injected externally, or it can be generated and refreshed by online training. If it is injected externally, a relevant algorithm injection update rollback interface needs to be provided.
(3) Execution layer: This is mainly responsible for execution, including computation and data transmission, as well as monitoring and statistical reporting of

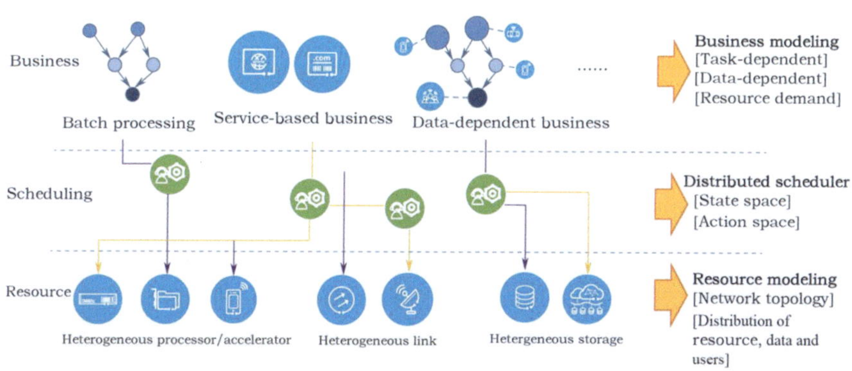

Fig. 15.6 Service system of XaaS

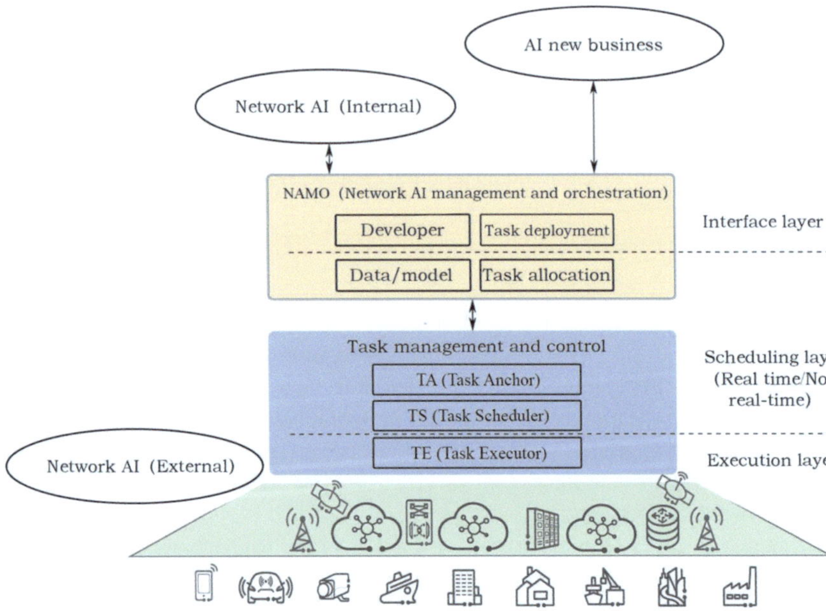

Fig. 15.7 Task orchestration scheduling and control architecture

network resources. After receiving the computing tasks allocated by the scheduler, the execution body carries out efficient execution according to the standard description form of the computing task (which could be a code snippet, image, or intermediate representation IR) and the actual local resource situation (including processor type, memory hierarchy, resource busy/idle status, etc.). At the same time, it monitors and reports changes in resource allocation, the completion of computing tasks, and timely reports to the scheduling layer, enabling the scheduling layer to map strategies more efficiently based on updated status information.

The logical functions involved are as follows.

(1) Network Artificial Intelligence (AI) Management and Orchestration (NAMO) function is responsible for the interface layer's developer interface, external application task deployment interface, orchestration layer's data/model management and non-real-time task allocation function. The NAMO function can run on the management plane, implementing task orchestration scheduling and control architecture and interaction with external systems and humans.
(2) Task Anchor (TA) function is responsible for the scheduling layer's task execution lifecycle management, completing task deployment, startup, deletion, modification, monitoring, etc., based on task QoS requirements, including the regulation of four essential resources to ensure task QoS. The TA function can

run on the control plane, acting as an interactive entity between the management plane and the control plane.

(3) Task Scheduler (TS) function is responsible for the scheduling layer's real-time control of tasks. During the deployment process of task instances, the TS establishes and maintains task-related context information, thereby controlling the tasks. The TS function can run on the control plane, ensuring the real-time nature of scheduling.

(4) Task Executor (TE) function is responsible for the execution layer's specific task execution, and for business logic interaction and execution status feedback. The workflow of the same service may be instantiated into multiple tasks, deployed to be executed among multiple TEs, hence there is business interaction between TEs.

15.4.3 Connection Function

From 2G voice services to 5G Internet of Things, the connection services provided by the connection plane have always been one of the most basic services of mobile networks. The connection plane includes the control plane (CP) for transmitting signalling and the user plane (UP) for business data. Based on network deployment requirements and to provide a better business experience, the control plane and user plane have evolved from initial unified deployment to current separate deployment. The future 6G network connection plane needs to meet the network requirements of new businesses and provide ubiquitous connection services, as detailed below.

1. Support for new business requirements

To support different connection businesses, mobile networks have designed specific QoS guarantees and network access functions. With the emergence of new businesses, the 6G network connection plane needs to provide connection service support for new hotspot businesses. For example, multi-angle live broadcasting, which provides users with the enjoyment of watching from different positions, zooming in and out of the screen, and 360° free viewing through high-definition video collection and processing at multiple positions on the scene. Cloud gaming, where the game screen is completely rendered in the server and transmitted to the user through the network, greatly reducing the data processing capability threshold of the user terminal. XR real-time interactive services, through AR/VR devices providing high-definition images and extremely low-latency interaction feedback, provide users with an immersive sensory experience in areas such as gaming and socialising.

New businesses have put forward new requirements for network connection services. For example, new application scenarios have temporary requirements for high-bandwidth and low-latency access services. To cope with this sudden connection service, distributed networks should support the special requirements of business for ultra-low latency and high bandwidth communication connections. To reduce the transmission and processing latency of data in the network, mobile edge computing services require mobile networks to connect data to the edge of the network, that is, closer to the user's location. The 6G network should support distributed MEC and application processing platforms to meet the communication requirements of different connection businesses.

2. Providing ubiquitous connection services

The future development trend of mobile networks aims to cater to multiple access scenarios, providing ubiquitous access services, as shown in Fig. 15.8. With the development of mobile networks, access to mobile networks is no longer limited to terrestrial cellular network access, but also includes fixed-mobile network fusion access, relay access, drone access, satellite access, and other types of access. 6G networks need to be able to provide access connection services for various types of access in the air, ground, and sea, to meet the wide coverage network deployment needs of the future. In addition, the 6G core network may need to provide a unified interface to flexibly cope with different access methods.

3. Flexible deployment of control plane/user plane functions

To support the above network services, the connection plane in 6G networks needs to support more flexible deployment, such as for different communication scenarios, the control plane and user plane can be separated or combined, fully utilising the advantages of different deployment modes. The combination of control plane/user plane simplifies the network form, allowing the control plane and user plane network functions to be deployed nearby, reducing unnecessary signalling interactions. The separation of control plane/user plane can meet the needs of a single user accessing

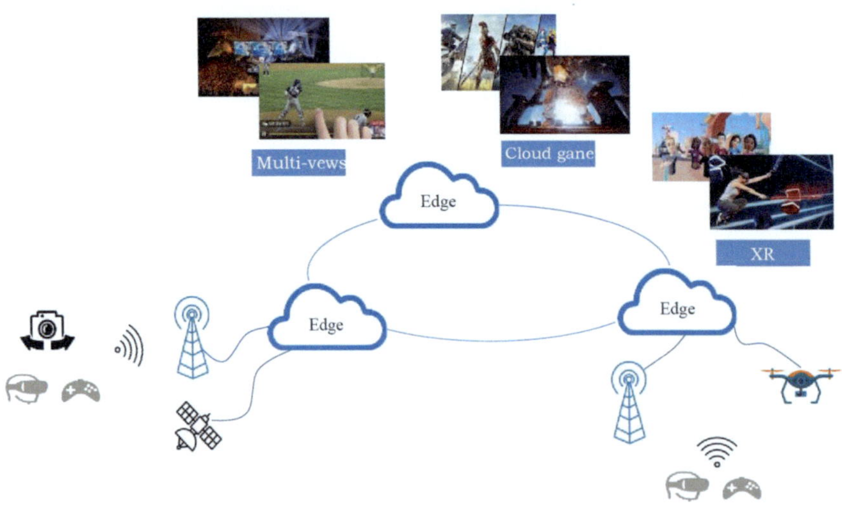

Fig. 15.8 Connection service scenarios

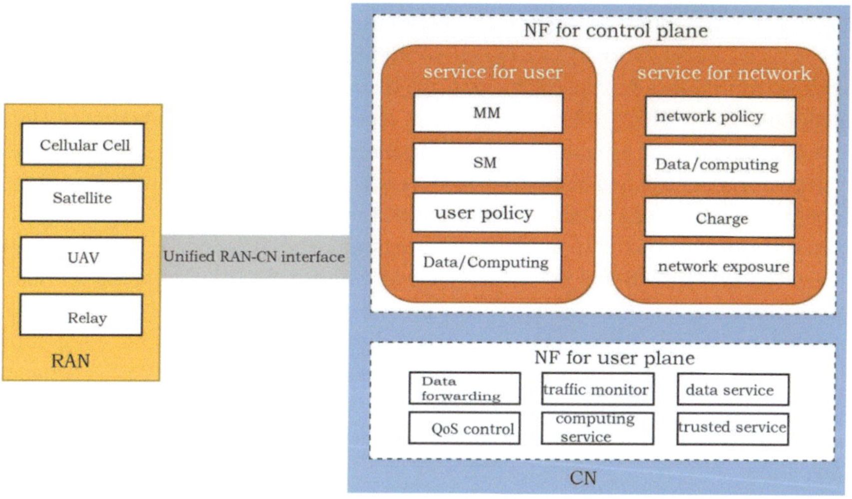

Fig. 15.9 Connection plane logical architecture

multiple network slices, which is beneficial for simplifying future operations, centralising control plane operations, and dispersing policies to the user plane. If the function architecture is considered during the deployment of the 6G connection plane, it is described in a combined manner as a priority, while retaining the possibility of separation to cope with various complex scenarios.

4. Network function redefinition

Existing control plane network functions include access and mobility management functions, session management functions, policy control functions, etc., while user plane network functions are relatively single. The network functions in future 6G networks need to be redefined or logically re-divided in conjunction with the design of the network architecture. For example, the connection plane needs to combine computing functions and data service functions, and may need to define new network functions to interact with other functions. In addition, for the sake of simplifying the network, the control plane functions of the connection plane may undergo further functional integration, simplifying the entire network's signalling process. For example, control plane functions can be divided into network functions for user signalling processing and network functions for network services. The logical architecture of the connection plane is shown in Fig. 15.9.

15.4.4 Computing Function

In 6G networks, computing resources will be distributed across various infrastructures including central clouds, edge clouds, network devices, and even terminal devices. 6G Computing as a Service (CaaS) is based on various distributed computing resources on the network to provide users with on-demand endogenous computing services, especially for terminals with limited computing resources or power, intelligent businesses with extreme performance requirements or high data security and privacy requirements, providing high-performance computing services. 6G CaaS will enable various distributed computing resources in 6G to "flow" on demand, breaking through the limitations of single-point computing performance and improving the overall performance of computing applications.

To realise 6G CaaS and better provide users with inclusive computing services, the 6G network architecture needs to natively support the fusion of communication and computing, achieve intelligent scheduling of ubiquitous network computing resources, and deeply coordinate with connection resources, enabling the 6G network to become a dual infrastructure of connection and computing. The following introduces the technical connotations included in the fusion of communication and computing.

(1) Integrated network architecture for computation and communication. Network elements and computing nodes form a schedulable cluster of computing and communication nodes, with their computing and communication capabilities integrated. This includes the following characteristics: endogenous computing power, distributed, heterogeneous computing power, and multi-level. The endogenous computing power characteristic specifically refers to network elements possessing both communication and computing capabilities, with the computing power generated by computing network elements and computing nodes referred to as endogenous computing power. The distributed characteristic specifically refers to deploying computing services to multiple nodes, allowing for more efficient use of communication, computing, and storage resources. The heterogeneous computing power characteristic specifically refers to nodes having different computing capabilities, storage capabilities, and types of computing power. The multi-level characteristic specifically refers to the integrated network architecture of computing and networking, which includes multiple network layers (end, edge, core network cloud), such as the cloud layer, fog layer, edge layer, and terminal layer, all interconnected and working together to provide services to users.

(2) Modelling, perception, discovery, and registration of heterogeneous computing power. Computing power perception specifically refers to the need for the 6G endogenous AI network to perceive computing power resource information, such as the type of computing power, the quantity of computing power resources, and the usage status of computing power resources. Computing power discovery specifically refers to the network perceiving new computing power resources during operation, which could be the discovery of new nodes with computing power resources, or the discovery of new computing power resources in existing nodes. Computing power registration specifically refers to the process of the network interacting with the node of the newly discovered computing power resource and integrating the new computing power resource into the network.

(3) Integrated control of computation and connection. The integrated control function specifically includes the scheduling and control of distributed heterogeneous computing power resources and connection resources. When scheduling resources, the integrated control function will consider information such as air interface status, delay, distribution of computing power, and heterogeneity of computing power, in order to achieve optimal performance or minimal resource usage while ensuring the QoS of computing services.

The logical architecture of the computing function is shown in Fig. 15.10. The logical architecture of the computing function includes the computing control part, the computing execution part, and the transmission part of the computing data (computing transmission part). The computing control part includes computing execution control and computing connection control; the computing execution part refers to the process of the node's computing execution function using the computing resources allocated by the computing control to execute computing tasks; the computing transmission part refers to the interaction of computing data between different nodes' computing execution functions using the computing connection, thereby achieving the completion of computing tasks through the collaboration of different nodes.

The deployment architecture of the computing function is shown in Fig. 15.11. The computing execution function can be deployed in terminals, next-generation

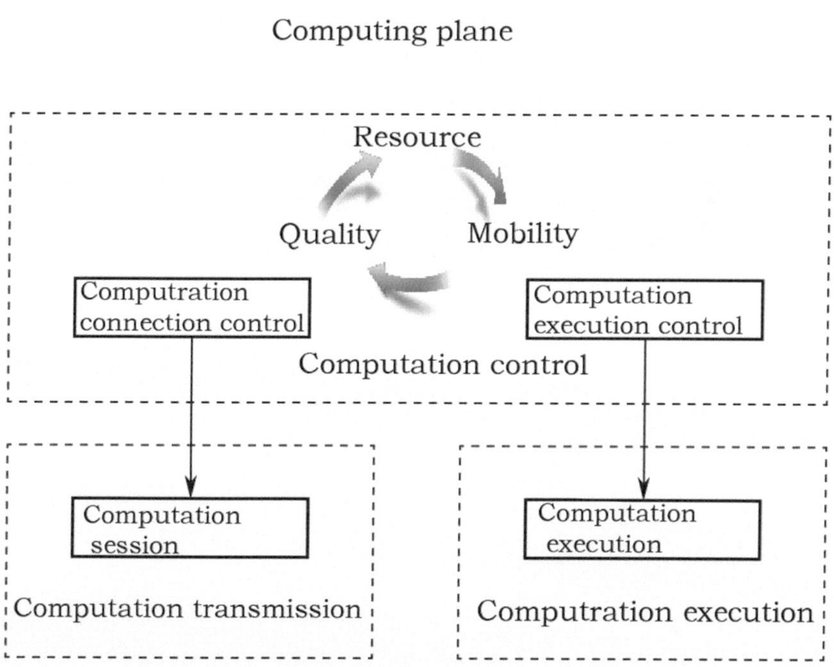

Fig. 15.10 Logical architecture of computing function

base stations (xNB), or core networks (xCN); the computing control function can be deployed in the core network or in base stations. Therefore, a computing session can include computing wireless bearers and computing bearers. The computing wireless bearer refers to the computing data transmission channel between the computing execution function of the terminal and the base station; the computing bearer refers to the computing data transmission channel between different base station computing execution functions, between different parts of a base station's computing execution function, or between the base station and the core network's computing execution function. The computing control (integrated computing execution control, computing connection control) integrated anchor point (communication-computing Converged Anchor, CA) deployed in xCN provides the following functions: computing control anchor point, computing power, address management; computing session management; access control of computing services, etc. The computing control (integrated computing execution control, computing connection control), CS (integrated scheduling) deployed in xNB (multiple xNBs form xRAN) provides the following functions: perception of computing power (computing resources) status and reporting of perception results; establishment/modification/suspension/recovery/release of terminal computing power resources; management of computing power resources when multiple computing power resources are aggregated, including addition/modification/release of computing power resources; triggering of the establishment of computing bearers.

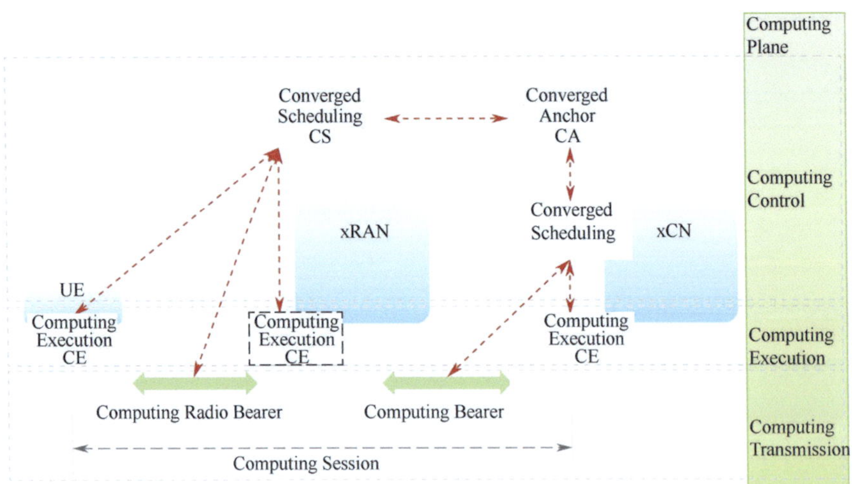

Fig. 15.11 Deployment architecture of computing function

15.4.5 Data Function

New technologies, businesses, and demands in the 6G network have brought about new requirements for data control and processing. Existing data control and processing based on cloud-based AI architecture, including network data analysis function (Network Data Analytics Function, NWDAF), the processing of data is first aggregated to the cloud, then pre-processed, stored, analysed, etc., and finally the analysis results are returned. Due to the introduction of communication perception and network AI, the future 6G network traffic model will change, with 70% of the traffic ending at the network edge, and the existing method of centralising to the cloud will no longer be applicable. We need to scatter the method of centralising data processing in the cloud into the network, and while the data is being forwarded at each node of its data pipeline, pre-processing, aggregation, analysis, etc., are carried out. This data carrying mode, which is brought about by distributed computing power and data, and which processes and forwards data at the same time, we call it data carrying. The edge-driven data flow has driven the distribution of data processing, thereby effectively supporting network AI, integrated sensing, etc.

The cloud-native trend poses new challenges to data collection. In a cloud-native environment, network functions can be instantiated at any time. Traditional hard collection requires hardware binding network elements and specific collection schemes. For the cloud-native environment, hard collection schemes are not up to the task; at the same time, the service-oriented network architecture makes the interface also service-oriented, encrypted, and traditional hard collection methods that collect data from the interface through taps will get encrypted data, which cannot be used. Under the cloud-native trend, the cost of hard collection has skyrocketed, and the method of built-in soft collection in network functions will become inevitable.

The control plane and user plane of the existing network are for session-oriented communication connection services. Although they have taken on the role of data carrying in some specific data-driven applications, such as Self-Organized Network (SON), Minimization of Drive Test (MDT) or Internet of Things (IoT) applications, it is obviously against the original intention of the control plane design to carry data on the control plane, and the carrying endpoints of the user plane carrying data are only UPF (from the perspective of PDU sessions) or base stations (from the perspective of wireless carrying), which cannot meet the scenarios of distributed data pipelines in 6G applications. In response to the foreseeable changes in data types and traffic models in 6G networks, it is necessary to add independent data functions in the 6G network architecture. The construction of data functions is to provide trustworthy data services for other functions in the 6G network, such as computing functions, trustworthy functions and various intelligent applications. Other functions are both producers and consumers of data functions. Data functions collect or collect data from other functions, process them accordingly and then output them to other functions or applications, achieving Data as a Service (DaaS). Through the analysis of numerous application scenarios and requirements, the data services that the data plane can provide are summarised into eight categories, as shown in Table 15.1.

Table 15.1 Data service classification

Service	Description
Raw data	Input the collected raw data into applications such as AI
Data pre–process	Preprocessing services such as data cleaning, filtering, aggregation, fusion, etc
Data storage	Provide centralized or distributed storage services on DA/ DSF/DLT
Data privacy and security protection	Provide end-to-end data privacy and security protection technology
Data sharing/trading	Trusted data sharing and trading
Data traceability	Technically compliant with regulations such as GDPR/ PIPL, with traceability/audit services, and distribution services for public keys, decentralized identities (DID), etc
Data analysis	Using AI, machine learning, big data, and other technologies for analysis and mining to provide intelligent services
Data dictionary	Wireless network characteristic data set Knowledge graph of 6G network

The data sources in the 6G network are diverse, and the types of data are numerous, and the processing procedures for different types of data are vastly different. Data services need to support the processing of different data types. Considering from as many orthogonal dimensions as possible, such as data type, data source, data consumer, etc., the data is divided into four categories, namely user data, network AI data, network data and IoT data. The four categories of data cover all the data in the network, as shown in Table 15.2.

To achieve programmable data functions, Data Orchestration (DO), Data Agent (DA), Trust Anchor Agent (TAA) and Data Storage Function (DSF) need to be introduced into the independent data functions. Data Orchestration translates data service requirements, i.e., it transforms the service requirements of data consumers into network requirements, and then orchestrates the Data Agents to form the pipeline and workflow for data processing. The Data Agent is the implementer of data processing, providing data collection, pre-processing, storage, analysis, etc., through Data Agents deployed on various network elements and terminals; providing data security protection, data privacy protection, etc., through the data security and privacy protection technology library. The Trust Anchor Agent enables users to have autonomous control over data, achieving data trustworthiness, auditability, traceability, and meeting PIPL/GDPR and other compliance requirements; through Data Orchestration, based on the orchestration and management of Data Agents, trustworthy data services are achieved according to the application's request. The logical architecture of the data plane is shown in Fig. 15.12.

Table 15.2 Data types

Types	Sources	Consumer	Notes
User data text)	Subscription data of users	Network functions,business operation support system,third–party applications	PIPL/GDPR compliance, privacy protection
Network AI data (dynamic)	AI models and parameters, AI meta data (number of layers, weights, etc.), training/testing datasets	AI/ML algorithm	Data security and privacy protection
Network data (Time series / Streaming)	UE (user equipment), network function (NF), location server, IT System; Digital twin network; RAN (Radio Access Network) nodes in ISAC (Integrated sensing and communication); Network metadata (type, category, source, quality indicators, correlation, location, etc.)	Network analysis, sensing algorithm	Data security and privacy protection
IoT data (Time series / Streaming)	Sensors, machines in IoT, connected devices or facilities in the Internet of Vehicles	IoT analysis; Autonomous driving or intelligent transportation systems	Compliance requirements, privacy protection

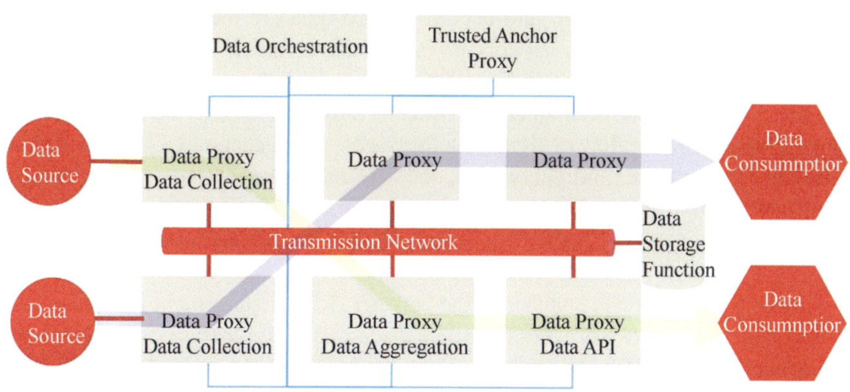

Fig. 15.12 Data plane logical architecture

15.4.6 Trusted Function

The trustworthiness of 6G networks needs to be driven by new trends, new scenarios, new technologies, etc., to build a unified security concept and security framework, forming an endogenous security system in 6G. The so-called “endogenous security” is manifested in the built-in integrated security capabilities for the entire network's security attributes on the one hand, and the inherent security capabilities of the 6G network can continuously evolve to dynamically meet the demands of network services on the other hand. 6G endogenous security guarantees the security, robustness, and continuity of the network and services fundamentally through its unique framework, components, and key technologies.

The main features of the 6G trusted function architecture are introduced below.

1. From Security to Trust

Traditional network security mainly refers to confidentiality, integrity, and availability (CIA). However, with the diversification of network value assets, the continuous improvement of attack methods, and people's demands for data rights, the trusted function of 6G networks has been extended from balanced security, persistent privacy protection, and intelligent resilience, reflecting the leap from security to trust.

(1) Balanced Security

One of the basic standards of 6G native trust is that the network can balance between the quality/user experience of the network and services and security capabilities, achieving balanced and appropriate security, dynamically weighing according to specific applications and usage scenarios.

(2) Persistent Privacy Protection

User identity, user behaviour, and user-generated data are the three types of data protected by communication network privacy. Only with user authorisation can the exposed user-related information be interpreted. Among them, the user identity and user behaviour of 6G have uniqueness, which is closely related to the unified definition of identity and the flow of signalling messages; user-generated data is not stored in the mobile network, and encryption, security management, and other technical means are mostly used in data processing and operation to protect it.

(3) Intelligent Resilience

Resilience focuses on risk analysis in the network. Risk management is divided into several stages. The network needs to be able to identify risks, analyse risks and threats based on big data, and take appropriate measures to avoid risks and reduce the impact of attacks. If the risk cannot be avoided, risk transfer is needed to quickly restore the network and control the risk impact at the lowest level. If the above measures cannot be achieved, it is necessary to ensure that only risks that do not cause fatal damage to the network are accepted.

2. Multi-modal Trust

Trust is a prerequisite for all parties in the network to exchange information. Establishing trust requires not only that all parties in the network can recognise each other, but also that the network has strong security and privacy protection capabilities to fully protect the entire network signalling and data flow. A multi-modal trust model can be built based on trusted functions.

Typical trust models include "bridge", "endorsement", and "consensus" modes. In the "bridge" mode, a centralised authorisation agency authenticates and authorises communication entities A and B point-to-point, thereby passing the trust relationship between A and B, and ultimately establishing a direct trust relationship between A and B. The "endorsement" mode relies on a third-party agency to make a trustworthiness judgment on the entity. In this mode, after the third party checks and evaluates one party A, it submits the result to the other party B, and submits the trustworthiness judgment conclusion of A to B. The "consensus" mode adopts a distributed architecture, transactions are distributed among entities, and the entities participating in the consensus mode can be a network element on the network, a subject on the supply chain, or an organisation in the industrial ecosystem. In this mode, transactions are verifiable, and responsibilities are shared by multiple parties.

Different trust models are suitable for different scenarios. With the development of new businesses such as computing, data, perception, and the emergence of spatially heterogeneous network deployment forms, 6G networks need to coordinate different trust models to jointly form a 6G multi-modal trust model, as shown in Fig. 15.13.

3. Combination of Centralised/Distributed Architecture

The trustworthy function of 6G greatly expands the boundaries of security compared to existing security technologies and introduces a new trust model, resulting in a completely new trustworthy function architecture. This trustworthy function architecture includes two major logical functions: trustworthy enabling units and trustworthy engines.

Trustworthy enabling units are trustworthy function enabling modules deployed on different communication network entities (such as core networks, RAN, and even UE and AF), providing the implementation capability of various trustworthy functions. Each communication entity can have different sets of trustworthy functions

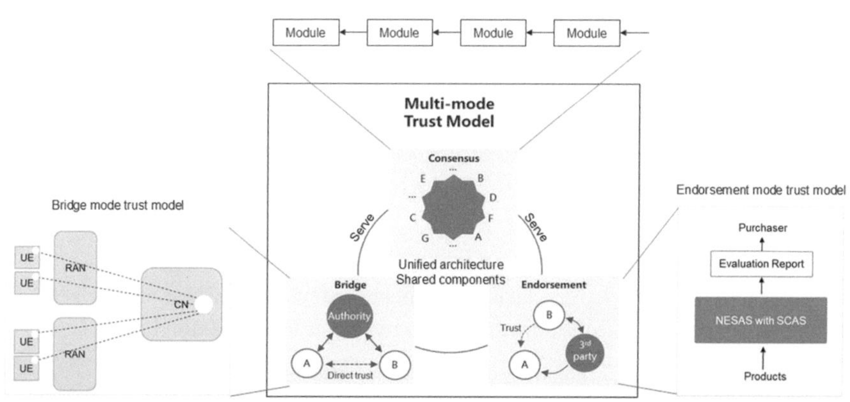

Fig. 15.13 6G Multi-modal Trust Model

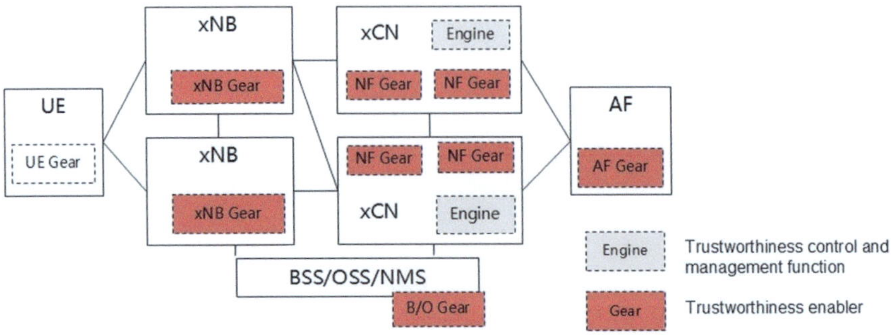

Fig. 15.14 A deployment mode of trustworthy enabling units and trustworthy engines

according to business requirements, such as cryptographic capabilities, blockchain capabilities, remote proof capabilities, etc. Trustworthy enabling units receive policies and configurations from the trustworthy engine.

The trustworthy engine is the central decision-making and management scheduling unit of network trust capabilities, usually located in the core network, and is a conceptually centralised, deployment-distributed management centre. The trustworthy engine manages, configures, maintains, etc., each trustworthy enabling unit, and also provides closed-loop transaction analysis of security situation awareness and other policies.

Figure 15.14 shows a deployment mode of trustworthy enabling units and trustworthy engines. Trustworthy enabling units and trustworthy engines work together to provide security for the 6G network itself and the services on the network.

15.4.7 Bearer Network

The development of 6G networks not only affects the functional architecture at the business level but also brings a lot of changes to the underlying transport network. These changes come from all aspects of the 6G network, affecting the traffic model in the network, and the overall change of the traffic model will also affect the architecture selection of the future bearer network from different angles. The changes in the traffic model are mainly reflected in the following three aspects.

(1) Overall increase in bandwidth demand: 6G introduces millimetre wave/terahertz/visible light communication, further increasing transmission bandwidth. In addition, from a business demand perspective, whether it is VR/AR/MR, holographic communication, or perception business, the amount of data transmitted in the network is also increasing rapidly. These surges exist not only in the backbone network and data centre network, but more importantly, in the access network, affecting both downstream and upstream traffic. This

brings about the requirement for ultra-large bandwidth in the mid-transmission/backhaul network.

(2) Edge-centric traffic distribution: In traditional mobile networks, backbone traffic is gradually aggregated from peripheral traffic, with the main pressure existing in the backbone network. With the increase in latency and security requirements and the emergence of 2B on-site business demands, the edge-centric trend of various applications has become a reality. This requires the bearer network architecture to not only solve the large-span data transmission of the backbone network but also pay extra attention to the local traffic surge in the edge network.

(3) Dynamic and widespread distribution of traffic initiation: The essence of existing mobile network services is to build a data tunnel to the application network for end users, but this single and static transmission mode will be broken in 6G. The data, computing, and perception services introduced by 6G will all regard core network functions, RAN nodes, and even UE as collaborative participants, and business traffic will be widely distributed among these objects and dynamically redistributed with changes in business requirements, user location, and network load. Therefore, a transmission network architecture that can adapt to this dynamism and breadth is needed.

In summary, the 6G bearer network may need to adapt to changes in the 6G traffic model, fully cross-layer collaborate with the access network and core network, form a new functional architecture and networking method, and meet the data communication needs under different business scenario requirements. The architecture of the 6G bearer network is shown in Fig. 15.15.

According to the 6G bearer network architecture, the following key technical features need to be considered.

(1) Widespread distribution: The 6G bearer network faces multi-dimensional deployment scenarios such as air, space, and ground, providing a unified and ubiquitous underlying transmission capability for mobile networks.

(2) Flexible and programmable: The 6G bearer network provides a routing forwarding mode driven by the business traffic model, supporting interconnection capabilities under central deployment, edge deployment, and multi-level complex deployment scenarios.

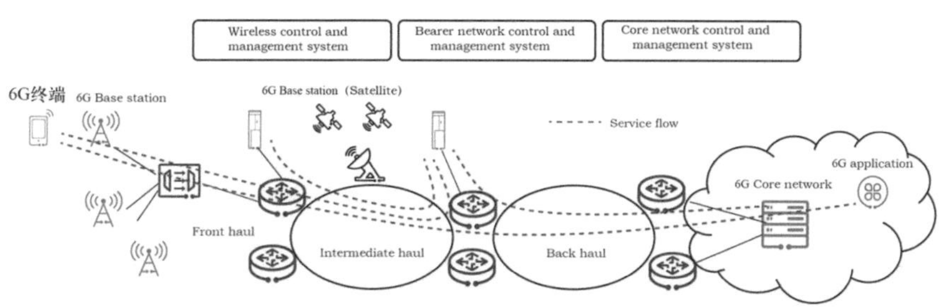

Fig. 15.15 6G bearer network architecture

(3) Business awareness: The differentiated requirements of applications need to be transmitted to the 6G bearer network through the mobile network, and end-to-end QoS guarantees are achieved through a consistent business awareness capability.

15.5 Key Technologies of 6G Network

15.5.1 Task-Centric Design Paradigm

In 6G, not only is the value of wireless connections important, whether it is providing XaaS-type services, or starting from the vision of 6G intelligent inclusiveness, providing a complete AI environment and service-based AI services within the network, the traditional session-centric design approach needs to evolve to a task-centric design mechanism of the 6G network, supporting the deep integration of connections, computing, data, and algorithms (four elements) required for XaaS, such as AIaaS business, that is, supporting real-time control at the unit of deep integration of the four elements, that is, task granularity control, and supporting task QoS guarantee mechanism. The corresponding 6G network architecture needs to achieve the following key changes in the design paradigm, to better describe, the following is an example of native AI business in the 6G network.

- Change 1: The control object in the wireless network system changes from “session” to “task”

Compared to the traditional “session”, “task” has two essential differences: one is the difference in technical goals; the other is the difference in technical means. Firstly, from the perspective of technical goals, traditional communication systems provide session-based services, with typical application scenarios providing session services between specific terminals or between terminals and application servers, with the ultimate goal still being the transmission of user data (including voice). However, the goal of network AI is different from that of a session, as the aforementioned network element intelligence and network intelligence aim to provide intelligent services for the network and improve communication network efficiency, while business intelligence aims to provide intelligent services at the App level for third parties. In summary, the goals of providing sessions and AI are different. Secondly, from the perspective of technical means, traditional communication services need to maintain user-granularity connection pipelines (such as UE to base station, base station to core network end-to-end tunnels) and lifecycle management and QoS guarantee mechanisms for connection pipelines in order to provide QoS-guaranteed data transmission services. AI belongs to data and computation-intensive services, which have the following differentiated features compared to sessions: on one hand, AI introduces new resource dimensions, including computing power (such as CPU, GPU and NPU), data (such as AI used and AI generated), and algorithms (such as neural network models, reinforcement learning), etc., therefore 6G networks need to introduce new resource management mechanisms; on the other hand, factors such as single-point computation bottlenecks, data privacy protection, and ultra-large model storage bottlenecks make it difficult for a single node to efficiently implement AI services, which can only be completed through the collaboration of computing power, algorithms, and data between multiple nodes, therefore 6G networks need to introduce new collaboration mechanisms between nodes. Based on the above two differences, it can be seen that the "session" system cannot support native AI, so a new "task" system needs to be designed to support the above new mechanisms (new resource management mechanisms and new collaboration mechanisms between nodes).

- Change 2: Control resources shift from connection resources to the four-element resources of connection, computation, data, and algorithm.

The "session" system establishes channels for users to transmit data and allocates corresponding connection and air interface resources, while the "task" system allocates four-element resources to complete AI tasks. Taking AI inference tasks as an example, the executor needs to first obtain resources information such as computation, data, and algorithms, and then execute related tasks. For example, computation information is the computation resource time slot or proportion corresponding to a certain task, data information is the data collected by the executor in real time or the data input from the outside, and algorithm information includes possible Graph Neural Network (GNN), Convolutional Neural Networks (CNN) and other AI models, or Reinforcement Learning (RL) and other AI algorithms. Taking federated learning tasks as an example, multiple executors cooperate to jointly train AI models, and during the training process, they need to use allocated connection resources to pass gradient information. In summary, due to the introduction of tasks, control resources have shifted from connection to the four-element resources of connection, computation, data, and algorithm.

- Change 3: Shift from "session control" to "task control"

Different from the function of traditional session control, the task control system in network AI mainly has the following functions: (1) decomposition/mapping from external services to internal tasks; (2) decomposition/mapping from service QoS to task QoS; (3) providing mechanisms for four-element collaboration and multi-node collaboration, orchestrating and real-time controlling the four-element resources of multiple nodes in the infrastructure layer, and ultimately achieving task-granularity distributed serial/parallel processing and its real-time QoS guarantee. Among them, for simple service requests, a service can correspond/map to a task; for complex services (such as the integration of multiple business flows or a service request with a huge amount of computation for only one business flow), it can be mapped to multiple nodes for systematic execution.

- Change 4: Shift from "session QoS" to "task QoS"

6G networks will no longer just serve as a conduit for traditional communication services. Different intelligent application scenarios will have different demands for

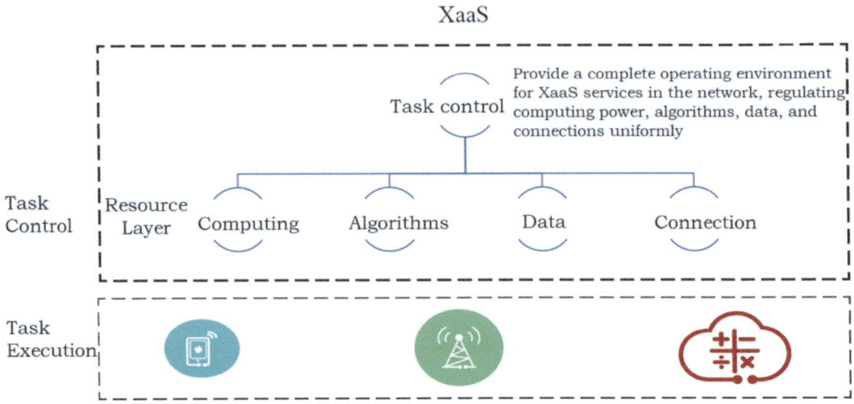

Fig. 15.16 Task-centric network

the quality of AI services, requiring a set of indicators to convey user needs and the overall effect of network orchestration control of AI elements (including connections, computations, data, and algorithms) in a quantitative or graded manner. The QoS of traditional communication networks mainly considers performance indicators related to connections such as delay, throughput, and transmission rate of communication services. In addition to traditional communication resources, 6G networks will introduce new resource dimensions such as computing power, algorithms, and data, and need to expand corresponding evaluation indicators. At the same time, with the implementation of "carbon neutrality" and "carbon peak" policies, the global intelligent application industry's general strengthening of data security and privacy concerns, and the increase in user demand for network autonomy, future performance-related indicators will no longer be the only indicators of user concern. The demands for cost, security, privacy, and autonomy will gradually deepen, becoming a new dimension for assessing service quality.

In a task-centric network (see Fig. 15.16), new features such as network AI orchestration, task control, and task resource layer have been introduced. The task control function is to control the multi-node (UE, base station, core network elements, etc.) and four-element resources in the resource layer in real-time through control layer signalling. Based on the task-centric network architecture, the efficient execution of AI tasks in the network and QoS assurance become possible.

15.5.2 New Non-access Layer Protocol

From the perspective of 6G network requirements, users can participate in the network as providers of capability services while enjoying the new services provided by network providers. The terminal held by the user itself can provide capability services for other nodes in the network, such as perception services, computing

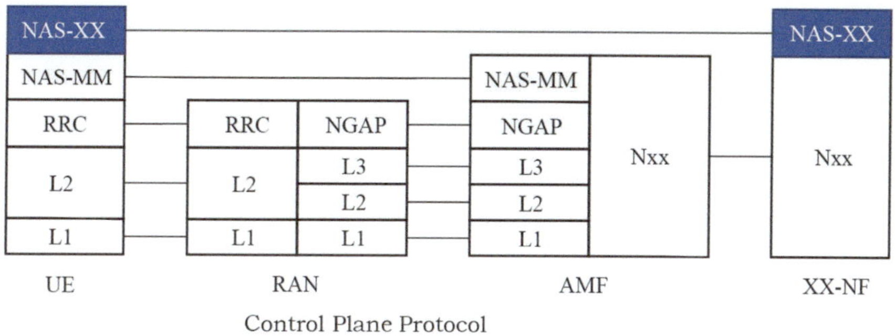

Fig. 15.17 Centralized NAS transmission method

services, etc. The resources of the user in the network can also be an important asset, controlled by the user, and open to other nodes or users in the network.

Therefore, the control plane channel between UE and CN is no longer simply connecting UE and the core network based on connection requirements, i.e., the user provides connection requirements to the network, and the network issues configuration parameters and instructions to the user. The future network needs to give users control over their own assets and provide users with richer interactive channels, specifically including:

(1) The network needs to obtain user authorization when calling user data or capability services;

(2) Users can manage and control user assets in the network.

In addition, for the transmission of control plane signalling, the existing 5G adopts a centralized (forwarded by AMF) NAS (Non-Access Layer) transmission method, as shown in Fig. 15.17. In the existing NAS transmission method, by adding a new service NAS container in the NAS message, signalling interaction between UE and new service function element (XX-NF) can be achieved. In the design process of the 6G network architecture, the transmission method of NAS messages may also have different evolution directions, such as distributed NAS (see Fig. 15.18), NAS over IP, etc. Through the protocol layer between RAN and CN (such as the NGAP layer), RAN can directly communicate with the NF (XX-NF) of the core network. Different NAS message transmission methods depend on the design of the future network architecture and the specific deployment method.

15.5.3 *Computational Wireless Bearer*

In order to achieve the integration of communication and computation in 6G networks, it is necessary to complete the mutual perception and coordination of computation execution and computation connection through a new computation protocol, real-time accurate computation resource discovery, flexible dynamic

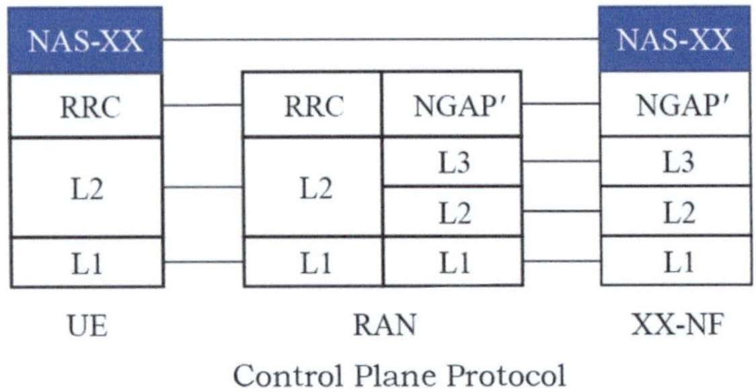

Fig. 15.18 Distributed NAS transmission method

computation resource and computation quality scheduling, providing ubiquitous computation services and connection services.

The computing function includes a set of new network logic functions, as well as the protocol interfaces between these logic functions and terminals, Apps, which will have a certain impact on the network architecture and network protocols. The control of the computing function includes computing execution control and computing connection control.

Taking the RAN architecture as an example, if the computing function is introduced into the RAN architecture (see Fig. 15.19 for the computing protocol stack), the air interface computing connection control can be implemented based on the existing RRC protocol mechanism, i.e., by modifying the RRC protocol or by invoking the basic functions of the RRC protocol to support the control of the computing connection; the computing execution control is implemented by the CRC (Computing Resource Control), which can be independent of the RRC or can be integrated with the RRC into xRC. The CRC is used to control the computing resources occupied by the computing execution function, the amount of computing operations, the quality of computing, etc. The computing connection control and computing execution control between base stations (Xn interface) can be implemented based on the existing Xn-AP mechanism (invoke or modify).

The user plane of traditional communication connects users with the Data Network (DN), and the transmission part of the computing function is used for the transmission of computing data between the computing execution functions of different nodes within the wireless network system, i.e., the data of computing is not transmitted to the DN, therefore, the design of the computing transmission mechanism needs to be distinguished from the traditional communication user plane. A possible computing transmission method is to introduce a new bearer method at the bearer level, such as the Computing Radio Bearer (CRB) for the air interface part, the Computing Bearer (CB) for the Xn interface part, and to introduce a new Radio Computing Session Protocol (RCSP) at the session level, at which point the computing session can also be referred to as an RCSP session.

RCSP is an efficient data communication protocol for computing resources within a wireless network, supporting the exchange of computing data between the computing execution functions of terminals, base stations or core network functions, thus supporting the computing collaboration between different nodes to jointly

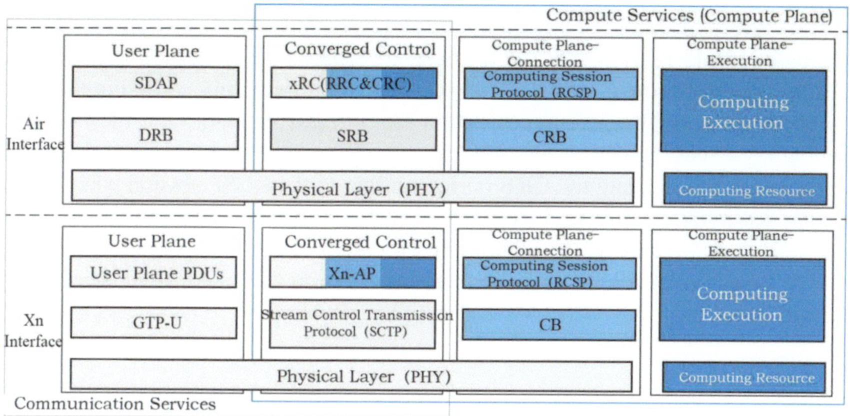

Fig. 15.19 Computing protocol stack

complete a computing task. The QoS of the computing task is determined by the QoS of the computing session and the QoS of the computing execution. Among them, the QoS of the computing execution is affected by the allocated computing resources, the amount of computation and the computing process, and the basic indicators include the computation time and the computation accuracy. The endpoint position of the computing session is determined by the node position where the computing execution function is located. For example, the RCSP session shown in Fig. 15.20a includes CRB, i.e., the exchange of computing data between the terminal and the base station; the RCSP session shown in Fig. 15.20b includes the Computing Bearer (CB), i.e., the exchange of computing data between different base stations, between the base station and the core network, or between the Distributed Unit (DU) and the Central Unit (CU); the RCSP session shown in Fig. 15.20c includes CRB and CB, i.e., the exchange of computing data between the terminal and the core network.

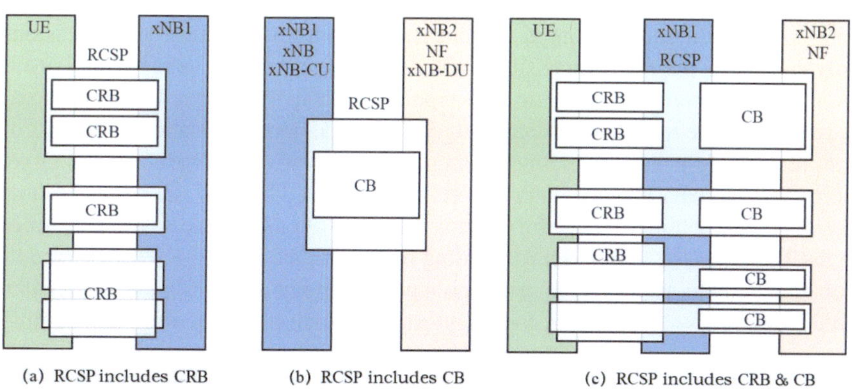

Fig. 15.20 Relationship between RCSP session and CRB, CB

xNB is responsible for maintaining the correspondence between CRB and RCSP sessions, its role is to enable the terminal to achieve the final data transmission by selecting the corresponding CRB when transmitting uplink computing data to the nodes of computing resources within the wireless network; and the nodes of computing resources within the wireless network can also achieve the final data transmission by selecting the CRB corresponding to the RCSP session when transmitting downlink computing data to the terminal. xNB implements data interaction between the terminal and the computing execution function deployed in the base station, based on RCSP, through the mapping between CRB and RCSP sessions in the middle.

15.5.4 Data Bearing and Forwarding

The 5G network is built based on sessions, and its user plane is used to bear session data. Within the mobile network, the user plane terminates at the UPF, while the potential data termination points can be located at the UE, wireless access network, core network, etc. As it cannot meet the "on-route computing" and "arbitrary topology" support required for 6G data bearing, the 5G user plane cannot bear the new data types and business requirements of the 6G network. Table 15.3 compares the differences between 5G user plane transmission and 6G data transmission from the perspectives of function, starting and ending points of data bearing, behaviour of data forwarding devices, data forwarding principles, topology structure, etc. The session connection of the 5G user plane implements the information exchange between two communication devices, specifically by the Protocol Data Unit (Protocol The Data Unit (PDU) provides an end-to-end user plane connection between user terminal equipment and the network.

6G data transmission consists of data collection, preprocessing, forwarding, storage, and analysis functions. User plane transmission is for communication connections between people or between people and machines, while data processing data is produced and consumed by machines/algorithms. 5G user plane sessions only implement packet transmission, while 6G data transmission needs to implement along-the-way computing. In the data pipeline, data is transformed and optimised to reach the state required for data analysis and intelligent applications. In terms of data forwarding behaviour, session data packets are forwarded based on the target address; in the data pipeline, data packets are forwarded based on data services and data pipeline identifiers. Data forwarding based on 5G user plane sessions belongs to the TCP/IP layer, while data forwarding belongs to the application layer. In addition, the topology based on the session is a point-to-point connection, while 6G data processing needs to support any topology structure (such as the tree structure required for data distribution and data aggregation).

Data bearing is used to transmit 6G data, which is different from the existing signalling plane bearing and user plane bearing in terms of QoS guarantee, scheduling priority, etc. In order to support the implementation of 6G data processing, a new data

Table 15.3 Comparison of **5G** user plane transmission and **6G** data transmission

Project	5G user plane transport	6G data transmission
Function	PDU sessions provide end-to-end user plane connectivity between user equipment and the network	Distributed data pipelines are composed of functions such as data collection, preprocessing, forwarding, storage, and analysis
The starting and ending points of data bearer (or data carrying)	UE and UPF	Any network element and terminal device
Behavior of data forwarding devices	Forwarding devices only forward data packets	It requires in-line computing: within the data pipeline, data is transformed and optimized while being forwarded, so that it can reach a state suitable for analysis and application
Principles of data forwarding	Data packets are forwarded based on their destination addresses	Data packets are forwarded based on data service and data pipeline identity
Topology	Point–to–point connection	Arbitrary topology

protocol is needed, hence the introduction of the Data Forwarding Control Protocol layer (DFCP layer). The data control protocol stack and data business protocol stack are shown in Fig. 15.21.

The DFCP layer is located above the PDCP layer and mainly completes the mapping of data services to data bearing, the routing and forwarding of business data, the parsing and recombination of packet headers, and the collection, analysis, and processing of data. The MAC layer adds the mapping of data bearing to logical

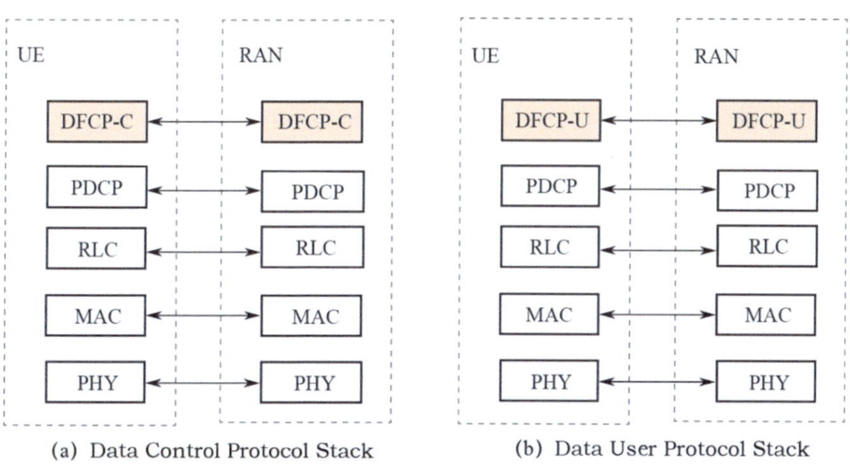

Fig. 15.21 Data control protocol stack and data business protocol stack

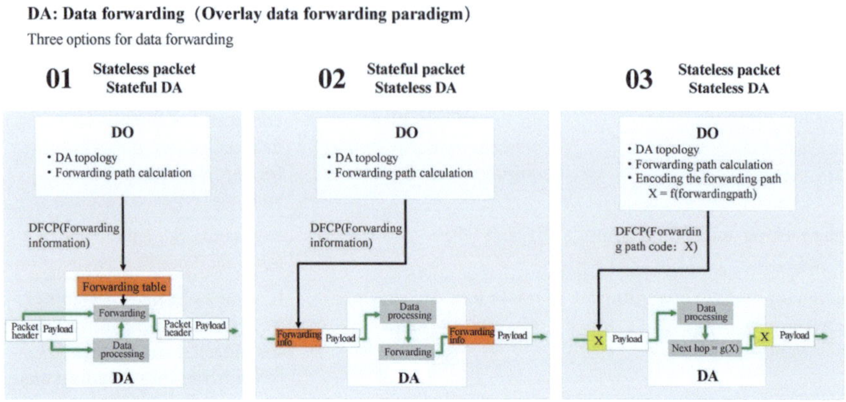

Fig. 15.22 Three wa data

channels, and the PDCP layer adds security and privacy protection functions to support data bearing.

6G data functions need to build a new data-oriented forwarding mechanism. Depending on the state of the data packet and the data forwarding entity (DA), the 6G network needs to consider different data forwarding technology solutions, as shown in Fig. 15.22. It should be noted that the data pipeline is directional, and a data service can contain multiple data pipelines.

When the packet is stateless and the DA is stateful: the Data Forwarding Control Entity (DO) arranges the data pipeline and its topology according to the data business requirements, and writes the data forwarding table items into the corresponding DA's data forwarding table. The DA forwards the data to the next hop according to the table items until the forwarding table items end. At the same time, the DA counts the number and bytes of the forwarded data packets and reports them to the DO as needed. After the data service execution ends, the data pipeline is deleted, and the DA deletes the data forwarding table items.

When the packet is stateful and the DA is stateless: the DO arranges the data pipeline and its topology according to the business requirements, and emits the data forwarding table items to the entrance DA. The entrance DA forwards the forwarding information as the data packet header information to the next hop. The DA in the forwarding path forwards according to the forwarding information carried by the data packet header and deletes the forwarding information involving this DA. The exit DA deletes the address/identifier information in the packet header and submits it to the upper layer application. The DA counts the number and bytes of the forwarded data packets and reports them to the DO as needed. The edge DA deletes the data forwarding table items of the given data service after the data service ends.

When both the packet and the DA are stateless: the DO arranges the data pipeline and its topology according to the business requirements. The DO encodes the data forwarding path corresponding to the data service and emits the encoding to the entrance DA. The entrance DA calculates the next hop of the data packet through decoding operation, and after completing the data processing, forwards it to the next node. The exit DA submits the packet to the upper layer application. The DA reports to the DO as needed based on the statistical data carried in the packet. The edge DA deletes the data pipeline after the data service ends.

Depending on different business requirement scenarios, the DO can choose to use different data routing schemes.

15.5.5 Trustworthy Distributed Network Technology

Distributed network technology has, to some extent, broken through the limitations of centralisation, driving the rapid development of internet services. The mainstream distributed technologies currently under study mainly include Distributed Ledger Technology (DLT), which shares, replicates, and synchronises databases among network members; the Inter Planetary File System (IPFS), which achieves decentralised peer-to-peer transmission for distributed data storage; the Distributed Hash Table (DHT), which realises distributed and rapid search and access of network functions; and blockchain technology, which combines various distributed technologies. Among these, blockchain technology, with its decentralised and immutable technical characteristics granted by its diverse integrated architecture, provides a trustworthy mechanism for information and value exchange in untrusted networks, solving trust and security issues in traditional centralised service architectures. Therefore, introducing blockchain technology thinking in resource sharing, inter-network collaboration, and network security can enhance network scalability, inter-network collaboration capabilities, and security and privacy protection capabilities. In addition, blockchain technology can provide high-performance and stable reliable data attestation services, ensuring the security, trustworthiness, transparency, and traceability of data. Drawing on the ideas of these distributed network technologies and integrating them into the design of future 6G network architectures will provide technical support for building network distribution autonomy, decentralised trust anchors, and distributed authentication, authorisation, and access control. It will also support user contract data autonomy and compliance with data protection regulations, reducing the risk of single-point failures and DDoS attacks. In distributed networks, a large number of diverse nodes (such as macro base stations, small base stations, terminals, etc.) are highly autonomous and have differentiated communication characteristics, caching capabilities, computing capabilities, and load conditions, etc. Therefore, it is necessary to coordinate different nodes to achieve distributed network resource complementarity and on-demand networking. However, as distributed network resources may belong to different enterprises, operators, individuals, or third parties, etc., it is necessary to establish a decentralised network security trust collaboration mechanism. Therefore, based on blockchain technology and thinking, achieving secure and trustworthy resource sharing, data security circulation, and privacy protection

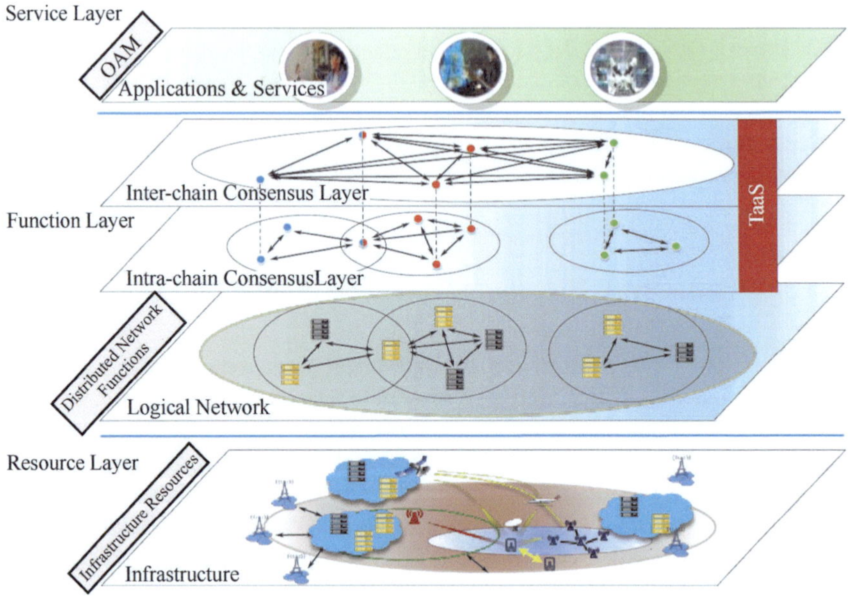

Fig. 15.23 Trustworthy distributed network technology

will become a new direction for future 6G networks to provide trust services. Individual distributed technologies have different shortcomings. By combining technologies organically, technological complementarity can be achieved, such as using DHT in combination with DLT to realise a user-centred network architecture, meet user-customised network functions, fine-grained personalised services, and provide decentralised Trust as a Service (TaaS) functionality. Trustworthy distributed network technology is shown in Fig. 15.23. User contract data and other data are stored off-chain by DHT to avoid blockchain bloating issues, and combined with authorised blockchain to protect user privacy, achieving deep integration of distributed technology and wireless communication, breaking down the trust barriers between "people – machines—things—networks", and improving the efficiency and security of wireless networks.

15.5.6 6G Blockchain

The decentralisation trend of the 6G network architecture necessitates the construction of a more inclusive multimodal trust model. The added distributed trust model, the entities participating in the consensus model can be a network element, a subject on the network, or an organisation in the industrial ecosystem. Blockchain is a typical decentralisation technology, and the integration of 6G and blockchain gives new meaning to 6G network security. 6G capabilities are inherent in 6G. BaaS. 6G blockchain is a redefinition of blockchain in the 6G scenario, with distinct 6G characteristics. It is different from traditional public chains like Bitcoin and Ethereum, and also different from various BaaS platforms and services based on the cloud. The 6G blockchain capability is inherent in the network through trusted enabling units: traditional public chains are based on the Internet platform, and any node can access the blockchain through the Internet; based on large cloud service providers' platforms, blockchain users such as government and enterprise, supply chain, finance and other industries purchase and use blockchain services from service providers like Huawei and Alibaba. The distinctive feature of the 6G blockchain is that its infrastructure is the underlying 6G network, with UE, base stations, core networks and other mobile communication network nodes as the infrastructure nodes of the blockchain, and the blockchain is centrally managed and scheduled through network functions. At the same time, the 6G blockchain serves 6G businesses, providing a secure mutual trust and sharing platform for upper-layer businesses based on the 6G blockchain, and businesses will need to reshape their business processes due to the introduction of the blockchain.

In certain scenarios, the 6G blockchain will have the following capabilities.

(1) Support for high-throughput transaction blockchain expansion capabilities. The current expansion of the blockchain can be achieved in three ways: optimising the data transmission protocol of the network layer and transport layer of the Open Systems Interconnection (OSI) model of the Internet architecture without changing the upper architecture of the blockchain, which is a performance improvement scheme that retains the original chain ecological rules; optimising and improving the basic structure, model, and algorithm of the blockchain data layer, network layer, consensus layer, and incentive layer, i.e., optimising and improving the blockchain's own system architecture to enhance blockchain performance, on-chain expansion schemes can be divided into block data related, sharding technology related, consensus protocol related, and new chain structure related; adjusting the contract layer and application layer of the blockchain, moving contracts and complex calculations off-chain, with the aim of reducing the on-chain burden to enhance blockchain system performance, off-chain expansion does not change the basic protocol of the blockchain. The 6G blockchain will have various characteristics of the chain, using different blockchain expansion technologies in different scenarios to increase transaction throughput, thereby meeting the transaction performance requirements of 6G.

(2) Support for GDPR and regulatory capabilities of editable technology. Editable technology is a technology that allows on-chain data to be safely modified and the modification records to be traceable through cryptographic algorithms, to meet the compliance with GDPR when there is personal information data on the chain, and the regulatory requirements when malicious information appears on the chain. A typical solution is to rewrite the block content using a chameleon hash function. Unlike traditional hash functions, the chameleon hash function is a keyed one-way hash function. Traditional cryptographic hash functions make

it difficult to find hash collisions, while chameleon hash functions can artificially set the next private key, and those who have it can easily find collisions, that is, those who have the key information can easily calculate the hash collision of any input data, thus allowing the input of the hash function to be changed arbitrarily without changing the output of the hash function. The key is shared among multiple validators, and the consensus process among the validators determines how to modify the block data.

15.6 6G Network Implementation Architecture

The functional architecture determines the completeness of the logic, and a good implementation architecture often brings benefits such as efficiency, reliability, openness, and agility. Therefore, we not only need to give the functional architecture of the 6G network, but also consider and suggest its implementation architecture.

In the 3GPP standard, the object described by the core network is network function, or it can also be considered as a form of microservice. The specific implementation of the core network network characteristics is described by dividing and defining the interaction of microservices.

The object described by the wireless access network is the node, and the specific implementation of the wireless access network characteristics is described by defining the function of the node and designing the protocol stack.

15.6.1 6G Core Network Implementation Architecture

The implementation architecture of the core network has basically completed the service-oriented evolution in 5G, but it still faces many problems, such as insufficient and unreasonable service-oriented decoupling, leading to coupling between network functions, making it difficult to achieve agile business launch and grey upgrade features, and the advantages of service-oriented cannot be fully utilised.

Full cloud-native is the consensus of the evolution of the 6G core network. Based on the existing SBA route, it provides further cloud capabilities to simplify the network and facilitate agile evolution. Here are a few implementation architecture design patterns that meet the requirements of cloud-native.

1. Microservice Decoupling Pattern

The 5G core network architecture adopts the SBA architecture, defining different entities as network functions (NF), each providing different services, and service registration discovery is carried out through the network repository function (NRF). Conceptually, this is a typical microservice pattern.

The 6G network can continue to use this pattern, carry out extreme decoupling of microservices for different functions and processing objects, and ensure that there

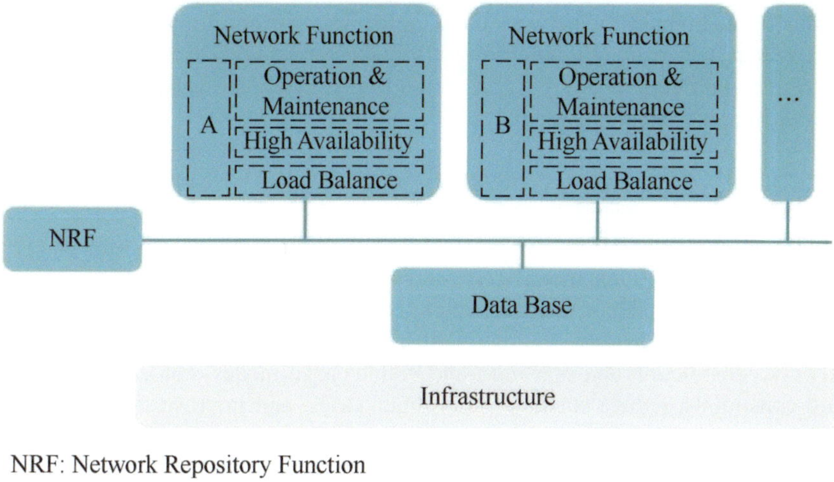

Fig. 15.24 Microservice Decoupling Architecture

is no mutual influence between different microservices. The functional coupling existing between the current NFs also needs to be decoupled and redistributed during the evolution process.

(1) Design Principle

Service-oriented reconstruction, first describe the logical architecture according to pure functions, and then determine the aggregation logic of NF based on the interdependence between functions (call relationship, data relationship, etc.), that is, high cohesion and low coupling.

(2) Architecture

The microservice decoupling architecture is shown in Fig. 15.24.

(3) Advantage Derivation

The advantages of microservice decoupling are:

(1) Services are fully decoupled, the network as a whole reflects the advantages of service-oriented, agile;
(2) Strong standard control, conducive to openness between partners;
(3) Closest to the 5G network service-oriented pattern, easy to evolve and interoperate.

2. Microservice Aggregation Pattern

With the flexible and diverse development of 6G network services, it is becoming more and more complex to define a complete business interaction process through standardisation; at the same time, not all couplings between NFs are suitable to be broken within the framework of the communication network, and it is necessary to consider how to make the network more agile under this constraint; in addition, no matter what, too many NF definitions will make the network increasingly difficult to operate and maintain, so it is considered to restructure the network through the microservice aggregation pattern.

(1) Design Principle

Service-oriented reconstruction, based on the existing NF division logic, select appropriate principles (network ownership, networking methods, etc.) for merging (in implementation, after merging, it can form super NF or network element, i.e., Network Element, NE), so that no matter how many logical functions increase, the number of super NFs is always controlled, and super NFs can converge networking, network construction and operation and maintenance modes, not subject to standard constraints, reduce standardisation boundaries, and improve integration and performance.

(2) Architecture

The microservice aggregation architecture is shown in Fig. 15.25.

(3) Advantage Derivation.

The advantages of microservice aggregation are:

(1) Complete decoupling does not rely on NF, and the agility advantage is provided by the vendor implementation within NE;
(2) The internal part of NE eliminates the performance overhead brought by service-oriented, with efficiency taking precedence;

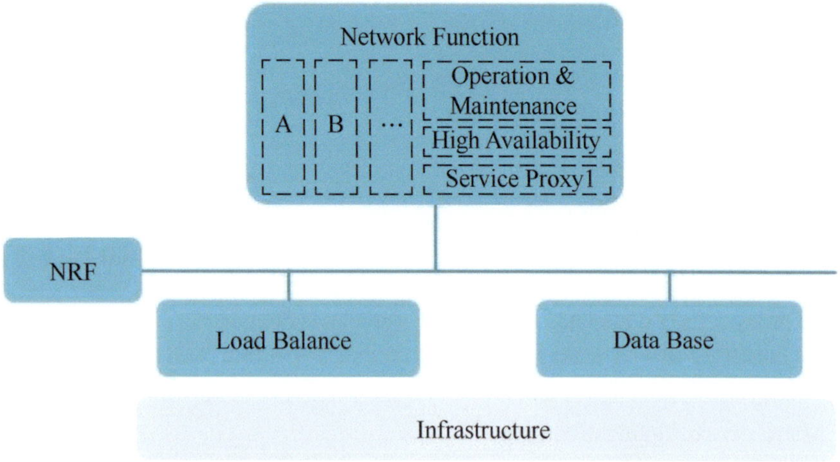

NRF: Network Repository Function

Fig. 15.25 Microservice Aggregation Architecture

(3) Reduces the complexity of network construction, network organisation, and maintenance objects;
(4) Standard control is weak, and partners can only collaborate on the NE boundary;
(5) It can evolve and interoperate with existing 5G NF.

3. De-microservice mode

With the continuous development of cloud-native technology, more breakthrough technologies have redefined the implementation of microservices. In the de-microservice work mode, the operation of the business is not bound to static execution bodies (virtual machines, containers, etc.), but is scheduled to run on demand by the platform. When a message or data triggers business execution, the platform schedules the execution logic to the appropriate execution body. At the same time, the platform provides basic support services to the business, ensuring the simplicity and scalability of business logic.

This de-microservice mode reflects the ultimate statelessness of the business and the full decoupling of the business and the platform.

(1) Design principles

Decouple common capabilities (service registration discovery, routing, high availability, operation and maintenance, etc.) from the business and decouple them into platform services. Platform services provide support functions for all businesses. The business logic is strongly controlled by standards; whether the platform logic is strongly controlled by standards has different implementation options.

(2) Architecture

The de-microservice architecture is shown in Fig. 15.26.

(3) Advantage derivation

The advantages of de-microservice are:

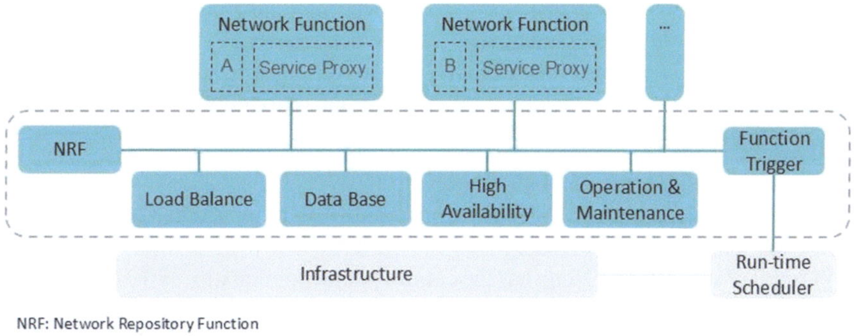

Fig. 15.26 De-microservice architecture

(1) The business platform is fully decoupled, the platform capability is simplified by common construction, and the complexity of business development is greatly reduced;
(2) The agility closest to IT.

15.6.2 6G Wireless Access Network Implementation Architecture

Based on the 6G network vision, technical requirements, and network function design, in addition to providing basic connection services, the 6G mobile communication system also needs to provide various new service capabilities such as computing, data, algorithms, and perception, effectively enabling 6GXaaS. Therefore, the 6G mobile communication system needs to build an endogenous integration and fusion of multi-dimensional heterogeneous resources to efficiently provide new service capabilities, which will drive the reconstruction of the 6G Wireless Access Network (Radio Access Network, RAN) architecture. The essential features of these new service capabilities require the collaboration of multiple terminals, multiple access nodes, and multi-dimensional resources (computing, data, algorithms, connections, etc.). However, the 5G access network is composed of a single gNB, which is a flat implementation architecture. In this implementation architecture, the efficiency of multi-node collaboration involving a large range is low. Therefore, from the perspective of the access network implementation architecture, it is necessary to introduce a centralized coordination cluster node (cluster Node, cNode) to achieve efficient and larger-scale collaboration within and between regions. Therefore, 6G RAN nodes can be divided into the following two types.

(1) Cluster nodes, providing regional centralized collaboration functions for multiple business service nodes, as well as collaboration functions between cluster control nodes across regions; within the cluster, providing anchor functions for collaborative tasks; on the air interface, not providing connection functions or only providing control functions for connections.
(2) Service nodes (serving Node, sNode), providing scheduling and execution functions for collaborative tasks; on the air interface, providing control and/or data functions for connections.

The 6G RAN architecture is shown in Fig. 15.27.

From the perspective of connection functionality, two options can be considered for the implementation of cNode.

(1) CP/UP separation: cNode has the control plane functionality of the connection, sNode has the user plane functionality of the connection. cNode provides the control function of the connection, responsible for CP functions, including MIB, SIB, SSB, Paging and RRC. sNode provides the data function of the connection, sNode provides data transmission function, physical layer control information (DCI, UCI) can be carried by sNode, or by cNode, thus achieving ultra-simple data carrier. Control and

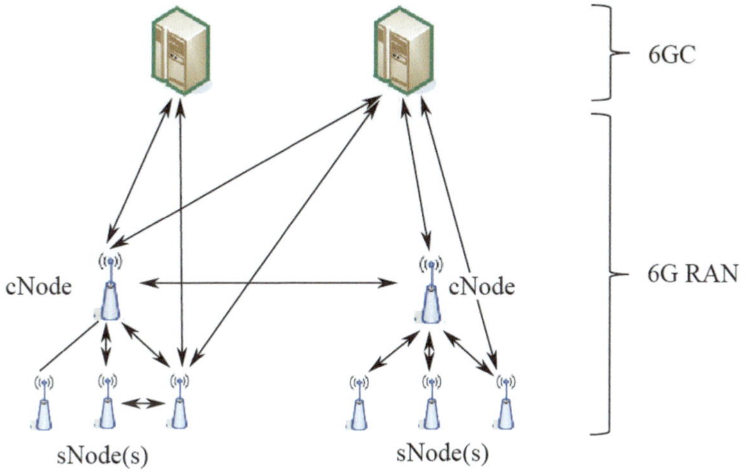

Fig. 15.27 6G RAN Architecture

data separation can enable high-frequency efficient transmission, network energy saving, enable multi-RAT extreme aggregation, etc.

From the UE perspective, in option (1), the communication between UE and RAN, the control plane communicates directly with cNode, and the user plane communicates directly with sNode; the communication between UE and CN, the control plane communicates with CN through cNode; the user plane communicates with CN through sNode.

(2) CP/UP non-separation: cNode has no connection function, sNode has both control plane and user plane functionality of the connection; from the perspective of connection functionality, sNode is equivalent to 5G gNB; cNode does not have connection functionality, only responsible for various new service capabilities introduced by 6G.

From the UE perspective, in option (2), the communication between UE and RAN, both control plane and user plane communicate directly with sNode; the communication between UE and CN, both control plane and user plane communicate with CN through sNode.

The above are based on the vision of 6G, the refinement of requirements and thinking, a kind of thinking about the functional and interface division of the 6G network architecture, for the industry and academia to think about.

References

1. Li Zhengmao, Wang Xiaoyun, Huang Yuhong, et al. TD-LTE Technology and Standards [M]. Beijing: People's Posts and Telecommunications Press, 2013.
2. Liu Guangyi, Fang Min, Guan Hao, et al. 5G Mobile Communication: Towards a Fully Connected World [M]. Beijing: People's Posts and Telecommunications Press, 2019.

3. Wang Xiaoyun, Liu Guangyi, Ding Haiyu, et al. 5G Technology and Standards [M]. Beijing: Electronics Industry Press, 2019.
4. ROSER C. How Managers Drove Boeing into the Ground[OL].(2022–03–22)[2023–02–22]. allaboutlean website.
5. Tong Wen, Zhu Peiying. 6G Wireless Communication New Journey: Crossing Human Connection, Object Connection, Towards Intelligent Connection of Everything [M]. Translated by Huawei Translation Centre. Beijing: Machinery Industry Press, 2021.
6. Liu Guangyi, Huang Yuhong, Cui Chunfeng, et al. 6G Reshaping the World [M]. Beijing: People's Posts and Telecommunications Press, 2021.
7. 6GANA. 6G Network AI Concept Terminology White Paper [R/OL]. 2022. http://www.6g-ana.com/upload/file/20220523/6378893017578935554735081.pdf.
8. 6GANA. 6G Network Native AI Technology Requirements White Paper[R/OL]. 2022. http://www.6g-ana.com/upload/file/20220523/6378893017730497434706068.pdf.

Abbreviations

1G	First generation mobile communication system
2G	Second generation mobile communication system
3D	Three dimensional
3G	Third generation mobile communication system
3GPP	Third generation partnership project
4G	Fourth generation mobile communication system
5G	Fifth generation mobile communication system
6G	Sixth generation mobile communication system
ACLR	Adjacent channel leakage ratio
ADC	Analogue to digital converter
AF	Application function
AI	Artificial intelligence
AI4NET	Artificial intelligence for network services
AIaaS	Artificial intelligence as a service
AMP	Approximate message passing
AOA	Angle of arrival
AOA	Azimuth angle
AOD	Departure angle
AP	Access point
App	Application
APSK	Amplitude phase shift keying
AR	Augmented reality
ARoF	Analog radio-over-fiber
ASK	Amplitude shift keying
BaaS	Blockchain as a service
BCD	Block coordinate descent
BCH Codes	Bose-Chaudhuri-Hocquenghem codes
BER	Bit error rate
BIC	Bit interleaved coded
BLER	Block error rate

G. Liu et al., *6G Migrates from Communication to XaaS*,
https://doi.org/10.1007/978-981-96-7579-1

BM	Beam management
BPSK	Binary phase shift keying
BS	Base station
BSC	Backscatter communication
CA	Communication-computing converged anchor
CG	Coarse grained
CGC	Coarse grained computing
CFO	Carrier frequency offset
CG	Configured grant
CIR	Channel impulse response
CJT	Coherent joint transmission
CLI	Cross link interference
CMLD-CFAR	Censored mean level detector-CFAR
CALDCFAR	Cell average level detector constant false alarm rate
CMOS	Complementary metal oxide semiconductor
CN	Core network
CNN	Convolutional neural networks
cNode	Cluster node
CoMP	Coordinated multi-point
CP	Cyclic prefix
CP/UP	Control plane/user plane
CPE	Customer premise equipment
CP-OFDM	Cyclic prefix orthogonal frequency division multiplexing
CPU	Central processing unit
CQI	Channel quality indicator
C-RAN	Cloud radio access network
CRB	Computing radio bearer
CS	Chirp scaling
CS	Compressed sensing
CSI	Channel state information
CSI-RS	Channel state information-reference signal
CSS	Chirp spread spectrum
CU	Central unit
CUDA	Compute unified device architecture
DaaS	Data as a service
DAC	Digital to analog converter
DAS	Distributed antenna system
DB	Database
DCI	Downlink Control Information
DC-RF	Direct current-radio frequency
DD	Delay-doppler
DDM	Doppler division multiplexing
DDoS	Distributed denial of service attack
FM	Frequency modulation
FMCW	Frequency modulated continuous wave

FME	Failure modes and effects
FMM	Fast multipole method
FMO	Flexible macroblock ordering
FMT	Filtered multitone
FN	False negative
FP	False positive
FPA	Focal plane array
FSK	Frequency shift keying
FSPL	Free space path loss
FSS	Fixed satellite service
FT	Fourier transform
FTTA	Fiber to the antenna
FTTB	Fiber to the building
FTTC	Fiber to the curb
FTTH	Fiber to the home
FTTN	Fiber to the node
FTTP	Fiber to the premises
FTTX	Fiber to the X
FLOPs	Floating point operations per second
FMCW	Frequency modulated continuous wave
FPGA	Field programmable gate array
FSK	Frequency shift keying
GBR	Guaranteed bit rate
GBSM	Geometry based stochastic channel model
GDPR	General data protection regulation
GEO	Geostationary earth orbit
GFDM	Generalized frequency domain multiplexing
GNN	Graph neural network
GNSS	Global navigation satellite system
GO-CFAR	Greatest of-CFAR
GPS	Global positioning system
GPU	Graphics processing unit
GSCM	Geometry based stochastic models
GSM	Global system for mobile communications
GTI	Global TD-LTE initiative
HA	High availability
HARQ	Hybrid automatic repeat request
ISD	Inter-site distance
ISI	Inter-symbol interference
ITU	International telecommunication union
MAC	Medium access control
MIMO	Multiple-input multiple-output
ML	Maximum likelihood
MMSE	Minimum mean square error
MRC	Maximum ratio combining

MS	Mobile station
MSE	Mean square error
NLOS	Non-line of sight
max-EE	Maximum energy efficiency
max-min Rate	Maximum minimum rate
max-SE	Maximum spectral efficiency
MBR	Maximum bit rate
MCMC	Markov chain monte carlo
MCU	Microcontroller
MDT	Minimisation of drive test
MEMS	Micro electro mechanical system
MEO	Medium earth orbit
MHCM	Map-based hybrid channel model
MIB	Master information block
MIMO	Multiple input multiple output
MISO	Multiple input single output
MiWEBA	Millimetre-wave evolution for backhaul and access
ML	Machine learning
ML	Maximum likelihood
MLP	Multilayer perceptron
MMSE	Minimum mean square error
mMTC	Massive machine type communications
mmWave	Millimeter wave
MP	Message passing
MPSK	Multiple phase shift keying
MRC	Maximum ratio combining
MRTD	Maximum receive timing difference
MSE	Mean square error
MUSA	Multi user shared access
MUSIC	Multiple signal classification
MVDR	Minimum variance distortionless response
NaaS	Network as a service
NAMO	Network AI management orchestration
NAS	Non-access stratum
NB-IoT	Narrow band internet of things
NCJT	Non-coherent joint transmission
NE	Network entity
NEF	Network exposure function
NET4AI	Network for artificial intelligence
NF	Network function
NGMN	Next generation mobile networks
NLOS	Non line of sight
NMSE	Normalized mean square error
NOMA	Non-orthogonal multiple access
NPU	Neural network processing unit

NR	New radio
NTN	Non-terrestrial networks
NWDAF	Network data analytics function
OFDM	Orthogonal frequency division multiplexing
OFDMA	Orthogonal frequency division multiple access
OM	Operation and maintenance
OMA	Orthogonal multiple access
OMP	Orthogonal matching pursuit
OOB	Out-of-band
OOK	On-off keying
OS-CFAR	Ordered statistics-CFAR
OSI	Open system interconnection
O-STBC	Orthogonal space time block code
OTFS	Orthogonal time frequency space
OFDM	Orthogonal frequency division multiplexing
PA	Power amplifier
PAPR	Peak to average power ratio
PAU	Power amplifier uncertainty
PBCH	Physical broadcast channel
PC	Partially-connected
PCA	Principal component analysis
PDF	Probability density function
PDMA	Pattern division multiple access
PD-NOMA	Power domain non-orthogonal multiple access
PDU	Protocol data unit
PER	Packet error rate
PIC	Parallel interference cancellation
PIE	Pulse interval encoding
PIN	Positive intrinsic negative
PIPL	Personal information protection law
PMI	Precoding matrix indicator
PN	Pseudo random noise
PON	Passive optical network
PR-ASK	Phase reversal amplitude shift keying
PSK	Phase shift keying
PSR	Pilot-to-signal ratio
PSS	Primary synchronisation signal
PAPR	Peak to average power ratio
PBCH	Physical broadcast channel
PC	Power control
PA	Pre-configured authorisation
PDCCH	Physical downlink control channel
PDSCH	Physical downlink shared channel
PHY	Physical layer
PL	Path loss

PMI	Precoding matrix indicator
PRB	Physical resource block
PSK	Phase shift keying
QAM	Quadrature amplitude modulation
QoS	Quality of service
Q-O-STBC	Quasi-orthogonal space time block code
QPSK	Quadrature phase shift keying
FBMC-QAM	Filter bank-base multi-carrier-quadrature amplitude modulation
QuaDRiGa	Quasi-deterministic radio channel generator
QAM	Quadrature amplitude modulation
QoS	Quality of service
RAN	Radio access network
RAT	Radio access technology
RB	Resource block
RC	Resistor-capacitance
RCS	Radar cross section
RSS	Received signal strength
RSU	Roadside unit
RT	Ray-tracing
RTT	Round trip time
RV	Redundancy version
RX	Receive
RAN	Radio access network
RF	Radio frequency
RLC	Radio link control
RRC	Radio resource control
RSSI	Received signal strength indicator
SDR	Software defined radio
SINR	Signal to interference plus noise ratio
SISO	Single-input single-output
SNR	Signal to noise ratio
SA	Standalone
SAE	System architecture evolution
SAGE	Space-alternating generalized expectation-maximization
SAI	System area identity
SBR	Shooting and bouncing ray method
SBS	Small base station
SC	Single Carrier
SCMA	Sparse code multiple access
SCA	Sequential Convex Approximation
SCDMA	Synchronous code division multiple access
SC-FDE	Single carrier frequency domain equalization
SCP	Service control point
SD	Service discovery
SDMA	Space division multiple access

SDN	Software defined network
SDR	Software defined radio
SDN	Software defined network
SE	Spectral efficiency
SF	Sensing function
SF	Spreading factor
SFC	Service function chaining
SFFT	Symplectic finite fourier transform
SFO	Sampling frequency offset
SGSN	Serving GPRS support node
SI	System information
SI	Self-interference
SI	Study item
SIB	System information block
SIC	Successive interference cancellation
SIMO	Single input multiple output
SINR	Signal to interference plus noise ratio
SISO	Single input single output
SL	Side link
SLA	Service level agreement
SLAM	Simultaneous localization and mapping
SLS	System level simulation
SM	Session management
S-MIMO	Single user multiple input multiple output
SMT	Space-time matrix
SN	Sequence number
SNR	Signal to noise ratio
sNode	Serving node
SO	Serving office
SO-CFAR	Smallest of constant false alarm rate
SOF	Start of frame
SON	Self-organizing network
SOR	Successive over relaxed
SP	Service provider
SPS	Semi-persistent scheduling
SR	Smart repeater
SR	Scheduling request
SRB	Signalling radio bearer
SRS	Sounding reference signal
SS	Synchronization signal
SSB	Synchronization signal block
SSB-ASK	Single sideband amplitude shift keying
SSC	Secondary synchronization channel
SSID	Service set identifier
SSS	Secondary synchronization signal

STBC	Space-time block coding
STC	Space-time coding
STF	Short training field
STO	Sampling time offset
SU	Single user
SUL	Supplementary uplink
SU-MIMO	Single user multiple input multiple output
SVC	Scalable video coding
SVD	Singular value decomposition
SW	Software
SWIPT	Simultaneous wireless information and power transfer
SX	Switch
TDD	Time division duplex
TTI	Transmission time interval
TA	Task anchor
TA	Time advance
TAA	Trust anchor agent
TaaS	Trust as a service
TAC	Tracking area code
TAE	Time alignment error
TAF	Traffic adaptation function
TAP	Test access port
TAU	Tracking area update
TBF	Temporary block flow
TCP/IP	Transmission control protocol/internet protocol
TC	Turbo code
TCH	Traffic channel
TCI	Transmission configuration indicator
TCP	Transmission control protocol
TD-CDMA	Time division-code division multiple access
TDD	Time-division duplex
TDL-A	Time-delay-line A
TDL-C	Time-delay-line C
TD-LTE	Time division long term evolution
TDM	Time division multiplexing
TDMA	Time division multiple access
TDP	Time division polarization
TDS	Time division synchronous
TD-SCDMA	Time division-synchronous code division multiple access
TE	Task executor
TE	Terminal equipment
TETRA	Terrestrial trunked radio
TF	Time-frequency
TF	Transport format
TFC	Transport format combination

TFS	Transport format set
TFTP	Trivial file transfer protocolTHP transmission HARQ process
TG	Test group
TIA	Telecommunications industry association
TIN	Temporary identifier
TIS	Total isotropic sensitivity
TMA	Tower mounted amplifier
TMN	Telecommunications management network
TMSI	Temporary mobile subscriber identity
TN	Terminal node
TOA	Time of arrival
TPC	Transmit power control
TPI	Test point interface
TPS	Transmission parameter signalling
TPU	Tensor processing unit
TR	Technical report
TRAU	Transcoder and rate adaptation unit
TRP	Total radiated power
TS	Technical specification
TSC	Training sequence code
TSG	Technical specification group
TSI	Time slot interchange
TSP	Time synchronization protocol
TTI	Transmission time interval
TTP	Trusted third party
TTS	Text to speech
TU	Typical urban
TV	Television
TVWS	Television white spaces
TX	Transmit
TXOP	Transmission opportunity
UE	User equipment
UAP	Upper address part
UARFCN	UTRA absolute radio frequency channel number
UAS	Unmanned aircraft system
UATI	Unicast access terminal identifier
UCI	Uplink control information
UCR	User centric routing
UDN	User defined network
UE	User equipment
UICC	Universal integrated circuit card
ULA	User link adaptation
ULCA	Uplink carrier aggregation
ULDPC	Uplink low density parity check
ULE	Ultra low energy

ULF	Ultra low frequency
ULP	Ultra low power
ULSCH	Uplink shared channel
UM	Unacknowledged mode
UMA	Unlicensed mobile access
UMB	Ultra mobile broadband
UMC	Universal mobile communications
UMTS	Universal mobile telecommunications system
UNI	User network interface
UP	User plane
UPC	Usage parameter control
UPI	User plane integrity
URA	UTRAN registration area
URLLC	Ultra-reliable low-latency communications
USAT	User services architecture team
USB	Universal serial bus
USF	Uplink state flag
USIM	Universal subscriber identity module
USRP	Universal software radio peripheral
UTRAN	UMTS terrestrial radio access network
UWB	Ultra wide band
V2I	Vehicle to infrastructure
V2V	Vehicle to vehicle
V2X	Vehicle to everything
VA	Virtual array
VBR	Variable bit rate
VDSL	Very high speed digital subscriber line
VDU	Virtualized distributed unit
VHF	Very high frequency
VHT	Very high throughput
VLAN	Virtual local area network
VLC	Visible light communication
VLR	Visitor location register
VM	Virtual machine
VNF	Virtual network function
VoIP	Voice over IP
VoLTE	Voice over LTE
VQ-VAE	Vector quantised-variational autoencoder
VPN	Virtual private network
VR	Virtual reality
VRRP	Virtual router redundancy protocol
VSA	Virtual spectrum access
VSAT	Very small aperture terminal
VSF	Virtualized storage function
VSWR	Voltage standing wave ratio

VT	Virtual terminal
VTC	Video telephony call
WAP	Wireless application protocol
WB	Wide band
WBA	Wireless broadband alliance
WCDMA	Wideband code division multiple access
WDM	Wavelength division multiplexing
WDS	Wireless distribution system
WEP	Wired equivalent privacy
WFQ	Weighted fair queuing
WG	Work group
WGF	Weighted graphic framework
Wi-Fi	Wireless fidelity
WiMax	World interoperability for microwave access
WLAN	Wireless local area network
WLL	Wireless local loop
WMAN	Wireless metropolitan area network
WMMSE	Weighted minimum mean square error
WMN	Wireless mesh network
WPAN	Wireless personal area network
WR	Windowing and restructuring
WRC	World radiocommunication conference
WS	Work station
WSN	Wireless sensor network
WWAN	Wireless wide area network
WWW	World wide web
X2AP	X2 application protocol
XaaS	Anything as a service
XCAP	XML configuration access protocol
xCN	next-generation core network
XDR	eXtended data rate
XG-PON	10-Gigabit-capable passive optical network
XML	eXtensible markup language
xNB	next-generation node basestation
XOR	Exclusive OR
XR	Extended reality
ZF	Zero forcing
ZFBF	Zero forcing beamforming
ZF-DFE	Zero forcing decision feedback equalizer
ZF-SIC	Zero forcing successive interference cancellation
ZIGBEE	Zigbee
ZF	Zero forcing
ZOA	Zenith of arrival
ZP	Zero padding

MIX
Papier aus verantwortungsvollen Quellen
Paper from responsible sources
FSC® C105338

FSC
www.fsc.org

If you have any concerns about our products,
you can contact us on
ProductSafety@springernature.com

In case Publisher is established outside the EU,
the EU authorized representative is:
Springer Nature Customer Service Center GmbH
Europaplatz 3, 69115 Heidelberg, Germany

Printed by Libri Plureos GmbH
in Hamburg, Germany